BRITISH CRUISERS

BRITISH CRUISERS

Two World Wars and After

NORMAN FRIEDMAN

Ship plans by A D Baker III, John R Dominy, Alan Raven and Paul Webb

Seaforth
PUBLISHING

Frontispiece: HMS *Dido* illustrates the configuration of her class by mid-1943, with the small lantern of the surface-search set atop the tripod foremast.
(This and all other uncredited photographs are from US official sources, by courtesy of the author)

Text copyright © Norman Friedman 2010
Plans © individual draughtsmen, as credited 2010

This paperback edition first published in Great Britain in 2022 by
Seaforth Publishing
An imprint of Pen & Sword Books Ltd
47 Church Street, Barnsley
S Yorkshire S70 2AS

www.seaforthpublishing.com
Email info@seaforthpublishing.com

British Library Cataloguing in Publication Data
A CIP data record for this book is available from the British Library

ISBN 978 1 3990 9791 8 (PAPERBACK)
ISBN 978 1 7834 6918 5 (EPUB)
ISBN 978 1 7834 6685 6 (KINDLE)

All rights reserved. No part of this publication may be reproduced or transmitted in any form or by any means, electronic or mechanical, including photocopying, recording, or any information storage and retrieval system, without prior permission in writing of both the copyright owner and the above publisher.

The right of Norman Friedman to be identified as the author of this work has been asserted in accordance with the Copyright, Designs and Patents Act 1988.

Typeset and designed by Ian Hughes, Mousemat Design Limited
Printed and bound in India by Replika Press Pvt. Ltd.

CONTENTS

	Glossary and Abbreviations	6
	Acknowledgements	7
CHAPTER 1	Introduction	8
CHAPTER 2	Protecting Trade	18
CHAPTER 3	Destroyer-Killers	36
CHAPTER 4	War Experience	74
CHAPTER 5	Treaties and Heavy Cruisers	96
CHAPTER 6	The 1930 London Treaty and its Cruisers	142
CHAPTER 7	The Slide toward War	200
CHAPTER 8	War	230
CHAPTER 9	Wartime Cruiser Design	252
CHAPTER 10	Post-War Cruisers	272
CHAPTER 11	The Missile Age	302
APPENDIX	Fast Minelayers	324
	Notes	334
	Bibliography	380
	Data List (specifications)	382
	List of Ships	410
	Index	422

GLOSSARY AND ABBREVIATIONS

A/S: anti-submarine
AA: anti-aircraft
ABU: automatic barrage unit
ACNS (W): Assistant Chief of the Naval Staff (Weapons)
ACNS: Assistant Chief of the Naval Staff
ADO: Air Defence Officer
ADP: Air Defence Position
ADR: Aircraft Direction Room
AIO: Action Information Organisation
AP: armour piercing
ARL: Admiralty Research Laboratory
ASW: anti-submarine warfare

BD: between decks (mounting)
BL: breech-loading (gun)
BOR: Bridge Operations Room

CDS: Comprehensive Display System
CID: Committee of Imperial Defence
C-in-C: Commander-in-Chief
CNS: Chief of the Naval Staff
CO: Commanding Officer
Commodore (T): Commodore of the Harwich Force (Commodore, later Rear Admiral, Sir Reginald Y Tyrwhitt)
COSAG: Combined Steam and Gas (turbine)
COW: Coventry Ordnance Works
CRBF: Close Range Blind Fire (system)

D of D: Director of Dockyards
D of P: Director of Plans
D of TD: Director of Torpedo Division
DA: direct attack (weapon)
DACR: direct attack close range (weapon)
DAS: Director of Anti-Submarine Warfare
DAW: Director of Naval Air Warfare
DBR: dive bomber reconnaissance (aircraft)
DCNS: Deputy Chief of the Naval Staff

DCT: director control tower
DDNC: Deputy Director of Naval Construction
DDNO: Deputy Director of Naval Ordnance
DDOD (M): Deputy Director of Operations Division (Mining)
DEE: Department/Director of Electrical Engineering
DFSL: Deputy First Sea Lord
DGD: Department/Director of Gunnery Division
DNAD: Director of Naval Air Division
DNC: Director of Naval Construction
DNE: Director of Naval Equipment (including arrangements for personnel aboard ships)
DNI: Director of Naval Intelligence
DNO: Director of Naval Ordnance
DOD: Director of Dockyards (same as D of D, above)
DPT: data link associated with CDS (*qv*)
DRC: Defence Requirements Committee
DRE: Director of Radio Equipment
DTM: Directorate of Torpedoes and Mines
DTSD: Director/Division of Training and Staff Duties (in 1945, Tactical, Torpedo, and Staff Duties; after 1945, Tactical and Staff Duties)
DTWP: Director of Tactics and Weapons Policy
DUW: Director of Underwater Weapons

ehp: effective horsepower
E-in-C: Engineer-in-Chief

F/R: fighter-reconnaissance (aircraft)
FDO: Fighter Direction Office
FKC: Fuse-Keeping Clock
FOST: Fleet Operational Support and Training (ship)
ft: foot/feet

GAP: Guided Air Projectile
GDR: Gunnery Direction Room
GDS: Gun Direction Systems

HA: high angle
HACP: High-Angle Control Position
HACS: High-Angle Control System
HADT: High-Angle Director Tower
HE: high explosive
HF/DF: High Frequency Direction-Finding
HMAS: His/Her Majesty's Australian Ship
HMCS: His/Her Majesty's Canadian Ship
HMS: His/Her Majesty's Ship
HP: high pressure

IFF: Identification Friend or Foe
in(s): inch(es)

kt(s): knot(s)
kW: kilowatts

LA: low angle
lb(s): pound(s)
LCS: Light Cruiser Squadron
LP: low pressure
LST: Landing Ship Tank

MCDP: Medium Calibre Dual Purpose (gun)
MoD: Ministry of Defence

NATO: North Atlantic Treaty Organisation
NDAC: New Design Armoured Cruiser
NID: Naval Intelligence Department
nm: nautical miles

oa: overall

PAC: Parachute and Cable (weapon)
PIL: Position In Line (rangefinder)
pp: between perpendiculars
PRO: Public Record Office

psi: pounds per square inch
PWQ Committee: Post-war Questions Committee

QF: quick-firing (gun)

RA (D): Rear Admiral (Destroyers)
RAE: Royal Aircraft Establishment, Farnborough
RAF: Royal Air Force
RCO: Radar Control Office
RDF: Radio Direction-Finding (i.e. radar)
RPC: remote power control
rpm: revolutions per minute

S of C: Superintendant of Charts
S/R: spotter-reconnaissance (aircraft)
SAP: Semi-Armour Piercing
shp: shaft horsepower
STAAG: Stabilised Tachymetric Anti-Aircraft Gun
STD: Simple Tachymetric Director

TIR: Target Indication Room
TIU: Target Indication Unit
TOM: Tachymetric One-Man (director)
TSR: torpedo spotter reconnaissance (aircraft)

UP: Unrotated Projectile (rocket)
USS: United States Ship

VCNS: Vice Chief of the Naval Staff

W/T: wireless telegraphy, i.e. radio
WA: warning air (radar)
wl: waterline
WS: warning surface (radar)

YARD: Yarrow-Admiralty Research Department
yd(s): yard(s)
YEAD: Yarrow-English Electric Admiralty Development

ACKNOWLEDGEMENTS

Above all I thank my wife Rhea, who has lived with this project and its forebears for many years, from the 1970s, when she first encouraged me to take vacation time to visit and revisit the Draught Room at the National Maritime Museum. She is a large part of how and why books like this get written. I have often enjoyed (and benefited greatly from) discussing the historical and policy issues raised in this work with her. She has always been very supportive, particularly at times when projects have seemed to me to entail walking through molasses. She has helped me adopt and to continue using photography to obtain copies of crucial documents, first using a film camera and tripod and more recently using a digital one. More than any previous book of mine, this one could not have been written without the digital camera, because the volume of documents consulted has been so vast. However, the book also benefited heavily from access to a considerable library of printed material. Rhea has often joined me in hunting through bookstores, here and abroad. I cannot thank Rhea enough for her loving support.

My research on British cruisers goes back to the 1970s, when it benefited enormously from help provided by the late David Lyon, at that time in charge of the Draught Room of the National Maritime Museum. Since his time crucial National Maritime Museum collections, including the Covers and the Constructors' Notebooks, have become available to the public at the Brass Foundry out-station of that museum. For critical assistance I thank its current chief, Jeremy Michell, his assistant Andrew Choong Han Lin, and their predecessors. I have to thank a former head of the Brass Foundry, the late David R Topliss, for alerting me to the value of the Constructors' Notebooks. I am also grateful to the Brass Foundry staff for helping me gain access to the trove of Vickers design material they hold. I am grateful to the staff of the Caird Library of the National Maritime Museum for assistance with the d'Eyncourt papers, and with the d'Eyncourt design notebook I have quoted. I much appreciate the considerable assistance provided by Captain Christopher Page (recently retired) and his staff at the Royal Naval Historical Branch and the Admiralty Library (to whose librarian, Jenny Wraight, I am much indebted). I have also benefited from access to the archive of the Royal Navy Museum in Portsmouth, arranged by Ms Wraight. I am, as always, grateful to the staffs of the Public Record Office (Kew: now The [British] National Archives) and the US National Archives and Records Agency, both downtown and at College Park. Given the close association of the Royal Navy and the US Navy, the ability to cross-reference materials from both sides of the Atlantic has been extremely valuable. I benefited greatly from relevant parts of the Thurston Notebook (export designs) provided by Stephen McLaughlin. Particularly for Australian ships, I am grateful to Dr David Stevens, head of the RAN Historical Branch, and to Paul Webb. Professor C M Woolgar of Southampton University and his assistant Karen Robson provided me with a microfilm of some of the pre-1914 Battenberg papers, which provided useful hints on British cruiser policy. In addition to those named, I benefited greatly from discussions with A D Baker III, with Alan Raven and with Dr Nicholas Lambert, whose knowledge of the workings of the Admiralty during the pre-1914 and First World War eras is remarkable. I am also grateful to Dr Eric Grove, whose insights into the post-1945 Royal Navy and into British defence policy in general have proven quite helpful. I would like to thank Dr George H Elder for supplying copies of some important Royal Navy documents. I would also like to thank members of the 'Steel Navy' web discussion board for their help, particularly in elucidating the final close-range armament of HMS *Exeter*.

My friend A D Baker III produced many drawings specially for this book, and provided invaluable advice and also many photographs. My friend Alan Raven kindly produced several drawings showing British cruisers at particularly interesting times during their wartime careers, drawn for this book, as well as much valued comment, particularly on wartime alterations. Paul Webb very kindly allowed me to use several of his drawings, and also provided important information and photographs. John R Dominy allowed me to use several of his drawings, all of which were based on official plans. Mr Baker in turn thanks several friends for their own assistance in his work: Robin Bursell, John Lambert, Darius Lipinski, Miles McLaughlin, Alan Raven, Paul Webb and the staff of the Brass Foundry, which holds plans on which many of the drawings were partly based (the list of those who helped Mr Baker suggests how much more his plans entail than simply tracing and simplifying the large originals in the National Maritime Museum).

For photographs, without which this book would have been empty, I am particularly grateful to my good friend Charles Haberlein, who recently retired as the curator of photographs at the Naval Historical and Heritage Command at the Washington Navy Yard, and to his assistants Ed Finney and Robert Renshew. Mr Haberlein's depth of knowledge extended well beyond finding the right photographs and properly identifying them. I would also like to thank Rick E Davis, Rich Gimblett, Dr David Stevens, Dr Jozef Straczek (formerly of the RAN Historical Branch, and of considerable assistance there), the staff of the photographic library of the US Naval Institute, and the staff of the photographic and cartographic branch of the US National Archives at College Park, Maryland. Bob Todd, the photo curator at the Brass Foundry, contributed invaluable insights and helped solve some major puzzles, particularly with regard to HMS *Caledon*, HMS *Delhi*, and ORP *Dragon*.

Although I greatly appreciate all the help given me, I am of course responsible for the contents of this book, including any errors.

CHAPTER 1
INTRODUCTION

It is difficult to define a cruiser in a way which embraces all the ships described in this book. The name implies a ship capable of cruising independently on a foreign station, which in the age of steam machinery entailed an ability to make running repairs far from home, as well as a long radius of action. Behind this name was the idea that the cruiser was smaller or more weakly armed than a battleship, yet still protected against enemy fire to some extent. The second, but not quite the first, attribute can be associated with the fast cruisers built during the First World War, which were to a considerable extent super-destroyers (which is why the big 'Tribal' class destroyers were candidate replacements in the 1930s). Although they were initially called armoured cruisers, I have not included the first battlecruisers; they are more closely associated with the last generation of large armoured cruisers, which I hope to describe in a later volume. I have included cruiser minelayers, and I have also included the escort cruisers and command cruisers (culminating in the *Invincible* class) of the 1960s and 1970s, because they were conceived very much for possible independent operations.

This book describes the British cruisers of the radio or wireless age. Radio changed naval warfare in profound ways, and cruisers operating far from home were changed more than most kinds of ships. British cruisers had three roles. One was to protect seaborne trade against surface raiders. A second was to support the battle fleet, both as scouts and by beating off enemy torpedo attacks. A third was to maintain order in the massive British Empire. During what might be called the pre-radio age, trade protection entailed large numbers of ships, covering convoys or occupying the focal areas through which most trade passed, and through which raiders, too, would most likely pass. Trade protection by either technique required large numbers of cruisers. In the late nineteenth century, likely enemies (France, Germany, Russia) all began building large, fast armoured cruisers and protected cruisers which might attack British trade. The Royal Navy built its own numerous fleet of large cruisers – each of which cost about as much as a battleship. Cruiser-building to protect trade was ruinously expensive because so many such ships were needed to cover so much trade. In effect the Royal Navy found itself building both a battle fleet and a cruiser fleet of similar or even greater cost. The French went so far as to write about an economic war ('guerre industrielle') in which the British would be defeated by being driven bankrupt. This war was deadly because the French (and Russians, and Germans) did not have to build large numbers of cruisers, while the Royal Navy had to place equivalent ships everywhere they might appear.

Radio changed trade protection. It became possible to envisage an intelligence system using radio reports of raider attacks to track the raiders.[1] On that basis, fast cruisers could be vectored to intercept them. Although the process was imperfect, it could deal far more economically with any raiders. Initially the expectation seems to have been that fast long-range ships (battlecruisers) would be held at readiness in home waters for despatch against raiders, but by about 1910 it was clear that groups of cruisers would be held on foreign stations awaiting radioed orders. This idea, which could not be discussed publicly, made it possible to imagine protecting British trade using an affordable number of cruisers. That number in turn shaped British cruiser design during the inter-war period, when most of the ships described in this book were built or at least conceived.

Cruisers also operated with the fleet. As scouts, they were expected to find the enemy fleet (and discover its disposition, course and speed)

Throughout most of the cruiser era, Royal Navy strategists were faced with a terrible problem: they had to defend the worldwide trade of the Empire, its lifeblood, with a single mobile fleet, supplemented by cruisers on remote stations. The encouragement of the Dominion navies was a partial solution, but the cruisers built for and operated by the Dominions were lumped with those of the Royal Navy under the inter-war naval arms-limitation treaties, particularly that signed in London in 1930 (which limited overall cruiser tonnage). HMAS *Australia* is shown between the wars. The two Australian 'Counties' differed from their Royal Navy counterparts in having taller funnels. (Photo by Allan C Green via State Library of Victoria)

INTRODUCTION

Above: The inter-war Royal Navy was dominated by the need to protect the seaborne trade which kept the British Empire alive. The big 'County' class cruisers were conceived largely to deter Japanese attacks on British trade by threatening Japanese trade during the weeks before the main British fleet could reach Singapore, its base for a war against Japan. The first five 'Counties' were therefore deployed to the China Fleet upon completion. HMS *Berwick* is shown in 1932, wearing tropical livery (white hull and buff funnels). Note the trolley, which indicates that she has been fitted with a catapult.

Above: HMS *Kent* emerged from a 1931 refit with a catapult, an important feature because an aircraft gave a cruiser a much better chance of detecting raiders. Within a few years the Royal Navy was intensely interested in the offensive potential of catapult aircraft, but the one shown here was limited to spotting duties. Aircraft were included in the design (the ships were completed with a foundation for a catapult) but initially were omitted to avoid exceeding the 10,000-ton Washington Treaty limit.

Above: HMS *Norfolk*, a later 'County' class cruiser, emerged from a 1937 refit with twin 4in anti-aircraft guns, part of a rearmament package proposed by the 1932 Naval Anti-Aircraft Gunnery Committee. The single centreline anti-aircraft director (aft) with which the ship was completed was replaced by a pair of such directors abreast the bridge. Other improvements were a pair of octuple pompoms, the most powerful existing light anti-aircraft weapons, abeam the after superstructure, and a pair of the new quadruple 0.5in machine guns on platforms abeam the gap between the first and second funnels. The inter-war Royal Navy invested more in anti-aircraft firepower than any other, and was air-conscious to the point that additional deck armour for capital ships was justified as a defence against bombing. The 4in gun was particularly important because it was assumed that enemy bombers would attack from medium altitude, hence could be engaged at a considerable distance and their formations broken up. Unfortunately the Royal Navy opted too soon for a fire-control system which depended on the controlling officer's estimate of the speed of the approaching aircraft. Just before the Second World War the Director of Naval Ordnance wrote that this had been the wrong choice; he was about to test a tachymetric (speed-measuring) system (which, probably unknown to him, the US Navy already used). He expected the new system to enter service in 1941. The ship's DF array is just visible near the juncture between the foremast and its topmast, just below the crow's-nest.

Right: Radio made it possible for the Royal Navy to set up a global ocean-surveillance system and to vector cruisers to hunt down surface raiders, something inconceivable before about 1908. In 1909 the Admiralty sought to convince the Dominions to create fleets which could secure Empire trade outside European waters. Only Australia and New Zealand responded, and only Australia created a full 'fleet unit'. When war came in 1914, HMAS *Sydney*, shown, participated in the hunt for the German raider *Emden*. The extent of the force required seems to have been an unpleasant surprise, and it inspired the post-war idea that British raiding cruisers could tie down significant Japanese forces. *Sydney* is shown pre-war; note the identifying funnel bands standard in the Royal Navy. (RAN Historical Branch)

while screening their own fleet from enemy discovery. If the British fleet was blockading an enemy port, British cruisers would operate off that port to raise the alarm when the enemy fleet sortied. To be viable, fleet scouts had to be powerful enough to survive in the face of the enemy's most powerful cruisers. In a pre-radio age, these scouts had to be backed by ships linking them to the main fleet, each within visual signalling range. With the advent of radio, the scouts could operate much further afield (a discussion in 1913 of fleet organisation mentioned scouts as much as 200 miles ahead) and the numbers in the cruiser force could be reduced dramatically. That became evident in post-First World War discussions of the number of cruisers the Royal Navy required.

In addition to their scouting role, the cruisers operating ahead of or with the fleet were expected to shield it from enemy torpedo craft. Before about 1910, the British expected the Germans to send out their destroyers (which they called seagoing torpedo boats) in hunting groups,

Radio also made it possible to limit the number of scouting cruisers required to operate with the fleet, since there was no need for linking ships to repeat the scouts' messages to the main force. By 1913 some British tacticians thought that scouts might operate as much as 200 miles ahead of the fleet, but the Royal Navy never had enough cruisers to form a useful screen at such distances. It did emphasise maximum radio range, achieved in part by lifting the 'flat-top' radio antenna as high as possible. This pre-war photograph of HMS *Bristol* shows her tall topmasts, their upper cross-trees supporting the roughly horizontal pair of wires of the 'flat top'. Neither the wires of the 'flat top' nor the vertical wire down to the radio room are visible. This photograph was taken before the Royal Navy introduced funnel recognition bands in 1913.

INTRODUCTION

Above: Radio made it possible to so reduce the number of scouts in the fleet that other cruisers could be assigned to beat off enemy destroyer attacks. After the First World War the ideal Royal Navy cruiser squadron consisted of five ships. By the early 1920s it was assumed that the battle fleet would have a two-squadron scouting line and two more squadrons to deal with enemy torpedo attacks, and to back up attacks by British destroyers. The seventy-cruiser force advocated from about 1924 on consisted mainly of ships assigned to trade protection, either as deterrents or to run down raiders based on ocean surveillance. The *Arethusas* (of 1913) and later small cruisers were conceived mainly as destroyer-killers. Later classes were given heavy torpedo batteries because, in attacking the enemy's destroyer force, they might find themselves in position to fire torpedoes at the enemy main body. HMS *Danae*, shown in 1930, was a mature example of this kind of cruiser, mounting four triple torpedo tubes – twice the battery of a destroyer, or a full destroyer battery on each side (she could not have mounted centreline tubes, hence could not use all of her tubes on either side). The main post-war modification was the addition of three 4in anti-aircraft guns, two abeam her funnels and one abaft No. 5 gun.

Below: The need for numbers of cruisers, mainly for trade protection, was a consuming requirement in inter-war British thinking. The Royal Navy pursued arms-control treaties to reduce cruiser size, hence cost, and its designers sought the minimum acceptable cruiser design. This ideal was reached in the *Arethusa* class. HMS *Galatea* is shown at Malta in April 1937, wearing neutrality stripes on 'B' turret to avoid attack during the Spanish Civil War. She had not yet been upgraded with twin rather than single 4in anti-aircraft guns. (Fahey Collection of the US Naval Institute)

so British anti-destroyer tactics were to blockade German torpedo boat bases using flotillas of destroyers stiffened by cruisers. About 1910 it was accepted that the Germans would take their destroyers to sea with their battle fleet, and the cruiser anti-destroyer (and pro-destroyer) roles with the fleet became important; the *Arethusas* and their successors were built for this purpose.

The third important cruiser role was protecting the Empire. It was complex partly because shadowing the formal British Empire was an informal one, consisting of close trading partners whose governments tended to benefit from British sea dominance. This informal empire was closely connected to the trading operations of the City of London, the financial centre of the United Kingdom and, before the First World War, the single most important financial centre in the world. The City financed world trade, and it well understood that free trade (free, for example, from anti-trade warfare) was key to British prosperity. It was understood that governments would favour Britain and the City if they understood that British sea dominance helped protect them. China, for example, was part of the informal empire, which explains why the Royal Navy maintained a large and expensive China Fleet through the inter-war period, far larger than the Asiatic fleet or squadron of any other European power. The extent of British investment helps explain why Japanese expansion into China in the 1930s was so threatening to the British. The informal empire seems to have been well understood in the British government, but rarely (if ever) explicitly discussed; it has surfaced in historical discussions only in recent years.[2] Yet the requirements of informal empire had profound implications for the British cruiser fleet. A cruiser was a particularly good package for colonial warfare: she combined a powerful gun armament with a substantial landing force of Marines and with command and control. In 1927, for example, HMS *Enterprise* and her Marines saved Kuwait from a Saudi attack. Kuwait was part of the informal empire, with a British resident, but was by no means a colony (a British amphibious carrier saved Kuwait again in 1961, this time from Iraqi attack). The great convulsion in the British cruiser force just prior to the beginning of this book

BRITISH CRUISERS

was the elimination of many small cruising craft on empire-protecting foreign stations, because neither the Foreign Office nor the Colonial Office was willing to pay for them, and in view of the new concept of trade protection they lacked a purely naval role. A few such craft (not true cruisers) survived, and an attempt to replace them, described in this book, failed just prior to the First World War because the demands of the main fleet were too insistent (the inter-war sloops, which are not described in this book, were the true successors to these small cruisers).

Informal empire could be expected to work as long as prospective partners could realistically expect Britain, which generally meant the Royal Navy, to help protect them. When someone wrote that 'trade follows the flag', what was often meant was that a country shielded by the Royal Navy would feel inclined to support that protection by

Above and below: British cruiser designers consistently sought to minimise the size, which they associated with the cost, of their ships. They therefore designed for minimum tonnage (weight-critical design practices), often not estimating the need for length or for space until late in the design process. That was possible because very experienced designers could estimate the needs of conventional designs. Unfortunately the situation changed quickly just before and during the Second World War, with dramatically increased needs not only for topweight (to accommodate radars and anti-aircraft guns) but also for internal space and for additional electric power. HMS *Manchester* is shown, newly completed, in 1938. She displays the first of the electronic devices which would soon proliferate, an HF/DF loop at the head of her foremast. This sensor was relatively common in the Royal Navy by 1939, but no other navy had it. Cruisers had HF/DF in order to detect and run down raiders beyond the horizon, and probably also to assist their aircraft in homing on them.

INTRODUCTION

Below: Early in the Second World War (this photograph is undated), *Birmingham* shows zarebas atop 'B' and 'X' turrets and on her quarterdeck for rocket launchers (UP projectors), a stopgap adopted in 1940 to make up for slow production of more adequate anti-aircraft weapons. Her main battery director shows the Type 284 gunnery radar, in its original form with separate transmitting and receiving antennas.

buying British, and by using British banks to float its loans. In a sense informal empire justified the cruiser squadrons maintained on foreign stations between the two World Wars. The stations were revived after the Second World War, but could not be maintained for long, as the war had destroyed too much of the British economy.

The Royal Navy ships in this book were designed by the Department of Naval Construction, headed by the Director of Naval Construction (DNC), who was officially adviser to the Admiralty Board on warship materiel (and as such was sometimes styled Deputy Controller). After a reorganisation in the 1960s, the design and construction organisation became the Ship Department and DNC became Director General, Ships; he figures only in the very last designs described in this book. Machinery was the responsibility of Engineer-in-Chief (E-in-C), and ordnance the Director of Naval Ordnance (DNO). During the inter-war period, as electrical machinery became more important, a separate Department of Electrical Engineering (DEE) was created. I have referred interchangeably to departments and to their chiefs. The DNC organisation designed ships up to the point at which bids could be invited, which for the purposes of this book meant to the point of mature designs. By way of contrast, E-in-C and DNO laid out specifications and evaluated designs (they also estimated weights and sizes so that DNC's designers could produce preliminary designs). DNO was responsible for fire control, but in 1941 a new Department of Gunnery and Anti-Air Warfare (DGD) was created, splitting DNO's role. There was also a separate torpedo and mine directorate (DTM). During the inter-war period an Anti-Submarine directorate, concerned with Asdic (sonar) was created, and ultimately this Director of Anti-Submarine Warfare (DAS) took over responsibility for torpedoes as well (as DTASW). Finally, between the wars a Naval Air Department (DNAD) was created; it was significant for cruiser design. Departments were reorganised after the Second World War on functional lines, creating, for example, a Director of Air Warfare (DAW) and a Director of Underwater Warfare (DUW). Neither is very important for this book. Director of Dockyards (D of D) was responsible for building and refitting ships, and provided information as to the capacity of the dockyard system.

Policy, including ship requirements, was set by the Board of Admiralty: First Sea Lord, his deputy Second Sea Lord (also responsible for personnel), Third Sea Lord (and Controller) and, at various times, Fourth Sea Lord (logistics) and Fifth Sea Lord (fleet aircraft). Before 1912 the Naval Intelligence Department (NID) functioned as both an intelligence organisation and a naval staff, evaluating various ship design issues, among other things. In 1912, as a result of controversy

Above: Photographed at Scapa Flow from USS *Wasp* on 17 May 1942, HMS *Manchester* displays the usual wartime modifications, which had to be squeezed into the ships. She had just completed a refit (18 March – 25 April 1942). Her masts carry the separate transmitting and receiving antennas of the Type 279 air-warning radar. The bridge carries the 'lantern' of the Type 273 surface-search radar, abaft the main battery director, which carries the separate transmitting and receiving antennas of a Type 284 gunnery-ranging set. Atop 'B' turret is a single Bofors gun. She had received one single Bofors and five single Oerlikons during a 16 January – 29 March 1941 refit, and another three single Oerlikons during a refit at the Philadelphia Navy Yard (23 September 1941 – 27 February 1942). Another two single Bofors were temporarily added for Operation 'Pedestal', the attempt to push a convoy through to Malta. During it she was sunk on 13 August.

Above and below: Photographed on 15 September 1943, *Birmingham* shows further modifications. She had been fitted with gunnery (Type 284) and air-search (Type 291) radars during a 1942 refit, but during a Devonport refit (23 April 1943 – 21 August 1943) she received a large-ship air-search set (Type 281B, with a single antenna for transmitting and receiving) and a Type 273 surface-search set atop her bridge. She was torpedoed on 28 November 1943 *en route* to Alexandria, and went to the United States for a further major refit.

concerning the navy's ability to staff war plans, a new War Staff was created, taking over the staff functions of NID. The staff functions became far more important during the First World War, and in 1917 a more elaborate staff organisation was formalised. By analogy with the army's staff, the war staff was given executive rather than advisory responsibility, and First Sea Lord was made Chief of the Naval Staff, with a Deputy First Sea Lord and Deputy and Assistant Chiefs of the Naval Staff (DCNS and ACNS), each of whom was responsible for parts of the naval staff. A new Naval Artillery and Torpedoes Division was created in June 1918 to decide weapons employment policy and also to develop weapon requirements. This division was also responsible for requirements for ship protection against weapons, which is why its chief became involved in discussions of the 'E' class cruiser design. This was Captain Frederic C Dreyer, who had been Admiral Jellicoe's Grand Fleet gunnery officer. When DNO (Director of Naval Ordnance) tried to shut down his new department, Dreyer argued successfully that DNO was far too involved with details to develop overall policy. In 1920 the new department was split into a Gunnery Department and a Torpedo Department; Dreyer became the first director of the Gunnery Department (DGD). A Training and Staff Duties Division (DTSD) was created in June 1918, initially to help organise the staff and also to consider conditions of entry into the Royal Navy; by way of contrast, the equivalent army organisation developed Staff Requirements, in effect deciding how new technology should be used to meet tactical and strategic needs. In 1918 the existence of numerous technical departments made such a development impossible, although to some extent Captain Dreyer's division filled them. A further reorganisation in 1920 made DCNS responsible for strategic policy and ACNS for tactical policy (including ship and weapon development); the office of Deputy First Sea Lord lapsed. On this basis ACNS was given a Tactical Section (he also had the Air Section).

By this time there was intense pressure to cut the staff as part of the post-war pruning of Royal Navy overheads. For example, DNO continued to see DGD as an unnecessary rival, and there was also a proposal to eliminate DTSD, among other divisions. For a time the new staff organisation survived due to memories of wartime disasters suffered because of inadequate staff work. In a further reorganisation in December 1928, the gunnery division was incorporated into DTSD, and the torpedo division into the tactical division (formerly the tactical section). Until 1939, Staff Requirements, at least for ships, were formulated by the Tactical Division. At that time it was folded into DTSD, which thereby gained full co-ordinating (never exclusive) responsibility for Staff Requirements. Note that these requirements were always a matter of discussion for all interested departments and divisions.[3]

First Sea Lord wore three hats. He was operational chief of the navy, a role made more important in the Second World War because he and his Admiralty staff had access to the ocean surveillance picture created on the basis of code-breaking and other sources of intelligence. He was also head of the Naval Staff, and he was also responsible for many decisions concerning materiel. In 1942 a new office of Deputy First Sea Lord (sometimes styled DFSL) created mainly to handle materiel. He was assisted by a new Assistant Chief of the Naval Staff (Weapons) (ACNS(W)); DFSL and ACNS(W) headed a new Future Building Committee, which largely but not completely shaped wartime ship policy. A Vice Chief of Naval Staff (VCNS) was also created. The Future Building Committee was considered successful. After 1945 it was succeeded by a Fleet Requirements Committee and a Ship Characteristics Committee, both of which were involved in the last cruiser designs.

Until the end of the First World War, Controller or, sometimes, First

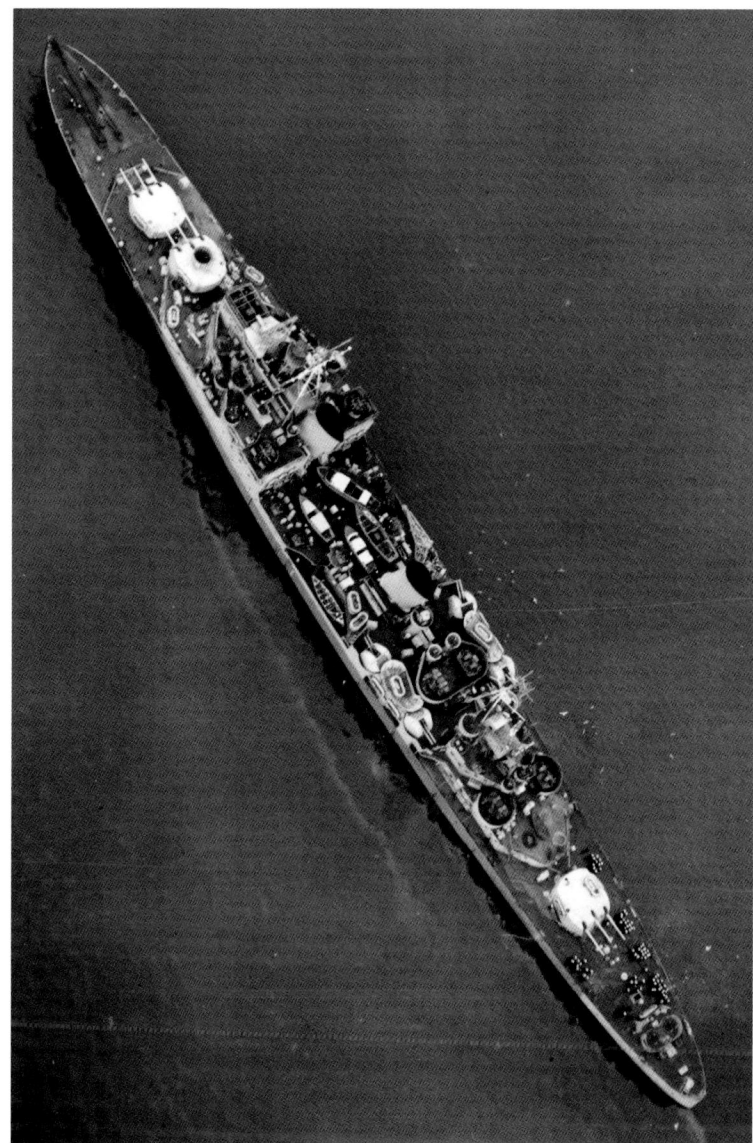

Above: By late 1943, drastic changes were needed to free topweight and space for further additions, particularly for more close-range weapons. HMS *Birmingham* is shown on 23 November 1944 at Hampton Roads after a refit in the United States (Norfolk Navy Yard, July 1944 – 28 November 1944), her 'X' turret having been removed. She retained the two quadruple pompoms atop her former hangar, but the refit added four quadruple Bofors aft plus five single and two twin Oerlikons. She already had two single Oerlikons (seven fitted during a Simonstown refit, 5 March 1942 – 1 April 1942, but five removed during the next refit) and eight twin Oerlikons (added during a Devonport refit, 23 April 1943 – 21 August 1943). Note the short depth-charge track right aft, a standard installation on board British cruisers from the First World War onwards.

Lord or First Sea Lord asked DNC for a sketch design to meet a very simple requirement, most of what would later figure in Staff Requirements being understood as conforming to standard practice. Later Controller generally formulated an initial set of requirements, DNC producing sketch designs to see what was practicable. Formal Staff Requirements typically reflected one such sketch design, although at times more general ones were formulated. Also, once formulated, Staff Requirements were debated within the Admiralty, as is evident in some of the cases described in this book. Controller was thus usually the key figure in defining what a cruiser should be, although he did not always succeed. The most obvious example is Rear Admiral Reginald Henderson's failure in 1936 to convince the Board to adopt a ship armed entirely with 5.25in guns as the 8,000-ton cruiser (the *Fiji*s had 6in guns). The only case in which a DNC took the initiative seems to have been the big cruiser eventually built as the *Hawkins* class. It may have been significant that the DNC involved, Sir Eustace Tennyson d'Eyncourt, came from a major private yard with its own design capacity (Armstrong) rather than from the ranks of the Royal Corps of Naval Constructors. All other DNCs rose through the ranks.

Many of the Controllers represented here later became First Sea Lords, and as such sometimes revived initiatives they had started as Controllers. The controversy over the internal arrangements of the rebuilt 'County' class is a case in point.

The Ministry of Defence (MoD) was created in 1940, but it had little effect on the Royal Navy before its minister Duncan Sandys conducted the 1956 Defence Review. Formal service unification came in 1959, and the Board of Admiralty was formally abolished in 1964 (it continued as a lower-level organisation). Ministry of Defence committees, such as the all-service Operational Requirements Committee, increasingly reviewed navy projects. These changes are peripheral to nearly all the ships described in this book, the only exceptions being the escort cruiser and its successor the command cruiser.

Given the set of Constructors' Notebooks preserved in the National Maritime Museum's Brass Foundry, plus the Ship Covers and papers in the Public Record Office, it has been possible to reconstruct (apparently) virtually all British cruiser designs prepared between 1920 and the last missile cruiser in 1956.[4] The Notebooks, particularly those left by Sir Charles S Lillicrap (head of the cruiser section in the late 1930s, and later DNC) provide insight into the way in which designs were prepared. The key design tool was the summary weight breakdown typically included in the Legend, the summary of ship characteristics presented to decision-makers. Typically the designer began with a target weight and with demands for particular armament, protection and speed. He could calculate (or estimate) armament weight, and therefore the weight of 'general equipment', which depended mainly on personnel and their stores. Displacement suggested overall dimensions, based on previous cruiser practice. Again, based on existing cruiser designs, the designer could estimate how much power was needed. E-in-C could estimate both machinery weight and the dimensions of the machinery box. The constructor could add up what he had and subtract from the total allowable displacement to give available protection weight. For much of the period covered by this book, protection meant a belt and deck over the machinery plus boxes covering magazines and (with reduced thickness) shell rooms. Hence machinery box dimensions gave armour weight or, for a given weight, available thickness. If the combination did not work, the constructor modified dimensions and tried again. Notebooks suggest that a few combinations of dimensions gave a practicable combination, on the basis of which more detailed work began.

The great strength of this technique was that, in the hands of an experienced designer like Lillicrap, it very quickly provided the basis for a cruiser design. It ruled out impractical alternatives. The weakness of the technique was that it did not explicitly account for ship volume. Experience was key, because a designer had a feel for what was wanted. Moreover, as long as ship designs were broadly similar, it was unlikely that a hull of reasonable size would fail to accommodate what was needed along the centreline of the ship: machinery and turrets (superstructure generally fit above a machinery box of reasonable size). If the ship seemed likely to be somewhat tight, the initial designer might add 10 or 20ft to its length, as a surrogate for adding deck space. This practice was unavoidable, because it was difficult at best to estimate deck areas and hence available space. That was done once overall dimensions and weights had been estimated, but it entailed far too much calculation for alternative layouts to be worked out.

Given experience and a long line of ships similarly arranged, designers generally found it unnecessary to work out the lengthwise arrangement of spaces. For cruisers the one exception, until 1939, seems to have been the unconventional aircraft-aft design investigated by Lillicrap late in 1936. However, lengthwise space analysis seems to have been the rule from 1939 on; it was certainly done for the wartime heavy cruiser designs.

British design practices worked because DNC split his organisation into sections, one of which specialised in cruisers (to some extent one might define a British cruiser as a ship designed by the cruiser section). Specialisation is obvious in the Constructors' Notebooks, which rarely show designs of multiple types (except as constructors moved from section to section). Those in the section worked on preliminary and detailed designs, and they also became aware of how the ships they designed performed. By way of contrast, the US Bureau of Construction and Repair was organised according to stages of design, the Preliminary Design section working on all types of ships (submarine design was somewhat more specialised). Contract Design, for example, was a separate organisation. In 1918 Stanley V Goodall, a British constructor (later DNC) seconded to the US organisation, delivered a lecture in which he argued that the British split according to type of ship made for better awareness of overall design issues.

The split by ship type encouraged a section to develop a style of design with implicit emphases. British designers favoured the tightest possible designs, with limited stretch for in-service modification. That became evident during the First World War, when, for example, extra generator power was wanted for more powerful searchlights. Early post-1918 designs, such as the 'County' class, seem to have had more stretch in them, but later inter-war designs were certainly quite tight. The *Fiji* class suffered particularly because they entered service just as major additions, such as radar and many more close-range anti-aircraft guns, were wanted. British cruisers designed after the First World War also seem to have suffered because E-in-C was more conservative than his foreign counterparts. Early reports of Italian practice seemed to justify his relatively bulky boilers and heavy turbines, but US designers produced roomier ships, probably because they had lighter and more compact machinery. During the Second World War DNC was forced to defend his design practices as British officers saw and admired many US designs. The British cruisers designed (but not built) late in the war were far larger than their predecessors, to an extent which shocked many of those defining requirements. The shock of growth was worsened because ships came to be described by their deep load rather than standard displacements, the difference amounting to several thousand tons in a large ship.

Opposite above and below: Off Guantánamo Bay on 17 November 1952, *Sheffield* shows relatively simple post-war modifications, in which light anti-aircraft weapons were partly standardised. At this time the ship had four twin 40mm and six single power-worked Bofors guns; she retained her original pair of quadruple pompoms. She had the standard end-of-war radar suite: Type 281B on the mainmast, Type 293 (target indication) on the foremast, Type 277 (surface search and limited height-finding) on a lattice tower before the foremast, Type 274 on the main battery director, and Type 285 on each of the 4in directors abeam and abaft the bridge. (US Navy photos courtesy of Rick E Davis)

INTRODUCTION

CHAPTER 2
PROTECTING TRADE

When Admiral Fisher took office as First Sea Lord in 1904, the British cruiser fleet included large armoured cruisers intended to work with the battle fleet or to deal with large enemy raiders, medium cruisers for trade protection and station work, and smaller cruisers intended for purposes ranging from Empire defence to linking scouts with the main fleet. Radio obviated many such functions, so that Fisher envisaged a fleet in which battlecruisers would scout and perhaps also form part of a battle line. Through 1909 the only cruisers he built were intended either as destroyer leaders (for independent flotilla operations blockading German destroyer bases) and as scouts for coastal defence destroyers. Because of their destroyer functions, these ships have been dealt within a previous volume devoted to British destroyers.

The *Bristol* class

Work on a 'new *Boadicea*' began late in 1907, six such ships being planned for the 1908/9 programme, including one to be built at the Royal Dockyard, Pembroke.[1] HMS *Boadicea* was essentially a destroyer leader, but the new cruiser was much more powerful. The principal role was understood to be to meet the new German Third Class Cruisers. That meant a variety of roles. For the fleet, it meant backing up blockading destroyers against a stronger German cruiser threat. Earlier, slower German cruisers would have been ineffective against fast British destroyers, but the newer ones could run down the British destroyers. It also meant trade protection, as the German ships could operate against British trade from the German colonies. During the First World War

Glasgow is shown with funnels raised, and with the pre-1914 arrangement of searchlights aft on a bandstand.

several of them did just that, *Emden* and *Königsberg* becoming famous in that role. Probably the ships involved were the first German turbine cruisers, the prototype *Stettin* and then *Dresden* and *Emden*, all armed with ten 4.1in/40 guns, displacing around 3,300–3,600 metric tons. Design speed was 23–24kts, increasing to 25.5–26kts in the next (*Kolberg*) class. Initial instructions (2 November 1907) were to design a 4,000-tonner capable of 25kts, armed with twelve 4in guns, with 50 per cent more fuel (coal and oil) than a *Boadicea* (the latest Scout), with a protective deck but no side armour, and with four months' stores. There was no apparent interest in higher speed to overmatch the latest German cruisers. DNC could meet these requirements on the desired displacement, with the same protection as *Boadicea* (½in deck throughout with 1in slopes over the machinery, and a 4in conning tower).[2]

The Board provisionally approved the 410-foot version of the 4,000-tonner, but DNC asked for more options with thicker armour decks: (A) with 1in flat and 1½in slope only over machinery and magazines (4,150 tons) and (B) with 1in flat and 2in slopes (4,300 tons). A detailed drawing showed a 420ft (pp) x 44ft x 14ft 9in ship (4,300 tons). The design showed two 4in guns side by side at each end plus three in the waist on each side, blocked from firing across the ship by the boiler casing. The new ships were rated as Second Class Protected Cruisers because they were powerful enough to fight the last British cruisers with that rating, the *Diana* class. These ships were too big to build at Pembroke (a ship had to be docked within six months of launching, and the yard had no dock large enough), so in January 1908 it was decided that one of the six

HMS *Newcastle* shows the effect of short funnels in this 19 August 1910 photograph.

1908/9 cruisers would be a smaller repeat *Boadicea*, the others being built at private yards.

It was proposed to replace the deck tubes with a submerged torpedo room.[3] That in turn cleared deck space for another two 4in guns, for a total of fourteen. These changes were decided early in January 1908. DNC sketched a 4,600-tonner with the desired heavier armour (1in flat, 2in slope), the larger gun battery, and the submerged tubes.

Controller considered a 4in battery on a ship this size weak; for a few more tons she could have 6in guns at the ends plus the eight broadside 4in. The 6in was considered the natural gun for a relatively small cruiser, because it was the largest whose shell could be handled by a single man, hence the largest which did not require a powered hoist and elaborate loading arrangements.[4] That also made the 6in the natural armament of armed merchant ships and other raiders which could be commissioned in significant numbers in wartime.

In mid January DNC ordered Legends prepared for this alternative, as well as for four 6in (guns paired alongside each other at the ends) and four 4in, and for six 6in and eight 12pdrs (3in guns: 6in paired at the ends, with another pair of 6in guns, one on each side abaft the break of the forecastle, and the 12pdrs on the broadside abaft them).[5] Although on 18 January the Board approved the version with single 6in at the ends (4,400 tons), Controller asked DNC to work out a slightly larger ship with two more waist 4in guns and a protected ammunition supply (3in tubes at the ends and in the waist). DNC thought that would add about 250 tons.[6] In addition to the 6in and 4in guns, the ship was to be armed with a Maxim machine gun (later increased to four).

On 17 February Controller told DNC (Philip Watts) to pursue this 4,650-ton design. On 13 May he added that all of the guns should protected with 3in shields (in previous designs the guns were not protected at all); estimated displacement rose to 4,700 tons. Further

BRITISH CRUISERS

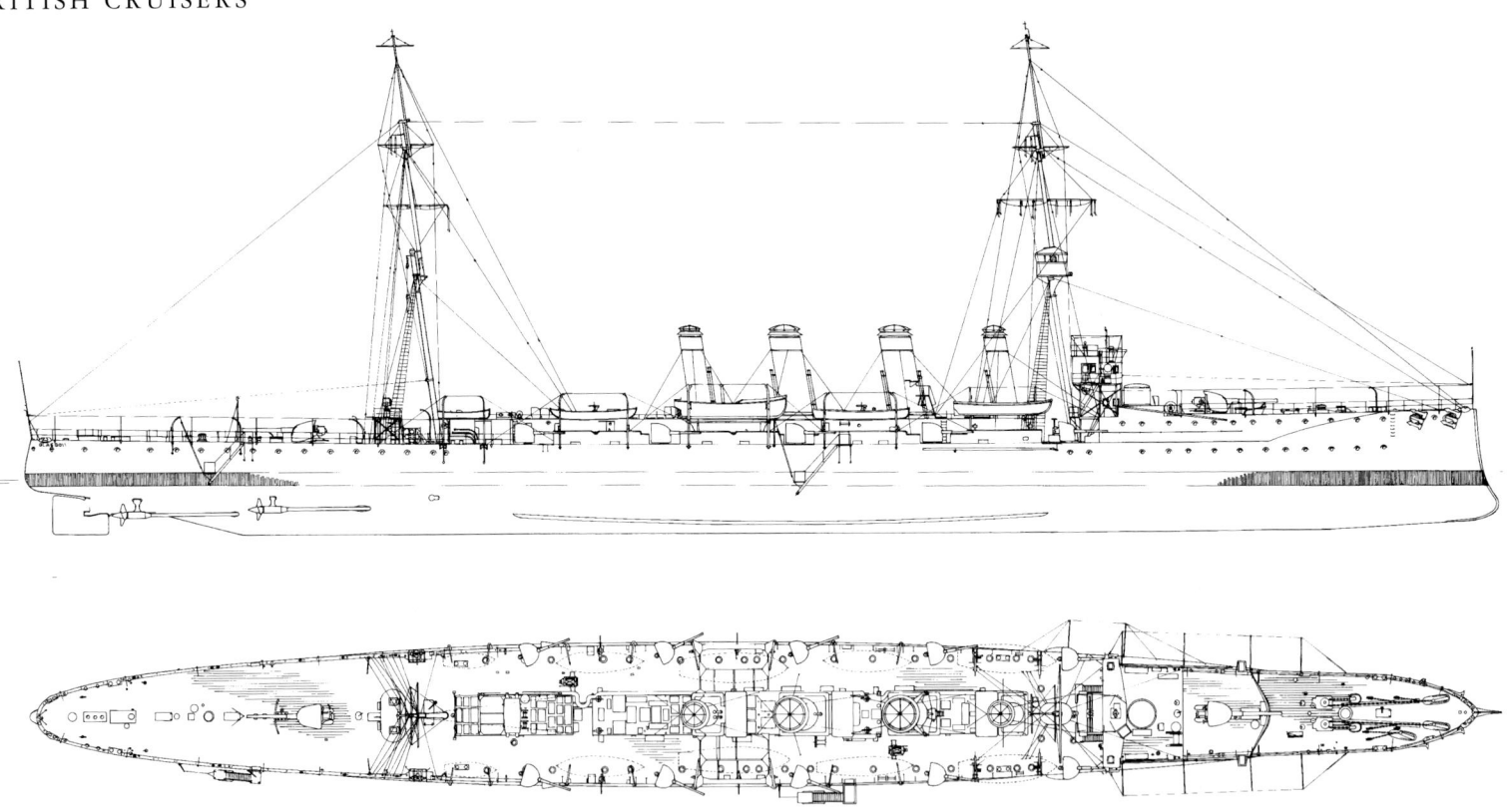

Above: Glasgow was a *Bristol* class cruiser. The 18in underwater broadside tubes were mounted well aft, abreast the mainmast, the starboard tube being mounted forward of the port tube. The elevation drawing was simplified somewhat by excluding the radio aerial rig, two multi-strand arrays rigged from the upper yards. It also omits the extensive coaling rig. The poles extending horizontally from the sides of the forecastle (and on the port side abreast the sick bay) supported 'sun screens'. When not in use they were stowed beneath the 35ft steam cutter. The two 16ft dinghies were stowed atop the 35ft cutter when the ship was at sea. The main changes during the First World War were slight enlargement of the fire-control platform (but the pole foremast was retained) and the addition of one 3in anti-aircraft gun. Several ships of this class had their main topmasts removed in wartime. In addition to the usual small arms for shore parties, the ship had fifty cutlasses stowed in the overhead of the passageway abreast the captain's cabin. (A D Baker III)

proposed detail changes would add another 120 tons. They included installing a 9ft rangefinder in a control position at the head of the foremast (with the guns having follow-the-pointer sights); fitting the 6in guns to have 1° rather than ½° depression; fitting the after 4in guns so that they could fire right aft; mounting four machine guns (Maxims) instead of one; adding a second searchlight (projector) on the after platform or engine room hatch; adding a 6ft screen (with open ports) between the upper deck guns, and across the deck in wake of the engine hatch; and installing magazine cooling (already provided in the *Boadicea* class). These changes would require another foot of beam. Watt particularly disliked the proposed screen, which he considered a possible shell trap which increased target area. The weight involved could be used instead to thicken the conning tower from 4in to 6in and also to thicken the deck over the steering gear from ¾in to 1½in. In rough weather the screens could trap water and this topweight would menace the ship. The screens were approved as a means of protecting guns on the off-side of the ship from the blast of guns when trained well off the beam; the gunnery school (HMS *Excellent*) estimated that without them the maximum training arc for broadside guns would be only 45°. Ultimately the conning tower was thickened to 6in. It was decided to save money by making the protective deck of nickel steel rather than non-cemented armour, experiments having shown no advantage for the latter.

Right: The *Bristol* class had funnels raised to reduce smoking; this also improved boiler draught. HMS *Gloucester* is shown shortly before the First World War, with identifying funnel bands. Note the marked difference between the large 6in guns at the ends and the broadside 4in guns. The objects visible on the compass platform are an open chart table and a rangefinder.

Watts considered the ships better subdivided than most unarmoured ships, and pointed out that all the main hold bulkheads were unpierced. Machinery spaces were redesigned between July and September 1908 for better subdivision. The engine space was divided into five compartments, three side by side forward of two spaces. The centre one of the three side by side contained the turbines driving the inner shafts, the turbines driving the outer shafts being in the two outer spaces. The other two compartments housed pumps and condensers. This arrangement protected the ship more than in the past against being disabled by a single shell penetrating the machinery spaces. The 'tween deck spaces above the armoured deck was much more subdivided than in the past, making it less likely that the ship would lose stability or buoyancy due

Above: HMS *Gloucester* was one of the initial group of 'Town' class cruisers designed partly to protect British trade, a reversion to an earlier cruiser function. She is shown as completed, with short funnels which smoked her bridge.

to riddling of her side. Of forty-one separate watertight compartments between the upper and protective decks, twenty were coal bunkers, three were offices and officers' cabins, six were crew accommodation, and twelve were washplaces, store rooms, etc. DNC argued that although any unarmoured ship was more vulnerable to loss of stability, he had considerably reduced that risk. By late September, weight saving in the detailed design had made it possible to increase conning tower protection from 4in to 6in, and to provide 2in over the steering gear.

The version with two 6in and ten 4in guns was reported to the Board as the 'New 2nd class Protected Cruiser', the Legend dated 30 May 1908 and submitted in June 1908. The new ship followed *Boadicea* in having engine rooms arranged so that either could operate if the other were bilged.[7] A different arrangement was being considered to this end. Required speed was 25kts, roughly that of a battlecruiser. At the outset, the Board clearly called for range, since it asked for 50 per cent more fuel (coal and oil) capacity than that of the earlier ship (about two and a quarter that of *Amethyst*, the last conventional small cruiser, and nearly four times that of the destroyer-leading Scouts). Compared to *Boadicea*, the new ship was given a block of coal stowage forward of the machinery, for extra protection. Similarly, her torpedo tubes were placed below the waterline, where they were considered protected, rather than unprotected above water. Gun armament was changed from an all-4in battery (and only six guns), suited to fighting or supporting destroyers, to a pair of 6in guns at the ends plus ten 4in guns along the sides. In contrast to *Boadicea*, all the guns were shielded. The new ship was far larger, 4,800 tons rather than 3,300 tons. As the designation applied, protection was limited to an armour deck. The Board approved this design on 7 July 1908.

Five of these *Bristol* class were built under the 1908/9 programme. They and their immediate successors were called the 'Town' class. Approval (on 16 January 1908) was subject to the demand that the cost of one repeat *Boadicea* and five of the new cruisers should not exceed that estimated for six improved *Boadicea*s. Estimated cost was £415,000, but shipbuilding conditions were bad, so builders bid low. These and the later versions of the design all had twelve boilers in three boiler rooms with four funnels, the middle pair being wider because they combined the uptakes from the after end of one boiler room with those from the fore end of the adjacent one. Each set of uptakes served two boilers set side by side, the stoking space in each boiler room being between two rows of boilers. Thus the foremost funnel was at the forward bulkhead of No. 1 boiler room, the aftermost at the after bulkhead of No. 3 boiler room. One of the five ships, HMS *Bristol*, had two-shaft Brown-Curtis turbines instead of the four-shaft Parsons turbines in the others. She had two engine rooms in tandem, each containing one turbine with its condenser on the other side of a longitudinal bulkhead (these bulkheads were on alternating sides, to suit the turbines driving the port and starboard shafts).

The *Dartmouth* class

The following year the Board asked that the 4in guns be replaced with 6in, for a uniform battery of 6in. On 28 January Controller asked for ships with six or eight guns, equipped as private ships or flagships, all of which were to have 21in rather than 18in torpedo tubes if possible. The eight-gun alternative was chosen; estimated displacement was 4,950 tons. Blast screens would be omitted. Legends submitted on 3 February showed a 430ft, 4,990-ton ship or a similarly-armed 440ft, 5,280-ton flagship; 22,000shp engines would drive each at 24.75kts. Watts proposed reducing fuel at deep load to 1,450 tons to compensate for extra armament weight (150 tons) without enlarging the ship. In each case armament included two submerged tubes for short 21in torpedoes. To get enough breadth for them, the tubes were moved to forward of the forward boiler room, the lower athwartships coal block (for protection) being eliminated. Watts considered the new cruisers the smallest which could accommodate the new torpedoes.

As the forward broadside 6in guns might be affected by spray, Watts relocated them to the lengthened forecastle. All three forecastle guns could fire right ahead, and the three after guns could fire right astern; the broadside was five 6in. More boilers (an additional boiler room) would be needed. In redesigning the ship, Watts lengthened the forecastle to provide sufficient space for all officer cabins, the lower deck being left clear for the crew. However, the Board had recently called for sufficient space for 15 per cent supernumeraries, and Watts wanted to know what complement he should allow for. In an accompanying note, Watts pointed out that to maintain the existing speed of 25kts, it would be necessary to add at least 400 to 600 tons. Alternatively, it would be far less expensive to retain the existing machinery and accept the loss of perhaps a quarter-knot. This might be no more than a nominal sacrifice, as many turbine ships were exceeding their rated horsepower and speed on trials (the state of turbine design was primitive, and reliable instruments, such as dynamometers, to measure output did not exist).

Controller proposed in mid-April that the eight-gun design be adopted for the 1909/10 ships. They would be fitted as flagships, accommodation being provided under an extended forecastle. Bunker capacity would be reduced by 150 tons as weight compensation. As Watts had recommended, the machinery would be unchanged, the slight loss of speed being accepted.

The completed design was submitted on 30 July 1909 and approved by the Board that day; Watts hoped to have building drawings ready for bidders by the end of August. Protection matched that of the *Bristol* class. Due to the loss of the lower coal block, fuel capacity was reduced from 1,600 tons to 1,500 tons, but 'this is still a very large fuel supply for this class of vessel'. Length was the same as that of the *Bristol* (430ft), but beam increased from 47ft to 48½ft and draft increased by 3in. At this stage displacement was given as 5,250 tons. On the basis of the cost estimate for the *Bristol*s, Watt estimated that the new ships would cost £440,000, but shipbuilding conditions were still poor and he expected some reduction (though not as much as for the *Bristol*s).

Four of these *Dartmouth*-class cruisers were built. As in the previous

The *Weymouth* class introduced an all-6in main battery. *Weymouth* is shown. Note her pronounced tumblehome in the waist.

PROTECTING TRADE

class, one ship (HMS *Yarmouth*) had two-shaft Brown-Curtis rather than four-shaft Parsons turbines. Although turbines were rated at 22,000shp, they often developed much more: three ships developed nearly 24,000shp, and *Falmouth* developed 27,900shp and attained 26.8kts. *Yarmouth* trials showed that on two shafts she had about the same performance as a four-shaft ship. Ships turned out lighter than expected: against the Legend figure of 5,250 tons, *Falmouth* displaced 5,040 tons, *Weymouth* 5,044 tons and *Dartmouth* 5,076 tons, with similar savings at deep load.

While these designs were being prepared, the British government was negotiating with major colonial governments in hopes that they would help provide naval forces for Imperial defence.[8] The British government would much have preferred that the colonies contribute to an Empire Fleet, but the Australian and Canadian governments held out for their own navies. The Admiralty pressed for each major unit in the Empire to create a 'fleet unit' which could help protect imperial trade, but which could also be combined with other fleet units to create a fleet for distant areas, i.e. for operations in the Pacific. It argued that the local defence forces the Dominions (particularly Australia) contemplated would provide only a very limited degree of mutual defence, because they could never affect the main threat of commerce raiding (the Australians originally contemplated a cruiser squadron to deal with raiders, backed by local defences). The fleet unit envisaged by the Admiralty comprised a battlecruiser, three *Bristol*-class cruisers, six destroyers, three submarines, and necessary supporting auxiliaries. At a meeting on 10 August 1909 Admiral Fisher explained the Empire or Pacific Fleet concept, and also argued that the core of any fleet unit had to be the battlecruiser, as without her the lesser units could not be very effective against a commerce raider. Ultimately only Australia bought a fleet unit.

No one explained how the battlecruiser and the cruisers would work together against commerce raiders, perhaps because it was so bound up with Admiral Fisher's intelligence-based scheme of operations. It is, however, reasonable to imagine that by 1909 he had accepted that a scouting line would be far more likely to find a raider than a single ship, the light cruisers detecting a powerful raider and falling back on the battlecruiser so that she could deal with it. The Australians opted for

HMS *Weymouth* is shown as built. This drawing omits the two eight-strand radio antennas slung between the upper yards. When coaling, the ship rigged a horizontal cable between the funnels and the masts to support coal-bag handling. When preparing for battle, additional stays could be rigged to support the foremast. Yards are omitted in this drawing to make it possible to indicate the complexity of the standing rigging for the masts. The two cruciform devices carried in cage racks on either side of the mainmast were quick-release lifesaving buoys provided with electric lights. During the First World War the foremast was altered to a tripod to support an enlarged fire-control top (ca. 1917). One 3in anti-aircraft gun was added in 1915 between the second and third funnels, a second being added on the quarterdeck in 1918, at which time the main topmast was struck. A flying-off platform was erected forward of the bridge in 1918 and removed in 1919. During a 1924–5 refit the compass platform (from which the ship was conned) was enlarged and extended forward, and the gun control platform equipped with a 'gun direction tower' atop an enlarged fire-control platform. *Weymouth* and her sisters retained their conning towers throughout their careers. Broader and slightly longer bilge keels were fitted at some point during the First World War. (A D Baker III)

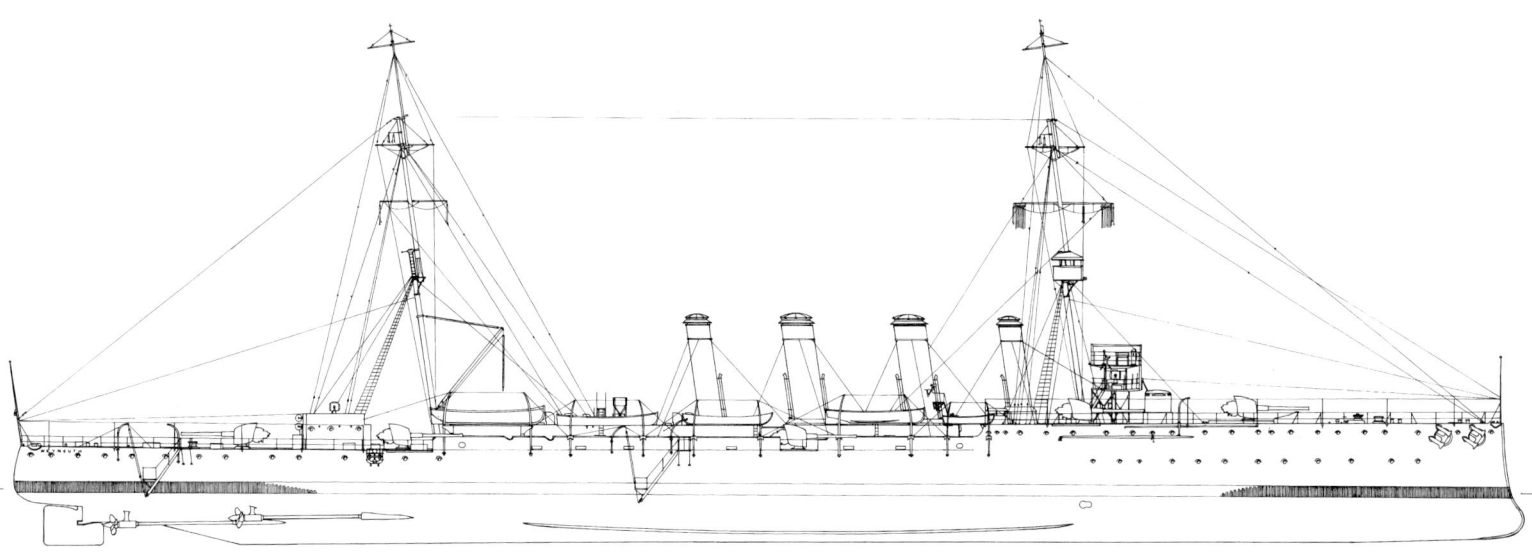

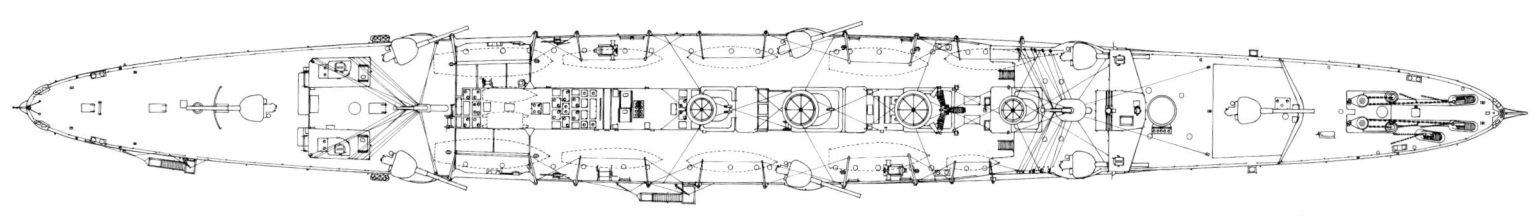

BRITISH CRUISERS

'Improved *Bristol*s,' which meant modified versions of the later *Chatham* class. Initially they hoped to complete all three by August 1912, which was when the *Weymouth*s were to be completed for the Royal Navy. Then the Australians decided to build two ships in the United Kingdom and the third in Australia, accepting the delay in order to develop their own industry.

The Canadians considered two alternatives: four improved *Bristol*s, a *Boadicea* and six destroyers; or three improved *Bristol*s and four destroyers. The Admiralty agreed to provide all possible assistance (April 1910), and the relevant specifications and drawings were provided in February 1911. Then the project died.

Right and below: Weymouth is shown in wartime, before the flying-off platform was fitted, with her mainmast cut down. These photographs were probably taken in 1917, after the tripod had been mounted but before the flying-off platform was added. Weymouth and Dartmouth had an additional 3in anti-aircraft gun on the quarterdeck, but it is not visible here. Weymouth served in the Mediterranean in 1915–16, when her mainmast was cut down, then with the Grand Fleet in 1916–17, then back in the Mediterranean in 1917–18, serving as flagship of the 8th Light Cruiser Squadron.

Above and below: Weymouth is shown in Malta in 1920, having been refitted in 1919–20 after having been torpedoed by the Austrian submarine *U-28* off Durazzo on 2 October 1918. During this refit her mainmast was restored to its previous height, presumably to provide her with increased radio range. The platform below the spotting top was for searchlights.

British Cruisers

The *Chatham* class

On 6 August 1909, while bidders' packages for the 1909/10 cruisers were being prepared, Controller issued instructions for the 1910/11 ships. He wanted them to have a main deck if possible (i.e. to be flush-decked) and to be good seakeepers without adding much size. Improved seakeeping was later explicitly connected with trade protection in distant seas. As in the previous year, horsepower should not be increased (speed should not fall below 23kts). The ships would not be fitted as flagships, and bunker capacity (i.e. range) should not be reduced. Displacement should not exceed 5,500 tons, and cost should not exceed £350,000. DNC pointed out in November that to hold down size he wanted to stay fairly close to the *Bristol* design. A flush-decked cruiser would come close to the old *Challenger*, but enlarged to provide 23kts or more. He offered designs for a ship with a forecastle deck extending for two-thirds of her length, carrying all but the after three 6in guns, and serving as the boat deck (to avoid a raised superstructure whose topweight would require increased beam). Armament, bunker capacity

Below: HMAS *Sydney* (*Chatham* class) soon after completion (about 29 August 1912). The obvious change from the previous class was the new bow profile (the below-water profile was the same).

Below: HMAS *Melbourne* displays standard First World War modifications, including a flying-off platform forward (in *Melbourne* and *Sydney*, but not in *Brisbane*), a tripod foremast topped by a director, and a long-base rangefinder abaft the fourth funnel. This undated photograph must have been taken soon after the war, since the flying-off platform (and conning tower) were removed when the ship returned to Australia. As completed, the ship had two 3pdr guns on quarterdeck level abaft the break of the forecastle, but by this time they had been raised to forecastle deck level. The searchlights amidships and aft were enlarged to 36in diameter.

Below: HMAS *Melbourne* is shown late in 1921, the flying-off platform and the conning tower beneath it having been removed. The 3in anti-aircraft gun in its bandstand is visible just abaft the long-base rangefinder, and the two 3pdrs are visible near the break of the forecastle. (Photo by Allan C Green via State Library of Victoria)

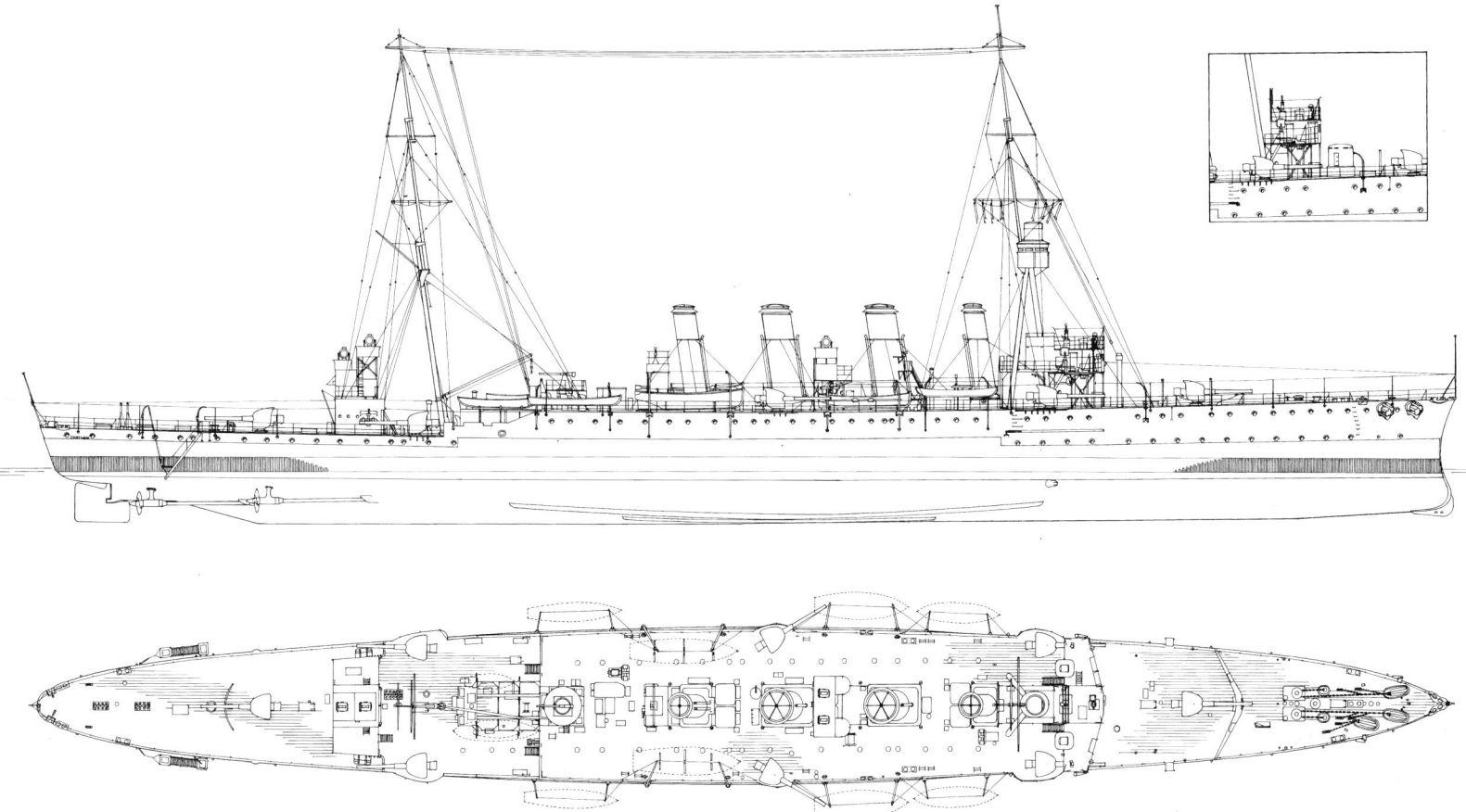

HMS *Chatham* is shown as fitted in January 1921 (the scrap elevation view shows her bridge and conning tower before the 1916 modification when she was given a tripod foremast to support an enlarged fire-control top forward). The director tower atop the fire-control top was added in 1920 when she was refitted to serve as flagship of the New Zealand Division of the Royal Navy (11 September 1920 through 1924). The after searchlight tower and the searchlight platform between the second and third funnels were added during a 1916 refit, and in 1917–18 a flying-off platform was erected forward of the bridge (removed 1919, when the ship went into reserve). Also in 1916 the bridge was extended aft to provide an admiral's sea cabin abaft the pilothouse, and a 24in searchlight was added on the centreline abaft the new cabin. The 3in gun was added in 1915. Four single Hotchkiss 3pdr QF guns were removed during the 1920 refit, when a portion of the after coal stowage was converted to additional oil fuel stowage. Also removed in 1920 were the ship's mine countermeasures paravanes and their handling gear. The accommodation ladder position amidships for non-rated personnel was moved to the port side aft. (A D Baker III)

and speed would match those of the *Bristol* and the subsequent class; estimated speed (with the same machinery) was 24.75kts. Alternative sketch designs showed officers' accommodation forward, as in the previous cruisers, or aft (Commodore (T) was reporting on his experience in HMS *Boadicea*, which had her officers forward). DNC favoured the *Boadicea* arrangement. DNC thought he could meet the cost limit.

The designs submitted in November were labelled 'New Colonial Cruiser', suggesting a ship intended to operate on a foreign station. Sketches embodying various modifications were dated 20 December 1909. They included a watch cabin with a sleeping berth on the bridge.

The experimental firing against HMS *Edinburgh* demonstrated 'once again' the value of thin armour against high-explosive (HE) shell, which were expected to be the main means of attack against unarmoured or lightly-armoured ships. The firing also showed the value of an armoured deck, a somewhat unexpected outcome since there had been no tests against armoured decks for many years. Well after this firing a target was used to compare the value of side and deck armour against 6in shells. This later experiment showed that side armour was much more valuable against HE shells, and that against powder-filled shells side armour gave no great advantage but was no disadvantage. Since almost all foreign navies had adopted HE shells, the HE experiments had priority. Watts therefore proposed developing the 1910/11 design on these lines, the planned ships for colonial navies (Australia and Canada) being of the same type.

It was then pointed out to DNC that by reducing the first and upper decks to 15lbs (⅜in), the ship could be given side armour, extending up from the lower edge of the protective deck to the edge of the upper deck over the length of the machinery, and from the protective deck to 3ft below the upper deck at the ends. On 5,300 tons the ship could have 1¾in side armour, and on 5,400 tons she could have 2¼in, for total thicknesses (including shell plating) of 2⅜in and 3in respectively (and 2in and 2½in at the ends). The displacement thus far put forward was 5,400 tons, but 5,300 tons was probably better for structural arrangement. Asked what thickness he could provide if the protective deck was thickened from ⅜in to ⅝in, DNC offered 2¾in rather than 3in, and 1¾in at the ends instead of 2in. Yet another possibility (considered in February 1910) was to thicken the lower deck only abaft the forecastle (adding 30 tons); ½in would be removed over the same length on the side. Watts liked the idea, because it would provide more protection where it was needed, further forward. The design finally submitted on 12 April 1910 showed a 3in belt extending to the upper deck over the whole length of the machinery spaces, with a 2½in belt forward to 3ft below the waterline, and a 2in belt aft instead of a thick protective deck. The lower deck would now be ⅛in thick, increased to ¾in abaft the engines, and to 1½in over the steering gear. In a history of recent British cruiser design of October 1918, DNC pointed out that the change to a thin belt had been of considerable wartime value. It led directly to the

Below: HMAS *Brisbane* is shown post-war. The searchlight platform below her fighting top distinguished her from her sisters *Melbourne* and *Sydney*. (Photo by Allan C Green via State Library of Victoria)

Bottom left: HMAS *Brisbane* outside Honolulu Harbor, 8 August 1928, late in her career. (US Navy photo courtesy of Rick E Davis)

Bottom right: HMS *Southampton* is shown in 1919. Note the concentration dials on the after sides of her spotting top and the after superstructure block, both facing aft so that other ships can see the range at which she is firing. (Perkins)

use of the side armour as part of the hull strength of the later small light cruisers of the *Arethusa* class.

Length between perpendiculars matched that of the earlier classes (430ft), but these ships had the above-water part of their bow raked rather than ram-form, retaining the below-water bulb (not really a ram). Displacement was 5,400 tons, a growth of about 150 tons compared to the earlier ship. Fuel matched that in the *Weymouth* class (650/1,500 tons), and rated speed was 24.75kts with power increased to 25,000shp.

The Royal Navy received three of these *Chatham* class cruisers; three more were built for the Royal Australian Navy (*Sydney*, *Melbourne* and *Brisbane*). Of the British ships, *Southampton* had Brown-Curtis turbines driving twin screws, the others Parsons driving quadruple screws. Of the three Australian ships, the last was built locally, at Cockatoo Dockyard. Construction was delayed by the late arrival of material from England.

The *Birmingham* class

Three more ships were included in the 1911/12 programme (*Birmingham* class). They were very similar to the *Chatham*s, but with nine rather than eight 6in guns (two guns on the forecastle side by side instead of one on the centreline, but a single gun aft on the centreline as in the earlier class). These ships introduced a new shorter, hence more easily manoeuvred, 6in gun (45-calibre Mk XII rather than 50-calibre Mk XI). The gun mounting also offered greater elevation. All had quadruple screws and were rated at 25,000shp. The Australians bought a fourth ship, which they built in Australia as HMAS *Adelaide*. Her hull and machinery were all made in Australia, the armament being imported from the United Kingdom.

In 1914 DNC wired the Australian government proposing a new design, the *Brisbane* design being four years old. *Brisbane* had eight 6in guns with 3in shields, but the newer *Birmingham* had nine, and the new light cruisers (*Arethusa* and *Calliope* classes) had spray shields rather than heavier protection to their guns.[9] Six 36in searchlights should replace the four 24in of the earlier design. Fire control and torpedo air compressor arrangements should be modernised (the earlier design lacked both a fire control platform aft and a gyro adjusting room for torpedoes, and its compressor produced 2,600psi rather than 3,000psi air). The outer thickness of the side armour should be worked longitudinally rather than vertically. The steering gear should be covered by a curved (turtle) deck rather than by a flat deck plus side armour. Parsons reaction turbines (four shafts) should give way to Parsons impulse turbines on two shafts with geared cruising turbines, and twelve boilers all burning coal or oil should give way to six dual-fuel boilers and four oil-burning boilers (in fact the ship had twelve boilers, like earlier 'Town' class cruisers).[10] There should be two sets rather than one of magazine cooling machinery. Fuel should be changed from 1,240 tons of coal and 260 tons of oil to 750 tons of coal and 600 tons of oil. This would meet Commonwealth requirements, but the change probably reflected both the greater efficiency of oil burning and the greater efficiency of geared cruising turbines. In view of experience, the captain's accommodation should be moved from right aft on the lower deck to further forward on the upper deck. Wooden lower masts should give way to steel ones which could carry searchlights. Overall, space and weight for machinery and fuel would be traded for equipment and armament. Displacement and form would roughly match those of *Brisbane*, but internal arrangements would be quite different.[11]

The 'Towns' were the last classic cruisers the Royal Navy built before

HMAS *Adelaide* in March 1939 after conversion to oil-burning. (Paul Webb)

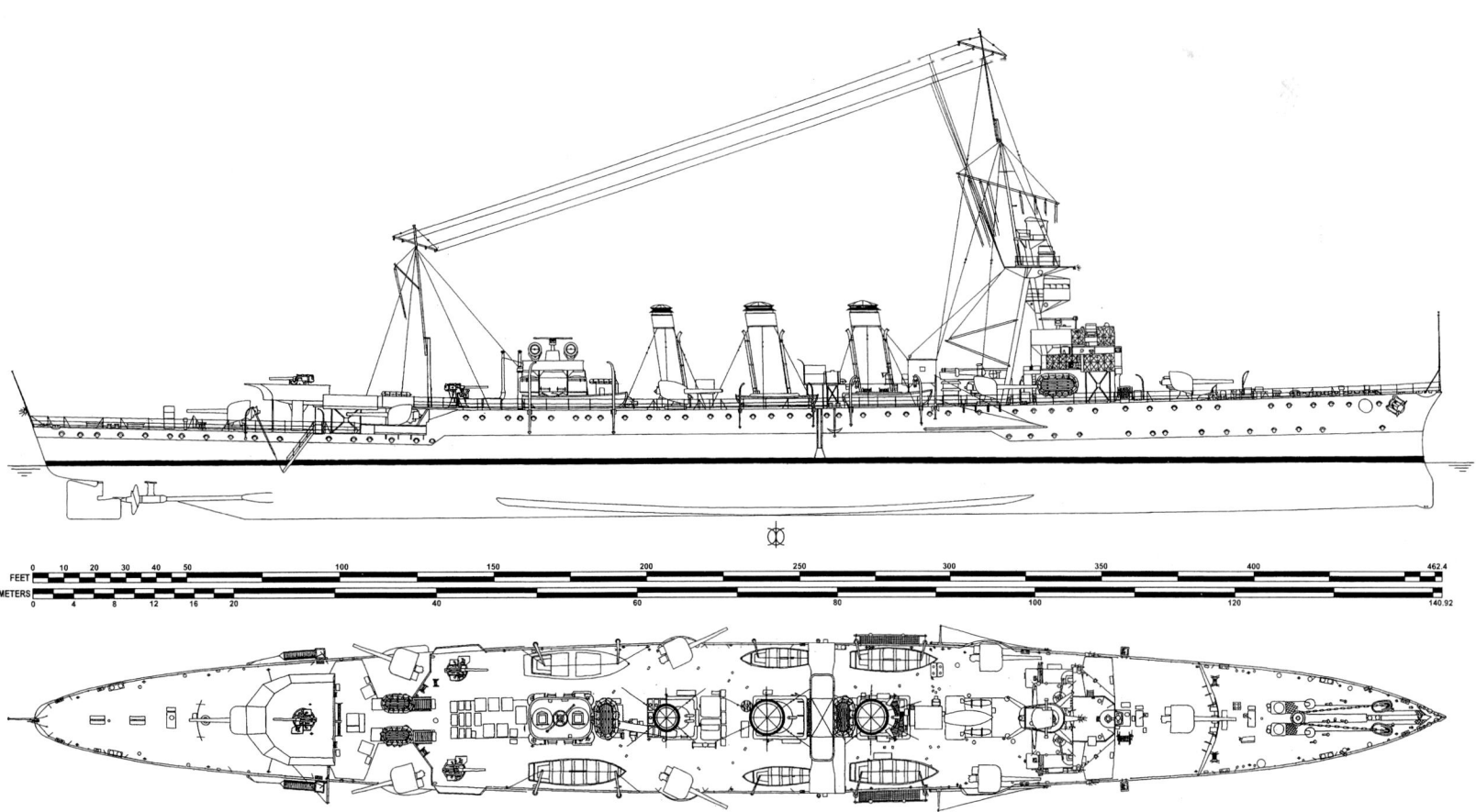

BRITISH CRUISERS

the end of the First World War, in the sense that they were intended for long-range independent deployment. During the war they served as the Grand Fleet's scouts, in the 'A-K Line' deployed ahead of the battleships. This scouting line became a fixed feature of post-First World War Royal Navy fleet formation. These large cruisers were succeeded by a series of much smaller cruisers best described as super-destroyers (with armour) or destroyer-killers.

Top left and right: HMS *Nottingham* (*Birmingham* class) is shown soon after completion. Note the paired forecastle 6in guns and the searchlight platform above the bridge.

Below: HMS *Lowestoft* was unique in the Royal Navy *Birmingham* class in being completed with a tripod. A similar mast was fitted to HMS *Birmingham* in 1916–17, but *Nottingham* was probably lost (19 August 1916) unmodified. All three were given 3in (20 cwt) anti-aircraft guns, abaft the after funnel, in 1915. Neither surviving ship was fitted with a flying-off platform.

HMAS *Adelaide*

As the Australians prepared to lay down their fourth cruiser in August 1915, Controller asked for armament alternatives using 9.2in and 7.5in guns. Controller thought, in view of 'recent trend of opinion' (presumably in connection with discussions of what became the *Hawkins* class) that the main alternatives (with the same weight as nine 6in) were three 7.5in and two 9.2in, of which the last could probably be ruled out. Controller asked DNC what four or five 7.5in guns would require, in terms of greater displacement, assuming all guns on the centreline and 300 rounds per gun (rather than 150), as the ships would probably fight at long range, with great waste of ammunition. DNC prepared tracings of both alternatives. A ship mounting four 7.5in guns would probably displace 1,100 tons more and would probably need another 2,500shp (but it seemed that no extra machinery weight would be involved). Adding another 7.5in gun would add another 460 tons, much of it for additional hull (240 tons). The four-gun ship would probably be 450ft long (vs 430ft for *Birmingham*), and the five-gun ship 465ft long. On 27,500shp both the enlarged ships would make 25.5kts. DNC's very rough sketches showed single mounts in 'A' and 'B' and 'X' and 'Y' positions. A fifth gun would have been worked in at the after end of the forecastle at roughly the same height as 'X' gun, which was on a free-standing barbette. Nothing came of this exercise, but in effect this was the first approach to what became the 'Improved *Birmingham*' or *Hawkins* class. The most striking difference from the *Hawkins* was the use of enclosed gunhouses (as in the B designs described below) rather than open shields. The Australian designs also offered considerably lower speed.

The Australian design was updated as she was being designed, so that in effect she reflected ongoing changes in British cruiser design practice. In June 1915 Controller decided that she would have a tripod mast and director control, then being fitted to new British cruisers, and he asked DNO to decide whether her 6in guns should have the 30° elevation then being used. It was clear that if the ship was to be laid down, as desired, in August 1915, she could not embody an entirely new design, but would have to be largely a repeat *Birmingham* brought up to date. Her hull could be modified to provide more oil fuel, director fire-control and the new way of working in side armour (which would be HT steel instead of the nickel steel used in *Birmingham*). As completed, *Adelaide* had the tripod foremast, with spotting top and director, added to British cruisers in wartime. E-in-C thought that the new boilers would provide full power using Australian coal. Machinery repeated that of HMS *Lowestoft*, except for some auxiliaries, the middle (of three) boiler rooms burning oil only. During design, the two coal-burning boiler rooms were lengthened (total 4ft), adding 20 tons. The ship had to be lengthened slightly, as machinery could not be shifted aft, and any shift forward would cause a trim by the bow. Coal was omitted from the upper bunkers over the after engine room and from the lower bunkers abreast the middle boiler room, stowage being provided for 860 tons (instead of 1,155 tons in *Birmingham*) and for 500 tons of oil (instead of 236 tons).[12] These were not quite the figures initially suggested. A 3in high-angle (HA) gun was added on the deckhouse aft, its magazine displacing the spirit room aft. The after control platform was enlarged to take a rangefinder, a provision made to fit the standard compass on the roof of the control platform.[13] Late in November 1915 DNC estimated that the new Australian cruiser *Adelaide* would displace 5,557 tons, compared to 5,441.8 tons for *Lowestoft*.

Adelaide was completed in 1922, and in November 1923 the Australian Naval Board decided to modernise her to burn oil fuel only, to have her guns on the centreline, and to have central storekeeping. By this time the Australians planned to build new oil-burning cruisers in the United Kingdom. Converting *Adelaide* during her planned visit to England would give Australia an all-oil burning squadron by 1931

HMAS *Adelaide*, the modified *Birmingham*, as completed, with a new-type bridge and a short mainmast. Unlike earlier cruisers, she carried her director below and before her spotting top, rather than atop it. Note the concentration dials on foremast and aft. Near-sisters *Birmingham* and *Lowestoft* also had their bridges rebuilt, but they carried their directors atop their spotting tops, with a searchlight platform below the spotting top.

BRITISH CRUISERS

Above: HMAS *Adelaide*, by then the last remaining British Commonwealth cruiser of pre-1914 design, as modernised in 1938–9 at Cockatoo Island. Her foremast has been rearranged to give her an HA director at its top, the 6in spotting top being brought lower down (the 6in director was not moved). Three 4in anti-aircraft guns were added, one on the centreline to superfire over the after 6in gun, and two at the after end of the forecastle. One of the two forecastle 6in guns was landed. The ship was converted to all-oil fuel, the two forward boilers and the forefunnel being removed. The sheet-anchor was removed, its hawse-pipe plated over. *Adelaide* recommissioned in this form on 13 March 1939, but the next month was laid up, her crew earmarked for the new cruiser *Perth*. She recommissioned in September 1939. (Photo by Allan C Green via State Library of Victoria)

Below: Adelaide at sea in wartime, before her 1942–3 Sydney refit, and before she was fitted with radars. She still had her original curved gun shields.

instead of by 1937, as originally planned. The Australians proposed replacing the existing twelve boilers with four larger ones. She would save considerably on complement during her remaining lifetime of fourteen years. The result would be more compact machinery spaces and reduced number and spacing of funnels. The Australians suggested that her armament could be rearranged on the centreline. The Australians hoped that the Admiralty could arrange favourable terms to justify having the work done away from Australia.

A slightly later proposed modernisation would retain six oil-burning boilers, the two end funnels being eliminated and the remaining uptakes trunked into one broad funnel and one narrow one. Armament would be reduced to six 6in guns arranged roughly as on a 'D' class cruiser, three guns being landed. To allow for the superfiring gun forward, the bridge would be rebuilt roughly as on a 'D' class cruiser, cruiser-type remote control and plotting arrangements being installed. The foremast would not be altered. The foremost 6in gun would have to be moved forward (further forward than in a 'D' class cruiser). It had recently been decided to strengthen the anti-aircraft armament of light cruisers, so the single 3in HA gun would be replaced by a pair of 4in HA guns amidships on the forecastle deck, with two 2pdr pompoms (singles) forward (as in the 'D' class). The 6in magazines would remain as before. The ship would carry eight torpedoes for her two submerged tubes.

Before this modernisation could be carried out, DNC produced a series of alternative schemes in which an aircraft catapult was placed between the new (fatter) after funnel and the searchlight platform, the former No. 1 and No. 4 funnels being eliminated (as before). The seaplane crane was stepped at the searchlight platform. The fatter after funnel was moved somewhat abaft of the original position of No. 3 funnel. In the simplest version, two 4in HA guns would have been mounted on the centreline, one in place of the forefunnel and one on a platform abaft the two after wing 6in guns. The single 6in guns and the bridge would not have been moved. In a more elaborate alternative scheme, 6in guns would have been mounted where the two 4in were, two forward wing 6in being landed; the two 4in guns would have been on a searchlight platform between the two remaining funnels. That would have given a broadside of six guns, compared to the original five. In a third scheme, a 6in gun was placed in a superfiring position forward (as in the original DNC scheme), and the two 4in HA guns were in the waist to either side of the gap between the funnels. A more elaborate version had a modified bridge and two 6in guns in the waist (total of seven), the 4in HA guns being to either side of the gap between the funnels (or, alternatively, on the searchlight platform). A final version retained the two waist 6in aft, but had the superfiring gun forward and the gun in place of No. 1 funnel. This one was unique in having the catapult aft, at the break of the forecastle deck abaft the two remaining broadside guns. DNC's sketches were dated 8 July 1926.

The project was evaluated at an Admiralty conference in December 1926. A key point was that the ship, which was slow by current standards, would be used in wartime for trade protection. Conversion to oil fuel made good economic sense. Moving guns to the centreline did not, because it did not add appreciably to the ship's fighting ability. However, a catapult and aircraft would be very valuable. The best position would be between the after funnel and the after control position (searchlight platform). It would also pay to modernise the ship's W/T systems, including installation of radio direction-finding. Late in 1927 the Australians decided to limit themselves to a conversion to oil burning, and even to retain the two forward boilers which had been earmarked for removal. Oil would replace coal previously stored in upper bunkers, these spaces becoming peace tanks.

Even this was not done; *Adelaide* was not modernised until 1938–9, when the two forward coal-burning boilers and the forefunnel were removed. The torpedo tubes were landed, the gun armament modified (the two side-by-side 6in forward replaced by one gun on the centreline and three 4in anti-aircraft [two in sided in the waist, a third superfiring over the after 6in gun] added, plus four 3pdr and eight Lewis guns), and fire controls rearranged. During 1942–3 the two waist 6in guns were replaced by four depth-charge throwers, one 6in replacing the 4in anti-aircraft gun superfiring over the after 6in (for a total of seven such guns). All 6in were fitted with new square shields. The two 4in in the waist remained, and the ship had six Oerlikons (she retained the Lewis guns).

Adelaide as seen from USS *Saratoga* on 28 April 1944. Note the new-type squared-off 6in shields. She had a mix of US and British radars: a US SC at the foretop for air search, but a British Type 272 in its flat-sided 'lantern' on the lattice tower forward of her forefunnel, and Type 285 atop the HA director. She had six Oerlikons: two in the flat-sided extensions to her bridge structure, two abeam the middle funnel, and two at the after end of the searchlight platform near the stub mainmast. The four depth-charge throwers were mounted, two on either side, behind the bulwark abaft the break of the forecastle. The major modifications were carried out during a 1942–3 refit at Sydney.

BRITISH CRUISERS

The *Chester* class

When war broke out in 1914, Cammell Laird was building two comparable cruisers for the Greek Navy. They were taken over in 1915 as *Birkenhead* and *Chester*. They resembled the *Chatham* class, but introduced a new 5.5in gun, which figured in some other British warships, including HMS *Hood* and the aircraft carrier *Hermes*. It was probably chosen for compatibility with the 5.5in secondary battery planned for the French *Bretagne* class battleship Greece was then planning to buy.

Below and inset: HMS *Chester* was one of two cruisers under construction for Greece in 1914, taken over in 1915 for the Royal Navy. They were built by the Coventry Syndicate, intended to compete with Vickers and Armstrong by linking an armaments firm (Coventry Ordnance Works) with three shipbuilders and steel makers (Cammell Laird, John Brown, and Fairfield). Unlike the very similar *Birmingham*s, she did not have paired guns on her forecastle. These ships also had differently-shaped gun shields. *Birkenhead* could be distinguished by her vertical mainmast. Note the bandstand (for a light anti-aircraft gun) before the bridge in the later photograph, which shows the 3in anti-aircraft gun aft, just abaft the after two waist guns. The ships were built with platforms, between the after pair of broadside guns and the aftermost 5.5in gun, for two 3in HA guns, but they were not available, and 3pdrs were mounted instead. *Chester* was unique among the 'Towns' in having all-oil-fired boilers; the others had mixed firing. The two ex-Greek ships could easily be distinguished from the other 'Towns' by their short mainmasts, only half the height of their foremasts.

These ships had ten 5.5in guns (two abreast at the bow, two on the centreline in tandem aft) and had a six-gun broadside. Like the 'Towns,' they burned coal and oil. Four shaft Parsons direct-drive turbines produced 25,000shp (25kts). *Chester* was modified while under construction to burn oil fuel only, for 31,000shp (26.5kts).

The Spanish cruiser *Reina Victoria Eugenia* is generally described as nearly an improved *Birmingham*. She was almost certainly a Vickers design.[14] Vickers also designed the next Spanish cruiser class (*Mendez Nunez*).[15]

Below: HMS *Birkenhead* in 1919, shows wartime modifications: a tripod foremast with a director atop the enlarged spotting top, a flying-off platform (with, unusually, wind protection for the aircraft), concentration dials, and large searchlights. Note the vertical mainmast.

The 'Atlantic Cruiser'

In 1912 the new DNC Sir Eustace Tennyson d'Eyncourt drew the Board's attention to the need to replace the large armoured cruisers then employed on foreign stations. At this time British policy was to match German cruisers on a two-for-one basis. The 1912 German Naval Law called for the construction of ten cruisers specifically for foreign service by 1920. If the Royal Navy built twenty such ships, it could have five each on the China, East Indies and Cape Stations, with five to spare for the West Indies or elsewhere as required.[16] Chief of War Plans Captain Ballard also pointed out that the Germans planned to equip ten of their largest and fastest merchant ships as armed merchant cruisers in wartime, probably for distant service (given their large coal capacity).

The 2 July 1913 report on the design of a new large cruiser (B3) was labelled (probably by Third Sea Lord) 'this design was got out as the result of rumours that the new German protected cruisers would be armed with guns of at least 6.9in (*sic*) calibre (probably larger). The First Lord [Winston Churchill] was anxious to have a design ready in case these rumours were true.' The ship was described as a Light Cruiser for Atlantic Service. The ship was armed with eight 7.5in guns. DNC pointed out that this was much more powerful than that of any foreign cruiser. He held down cost by limiting displacement to 7,500 tons, on which he was unable to provide more than 4in of belt armour amidships (3in at the ends) with a 3in upper strake amidships. To achieve the desired speed (26kts) on moderate power (as on the new light cruisers, but using heavier machinery for greater reliability), he had to make the ship rather long (500ft); he also gave the ship relatively deep draft (20ft mean) for good seakeeping performance. Unlike existing cruisers, this one would have mounted her heavy guns in turrets: single ones at the ends, and the others in single turrets paired port and starboard. She would have had two underwater torpedo rooms (two tubes each), forward and abaft her machinery spaces.[17]

Third Lord was impressed by her power and good freeboard – 'compared with the County class, it is remarkable what a powerful ship she would be for her size'. He also noted that if the existing 6in gun were retained, there would be little scope for improvement over the current *Birmingham*, although greater length and oil fuel might add another knot and perhaps two more 6in guns or two more torpedo tubes could be mounted.

Churchill was unhappy with the size and cost of the ship; on 4 August he wrote to First Sea Lord that 'I question whether it does not go beyond anything required by German cruiser construction. I do not like the expression "for Atlantic service."' He asked DNC for a second design for comparison: smaller (6,500 tons) but faster (27.5kts on oil fuel), so in August DNC offered B4. It retained the end 7.5in guns but substituted six 6in for the rest: two in 'B' and 'X' positions, two abreast the bridge, and two on the centreline abaft the funnels. Length was 510ft to achieve the desired speed on the limited power. Alternatively, a 7,000-tonner could burn coal and oil as in the *Birmingham* class. Given the lower energy content of coal, she was larger (7,000 tons) and slower (26.5kts). As desired, the new design was expected to be considerably less expensive (£550,000 rather than £700,000, in each case exclusive of guns; the mixed-fuel ship would cost £590,000).

Nothing happened for the moment, presumably because the new destroyer-killer (see the next chapter) was more urgently wanted, but design work was approved. In concept, the big cruiser became the basis for the *Hawkins* class built during the First World War, the justification for which was almost exactly what DNC had written a few years earlier. Despite the German Fleet Law, the two for one policy was applied to produce the small fleet cruisers described in the next chapter.

There was also interest in a new low-performance 'colonial' cruiser, in effect the ancestor of the inter-war sloops.[18] It died because British finances were badly strained simply to match German cruiser construction as desired. If the colonial cruiser was included in the two-for-one numbers, it detracted from effective British cruiser strength. If it was counted outside those numbers (which might be difficult to explain to Parliament), it stretched the badly extended budget. In 1912 the financial problem was bad enough that First Lord Winston Churchill was interested in an arms control agreement with the Germans, who were also in considerable trouble (but it could not be negotiated).

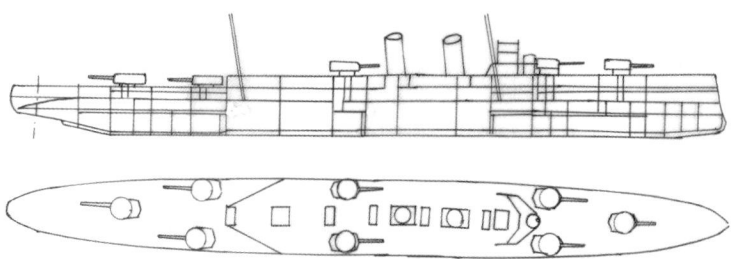

Above: Design B3. The first sketch design (July 1913) showed a 7,400-tonner (500ft x 52ft x 20ft) with 26ft freeboard forward, armed with eight single 7.5in guns, four HA guns and four submerged torpedo tubes. A 30,000shp powerplant would have driven her at 26kts, and she would have burned only oil. Like later British First World War cruisers, she had only side armour, in this case 4in and 3in amidships. B3 was this size, with the same armament of eight 7.5in in single turrets – not the open mountings of the later *Hawkins* class, in effect continuations of this theme. The three boiler rooms were separated from two engine rooms by oil fuel stowage. (Norman Friedman)

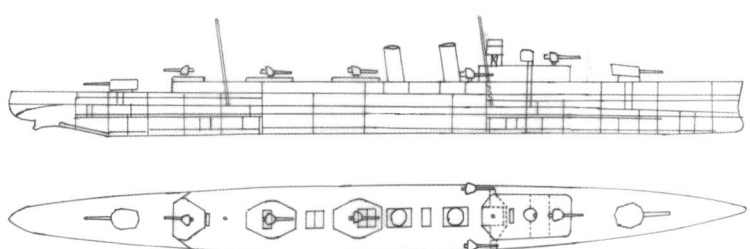

Above: Design B4 (August 1913). This version of the Atlantic cruiser was somewhat smaller (6,500 tons, 510ft x 53ft x 17ft 6in), armed with two 7.5in guns in turrets (like an armoured cruiser) and six open shielded 6in guns, like a light cruiser. Armour matched that of B3. B4 would make 27.5kts on oil fuel. An alternative version burning coal and oil would be larger (7,000 tons, 510ft x 54ft x 17ft 6in) and slower (26.5kts on 28,000shp). As in B3, each 7.5in gun would have 150 rounds, and each 6in would have 200, plus 300 per HA gun. (Norman Friedman)

CHAPTER 3
DESTROYER-KILLERS

The *Arethusa* class

In about 1907 the Germans became interested in taking destroyers to sea to support their battle fleet, but by this time the Royal Navy considered destroyers operating in direct support of the battle fleet a danger both to themselves and to the capital ships. This departure from what had become standard practice in the Mediterranean Fleet can probably be traced to increasing gunnery range, which made it far more difficult to insure against friendly-fire accidents. As Mediterranean Fleet commander, Admiral Sir John Fisher had in effect invented tactics in which British destroyers worked directly with the fleet, but as First Sea Lord he abandoned them in favour of distant blockade of German destroyer bases coupled with the use of destroyer flotillas for home defence. He became less and less convinced that the battle fleet could

operate effectively in the North Sea, to the extent that he preferred to buy battlecruisers, which had an important trade-protection role in distant waters. This was by no means a widely-held view. When Fisher's successor as Mediterranean commander, Admiral Sir Charles Beresford, took over the Channel Fleet, he objected to Fisher's centralisation of torpedo craft command, arguing that he had to have destroyers as part of his fleet. Beresford went so far as to conduct exercises in 1908 to demonstrate that without them he could not fight a fleet with its own integral destroyer force. The Admiralty staff (i.e. Fisher) dismissed his account of the exercises with the bald statement that the Germans, like the British, would be operating their destroyers separately from the fleet. This claim may have been mirror-imaging, or part of Fisher's ongoing war with Beresford. To some extent, also, Fisher's belief that neither battle fleet could survive in a narrow sea in the face of masses of torpedo craft, including submarines, rendered such exercises pointless.

Once Fisher left the Admiralty in January 1910, battle fleet operations in the North Sea were taken far more seriously. Admiral William H May conducted exercises to examine the consequences of the newly-perceived German practice of operating destroyers with their fleet. Having completed the experiments early in 1911, May concluded that light cruisers were the appropriate counter to German fleet destroyers. In the blockade role destroyers were necessary, because they needed the speed to run down German destroyers as they appeared. They were poor gun platforms, but they would have time to sink the German ships as they ran with them, but when trying to deal with German destroyers rushing the British battle line, they would not have enough time to make up for their poor gun platform characteristics. This conclusion seems not to have had an immediate impact. First Sea Lord Sir Arthur K Wilson apparently planned to attack the German destroyers and submarines in their ports (later tacticians would call this 'attack at source') because he doubted that any form of blockade would succeed. He seems not to have accepted the new view of how the Germans would use their

HMS *Galatea* is shown in 1919, with typical First World War modifications. She, *Aurora*, *Inconstant*, *Phaeton* and *Royalist* were all fitted with two 3in guns in 1917, replacing the former single gun (mounted in 1915). However, *Penelope* and *Undaunted* both had a single 4in anti-aircraft gun forward of the after 6in gun. (Perkins)

BRITISH CRUISERS

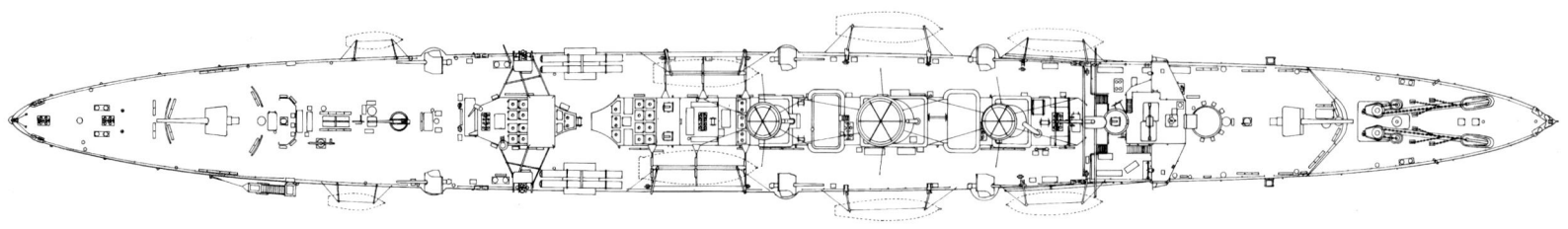

HMS *Arethusa* as in March 1914. The forecastle and upper (main) deck were planked, as was the bridge, but the deckhouse amidships (in effect the boiler casing) had steel decking. A reload torpedo magazine was located on the first platform deck, the reloading hatch being just abaft the aftermost deckhouse (six reloads could be carried). Torpedoes were reloaded via overhead travelling lift gear fitted between the centreline deckhouse sides and the officers' and warrant officers' washroom deckhouse. Folding platforms outboard the 4in mounts were removed during 1915. Surviving ships had two more twin torpedo tubes added during 1917. In 1918 conning towers were removed and a third 6in gun replaced the two after 4in guns. The foremast was given tripod legs to support a fire-control top, and a flying-off platform mounted forward of the bridge. Four of the surviving ships (*Galatea, Phaeton, Royalist* and *Undaunted*) could tow kite balloons. In 1918 all were fitted to lay mines. (A D Baker III)

destroyer force. Wilson described his war plan at an informal meeting in 1911 in Whitehall Gardens, called because war seemed imminent due to the Agadir crisis. A horrified Prime Minister Asquith decided that Wilson was a fool (he had failed adequately to explain his reasoning), and appointed Winston Churchill First Lord of the Admiralty (roughly equivalent to a US Secretary of the Navy) to push the Royal Navy towards saner operating concepts.

Churchill had no naval experience whatever, but he was nominally responsible for the main features of the ships of the 1912/13 programme, including the *Queen Elizabeth* class battleships and the *Arethusa* class cruisers. It is not clear who advised Churchill, but he was close to the former First Sea Lord Admiral Sir John Fisher. Both classes (and also the destroyers of this programme) emphasised speed. The new 'Town' class cruisers were not fast enough to deal with German destroyers, and so early in 1912 Churchill verbally instructed DNC Philip Watts to develop a new fast light cruiser to deal with German destroyers. The DNC Department First World War cruiser history, dated October 1918, attributes the new ship to the considerable interest aroused in the autumn of 1911 by the new Italian cruiser *Quarto*, which achieved 28kts although she was about the same size as the earlier British *Boadicea* (25kts). In effect *Quarto* demonstrated that a small ship could combine the performance of new seagoing destroyers with cruiser protection and firepower, something not previous achieved. It was also reported that the newest German cruisers (the *Breslau* class) were faster than British light cruisers, and better protected, though not as well armed. It appeared that these fast new cruisers achieved their speed by using oil fuel and by adopting faster-running machinery closer to destroyer standards. The *Arethusa* Cover does not mention any of this.

The Cover is marked 'New *Fearless*', so presumably it was conceived as an upgraded version of that ship, which in turn had been (in effect) a destroyer leader. An initial sketch submitted on 10 January 1912 showed a 3,500-tonner (30kts) protected with 2in side armour over the machinery spaces, 410ft x 42ft, carrying 300/800 tons of fuel. She was about the size of the earlier cruiser *Active*. Armament was two 6in guns, four 12pdrs, four machine guns (Maxims) and two 21in torpedo tubes, whose total weight was the same as the armament weight of the *Active*. DNO preferred 4in guns to 12pdrs (the same shift was occurring in contemporary British destroyers), and wanted two more torpedo tubes. DNC submitted the resulting revised design on 16 January, commenting that no modification would have been required had the after 6in gun been replaced by a 4in. At this stage the ship was called a Third Class Cruiser. An earlier but undated design (which survives in the Cover) had an all-4in gun armament, including two guns side by side on the forecastle and the quarterdeck (and four single 4in on each side in the waist). This sketch may merely be an updated version of HMS *Active*, which was armed with ten single 4in guns (although the sketch is not marked as

Above and below: Nearly all British First World War light cruisers were derived from the pre-war *Arethusa* class. HMS *Aurora* is shown newly completed (the towers aft were presumably temporary trials fittings). The reduction to one mast reduced radio range but it also made it more difficult for enemy ships to estimate the cruiser's course.

such). Note that *Active* lacked the side armour of the new design. The new cruiser would be longer than *Active* (410ft rather than 385ft between perpendiculars) and much faster (30kts rather than 25.3kts, on 40,000shp rather than 18,000shp).

There may have been some interest in a cruiser based on the big destroyer *Swift*; the Cover includes estimates of steaming radius for both the Super-*Active* and the Super-*Swift*. However, that comparison is the only reference to the Super-*Swift* in the Cover, and it is difficult to see a Super-*Swift* as a heavily-gunned destroyer killer. The new cruiser had far more endurance, making her much better suited to operating as an integral part of a battle fleet (4,400nm at 15kts compared to 2,400nm at 16kts). Because direct-drive turbines offered poor efficiency at low speeds, the ships had cruising turbines. Turbines were rated at 7,500shp each, with 10,000shp overload rating for a short time (at 650rpm). Boiler pressure was 235psi. The engines were in two engine rooms, and the eight boilers in two boiler rooms. All had Parsons Impulse-Reaction turbines except for *Arethusa* and *Undaunted*, which had Brown-Curtis. As in a destroyer, the uptakes from the after boilers in the forward room and from the forward boilers in the after room were trunked together to form a fatter second funnel between two narrower ones. Late in the design process it was decided to rake the mast and funnels because the mast did not support a boat derrick, hence did not have to be vertical. While the ships were being built, it was pointed out that the high power (10,000shp per shaft) would probably necessitate experiments with alternative propellers, but that proved impossible due to the outbreak of war, and the ships actually made 18.5 to 29kts.[1]

An internal DNC memo dated 29 February 1912 instructed the DNC staff to push the 'New *Active*' design as quickly as possible, with an armament, if possible, of twelve 4in guns – which might be difficult to fit in. Submerged torpedo tubes should be adopted to leave more space (for the guns) on the upper deck. Space was a problem because, DNC wrote, 'it is inexpedient to lengthen ship'. A sketch design for the all-4in ship was submitted in March 1912. In March the First Sea Lord asked for alternative ten- and eight-gun arrangements. DNC pointed out that weight saved that way could go into side armour over the machinery: an extra quarter-inch with ten guns, an extra ½in with eight guns. One of the ten-gun arrangement showed the two after guns *en echelon*, well separated lengthwise, to give a six- rather than five-gun

broadside (as in the twelve-gun ship). One of the eight-gun arrangements showed single centreline 4in guns at the ends, to give the same broadside as a ten-gun ship with paired guns fore and aft. The result might seem comparable to *Active*, but a March 1912 note on complement observed that the earlier ship provided enough crews only to man one broadside or the other, whereas the new cruiser would have enough to man all her guns simultaneously – which would make sense, as she would be engaging German destroyers rushing past her to attack the British battle line. In addition, it appeared that she would be armed with faster-firing QF (quick-firing, i.e. fixed ammunition) rather than breech-loading (BL, i.e. separate ammunition) 4in guns (but ammunition allowance per gun, 250 rounds, would not change).

To save hull weight, the armour was incorporated into the hull strength, a 2in outer layer covering the 1in hull plates, which had not been done in the 'Town' class. To this end strakes were worked longitudinally rather than vertically; there was later some fear that hits would break up the sandwich of armour involved. Total thickness was 3in over the machinery (and fuel tanks) amidships. That was quite respectable. After her action with *Emden*, the Captain of HMAS *Sydney* wrote that his 2in side armour had proven very valuable, since it defeated the standard German cruiser gun (4.1in) at 8,000yds, and probably at much shorter ranges. Fortunately the German shells rarely burst (had they done so, the ship's fore and aft controls would have been put out of action in the first few minutes). They also fell steeply enough not to ricochet. The Captain concluded that at least one control position should be behind armour. He rejected the existing conning tower.

DNC planned to extend side armour to the ends (1¼in thick); Churchill preferred protective decks at the ends. However, the extended side plating was important for structural strength; more structure would be needed to make up for the discontinuities at the ends of the belt. The weight saving was vital if the ship was to make the desired speed. Also, First Lord's preferred armour decks would complicate internal arrangements in a ship in which space was at a premium. DNC suggested a compromise, extending the belt armour about 60ft fore and aft of the machinery spaces (but to reduced height). That would provide 2in sides instead of the earlier 1½in sides, and a 1in arched deck over the steering flat instead of the previous quarter-inch deck. The extended belt would cover the lower conning tower, magazines, and shell rooms. The original 1½in plating was extended all the way to the bow, but 2in plating was extended only 30ft further aft and to a reduced height. All of that would cost about 40 tons, but about 20 tons would be gained back by eliminating the 1½in side over the steering compartment. Other detail cuts might save another 10 tons (for example, the lower conning tower side could be cut by an inch since it would be behind thicker armour). The new arrangement had the additional virtue of improving internal arrangements. Churchill approved. Ultimately ships had 1in to 1½in additional armour forward of their machinery, extending from 3ft below the upper deck to 2ft 6in below the load waterline. Aft of the machinery spaces they had 1½in extra armour back to the rudder head, from 3ft below the upper deck (cut down to 5ft 6in aft) down to 2ft 6in below load waterline. Hull plating itself was 1in to ½in thick. The machinery was protected from aft by a 1in after bulkhead, presumably to resist the fire of a pursuing enemy cruiser. *Arethusa* had a 6in conning tower with a 4in tube, ultimately replaced by a 3in conning tower with a 2in tube in these ships. The Legend showed no protective deck other than plating over the steering gear (side and deck were considered equivalent to 1½in). The continuous part of the upper deck was 1in thick for strength rather than for protection. Its role was made clear when the Captain of *Arethusa* (lead ship of the 1912/13 cruiser class) complained after the Heligoland Bight battle (28 August 1914) that parts of the superstructure acted as a shell trap, and that therefore the forecastle should have 1in armour to protect spaces below from fragments, but that would have added too much weight. Much of the lower deck had to be cut away to accommodate the large boilers. Because the ships burned oil to achieve their desired high speed, they lacked the coal bunkers whose bulkheads transversely stiffened earlier ships. It therefore became important to stiffen the ships transversely around their boiler rooms, by continuing the deep web frames in the machinery spaces above the lower deck. The structural arrangement incidentally doomed the idea of lower-deck torpedo tubes, because they entailed cutting the ship's side just where the heavy plating was. Moreover, there would be insufficient deck height to handle torpedoes freely, and tubes set low in the ship's side

HMS *Royalist* shows typical First World War modifications: a flying-off platform, the conning tower removed, a tripod foremast with a spotting top (but no heavy director; these ships had limited reserves of stability), a 3in anti-aircraft gun aft (actually one of two, to either side). One pair of 4in guns was replaced by a 6in gun on the centreline, forward of the after conning tower (barely visible here). The two pairs of deck torpedo tubes are not visible. (E Hopkins of Southsea courtesy of Josef Straczek)

would fire their weapons directly into the waves created by the ship at high speed. The combined structural and protective arrangements became standard for the First World War British light cruisers derived from the *Arethusa* class.

Estimates of the required complement showed that the ship was not large enough to accommodate it; manning the guns and the submerged torpedo tubes was apparently the problem. In April, Churchill proposed to solve the problem by cutting to ten guns, with crews for seven of them (saving thirty men) and by adopting deck torpedo tubes, which saved another three men. However, the twelve-gun alternative was chosen by the Board, with QF guns of higher velocity than those being adopted for destroyers (45 calibres rather than 40 calibres). In May 1912 the Board sought to solve the weight problem by limiting each gun to 140 rounds, rather than the 250 earlier envisaged. That still left the space problem.

The Board approved the design on 7 July 1912 (DNC submitted it on 1 July). This version showed ten 4in QF guns and two upper-deck single 21in torpedo tubes; it is not clear when the Board reversed itself and approved a ten-gun armament. The Legend showed 200 rounds per gun.

Endurance was clearly considered important, because in the submission to the Board Watts pointed out that the ship could make 5,000–5,500nm by running on two shafts and trailing the other two. Estimated radius with four shafts on line was 4,000nm. In Legend or trial condition, ships carried 300 tons of oil fuel. Deeply loaded they carried 800 tons, including 140 tons in peace tanks (so called because, being above the waterline, they could not be filled in wartime). Later the total was given as 810 tons, a great deal for a 3,500-ton ship. As more and more topweight was added in wartime, stability in the light condition became less and less satisfactory – but it was possible to remedy that by flooding empty fuel tanks with sea water. Arrangements to this end were ordered in August 1916. DNC considered this modification a prerequisite for many newly-required improvements, including a tripod mast for fire control. The 1912/13 programme included eight ships of this *Arethusa* class.[2]

About November 1912 the First Lord (Churchill) decided to substitute single 6in guns for the paired (abreast) 4in guns at the ends of the ship, despite the considerable modifications involved; for example, the conning tower and bridges had to be raised about 3½ft to avoid gun smoke interference with the rangefinder on the bridge roof, and the funnels had to be raised similarly. The ships were given additional flare (to reduce wetness on the forecastle) and high spray shields fitted to reduce spray over the forward guns. The ships had already been ordered, so this change was hardly inexpensive. The rapidly-fired 4in gun was considered ideal as a destroyer-killer, but the 6in was wanted to deal with enemy cruisers which might support destroyer attacks.[3] The change left the ships with two 6in and six 4in, the latter in the waist at upper deck level.

There was, however, a problem: the ships were lively, and the 6in gun, which was manually trained and laid, was heavy. In 1913 trials on board HMS *Falmouth*, which was larger and hence not as lively as an *Arethusa*, showed that the existing 6in mounting could not be laid and trained quickly enough to deal with the motion of the ship. The metacentric height (in deep condition) in the new ships was therefore deliberately reduced. (and a squarer-section hull and deep bilge keels also adopted, to reduce motion). Goodall wrote that these changes could be made without detriment to stability because of the shift from coal and oil to oil fuel only, since in the deep condition the centre of gravity of the fuel was low rather than high. Much the same thing was done at this time in the design of the 'R' class battleships.[4] Coventry Ordnance Works (COW) proposed a power-worked solution. The alternatives were the Vickers experimental N mount and the existing P.VI pedestal mounting. The Vickers mount seemed most promising.

Ships had little or no centralised fire control, which made sense for a ship expected to engage several destroyers more or less simultaneously. The anti-destroyer guns were grouped for a degree of control, and the ship had two standard 9ft rangefinders. After she destroyed the German raider *Emden* in 1914, HMAS *Sydney* turned in a report that the rangefinders became useless once fire was opened, hence that the new technique of 'rangefinder control' was out of the question.

In March 1913 DNC was told to work out how to replace the two single deck torpedo tubes with twin tubes.[5] It would not be very expensive, but the ships had already expended their Board margins (for growth during construction), and the 5 tons involved might turn out to be the last straw. If it could not be accepted for the *Arethusa*s, surely it should be accepted for the 1913/14 ships. The First Sea Lord accepted the loss of speed. He could imagine occasions when 'these very fast ships could make night torpedo attacks on a battle squadron or division, which could not be avoided or resisted and which probably would not mean much risk to the smaller vessels'. The new cruisers were in effect super-destroyers, intended both to kill destroyers and to conduct their own torpedo attacks. The Royal Navy seems to have been unique at this time in seeing its light cruisers as torpedo craft. The idea apparently lapsed for a time, but it certainly returned strongly during the First World War.

Finally, in 1913 installation of single 3in HA (high angle, i.e. anti-aircraft) guns was proposed for light cruisers. It was not approved until 1916. *Galatea* was completed with one 6pdr, *Inconstant* and *Phaeton* with one 3pdr, and *Undaunted* with one 1½pdr. The ships were completed with a single 0.303in Maxim (Vickers) machine gun, which was also standard in contemporary destroyers.

The first three ships entered service soon after war broke out in 1914. The broadside positions of the 4in guns were so wet that, at least in HMS *Aurora*, they were nearly useless. The ship's captain interpreted his experience to mean that 4in guns were useless compared to 6in, but it seems fairer to say that low-lying guns necessarily nearly at the edge of the deck were much wetter than guns on the centreline. The position nearly at the deck-edge had other unfortunate consequences. Layers and trainers had to be strapped into their seats, because on some bearings they were actually over the sea. Shells striking the belt armour created splinters which, on one occasion on board HMS *Arethusa* in 1914, killed a gunlayer.[6] DNO suggested building a light sponson under the guns. Although DNC protested that it would create more spray, it was added.

The *Caroline* or *Calliope* class

Eight more ships, slightly modified, were ordered under the 1913/14 programme as the *Calliope* class (later called the *Caroline* class). Compared to the *Arethusa* class, these ships had two more 4in guns. Two 4in were abreast forward of the bridge instead of a 6in gun; the two 6in were superimposed aft in 'X' and 'Y' positions. This unusual arrangement seems to have reflected the destroyer-killer logic: the ships would chase destroyers, against which their rapidly-firing (actually QF) 4in guns were their main weapons. Having four of them on the forecastle would provide more fire in the desired direction, and these guns would be dry in almost any weather. The 6in guns were intended to be used against enemy cruisers chasing the destroyer-killer. The ships were designed at about the same time that trials showed that existing 6in guns were unlikely to be very effective from such lively platforms.[7] According to the DNC First World War cruiser history, an alternative

HMS *Caroline* as completed in 1914, with four single 4in guns forward and two 6in aft, plus deck torpedo tubes as in *Arethusa*. (John R Dominy)

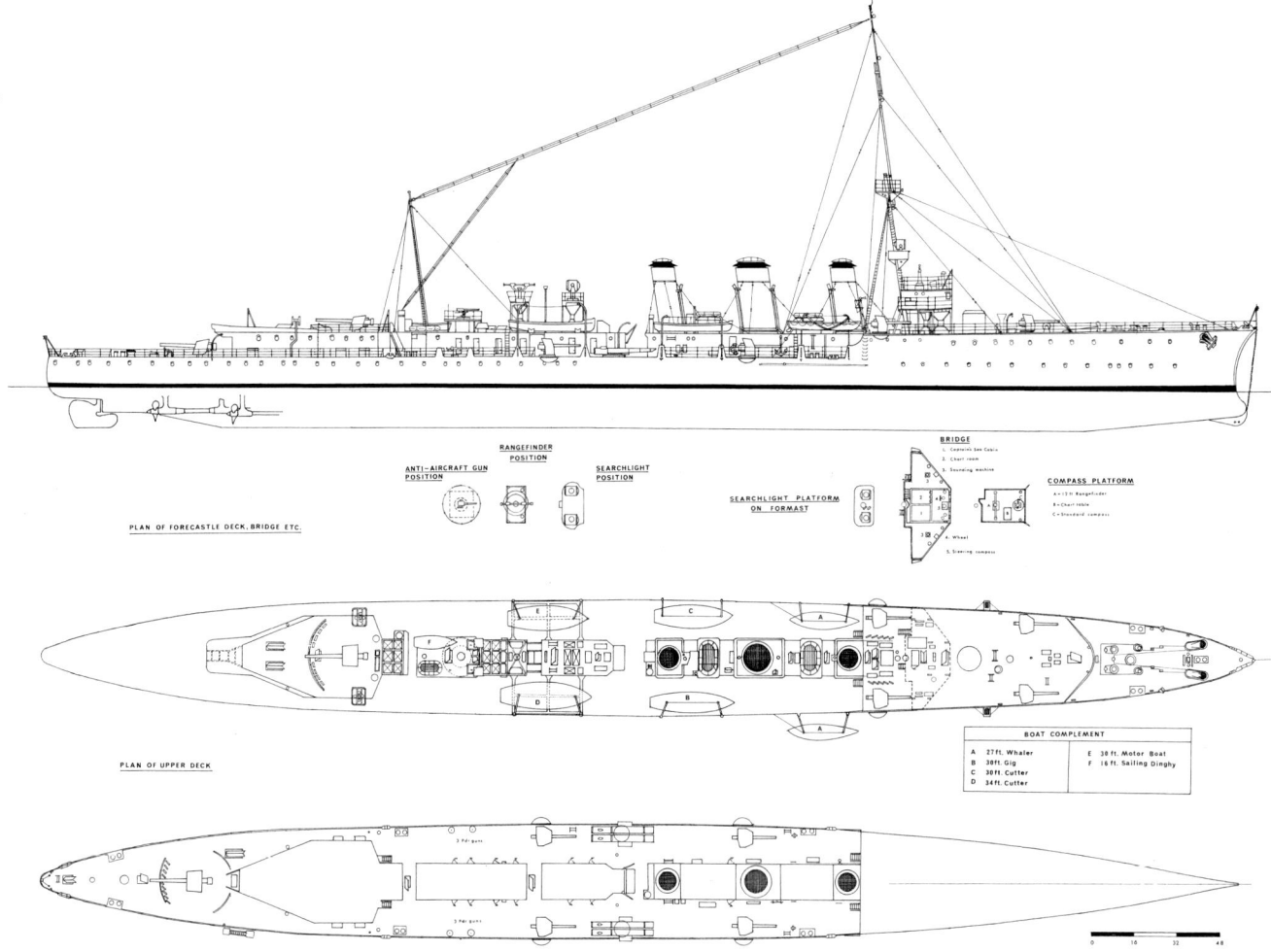

Caroline and *Carysfort*. *Caroline* is shown as in May 1917. *Carysfort* is shown as partially re-armed in 1918, with additional torpedo tubes in place of one of her waist 4in guns. Note her covered fire-control top, with (empty) searchlight platform below it, and the semaphore on the signal platform on her bridge structure. (John R Dominy)

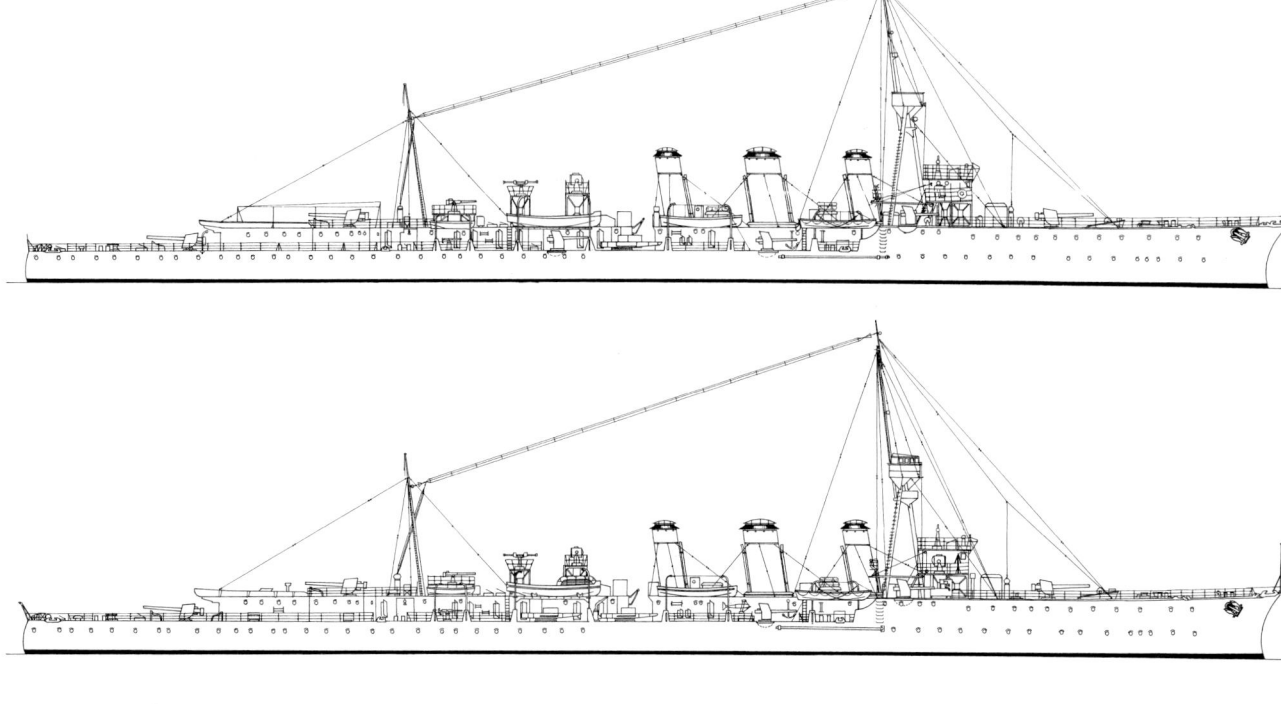

Cordelia (1918) and *Cleopatra* (1919). *Cordelia* is as rearmed with four 6in guns, one 4in HA and eight deck torpedo tubes. Note the derrick for handling boats. *Cleopatra* shows new large gun shields for her 6in mountings. She is shown as she appeared in the Baltic fighting the Bolsheviks. (John R Dominy)

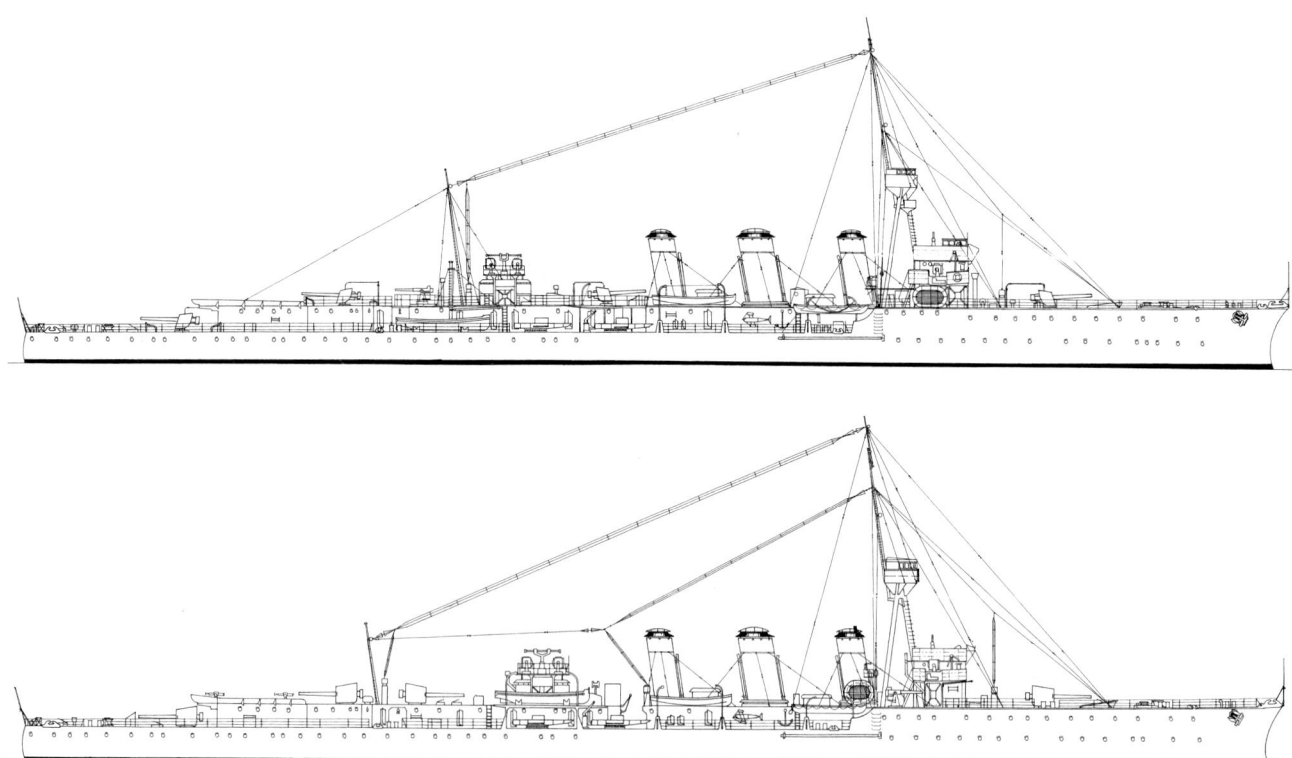

HMS *Calliope* is shown in 1916 (top) and in 1918. In the upper drawing the main difference from a *Caroline* is the absence of deck torpedo tubes. Note also the stowage of the motor cutter. The lower drawing shows her as rearmed with four single 6in guns, two 4in HA guns and two twin deck torpedo tubes in addition to her submerged tubes. (John R Dominy)

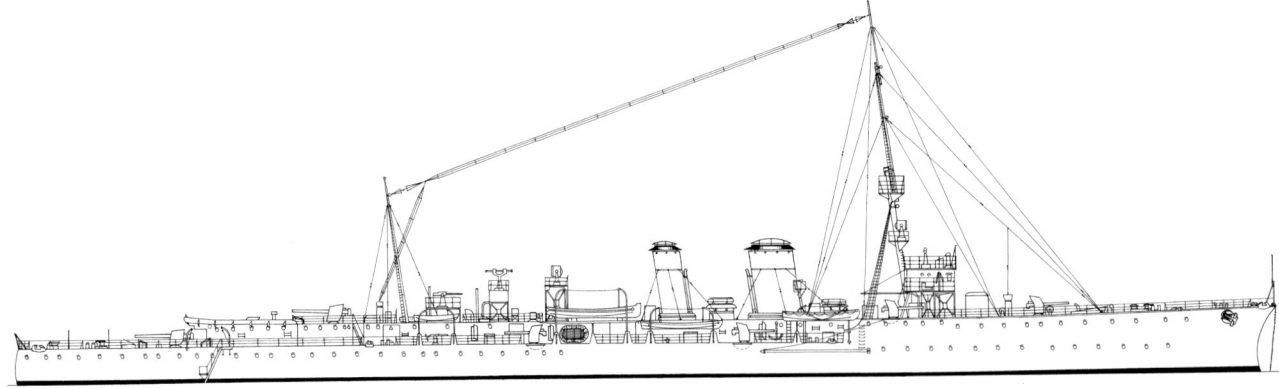

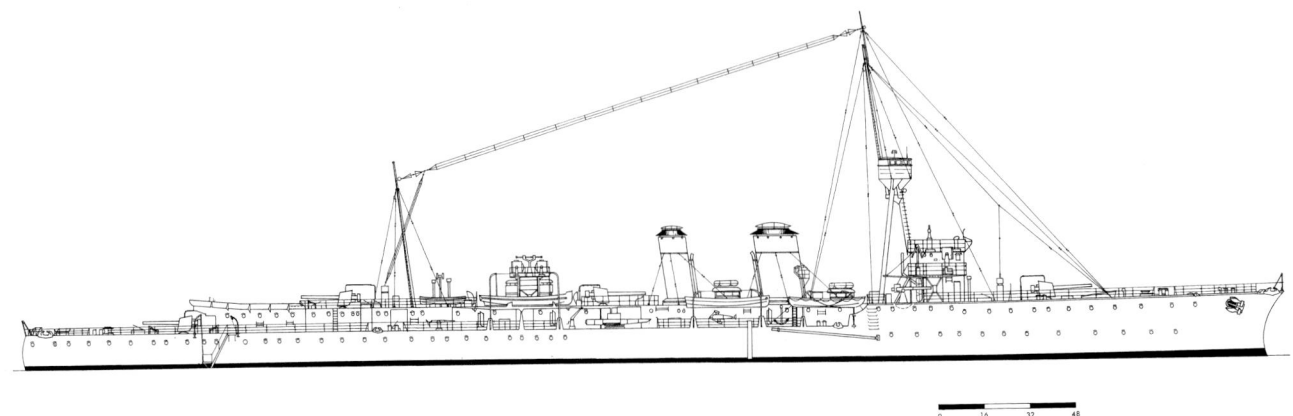

HMS *Calliope* is shown in 1925. She was refitted from November 1919 to March 1920 after a serious oil fuel fire. Her above-water torpedo tubes were removed, the after control simplified and two 2pdr (single pompoms) were added. The two 4in HA were replaced by 3in HA.

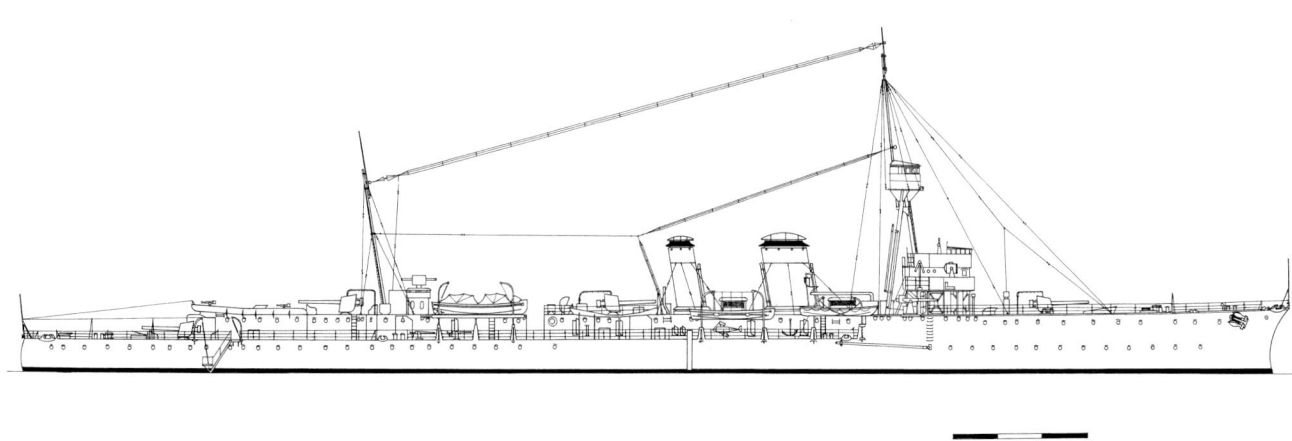

HMS *Champion* is shown in 1919, with a short 'H' mainmast to take her wireless antennas. Wartime and early post-war modifications include the deck torpedo tubes, the large shields for the 6in guns and the single 4in HA gun. Later she was given two single pompoms and a short pole mainmast. She retained the after control position shown.

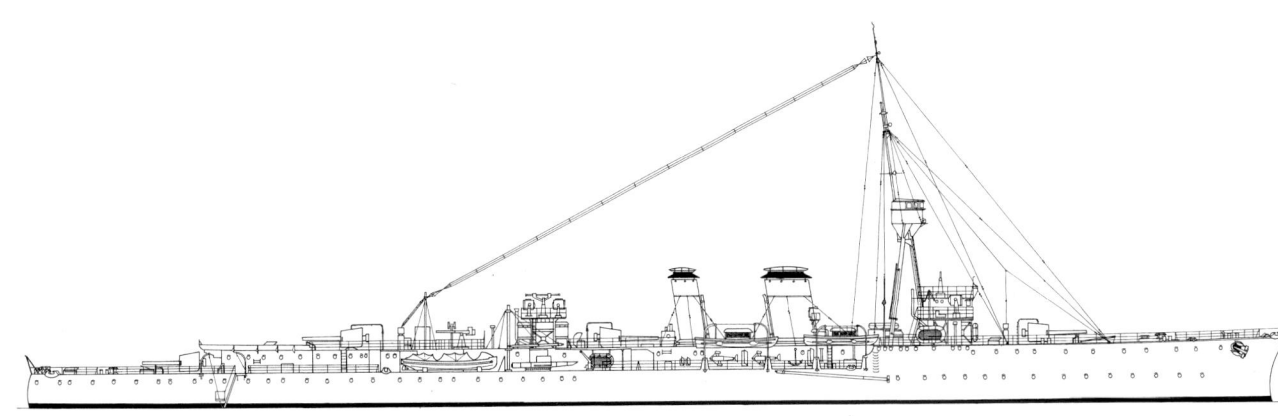

Comus (1927) and *Conquest* (1928). *Comus* has had her after controls and extra torpedo tubes removed, a hood added to her rangefinder and some boats repositioned. *Conquest* is shown as leader of the 1st Submarine Flotilla. Surface ships were used to lead groups of submarines attached to the fleet because they could maintain communications and could scout for the submarines, much as light cruisers had been envisaged as the 'eyes' of pre-1914 destroyer flotillas. In 1925 the amidships 6in gun in *Conquest* was replaced by a deckhouse as shown. (John R Dominy)

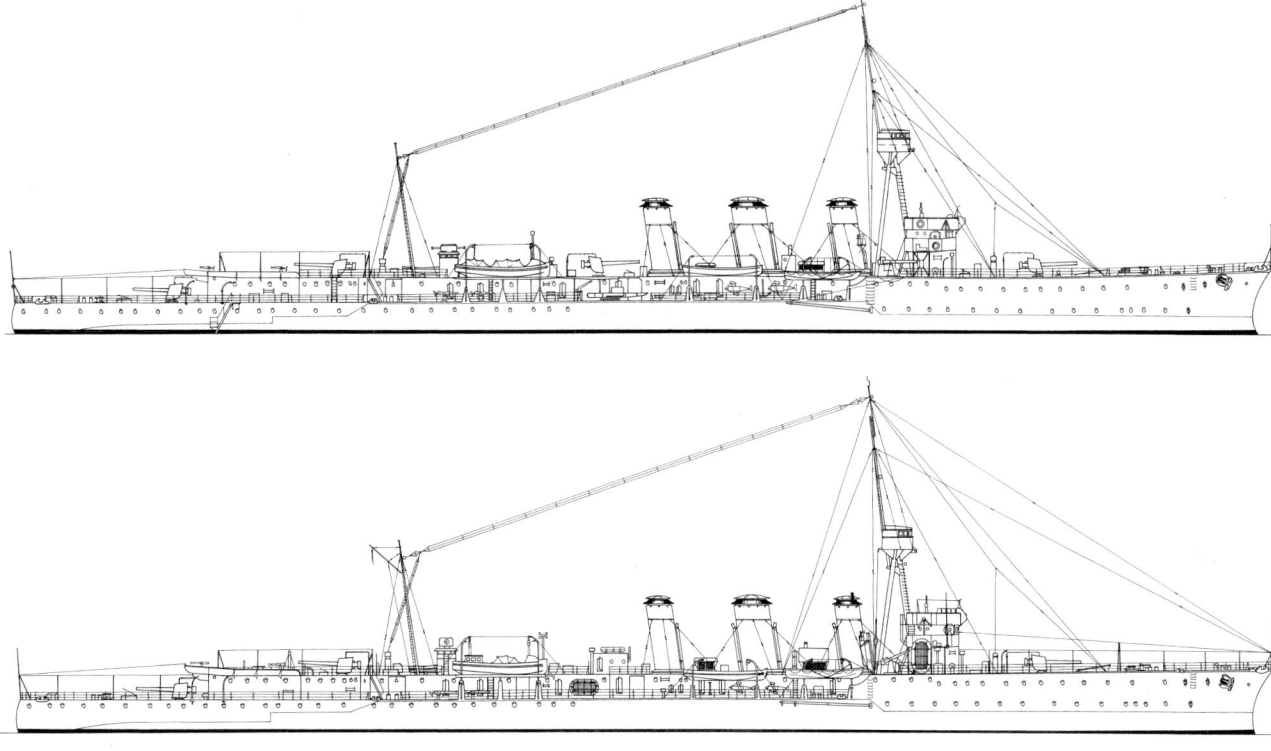

Castor as completed in 1915, with her original gun battery of four 4in forward and two 6in on the centre-line aft. She had no deck torpedo tubes. (John R Dominy)

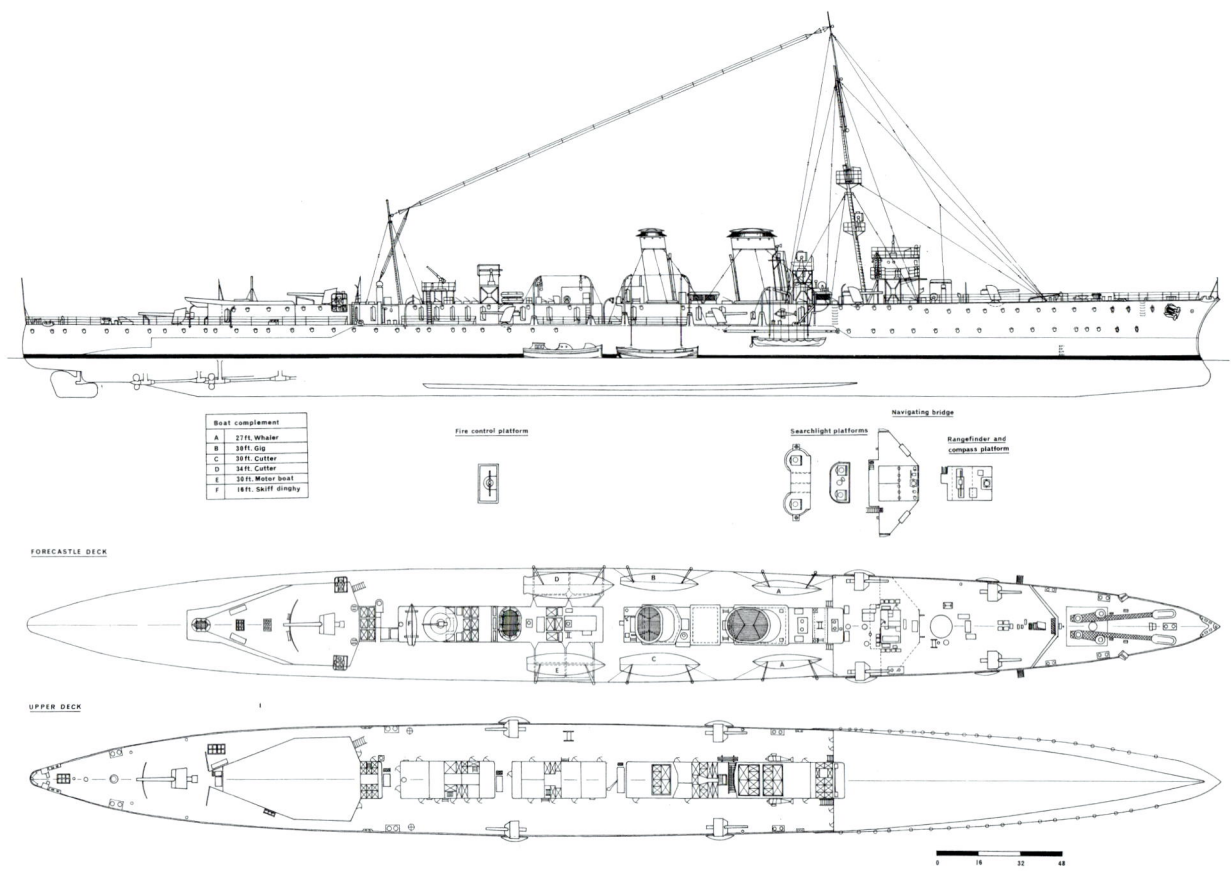

Cambrian (1916) and *Canterbury* (1917). *Cambrian* is shown as she appeared on joining the Grand Fleet. *Canterbury* shows initial modifications: her after waist 4in guns have been replaced with torpedo tubes and she has a prominent searchlight platform aft. (John R Dominy)

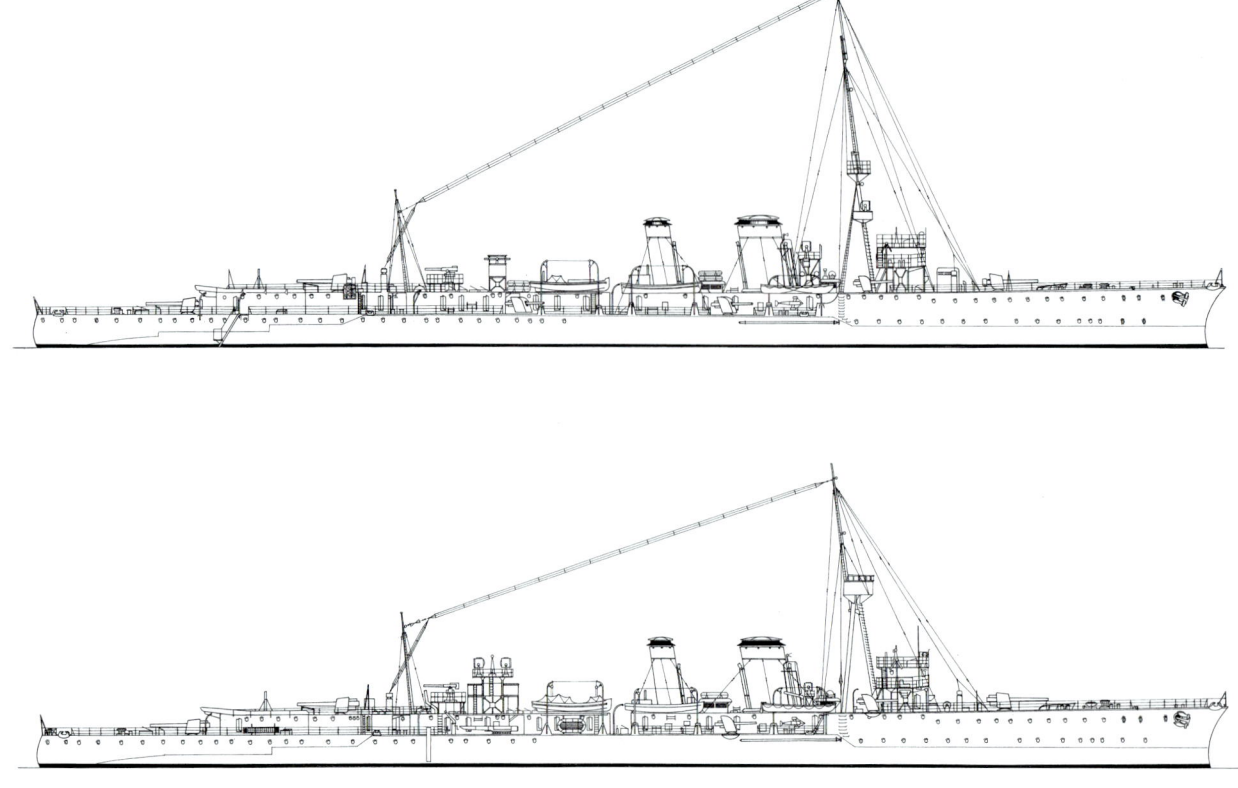

DESTROYER-KILLERS

armament of five (later six) 6in guns was proposed and rejected, but it does not appear in the Cover. Both versions showed 3,700-ton displacement (200 tons more than *Arethusa*) and 3in side protection carried 28ft further. Peace tanks were omitted, but it was expected (according to the DNC history) that more economical machinery would give a radius of action of 7,000nm on 900 tons of oil. This was wildly unrealistic.

The forecastle was carried further aft and the two forward waist guns moved up to the forecastle. Two ships, one building at Chatham (*Calliope*) and one at Hawthorn Leslie (*Champion*) were built with geared turbines and six small-tube boilers rather than the eight of the other ships. They had reduced power because their slower-turning screws were more efficient (37,500shp rather than 40,000shp in *Calliope*).[8] The eight-boiler ships all had Parsons Impulse Reaction turbines except for *Carysfort*, which had Brown-Curtis. *Champion* had two rather than four shafts. They therefore needed no narrow forward funnel. Machinery

The *Caroline*s were essentially repeat *Arethusa*s with a new main battery arrangement, including four 4in on the forecastle. HMS *Cleopatra* is shown, completing.

spaces were materially shorter, and it was possible to enlarge the compartment between boiler rooms and engine rooms for submerged (rather than deck) torpedo tubes. Controller (Rear Admiral Archibald G H W Moore) much favoured this change, the only disadvantage being the existing problem (which he was sure would soon be solved) of discharge at high speed. The First Lord (Churchill) welcomed this development, which he had long espoused, and wanted it extended to the rest of the class – which was impossible. The change was particularly attractive because the twin deck tubes limited the firing arcs of the aftermost 4in guns. However, the chief advantage was better protection. It was accepted that in peacetime the upper deck tubes would enjoy a higher rate of fire, and that in wartime the upper deck tube might be better placed to take advantage of a fleeting opportunity because a submerged tube would require more preparation time and in view of the difficulty of communication between the firing officer and the tube. Also, although the tube itself might be better protected, it still depended on a vulnerable firing position on deck. The decision for below-deck tubes was firm by October 1913. Below-deck installations had seven torpedoes, upper deck ones ten.

The changes added weight. The Board approved a 3,750-ton ship with an expected speed of 30kts. Raising two 4in guns and extending the forecastle added 17 tons, submerged tubes added 36 tons, oil fuel heating added 20 tons, an increase in galley coal added another 20 tons, and the Board decided that the margin in future ships should be 1 per cent rather than 20 tons (another 20 tons) – a total of 113 tons. Machinery changes added another 19 tons, and the 132 tons added since design would add 2in of draft and cost half a knot. The only way to get back to the Legend figures initially approved would be a drastic cut in fuel oil, which was unacceptable.

The 1914/15 programme was reduced to four light cruisers (*Cambrian* class), repeat versions of *Calliope* herself with two funnels and submerged torpedo tubes. All had six boilers but not the geared turbines of the *Calliope*s. It is not clear why the annual programme was cut, because when war broke out in 1914 the planned 1915/16 programme was eight cruisers.[9] In fact only two ships were ordered during the autumn of 1914, initially as repeat *Calliope*s (*Centaur* and *Concord*).

For machinery, E-in-C asked for independent Parsons or Brown-Curtis turbines on four shafts with geared cruising turbines and small-tube boilers. Most firms offered alternatives. One was combined turbines, high-pressure (HP) turbines on the outer shafts exhausting to low-pressure (LP) on the inner shafts (which also had the geared cruising turbines). It offered about 6.5 per cent better economy, but each set of main engines would require a separate astern turbine, which would tend to cramp the restricted engine rooms. Another was all-geared Parsons turbines on two or four shafts. It was rejected pending experience with the geared turbines of HMS *Calliope* and HMS *Champion* and with destroyers. The third alternative, the German Fottinger hydraulic coupling (between turbines and shafts), was rejected on the grounds that its claimed efficiency was overstated, particularly at low power. The ships had Parsons Impulse Reaction turbines except for *Canterbury* (Brown-Curtis).

By this time it was clear that generator capacity (two 500-amp) was inadequate due to the introduction of 36in searchlights and installation of twenty-two electric radiators. The two generators in a *Calliope* could barely power two such searchlights, without the radiators or workshop pumps or 12½in ventilator fans. In April 1914 DEE wanted enough power for four such searchlights and for two 12½in fans in any new cruiser. His plea was rejected because no additional weight could be added.[10] However, in 1915 it was decided that future light cruisers would have more powerful generators (two of 800 amps working at 105 volts), and provision of a third 500-amp generator was being considered for existing light cruisers and for ships under construction, for which provision of the larger generators would involve delay.

The *Centaur* class

Once war broke out, the two vital issues were gun and torpedo armament. Experience at Heligoland Bight convinced Captain Nicholson of HMS *Aurora* that his 4in guns were useless, hence that future light cruisers should be armed only with 6in guns.[11] In November 1914 he sent a paper to this effect via an enthusiastic Commodore (T) Tyrwhitt of the Harwich Force (at that time the First Fleet Flotilla). He sketched an arrangement offering a total of five, all on the centreline, one gun firing forward and one right aft. The Controller (Rear Admiral F C T Tudor), who had lived through the *Calliope* design, was less than enthusiastic. These were special-purpose ships, intended to operate with and against destroyers and other light craft, and likely to have to fight several ships at the same time. They needed heavy ahead fire from several guns and the heavy astern fire provided in the design. Tudor was undoubtedly aware that his preference for a mixed armament might make it appear that he opposed single-calibre battleships (dreadnoughts) – Fisher had recently returned as First Sea Lord, and would hate to be reminded of the opposition to his great innovation of less than a decade earlier. Thus Tudor pointed out that he personally favoured one-calibre armament for ships intended to fight either single-ship or fleet actions, as that simplified both fire control and ammunition supply. Tudor saw Nicholson's proposal as one of a series of cycles in cruiser design in which single large-calibre armament superseded mixed armament, the latest being the shift from the *Bristol* to the *Birmingham* classes. Each time design reverted back to mixed armament because it became obvious that a heavier gun could not shoot effectively from a lively cruiser. The 1913 tests of 6in gunnery aboard HMS *Falmouth* seemed to prove exactly that. Tudor felt that a 4in gun could be manoeuvred rapidly enough to make up for the ship's motion, but a 6in could not. He suspected that the dividing line between effective and ineffective was half-way between, meaning a 5in or 60pdr gun. HMS *Aurora* had fought in fine weather, which was deceptive, with two of her six 4in out of action before the battle began, and the rest more or less quickly jammed. He had also observed the motions of such guns in bad weather but not in action. Tudor particularly disliked cutting the number of guns, hence the number of targets the ship could engage, and also

HMS *Centaur* as flagship of Commodore (D) of the Atlantic Fleet. She was given an enlarged bridge structure to house the necessary staff, and No. 2 6in gun was landed. She and *Concord* could be distinguished from the very similar *Caledon* class by their clipper bows.

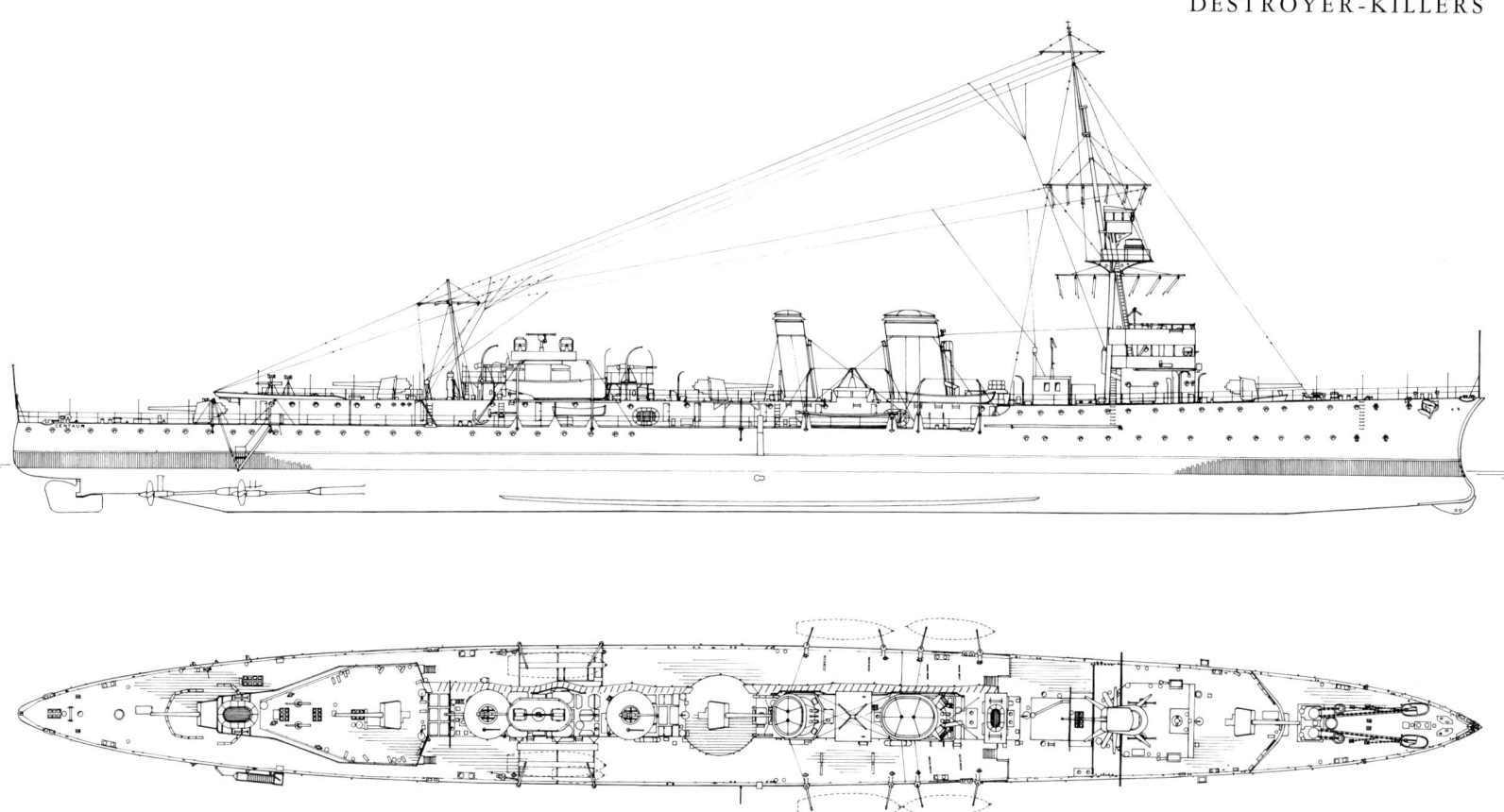

halving her astern fire (it was necessary to cut down the deckhouse aft to accommodate the guns on the centreline).

What Tudor did not say was that, if the guns could be manoeuvred quickly enough, the new method of director firing would go a long way towards counteracting the ship's motion, particularly when she was firing broadsides. Director firing in turn was most effective for a single-calibre main armament – engaging a single target. The question, which was not raised again, was whether Tudor's vision (doubtless not his alone) of a single cruiser fighting many destroyers at once was particularly realistic.

There was no question of altering ships under construction, but the two repeat *Calliope*s just ordered could be modified. As a compromise, Tudor offered to substitute a 6in gun for the pair of 4in on the forecastle forward of the bridge, to give an armament of three 6in and six 4in, with an ahead fire of one 6in and two 4in. The three 6in would give a powerful broadside for use against light cruisers in bad weather, when only broadside fire would be effective, and a large volume of fire against destroyers under favourable conditions.

Fisher and Churchill both strongly favoured the all-big-gun armament, and it was adopted for the two repeat *Calliope*s, HMS *Centaur* and HMS *Concord*.[12] They retained the superimposed arrangement aft. The bridge was moved forward so that a second 6in could be inserted between it and the forefunnel, and a third 6in was inserted abaft the after funnel.[13] As Nicholson had suggested, the heavy conning tower was eliminated to reduce weight in the bow due to the 6in gun. The 6in guns were modified to increase their elevation to 30° and hence their range. Beam was increased 6in, to 42ft. The new arrangement became the basis for the next class of light cruisers. Tudor's comment that a 60pdr might be ideal (the 6in was a 100pdr) aroused Churchill's interest. It seemed that six such guns could be accommodated. Nothing could be done for the moment because no such gun existed, and because a prototype would have to be built and tested. The prototype was

Centaur is shown in March 1925 after her 1924–5 refit at Rosyth to serve as flagship of the Commodore (D) commanding Atlantic Fleet destroyers (she recommissioned in April 1925). The 6in gun abaft the bridge structure was replaced by a 'recreation' deckhouse, and the bridge was greatly modified and enlarged, with the signal flag lockers enlarged and moved to the forecastle deck forward of the new deckhouse (which had a single semaphore atop it; note also the semaphore atop the bridge). The conning tower seems to have been removed during the same refit. The *Centaur*s were the first Royal Navy cruisers to have director control for their 6in guns; the secondary control position with two 36in searchlights atop it was added during the 1924–5 refit, and at the same time the lattice searchlight support structure just forward of the forward funnel was removed. Note the 'elevation rails' to prevent the 3in HA guns from firing into the ship's structure. The ship retained her paravane fitting in her forefoot, but not the paravane gear. The torpedo spaces were located just abaft the after funnel on the lower deck. *Centaur* and her sister were sometimes incorrectly described as ex-Turkish. In fact light cruisers were included in the Turkish Vickers naval programme which produced the battleship *Reshadieh* (which became HMS *Erin*); Vickers built the two *Centaur*s. The company's records show two designs for fast protected scouts (No. 771 of 27 January 1914 and No. 771A/B of 19 February 1914). Design 771B was relatively small at 3,558 tons (400ft pp, 423ft loa x 39ft 6in x 23ft 6in x 13ft 6in), and was expected to make 27kts on 22,000shp – which means that the Turkish ships, if they were ordered, did not provide the turbines used in the *Centaur*s. Armament would have been eight 4in/50 without shields, six 3pdr with light shields and two 21in deck tubes (probably singles). (A D Baker III)

ordered, emerging later as the 4.7in mounted in flotilla leaders.[14] Both ships had six boilers and four shafts driven directly by Parsons Impulse Reaction turbines.

The issues for torpedo armament were means of control and loading arrangements. The torpedo directors in conning towers could rarely be used because of spray, so an alternative position was provided; by June 1915 this modification had already been approved for the 'Towns' and was about to be extended to the new light cruisers. Torpedo tactics were moving towards the use of 'browning' shots at longer ranges, and that in turn required dedicated rangefinders (typically a third rangefinder in a cruiser). It was placed on the same platform as the torpedo director.

It became obvious that upper deck torpedo tubes could not be reloaded at sea from below, hence that reloads were useless in battle. At a conference on board HMS *Lion* in June 1915, two alternatives were raised for the new light cruisers: either a second set of twin tubes could be added, or a box for reloads could be built on deck in line with each set of tubes. Either could be provided in an *Arethusa*, but the *Calliope*s lacked space for additional tubes.

The *Calypso* class

The war programme assembled in June 1915 included no light cruisers, although an appended note did call for ships for overseas service, which eventually became the *Hawkins* class (see below). All available slips may have been occupied by the six ships ordered in 1914. In November, however, a new six-ship programme was adopted. In December the Admiralty offered immediate contracts to four builders (Beardmore, Hawthorn Leslie, Scott and Vickers) provided they submitted firm fixed-price bids, and were prepared to accept a final Admiralty decision as to the amount to be paid. Cammell Laird built the ship planned initially for Beardmore. These ships were named *Caledon*, *Calypso*, *Caradoc* and *Cassandra*. In March 1916 it was decided to order another ship (*Curacoa*) from Pembroke Dockyard and a sixth (*Ceres*) by contract (ultimately to John Brown). Three more were soon ordered, from Fairfield, Swan Hunter, and Vickers (*Caprice*, later renamed *Cardiff*, *Corsair*, later renamed *Coventry*, and *Curlew* respectively).[15]

This *Calypso* class (1915/16 programme) was conceived as a slightly modified repeat *Centaur* with new geared-turbine machinery. Controller resisted changes, but a few crept in. The 6in guns were raised a foot above the deck, presumably for easier loading at higher elevation, and thirty rounds of ready-use ammunition per gun was provided on the upper deck or forecastle. These ships were also provided with two 3in HA guns each (250 rounds per gun plus fifty rounds of practice ammunition). Beam was increased 9in and the hull form slightly altered, the stem being modified to match that in the much larger Ocean-Going Cruiser (*Hawkins* class) then being designed, in hopes of reducing spray over the bow gun. Initially the anti-aircraft armament would have matched that of the earlier cruisers: two 3pdr Vickers QF guns with 300 rounds per gun and one 0.303in Maxim machine gun with both ship and field mountings. However, in February 1916 DNO proposed replacing the 3pdrs with 3in 20cwt HA guns (some of the additional weight would be saved by eliminating the usual pair of 3pdr saluting guns). The guns were characterised as both anti-aircraft and anti-submarine. The change (and others proposed by DNO) was rejected by Controller in the interest of speeding completion.

Below and opposite: HMS *Calypso* (*Caledon* class) as completed. Note the director below the spotting top, an improvement not made in earlier classes, and the two sets of twin deck torpedo tubes visible on this side.

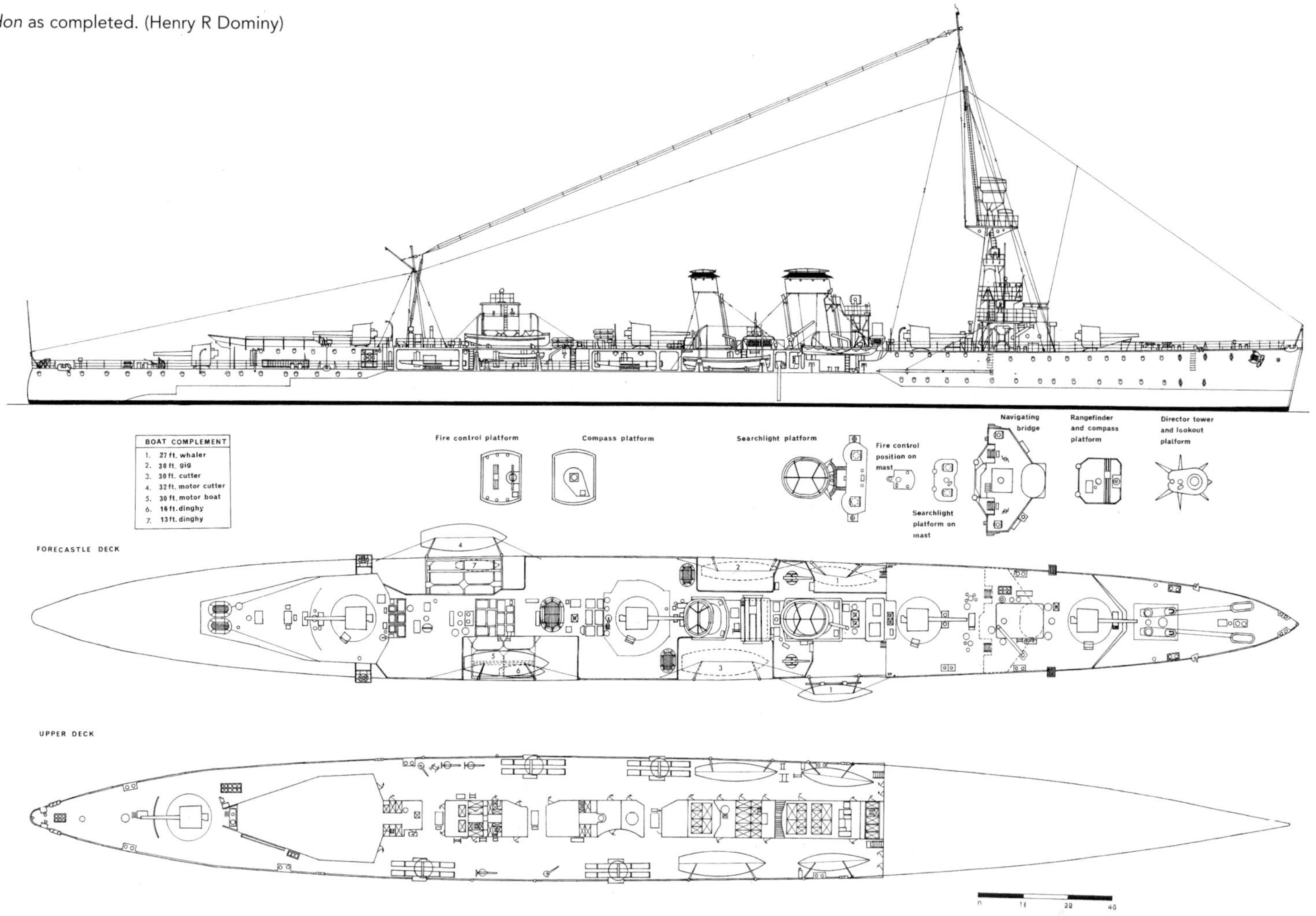

DESTROYER-KILLERS

Caledon as completed. (Henry R Dominy)

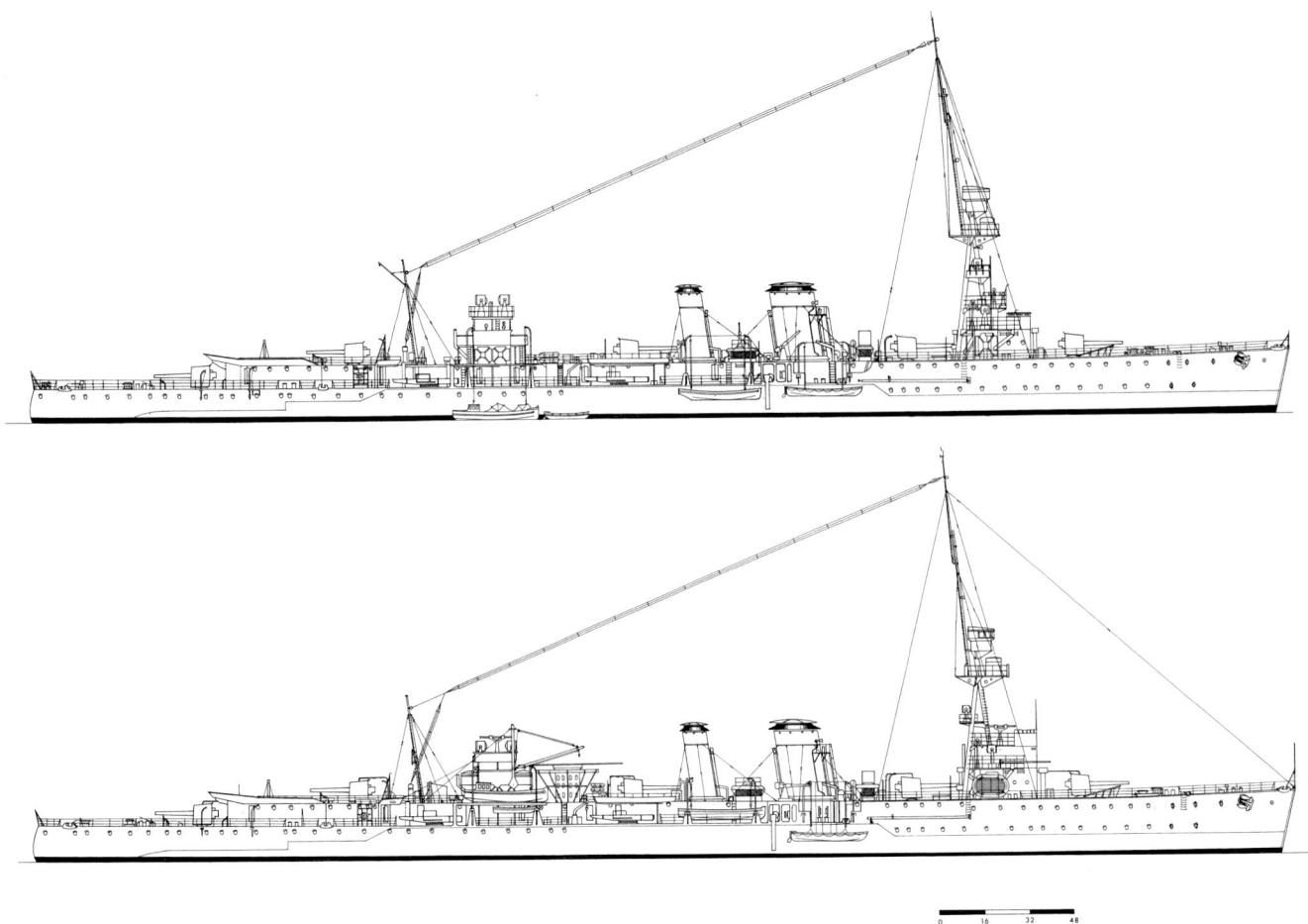

Above: Caledon is shown in 1917 (top) and in 1923 (bottom). The 1917 view is as she appeared after a May 1917 refit. The 1923 view shows a rotating flying-off platform forward of No. 4 gun, the conning tower removed, the bridge enlarged (the compass platform, from which the ship would be conned following late wartime practice, was built up), and the after control and searchlight positions modified. She now had two rangefinders (fore and aft). The lines extending into the 6in shields carried director data to the dials on the mounts, which gunners matched to lay and train their guns. (John R Dominy)

Below: Curacoa as in February 1918. This drawing was based on a detailed model plan prepared by the late Norman Ough in 1953 using original Admiralty as-fitted drawings. This version omits the ready-use racks for the 6in guns. The 15ft rangefinder at the fore end of the second superstructure platform was removed during the 1920s and the position plated in. Capetown and Carlisle initially had aircraft hangars built into their bridges, like that of Dauntless and Dragon. The name Curacoa recalls the capture of that Dutch island (now spelled Curaçao) in 1806. (A D Baker III)

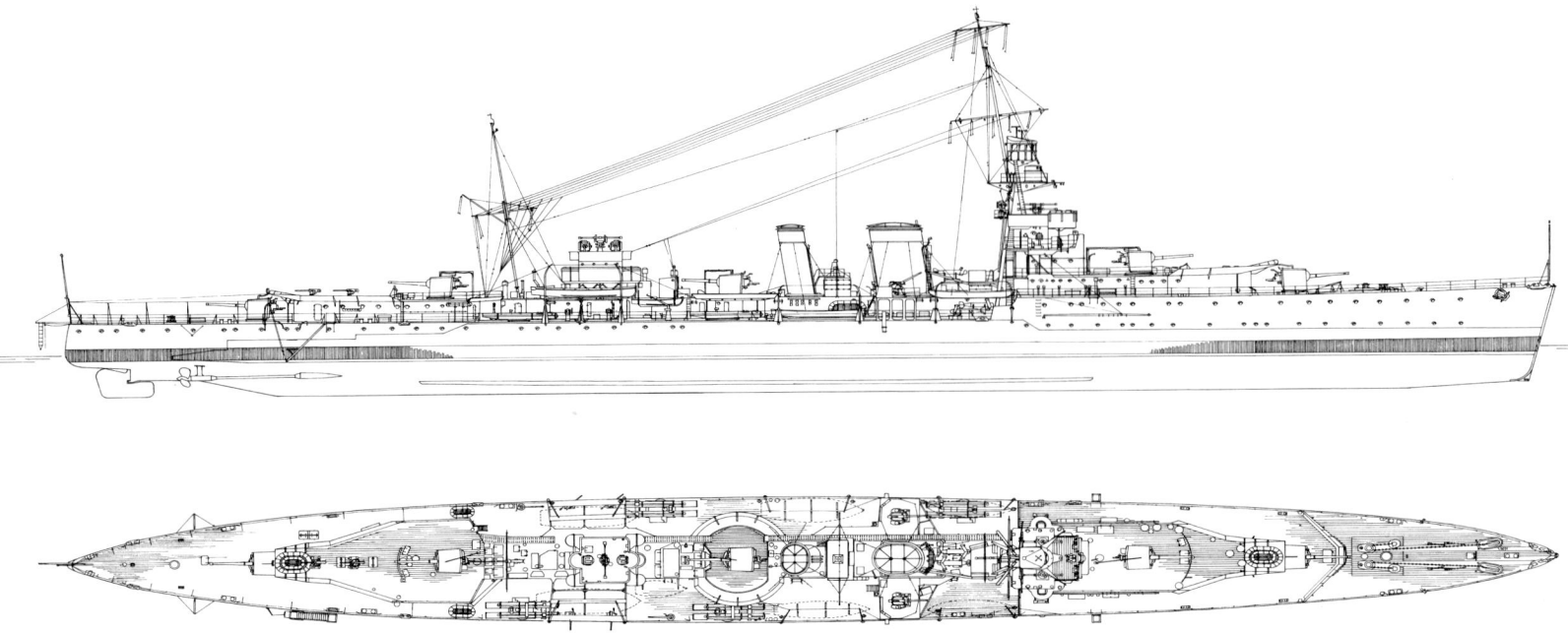

DESTROYER-KILLERS

HMS *Caradoc* in 1941 with limited initial war modifications. She has a Type 286 air-search radar atop her foremast and a Type 271 surface-search set aft. She has a 3in anti-aircraft gun abeam the gap between her two funnels, and the shields of two Oerlikons are visible abaft the shield of No. 4 6in gun. Another Oerlikon is visible on the bridge wing, behind a bulwark. In 1938 the ship had two 3in anti-aircraft guns and two single pompoms. The date written on the photograph (by the US Office of Naval Intelligence) is something of a puzzle, as it appears that the Oerlikons and radars were all installed during a refit at the New York Navy Yard, 21 October 1941 – 3 March 1942 (the ship received a total of five Oerlikons). *Calypso* was converted into a gunnery training ship in mid-1943.

The *Ceres* class set the pattern for later British First World War cruisers, with their super-firing main battery providing more ahead fire. HMS *Ceres* is shown upon completion.

The problem of firing submerged torpedo tubes at high speed had not, it turned out, been resolved; at a conference, representatives of HMS *Vernon* said that good running could not be guaranteed above 24.5kts, so the new Improved *Centaur* had to have above-water tubes with their centres not less than 10ft above water. Tubes on deck should be strong enough or roomy enough to resist splinters. Ideally they should be placed behind some form of armour, but that was impossible on weight grounds. It seemed best to fit two sets of double revolving 'slack fit' improved *Arethusa*-type tubes, ideally not exactly opposite each other.

Admiral Jellicoe, the Grand Fleet commander, had been pressing for two twin above-water tubes on each side as early as December 1915. He had little faith in any attempt to reload at sea, and with two sets of tubes on each side all torpedoes would be ready for immediate firing. Jellicoe pointed out that his Grand Fleet Battle Orders placed particular stress on the duty of light cruisers to attack the enemy battle line with torpedoes. 'In attacking enemy torpedo craft or supporting our own

BRITISH CRUISERS

torpedo boat destroyers, light cruisers are very likely to pass through positions from which torpedoes can be successfully fired at the enemy's battle line.' Once again, the line between light cruisers and destroyers (as torpedo craft) was clearly very thin. 'The attack with torpedoes on the enemy's battle line by light cruisers is the more important because many of our destroyers may be absent from the general action owing to shortage of fuel or bad weather encountered off the Northern Base, and the destroyer duties must then be performed by the light cruisers.' Conversely, Jellicoe often demanded more destroyers on the grounds that he did not have enough light cruisers, and the demand for destroyers (which could be built much more quickly) presumably reduced building resources available for these destroyer-like light cruisers. By this time the Royal Navy accepted that long-range shots aimed at individual ships were unlikely to hit, given the long running time of the torpedoes and various inaccuracies. It also accepted that neither destroyers nor light cruisers were likely to get close enough to fire short-range aimed shots. Instead the Royal Navy emphasised 'browning' shots at groups of enemy ships. The more torpedoes fired, the better the chance of hitting. Just as British destroyers had pairs of twin tubes, Jellicoe wanted two twin tubes on each side for his cruisers. By March 1916 Controller had acceded; the operational Commander-in-Chief (C-in-C) should get what he needed, despite the effect of any changes. First Sea Lord (Admiral Jackson) approved the pair of twin tubes on each side. They added about 3 tons to the weight of a pair of tubes plus reloads. A proposed alternative using fixed tubes was rejected because they entailed too much additional structural weight, including extra length.

Right and below: HMS *Dauntless* is shown in April 1930, little modified since the First World War. Note the considerable difference between her bridge and that of the 'C' class cruisers; hers is much closer to the ideal bridge developed late in the First World War, which provided officers on the compass platform the maximum view on the basis of which to conn the ship.

The 'D' *(Danae)* class

In about March 1916 DNC was asked to develop a new light cruiser for the programme planned for that May. The starting point seems to have been reports of new German light cruisers armed with ten or eleven 5.9in guns. DNC pointed out that the Germans were still following orthodox cruiser design practice, with guns on each broadside, so that the battery reported would give them a broadside of no more than six guns.[16] That would outclass the new British cruisers. DNC added that it was almost impossible to produce a light cruiser combining a displacement much greater than that of a *Centaur* with the same speed without going to something like the new overseas cruiser *Effingham*. He presented his solution on 17 March 1916, adding another 6in gun.[17] He could do so at minimum cost in additional length (and displacement) by making No. 2 gun superfiring and moving the bridge back towards the funnels. Two more centreline guns were mounted, as in the latest 'C' class, before and abaft the two funnels. In his view, the 'C' class, designed for 30kts, had reached the limiting speed of 29.5kts for such small ships. He had to accept much greater displacement in the *Effingham* to regain that speed, and for so large a ship the 6in gun was too small. The new sketch design traded about half a knot of speed for an additional gun. Moreover, superfiring meant that two guns could fire

DESTROYER-KILLERS

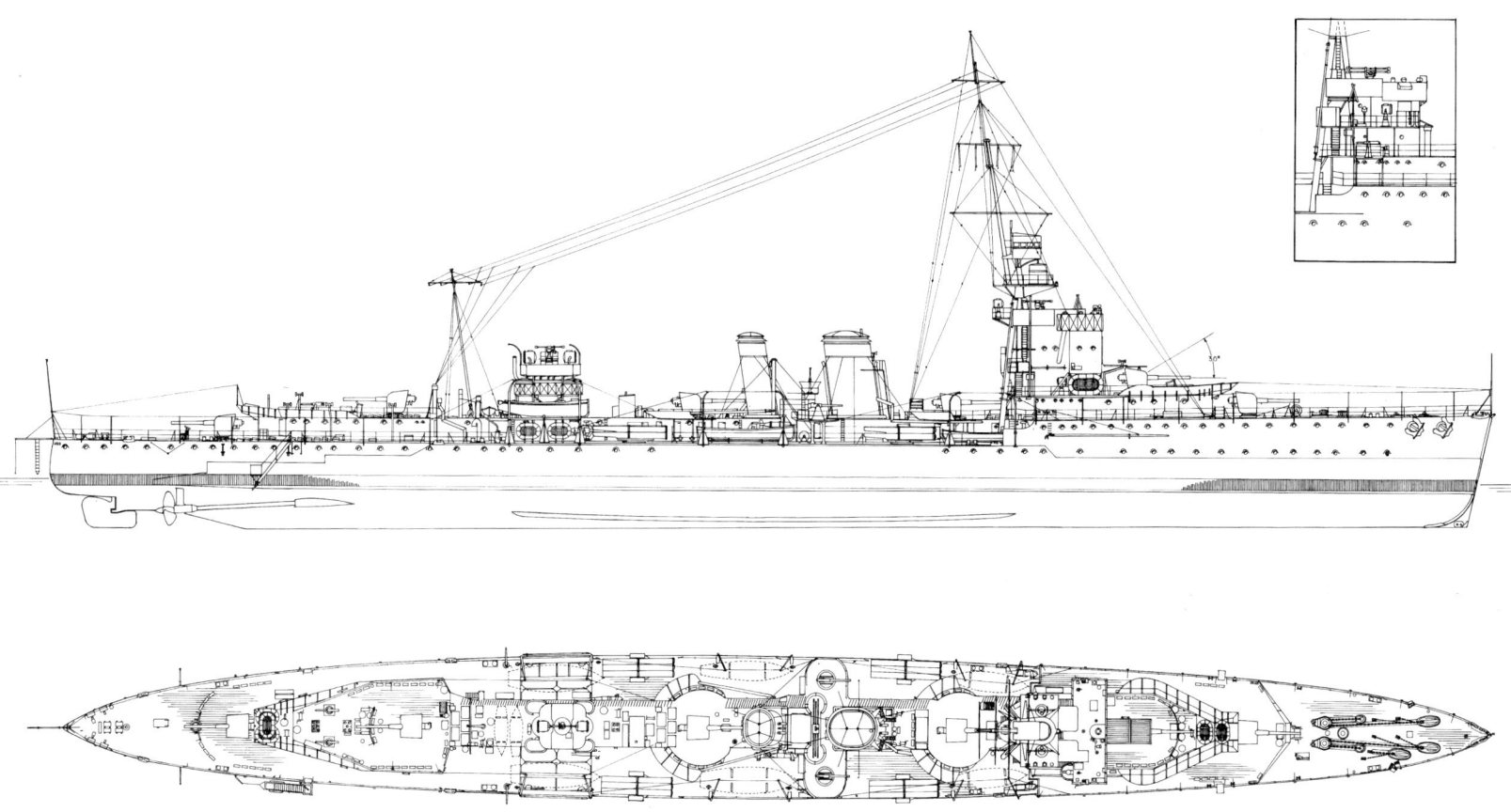

Dauntless as completed in February 1919, with a hangar built into her bridge structure (it was never used). *Dragon* had an identical hangar. The section of the flying-off platform which would normally lie atop 'B' gun could be folded back over the fixed portion. It was removed in 1920, and the fixed part used to mount two 24in searchlights (the starboard light was about 4ft abaft the port). The hangar face could be closed by doors, but the sides were open. The aircraft was stowed ready for take-off, its tail resting on a railed support within the narrower extension of the hangar. An aircraft loading (and wreck recovery) boom could be stepped on the forward port corner of the hangar. The captain's sea cabin was to starboard and the charthouse to port on the open bridge. The X-crossed panels on the sides of the open bridge and the weapons control station aft represent splinter protection mats, which were removed during the 1920 refit. Note that there was a steering wheel and engine order station on the bridge, to starboard of the compass platform. By 1930 *Dauntless* and her sisters *Danae* and *Dragon* had three 4in HA in place of the two 3in mounts, while the two single pompoms were moved to new platforms abreast the bridge structure. The 'D' class was the only Royal Navy steam-turbine light cruiser class designed with two rather than four shafts (*Undaunted* also had two shafts). (A D Baker III)

dead ahead. DNC thought the change would cost a few months of building time. On 20 March the Sea Lords approved the new design in principle for some of the cruisers of the May 1916 programme. In addition to the sixth 6in gun, the new design showed two 3pdr HA and one Maxim machine gun. As the design developed, No. 2 gun and bridge were moved 14ft further from the bow, and the bridge and conning tower (4in with 3in tube) raised. Although horsepower would match that of the *Calypso* class, revolutions would be reduced from 300rpm to 275rpm at a cost in weight and engine-room length (about 2ft). Some 2in armour was worked in aft. The design was approved in June 1916 (Board Stamp 30 June) with slightly more protection: magazine crowns were thickened to 1in, and gun shields were to provide protection from splinters as well as from spray.[18] By this time, the design also showed triple rather than twin torpedo tubes.

A slightly later attempt to cut weight by reducing ammunition from 200 to 150 rounds per gun was rejected on the grounds that 'the new method of ranging will lead to increased expenditure by the main armament. These ships will also be able to open fire at very long ranges as they will have 30° mountings and Director firing; and more opportunities occur for firing when using Director.' DNO offered only that the usual sixteen practice rounds could be included in the 200 rounds per gun.

By May 1916 the new ships were designated the 'D' class, three being ordered in September 1916. Three more were ordered in July 1917, and another five in March 1918 (four were cancelled at the end of the war). Of this class, *Dauntless* and *Dragon* were fitted to carry an aircraft.

It was decided not to modify the six 'C' class ships ordered in December and March 1916, so DNC asked whether the last three ships (ordered in April 1916) should be modified. Proposed modifications were substitution of 3in HA for 3pdrs (12 tons), substitution of armour for spray shields on 6in guns (23¾ tons plus increased beam and hull weight), and modified ammunition supply to the 6in guns.

The Improved *Calypso* class

With the 'D' class design proceeding in April 1916, Controller planned that they should embody all improvements, and that the new 'C' class cruisers just ordered should be repeat *Calypso*s. However, he was well aware of the value of superfiring mounts, which had just been arranged for in new destroyer leaders. Could the same thing be done in a 'C' class hull, moving bridges further aft? On 4 May DNC submitted just such a plan – 'the improvement is so marked that it is thought to be well worth carrying out.' The space between boiler room and engine room was reduced by 18ft, the boilers being moved aft this distance. Beam was increased by 9in, at a cost in weight and thus in speed. The other proposed changes were not adopted. Controller agreed that the improvement in bow fire and in the position of bridge and conning tower were so great that they were well worth a slight loss of smooth-water speed. The First Sea Lord and the First Lord of the Admiralty (now Arthur J Balfour) initialled the change on 6 May 1916. The one change that was carried out was substitution of 3in HA for 2pdrs, which was justified by the need for more powerful anti-aircraft weapons. It was urgently requested by the Grand Fleet, and had already been approved for *Centaur* and *Concord* on that basis.

The question was now which of the *Calypso* class under construction could be modified with minimum delay. A naval constructor was assigned to visit the yards to see what could be done. He concluded that the first four were all too far along, even the least advanced *Calypso*. That left the five most recently ordered ships, all of which were completed with superfiring guns forward. They were described as the Improved *Calypso* class. They had two-shaft geared Brown-Curtis turbines. According to the report of gun trials by *Ceres*, the first of this modified 'C' class, the absence of vibration in way of the bridge was noticeable, appreciably less than in *Champion* with similar machinery. It was attributed to the position further aft and to the absence of bridge wings, 'which readily take up any vibration in the remainder of the structure; and to the solidity of the bridge structure to which special attention was paid in this ship'. The director control tower (DCT) vibrated worse than the control top immediately above it, probably because the former was overhung forward of the mast. The thick side plating and the shell plating were both faired into the skin of the ship to leave a smooth surface, spray being due almost entirely to the paravane chains.

Another eight light cruisers were ordered in June–July 1917. Plans may initially have called for three repeat 'C' class and three repeat 'D' class. At one point six of the eight were to have been 'D' class modified with greater endurance for overseas service, but instead the order was split between three 'D' class and five 'C' class. Both groups were given raised ('trawler') bows for better sea-keeping. All of the repeat 'C' class had geared Brown-Curtis turbines except for *Curlew*, which had Parsons. Five more 'D' class cruisers were ordered in March 1918, identical to the 1917 ships except for 6in more beam.

In the autumn of 1916 the Board decided that six of the eight light cruisers of the November 1916 programme should have a 7,000nm radius of action (compared to 5,000nm for a 'D' at 13.5–14.5kts) for use on foreign stations (unlike the *Hawkins* class, however, they were not to burn coal for convenience in foreign ports).[19] The Controller issued instructions in a Minute dated 27 September 1916. On 24 November DNC submitted a design differing from a standard 'D' class cruiser only in having greater displacement to carry more oil. Two weeks later the Controller (Rear Admiral Tudor) directed DNC to develop full working drawings as soon as possible, so that orders could be placed. DNC increased length from 445ft to 460ft, buying 16ft more (for fuel) between engines and boilers, but limiting increased length by shortening the part of the ship forward of the foremost fuel tanks. He also provided peace tanks between the platform and lower decks (capacity 100 tons), for a total capacity of 1,325 tons to give the required radius at 13–14kts. The design showed the recently-adopted triple torpedo tubes. Machinery was 10 tons heavier than in a 'D', presumably mainly due to longer propeller shafts. Deep displacement was given as 5,250 tons, about 600 tons more than that of a 'D' class cruiser as completed. The loss of speed was not indicated. Jellicoe, now First Sea Lord, approved the design on 20 December (it received the Board Stamp on 3 January), but no orders were given, and the design died.

As recently-appointed First Sea Lord, Admiral Jellicoe was very much concerned with the wetness of British cruisers and destroyers, having operated with them in rough weather. On 9 December 1916 he asked Controller to look into a design with the bridge moved further back by interchanging bridge and mast with No. 3 gun, leaving a gap between forecastle and bridge (he was interested in a similar scheme for destroyers). He was probably unaware of how far back the bridges of the ships with superfiring guns were: 136ft in a modified 'C' and 132ft in a 'D', compared to 92ft in a *Calypso*. DNC pointed out that he would have to add 7ft more length, because it was undesirable to move Nos. 1 and 2 guns closer to the bow. That would entail greater beam and hull depth (for strength) and displacement, and more length might be needed because the space between the end of the extended forecastle (carrying the bridge) might be too limited to accommodate boats, four triple torpedo tubes, and two 3in HA as then to be carried by a 'D' class cruiser. No. 2 gun would need an additional blast screen to protect it from No. 3, and the screen in turn would limit the elevation of No. 3 gun on extreme forward bearings. Raising No. 2 gun would enlarge the target represented by the ship. The idea died.

Bridge Design

In March 1917 a special committee under Commodore C F Lambert was convened on board HMS *Southampton* at Rosyth to decide future cruiser bridge policy on the basis of wartime experience. The discussion helped shape the bridges of post-war British cruisers. All present agreed that it was not necessary for the bridge wings to extend all the way from side to side. However, in that case the view of signal lights might be blanked off by the ship's broad forefunnel. There was strong interest (as in destroyer bridges) in streamlining the superstructure and also in shaping it to resist the impact of heavy waves. The bridge should be as far aft as possible, limited only by the position of the tripod mast, which had to be forward of the boiler room. The captain, navigator and officer of the watch would occupy a navigational platform incorporating the compass platform, with the best possible view, with the captain on the compass platform. By this time standard British practice was for officers on an open compass platform to pass steering orders to a helmsman below them. Ideally the chart house would be directly below the navigational platform, with a steering position built onto its fore end, with only wheel fittings and room for a helmsman to stand. To keep the navigational platform as clear as possible, gunnery and torpedo controls and signalling should be moved away. Ideally the bridge should incorporate a weather-proof and splinter-proof compass platform and a weather-proof position for the chart house, signal house and officers' and captain's sea cabins. Weight could be made available by eliminating the more-or-less useless existing conning tower. The space below the chart house would be divided into a signal house and a captain's cabin. It should be possible to enter the bridge from the main deck in foul weather. The latest 'C' class bridge embodied most of these ideas, but it had proven impossible to find other locations for fire controls.[20] The 'C' class bridge was being fitted to all light cruisers under construction.

Existing steering positions were protected by 3in HT plating, but to avoid disturbing the magnetic compass, the upper bridge (and compass platform) was protected by wire mattresses. Weight saved by eliminating the conning tower would go into these mattresses.

Above: In this 1930 view, *Calypso* shows her new block bridge structure, similar in concept to the bridges then being built for new cruisers. The helmsman on the level below the compass platform has only a few deadlights rather than the previous row of windows, because the ship was conned from the compass platform. These ships were excluded from early plans for anti-aircraft conversion because they would have required extra design and construction work due to their lack of superfiring guns forward.

Below: Bridge details of HMS *Caradoc* (*Caledon* class), probably in the 1930s.

Wartime Rearmament

The changes in policy which produced the succession of classes affected the existing ships. Because they were so small, standard additions of weight had large consequences. From the *Centaur* class on, ships were completed with tripod foremasts supporting spotting tops and directors. Similar facilities were proposed for the 'Towns' and the earlier light cruisers, but the ships were badly needed, and it was considered unwise to withdraw them for such extensive work. As a stopgap they received lightweight directors or even had one gun designated the directing gun. By August 1916 tripods had been tentatively approved for the *Arethusa*s (they were fitted to all ships after a successful inclining experiment in HMS *Penelope* in 1917). Ships were also fitted to carry and drop Leon drifting oscillating mines (it was believed that the Germans planned to drop such drifting mines in the path of the British fleet).

In September 1916 a simplified system, not involving a tripod, was approved for HMS *Cleopatra*. A simple director was installed in her fore top. It was described as a complete director sight as far as training was concerned; the guns would be laid for elevation by a pointer from the sights, and fired from the sights. In this way advantage could be taken of the roll to gain range for the guns. When the tripod issue was revived in 1916, it was pointed out that in light condition the ships would have to have water ballast. The first ship so fitted, HMS *Caroline*, was inclined in March 1917. Weights proved somewhat greater than expected. More generally, by mid-1917 the weight situation was so bad that nothing could be added unless compensating weight was landed. That issue was raised when DTM asked for remote searchlight control (rangefinders could not be used when searchlights were on because they would blind the operators).

In March 1916 Admiral Jellicoe proposed adding two more twin tubes at the expense of the after 4in on each side (and eliminating the deck reloads). In the two modified ships (*Calliope* and *Champion*) stability problems might preclude the twin tubes, but Jellicoe hoped that single deck tubes could be fitted. Alternatively, the addition of twin tubes

Left: HMS *Conquest* shows typical war modifications; by the time this photograph was taken, she no longer had a flying-off platform. The forecastle 4in guns was replaced by a single 6in on the centreline in 1916–17. When the conning tower was removed and the tripod stepped, the remaining pair of 4in guns in the waist gave way to a fourth 6in, abaft the funnel. Note that by this time she had reverted to a single pair of deck torpedo tubes on each side. Ships were initially fitted with a single 13pdr Royal Horse Artillery anti-aircraft gun. Later it was replaced. *Conquest* had one 4in anti-aircraft gun and two single 2pdr pompoms. Others in the class varied. *Caroline*, *Carysfort* and *Comus* each had two 3in/20 cwt; *Cleopatra* had two 4in and two single pompoms; and *Cordelia* had one 4in and no pompoms. This photograph shows the ship without No. 2 6in gun, which was replaced by a deckhouse in 1924.

Right: HMS *Caroline* with wartime changes, including the shift to an all-6in main battery. (Josef Straczek)

Below: HMS *Cambrian* post-war, showing her all-6in main battery and her tripod foremast.

might be compensated for by reducing coal stowage (culinary, on the upper deck) and landing some fittings. DNC agreed (in October) that the additional weight would be limited, but in view of the many additions already made to the ships, he wanted a constructor to visit Scapa Flow to discuss what could be done. Unfortunately most of the weights added since commissioning had been high in the ship.[21] Director of Naval Equipment (DNE) suggested approving C-in-C's proposal for the *Arethusa* and *Calliope* classes (except for the up-gunned *Centaur* and *Concord*). By this time the proposal was to add the tubes to the *Arethusa* class and to add them to the 'C' class in place of the after 4in gun on each side. DNC agreed to these changes except for ships with submerged tubes, in which other weights might have to be surrendered. For the two up-gunned ships with submerged tubes, extensive blast screens would be needed if deck tubes were fitted.

In connection with the proposal to replace 4in guns with additional deck tubes, Jellicoe asked for proposals to save topweight. The list was headed by the comment that it was pointless to leave two 4in guns on each side, whose value would be negligible compared to 6in guns. Better to replace them with another 6in gun. Jellicoe rejected the idea on the grounds that even if adequate gun support could be arranged it would be impossible to find sufficient compensating weight.

In May 1916 Admiral Jellicoe asked that the two foremost 4in guns be replaced by a third 6in gun on board the mixed-battery 'C' class cruisers. *Conquest* and *Cambrian* were modified in advance of Board approval (June 1916), and work was done on a not-to-delay basis. The Board also ordered that a 4in HA gun replace the existing 3pdr Vickers gun; ultimately two 4in HA guns would be installed. The Board rejected Jellicoe's proposal to save weight by removing the conning tower and its tube (in *Conquest* the 6in gun had been added without interfering with the conning tower). Deck reloads would be carried, the ships retaining their original pair of twin deck torpedo tubes (except for those with submerged tubes). At about the same time deck reloads and 4in HA guns were authorised for the *Arethusa*s.

Before a decision was made, Commodore (T) (Tyrwhitt of the Harwich Force) was asked his views on these ships under his command, as their role was rather different from that of similar cruisers in other Light Cruiser Squadrons. In November he heartily agreed with the idea of adding 6in guns; in his view, the ships were lamentably underarmed with torpedoes. He also repeated his earlier argument for more 6in guns. Jellicoe agreed with Tyrwhitt (and said that he had opposed the mixed batteries when the ships were being designed) but feared that the work involved in rearmament would take the ships out of action for too long. Nothing was done. However, the additional pair of twin tubes on each side was ordered installed in place of the after 4in gun in all mixed-battery 'C' class cruisers, including those with underwater torpedo tubes. Manufacture of the tubes was approved late in 1916, with first deliveries in December. By this time it was accepted that underwater tubes were unsuited to fast light cruisers, but they could not be surrendered from a stability point of view.

The Board later reversed itself regarding the *Caroline* and *Cambrian* classes, probably at the instance of Tyrwhitt. In a note dated 2

September 1917 he summarised wartime lessons of his force, which often fought German light forces in the North Sea. In most cases he felt a need for more ahead fire; minor actions 'invariably resulted in a chase'. The new *Ceres* class (two superimposed guns forward) were an improvement, but surely more could be done. Surely a ship could have two guns on the beam in addition to one on the centreline. The resulting ship could have a five-gun broadside and could fire four guns ahead and astern on most bearings. Tyrwhitt also rejected submerged torpedo tubes (which had been dropped in light cruisers).

Tyrwhitt's letter revived the earlier idea of replacing 4in guns with a pair of 6in guns on the sides of the ship. In about September 1917 DNC approved the idea. The C-in-C of the Grand Fleet, Admiral Beatty, had already endorsed the idea of adding 6in guns to the light cruisers at a 15 August 1917 conference with the Third Sea Lord. The work was justified by the heavy armament of new German light cruisers, the same issue which had led to the design of the 'D' class. As initially proposed, the guns could be mounted in an *Arethusa* at the expense of the two forward pairs of 4in guns, and in 'C' class cruisers at the expense of these guns plus moving the 4in HA gun to a lower position (the quarterdeck) plus elimination of the conning tower. DNO pointed out that 6in guns on the sides would create considerable blast on some bearings. He was also less than enthusiastic about available numbers of spare guns for rearmament. Further complicating the issue was a decision by the Grand Fleet Operations Committee that all light cruisers should carry fighter aircraft. Normally they were stowed on one side of the forecastle, fouling the position in which Tyrwhitt hoped to put a 6in gun. Several of those commenting on Tyrwhitt's proposals suggested that they really applied to new construction. The outcome (see below) was the design of the 'E' class.

In December 1917 DNO proposed rearming the *Arethusa*s as Tyrwhitt proposed, with five 6in (two on the forecastle abaft the existing gun, one superfiring over the existing after gun, the 4in HA gun being moved to the forecastle deck above the torpedo tubes), giving a broadside of four guns. Compared to DNC's recent proposal, DNO offered an additional centreline 6in gun aft instead of the two 4in DNC would have retained. DNO offered only one additional 6in firing right aft, and to provide it he would have to land or relocate the after torpedo tubes. It also turned out that blast from No. 3 gun would affect the crews in any position further forward (although crew shelters could be provided, as in the *Caledon* and *Danae* ['D'] Classes). They were too important to eliminate. DNC was asked to submit further designs after a conference in January 1918. In one version, the three forward 6in were retained and an additional 6in gun was mounted on the centreline aft, just forward of the original after 6in gun, the after sets of twin torpedo tubes and all 4in guns being landed. The alternative was a superfiring gun forward, the after guns being as in the first alternative. As weight compensation, the conning tower and tube would be removed altogether (in mid-December 1917 C-in-C Grand Fleet agreed to the removal of all light cruiser conning towers). There was very little space between the forward 6in gun and the tripod foremast. The mast could be moved aft a few feet (otherwise its fore leg would pass through the charthouse). Even so, there was very little space between No. 2 gun and the charthouse. A drawing of the proposed bridge structure shows the compass platform built out over the forward end of the chart house, its own forward end supported by a strut extending up from No. 2 shield. Controller preferred this arrangement, even though it precluded the desired flying-off platform for a fighter. Before agreeing to rearmament, C-in-C Grand Fleet wanted to know the earliest date a ship could be taken in hand and whether rearmament could be done at the ships' normal refit ports. He also wanted to know how long rearmament would take. Of the seven ships in the class, HMS *Undaunted* was refitting as a minelayer. The others were due for refits which would begin from 27 April up to about July 1918. Work would probably take five weeks, once material had been received. However, discussion of arrangement (for example, the position of HA guns, and whether the ships would have one 4in or two 3in HA) continued through the summer of 1918.

In July, HMS *Penelope* received an interim refit in which an additional 6in gun was mounted aft (making three 6in in all), but the superfiring structure was not built. She surrendered two (of six) 4in low-angle (LA) guns and her 4in HA gun and added two 3in HA guns. The rest of the 4in would go when she received her fourth 6in gun. She conducted gunnery trials on 16 August 1918 (and again, defects having been made good, on 19 October). *Inconstant* was next, and in September 1918 *Phaeton* was ordered modified in the same way during her upcoming refit. Work continued after the Armistice, *Galatea* being modified during a December 1918 refit. *Royalist* may also have been refitted.

Initially the Board favoured a four-gun armament for the mixed-battery 'C' class, of which two would be mounted forward, one superfiring. 'C' class stability was sufficiently tender that not even the lightweight conning tower could be added in this case. A new position was settled at an August 1917 conference: a fourth 6in gun replaced the 4in HA gun previously mounted on a bandstand (to clear an engine-room ventilator) atop the casing. The two remaining 4in guns on the forecastle were converted to HA. After arguments were raised against this arrangement (initially by the Captain of HMS *Constance*), it was reconsidered (in April 1918) and changed to retaining the existing single 4in HA gun and mounting the fourth 6in gun further aft, just forward of the after control position. *Cleopatra*, the first ship rearmed, was apparently the only one to have the two 4in HA guns. The conning tower was removed (a 1in splinter-proof tube was substituted). The big refit also involved replacing the earlier 15° elevation mounts with 20° mounts, and there was some difficulty over the supply of the modified gun mounts, fourteen of which had been ordered from Coventry Ordnance Works. Given the problem, DNO proposed simply to replace the three 6in on board the first four ships during their ordinary refits, on the theory that they could not be spared for long refits in any case, and this work would have to be done as part of the larger refit planned. All ships, including those with submerged tubes, would be armed with a total of eight 21in torpedo tubes. Topweight was a problem; a telegram to HMS *Calliope* (dated 2 April 1918) approved retention of the two forecastle 4in guns (as HA) only if the planned amidships 6in gun was not fitted and if the 4in HA gun and its bandstand were landed. The two 4in guns were to be removed as soon as the fourth 6in mounting was fitted. At this time C-in-C Grand Fleet was unwilling to surrender ships for the five-week refit involved, but Tyrwhitt (now Rear Admiral Harwich Force) was anxious to have the extra gun.

Also in November 1917 it was decided to provide ships of the Dover Patrol (except destroyers) with two single 2pdr pompoms to defend against German motor attack boats (CMBs and DCBs, the latter Distance Controlled [i.e. remote-controlled] Boats used off the Belgian coast). Cruisers involved were *Centaur*, *Cleopatra*, *Concord*, *Carysfort*, *Conquest*, *Canterbury*, *Penelope*, *Aurora* and *Undaunted*; this addition was also planned for the leaders *Shakespeare* and *Valkyrie*.

In July 1918 the Captain of HMS *Conquest* raised another point. As rearmed the ships would have no guns suited to engaging surfaced U-boats, other than the bow 6in, which he considered too slow-firing (no gun would have much time to fire before a U-boat surprised on the surface dove). The two 4in on the forecastle would have been ideal, but under the revised rearmament scheme they were gone. Rear Admiral Harwich Force proposed 3in guns on the forecastle as anti-submarine weapons in addition to the 4in HA gun aft. DNO much liked the idea, suggesting simply substituting the 3in guns forward (where they could bear on submarines) for the 4in HA aft. For all-round fire the ends of the bridge would be clipped off (as in HMS *Cleopatra*, the only ship with 4in guns in these positions). The deck would be stiffened for anti-aircraft fire. Vice Admiral Light Cruiser Force (of the Grand Fleet) recalled that in May the idea of substituting two 3in for one 4in HA had been accepted in capital ships; it was time to extend it to the cruisers. The single 4in was clearly superior for attacking aircraft at considerable altitude (interpreted as offensive anti-aircraft fire). However, fighters operating with the fleet now dominated the offensive anti-aircraft role. At long range the 6in gun (20° elevation) could be effective against aircraft. When attacking a ship aircraft had to descend to altitudes at which 3in guns would be an effective defence. Also, only HA guns could fire star shell, essential for night fighting. As such they needed good ahead bearings – which the amidships 4in lacked. He therefore agreed with Tyrwhitt, but he wanted the 3in guns on the forecastle to have HA mountings. They should replace the 4in amidships. C-in-C Grand Fleet added that the recent experience of HMS *Galatea* under air attack showed that ahead fire was essential for engaging aircraft. HMS *Caroline* seems to have been the first ship rearmed with 3in rather than 4in anti-aircraft guns (in November 1918).

Some ships became minelayers. In January 1917 the First Sea Lord asked that the 'C' class be considered for conversion, but the *Aurora*s were actually better: *Penelope* could carry 100 mines. Of the 'C' class, removing a 6in gun from a *Caledon* would allow her to carry fifty mines, and a *Curacoa* could carry eighty. *Aurora* was the first to be converted (while under refit, completed May 1917). Ultimately all seven surviving ships were fitted for minelaying. Ships could carry seventy or seventy-four mines. Mines were laid by *Aurora* (212), *Galatea* (220), *Inconstant* (370), *Penelope* (210), *Phaeton* (358) and *Royalist* (1,183), but not by *Undaunted*. On 31 December 1918 removal of minelaying fittings was ordered; when queried by the Naval Staff in November 1919 the responsible Constructor did not know whether that had actually been done.

At the end of the war ships were so overweight that when refrigerating machinery (for magazines) was wanted for service in the tropics, DNC said they had to surrender either a 6in gun or a pair of twin torpedo tubes. The choice was No. 2 gun (in 1923). Probably this was done only in HMS *Conquest*.

Aircraft

Interest in flying-off platforms carrying fleet fighters dated from at least 1916, when Director Air Service (DAS) proposed fitting a landing and taking-off platform to 'C' class cruisers.[22] DNC rejected what he called 'a forecastle deck above the forecastle deck', but suggested a temporary platform which could be unshipped; he preferred the hydroplane lighter then being designed for destroyers. The idea survived because fleet guns were unlikely to destroy German reconnaissance Zeppelins quickly enough, or at sufficient range. The British were almost certainly aware that the Germans had escaped from what amount to a trap in August 1916 because their Zeppelins spotted British warships (actually not the

Yarmouth had the prototype cruiser flying-off platform. Other major wartime modifications included the tripod foremast carrying an enlarged spotting top with a director above it (fitted in 1917 to all but *Falmouth*, lost on 19 August 1916 after being torpedoed by two different U-boats), a 3in anti-aircraft gun between the second and third funnels, and searchlight control platforms aft. Note also the splinter mattresses on the bridge. Additional structure was added abaft the original bridgework. After the First World War *Yarmouth* had her after 6in gun removed and replaced by quarterdeck deckhouses.

Grand Fleet itself). At a 15 August 1917 conference, C-in-C Grand Fleet asked that one light cruiser in each squadron have a flying-off deck for an anti-Zeppelin machine. He selected *Caledon* (1st LCS), *Dublin* (2nd LCS), *Yarmouth* (3rd LCS) and *Cassandra* (6th LCS), *Yarmouth* being the prototype. Additional light cruisers would be converted if the idea proved successful. Because it was not in the 'A-K line' in the van of the fleet, 4th Light Cruiser Squadron would not have any of its ships fitted. Late in October 1917 the Third Sea Lord's Committee on flying-off platforms chose a pair of troughs (with a platform between them) along which the aircraft's wheels could run. Further ships were also fitted.[23] Separate drawings were prepared for the light cruisers (*Caroline*, *Cambrian*, *Aurora*, *Chatham* and *Weymouth* classes); for *Birmingham* and *Lowestoft*; and for *Birkenhead* and *Chester*. Except for *Birmingham*, *Birkenhead* and *Chester*, the forward end of the platform had to be held up by a support fixed to the forward 6in gun shield (as already fitted in *Yarmouth* and *Cordelia*). *Bristol* and earlier light cruiser classes were not to be fitted. Given the critical weight situation, the captain of HMS *Constance* (scheduled to receive a flying-off platform) suggested that the aircraft would have to be landed before a fourth 6in gun was mounted.[24] The Grand Fleet would lose aircraft from five of its twenty-four light cruisers. A conference in April 1918 between the Third Sea Lord and representatives of the departments agreed that it was most unwise for these ships to have flying-off arrangements plus the extra gun, and that the C-in-C Grand Fleet had to be asked about the problem. Erroneous references in correspondence about *Constance* to a revolving platform indicate that it was considered a solution to the problem of a platform built out above No. 1 gun.

Ships with superfiring guns forward required something far more elaborate, because the platform had to clear both guns and the hangar was built under a raised bridge. HMS *Carlisle* was the first ship so fitted. Rear Admiral Harwich Force commented in October 1918 that 'these magnificent ships are completely ruined by their present form of bridge, which completely disorganises the control of the ship and renders the life of the personnel one of misery'. This was despite the great advantage of her 'trawler' bow; in December 1918 she ran into a strong SSW gale at 16–24kts for eleven hours. Despite the short steep sea running, she shipped no water, and her mess decks were dry except for a few minor leaks. However, because the hangar raised it so high, the fore bridge was close under the fore top, hence extremely drafty and cold. 'Everything was blown about in a most distressing fashion. The noise of the wind round the mast and rigging makes the efficient use of voice-pipes and telephones in the Fore Control position most difficult.' *Capetown* was similarly fitted, as were *Dragon* and *Dauntless*. By this time removal of the awkward hangar had already been approved (for the three *Carlisle* class and 'D' class cruisers under construction), the fixed flying-off platform to be replaced by a revolving one on the fore side of the after control position (which had to be moved about 9ft aft). A new bridge would be installed, similar to that in non-platform 'D' class cruisers. Unfortunately for *Carlisle*, she was urgently needed on the China Station, and there was no time to rebuild her bridge before she left (there was hope that it could be done at Hong Kong). Ironically, she was not expected to carry any aircraft while on the China Station. *Caledon*, *Delhi*, *Despatch*, *Dunedin* and *Durban* were all fitted with a revolving flying-off platform aft in 1919. By 1928 only *Caledon* retained the platform.

Flying-off ended with the end of the First World War, although the unwieldy hangar-platform arrangements survived into the 1920s (but were not used). When it was revived about 1926, only the two new 'E' class cruisers and *Caledon* had revolving flying-off platforms, those on board the two larger ships surviving to 1935–6. In all, thirty-two light cruisers had some type of flying-off platform.[25]

The *Emerald* class

The final wartime light cruiser, the 'E' class, apparently began as a result of the comments on existing light cruisers by Commodore (T) Tyrwhitt of the Harwich Force. At about the same time (the week of 20 October 1917) Controller initiated a study of increasing the speed of light cruisers. In November 1917 Controller (Rear Admiral Lionel Halsey) asked DNC and E-in-C for a completely new cruiser design, with greater speed combined with gun power and if possible three guns capable of firing ahead: 'It appears to me that it would be possible to retain the foremost superimposed gun and to take away the foremost forecastle gun.'[26] Two sketch designs were completed in December 1917.[27] The first was Tyrwhitt's ship, with four 6in guns (presumably three on the forecastle), using 60,000shp to reach 31kts deeply loaded on 4,500 tons. As DNC had written in connection with the 'D' class, the alternative was a much larger ship, a 510-footer (six 6in guns) using 80,000shp to make 32.75kts deeply loaded (34kts light). Both designs used lightweight (destroyer leader) machinery (note that a 'D' class cruiser needed almost as much weight as Design A for two-thirds the power). The larger ship offered the standard ammunition stowage (200 rounds per gun) rather than the 150 rounds of the smaller design, and it also offered better protection. It became the basis for further development. Although the design was incomplete, three ships were ordered in March 1918, at the same time as the improved repeat 'D' class. One, *Euphrates*, was cancelled after the end of the war.

Unfortunately no drawing of either alternative has survived, but it seems likely that, like Design A, Design B included two beam guns. Even if they were well abaft the bridge, the ship would have three guns firing over nearly all forward arcs, except for dead ahead. The fleet wanted more space between the machinery spaces, presumably to reduce vulnerability to single hits, and it also wanted better ammunition arrangements. A design was circulated in March 1918. The armament was considered satisfactory, but not the protection. Diagrams of new German light cruisers in the intelligence handbook showed a protective deck over their vitals, but the new British light cruisers had only a protective upper deck (i.e. the deck below forecastle deck level) and in the new ship even that did not extend over the boilers. A British shell hitting the side armour of a German light cruiser would burst, its fragments kept out of the vitals by the deck. Fragments from a German shell hitting the high side armour of the new cruiser would encounter no deck at all, or else would likely pass under the protective upper deck. In 1918 the British were finding it difficult to design a 6in shell which would defeat German-style protection. Why couldn't the new cruiser be as well protected? It seemed that the existing arrangement actually increased the chance that a shell would burst in the ship's vitals.

Naval Intelligence was forced to admit that deck thickness was the most difficult of all data to obtain. Its best hope was that the survivors of the German cruiser *Breslau*, recently sunk in the Mediterranean, would provide information, but their passage (by convoy) to England had been delayed, and there was no hope of obtaining information in time. Intelligence was sure that the Germans had rearmed some cruisers with seven 5.9in guns each, which made the new more heavily-armed British cruiser more important.[28]

DNC could do nothing about protection, because it was really incidental to the structural designs of all the existing fast light oil-burning cruisers. High speed demanded great length, and to minimise structural weight he had integrated armour into the hull – conversely, where there was no hull steel there was no protection. A German-style deck lower in the ship would contribute almost nothing to strength (girder strength depended mainly on the uppermost continuous deck and on the

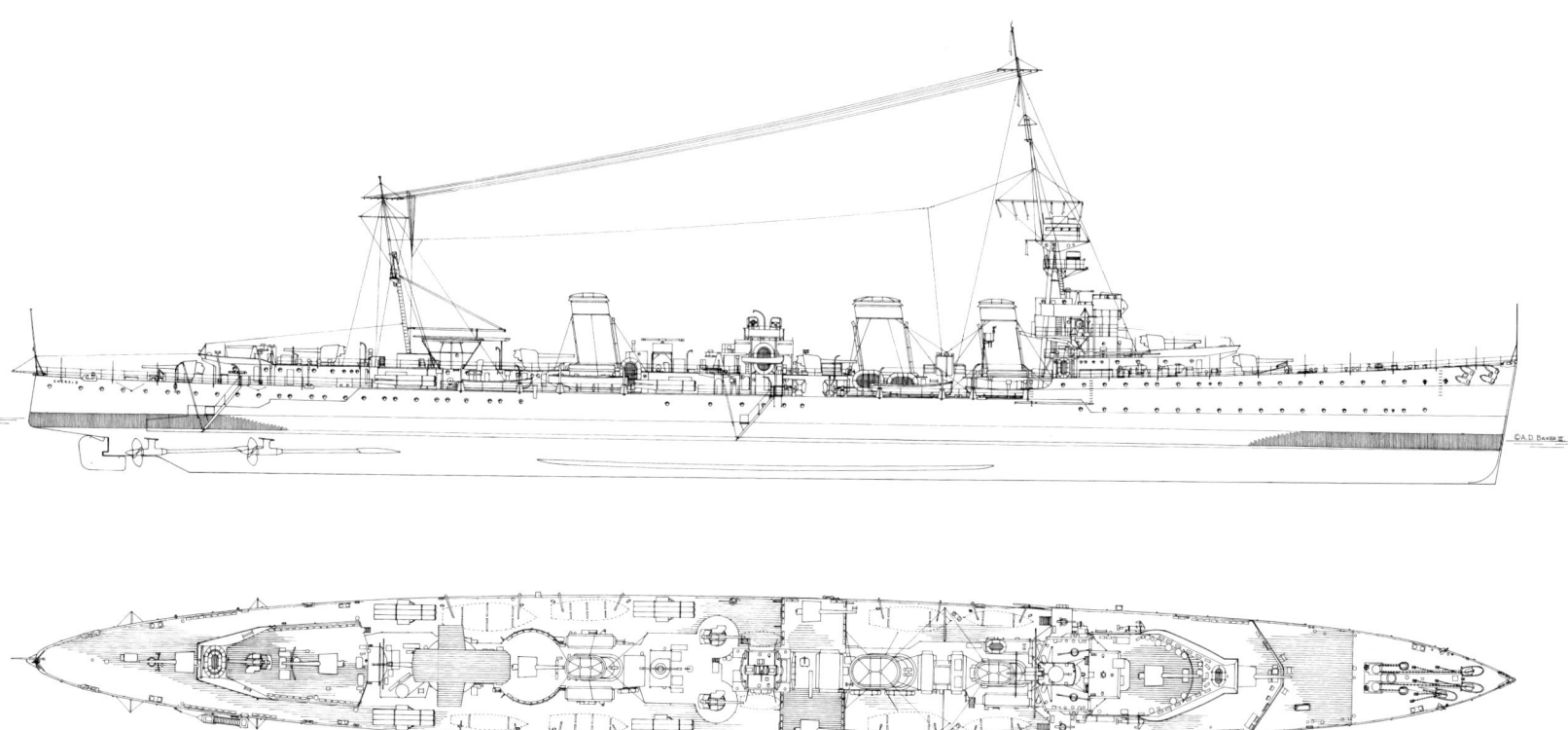

Above: Emerald in March 1926, as completed, with a rotating flying-off platform and triple torpedo tubes. The triples were replaced by quadruple tubes during a 1928–9 refit, with maintenance sponsons added outboard, and with 'torpedo shelter' decks added over the mounts. A 'native hut' was added abaft the second funnel (atop the engine room ventilation trunk). During a refit at Chatham in 1934 the funnels were raised, the flying-off platform replaced by a 44ft catapult, the mainmast moved to forward of the second funnel, an aircraft and boat crane added abaft the catapult, the 15ft rangefinder amidships replaced on a larger platform by a Mk I HA director for the 4in guns, and the 30ft gig to starboard aft moved to the port side (replacing a 30ft cutter) and replaced by a 36ft motor cutter. A depth-charge rack was added to port at the stern, but no Asdic was fitted. The elevation drawing omits the ready-use stowage racks for 6in ammunition (they appear as small rectangles in the plan view). Each rack held one projectile and was tilted slightly so that the projectile point was higher. (A D Baker III)

Below: The 'E' class was the ultimate development of the wartime British light cruiser, reverting to side-mounted single 6in guns to gain ahead fire. HMS *Emerald* is shown as completed, with a rotating flying-off platform aft and with triple torpedo tubes (although the decision had already been made to fit quadruple tubes).

Above: By 1931, *Emerald* had her planned quadruple torpedo tubes, but she (and her sister) retained the old rotating flying-off platform.

Below: Emerald shows her new catapult and crane in this 1933 photograph. She and her sister retained their catapults longer than other British cruisers.

keel, with contributions from the ship's side plating). He could provide vertical armour over their boilers, but the boilers were so large that it was impossible to cover them with an effective armour deck. For that matter, eliminating side armour was unwise because in that case the waterline could be riddled and stability lost in battle. He considered the cruisers virtually large destroyers, which had to depend on speed for protection. He soon repeated this view during discussions of what became the 'County' class heavy cruisers. Moreover, very little could be changed, because in hopes of getting ships rapidly DNC had promised the builders information as early as possible, probably well before the design had formally been approved. Some information had already been supplied, and any basic change would require an entirely new design. DNC's comments were dated 30 March 1918. First Sea Lord accepted these arguments (about 10 April), but wanted the issue kept in mind for future ships.

The design presented to the Board on 2 May 1918 was stretched to add a seventh 6in gun, on the centreline.[29] Protection broadly matched that of earlier light cruisers, the side plating (3in over machinery) being integrated with the hull, covering the machinery and the midships magazines from upper deck to 2ft 6in below the load waterline. The belt extended fore and aft, 2½in thick over end magazines and oil fuel tanks and 1½in to the bow and 2in to the stern beyond. Amidships, the belt extended from the upper deck (weather deck level aft) down to the platform deck. It sloped down fore and aft of the machinery spaces, more sharply aft than forward. In contrast to earlier designs, DNC added 1in decks (at platform deck level, i.e. at the lower edge of the belt) over the magazines and 1in over the machinery at lower deck level (below the upper deck). Because the magazine sides were entirely below the belt, ½in was added to their sides and ends to a depth of 3ft. The 1in deck over the engines, but not the boilers, was described as an improvement on previous ships. Unlike previous cruisers, this one lacked an armour bulkhead over the after end of the machinery.

The machinery arrangement was unusual, spaces being separated for survivability. There were four boiler rooms (two boilers each), the forward engine room being abaft the three forward engine rooms, with No. 4 boiler room abaft that and then the after engine room. The midships 6in magazine and shell room (and a crew space above them) separated the three forward (contiguous) boiler rooms from the forward engine room. The after boiler room was immediately abaft this engine room, but the magazine and shell room for No. 5 gun separated it from the after engine room further aft. The forward engine room drove the two outer shafts. Boilers in Nos. 1 and 4 (end) boiler rooms were in tandem, with oil bunkers alongside, an arrangement revived in the 1930s. The boilers in each room fed a funnel. The boilers arranged side by side in Nos. 2 and 3 boiler rooms all fed the same fatter funnel. The ship used two sets of *Shakespeare*-class (destroyer leader) machinery. As in other light cruisers, the foremast and bridge were right up against the tripod. With so much power on a relatively light hull, they enjoyed high speed; even in June 1944 HMS *Enterprise* could outpace much newer US cruisers (USS *Tuscaloosa* and *Quincy*) off Normandy. DNC could not provide a bulge as in the *Hawkins* class without sacrificing speed, but he argued that the unusual machinery arrangement provided much the same survivability against underwater damage. On this much length, it was not difficult to provide length for a revolving aircraft take-off platform, which both ships carried until 1935–6, when catapults were fitted.

The Board Stamp was applied on 9 May 1918. In August 1918 two versions of the 'E' class were sketched with 7.5in (as in *Hawkins*) rather than 6in guns as responses to a reported German cruiser armed with 8.2in guns (this information was considered dubious).[30]

The *Emerald* class became the basis for initial post-war cruiser thinking. In 1920 replacement of the triple tubes with quadruple ones was approved, although the change was not carried out until 1929, when the ships had their first major post-construction refits. *Emerald* was completed to the original design, but *Enterprise* had the prototype twin 6in turret and DCT.[31]

Construction was drastically slowed by the 1918 Armistice. In the 1923/4 Estimates Controller argued for completion of these two ships rather than the beginning of work on the new *Kents* on the grounds that they reflected advances which might be incorporated in the later ships (presumably referring to the turret and DCT).

The *Hawkins* class

The only other First World War cruiser design was DNC's favourite large overseas design, which became the *Hawkins* class. He repeated his proposal for these ships in a memo dated 12 October 1914 to the Admiralty Board. This time his argument was that for commerce protection something like a *Birmingham* was needed, but larger and faster, with a good steaming radius, and adapted to burn coal and/or oil, so that they could obtain fuel easily abroad. In his view it was an open question whether they needed sufficient protection to resist 6in fire, in which case they would have to be considerably larger than a *Birmingham*.[32] Design work was ordered after a Sea Lords meeting on 9 June 1915.[33] The ships were described as improved *Birmingham*s capable of 30kts, with ten or more 6in guns mounted as far as possible on the centreline. Their machinery should be arranged so that one-fifth power could be achieved with coal, 'a great convenience, if it is not an absolute necessity, in ships intended for use abroad which may have to pick up supplies in all sorts of inaccessible places.' The new 5.5in gun had been discussed as an alternative to the 6in, but it had been rejected in view of reports that new German light cruisers would have 5.9in guns (however, given its lighter shell and therefore faster rate of fire, 5.5in might be the ideal gun for future capital ship secondary batteries – as it was in HMS *Hood*). In a Minute for the Board (and, presumably, the Cabinet) on the new construction programme for 1915/16, the First Lord (Balfour) repeated DNC's argument: 'The rapid deterioration of our older cruisers will necessitate before very long a replacement of cruisers suitable for foreign service; and the Board consider that preparation should be made for six

HMS *Hawkins* is shown dressed overall for a naval review, possibly in 1935. At this time she retained all seven of her 7.5in guns.

BRITISH CRUISERS

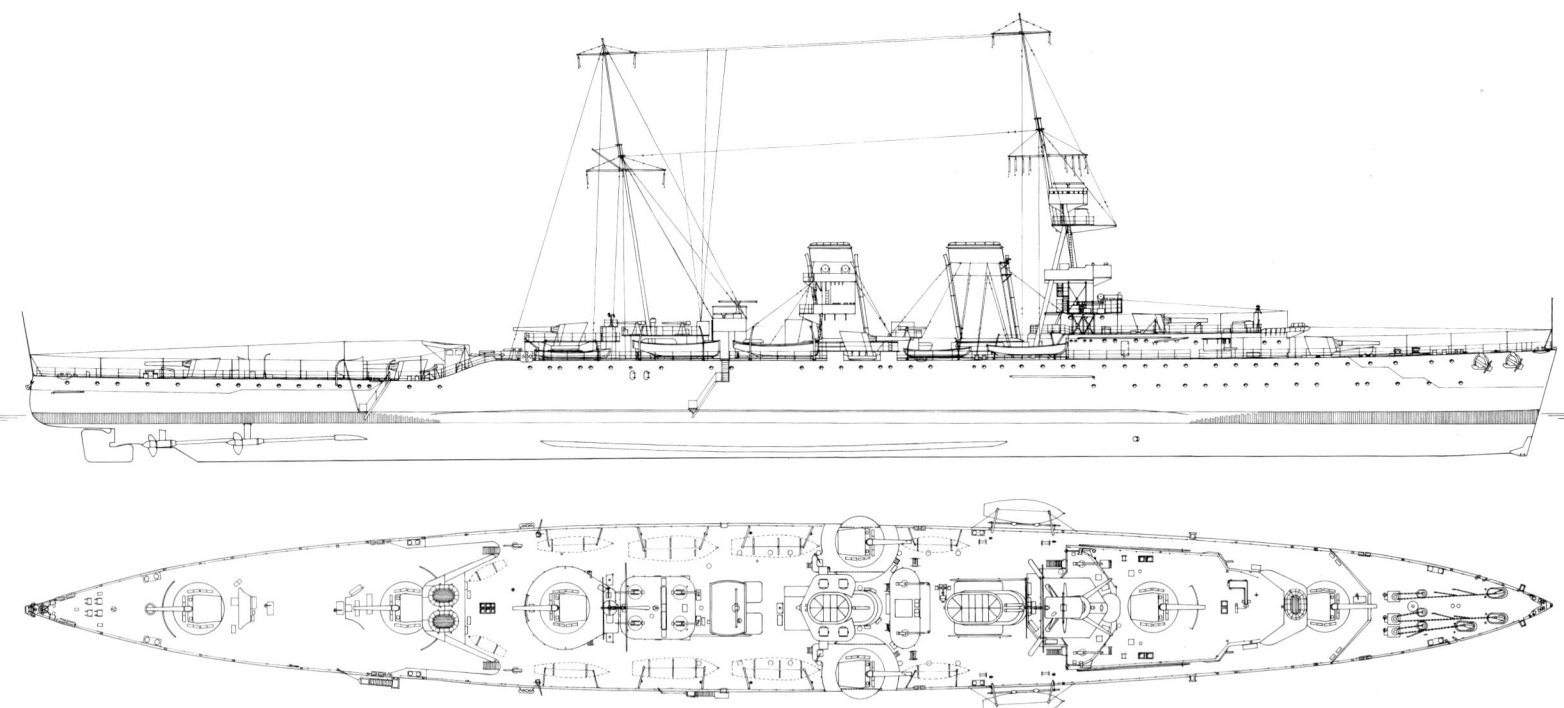

Above: Raleigh as completed in September 1920. The sides amidships had considerable flare above the bulges. Except for the two guns abreast the after funnel, the 7.5in guns were mounted in pits and had a circular working area that revolved with the mounting. The amidships pair had folding semi-circular working deck platforms inboard and outboard of the mountings. Wooden decking was not shown on the as-fitted drawings, but it may have been fitted. Annotations on the plan indicate that the bridge was modified as on later units of the class prior to the ship's loss. She ran aground at high speed at Point Amour, Forteau Bay, Labrador, on 8 August 1922. She was stripped of useful equipment and used as a gunnery target until blown up by a party from HMS *Calcutta* during September 1920. (A D Baker III)

Below: Frobisher as rearmed in February 1942. She had served pre-war as a cadet training ship, and she reverted to that role in May 1945. As shown she had Type 281 air-warning radar (two antennas at her mastheads), a Type 273 surface-search radar in a 'lantern,' and two Type 285 radars atop her 4in directors. She had passive acoustic intercept gear near her bow. The pompom directors lacked radars. The 7.5in guns were controlled by a single foremast director, without any associated radar. In May 1944 at Lyness she was fitted with eight additional Oerlikons: two Mk IIIA mounts slightly staggered abaft the existing centreline mount on the forward director, and six lightweight Mk VIIA (four atop the flag officer deckhouse forward of the break of the forecastle, two abreast the 4in gun on the quarterdeck). At the same time her four fixed torpedo tubes were apparently removed. As a cadet training ship (conversion completed 1938) she retained a single 7.5in gun (the aftermost mount on the quarterdeck) but retained the forward pair of 4in HA mounts. On the platform formerly occupied by 'B' 7.5in gun a two-level deckhouse was erected, and the other 7.5in on the quarterdeck was replaced by a crane to handle a floatplane stowed at the after end of the forecastle deck. Annotations on the 1942 plan show that it was intended to remove the foremost 7.5in mount on the quarterdeck and to add two more Oerlikons. However, the ship emerged in May 1945 with only three 7.5in guns ('A' gun, the superfiring gun aft, and the foremost quarterdeck mount). An open 6in mount replaced 'B' gun. Of the 4in guns, only the quarterdeck mount was retained. The ship also had eleven Oerlikons, including two in the circular tubs high in the superstructure formerly occupied by pompom directors. The depth-charge rack was removed, and the aftermost 7.5in gun replaced by a 21in torpedo tube (probably a twin). All splinter shielding around the 7.5in and 4in guns was removed, and the Type 281 radars replaced by a single antenna at the top of the foremast (probably Type 291). (A D Baker III)

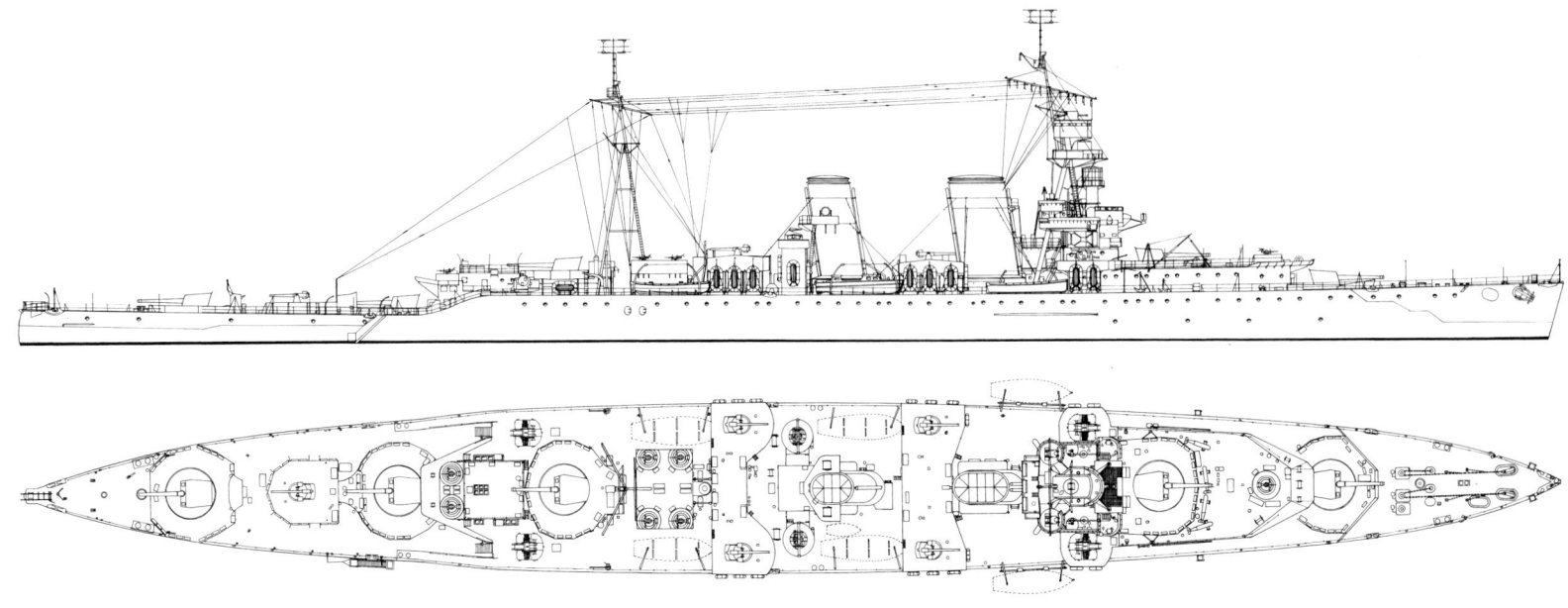

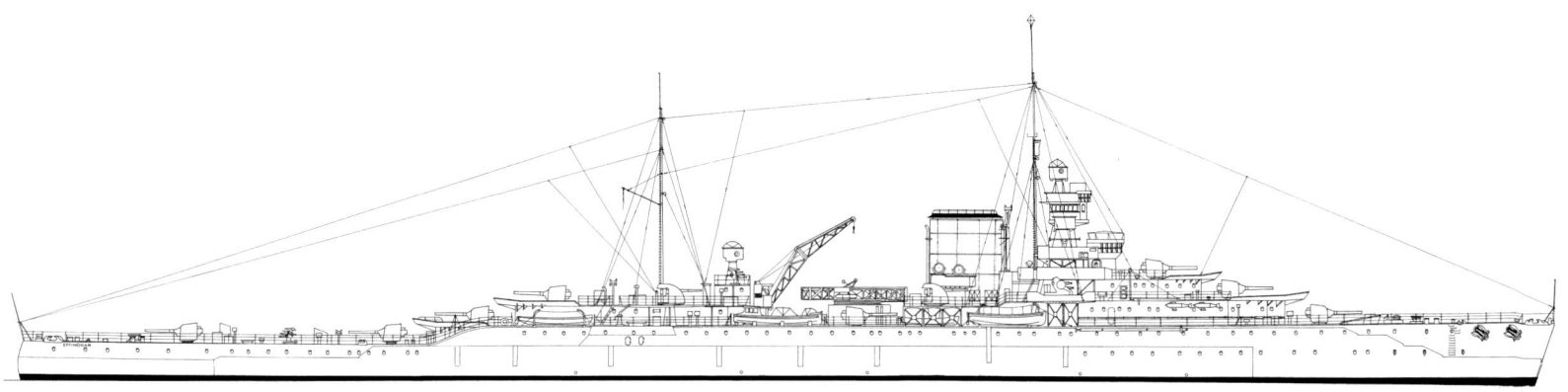

Above: *Effingham* as rebuilt and rearmed, 1939. (John R Dominy)

Below: *Vindictive* as a repair ship, 30 March 1940. She was armed with six 4in HA guns, two quadruple pompoms and two single depth-charge release gear (total of six depth charges). The upper masting and rigging arrangement is estimated from the one available photograph, and is incomplete. Note the unusual height of the boot topping, shown as the ship was painted in 1943. By that time she had received six single Oerlikons, three on each side atop the large workshop structure abaft the funnel. Also by that time the forward-most of the two large Carley floats had been replaced by a smaller one. She did not carry any radar early in 1943. However, she seems to have been fitted with Type 286 by August 1943 and with Type 291 January 1944. By August 1943 she also had Type 285 for 4in control. As a demilitarised training ship she had already been reduced to six operational boilers, and the inboard turbines had been removed; but the inboard shafts, minus propellers, were retained, as was the hull armour.

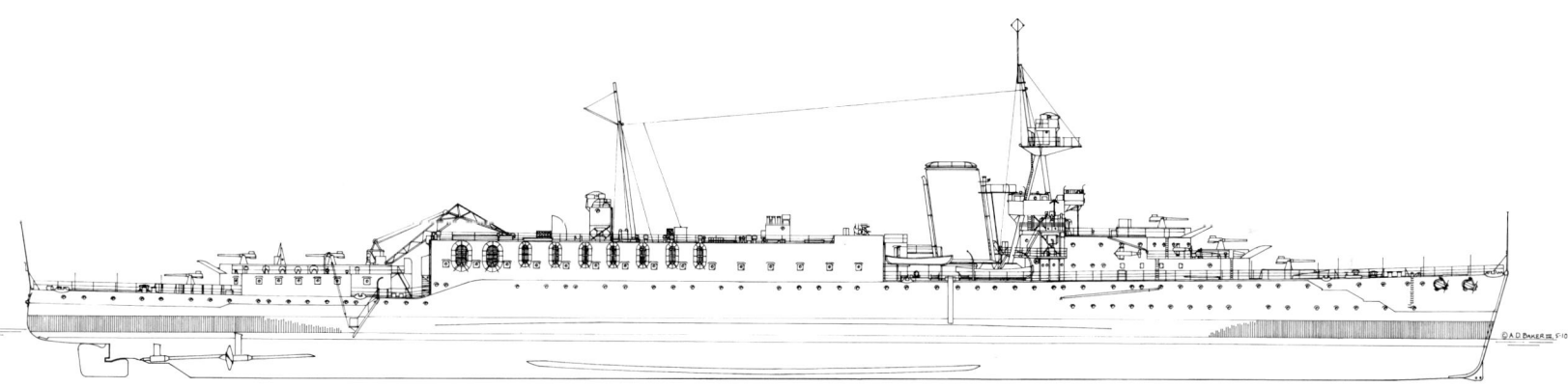

such vessels later in the year.'[34] Alternative designs were already being prepared. Current knowledge of the German shipbuilding programme made it unnecessary to order any more North Sea light cruisers. Once a design had been chosen and orders placed, ships might be delivered in eighteen months.

DNC looked at five alternative batteries: eight, twelve and fourteen 6in; two 9.2in and eight 6in; and eight 7.5in QF. The chosen design was armed with seven 7.5in plus ten 3in (four of them HA). In the proposals for HMAS *Adelaide*, the 7.5in guns were in gunhouses (at least as sketched), but in the ship finally chosen they were in central-pivot mounts like those of 6in guns. Five were mounted on the centreline (two were superimposed forward and two aft, with another after gun on the same level as the lower of the superimposed pair), and two on the sides amidships, roughly abeam the second funnel. It is not clear in retrospect why the ships had a secondary 3in (12pdr) armament, but these guns may have been wanted to beat off attacks by small fast craft, the 7.5in firing too slowly for that. Two were mounted in embrasures in the forward superstructure which supported 'B' 7.5in gun, two on the bridge wings, two on a platform abaft the forefunnel, and two abaft (and to either side of) the superfiring 7.5in gun aft. All were subject to blast from the 7.5in guns. The four 3in HA guns were concentrated on a platform just forward of the after superfiring 7.5in gun. While the ships were being designed, underwater torpedo tubes were rejected for the light cruisers in favour of deck tubes. The same reasoning ultimately led to the addition of four above-deck tubes just forward of the superfiring 7.5in gun aft.

The most unusual feature of the design was the combination of inward-sloping sides and bulge similar to that chosen a few months earlier for the 'large light cruiser' (or light battlecruiser) HMS *Furious*, recently designed (a similar form was chosen for HMS *Hood*). DNC claimed later that subdivision provided a two-compartment standard of protection against underwater damage, and that the sloping sides contributed substantially to the ship's girder strength. The slope (10°) was chosen so that the beam at the waterline matched that of the outward-flaring forecastle deck. The limited draft and high freeboard provided the ship with great reserve buoyancy. Belt armour extended from the forecastle deck down to the platform deck. Machinery occupied three adjacent boiler rooms separated from the two engine rooms by a 7.5in magazine and shell room which served both the two wing 7.5in guns and the foremost of the after centreline mounts. The coal bunker was at

BRITISH CRUISERS

Above and below: HMS *Frobisher* is shown in June 1930, the after superfiring gun having been removed to provide space for a floatplane. It was served by a mobile crane, but the ship had no catapult, at least at this time. Neither crane nor floatplane is present here.

Left: At a naval review on 15 July 1935, *Frobisher* displays her aircraft arrangements. By this time both the after superfiring gun and one of the two quarterdeck guns had been landed.

Below: Under the terms of the 1930 London Naval Treaty, the Royal Navy disarmed its *Hawkins* class cruisers in return for ending construction of heavy cruisers by other navies. HMS *Frobisher* is shown as a nearly disarmed training ship.

the after end of the midships boiler room. Uptakes from the forward boilers and from the forward boilers in the midships boiler room were trunked into the forefunnel, and from the after boiler room and the after boilers of the midships boiler room into the after funnel. Four boilers occupied each engine room, with four coal-burners in the amidships room and four oil burners in each of the others. As a measure of the lower efficiency of a coal-burning boiler, it took four coal-burning boilers to produce a sixth of the total machinery output.

By this time the existing 3in and 4in anti-aircraft guns no longer seemed adequate. First Sea Lord asked whether the 'E' class could have two of its 6in guns on dual-purpose mountings. This seems to have been the first proposal for such weapons.[35]

Although six ships were planned, four (*Raleigh*, *Effingham*, *Frobisher* and *Hawkins*) were ordered in December 1915. A fifth (*Cavendish*) was ordered in April 1916. In 1917 *Cavendish* was ordered converted into a carrier (and her construction expedited), and in 1918 she was renamed *Vindictive* to commemorate the ship which had served heroically as a blockship at Zeebrugge. In November 1917 the Board ordered the three least advanced (*Raleigh*, *Effingham* and *Frobisher*) converted to burn only oil fuel, their rated power increasing to 70,000shp. Their four amidships coal-burning boilers were replaced by Yarrow small-tube boilers like the eight in the other two boiler rooms. *Frobisher* and *Effingham* received only two new oil-burning boilers, and were rated at 65,000shp.[36] When *Hawkins* was converted in 1929, the coal-burning boilers were removed without replacement, but the remaining boilers were boosted to give an output of 55,000shp.

The ships were criticised for their gun arrangement, the captain of HMS *Hawkins* pointing out that the two guns aft on the quarterdeck were very wet with any sea running, or with the ships steaming at high speed, and that because the berthing rails were taken down when the guns were cleared for action, their crews, particularly at the after gun, were uncomfortably close to the ship's side. Controller asked the captain of the other ship of the class in service, HMS *Raleigh*, whether he agreed. He pointed out that after guns in all light cruisers were wet (and not from speed alone), but that No. 1 gun was worse. Also, the after gun suffered from extreme vibration at high speed; 'this is so bad that I consider it would adversely affect the nerves of the men stationed at the gun for any length of time under these conditions.'

He considered the 7.5in gun unsuitable for light cruisers due to its low rate of fire (two-thirds that of a 6in under favourable conditions, offering 30 per cent more weight of metal but worse control due to more time between shots); and the large gun crews required to supply four rounds per minute ('absurdly out of proportion to those required in, say, a 15in turret'). They amounted to twenty-five men for No. 1 gun, but to forty-three for the superfiring No. 2, and to thirty-nine and forty-four for the two waist guns, which shared one magazine and one shell room. Ammunition for No. 2 gun had to be passed along the deck between dredger hoist and hand-up to the gun. It would be a vast improvement if magazine and shell room could share a single enlarged dredger hoist (which had to be large enough to take the protective cases over the bagged powder). DNO saw proof that any new heavily-armed cruisers should have power-worked shell and cordite hoists directly under the

Above and above right: HMS *Effingham*, the third *Hawkins* class cruiser, was rearmed with 6in guns, preserving her as a combatant ship under the 1930 and 1936 treaties. Had war not come in 1939, the other two ships would also have been rearmed, though possibly not with single open 6in guns. Presumably single 6in guns became available as 'C' class cruisers were converted into anti-aircraft ships. *Effingham* was lost when she struck a rock during the 1940 Norwegian campaign.

Above and above right: HMS *Frobisher* is shown in April 1942, newly rearmed, just before joining the Eastern Fleet. She had been laid up at the outbreak of war for rearmament, to follow *Hawkins*, but work was slowed, presumably by the press of more urgent projects. She was to have had the same armament as *Hawkins*, but two more multiple pompoms replaced the two waist 7.5in guns. The rearmament refit, at Plymouth, lasted from 5 January 1940 through March 1942. *Frobisher* commissioned on 10 February 1942, ran trials, and sailed for the Clyde (where these photographs were probably taken) on 4 March. Note that the two waist 7.5in guns were not remounted; she was reduced to five such guns. Other armament at this time was five 4in HA, four quadruple pompoms, seven Oerlikons, and two Lewis guns. She also had four single fixed above-water torpedo tubes. Note the Oerlikon just forward of No. 2 gun, and the 4in gun just abaft one of the quarterdeck 7.5in guns.

guns.[37] He was looking forward to the 8in cruisers then being discussed (see below). Although the point generally was not made, a cruiser had relatively few potential magazine positions. If 8in guns had to be directly above each such position, and if more than three or four were desired, they had to be in multiple turrets or gunhouses – as was done in the next class of large cruisers, the *Kent*s (whose earliest stages coincided with this discussion). DGD defended the 7.5in gun as far more formidable than a 6in, given its much heavier shell (200lbs) and greater stopping power (as demonstrated in trials with the ex-German cruiser *Nürnberg*). It offered longer range, and it could be controlled at much longer range because the splashes from its shells were much more visible. DGD noted that *Raleigh* had carried out relatively little firing, hence these virtues had not become evident (she would soon be wrecked). These were important arguments for DGD, who strongly espoused 8in guns for future cruisers.

A preferable armament would be six 6in guns (in positions now occupied by Nos. 1, 2, 4, 5, 6 and 7 guns), the two waist guns being replaced by one on the centreline. A 6in gun should replace the after control position on the centreline, at the same height (on the forecastle) as No. 4 gun. Two 4in HA guns could replace the waist 7.5in. The HA guns were poorly placed, blocked by rigging (including wireless aerials).[38]

Although in 1922 it seemed impossible to imagine rebuilding the *Hawkins* class, exactly that idea was raised in 1925. The advent of the

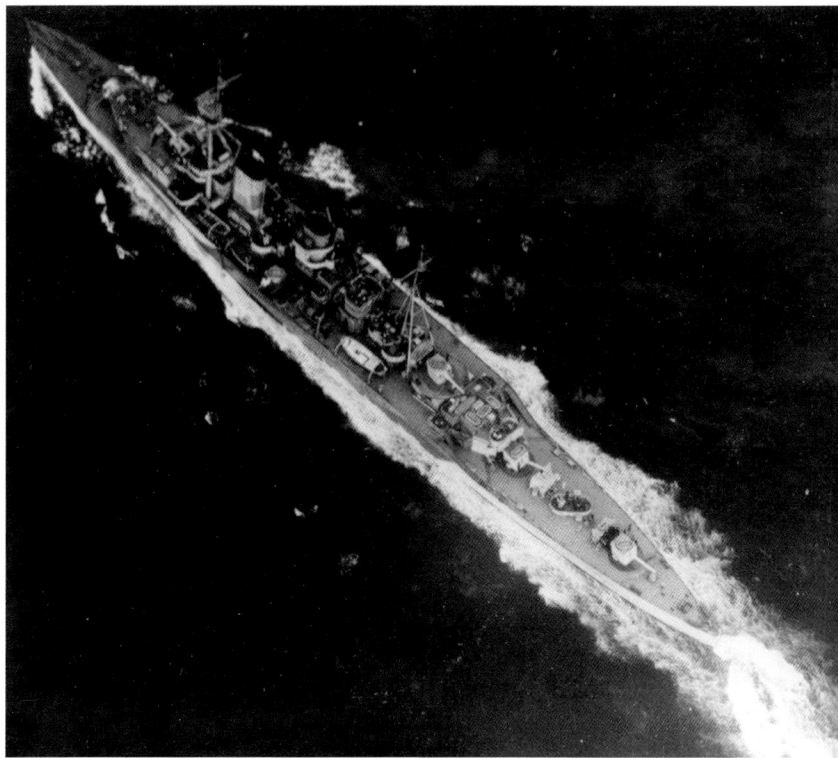

Above and right: As rearmed, HMS *Hawkins* differed from *Frobisher* in having all seven 7.5in guns (as refitted at Portsmouth, 4 December 1941 – 7 May 1942 she had two rather than four quadruple pompoms). Hawkins also had two single pompoms, seven Oerlikons and two Lewis guns, plus four fixed above-water torpedo tubes. Radar was installed (Types 281, 273 and 285). These photographs were taken in June 1942. *Hawkins* had been brought back into service in 1940. At that time she had four 4in HA guns, four single pompoms and two machine guns. The quadruple pompoms were replaced by octuple pompoms during a further Portsmouth refit, 8–23 August 1944, and two single Oerlikons were added.

Below: HMS *Hawkins* late in the Second World War, presumably after the 1944 refit. Note the Oerlikon atop the shield of No. 2 7.5in gun.

8in cruisers led to interest in rearming the four surviving ships with three 8in turrets each. In March 1925 DNC produced a sketch design for *Frobisher* showing two turrets forward and one aft, using the same turrets which had been selected for the *Kent* class. This design would also apply to *Effingham*, and to *Hawkins* if the ship were converted to all-oil fuel (*Vindictive* was a more complicated proposition because of the catapult in 'B' position). DNO pointed out that although there was sufficient space forward, the side bulkheads of the magazines and shell rooms aft, which were also the sides of the shaft passages, could not be moved to provide sufficient space, and the vertical height was also insufficient.[39] A new twin 8in design would be required, and it was not clear that it could provide a sufficient rate of fire. Ammunition capacity would be about 120 rounds per gun, rather than the 150 in a *Kent*. Worse, rearmament would add weight, and the *Hawkins* class was already close to the 10,000-ton limit. Not only would the new armament weigh more (although crew would be cut somewhat), but they would need an additional generator. Even with only 100 rounds per gun, the ship would gain 170 tons; she was already only 150 tons below the limit, and, DNO wrote, 'great importance was attached by the last Cabinet and Foreign Office to keeping within [the 10,000-ton] limit'. Rearmament would slightly reduce metacentric height (by an acceptable 6in), and the ship would trim slightly by the bow (which could be offset by using oil fuel tanks in the correct sequence). About 100 tons of oil would be sacrificed (300nm at 12kts). The arrangement of HA guns would improve. The cost would be high: £517,500 for guns, mountings, and ammunition and about £170,000 for the reconstruction; converting three ships would buy a new ship of about the same size. Moreover, the *Hawkins* class had poor magazine protection compared to a *Kent*. It would take about nine months to prepare a design for the after mounting and another eighteen months to deliver the first such mounting, plus nine to twelve months to rebuild the ship and run trials. Against all of this, the six 8in power-operated guns offered considerable advantages both in rate of fire and in weight of broadside (6,144lbs vs

Above and below: Frobisher is shown on 1 May 1945 in the Firth of Forth, having been converted into a training ship with a wide variety of weapons. Note that her air-search radars, but not her other radars, have been removed.

3,500lbs per minute); and the 8in shell offered more explosive, weight for weight, than a 7.5in shell. Turrets themselves offered considerable advantages, including blast protection from neighbouring guns. Controller noted all the disadvantages cited by DNO and cited another: other powers would notice what was being done, and might accelerate their own cruiser programmes accordingly. The improvement was not worth the money. ACNS agreed, pointing out that the armour protection of the *Hawkins* class was outdated, their magazines unprotected against long-range plunging fire from 8in guns.

ACNS had another way to look at the issue. The four *Hawkins* could be seen as individual counters to four Japanese 33-knot cruisers armed with 5.5in guns, superior to them in everything but speed. Laying the four British ships up for eighteen months to three years to become counters to the four *Furutaka*s would leave the British with the problem of commissioning four 6in cruisers from reserve – ships of which the Royal Navy was short. Four such cruisers would have to be built, their cost added to the cost of rearming the *Hawkins* class. ACNS assumed a modern 6in cruiser would cost £1.3 million, compared to £2 million for a *Kent*, the difference of £700,000 being about what it would cost to rearm a *Hawkins*. On this basis rearming the *Hawkins* class and building the required new cruisers would cost as much as simply building four *Kent*-class cruisers. Against this, DGD suggested that the Royal Navy risked entering the next arms control negotiation with fewer 8in cruisers than the Japanese. DCNS agreed with ACNS that rearmament was not worthwhile. The Sea Lords decided not to pursue the idea. ACNS also asked DGD about simply modifying *Hawkins* class magazines.

Since the London Naval Treaty of 1930 limited the number of British cruisers with guns of more than 6.1in calibre (see below), DNC was asked whether the four large *Hawkins* class cruisers could be rebuilt with the standard light cruiser battery of four twin 6in guns.[40] They would be somewhat large for this battery, but they had reasonably modern hulls. Turrets and ammunition (200 rounds per gun) would weigh 466 tons, compared to 460 tons for the 7.5in guns (150 rounds per gun), so main armament weights would balance out. The only major addition would be protection to the ammunition lobbies and gun supports. *Frobisher* was rated at 9,860 tons standard and *Effingham* at 9,770 tons, so both offered sufficient weight. Existing magazines and shell rooms were large enough, and could be adapted; the ammunition lobbies would displace some living space, but on the other hand complement would be reduced. Adding 3in to the magazine crowns would cost another 250 tons (1½in would cost 125 tons). Adding a catapult crane, and seaplane would add another 50 tons, and with extra magazine armour would bring displacement somewhat above 10,000 tons. DNC attached an outline profile to his report to Controller on 1 April 1930; it has been lost.

DESTROYER-KILLERS

Above and above right: The last ship of the class, HMS *Cavendish*, was renamed *Vindictive* to honor the cruiser used as a blockship at Zeebrugge in 1918, and was completed as a carrier. She is shown soon after the First World War. In this guise she retained four of her seven 7.5in guns: one right forward, one right aft, and the two waist guns. She operated in the Baltic during the war of intervention against the Bolsheviks, ran aground, and was severely damaged. Repairs at Portsmouth continued into 1921, and she was then laid up (used at times for trooping). She was converted back into a cruiser in Chatham in 1923–5. (View from aft from RAN Historical Branch)

Nothing was done. The treaty required that navies reduce their cruiser forces to the allowed numbers by 31 December 1936. By that time the British had invoked an escalator clause, which allowed them to retain some smaller cruisers which would have been scrapped. However, they much valued the limit on cruisers with larger guns, and it was retained in the 1936 treaty which superseded that ratified in 1930. Something had to be done with the four *Hawkins* class. *Frobisher* and *Vindictive* were reduced to training ships in 1932, *Frobisher* being reduced to five 7.5in than then to one 4.7in in 1936. As a training ship *Vindictive* was reduced to two 4.7in guns and her original pair of submerged torpedo tubes. She also lost half her boilers. *Hawkins* was disarmed and laid up in 1937.

In 1937–9 *Effingham* was rebuilt with nine 6in guns in single rather than twin mounts, plus the usual four twin 4in HA guns, two octuple pompoms, and the usual pair of quadruple 0.5in machine guns. She was given an E.IV.H catapult. Her ten boilers were reduced to eight, the after boiler room being converted into oil fuel tankage. That made it possible to trunk the remaining uptakes into one large funnel. Power was reduced from the original 66,000shp to 61,000shp. About June 1938 it was proposed that the other two ships be rearmed with six twin 5.25in guns and fitted with heavy catapults as in current cruisers and the rebuilt 'Counties'.[41]

Instead, *Hawkins* and *Frobisher* were rearmed with 7.5in guns after war broke out. *Hawkins* emerged in 1940 with seven 7.5in, four 4in HA, two quadruple pompoms and seven Oerlikons, plus four fixed above-water torpedo tubes. For *Frobisher*, a planning conference was held on 27 August 1939, the ship to be taken in hand in September. At this time she was to be fitted with four single pompoms, two of which would later be replaced by quads. The submerged torpedo tubes would be removed.

Above and below: Vindictive is shown as a cruiser in 1927. She had the first Royal Navy catapult, mounted athwartships atop the hangar retained from her service as a carrier. The prototype Carey catapult was driven by compressed air via a system of pulleys, the air piston travelling 20ft to move the catapult cradle 60ft. Capacity was 7,000lbs at 45kts. This catapult first launched aircraft (Fairey Flycatcher and IIID) at sea in October 1925.

The ship was completed in 1942, with five 7.5in (seven were originally intended; the two waist guns were not re-installed), five 4in HA, two quadruple pompoms (increased to four in 1941), two single pompoms (removed in 1941), three Oerlikons (seven in 1941), and four upper deck torpedo tubes. The support for No. 5 7.5in gun had to be replaced (all the other gun supports remained in the ship), the above-water tubes had to be replaced, the crane and seaplane gear had to be removed, and 4in guns and magazines re-installed.

In August 1944 a conference was held to decide how to convert *Frobisher* into a training ship for 150 cadets. She was then being repaired at Chatham after battle damage, and was to be taken to Rosyth for conversion beginning in September. The 7.5in main armament was not required, nor was a speed over 15kts (for which two shafts would suffice; the after boiler room, with its two boilers, would be shut down). The pompoms would be removed, the Oerlikons retained. It was decided to replace No. 2 gun with a 6in or 4.5in gun, and No. 5 gun with, if possible, an above-water torpedo tube; if possible No. 4 gun would be removed to add clear space on the quarterdeck. All of the 4in guns would be landed. As a cadet training ship in 1945, *Frobisher* had three 7.5in guns, one single 4in HA gun (in 'B' position), thirteen single Oerlikons,

DESTROYER-KILLERS

Like her two sisters, *Vindictive* was largely disarmed and reduced to training duties under the 1930 London Naval Treaty. However, she also had her after boilers removed, permanently reducing her speed and making it unprofitable to rearm her like her sisters.

a set of quadruple torpedo tubes, and two light machine guns. She was ultimately replaced in this role by the newer heavy cruiser *Devonshire*.

Once war broke out *Vindictive* was taken in hand at Devonport for a conversion similar to that of HMS *Effingham*. This work had low priority, so little had been done by early October. Given her reduced speed, a simpler proposal was made to use equipment like that in an armed merchant cruiser.[42] The minimum conversion would provide four 6in on her centreline; another two could be added, one on each side of the old 7.5in platforms. In addition, she could have three single 4in HA guns (with a HA computer position containing a destroyer-type Fuse-Keeping Clock[FKC]), and four single pompoms. The computer position would be placed on the centreline in the existing superstructure. E-in-C was to consider converting the after boiler room (now a laundry) to oil fuel stowage, giving a total of 2,640 tons rather than 1,000 tons, and a radius at full speed of 3,470nm rather than 2,240nm. Added weight would decrease the ship's maximum speed from 24kts to 23kts. The earlier above-water torpedo tubes would not be replaced. Instead, *Vindictive* was converted into a repair ship.

Below: Vindictive was converted into a repair ship in 1939–40, completing on 30 March 1940, in time to serve during the Norwegian campaign. She was assigned to the South Atlantic in 1940–2 and to the Mediterranean in 1943–4. She is shown entering Mers-el-Kebir in 1943. At that time she was armed with six 4in HA guns and two quadruple pompoms; by April 1944 she was also credited with six Oerlikons (at least two of which are visible here).

CHAPTER 4
WAR EXPERIENCE

If Britain had entered the First World War because the rise of German sea power was a mortal challenge, in 1918 the two rising sea powers, the United States and Japan, were potential future enemies. British naval rhetoric, both within and beyond the Admiralty, often referred to both. However, from the beginning, it seems to have been accepted that war against the United States was unlikely to the point of being nearly impossible, whereas war against Japan was quite possible. The requirements of such a war shaped the inter-war British fleet, much as the requirements of war against Japan shaped the inter-war US fleet. For both the Royal Navy and the US Navy, the stated possibility of war against the other had important advantages. Success against Japan would demand a considerable edge in naval power, at the least because of the vast distances involved, but after the First World War, the British and US governments and populations had little taste for naval expenditure. Given a declaration that the navy was intended mainly or solely to deal with Japan, the government of the day could demand that the navy cut down to parity with the smaller Japanese navy – which would be fatal in the event of war. Also, having a nominal opponent whose capabilities were relatively openly known made it far easier to guess what had to be done to maintain a fully modern fleet. Japan was far too secretive for that.

From a strategic point of view, the greatest surprise of the First World War was the enormous dislocation caused by a handful of German light cruisers operating against British trade in 1914–15. The war showed that seaborne trade was the single most important naval target in the British Empire. The two threats were surface cruisers and U-boats. U-boats had been effective mainly near ports, because they had limited mobility, hence could not easily search the broad oceans (in the Second World War the Germans would solve that problem with radio intelligence and wolf packs). In the Pacific the likely main future threat to British Empire trade would be cruisers and other surface raiders. The British tried and failed several times simply to outlaw submarines. Many believed that by violating the pre-1914 rules of blockade when adopting unrestricted submarine warfare, Germany had in effect committed suicide by drawing the United States into the war, so when planning a war against Japan in the 1920s, American officers rejected a strategy of

WAR EXPERIENCE

By 1918 the Director of Naval Ordnance was convinced that cruiser guns should be enclosed and power-operated. His first step in that direction was the prototype enclosed mounting on board HMS *Diomede*, shown on 10 July 1929. (Photo by Allan C Green via State Library of Victoria)

such warfare for fear that it would bring Britain, with the largest merchant fleet in the world, into the war (this reason evaporated in 1941, and unrestricted submarine warfare played a major role in defeating Japan). These considerations made cruisers a primary threat to trade, and focussed inter-war British cruiser thinking heavily on the needs of trade protection. The British did not dismiss the possibility of war in Europe, but it was clearly a lesser problem.

Against the new problem of war against Japan, Britain after 1918 clearly lacked the financial resources she had enjoyed in 1914. In August 1919 the planning assumption that there would be no major war for ten years was introduced: the 'Ten Year Rule'.[1] The Admiralty interpretation was that the fleet had to be modernised by 1929. Because the existing fleet had been designed for a North Sea war, that implied a major new construction programme, particularly in cruisers (see the next chapter). The British government had nothing like that in mind. It tolerated Admiralty calls for new cruiser construction, but it avoided long-term programmes designed to build up the large cruiser force envisaged.

The great wartime tactical surprise was that actions so often turned into chases. Before 1914 it was assumed that the enemy fleet would seek battle. However, the German fleet, and German naval forces in general, fled if they did not enjoy an overwhelming advantage. After Jutland, for example, the British sought a new kind of battleship shell which could stop a typical German battleship, that being the first step needed in forcing any sort of decision. The last cruiser design of the war, the 'E' class, was conceived to fight a chasing action, with a combination of maximum speed and maximum ahead fire.

Actions would normally be fought at maximum range, far beyond what had been envisaged for cruisers before the war. That happened in 1914 at the Falklands. The maximum range argument also applied to chases, because longer gun range could make up for any slight deficiency in speed. Giving cruisers longer gun range meant giving them the same sort of sophisticated fire-control systems which were being developed for new battleships.

The Post-war Cruiser Force

Pre-1914 discussions of a 'one power' or '1.6 power' standard had always omitted the US Navy. The US entry into the First World War in April 1917 suggested that the United States could no longer be omitted from British strategic calculations, and by 1919–20 the Admiralty was promoting a 'one power standard'. Whether or not it had any expectation of fighting the United States, this approach guaranteed an essential margin over the more likely enemy, Japan. Thus there could not be enough navy to send a fleet to the Far East and to retain a sufficient deterrent fleet in European waters – exactly the problem which would bedevil the British in 1939–41. The Admiralty therefore promoted a strategy of 'naval mobility'. It demanded wholesale adoption of oil fuel, and the build-up of sufficient reserves throughout the Empire. Calculations of the necessary reserves were a staple of Admiralty planning throughout the inter-war period.[2] They focussed on the requirements of war in the Far East. The shift to oil fuel also required that the Royal Navy protect the routes from the sources of oil, hence that the fleet had to be larger than in the days of coal produced in the British Isles. In the 1930s there was some public interest in reversion to coal, and DNC and E-in-C had to demonstrate (for internal consumption) that ships so fuelled would

Converted to a gunnery training ship, *Diomede* no longer had any need for her special enclosed gun mount, so it was replaced by a standard open 6in gun. The ship is shown in the Firth of Forth, 25 October 1943. Note the twin power Oerlikon at shelter deck level under the bridge.

have neither the speed nor the endurance of those burning oil. Because it had a global chain of bases, at which ships could fuel, the Royal Navy had little interest in fuelling at sea.

Calculations of fleet shape assumed that cruisers would be organised in five-ship squadrons, and usually a 25 per cent margin was allowed for refits. The earliest estimate was developed by the Admiralty Reconstruction Committee, formed in February 1918 to decide the appropriate shape of the future Royal Navy. Early in 1919 it envisaged three main fleets (Atlantic, Home and Mediterranean) plus a China Fleet built around a single battlecruiser. One cruiser squadron would serve with each fleet, and the Atlantic and Mediterranean fleets would have one extra cruiser to serve as the destroyer commander's flagship. In addition, each of four stations (East Indies, Cape of Good Hope, South America, and Western Atlantic) would have one cruiser squadron. That made eight squadrons, forty (actually forty-two) ships, requiring a total of fifty active ships; the committee wanted another ten in reserve. Given the wartime needs of the Grand Fleet, the allocation per fleet was quite small. This calculation probably explains an offhand remark by Admiral Beatty at the Washington Conference (1921) that the Empire needed about fifty cruisers. After the desired number was set at seventy a few years later, as Chancellor of the Exchequer Winston Churchill used Beatty's lower figure as a debating point.

Attention shifted East. In 1918 the Dominion Prime Ministers invited Lord Jellicoe, who after commanding the Grand Fleet had been First Sea Lord in 1916–17, to tour the Empire to advise them on naval defence. Jellicoe left with his staff aboard HMS *New Zealand* on 21 February 1919, returning home that autumn. The First Lord of the Admiralty Sir Eric Geddes omitted strategy from the sorts of recommendations Jellicoe was to provide, but inevitably he developed a strategic framework. He considered Japan the only likely threat to the British Empire in the East, citing Japanese acts and statements during the First World War. Jellicoe laid out the basis of inter-war British naval strategy: a main fleet based in the East. No base had the infrastructure (or the specialist personnel) to support such a fleet in peacetime, so for a time between the outbreak of war and the arrival of the fleet the Japanese fleet would be more or less free to act, for example to attack Australia. The long Eastern trade routes would require convoy protection. Jellicoe identified Singapore as a potential fleet base, but pointed out that without fortifications it could be taken by a large Japanese landing force. He envisaged creation of an Empire Pacific fleet including ten cruisers to form an 'A-K Line' plus another fourteen (and thirty-seven armed merchant ships) for convoy work. Given his wartime obsession with enemy torpedo attack, and the role of cruisers in dealing with it, Jellicoe seems to have understated what the future main fleet needed, perhaps to limit the cruiser requirement.

From the Dominions' point of view, Jellicoe's most important conclusion was that while the United Kingdom would provide the battle fleet (including its cruisers), the Dominions should provide the trade-protection cruisers. In fact there was no real prospect that the Dominions would provide the numbers Jellicoe stated, let alone the much larger ones later proposed. The Canadians showed little interest in a trans-Pacific threat, and the Indian Government concentrated on the Russian threat from the north. Australia alone was unlikely to build a large enough fleet, and New Zealand had much smaller resources. South Africa had only a limited interest in building a navy.

The Admiralty professed itself unhappy with Jellicoe's report, but it reached much the same conclusions. Possibly Jellicoe had been too explicit in dismissing any American threat, hence making it difficult for the Admiralty to explain to British politicians why the Royal Navy should be much larger than the Japanese. The Admiralty certainly did adopt the idea that trade-protection cruisers should be provided, at least for the Far East, by the Dominions there. Thus post-war discussion of the trade-protection cruiser were headed 'Colonial Cruiser'. Although Admiral Jellicoe himself soon left to become Governor-General of New Zealand, his Chief of Staff Captain F C Dreyer resumed office as Director of the Gunnery Division (DGD) of the Admiralty. In that capacity he was deeply involved in formulating the characteristics of the post-war cruisers.

In 1921, when the 'Colonial Cruiser' (soon to become the 10,000-ton 8in cruiser) was being discussed, the idea was raised that the threat of such ships against Japanese seaborne trade could help tie down Japanese naval forces until the main fleet arrived. The Germans had demonstrated exactly that possibility with their *Emden* and other small cruisers in 1914–15. The raider threat presented by the British would also help protect their own trade. Within a few years planners apparently saw cruiser trade protection mainly in terms of running down raiders; almost all convoy escorts would be armed merchant ships taken up from trade. By 1924, the naval staff had concluded that to fight a war in the East the Royal Navy needed seventy cruisers. Unfortunately no credible explanation has survived, but at times it was stated that twenty-five were for the fleet and forty-five to protect trade.[3] Each figure included the 25 per cent refit margin (and no explicit margin for losses). On this basis an explanation can be deduced. The fleet required the 'A-K' Line (ten cruisers, two squadrons) and two more squadrons for pro- and anti-destroyer work, a total of twenty ships (with the margin, twenty-five). The Royal Navy stationed two cruiser squadrons in China, both for security and, given the new idea of pro-trade raiding, as a sacrificial force to hold the area until the main fleet arrived. Other squadrons were needed in the South Pacific (Australia and New Zealand), in the Indian Ocean (Africa Station), and in the Atlantic (North and South as separate stations) – a total of seven cruiser squadrons, thirty-five ships. Probably a thirty sixth was added as flagship of the Australia-New Zealand station. With the 25 per cent margin, that made forty-five ships.[4] The seventy-cruiser figure became a driving force in inter-war British cruiser thinking. It included not only the Royal Navy but also the Dominion navies.

Lessons Learned Committees

Two committees were formed to collect lessons on future material: first the Fire Control Requirements committee and then the Post-war Questions (PWQ) Committee.[5] The Fire Control Committee saw light cruisers primarily as gunnery ships, though carrying a powerful torpedo armament. Thus no increase in torpedo armament could be accepted at the expense of gun armament. It was pointless to use cruisers for torpedo attacks when larger numbers of destroyers would be more likely to succeed. It also seemed unlikely that aircraft carriers would relieve cruisers of their scouting role, which in turn would require them to carry scouting aircraft. In order of priority, cruiser requirements were numbers, speed, guns and torpedoes, and the 'D' class seemed to carry the right armament, although the 'E' class offered better speed and sea-keeping. An important pointer to the future was the recommendation that capital ships use dual-purpose guns as their secondary armament, preferably in dual mountings (to reduce personnel) using QF weapons. This reasoning affected light cruisers, with their anti-destroyer mission, as much as capital ship secondary batteries. Eventually a 5.1in and then a 5.25in gun was chosen for battleship secondary armament – and thus as a cruiser main battery.

The Committee considered the number of main battery fire control

BRITISH CRUISERS

positions in modern capital ships and cruisers excessive (though secondary control might be deficient); torpedo control positions could be reduced. Ships should have a plotting room below and behind armour containing their fire-control computer (Dreyer Table or the planned automatic computer); it would also provide torpedo control. The Committee proposed a director control tower (DCT), to carry the director, its personnel, the spotting and rate officers, and a rangefinder, as the only way to simplify above-decks control arrangements.[6] The DCT would be mounted atop the bridge, eliminating the usual heavy tripod carrying a fire-control top and a director. It combined all the fire-control instruments in such a way that it was impossible to train any of them on anything but the correct target. The Committee envisaged a cruiser bridge stiffened to carry the DCT atop an extra tier at its after end, the rest of the platform carrying the night defence fire-control instruments formerly carried in the foretop. The torpedo control and firing position should be on the level of the navigating bridge and abaft

Left: For the Royal Navy, the First World War was a radio war; wireless (radio) intelligence made an enormous difference, providing both ocean surveillance and an ability to detect enemy ships beyond the horizon. The Royal Navy distinguished between navigational (safety of ships and aircraft) and strategical (location of own and enemy ships and aircraft) direction-finding; the former required much greater accuracy than the latter. However, the latter required something more, rapid search and quick sense finding. After the First World War the Royal Navy seems to have led all others, by a wide margin, in providing its ships with radio direction-finders. By the mid-1930s, sets were designated by pairs of letters, the first indicating the type of aerial and the second the frequency band. First letters were A (Adcock), F (fixed frame), L (loop), and R (rotating frame coil). Second letters were A (alternative HF and MF), C (common aerial, usable for HF or MF, with a receiver which could handle both bands using two aerials at once), H (HF only), M (MF only), and S (simultaneous use on HF and MF), of which the S series did not yet exist. Numbers indicated different sets within a series. Although navies had adopted high-frequency communication, they generally still used medium frequency (MF) as well. This one was aboard HMAS *Perth*, photographed (probably in New York) on her delivery voyage to Australia in 1939. The rotating coil (its motor is visible) served D/F Outfit LM. *Perth* and her sisters were scheduled, in 1939, to be fitted with Outfit RA, already installed on board HMS *Apollo* (*Hobart*). Outfit FC was planned for all the new cruisers of the *Dido* and *Fiji* classes and *Liverpool*, *Gloucester* and *Manchester*, *Belfast* and *Edinburgh*. It also equipped *Kent*. Further installations were planned for *Dorsetshire* and *Norfolk*, for other heavy cruisers (except *Cumberland*, assigned an FH), and for *Frobisher* and *Hawkins*. The earlier *Southampton*s were credited with RC. FA equipped *Aurora* and *Ajax*. It was to be installed on board the other *Arethusa*s, *York* and *Exeter*, the *Leander*s, and *Adventure*. Some ships (*Leander*, *York*, *Exeter*, *Dorsetshire*, *Suffolk*, *Berwick*, *Australia*, *Canberra*, *London*, *Devonshire*, *Shropshire*, *Sussex*, *Adventure*, *Hawkins*, *Frobisher*, *Vindictive*, the *Danae* class, *Coventry*, *Curlew*, *Cairo*, *Calcutta*, *Curacoa*, *Cardiff* and *Adventure*) had Outfit SD, which used the triatic (mast) stays as antennas (it was between the funnels in *Coventry* and *Curlew*). SD and LM operated at 60 to 600 kHz. FH operated at 700 kHz to 20 MHz, and the other sets at 60 kHz to 20 MHz. The wartime HF/DF sets, which were not installed on board cruisers, were designated in the FH series, and a new VHF series was created (FV1 etc). In 1937 HMS *Cornwall* was credited with an FH at her foremast head, plus an LM (*Berwick* had SD and FH). *Penelope* had FM at her foremast head. LM was installed on board *Arethusa* and *Galatea*; *Achilles*, *Neptune* and *Orion*; *Amphion* and *Sydney*; *Cornwall*; and *Cumberland* and *Suffolk*.

Above: The new cruiser *Gloucester* in 1939, showing her FC masthead array. The object under the searchlight, under the bridge, is a tactical and torpedo rangefinder.

Above and below: Direction-finders were retained after radar was introduced, creating some problems (the best place for the direction-finder, the top of the mast, was also the best place for the radar). *Euryalus*, shown as completed in 1941, illustrates a standard solution for the FC coil on her mainmast. She retained the coil as late as 1943. Note that she had a ranging radar (Type 285) for her after HA control system but not for the forward one.

Above and inset: The converted anti-aircraft cruiser *Coventry* had an FA2 coil, initially on a tall mainmast and then, as here (in 1940), below the radar transmitting aerial on her foremast. She was converted because, by the mid-1930s, the Royal Navy was acutely aware of its limited anti-aircraft firepower. A 1921–2 naval anti-aircraft gunnery committee discussed the creation of specialised anti-aircraft ships, but that idea was dismissed on the ground that it was better for all major units to be self-defending. No such idea was even brought up when a further anti-aircraft gunnery committee met in 1931. However, the conversion of the two cruisers *Coventry* and *Curlew* seems to have been part of the anti-aircraft rearmament programme developed in the wake of the 1932 report.

it, with searchlight control from the torpedo control positions. A second DCT would be mounted aft, as well as duplicate torpedo controls. By the time the committee reported, a mock-up DCT was being built at Portsmouth. The DCT became a key feature of post-war British warships. The prototype was installed on board HMS *Enterprise*. In later ships the DCT was associated with the new automated fire-control computer, the Admiralty Fire Control Table, but not in *Enterprise*, which had the earlier Dreyer Table. The Committee also wanted a 12ft torpedo rangefinder (for long-range 'browning' shots) on each side of the bridge. The foremast should be well abaft the bridge, to avoid transmitting vibration to the bridge, and to reduce interference with the director.

About the time the final report of the Fire Control Requirements Committee was ready (March 1920), the PWQ Committee reported. For fleet work, the light cruiser had to drive in the enemy's screen in the course of her screening duties, beat off enemy light craft trying to attack the battle line with torpedoes, and press home her own torpedo attacks on the enemy's battle fleet. For ocean (i.e. trade protection) warfare, a light cruiser had to deal with other light cruisers, with disguised raiders, with cruiser submarines operating on the surface, and with other enemy craft. During the recent war, superior gun power was decisive in all cruiser actions fought to a finish. That the Germans rearmed their cruisers with 5.9in guns was taken as evidence that they learned this lesson. The *Hawkins* class outgunned enemy light cruisers, but were expensive, and could easily be outgunned in turn, e.g. by ships with 9.2in guns. It might be more efficient to reinforce cruiser squadrons abroad with the older battlecruisers than to buy more large cruisers. War experience 'having amply proved' the 6in gun ships, they should be developed further.[7] The light cruisers had been quite successful, but although valuable in the North Sea and in the Adriatic, the destroyer-killers could not have performed the ocean work done by the 'Towns'. Something larger was needed.

The Committee assumed that the same gun and mount would arm light cruisers and would become the capital ship secondary weapon. The Fleet favoured hand-worked guns, the largest of which was the 6in, with its 100lb shell. It was too light to get the best ballistic results out of such guns, but during the action of 21 November 1917 crews became exhausted loading them into guns high enough above the deck to give 30° elevation. Loader trials at HMS *Excellent* suggested that 90lbs was the heaviest projectile which could be continuously and rapidly hand-loaded (in July 1920, however, the Naval Staff chose the 6in over the 5.5in as the future light cruiser and capital ship secondary weapon). It accepted a heavier (112lb) projectile. Elevation should be at least 40°, to get sufficient range.

DNO argued that any future light cruiser should be armed with enclosed 6in mounts. The open single mounts in use had been conceived

A major lesson of the First World War was the need for tactical situational awareness, using a pair of plots of the current situation (tactical and strategical). Admiral Jellicoe instituted such plots in 1914, but the idea was first really tested at Jutland (it enabled him to deploy effectively, but it did not enable him to follow the course of the battle effectively). By 1918 the Royal Navy understood plotting, and it developed the technique further between the wars. Tactical plots made possible complex tactics, including night fighting. Navies allied to the Royal Navy during the First World War all learned how to use tactical plots (the Germans were apparently totally ignorant of the idea). Plotting in turn required that a ship be able to measure the ranges to other ships with which she was not engaged. Meanwhile the Royal Navy became interested in long-range torpedo firing, which also required accurate ranges (allied navies, including the Japanese, surely picked up the idea during the First World War). Post-war, many ships, including cruisers, were fitted with tactical rangefinders outside their fire-control systems. HMAS *Australia* is shown on a visit to New York in the 1930s. One of her two tactical rangefinders is visible next to the base of the square tower supporting her director.

before the advent of director control, and were therefore designed for individual gunlaying. With director firing the primary means of control, visibility from the gun mounting was no longer very important. Cruisers were inherently wet, and they had limited centreline space. Length limited not only space for gun mounts, but also internal space for magazines – which, for efficient supply, should be directly under the guns. DNO began work in the spring of 1918 on a single enclosed 6in gun to replace the forecastle guns in 'C', 'D' and 'E' class cruisers with minimum alteration. DNO saw such a mounting as a way to keep the bow guns of the 'E' class dry despite their high speed. The single mounting was installed aboard HMS *Diomede*.

Single open mountings were ill-adapted to the high angles of elevation required for long ranges, which required higher trunnions. The loaders on the ship's deck had to lift shells and cordite much higher. In an enclosed mount, shells and cordite were lifted by power onto a platform placed closer to the trunnions. This sort of arrangement was far heavier than the earlier open mounting, so it had to be power-trained (with manual backup). It was hand-elevated with power backup. It was hand-loaded. The weight of the electro-hydraulic motor in the gun house could counterbalance the weights of the rear numbers in the crew and of the rear overhang of the gunhouse required to provide the necessary space.

BRITISH CRUISERS

From the beginning DNO considered that only multiple mounts could make the most of a cruiser's length. DNO preferred twins to triples because ammunition supply to the third gun would be difficult (a single hoist could easily supply two guns alternately), because the mounting would be too heavy for manual back-up training, and because a nine-gun salvo was too large for control (double salvos would be five and four guns, and a three-gun salvo would be insufficient). Unless the ship had twelve guns, the triple would not be worthwhile.[8] Twin mountings (eight or ten guns) would give four- or five-gun salvos alternating left and right guns. Four mountings would suit a successor to the 'E' or *Hawkins* class (in their case, with 7.5in guns), and three would suit a 'C' or 'D' class successor. Unhappy experience with the old 'County' class armoured cruisers made twin mounts unpopular, but the problems of the past could be attributed to cramped design, 'most unsatisfactory' electric training and ammunition hoists, and poor gunsights. These guns were also in one double cradle. A new gun mount would repeat none of these errors. The PWQ Committee liked twin mounts because they saved space, simplified ammunition supply, and reduced personnel (fewer supply parties). At Jutland personnel in the open batteries of HMS *Chester* and *Southampton* suffered badly, though the guns themselves were little damaged. Gun positions should therefore be isolated and splinter-proofed.

Once loading arrangements for the enclosed gun had been thoroughly mocked-up, the two gun mounting firms (Vickers and Elswick) designed twin mounts, whose rear portions were mocked up and tested at Whale Island, with satisfactory results. The base requirement was that the two guns should be capable of being loaded and fired as rapidly as two separate single mounts. The design adopted had a short revolving trunk suspended from the turret turntable, supplied with ammunition from a fixed flashtight 'tween deck compartment (ammunition lobby) immediately below it, which in turn was supplied by a fixed hoist from the magazine and shell room. This short-trunk mounting equipped British cruisers until the *Belfast* class. HMS *Enterprise* was completed with the prototype twin 6in mount, which was also the secondary weapon of the *Nelson* class.

The Committee recommended an improved 'E' class light cruiser mounting four twin 6in, with a maximum range of 20,000yds, two turrets superfiring forward and two aft, offering a heavier broadside and heavier ahead and astern fire.[9] The ship should also have two 3in anti-aircraft guns (or whatever calibre was chosen for battleships) plus machine guns arranged so that three could fire on any bearing. Because the Committee recommended specifically against a specialised torpedo cruiser, it called for the maximum number of tubes for the conventional cruiser, choosing two quadruple tubes on each side. By this time quadruple torpedo tubes had provisionally been arranged for the 'E' class. Since the Committee considered the 'E' class the limiting cruiser size, it rejected any increase in protection. No bulge would be suitable, but a partial torpedo bulkhead might be worked in over the vitals. The ship should carry as many aircraft as possible, and one set of davits on each side should be capable of hoisting the smaller type of Coastal Motor Boat (motor torpedo boat).[10] She should have sufficient endurance for Atlantic and Pacific operations (6,000nm at 14kts, 15 per cent less than actual endurance). Speed would be reduced from 32kts to 30kts to provide the extra oil stowage required.

The step beyond the single enclosed mount was a twin, which in the early 1920s was seen both as a battleship secondary weapon and as a cruiser primary weapon. It was tested on board HMS *Enterprise*. The cruiser also tested the new director control tower (DCT), which combined the functions of a director with those of a gunnery-control position. Without such a combination, it was possible for a control officer to order corrections based on splashes surrounding the wrong target. The problem returned for air spotters, and it caused real difficulties during the battle of the River Plate in 1939. In HMS *Enterprise* the new DCT was linked to the existing semi-manual Dreyer Table computer, but later ships linked it to the new automated Admiralty Fire Control Table, the DCT providing the table with feedback, and vice versa.

WAR EXPERIENCE

While the PWQ Committee was still deliberating, in February 1920 DNC assigned constructor Charles S Lillicrap (a future DNC) to estimate the characteristics of the proposed ship.[11] Lillicrap assumed that the vitals would be protected against 6in shellfire at 7,000yds (the origin of the 7,000yd requirement is unclear). That would require 200lb (5in) cemented armour on the ship's sides and a 50lb HT steel deck, based on sketchy information about the 6in armour-piercing (AP) shell. An 'E' class cruiser had an 80lb HT side included in the ship's hull strength (80lbs for the two upper strakes for sagging and 60lb for hogging). A total of 250 tons could be saved by removing the lower strake and by moving some of the plating to the deck. The resulting deck would more than meet the requirement (Lillicrap omitted to point out that deck area was quite limited). Required additional side protection would amount to 770 tons (total 1,070 tons), hardly a trivial amount. To make the desired endurance, a new cruiser burning oil at the same rate as an 'E' class cruiser would need 2,600 tons of oil, giving a new deep displacement of 10,060 tons rather than 8,940 tons, this figure including 120 tons more for the additional 6in gun, ammunition, turrets, new controls fore and aft, and additional torpedo tubes. Against that, the reduced speed would cut required power from 80,000shp to 60,000shp, so one boiler room (165 tons) could be given up. That would roughly balance weight added in armament and control. Lillicrap concluded that on 9,250 tons the ship could carry another 300 tons of endurance oil, for an endurance of about 5,100nm. In effect he showed that added endurance and anything like serious protection could be very expensive in terms of ship size. DNC must have remembered his repeated wartime comments that cruisers should not depend on armour for survival.

DNC commented that to provide each pair of guns with its own magazine and shell room aft would require about 40ft more length abaft the machinery than in the 'D' and 'E' classes, due to the difficulty of arranging these spaces in the narrow after end of the ship. More space would also be needed forward of the machinery for the magazines and shell rooms and also for the larger fire-control spaces and for the larger bridge superstructure above decks. Extra length would be needed in any case to balance the greater length aft. Reduced power required might make it possible to mount all eight guns and the new control and bridge arrangements on 'E' class length.

A Sea Lords' conference on 24 March 1920 concluded that a single type of light cruiser would serve for both fleet work and foreign service. It should not exceed 'E' class tonnage.[12] Future cruiser guns should defeat 3in side armour and a 1in deck. The choice of gun calibre depended partly on elevation, because at higher elevation a shell would carry further and would be more effective against enemy deck armour. Higher elevation required enclosed gun mounts. The lightest gun which would defeat 3in side armour at 20,000yds was the 9.2in (which could

Enterprise was modernised in 1943, the most obvious consequence being removal of half her torpedo tubes. She began the war with seven 6in guns (the twin mounting forward, two in the waist, one on the centreline amidships, and two aft), three single 4in HA guns, and two single pompoms. She emerged from a large repair at Colombo (11–18 March 1941) with two 6in removed but one quadruple pompom added. By October 1942 she also had four single Oerlikons. During her major refit on the Clyde, 25 December 1942 – 31 October 1943, the two 6in were restored and the single pompoms and Oerlikons removed. A second quadruple pompom was installed, and she received six twin Oerlikons. At the same time she was fitted with major radars (Types 281, 272, 282, 284 and 285). During a short Devonport refit (3 – 29 February 1944) her catapult was removed and four single Oerlikons added. Type 650 missile jamming gear was installed during a short availability at Devonport (27–31 March 1944). The main battery ranging radar (Type 284) apparently did not work very well, and the ship was badly shaken up during a December 1943 engagement with German destroyers in the Channel. She was repaired only sufficiently to conduct shore bombardments during and after D-Day; once the army had advanced beyond the range of naval gunfire she was withdrawn and quickly reduced to reserve.

defeat it out to 22,000yds); a 7.5in gun could defeat it at 16,000yds. A 6in gun could defeat both side and deck protection at 14,000yds, while the 5.5in would probably defeat the side armour at 12,000yds.

DNC and DNO argued against thinking in terms of armour protection. Jutland seemed to show that protection was a matter of chance: it took hours of 12in shelling to sink *Scharnhorst* and *Gniesenau* at the Falklands in 1914, but the three battlecruisers at Jutland, which had better armour, were all blown up by single salvos.[13] On the other hand the light cruiser *Conquest* survived a salvo of heavy shells, all of which easily penetrated her armour. No modern British light cruiser was sunk by gunfire. Although 6in fire destroyed the German light cruiser *Mainz*, 6in fire by HMAS *Sydney* neither sank nor stopped *Emden*. Prolonged gunfire did sink the German light cruisers *Nürnberg* and *Leipzig* at the Falklands. Overall, 'no absolute conclusions as to the stopping power of guns of a given size against such vessels can safely be drawn'.

Alongside the conventional cruiser, the Committee envisaged a convoy cruiser to deal with enemy cruiser submarines operating on trade routes or blockading British ports. It should have superior or equal gun power, slightly superior surface speed, equal endurance, good seaworthiness, and superior habitability. German U-cruisers devoted about 10 per cent of their displacement to machinery and fittings for underwater operation; that suggested to the Committee that a ship of about the same displacement, powered by diesels like a submarine, could enjoy superior gun power and slightly superior speed. It could be expanded to the point of being able to fight a light cruiser, though with her diesel engines she would be slower. She would have her cruiser guns on the centreline, a triple anti-submarine torpedo tube on each beam, a small seaplane scout, 'the best submarine detection apparatus' (with two-mile or better detection range), long-range wireless and depth charges. She could exploit her submarine detector by remaining stopped while her seaplane scouted. She could also relieve armed merchant cruisers, which during the war had been badly needed on trade routes. Nothing came of this proposal.

Anti-Aircraft Weapons

A Naval Anti-Aircraft Committee was formed in 1919.[14] Its 1921 recommendations helped shape inter-war cruisers. Probably its most important was to develop an automated form of HA fire control, similar in concept to the new automated form of surface fire control, one director and its computer controlling multiple guns.[15] Existing methods were manual.[16] Heavy-calibre fire control required both that the shell be brought close to the air target, and also that its fuse be set ('cut') so that it exploded within lethal range (few if any shells would actually hit, but a near miss would shatter an aircraft). The faster the target, the more such systems were affected by inherent errors, such as delays in passing information. In 1919 fuse timing was set by a powder train whose length could be varied, but the next step was a mechanical time fuse. It could be set by a fuse-setting machine controlled by a fire-control system (such fuses were again recommended by the 1932 Naval Anti-Aircraft Gunnery Committee, and they were adopted shortly afterwards).

Like its surface counterpart, the automated British High-Angle Control System (HACS) combined a director (HADT, HA director tower) and a below-decks computer (HACS Table) in a High Angle Control Position (HACP) corresponding to the DCT and computer in a transmitting station of the LA battery. The British configuration differed from that in the contemporary US and Japanese navies, in that it separated the relatively small HADT from its below-decks computer. Initially the HADT and anti-aircraft rangefinder were separated, but they were soon combined, to preclude ranging on the wrong target (the Japanese were to have just this problem during the Second World War). Because of the connection between HADT and HACP, to increase the number of separate targets a ship could engage required not only more topside space for HADTs but also more internal space for additional HACPs. On the other hand, because the anti-aircraft calculating function was in the HACP below decks, it was possible to rig an aloft director for both LA and HA fire, an important consideration in cruisers with limited topside space. HADTs first appeared aboard the 'County' class cruisers and modernised battleships. Probably because the Royal Navy worked so intensively on its HACS so early, the system was badly flawed, compared to later US systems.[17] As a consequence, in 1941 the Royal Navy was interested in buying the US Mk 37.

The only forms of air attack in sight were level bombing, torpedo bombing (the Royal Navy led the world in this), and strafing attacks against ships' bridges and upperworks. Dive bombing was not considered because it did not yet really exist (except as a variation on low-level bombing). Fleet anti-aircraft fire was also expected to be important as a way of driving off enemy spotters (who would make enemy surface fire more effective) and as a way of driving off reconnaissance aircraft. During the First World War, such fire was sometimes called offensive anti-aircraft fire.

The fleet already had 3in and 4in HA guns which, it was understood, would keep attackers at such a height that bombing would be inaccurate, and would also prevent enemy spotters from staying over British ships. A ship armed only with lighter guns could be bombed effectively. It also seemed that airmen were 'more affected by a few large shell bursting close to them than [by] a greater number of smaller shell'. Thus maximum calibre should be set only by the point at which shells became so heavy that they could not be fired quickly enough. Volume of fire was key: the amount which could be fired as long as the aircraft was in range but was not yet in attacking position. Four experimental 4.7in anti-aircraft guns were on order. In 1921 the Committee recommended that four of the heaviest possible, i.e. longest-range, guns (4.7in) bear on any part of the sky, to produce a sufficient volume of fire.[18] HMS *Excellent*, the gunnery school, considered the 4.7in fixed round the largest which could be manhandled and loaded at all elevations, although the Committee was also interested in power-loaded 5.5in and 6in guns, particularly for battleship dual-purpose secondary batteries. There was some feeling that the 4.7in would fire appreciably more slowly than the 4in, so that it might be a step beyond what was worthwhile. The 4.7in gun was mounted on board the *Nelson*s and the minelaying cruiser *Adventure*, but other new cruisers all had single 4in guns instead. All guns had to have clear sky arcs, so that four guns could be mounted in such a way that all of them could cover nearly the whole sky. The four-gun battery became standard in large cruisers. The committee hoped that a dual-purpose gun could be developed that would be suitable for capital-ship secondary batteries.

Meanwhile the Board approved increased fleet HA armament, primarily substitution of 4in for 3in guns.[19] In 1920 light cruisers had first priority, the priority within that category being the *Hawkins* and then the 'D' class (those without 4in HA guns). Battlecruisers came next, then battleships. In January 1922, however, DGD asked that the subject be reviewed. At that time he planned to provide *Hawkins* with the same HA armament as her near-sister *Effingham* (three 4in). *Vindictive* would also have three 4in guns. 'D' class cruisers would have two 4in (instead of 3in) guns each, and DNO proposed the same for the 'E' class. More generally, DGD argued for the most powerful possible HA armament for light cruisers. Even if a 'D' could not accommodate more than two 4in guns, perhaps an 'E' could receive three of them.

WAR EXPERIENCE

Only a machine gun or machine cannon could fire rapidly enough to deal with either a nearby aircraft or a fast boat. By the end of the First World War ships had both machine cannon, such as the single 2pdr pompom, and lighter machine guns such as the Lewis gun. Pompoms were considered more effective against low-or intermediate-level bombers (as would be encountered when the ceiling was low); against torpedo bombers, which had to come close before attacking; and also against the torpedo boats and remote-control boats the Germans used to protect the Belgian coast from seaborne attack. These boats seem to have made a strong impression, because the threat of them was frequently cited during the inter-war period; pompoms were currently carried particularly because of their value against such targets.

The Committee saw light weapons as the last-ditch defence against aircraft which had penetrated the main (larger-calibre) defence. They should provide 'almost certain defence' at 2,000–2,500yds. The Committee favoured a multiple pompom, which offered volume of fire, since the available shooting time was so short. The 2pdr turned out to be the best gun to counter torpedo bombers and small boats.[20] A six-gun mount was built at Portsmouth for further experiments and tested on board HMS *Dragon* in 1921–2. It was not, however, a prototype for a later service weapon. In November 1920 the Committee offered DNO its specification for the service weapon, for circulation among potential developers.[21] Vickers won the contract over Armstrong (Elswick) with an octuple multiple pompom, provision for which was made in the British cruiser designs of the 1920s. The mockup was completed in 1923, and HMS *Tiger* carried out sea trials in 1928. The first production unit was mounted in the battleship *Valiant* in 1930. By that time a lighter quadruple mount (Mk VII rather than Mk V/VI) was planned for cruisers.

Lewis guns were intended specifically to deal with strafing aircraft. The Committee decided that they should be replaced because future

The multiple pompom was developed as a direct result of First World War experience. It was the main British heavy naval anti-aircraft machine cannon of the Second World War. This octuple pompom (known as a 'Chicago piano') was photographed aboard the Australian cruiser *Shropshire*. Visible above it is the pompom director. Because such directors needed accurate rangefinding, but could not make good use of optical rangefinders, the development of the ranging radar visible here, a Type 282, had higher priority than similar work on long-range HA radars and on main battery fire-control radar. (RAN Historical Branch)

BRITISH CRUISERS

Above: The most visible change in cruisers during the First World War was the advent of shipboard aircraft. This Sopwith Camel is aboard HMAS *Sydney*. (RAN Historical Branch)

Above: It was difficult enough to launch an aircraft over a deck gun, but the advent of superfiring guns made matters far worse. HMS *Dragon* (about 1921) shows one solution, a hangar and flying-off platform raised above the gun mount, closed by shutters. Aircraft took off in such short distances that it might be possible to dispense with a platform over the gun itself. However, the arrangement badly affected the ship's bridge, and it seems not to have been used in practice (it survived only because ships did not go into refit).

aircraft might well be armoured. A quadruple 0.5in machine gun was developed both to supplement the multiple pompom and as an alternative where it could not accommodated. It had a much higher rate of fire (600 rounds per barrel per minute, or 2,400 rounds total, compared to 720 rounds total for the octuple pompom), but the rounds were far less destructive and the effective range shorter. Effective 0.5in range, as evaluated in 1932, was about 1,000yds (out to 1,800yds time of flight was less than for the 2pdr, but bullets were not densely grouped enough

Above: HMS *Caledon* shows a preferable solution, a rotating flying-off platform, in this case just forward of her searchlight platform.

to do much damage). The gun was far inferior to the multiple pompom, but also far better than a single 2pdr. Ships also retained twin Lewis gun mounts, which were set up on a temporary basis.

The Committee raised the possibility that a specialised fleet anti-aircraft ship was needed. A 1919 Atlantic Fleet report pointed out that the fleet should be able to concentrate anti-aircraft fire on an approaching group of aircraft, but that existing arrangements for concentration fire were 'strained to the limit' to deal with much slower surface targets. If concentration among battleships was impossible, perhaps a special anti-aircraft ship armed with twenty guns should be built, the equivalent of at least eight battleships. Such a ship was proposed in 1925, but nothing came of the idea.[22] The steady preference through the inter-war period was to reinforce the defences of individual ships.

DGD argued that until it was possible to combine HA and LA functions in the same weapons, cruiser armament should continue to emphasise LA fire. Since cruisers were far less important than the battleships they often accompanied, it seemed unlikely that the enemy would concentrate high-level bombing attacks on them. However, cruisers were fleet screening ships, and in this capacity were intended to beat off attacks by enemy light forces – including enemy aircraft. That made their long-range HA armament important. Hence the interest in 4in HA guns, of which two should, ideally, fire on any bearing. On the other hand, the new multiple pompom was intended to defend an individual ship against attacks by aircraft and boats. It would certainly be more effective than a single pompom, but it and its ammunition would also be a good deal heavier; the added weight would better be devoted to main and heavy-calibre HA armament. Controller approved this policy on 10 February 1923, and CNS (the First Sea Lord) concurred.

Cruiser Aircraft

Despite having lost nominal control of naval aircraft to the Royal Air Force in 1918, the Royal Navy was well aware of their value. It had invented the aircraft carrier, and at the Washington Conference it pressed for a large allowance for future carrier construction. However, British carrier operating practice drastically limited the number of aircraft any one carrier could operate. It therefore became important for

WAR EXPERIENCE

The better solution was a catapult. After successful trials, in 1928 the Admiralty decided to fit catapults to all capital ships and cruisers (the 'C' and 'D' classes were later excepted). At San Diego in September 1934, HMS *Norfolk* shows the standard catapult aircraft of the period, a Fairey IIIF. This aircraft first flew in 1926; 379 were built for the Fleet Air Arm, and another 243 for the Royal Air Force, making it the most-produced British catapult aircraft. It could lift a 500lb bomb load, but was considered mainly a spotter and reconnaissance aircraft.

surface combatants to contribute combat aircraft to the fleet. To some extent this could be seen as a simple extension of the wartime placement of fighters, for fleet air defence, on board capital ships and cruisers (atop capital ship turrets and on flying-off platforms forward of cruiser bridges). The two 'E' class cruisers were completed in 1926 with turntable flying-off platforms. Similar platforms had been installed on board several 'C' and 'D' class cruisers, but by 1927 only that on board HMS *Celedon* remained. Revolving platforms were considered preferable to fixed ones because they did not require the ship to turn into the wind, but they took up more of a ship's length. Only fighters could use them. Catapults were better because they could be used in still air and because they offered valuable freedom of manoeuvre.

The first catapult on board a British warship was installed on board HMS *Vindictive* when she was converted back from a carrier into a cruiser in 1923–5. Her fixed athwartships catapult occupied 'B' position, the superstructure below it forming a hangar for three aircraft. *Vindictive* made the first British catapult launches at sea in October 1925, with Fairey IIID and Flycatcher aircraft. Her prototype Carey catapult was removed in 1926 for installation on board the battleship *Resolution*. As of early 1927 a heavy Farnborough catapult (the prototype F.I.H) was about to be ordered for *Frobisher*; apparently it was carried on board only during 1927 and possibly into 1928 (*Frobisher* was credited with a catapult and one spotter/reconnaissance aircraft in the April 1930 register). HMS *York* was the third cruiser to have a catapult, making the first British cordite catapult shot at sea in May 1928.

Policy was set out in a 1926 Admiralty memorandum.[23] The limited carrier force allowed by treaty could not support detached squadrons or help protect shipping. Within a fleet, cruisers and battleships could provide supporting fighters and reconnaissance aircraft, freeing limited carrier capacity for the torpedo bombers only they could operate.

Norfolk hoists in her Fairey IIIF, 1935.

Outside the fleet, cruisers would find aircraft useful in trade route protection and they could supply aircraft to support detached squadrons. Capital ships and cruisers could operate both three-seat spotter-reconnaissance (S/R) and two-seat fighter-reconnaissance (F/R) aircraft. Light catapults (for loads of up to 4,000lbs) were designed for the F/R, heavy ones for the S/R. In 1928 official policy called for installation of catapults on board all capital ships and cruisers, but the smaller First World War cruisers were never fitted.

In a fleet action, ideally two S/R would operate continuously in support of each division of heavy ships (requiring a total of five); others would support detached cruiser squadrons (*Vindictive* carried three). As envisaged in 1926, one S/R would be assigned to each heavy cruiser catapult. The F/R would replace and supplement S/Rs once contact was made with the enemy. For example, they would drive off enemy spotters. Plans called for eighteen of them to work with the fleet, carried on board capital ships and the cruisers of the 'A-K' scouting line, plus another eighteen for cruisers on detached duty. These were in addition to at least fifty fighters assigned to the fleet to attack enemy aircraft and to protect British torpedo bombers and ships (plus others assigned to the small carriers *Argus* and *Hermes*, intended for detached duties). A cruiser in the scouting line should carry an F/R. The rest of the cruisers in the fleet should carry fighters to help defend against air attack, just as their guns would help break up enemy destroyer torpedo attacks. A cruiser on trade-protection duty needed a reconnaissance aircraft and, if it could be developed, a catapultable torpedo bomber.

The 1926 memo was written as the Naval Staff was deciding that the Royal Navy could not afford separate fleet and trade-protection cruisers, so that all cruisers should be able to operate all three types of aircraft (fighter, F/R, S/R). Aircraft requirements had enormous impact on interwar British cruiser design, because catapults and aircraft stowage (if provided) demanded so much space. Further space was consumed by magazines (bomb rooms) for air ordnance. It had been assumed that cruisers would carry only one aircraft each, if that, but the new *Kent*s had space for two light catapults, albeit probably weight for only one. It was hoped that later classes (beginning with HMS *York*) would have one light and one heavy catapult, and thus could accommodate one three-seat S/R and one two-seat F/R. That proved difficult, because the only available catapult positions were the usual one amidships, the top of 'B' turret and the quarterdeck. Experience with the quarterdeck catapult on board HMS *Hood* eliminated that position, while experience with HMS *York* showed that it was impractical to place a light catapult atop 'B' turret. Ships therefore continued the previous practice of placing one rotating catapult amidships, *Exeter* being the sole exception.

No F/R yet existed in naval service; the only fighter was the single-seat Fairey Flycatcher, which could have a wheeled or float undercarriage. The S/R was the Fairey IIID. The Fairey IIIF, which had wireless

When the big 'Counties' were first completed, it was not at all clear that enough weight remained, within the Washington Treaty limit of 10,000 tons, for a catapult. HMAS *Australia* carried this Supermarine Seagull III on the pedestal that would later accommodate her catapult. The ship was fitted with a catapult in September 1935 (her sister followed in April 1936). (Photo by Allan C Green via State Library of Victoria)

as a standard fitting, was in service a year later. The first F/R, the two-seat Hawker Osprey, entered service in 1932. By 1935 the Fairey IIIF was being superseded by the Fairey Seal, another three-seat biplane.

Catapults were designated by type letter, sequence number (Roman numeral) and a suffix indicating light (L) or heavy (H). Type letters for cruiser catapults were C for Chatham Dockyard, D for double-acting (fixed, athwartships), E for a rotating extendable unit, F for Farnborough (RAE), and S for a slider (Admiralty design, cordite-operated). C.IV.H was the Carey unit in HMS *Vindictive*.[24] *York* had the prototype E catapult, an E.I.H. Three E.II.H were ordered in 1928, for two *London* class cruisers and either a heavy cruiser of the Atlantic Fleet or a third *London*, all to be installed when the ships came home for to recommission in 1931 or early 1932. No further catapults were immediately (1929) ordered for *London* and *Norfolk* class cruisers pending tests of other designs. E.III.H was designed to a 1930 specification, one being earmarked for HMS *Hawkins*. At that time no other cruisers, besides HMS *Vindictive*, had catapults.

It appeared that the 'D' and later classes could carry catapults, the *Birmingham*s and 'C' class being limited at best to a revolving platform. In fact the 'D' class never received catapults, and the *Birmingham*s were stricken before many catapults were made. Potential shipboard aircraft strength as projected forward to 1932 was 105 aircraft, compared to 238 on board the carriers (including one projected ship, which was not built).

Initially many new cruisers had aircraft but not catapults. In April 1931 the new heavy cruisers *Dorsetshire* and *Norfolk* each had one S/R but no catapult. *Cornwall* had a light S.II.L catapult (48kts) and carried one fighter (a Fairey Flycatcher floatplane); *York* had a heavy catapult (39kts) and carried one S/R. A year later *London* and *Sussex* carried one S/R each, without any catapult. However, catapults were being installed on board the other 'County' class cruisers: *Exeter* had two heavy E.II.H catapults (50kts) and carried two S/R; *Kent*, *Suffolk* and *Cumberland* each had a light S.II.L (*Kent* originally had an F.I.L) catapult (48kts) and one fighter; and *Norfolk* and *Shropshire* each had a heavy (50kt) E.II.H catapult and carried one S/R. A year after that (April 1933) all the 'County' class cruisers had either heavy or light catapults. *Cornwall* and *Kent* had light (48kts) S.II.L catapults and each carried a fighter. *Suffolk*, *Cumberland* and *Berwick* all had light (50kts) S.II.L catapults and each carried a fighter. *Norfolk*, *Shropshire*, *Dorsetshire*, *London*, *Devonshire* and *Sussex* each had a heavy E.II.H catapult (50kts) and carried one F/R. The new *Leander* had a more powerful E.III.H heavy catapult (56kts) and also carried one F/R. Similar catapults were installed in the rest of her class and on board HMAS *Sydney* as well as the two Australian 'County' class

HMAS *Sydney* launches her Supermarine Seagull V (Walrus) from her extending-type E.III.H catapult (note the extension). The aircraft changed cruiser catapult operations because it was designed to land in an ocean swell, thanks to its structural strength and its relatively low landing speed. It was designed to an Australian specification, then adopted by the Royal Navy as the Walrus. Total production was 746 aircraft. In effect the slow but effective Walrus became the alternative to the combat aircraft which the Royal Navy sometimes planned to operate from its cruisers. (RAN Historical Branch)

cruisers *Australia* and *Canberra* (S.II.L). *Emerald* and *Enterprise* both had light catapults, S.II.L and S.III.L, respectively.

The first four *Leanders* all had heavy catapults. In May 1932, however, both major fleet commanders suggested that light catapults would suffice for fleet service (for either fighters or for light reconnaissance aircraft). DNAD was developing both a light reconnaissance aircraft (5,000lbs) which could be launched from a light catapult (the Fairey Seafox), and a heavier torpedo spotter reconnaissance (TSR) aircraft, which became the Fairey Swordfish. The TSR was expected to be at least as large and heavy as the Fairey IIIF, which had proven cumbersome for cruisers. The advent of the TSR probably explains the shift to the big D-series athwartships catapults and to large hangars. Cruisers equipped with TSRs typically carried three of their 18in torpedoes. The light reconnaissance aircraft carried 100lb anti-submarine bombs (to which 250lb bombs were added about 1936). DNAD argued against the light catapult, on the grounds that aircraft were becoming heavier and heavier.

Since weight was critical, saving even 10 tons by changing to light catapults was attractive. However, the same cruisers might also be needed for trade protection. DNAD argued that therefore every effort should be made to accommodate the heavy catapult. Director of Plans and Director of Tactical Duties (i.e. of Staff Requirements) supported DNAD in arguing that the heavy-catapult policy for *Leanders* should stand, but that proved impossible for the Modified *Leanders* (*Amphions*) and for the *Arethusas*. These ships were intended to receive S.II.Ls, although that decision was reversed for the *Amphions* (*Sydney* had an E type catapult and *Perth* never received hers).

By 1934 it was standing policy to fit all cruisers from the two 'E' class onwards with catapults by the beginning of 1935. In 1936 all the modern cruisers, except *Aurora* (which was scheduled for fitting, but never was fitted) had catapults. *Hawkins* was being considered for fitting at her next large repair (this was not done), and *Frobisher* and the 'C' and 'D' classes lacked catapults (the two 'E' class had recently had catapults installed). The great question was whether ships should be fitted to launch the next-generation Fleet Air Arm aircraft, weighing 12,000lbs with a 50ft span. The only projected aircraft in this category were the new dive bomber reconnaissance (DBR) and the replacement TSR, the Blackburn Skua and the Swordfish, both primarily for carriers. Both weighed about 9,000lbs and were expected to enter production about mid-1936. They would be the first cruiser catapult aircraft with real anti-ship striking power. Plans called for the TSR to be replaced by two-seat fighters (Blackburn Rocs) and possibly by three-seat fighter spotters (which never materialised). The fighter would be somewhat smaller than the TSR: 46ft span, 8,500lbs on floats. Alternatively, ships could be equipped with a large but lower-performance reconnaissance aircraft, the Fairey Seagull.

Adapting ships for the new generation of heavy aircraft entailed considerable efforts. D.I.H (fifteen built) was introduced in the 'Town' class cruisers and equipped them and most rebuilt 'Counties'. D.IV.H (seventeen built) equipped the *Fiji*s.[25] These fixed catapults used below-decks machinery which affected accommodation. The design of the compact *Fiji* class in particular raised the question of whether aircraft were really worth the considerable sacrifices involved. Catapult and hangars took up about a sixth of the space between the main gun mountings, and they and the aircraft accounted for 190 tons in the new 10,000-ton *Belfast*. The 1936 London Naval Treaty removed the limit on total carrier tonnage, hence made it possible for the Royal Navy to imagine doing away with cruiser aircraft as substitutes. That left the very different mission of trade protection, for which cruiser aircraft certainly were worthwhile. For that purpose the F/R and S/R ideas were not too important. Instead, the ships could have either an aircraft designed specifically to land in an ocean swell (the new Supermarine Walrus) or an autogyro (a predecessor to the helicopter) which could land on the ship's short deck. For cruisers which could not accommodate the Walrus, a small two-seat spotter reconnaissance (light reconnaissance, or L/R) aircraft was being developed, the Fairey Seafox. Like the Walrus, it was designed to land at low speed so that it could tolerate worse sea conditions. Unlike previous cruiser aircraft, it had no strike capacity at all. The Staff considered the autogyro so promising that they were willing to spend heavily to perfect it. DNAD, DTSD and Director of Plans all argued that the big athwartships catapults entailed too great a sacrifice in anti-aircraft firepower. Instead, ships might well revert to either a single revolving catapult or they might rely on aircraft hoisted over the side to take off as well as to land. The best way to provide the fleet with fighting aircraft was to accelerate the carrier programme. ACNS agreed. Perhaps surprisingly, Controller, who had invented the new armoured carrier, and who was certainly pressing for a larger carrier programme, disagreed. Controller estimated in 1936 that an armoured carrier cost more than a battleship, so large numbers would be difficult to build. She carried only thirty-six aircraft, so there was little prospect of distributing such ships around the world for trade protection. Catapult aircraft would still be needed in some numbers. The *Fiji*s not only had the same hangar arrangements as their predecessors, to accommodate TSRs, but they had even more powerful catapults. The senior aircraft commander, Rear Admiral (Aircraft Carriers) wanted the battle fleet to be self-contained in spotting and action observation, so that all essential air requirements could be met even if a proportion or all the carriers were destroyed. The aircraft stayed.

Catapults and aircraft cost considerable space and weight, and thus were early candidates for removal in favour of additional anti-aircraft guns. *Aurora* was completed as a flagship for Rear Admiral (Destroyers), her catapult space taken over for the extra accommodations required. By May 1941 the other three *Aurora*s, the smallest modern cruisers, had had their catapults landed. In other classes, replacements of catapults by anti-aircraft weapons began in 1941. Work was usually done when ships became available for major refits.[26]

The Fairey Seafox was the smaller counterpart to the Walrus, also designed (to an earlier specification) to land in ocean swells. This one is shown aboard HMS *Orion*, at a southern US port in 1937. The catapult extensions have been retracted. A Seafox from HMS *Ajax* spotted successfully during the battle of the River Plate.

The Big Cruiser

An Imperial Conference was scheduled for July 1921. The Admiralty revived Jellicoe's idea of Dominion-sponsored cruiser construction. It argued that light cruisers were particularly well adapted to Dominion navies because they were 'the smallest and least expensive vessels in which officers and men can adequately be trained and given general seagoing experience in peace'.[27] For this reason the big trade-protection cruisers were initially called 'colonial cruisers'. There was no attempt to lay out total numbers, possibly because they would have been too frightening. Plans Division pointed out that numbers would be needed above all, so that individual cost should be kept as low as possible (a recurring theme throughout the inter-war period), that the cruisers should burn oil, and that endurance should be 5,500nm at 16kts under war conditions. This was far beyond what existing small cruisers could do.[28] Plans Division added the new idea that cruisers already present could exert enormous influence as long as they survived, during the six weeks or two months it would take the main fleet to arrive in the East. As before, both Japan and the United States were listed as potential enemies, but planning clearly concentrated on Japan.[29]

DGD, Rear Admiral F C Dreyer, had been thinking about what sort of cruisers the Empire needed while he had been Jellicoe's Chief of Staff less than two years earlier.[30] He was struck by the trend towards much larger cruisers. The US *Omaha* class might be seen as a copy of the 'E' class (it was not, having been designed earlier), but after the Armistice the US Navy became interested in 10,000-ton cruisers armed with 8in guns.[31] In January 1920 in Washington, the American Admiral Mayo explained to Dreyer that in his opinion nothing smaller could carry enough fuel to be of much use in the Pacific. The General Board, responsible for formulating US Navy programmes, recommended that year that thirty such cruisers, armed with 8in guns, be built over the next three years. The US government was unenthusiastic, and even a scaled-down plan for five cruisers in 1921 failed, but clearly future US light cruisers would be 10,000-tonners. The US Navy was reportedly planning to mount ten 8in guns in the planned ships (which seemed to be too much on that displacement). The Japanese had already announced plans for four cruisers of over 7,000 tons (the *Furutaka*s). No details were known.

Dreyer suggested that former German officers and the French gave some pointers towards the future. To the Germans, their wartime *Köln* class (about 5,500 tons, eight 5.9in guns, 28.5kts) was too slow and too large for the fleet and too small for foreign service. They violated the cardinal rule that ships of inferior fighting power should be fast enough to escape superior ships. The first requirement for a 'foreign service' cruiser stated in 1917 was the ability to keep the sea. Speed should be 25kts for long periods and 26–29kts for short ones, to run down and examine fast merchant ships and to avoid the enemy. Guns should be 6.7in or 7.5in; the proposed armament was eight such guns in twin turrets. Torpedoes were desirable. The ship should be armoured against 6in fire. The original proposal was for 12,000 tons, but the Kaiser considered that too small and recommended 14,000 tons.

According to the Naval Attaché in Paris, Admiral Grasset argued that since the Versailles Treaty limited the Germans (still the main enemy) to 10,000 tons, France should go one better with 10,000 tons and 7in or 8in guns. No such ships had yet been ordered. Dreyer considered the French reasoning vicious, because it would start an upward spiral of cruiser development which would prevent the Royal Navy from building enough such ships (he regretted the *Hawkins* class, which had started the process). Overall, it was clear that cruisers were tending towards 10,000 tons.

Ideally the Royal Navy would build somewhat smaller ships in

The great problem of catapult installation was always the centreline space a rotating catapult required, even if it had retractable extensions. HMS *Exeter*, shown, solved the problem by using two fixed catapults, angled outward. Their success inspired the developed of the fixed athwartships catapults installed on board *Southampton* and *Fiji* class cruisers.

larger numbers. The places to cut would be torpedo tubes (not needed in a trade-protection cruiser) and side protection against 6in guns (it would suffice to provide a protective deck of moderate thickness).[32] In any case, enough armour to defeat 7.5in or 8in guns would add prohibitive weight. Presumably torpedo tubes could be fitted if the cruiser was needed for fleet work.

As a gunner, Dreyer advocated the 8in gun because engagements would probably be fought at extreme range, and because effective range depended on the ability to observe the fall of shot. Although a 6in gun could range out to 20,000yds, only the splashes of the larger 7.5in and 8in shells could be spotted reliably at such ranges.[33] Recent trials suggested, moreover, that a well-designed enemy light cruiser could not be stopped by 6in fire. Dreyer preferred the 8in gun to the 7.5in because it offered superior penetration and bursting effect for a small increase in weight. He hoped that a power-operated 8in mounting could fire five rounds per gun per minute. Ships would have no secondary LA armament, but should have four 4in HA guns for anti-aircraft and star shell. They would also need automatic weapons to counter torpedo planes and distance-controlled boats, both of which Dreyer claimed the US Navy was developing.[34] Two of the multiple pompoms then being proposed by the Naval Anti-Aircraft Committee seemed adequate. Any such ship should carry one or more amphibious aircraft (Dreyer recalled the wartime German raider *Wolf*, which had one such aircraft). Dreyer envisaged a revolving flying-off platform for an amphibious aircraft and a crane to hoist it in. The aircraft would be used for both reconnaissance and spotting. Since the cruisers would operate mainly in the tropics, they should have improved ventilation arrangements and a magazine cooling plant.

On this basis Dreyer suggested five alternatives:

A: 10,000 tons, 31kts, eight 8in twin splinter-proof on centreline.
B: 8,500 tons, 32kts, five 8in single splinter proof on centreline.
C: 7,500 tons, 35kts, four 8in single splinter-proof on centreline.
D: 7,500 tons, 32kts, four 8in single splinter-proof on centreline.
E: 7,500 tons, 25kts, four 8in single splinter-proof on centreline.

Design A would counter the projected US 10,000-ton cruisers, if British finances permitted (the 7,050-ton *Omaha*s would, however, outrun them). It would not be desirable to go below 7,500 tons, 'as this is the smallest size now advocated by other countries'. Director of Plans protested that existing Japanese cruisers were much smaller, but for Dreyer the problem was what was coming, not what already existed. No 6in cruiser could effectively fight an 8in cruiser. Dreyer preferred a 7,500-tonner armed with four centreline 8in guns (he was willing to accept 7.5in if DNC could not provide power hoists and power ramming while providing the desired endurance, maximum speed, and other items on the tonnage [i.e. on a limited cost]). The armament decision seemed urgent, if a concrete plan was to be presented to the Imperial Conference. Dreyer particularly cautioned that the Dominion governments should not be misled into imagining that they were being asked for nothing more than the wartime fleet cruisers. However, he also feared that buying cruisers comparable to the largest ones being planned abroad might (as with the *Hawkins* class) lead other navies to build even larger ships. Hence his preference for the four-gun 7,500-tonner. He also warned that, given his own experience over the last seven years, it might be some years before any Dominion ordered a new light cruiser. He did not make the implication explicit: some or all of those trade protection ships would have to come out of Royal Navy funds. Of the alternatives listed, C to E differed in endurance and protection. Cruiser E was a minimum ship for convoy protection, but she would be unable to attack or run down enemy cruisers. In effect Dreyer had described the next step in cruiser development.

DCNS agreed that any new trade protection light cruisers would have to be armed with (at least) 7.5in guns, and would probably be comparable to the big *Hawkins* type. He doubted that a ship of smaller displacement could combine sufficient radius of action and armament.

The problem was numbers. In July 1918, when practically all trade between North America and Europe was being convoyed, as well as a proportion of vessels outward bound to North America, and ships operating between Great Britain and Sierra Leone and Dakar, convoys required no fewer than seventy ocean escorts, including cruisers, armed merchant cruisers and commissioned escort ships. A worldwide convoy system would have required about 150 ocean escorts (apart from ASW ships in local escort groups). The most powerful potential Japanese raiders were the four *Kongo* class battlecruisers, which could be contained only by their British equivalents. For this reason the Royal Navy periodically considered stationing some or all of its battlecruisers in the Far East (a plan to this effect was nearly put into effect in 1929). However, a smaller number of unusually powerful British cruisers working with convoys could make Japanese attacks on convoys too risky. To attack Empire commerce, any Japanese cruisers would have to operate far from their bases; even limited damage might prove fatal (as was the case with the German *Admiral Graf Spee* in 1939). The wartime 'large light cruisers' (*Courageous* class) might be a useful model for future construction. In the past cruiser size had been held down to make it possible to build such ships in quantity, particularly for fleet operations. However, the fleet might need fewer cruisers if the promise of carrier-borne reconnaissance aircraft was realised. Five *Courageous*-class cruisers would cost about as much as eight *Hawkins*.

ACNS suggested (and DCNS agreed) to ask DNC to consider two alternatives. One would be a 33kt 10,000-tonner armed with 7.5in or 8in guns, without torpedo tubes, and otherwise as Dreyer had proposed. Endurance would be 5,500nm at 16kts, and in contrast to the wartime *Hawkins*, the ship would burn only oil fuel. The second would have much the same characteristics, but with more powerful (preferably 10in) guns, and magazines protected against 8in fire. Maximum displacement would be 15,000 tons. DCNS added that the term Commerce Protection Cruiser should be dropped in favour of some alternative, preferably Station Cruiser – which would recall the much earlier practice of keeping powerful armoured cruisers on the foreign stations, for presence as well as for trade protection. The Dominions should want a ship which could go anywhere and fight anything short of a battlecruiser.

DNC could not produce the desired pair of designs, because his department was fully occupied producing the new battleship and battlecruiser designs as well as other vital work (including the cruiser-sized minelayer described in the Appendix and the flush-deck carrier conversion of HMS *Furious*), but he produced some quick estimates.[35] His main conclusion was that the Staff had grossly underestimated what was needed to achieve either the desired speed or the desired endurance. For example, using lightweight ('E' class) machinery, an enlarged *Hawkins* (11,000 tons) might make 31kts. To achieve the desired endurance, the ship would have to be lengthened to about 600ft (about 12,000 tons). To make 33kts, she would need about 30 per cent more power (using lightweight machinery, about 12,500 tons). To provide deck space for the amphibian, she would have to concentrate her armament (six rather than seven 7.5in) in three twin turrets; without the amphibian she could probably have another pair of such guns. The proposed 10in ship would probably be about the size of HMS *Courageous* (19,000 tons).

CHAPTER 5
TREATIES AND HEAVY CRUISERS

The Washington Conference

In November 1921, before any further conclusions could be drawn, the US government invited the world's major seapowers to what became the Washington Naval Conference. The United States proposed to stop the construction of warships of over 3,000 tons (and 15kts speed) for a decade.[1] Only the replacement of overage cruisers (over fifteen years old) would be permitted. With the world's largest fleet of underage cruisers, Britain would have been banned from building new ones for several years. With a largely outdated cruiser fleet, the United States would have been free to build many of the desired 10,000-ton cruisers – which would, incidentally, outclass nearly all existing British cruisers. Because the Imperial Japanese Navy included several old armoured cruisers it too would be able to begin new construction of large cruisers. Britain also could not accept the idea because it would have wrecked an important source of national power, the world's largest private naval shipbuilding industry. The British delegation answered that 'had it been the intention of the authors of this proposal to abolish naval forces, this . . . holiday would have been a fitting preliminary'.[2] New construction should continue, though on a limited scale. Within a few weeks, according to a British account, the US delegates understood that the ten-year total building ban was a mistake, but they could not admit as much due to US public opinion. American newspapers had found the ban far too attractive.

The unexpressed essence of the US treaty plan was equality with the Royal Navy. Without such equality, the treaty was unlikely to be accept-

TREATIES AND HEAVY CRUISERS

The Royal Australian Navy bought two *Kent* class cruisers, at least initially with taller funnels than their British sisters. HMAS *Canberra* is shown on 24 October 1937. Minor initial modifications were enlargement of the top of the fire-control tower, enclosure of the after AA rangefinder, extension of the machine-gun platform abaft 'B' shelter deck, and addition of a light pole to the mainmast. (Photo by Allan C Green via State Library of Victoria)

able at home. As for cruisers, the British felt that, with by far the world's largest merchant fleet, they had a special need for cruisers for trade protection over and above what every other navy needed for its fleet. This idea of splitting trade protection from fleet cruisers recurred in later treaty negotiations. At the end of November 1921 Lord Lee, First Lord of the Admiralty (and a delegate) proposed both that cruisers be limited to 10,000 tons and that the 5:5:3 ratio then being proposed for battleships apply also to cruisers – but that the British be provided with additional cruisers above the ratio for trade protection. The 10,000-ton limit would cap cruisers at *Hawkins* displacement, precluding the vicious spiral Dreyer had foreseen a few months earlier. The Royal Navy might hope to squeeze tonnage further down to make cruisers affordable in the numbers it needed, but this was a start. The idea of further capping recurred in later negotiations.

The cruiser issue was little discussed; energy at the Washington Conference went mainly into questions of capital ship numbers and treaty arrangements in the Western Pacific, including clauses preventing fortification of many places. The attempt to limit total numbers of cruisers collapsed, but the British proposal for a 10,000-ton limit survived. The cruiser position provides an interesting contrast with the capital ship position. At 41,200 tons, HMS *Hood* exceeded the maximum tonnage for future battleships (35,000 tons) and, at least at the time, uniquely combined battleship protection and battlecruiser speed. On the other hand, the treaty allowed other navies to build ships as large as the *Hawkins* class, with more powerful (8in) guns. The British were already aware that the US Navy would not lightly accept any lower limit, and they counted on their ability to build large numbers of cruisers rather than on any intention to build (or retain) individually more powerful ones. The two 'large light cruisers', which a few months earlier had seemed to be useful counters to enemy heavy cruisers, were to be converted into aircraft carriers. The Washington Treaty also banned the sale of existing warships (the sale of a cruiser to Venezuela was cancelled as a result). The rationale was that a signatory might salvage ships otherwise condemned to be scrapped by bogus sales, but the effect was to provide national shipbuilding industries in the major naval powers with at least potential further orders, to help keep them alive while reducing pressure on governments for new orders outside the categories of warships limited by treaty. This clause precluded a flood of surplus warships onto the world market like that which was to follow the Second World War.

A new 'standard displacement' was defined. Because navies differed enormously in the endurance they demanded of their ships, it omitted fuel and reserve feed water, both of which were set by required endurance. Longer range still affected ship design, because extra tankage affected hull weight, but the difference was more subtle. Another subtlety was that existing ships were credited with a standard displacement equal to their pre-treaty 'normal' displacements – which included fuel and reserve feed water. Because replacement was on a ton-by-ton basis, replacements could be somewhat larger than older ships. The treaty recognised that most existing capital ships had been built before air and submarine threats had been either well known or well countered by new kinds of protection (such as bulges). It therefore allowed 3,000 tons for additional protection against air and underwater threats – but nothing at all for cruiser reconstruction. The tacit assumption, at least within the British Government, was that any modification to an existing cruiser could not be allowed to increase its displacement above 10,000 tons. That applied particularly to the *Hawkins* class.

The 8in Gun Cruiser

The Washington Treaty defined the 10,000-ton cruiser with 8in guns which DGD (Dreyer) and others had discussed the previous year. On 6 February 1922 DGD called for information about both the 8in gun and its shell (and their performance) and the size and weight of twin and single 8in gun mounts, against the contingency that a 10,000-ton cruiser might be required. Figures on 8in AP shell performance would be essential if a future cruiser were to be protected against the fire of the new kind of cruiser. At this time the *Hawkins* class did not have 7.5in AP shells, having been designed to fight lightly-protected raiders. Dreyer argued that these data would be the determining factors in the time required to complete a sketch design. One side benefit might be the design of a 7.5in AP shell, which could equip the *Hawkins* class in an emergency (so they could fight foreign heavy cruisers of the new type). DGD wanted figures for ranges of 8,000yds, 12,000yds, 15,000yds, 20,000yds and 25,000yds, which he suggested could be based on army 8in howitzer data scaled up for higher muzzle velocity (2,800ft/sec) and a 256lb shell (based on the weights of smaller-calibre shells) for various shell shapes (4 and 8 calibre radius head). Wartime cruiser actions had not been fought broadside to broadside (firing at right angles to target's course), so the future 8in shell should probably be uncapped.[3] DNO apparently stalled. In June he replied that although he had no difficulty investigating gun designs, he had to contract for turret designs, and the two firms involved (Armstrong [Elswick Ordnance Co.] and Vickers) would not be willing to develop them unless they had a good chance of receiving orders.[4] However, sketch designs could be relatively inexpensive. DNO suspected that DNC would much prefer a twin to a single. Unless such a mount had a thick (i.e. heavy) barbette, cordite charges had to go up to the guns in Clarkson's cases (as in the new twin 6in mount). The necessary arrangements to send those cases back down to the magazine would make for a larger mount, and for larger working spaces around the magazine.[5]

DNC could not design a ship without knowing the desired balance between protection and other features, particularly speed. In 1920 ACNS had called for protection sufficient to keep 6in AP shell out of the vitals of a ship armed with 6in guns. Assumed fighting range was 10,000–15,000yds at an average inclination of 67½° (the angle between the line of enemy fire and the ship's centreline). No wartime light cruiser met this standard. DNC estimated that it would require 4½in homogenous side armour and an 80lb deck (of the new type of deck armour), amounting to another thousand tons in an 'E' class cruiser. That would cost a knot at full power but, more importantly, it would entail an unacceptable loss of stability, meaning that the ship would have to be considerably larger. To provide similar protection against 8in shell would require a 6in belt and a 2½in deck.[6] DNC reminded the Staff that he still regarded extensive armour on a light cruiser as a serious and costly mistake; the PWQ Committee had not recommended extra armour for an improved 'E' class cruiser. As he wrote in 1918, these ships should be included in the destroyer category, in the sense that they depended mainly on speed for protection. During the war no British light cruiser had been sunk by gunfire, even though their armour was penetrable by the enemy's 5.9in shells. DNO disagreed. If the ships fought at long range, their magazines might be penetrated by plunging fire (at short range they were safe because they were below the waterline). The memory of the explosions at Jutland was obviously still raw.

DNC had his cruiser designer, Lillicrap, prepare rough data.[7] He tried an armament of three twin 8in guns, four quadruple torpedo tubes (as in the 'E' class), four 4in HA guns, and two multiple pompoms – essentially a scaled-up 'E' class armament with the desired anti-aircraft battery.

DNC pointed out that the ship's centreline would be badly congested, given other demands such as aircraft and main and secondary control towers and searchlights (preferably on the centreline), so that it would be difficult to arrange even twin turrets. The maximum might well be three. The ship would be bulged. Lillicrap estimated that each twin 8in mount would weigh 150 tons, based on the 75 tons of the twin 6in. That gave him an armament weight of about 800 tons (he used 820 tons), including 150 rounds per 8in (as in earlier 6in ships; but during the war the standard for 6in guns had been raised to 200 rounds per gun). He tried two design alternatives, one based on the new cruiser-minelayer *Adventure* and one based on *Hawkins*. In each case, he subtracted the expected weight of armament, the estimated weight of equipment (including crew), and the estimated weight of armour (belt depth 9ft [to extend 2ft below the waterline], length 400ft, to cover the vitals) from the available 10,000 tons, then used what was left for hull and machinery. Given minelayer lines and 'E' class lightweight machinery, he had enough machinery weight for 60,000shp, which gave him about 28kts. Scaling from *Hawkins* gave him another set of estimates. On 23 May 1922 he sent DNC a sketch: 525ft (pp) x 63ft x 19ft (10,000 tons standard) with 15ft freeboard amidships, and a speed of 27kts (54,000shp), with an estimated fuel capacity of 1,800 tons. Including the 100-ton Board Margin, estimated displacement was 10,020 tons, but the 20-ton excess could be pared down during detailed design. DNC had proven again that light cruisers should not be armoured. Low speed was unacceptable. ACNS and DCNS (Admiral Roger Keyes) agreed.

In February it seemed that new British cruiser construction might be deferred, since the Royal Navy already had the five *Hawkins* class heavy cruisers, but in June ACNS pointed out that the Royal Navy would face a bloc obsolescence problem because its entire light cruiser fleet had been completed in roughly a ten-year period. The oldest surviving ships, the *Southampton*s, had been completed in 1912, and replacements would normally have to be laid down in about five more years. Mass replacement over a ten-year period might be financially impossible, and new construction would have to begin in two or three years. In evaluating US arms control proposals the Admiralty had proposed to the Committee on Imperial Defence that cruiser lifetime be reduced to eight years for ships which had participated in the recent war, which made replacement urgent.[8] Age, however, was elastic. The Admiralty felt compelled to extend ships' lives as it became clear that successive Governments would not buy replacements quickly enough. By 1926 cruiser replacement age was twenty years.[9]

Much depended on foreign programmes, so in September Director of Plans (Dudley Pound, later First Sea Lord) asked Naval Intelligence to learn what it could of the projected Japanese 10,000-ton cruisers (which became the *Myoko* class). The Japanese programme called for four 7,000–7,500-ton ships and four 10,000-tonners, all to be completed in 1928–9. In October, with no more information available, Keyes called for a sketch design for the most powerful light cruiser which could be built within the treaty limit. The first such ship could be included in the 1923/4 programme, nominally to replace HMS *Raleigh*, wrecked in Canadian waters on 8 August 1922. On 19 October Controller ordered DNC to proceed as soon as the new *Nelson* design was complete, which should be very soon. However, he rejected including the new ship in the 1923/4 programme, on the grounds that he wanted the money spent to complete the two 'E' class cruisers, which he described as a departure in cruiser construction (presumably because of their lightweight machinery).[10] DNO had already contacted Vickers and Elswick, which expected to have sketches ready in about January 1923. They would be power-worked throughout, including their shell rooms, and would have 1in shields. It was reported (incorrectly) that the Japanese intended to use triple mounts, so DNC proposed (and DNO agreed) to ask the firms to make their priorities triple, twin, and single mounts, in that order. Without working out designs DNO could not know whether a triple was practicable for 1923/4, but that it would be worth studying in any case.

Pound suggested estimating endurance on the basis of the new mission of harrassing Japanese shipping while the main fleet was *en route* East. At least one such ship would have to operate continuously somewhere between Hong Kong and Japan, the ends of this notional cruiser patrol line 1,920 and 2,300nm from Singapore. Hong Kong probably would not be available for fuelling. At 16kts a cruiser would steam from Singapore to the northern end of the line in six days. She would spend five days on patrol at 16kts, then return to Singapore, steaming a total of 6,520nm over seventeen days. She would spend at least two days between patrols at Singapore, fuelling and making good defects, so she would spend five days on patrol out of every nineteen. More realistically, she would probably patrol at lower speed, spending seven or eight days on station out of twenty-one: it would therefore require a total of three cruisers to have one permanently on patrol. Endurance should therefore be 6,500nm at 16kts – an unusually high figure for the time. Survival would require that the ship be able to outrun existing Japanese battlecruisers (27.5kts), cruisers (33kts) and destroyers (34kts); Pound suggested aiming for 33kts.

Pound suggested that the ship carry torpedo tubes because ships protecting enemy commerce might well use them. They should be included unless they carried a serious cost in speed, endurance, or gun fire. Furthermore, not only would aircraft be valuable for reconnaissance, they might be worth some guns because spotting greatly improved gun performance.

The *Kent* class

DNC gave Lillicrap new orders.[11] He wanted an entirely fresh design, not bound by previous cruisers ('E' and *Hawkins* classes). The ratio of hull depth to length should be increased to save weight.[12] He might use a flush deck (which was adopted). Hull weight could also be reduced by using new steel with greater tensile strength, and this key choice made

HMS *Suffolk* enters Malta. She had recently been completed, and as yet had no catapult. As one of the first five British heavy cruisers, she was assigned to the China Fleet, and this photograph presumably showed her *en route* to the Far East (had she been returning, she would have been flying a paying-off pennant). Sunlight falling on her side makes the top of her bulge visible. (RAN Historical Branch)

BRITISH CRUISERS

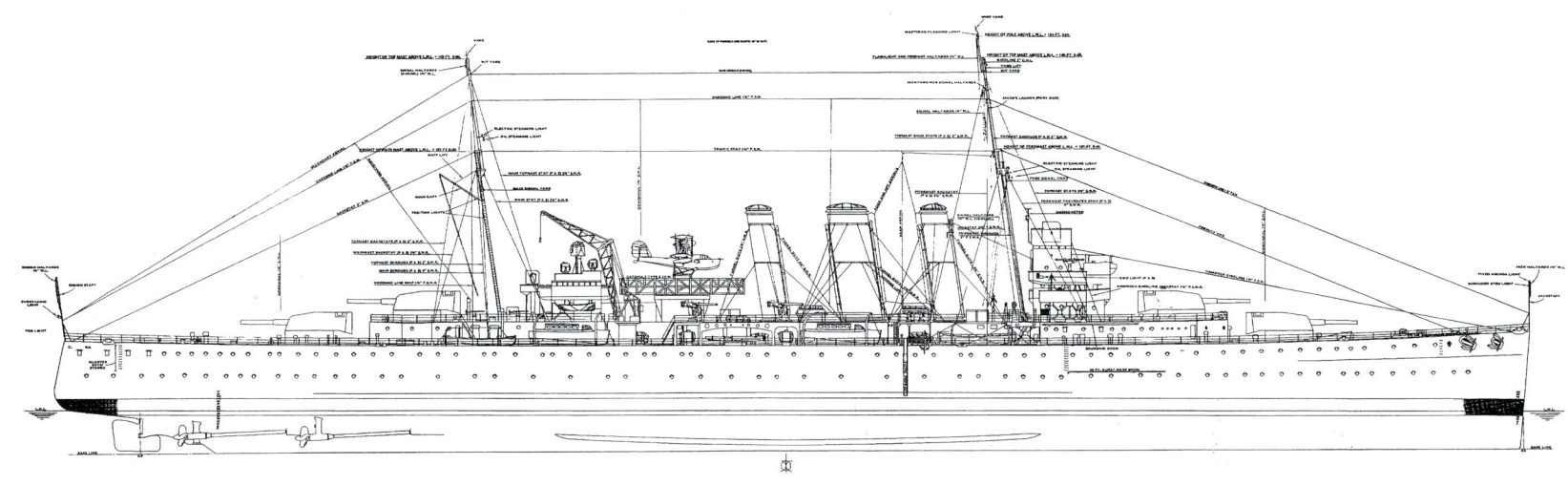

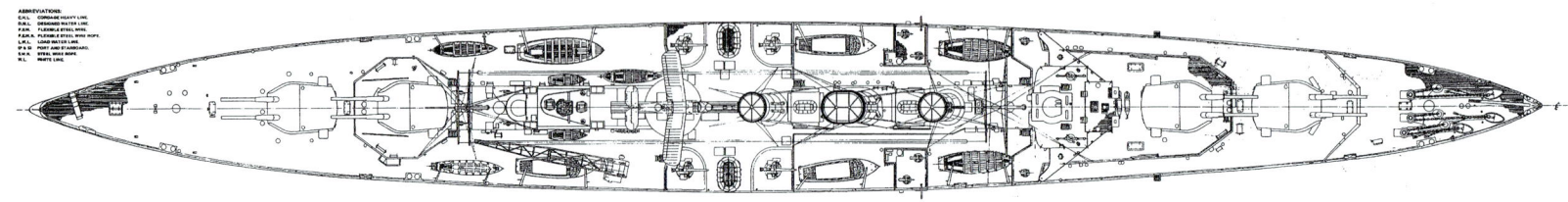

HMAS *Canberra*, typical of *Kent* class cruisers as built. (Paul Webb)

HMS *Kent* off Wei-hai-wei, China, on 11 July 1928. As yet she had no catapult.

the new cruisers unusually roomy, which made them desirable as flagships even after the Second World War. Armament should be four twin 8in turrets and speed 33–34kts. However, DNC wanted three-turret alternatives offering centreline space for hangars for two amphibious aircraft. The turrets could all be grouped forward, as in *Nelson* (with two 5.5in HA aft) or with one twin 8in aft. The ships would also have two quadruple deck torpedo tubes. Protection would be limited to a

HMS *Kent* emerged from a refit in 1931 with the planned catapult aboard.

bulge (as in the monitor *Gorgon*, perhaps) and, if possible on the weight, deck armour over the steering gear, magazines, etc (i.e. not over the machinery). The bulge was essential because a torpedo hit could slow or stop the ship and thus end the protection her speed offered. Later, in connection with a proposal to rearm the *Hawkins* class, ACNS remarked that the main idea of cruiser protection was to guard against total destruction of the ship by one lucky hit in a magazine – i.e. against the Jutland battlecruiser disasters. The emphasis on long-range 8in fire may also have been an attempt to preclude underwater damage, as gun range considerably exceeded torpedo range. High speed would make submarine attack difficult. DNC suggested a 100,000shp powerplant, using eight 12,500shp boilers. Lillicrap thought he would need 580ft between bow and after end of magazines, and perhaps another 40ft abaft that, for a total of 620ft. In HMS *Adventure* some length and weight had been saved by cutting off the usual cruiser stern into a transom, in which case 20ft could be saved on the cruiser, for a total of 600ft.

Lillicrap estimated the power he needed by scaling up the minelayer design: 100,500shp for 33kts or 117,000shp for 34kts.[13] DNC suggested trying sloop and *Courageous* hull forms (he later decided to use lines similar to those of *Courageous*, which had a shallow bulge).[14] To reduce resistance, he proposed a long forebody (but in the end accepted a fairly conventional hull form). Reviewing the design in November, DNC pointed out that speeds above a speed-length ratio of 1.2 (29.4kts for a 600ft ship) were very expensive. He reminded Lillicrap that the ship should have a protective deck over her magazines, plus a deck or sides amidships to protect her stability. It was not clear that she could have a hangar.

DNO was specifying 8in/50 guns firing four 280lb shells per minute with maximum elevation of 45° and maximum depression of 3°. Guns would load at a fixed angle of elevation (10°). Each turret would require its own power. If possible, the use of protective cases over the charges should be avoided, so the conveyors between handing room and gunhouse would have to be completely flash-tight, leaving no dangerous train of cordite between gunhouse and handing room. By the time the ships were being built, DNO had decided on 70° maximum elevation, providing the ship's main battery with some anti-aircraft potential.[15]

Reviewing the design at the end of January 1923, DNC proposed that Lillicrap work to a deep displacement of 12,000 tons and 32kts to get 33kts at standard displacement (10,000 tons). Maximum power should be 100,000shp. The machinery would be protected by a low flat deck with a short belt, to protect buoyancy against riddling by numerous small-calibre hits or by splinters. Magazine protection would be separate, intended to protect against cruiser fire. This basic idea survived for about a decade. Based on tank tests, assuming a propulsive coefficient of 0.5 (slightly conservative), Lillicrap estimated that he could drive the ship at 33.8kts on 10,000 tons with 100,000shp. Even 10,000 tons was not much, when so much power was needed. Lillicrap estimated that armament would consume 1,000 tons, and the 100,000shp powerplant another 2,100 tons.[16] He estimated 675 tons for equipment (including crew) and 5,300 tons for the hull (which he suspected was too low).

That left only 820 tons for protection. At 20,000yds it would take 4in side armour and a 3in deck to stop an 8in shell. There was nothing like enough weight to protect machinery, but small boxes this thick could be built around the magazines. Scheme A has been lost, but Scheme B (July 1923) envisaged a 2in belt and 1in deck over the machinery spaces, with 2in sides and deck over the magazines, and 1in plating aft. The amidships magazine for 4in HA guns was given 1in deck and sides. Scheme B consumed 830 tons, but it bought very little. More than half the weight (314 tons of deck and 154 tons of belt) went into machinery protection; this was surely the most that could be provided. It offered only protection against destroyer guns (4in, not 4.7in) – but given their flat trajectories at the sorts of ranges at which destroyers could hit, shells would generally pass through the ship anyway and do little if any damage – and surely destroyers would not get close enough to do such damage. The deck was inadequate against bombs. Lillicrap apparently did not take seriously the argument that even a thin belt would protect the ship's side against being riddled by small-calibre hits and by fragments.

Lillicrap argued that the limited armour weight should be concentrated around the magazines. He offered three more schemes. In Scheme C, the magazines and shell rooms were well underwater in the hold (the platform deck was raised 1½ft), hence needed only deck protection (3in NC and 3in C end bulkheads, since shells could enter the ship above water to hit the ends of the magazines). That consumed 735 tons, and another 60 tons went into protected trunks between magazines and platform deck. A turtle deck would protect the steering gear. Stability would be protected by filling the upper part of the bulge with balsa wood, cork, etc. In Scheme D (820 tons) the magazines were totally enclosed, with 4in sides as well as the 3in deck and ends (695 tons of magazine armour). In addition, 3in rings enclosed the parts of the hoists emerging from the magazines (otherwise there would have been a gap in the deck armour: 50 tons). There was also a 1½in turtle deck over the steering gear. Scheme E (800 tons) further protected the magazines by placing the shell rooms above them (armour thicknesses were the same as in Scheme D). Placing the shell rooms above rather than alongside the magazines also freed more length for fuel oil below the platform deck (but oil could be carried above the platform deck). DNC chose Scheme D. He rejected pure deck protection (Scheme C) because a shell could reach the magazine if the ship were rolling badly or had a bad list. Scheme E was objectionable because the shell rooms were not as well protected, being higher in the ship, and shell rooms could not be flooded from the sea (as was now considered preferable).

In October 1923 DNC offered the Board three alternatives: X (eight guns, 130 rounds each), Y (six guns, 150 rounds each), and Z (eight guns, 150 rounds each), each with four single 4in anti-aircraft guns, two Mark 'M' pompoms (multiple pompoms, then under development), and two quadruple 21in torpedo tubes. Each had the same 626ft 9in (overall) x 69ft x 16ft 3in hull displacing 10,000 tons, each with the same hull depth of 43ft 6in amidships. X and Y each had a 100,000shp powerplant, for a speed of 32kts in deep condition and 33kts in light condition; 2,800 tons of fuel oil would give a radius of 7,000nm at 12kts. Z had a 75,000shp powerplant (30/31kts) and 3,200 tons of fuel (8,000nm at 12kts). A sketch of Design X showed much the form the British 'County' class eventually took, but with a tripod foremast.

Because the DCT had not yet been adopted when these ships were designed, they had a small director forward topping a square-sided windowed tower containing the fire-control officers, with the plotting room at its base. The tower in turn created unacceptable drafts on the open bridge surrounding it. This problem in turn led to the design of the streamlined bridge structure first tried on board HMS *Exeter*. This is HMS *Cornwall*.

The Board chose to trade reduced power (fewer boilers: 75,000shp) for machinery space protection: a 1in deck and 2in side.[17] DNO supported this addition on the grounds that it would keep out ricochets and splinters of shell bursting short – as had been shown at Heligoland Bight in August 1914. DGD pointed out that when attacked from abeam, the target presented by the ship's deck was more than twice as large as that presented by its side at 16,000yds. Given that and also the likely future threat of air attack, DGD suggested emphasising deck protection, but there was not enough weight available (giving up side armour altogether would buy only ½in of deck). The 1in deck would likely be effective against 8in fire at 16,500yds (angle of descent 20°) and against 6in fire at all but extreme range. DGD also pointed out that the planned 4in magazine side would likely be ineffective at decisive gun range (under 10,000yds). He wanted at least 5in, but that was impossible on weight grounds, and it was argued that liquids and the ship's structure provided added protection. DGD also wanted the magazines brought below the shell rooms, so the latter could offer them some protection. The design submitted to the Board on 29 October 1923 (and approved by First Sea Lord on 6 November) showed the 1½in deck and side armour halved to 1in (but made more effective because the ship's sides sloped), plus box protection for the magazines.[18]

The 8in gun had been chosen because it offered greater range, so naturally the ships received the new Admiralty Fire Control Table (analog fire-control computer) also being installed on board the *Nelson* class battleships. It embodied the automation espoused by the Fire Control Requirements Committee. Unfortunately the DCT, which was intended as part of the new fire-control system, was not ready in time. Thus the ships had simple power-worked directors (carrying rangefinders) atop a square-faced tower which accommodated spotters and the new plotting space. The tower was needed so that its windows (on two levels) could face in all directions. The heavy tripods of previous cruisers were abandoned, presumably to save weight, despite earlier arguments that such high positions were needed for extreme range. The tower or island in turn created an air flow which blew smoke from the forefunnel back onto the bridge, so the funnels all had to be raised after completion.[19] This was an unpleasant surprise.

The bridge itself was modelled on the successful structure in HMS *Frobisher*, whose officers considered it the ideal for future ships. No one seems to have realised that placing the tower on top of this bridge completely changed the air flow over and around it. The problem was discovered during the trials of HMS *Cumberland*. Modifications to the bridge of HMS *Cornwall* helped somewhat, but not nearly enough. Writing in April 1928, in connection with the next major design (HMS *York*), DNC doubted that any simple remedy could be found. The solution was twofold: the tower was eliminated when the DCT appeared, and ships from HMS *Exeter* onward had a new streamlined enclosed bridge structure, but there was little hope of installing either on board existing cruisers.

At each end of the ship, the magazines and shell rooms were arranged around a handing room from which shells and cordite charges were sent up to the turret. The magazine was on one side of the handing room, the shell room on the other. At least in the version of the design approved by the Board in 1923, there was a series of five spaces at each end: a magazine, a handing room, a shared shell room, then another handing room (for a second 8in mounting), and finally a second magazine. Magazine and handing room both had to be given maximum protection. At this time it was assumed that shells were relatively inert, to the point that placing shells above a magazine was a form of deck protection.

DNC had certainly protected the ships against catastrophic damage, but their fighting ability depended on some key spaces that were unprotected: the transmitting station (housing the fire-control computer, key to long-range gunnery), the low power room (for transmitting data between elements of the fire-control system and the guns), and the telephone exchange. This problem seems to have been raised about mid-1924, too late for the *Kent*s.

The reduction in power did not save as much weight as might be imagined, because at about the same time the lightweight 'E' class type machinery was abandoned in favour of a heavier and, it was hoped, more reliable type.[20] DNC thought that foreign navies achieved their high powers and high speeds by using lightweight (i.e. unreliable) destroyer-type machinery.[21] Lillicrap felt that the heavier machinery made much less difference in a 10,000-ton cruiser than in, say, a destroyer or a small cruiser. He also claimed that in some cases the greater length of the British treaty cruisers made up for their lower power output (he cited USS *Indianapolis* as a case in point). At a Controller's conference on 29 October 1923, E-in-C said that he could provide 80,000shp in the space allocated for 75,000shp, and that rating was adopted, using eight boilers in two boiler rooms. As a survivability feature, the after boiler room and the forward engine room formed a single unit which could operate independently of the other boiler room and engine room. Also for survivability, each pair of boilers, port and starboard, fed port and starboard engines independently, in what the Royal Navy called the unit system. This arrangement was considered insurance against trouble due to small splinters damaging a main steam pipe. However, there was no insurance against the loss of all four boilers in a room due to a single penetrating hit. War experience had shown, however, that running a bulkhead down the centreline of the boiler room, which would provide exactly such protection, would sink the ship if one side flooded. As in earlier British cruisers and destroyers, the forward uptakes of the after boiler room and the after uptakes of the forward boiler room were trunked together into a wider funnel between two narrow ones. The two engine rooms were separated from the two boiler rooms by the 4in (HA) magazine.

On the basis of the 'E' class, DEE asked for four 150kW generators. DEE wanted them dispersed, two in engine rooms and two in a separate dynamo room, but E-in-C preferred to place them all in the engine rooms, where they could be tended by engine-room personnel. It turned out that the action load was about 700kW, so at the least the ship needed 200kW generators.[22] Ratings were later raised to 250kW and then to 300kW.

The desired endurance, 6,500nm at 16kts, was a problem. Estimation of hull volumes, such as oil fuel stowage, entailed laborious calculation. Lillicrap estimated that the ship would carry 1,900 tons of oil, giving an endurance of 4,600nm at 16kts (based on the oil-burning *Raleigh*) or 5,200nm at 12kts (based on the 'E' class). He thought he could squeeze in another 140 tons to manage 4,500nm at 16kts or 5,600 at 12kts – still well short of what the planners wanted. The design sent to the Board showed much more oil (3,200 tons) and an estimated endurance of 8,000nm at 12kts.

Drawings and a Legend were taken to the Sea Lords on 1 December, the drawings being approved on 13 December 1923 with minor changes, including a trawler bow and provision of a tube so that Asdic could later be installed.[23] Initially the winning gun mount design (by Elswick) was expected to weigh 130 tons (125 tons without guns), which was less than Lillicrap initially guessed. The detail design submitted in May 1924 showed 164 tons of revolving weight (140 tons without guns). Elswick ascribed extra weight to heavier than expected guns and to changes demanded by the Admiralty. This soon grew to 155 tons per mounting (without guns), but on average actual weight was 206 tons (including guns). Cuts were difficult to make. A proposal to reduce handing-room machinery was rejected because it was needed to achieve the desired high rate of fire, eight rounds per gun per minute.

Some weight was saved by drastically reducing protection to the ammunition hoist. There was no barbette, the hoist being an open skeleton structure rising out of a protective trunk just above the magazine, carrying flash-tight containers of cordite. The situation was so bad that the peacetime load was reduced from 150 to 130 and then to 100 rounds per 8in gun (there was space, however, for 150).[24] As the ships were being built, it turned out that the cases used to hold powder charges (five half-charges per box) were substantially overweight, adding as much as 14 tons for 800 rounds in the *London* class.

The mountings proved unreliable. In June 1930 CO of the First Cruiser Squadron (on board HMS *London*) wrote that sustained firings (twenty rounds per gun) were disappointing, as they averaged six rather than the desired eight rounds per minute. Torpedo firings were also a problem, with five lost out of fourteen runs (problems could not be diagnosed because runs were made in deep water, the torpedoes going to the bottom and so being impossible to retrieve).

The design called for a pair of octuple ('Mk M') pompoms. By 1928 a lighter quadruple pompom was being planned for British 8in cruisers. There was also a new quadruple 0.5in machine gun in prospect. Thus as of January 1928 the approved anti-aircraft battery was either two quadruple pompoms; or two quadruple 0.5in machine guns; or four single pompoms. Ships were to be strengthened to take the two pompoms. DGD pointed out that the thinner stream of shells from the quadruple pompom would take longer to bring down an aircraft, and also that it would probably be useful against destroyer attacks. He therefore suggested that instead of the 1,000 rounds per barrel formerly planned for the pompom, the allowance should be 1,500 (later it was increased to 1,800). Ships were all completed with four single pompoms, the only weapons of the three which were available. The two small 8in cruisers *York* and *Exeter* (see below) had limited deck space, hence were armed with only two single pompoms.

DTM wanted the ships to have his new E ('Enriched Air') type oxygen torpedo, considerably heavier than the current type, which also required two large oxygen-generating compartments in the ship. The only space they could occupy was at the fore end of the ship, some of it at the expense of fuel (hence range). This was initially accepted, as of November 1923, but the oxygen torpedo was not introduced until the *London* class. Requirements for cruiser torpedoes were formulated in 1927, the oxygen-enriched Mk VII being chosen as interim weapon, and supplied to the *London* and later classes and to HMAS *Australia* and *Canberra*.[25] The first cruiser tests were made from HMS *Berwick*, but this weapon was not supplied to the *Kent*s. DTM also disliked deck torpedo tubes, because they entailed so long a drop when torpedoes were fired, but instead the Mk V torpedoes supplied to the *Kent*s were strengthened after trials from HMS *Berwick*.

The weight situation so worried DNC that he was unwilling to allow installation of catapults. Eventually it became clear that ships were slightly underweight, so that single light (4,000lb load) catapults could be installed (in 1931–2), even though the trade-protection mission favoured heavy ones capable of launching the S/R aircraft such a cruiser would use to search the area around her. *Australia* received her catapult in 1935, *Canberra* in 1941. The Australians accepted light catapults in their two *Kent* class cruisers, but were prepared to switch back to the heavy catapult in the interest of interchangeability with the Royal Navy.

Numbers and An Abortive Treaty

By mid-1923 the Board was calling for a seventy-cruiser fleet, ten of which could be overage (in reserve at home and in the Mediterranean), to be ready by the 1929 target modernisation date. Director of Plans proposed a building programme.[26] He envisaged laying down eight 10,000-ton cruisers in each of 1924, 1925 and 1926, and four in each succeeding year. He assumed, also, a three-year building time. That would give twenty-eight large cruisers in 1929, and a total of fifty-nine under-age ships at that time. On this basis, in 1923 the Admiralty proposed the first year's programme, including the eight large cruisers, a second cruiser minelayer, a carrier, and two prototype destroyers. Three of the cruisers would be built at Royal Dockyards in southern England. The private shipyards had little work, and this programme was justified partly as a way of relieving unemployment in Scotland. Conversely, without some new construction programme employment under existing programmes would collapse beginning in 1925/6. The government was not particularly interested in expanding the Royal Navy, so the cruisers were all explained as replacements: the 10,000-tonners for old (discarded) 'County' class ships for trade protection, and the minelayer as a replacement for the *Princess Irene*. Probably to help him with his Cabinet, the Admiralty had to provide justifications directly to Prime Minister Ramsay MacDonald, the first Labour Prime Minister and no enthusiast for military spending. For example, it had to head off criticism that the British had somehow triggered the Japanese and US naval programmes (to which the Admiralty was responding), or, for that matter, those of the French and the Italians. The Admiralty pointed out that in 1919 the Royal Navy had eighty-four cruisers, of which thirty-six, including the 'Counties,' had since been scrapped. Of the remaining forty-eight ships, ten were already or were about to become overage. The Admiralty also emphasised the unique need of the British Empire for numbers of cruisers for trade protection.

In January 1924 the First Lord offered a reduction to four cruisers in 1924/5 and to five in 1925/6, provided the reduction was made up in later years. By February the Cabinet had decided to buy five in 1924/5, presumably to help relieve unemployment. They became the *Kent* class. The Commonwealth of Australia ordered another two, HMAS *Australia* and *Canberra*. The minelayer was dropped. Given the decision for five cruisers in 1924/5, four follow-on cruisers (*London* class) were bought under the 1925/6 programme. It was impossible to get eight large cruisers each year, so the Admiralty rethought and adopted a twenty-year cruiser lifetime in 1926. It sought a steady annual programme including three cruisers. That would eventually give the desired sixty underage cruisers, another ten making up the seventy. The first such programme (1926/7) provided the two *Norfolk*s and the smaller *York*, which was described as a trade-protection cruiser.

The government was clearly nervous about the cost of the Admiralty's programme, which included development of the Singapore base required if the fleet were to operate effectively in the Far East. Moreover, in peacetime Singapore was expected to support a substantial deterrent fleet.[27] A new round of disarmament could be substituted for the naval build-up. On 20 February 1924 the Cabinet formed a committee to consider the replacement of units other than capital ships (as their replacement had been deferred by the Washington Treaty) and also to reconsider plans for the Singapore base.[28] Prime Minister MacDonald's statement that failure to 'improve world conditions and reach an agreement on armaments' would require that the base go ahead could be read as a call for a fresh round of arms-control talks. He announced that the Singapore base would not go ahead. This was consistent with the idea current in the Labour government that British initiatives were causing the Japanese to arm. With Singapore, and therefore the threat of a British fleet movement to the East, removed, surely the Japanese would agree to drastic naval arms cuts. The Dominion governments affected by the decision disagreed, but that did not matter. The

First Sea Lord Admiral Beatty argued that the tonnage ratios Britain had accepted at Washington were predicated on the existence of a Singapore base (which was not the case), and the Admiralty position was that the goal of any negotiation with Japan should be the elimination of the Japanese battleship force – which was impossible. No negotiations followed.

However, in April 1924 the Cabinet directed the Admiralty to prepare a memorandum on further naval arms-reduction measures.[29] By this time it was clear that the large cruisers permitted under the Washington Treaty would be expensive. Limits were also suggested for other types of ships. DNC seems to have foreseen the demand for a smaller cruiser; discussing the design of the big cruiser on 9 April 1923, DNC ordered Lillicrap also to work on an Improved 'E' class cruiser with 6in guns. DNC provided the Board with a sketch design for a 6in cruiser about April 1924, presumably to support the reduction proposal: a scaled-down *Kent* displacing 7,500–8,000 tons, armed with four twin 6in. He could provide no protection at all, he had to use destroyer-leader machinery, and endurance would be half that of the larger cruiser.[30] Later the Admiralty would discuss attempts to 'cap' overly-expensive ship developments, and the idea of new qualitative limits on cruisers and lesser warships would be revived in 1927.

The word 'cruiser' covered far more than medium surface combatants. The *Kent* class Covers refer to 'Light Cruiser A,' a cruiser-sized carrier.[31] It counted as a cruiser because the Washington Treaty did not affect aircraft-carrying ships displacing less than 10,000 tons. Its existence reminds us that small carriers used about the same building resources as cruisers, a point of some significance during the Second World War.

HMAS *Shropshire* in December 1945. She retained the square tower and director of her class. (Paul Webb)

The *London* class

Sir Eustace Tennyson d'Eyncourt retired as DNC at the end of 1923, being succeeded by William J Berry (who had led the 'County' class design team). As he entered office, Berry wondered whether he could have produced a better treaty cruiser, for example using triple turrets. He assigned Lillicrap to investigate an alternative design with three triple turrets instead of four twins, which DNO estimated would weigh about the same, or with four triples (as it was rumoured the US Navy was planning). Four triples would weigh about 800 tons rather than the 600 tons then estimated for the twins, and at 130 rounds per gun extra ammunition would add another 105 tons. Magazines, shell rooms and handing rooms would be larger, so the armour covering them would weigh more (and would cost fuel space). Lillicrap set that at 95 more tons, for a total increase of 400 tons. He could claw almost all of that back by shaving protection over the machinery and by eliminating the torpedo battery and the steering-gear protection.[32] Unfortunately the wider magazines would run up against the shafting aft. They could be made narrower but longer, adding some armour weight (which could be offset by further cuts in machinery protection). Lillicrap also suggested shrinking the machinery spaces by cutting back to six boilers (60,000shp). That would make up for oil tankage lost to enlarged magazines, so that the ship would carry 3,290 tons of oil, and she could still make the desirable 8,000nm at 12kts.

Lillicrap's study seems to have been the basis for DNC's first approach to designing the next cruiser class (for 1925/6). At the end of July 1924 DNC suggested, for the 1925/6 cruisers, an entirely new design with four triple 8in, but with no protection and no bulge, and a speed of at least 34kts (attainable, perhaps, on 100,000shp). Armed with his earlier study, Lillicrap estimated that armament weight would increase by about 300 tons. Omitting the bulge would save 150 tons, partly balanced by heavier turret foundations and larger magazines; the net saving would be 65 tons on the 5,600-ton *Kent* hull. That saving would roughly balance increased hull weight due to the larger power-

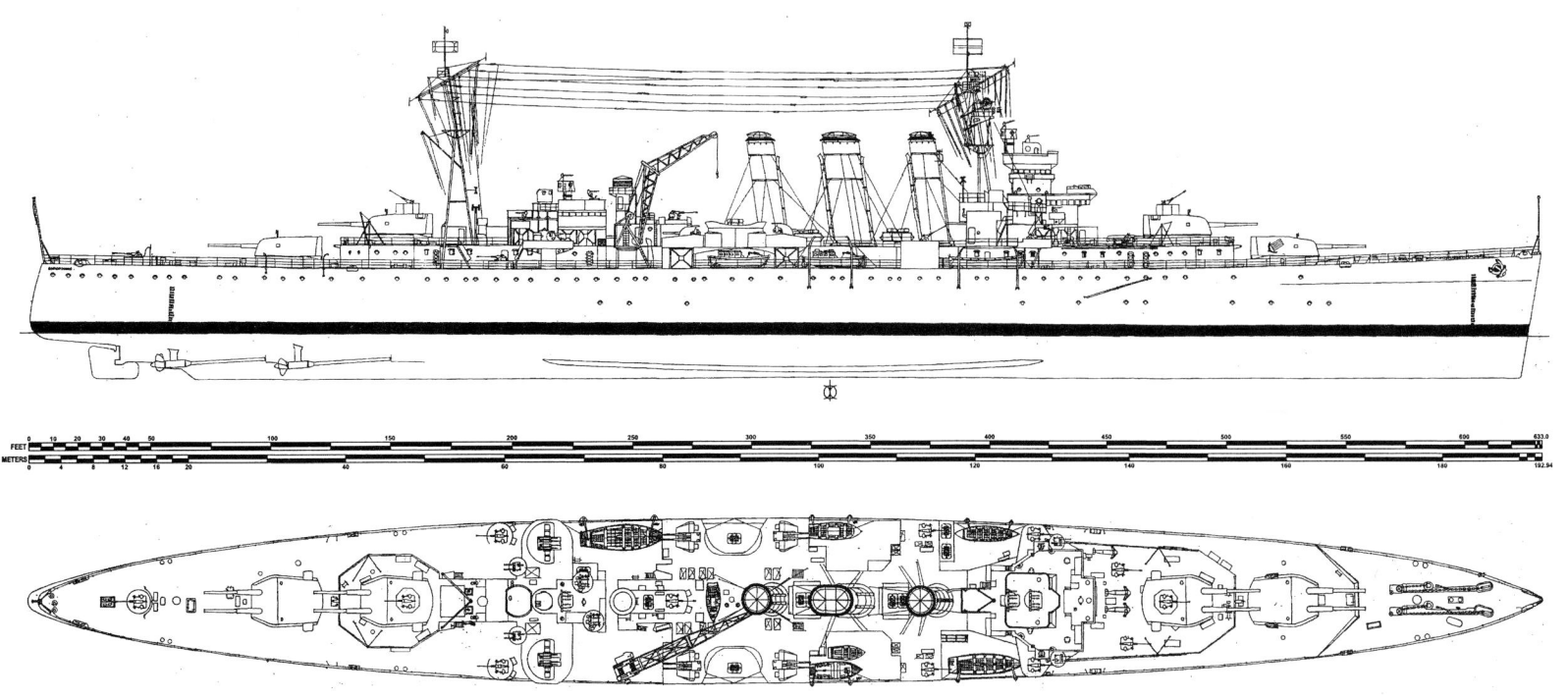

Above: The *London* class in effect packaged a *Kent* in a new hull without bulges. The 4in HA guns were moved forward to a boat deck (although the HA DCT remained aft). HMS *London* is shown at Portsmouth, February 1937.

plant. By this time more precise estimates were at hand for *Kent* weights, so Lillicrap could be confident that her machinery weighed 1,686 tons (he took 1,700 tons). He took the same equipment weight as in *Kent*, knowing that it could be cut somewhat. Most of the armour weight of the earlier ship could be applied to machinery, which could weigh as much as 2,530 tons (about 2,340 tons for E-in-C machinery). Lillicrap thought that would easily provide him with the 110,000shp needed to achieve 34kts (Haslar curves showed 55,000ehp to reach 34kts at 10,000 tons).[33] Lillicrap also considered a 120,000shp plant. The new ship would have a third boiler room.

Detailed estimates completed in mid-September 1924 showed that 110 tons would be left over for protection, enough for very limited deck armour over the magazines.[34] Since that was unacceptable, DNC was driven back to four twin 8in guns. In mid-October he asked how much protection that would buy (keeping the high power). He found that he could get good magazine protection, which was much the same conclusion reflected in the previous year's X, Y and Z designs.[35]

At the end of October, DNC asked Lillicrap for two alternatives. Design A would be based on *Kent* but would have the bulge eliminated to improve her lines and hence her performance. Weight saved would go into a catapult, crane, and other aircraft arrangements. The weight used to protect the 4in magazine would go instead into the lower conning tower, transmitting station and telephone exchange. With *Kent* machinery, the new lines offered another three-quarters of a knot.

Design B was the higher-speed unprotected version. DNC wanted a sketch showing torpedo tubes (which could be triples rather than quads, for oxygen torpedoes) in place of her catapult. Lillicrap added a 1in deck over the transmitting station, telephone exchange, and lower steering position (lower conning tower), a total of only 33 tons. He also added 24 tons of ready-use ammunition lockers. A developed version used 110,000shp machinery. Controller chose Design A by 2 January 1925, with torpedo tubes and a catapult.

DNC considered rearranging the machinery to improve subdivision. The existing eight-boiler plant lacked space to withdraw superheaters. Providing that in four rather than two boiler rooms would add 24ft (each boiler room would be 28ft long) and would add about 400 tons to displacement. Similar subdivision could be provided by a bulkhead cutting each existing boiler room in half, but in that case there would be no space to withdraw superheaters, unless holes were cut in the new bulkheads. Alternatively, the ship could be powered by nine 9,000shp boilers, three abreast in three 20ft boiler rooms. They could fit within the ship's beam (the double bottom would have to be somewhat

Below and opposite: HMS *Norfolk* (shown, in 1932) and HMS *Dorsetshire*, the last of the 'Counties', finally had the DCT, hence did not have towers atop their bridges.

shallower under the two forward boiler rooms). However, this arrangement could not be adapted to unit operation, as in the *Kent* class; it offered less security against local damage. By this time the Royal Navy was associating a particular length of damage with a hit: a single torpedo hit at the centre of a machinery space would damage boundary bulkheads 40–50ft apart beyond the ship's ability to cope with the inflow of water. In the worst case three such compartments would be opened up. In the nine-boiler design the length opened up would be 84ft instead of 112ft in the existing design. The existence of a midships magazine separating engine and boiler rooms became a valuable survivability feature. On the other hand, it could be assumed that an 8in shell burst could be confined between two sets of bulkheads, so with three separate boiler rooms one-third instead of half the boilers would be put out of action. Against that, the nine-boiler arrangement was more sensitive to local steam-pipe damage. Ultimately the new machinery followed that of the earlier class, although it was heavier, with a different steam pipe arrangement.

As the two main alternatives were being evaluated, Lillicrap was ordered to see whether a smaller cruiser might be feasible. He started by scaling up from the 'E' class. Using new higher-strength steel he could save 410 tons on the hull weight. He assumed *Kent* machinery, armament, equipment and protection weights. Without protection, the ship would displace 7,340 tons (standard, not the Legend condition at which an 'E' displaced about 7,500 tons). With *Kent* protection, the ship would displace 8,340 tons standard, but that was only part of the story. For example, she needed much more magazine volume, so to retain sufficient oil tankage she would have to be lengthened. Size would spiral up towards that of the real *Kent*. In effect Lillicrap was proving to DNC's satisfaction that the dimensions chosen were about the best that could be obtained.

The four 1925/6 cruisers became the *London* class. Except for the lack of bulges and the vertical rather than sloped hull sides, they appeared to duplicate the *Kents*. Unlike the *Kents*, they were fitted with single heavy catapults appropriate to their trade-protection mission. They carried enriched-air Mk VII rather than conventional Mk V torpedoes.

An October 1927 tabulation of Legend vs actual (weighed) weights for *London* showed a saving of 80 tons on the hull (originally 5,480 tons); a late tabulation showed that the hull actually weighed only 5,153 tons. By way of comparison, as built, *Suffolk*'s hull weighed 5,250 tons. Protection was held to 932 tons (compared to 960 tons as designed, and to 950 tons in *Suffolk*). Machinery weight was 1,570 tons. On the other hand, equipment was overweight (635 tons, compared to the estimated 570 tons, and to 530 tons in *Suffolk*), and armament was 1,220 tons (1,004 tons estimated, 1,200 tons in *Suffolk*). Standard displacement was thus 9,710 tons (9,820 tons for *Suffolk*). Thus *London* was within the 10,000-ton limit, but with little to spare. This displacement included only 100 rounds per 8in gun, but it also included stores for six months, provisions for ninety war days, and 8 gallons of fresh water per man for three days. There was enough weight to increase the load-out of 8in shells and cordite back to the 150 rounds per gun originally desired (another 70 tons), and also to add desired features such as a motor boat as Admiral's barge. On completion *London* did not have the planned pair of multiple pom-poms, because they were not yet ready; instead she received four single 2pdrs, which were much lighter.

By March 1926 the Board had decided that the two 10,000-ton cruisers (*Dorsetshire* class) of the 1926/7 programme should duplicate the four *London*s of the 1925/6 programme. Any major design changes were ruled out, because tenders for one of the two cruisers had to be invited by 1 August 1926 (for the other, to be built in a Royal Dockyard, tenders would be invited on 14 October). New 8in mountings were expected to save 76 tons, but increased armour would add about 140 tons, the net increase of 65 tons to be offset as far as possible by taking advantage of rolling margins in the steel (i.e. by a slight decrease in scantlings) as in *Nelson* and *Rodney*.

However, Controller was still interested in making protection more appropriate. Data were correlated beginning late in May 1926. What would fighting range be? Based on war experience, Heligoland Bight was probably the low end (7,000–8,000yds). Jutland was the high end: 15,000–19,000yds (the new large cruisers would probably be used like

the old battlecruisers). Rolling would expose spaces protected by water when the ship was on an even keel, and ships would not fight dead abeam. fight at a bearing angle. The 1926 analysis was based on an enemy bearing 40° abaft the ship's beam, and on 5° and 10° roll angles. Only magazine protection was considered, as there could not possibly be enough weight to protect machinery against 8in fire.

The greater the range, the steeper the trajectory of the enemy shell, and the larger the side area of the magazine which it could strike after passing through part of the ship after entering above water. At 10,000yds only 0.3 per cent of the side of the magazine of a fully-loaded ship could be hit, but that grew to 12 per cent at 13,500yds, and to 20 per cent at 15,000yds. At the after end, where the shell could strike at the greatest depth (due to the ship's wave pattern), at 15,000yds a shell could reach a third of the depth of the plate. In half-oil condition a ship drew 2ft less, so the figures would increase to (respectively) 12, 32 and 40 per cent, and the greatest depth reached would be 0.53 of total depth. The 4in magazine side could be penetrated by 8in fire out to 20,000yds; it took 5½in to stop an 8in shell at 13,500yds. DNC argued further that it was so unlikely that a shell would strike just at the waterline at 15,000–20,000yds that there was little point in special protection. Better to give the upper part of the magazine adequate protection and reduce the thickness of the lower part. He offered 5in plate on the upper part of the magazine and two ¾in D quality plates on the lower part for the same weight currently required by a uniform 4in plate. ACNS disagreed: in the likely event of a stern chase, the ship could be hit in the bow. A shell falling steeply through her decks (at long range) could get deep enough in the ship to explode near the lower (thin) part of the magazine end or side (if it arrived obliquely enough). Surely it would be wiser to use uniform 5in plating. DNC countered that this shell would have passed through several decks, and probably could not penetrate the final 1½in magazine side. However, the 1½in would resist its splinters, if it burst alongside the magazine. Dreyer favoured sloping magazine armour. However, if the floor area of the magazine was fixed in size, sloping the sides outward would enlarge the magazine crown and would cost fuel stowage. It was also questionable whether any such slope would matter for shots fired from 40° off the beam. At up to 60° off the beam, a shell passing through decks might travel a shorter distance than the fuse delay, hence might penetrate the magazine from the side.

DNC, DNO and DGD met on 23 July 1926 to decide the best magazine protection. The existing armour offered no protection at any range out to 20,000yds. Minimum immune range for the 5in/1¼in arrangement was 13,500yds. Sloping armour was effective at 12,000yds, but it entailed crippling practical problems. Any attempt to solve the problem of shells passing through the ship to explode entailed providing magazine crown armour all the way to the side of the ship, adding unacceptable weight. DNC concluded that it would be best simply to retain the earlier uniform 4in magazine side plating. Controller and then CNS (First Sea Lord) agreed. ACNS gave instructions on 4 August: the 8in magazines and shell rooms would be arranged as in the new *York* (see below), with the same protection, rather than as in *London* (DNC had prepared a submission to Controller in this regard).

The Board approved the Legend and drawings on 28 October 1926. The modified *London* became the *Dorsetshire* class in February 1927. ACNS proposed installing the new DCT on board the new ships (approved 8 February 1927). That May DNC's proposal that protection match that of the new *York* class described below was approved. A proposal to double the torpedo battery (and to carry the usual spare for each tube) was rejected because the extra weight involved would have required cutting 8in ammunition to sixty rounds per gun and halving 4in and pompom ammunition. Controller (Chatfield) approved the design on 1 June 1927. The visible change from the *London*s was the DCT atop the bridge, the tower of the two earlier classes being eliminated.

The Smaller Cruiser

In February 1925 DNC assigned Lillicrap to return to work on the smaller cruiser with eight 6in guns. He started with a big fast hull chosen by DNC: 570ft x 60–62ft x 15¾ft, 8,500 tons, 100,000shp for 34.5kts. Machinery weight was scaled up from that of *Kent* (i.e. it was not a lightweight plant), and the ship was given *Kent*-class magazine protection (with no protection over her machinery) and secondary armament, including a pair of quadruple torpedo tubes. A trial Legend suggested that this combination was feasible. However, it might not carry enough oil for the desired 8,000nm at 12kts. On the basis of a 575-footer worked out in detail, Lillicrap estimated that the ship could carry about 2,600 tons of oil. To reach 8,000nm, an 'E' class hull would need 2,800 tons and a *Kent* 3,200 tons; Lillicrap estimated that his ship could make 6,900nm at 12kts. A further estimate showed 2,800 tons of oil below the platform deck. DNC told Controller that he could manage 7,000nm at 12kts on the usual tankage below the platform deck. He could do even better if he stowed 400 tons in above-water 'peace tanks', which would of course have to be emptied before the ship went into action.

Lillicrap also estimated details of a smaller 6in cruiser of about 6,000 tons (495ft x 50.4ft x 15.25ft as an initial guess), with 60,000shp or 80,000shp machinery. As in a large cruiser, he could provide magazine, handing room and shell room protection as well as protection to the steering gear. By the end of March 1925, this project had grown into a 6,630-ton design. This was a serious project; ACNS set the level of protection: splinter-proof shields and ammunition trunks for the 6in guns and the lower conning tower, and a 2in forward boiler room bulkhead, presumably plus the usual box magazine protection. The ship should have a maximum speed of 33–35kts (31kts was considered too slow) and an endurance of 8,000nm at 12kts (as in a *Kent*). Deputy Director of Naval Construction (DDNC) told Lillicrap to use 'E' class length and get what speed he could on 80,000shp ('E' class or *Kent* power). The hull should be flush-decked (like *Kent*) but without bulges. Displacement should be about 7,500 tons. Main armament should be four twin 6in guns, torpedo tubes were necessary (preferably two triples as a minimum, or one triple on the centreline), and a catapult was considered essential. Lillicrap reported an 8,500-tonner; DNC made 7,500 tons an upper limit.[36]

Lillicrap also looked at an alternative main armament of three twin 8in guns, giving a total armament weight of about 850 tons, compared to 675 tons for the ship with 6in guns. He estimated that this ship would displace 8,760 tons and would make 31.75kts on the 80,000shp plant used for the larger cruisers. The second-rate cruiser might be able to deal with the emerging class of 8in cruisers, or else dominate the majority of the world's cruisers, with their 6in guns.

At least at first, DNC was much more interested in a smaller 6in cruiser. He suggested using lightweight destroyer machinery (two 40,000shp plants, as in the new *Amazon* and *Ambuscade*). Two sets of Thornycroft machinery would weigh about 1,100 tons; auxiliaries such as generators would add another 120 tons, for a total of 1,220 tons instead of about 1,600 tons. Initial estimates gave 6,795 tons, although Lillicrap later worked on the basis of 7,500 tons.[37] He was still interested in the alternative design with six 8in guns, the displacement of which he now estimated would be 7,710 tons. E-in-C was unwilling to approve destroyer machinery for a cruiser, so Controller asked at the end of June

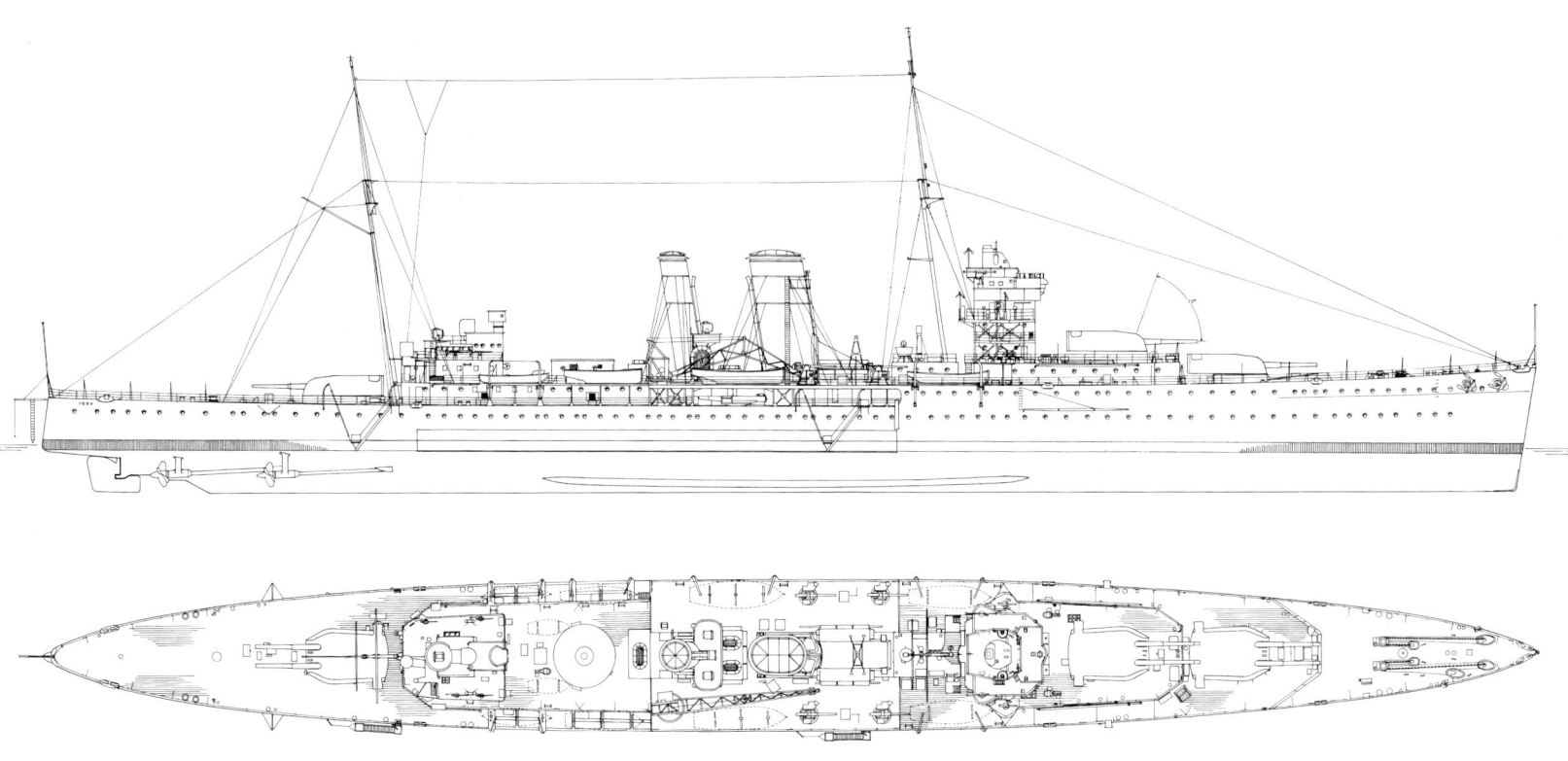

HMS *York* as fitted in 1930. During a 1931–2 refit she received a Type E.II.H catapult, and the side plating was extended aft to just abaft the after pair of 4in guns. During her 1934–5 refit her 2pdr AA guns (single pompoms) were replaced by two quadruple 0.5in machine guns. Her signal deck was extended aft and the sidelights were mounted on new small bridge wings. In 1938–9 the catapult was replaced by a Type S.I.H and the single Fairey IIIF replaced by a Supermarine Walrus, stowed on the catapult. A five-round depth-charge rack was added on the port quarter in 1938, with two reloads on deck just forward of it and three more atop the after superstructure just forward of the mainmast. No Asdic was installed. The only Second World War armament improvement was the fitting of single Oerlikons atop 'B' and 'Y' turrets. Most of the lower deck portholes were plated over, and a bulwark was erected around the 4in guns. She was fitted with a Type 286 radar some time before her loss in May 1941. In 1929 the Chilean government wanted to buy a modified *York*, for which it negotiated with Vickers. Vickers' files dated 9 May 1930 describe work on an improved and slightly enlarged ship intended to achieve 32kts on the same 80,000shp power-plant. Proposed dimensions were 550ft rather than 540ft pp, 580ft rather than 570ft lwl, and 58.5ft rather than 57.5ft beam. The *York* project collapsed, but Chile decided to revive it in 1938, after a project for a smaller cruiser (to be bought from Italy) was raised (the design involved was probably the one sold to Siam). That caused considerable problems, since the project threatened to upset the delicate balance of cruiser limitation. Ultimately the Chileans seem to have been stopped by financial problems; in 1939 Vickers was negotiating to build what amounted to a somewhat enlarged four-turret *Dido*. As of 1929 Chile had already expressed interest in a heavy cruiser; Vickers offered Design 1208 about October 1925. It was comparable to the new British cruisers and to the cruiser being designed for Spain: 10,750 tons (600ft pp, 635ft lwl, 639ft loa x 64ft over bulges, 60ft at wl x 37ft x 17ft 3in). Armament was four twin 8in/50, six single 4.7in HA and four triple torpedo tubes. There would have been virtually no armour (the design book lists 1in barbettes, above the weather deck only). The ship would have made 35kts on 120,000shp (ten Yarrow boilers). The crude sketch in the design book shows a forecastle extending most of the length of the ship, back to abaft 'Y' turret, but not the whole length. The torpedo tubes are recessed below the forecastle deck. The ship has three vertical funnels, a tripod mainmast, and a director forward on a tall pylon (with a second director aft, forward of 'X' turret). A rotating flying-off platform was set abaft the third funnel, with a separate aircraft stowage platform (aircraft athwartships) forward of it. This was a considerable step up from the previous post-1918 Vickers design series for Chile, Design 758 (estimate dated October 1919), with two twin and three single 6in (525ft x 51ft 6in x 28ft 7in x 15ft 6in, 7,000 tons, 43,000shp [coal and oil] for 29kts). Other armament was two 4in HA and three triple torpedo tubes; an alternative 758A had four twin 6in and four triple tubes. Vickers also offered cruisers to Argentina's rival Chile. Design 1084 (7 March 1924) would have displaced 6,600 tons (510ft pp, 538ft lwl, 540ft loa x 51ft 6in x 28ft x 15ft 6in). She would have made 32kts on 68,000shp (six boilers), with a radius of 6,000nm at 15kts. Armament would have been seven 6in (three twin and one single, all on the centreline), four 4in HA and four triple 21in torpedo tubes. A slightly later Design 880 (6 February 1925) would have displaced 6,000 tons (490ft x 51ft x 28ft x 15ft), armed with seven 6in/50 (two twin, three single), four 105mm HA, two single pompoms and one depth-charge thrower (apparently no torpedo tubes). The ship would have made 29kts on 42,000shp (six Yarrow oil-fired boilers). Design 1287 (16 March 1927) resembled a scaled-down 'County' (with the same three funnels). She would have displaced 8,500 tons normal (570ft pp, 605ft oa x 56ft 4in ext x 16ft 6in), with 80,000shp engines for 33.5kts. Armament would have been six 7.5in/52 (twin mounts), seven twin 4in HA, four single pompoms and six torpedo tubes. It was rejected in favour of the Italian design which was built as the *Almirante Brown* class. Vickers-Armstrong succeeded with Design 1076, which became the training cruiser *La Argentina*. Like the others, it had no direct Admiralty equivalent, although it seems most comparable to the *Apollo* (later *Sydney*) class or to an enlarged *Arethusa* with triple rather than twin turrets. At the design stage it was expected to displace 6,850 tons. (A D Baker III)

how much the ship had to grow to get the same features. Alternatively, how much power would she lose if her displacement could not be increased? E-in-C offered 72,000shp on the available weight, which would give the ship 32kts at standard displacement and 31kts fully loaded.[38]

In August Lillicrap was told to concentrate on the 7,500-tonner armed with six 8in guns. DNC asked for 3in side and deck armour over the 8in magazines and handing rooms, which pushed estimated displacement up to 8,000 tons. The ship was lengthened.[39] By this time the flush deck had apparently been abandoned. If 'X' turret was mounted on the upper deck, there would be insufficient space between it and the magazine below water. However, raising 'X' turret onto the superstructure would add too much topweight (in the end, it was mounted on the upper deck). The reduced height required a new mounting design.

The Board decided that the one of the three 1926/7 cruisers would be the smaller type Lillicrap was sketching. She was called the B Cruiser. Plans Division saw the ship as a specialised trade-protection cruiser capable of attacking enemy 10,000-tonners, although Trade Division (responsible for protecting trade) doubted that a B Cruiser should be pitted against an A (10,000-ton) Cruiser. Operations went further: whatever juggling was done, an 8,000-ton ship could not be built to take on a 10,000-tonner. Plans Division split the 10,000-ton cruisers into those intended for the fleet and eleven intended to back the B Cruisers on the trade routes.[40]

The development of Staff Requirements was put off until the autumn of 1925 pending design work. DNC pointed out that, because of her reduced size, she would not be able to keep speed at sea as well as a larger cruiser, so Legend speed was not reduced. With a shorter hull, she needed the same power to drive her at the same speed. Aside from one twin 8in mount, the only reduction in armament compared to a 10,000-ton cruiser was substitution of triple for quadruple torpedo tubes. DNC claimed that he had abandoned the flush-deck hull of the larger cruisers in order to save weight, but presumably he was also losing the advantages of the deeper hull girder for strength with minimum weight. He pointed out that reduced hull volume would reduce many features to the level of the 'D' and 'E' classes.

Initially DNC tried simply to duplicate *Kent* class protection, on much the same protection weight. Because the ship was smaller, he was able to provide a 1½in rather than 1in deck over the machinery. The Staff pointed out that in her most likely role, trade protection, the B Cruiser would have to fight enemy 10,000-tonners, hence needed all the gunnery help she could get. A second catapult, perhaps on a turret top,

Below and opposite: HMS *York* was, in effect, a smaller version of *Norfolk*, with the same new DCT. She is shown in 1930.

TREATIES AND HEAVY CRUISERS

would be valuable, launching an F/R or fighter.[41] The following year it was decided that all cruisers should have both light and heavy catapults. Director of Torpedo Division (D of TD) asked whether DNC could estimate the cost of adding a fourth turret.[42] Like the larger cruisers, this ship was criticised for its thin side armour. Since she was well below the treaty limit, she could be more heavily armoured. It would take a 7in side to protect against 8in fire, but 3in would do against 6in fire and 1½in against 4in. DNC speculated that the ship would be far more likely to encounter the numerous 6in cruisers than the few 8in. The US *Omaha* class and probably the Japanese *Furutaka* had this level of protection (he was right about both). The 1½in deck had been recommended for the 10,000-ton cruisers, but insufficient weight had been available. In effect the additional protection compensated for the greater damage a cruiser with more 8in guns would probably inflict on the B Cruiser. There was some interest in arming the ship with higher-velocity guns, but DNO pointed out that they would need significantly larger magazines, and that would have been difficult (DNC pointed out that oil stowage would shrink substantially).

The thicker armour was incorporated in a Legend DNC submitted on 22 December 1925. The ship was shorter and somewhat beamier than what Lillicrap had proposed, and was expected to displace 8,200 tons. Compared to a conventional 10,000-ton cruiser, she had much more side armour and slightly less magazine crown armour.[43]

Endurance was reduced from the 12,000nm now expected of HMS *Kent* to about 7,000nm. She was expected to cost £220,000 less than a 10,000-ton ('A') cruiser, which in 1926 was expected to cost about £2 million. The sketch design submitted to Controller on 11 February 1926 was approved by the Board on 25 February.

It seemed that there was no space aft for a second main battery director, but space became available because the HA director did not need a separate rangefinder. The after DCT had to be 14in off the centreline to starboard, so that the angles at which it was blocked (wooded) by the after searchlight were equal. A new twin turret design was ordered from Armstrong (Elswick); the company hoped to reduce weight by about 20 tons per mounting.[44]

The final Legend received the Board Stamp on 23 July 1926. The ship was now expected to displace 8,400 tons, including her multiple pompoms. She was somewhat longer than had been envisaged.[45] The most obvious change was that decks over magazines and shell rooms were 3in thick (as in the larger cruisers) rather than 2½in.

As yet there was no suitable light catapult design for a turret top, so the designers had to estimate its size. As estimated size and weight grew, the compass platform had to be raised 3ft 6in higher than originally expected.[46] That made the weight of the new DCT critical, so its protection was cut. The extra weight (but not so much the topweight) involved was acceptable because the ship was nowhere near the 10,000-

York had a high bridge structure because she was conceived to carry both heavy catapult aircraft aft and a light catapult aircraft on a catapult atop 'B' turret. However, no such catapult was ever designed (there were questions as to whether it was practicable).

Like *Norfolk*, *York* had her anti-aircraft guns forward (in her case, on the forecastle deck). They are barely visible around the boat roughly abeam her forefunnel. The ship was photographed at Coco Solo, in the Canal Zone, on 28 February 1939. She never received twin 4in guns. The only early-war addition to her battery seems to have been four single pompoms.

ton treaty limit. DNC pointed out that the catapult would be subject to muzzle blast from the turret guns.[47] As the ship neared completion in 1929, there was still no light catapult design, so the idea was abandoned, and the derrick, canvas cover to catapult, forward petrol tank, etc were all ordered omitted.[48]

HMS *York* (as the new ship was named) seems to have been the first post-war cruiser subjected to wind tunnel experiments (at the National Physical Laboratory), in May 1928. The question, raised in February, was whether the newly-designed raised bridge would be subject to smoke interference, hence whether the funnels should be raised (as they were being raised in the *Kent* class). She was designed with the same three-funnel arrangement as the other British 8in cruisers. Smoke from the funnels was driven down onto the deck amidships, continuing aft along the after control station and the quarterdeck. DNC recommended raising the funnels by 10ft, but even then the after control positions would not be clear of smoke when the ship ran right into the wind (conditions at half speed were worse than at full speed). DNC attributed the problem to eddies created in the wake of the bridge structure, and suggested remedies such as removing bulwarks from the sea cabin deck (the lowest level of the bridge structure). Rounding off bridge projections made no great difference. As long as the bridge height was maintained, it seemed impossible to eliminate the eddies. Experiments resumed in January 1929 showed that to get any improvement the forefunnel had to be raised 15–20ft. That was impractical, so a modified model had the two forward funnels trunked together. If they were raised 10ft, most of the smoke cleared the after control station, presumably because most of the eddies dissipated before they reached the new trunked funnel. If the funnels were trunked and the bridge lowered (up to 10ft), the (raised) funnels could be lowered by a similar amount. *York* was completed with trunked raised funnels. It was too late to lower her bridge now that there was no turret catapult.

Like other cruisers, *York* suffered from increased weight during construction, mainly 90 tons in her 8in mountings and 25 tons due to modifications to her bridge and DCT. Unfortunately the additional weights were both high in the ship and mainly forward, so that the ship lost stability (metacentric height) and trimmed forward when fully loaded. According to a sheet dated 28 November 1927, as designed *York* had been expected to displace 8,425 tons, but as laid off she was expected to displace 8,434 tons and possibly 8,549 tons in standard condition. Her designed metacentric height in standard condition had fallen from 3.4ft to 2.9ft – and she would not be completed for another two years.

York must never have been completely satisfactory, because in 1937 reconstruction of her bridge was proposed, presumably on the lines of HMS *Exeter*, in connection with anti-aircraft rearmament. Unfortunately no drawing was preserved, but a paper rejecting the idea mentions a vertical mast. The whole bridge above 'B' gun deck would have been dismantled. It was understood at the time that the ship would be due for reconstruction in 1940, at which time the bridge structure could be rebuilt.

Back to Treaty-Making: Geneva 1927

The British government continued to see treaties as a way of cutting the cost of the Royal Navy. The Labour government fell in October 1924, but its Conservative successor was no less aware of the country's difficult financial position. Pressure for further naval arms limitation also came from the League of Nations, whose Covenant linked arms reductions to international security, and from the United States, whose government had called the Washington Conference in the first place. In August 1924 President Calvin Coolidge announced that he would call an arms-control conference as soon as European conditions justified one. A League of Nations Preparatory Commission for a new world disarmament conference met for the first time in May 1926. That December three of the Sea Lords (Admirals Dreyer, Field and Chatfield) proposed limiting the size of warships in all categories to keep them affordable.[49] The life of capital ships should be extended (i.e. the building holiday should continue past 1931), the 5:5:3 ratio should be extended to 8in cruisers 'until they become obsolete', destroyer numbers should be reduced (if submarines were limited) and torpedoes should be limited in calibre. The last was essential because smaller capital ships were disproportionately vulnerable to underwater hits. The British draft presented to the Preparatory Commission embodied these points.[50] President Coolidge proposed his new conference in a February 1927 address to Congress and it met at Geneva in June 1927.

The planned 1927/8 programme included one large and two smaller 8in gun cruisers. Single A Cruisers were also planned for the 1928/9 and 1929/30 programmes. From the British point of view, abolition of 8in cruisers would be so advantageous that it would pay to abandon this programme so that none of the participants in the coming conference could imagine that the British hoped to build a last few such ships before the conference killed them off. It was probably inconceivable that nothing would come of a second naval disarmament conference, given the success of the Washington Conference and widespread belief in peace through arms control. Among the British objectives was the end of 8in-gun cruiser construction (the maximum cruiser calibre should be 6in) and a reduction of maximum cruiser size to 7,500 tons, the figure Lillicrap had worked to the previous year. The British also sought an agreement on the number of 8in gun cruisers each major power could have, as well as the total size of its cruiser fleet. British logic was the same as at Washington in 1921: other powers needed cruisers for their fleets, but the maritime Empire needed many more to protect its long trade routes. The British still wanted a total of seventy cruisers; they estimated US and Japanese requirements as forty-seven and twenty-one respectively.

The US General Board had secured authorisation for eight large cruisers in 1924 after several years of failure, but it required many more to execute the war plan against the most likely enemy, Japan (which had a larger modern cruiser force). It appears in retrospect that the US Navy delegates to the Geneva conference deliberately constructed a cruiser policy (of equality and reduction) they knew the British would reject, in order to sink the conference. On several occasions during the conference the British thought they had secured US agreement, only to find the US Navy delegates reversing themselves abruptly. US arms-control advocates were so angry that they tried to prevent any naval delegates from attending the next conference in 1930. From the US Navy's point of view, it was easy to stir anti-British feeling at home to justify the failure of the conference. The conference broke up in August 1927.

However, some of the final British proposals pointed to the agreement that would be reached in 1930. Cruisers should be split into two categories, the 10,000-tonners with eighteen-year lifetimes, and smaller ships (with a 7,500-ton limit and 6in guns) with sixteen-year lifetimes. Countries would be allowed a total tonnage of underage ships plus 25 per cent in overage ones. Existing ships between 6,000–10,000 tons would be spared: for the Royal Navy, the four *Hawkins* class, *York* and the two 'E' class; for the United States, the ten *Omaha*s; and for Japan, the four *Furutaka*s. The British tacitly accepted that the 10,000-tonners were strategic scouts, as the US Navy perceived them, so it accepted equality in that category (twelve each for the United States and Britain, eight for Japan). The US delegates proposed another idea incor-

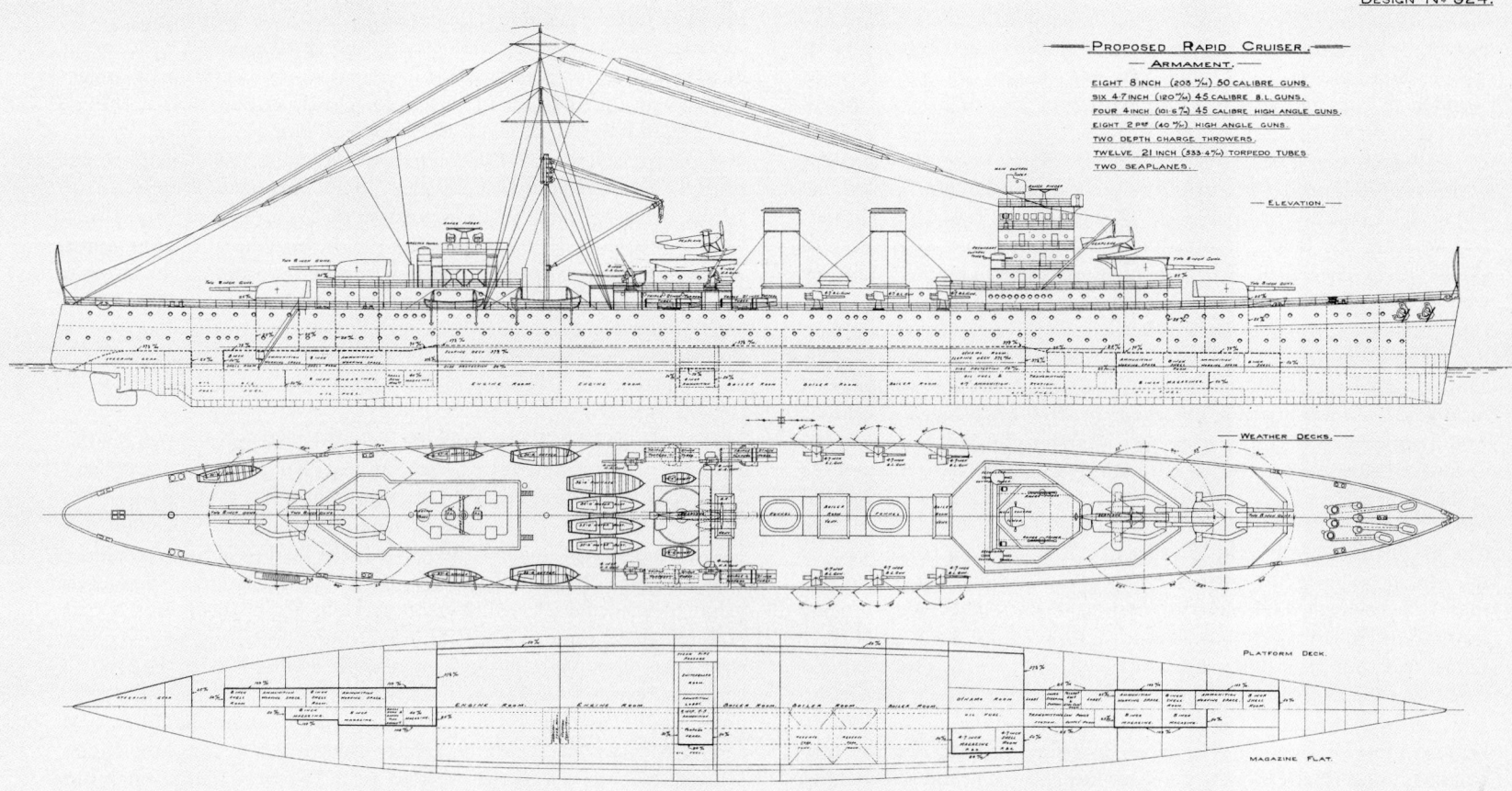

Vickers Design 924 was the basis for the Spanish *Canarias* class. It was Vickers' only inter-war export cruiser success prior to the training cruiser *La Argentina* in the mid-1930s. Note the two seaplanes, atop 'B' turret and amidships. Unlike *Canarias*, this ship carries her torpedo tubes in triple deck mounts. When the Spanish government decided to modernise its navy with a 1908 Navy Law, it created a national shipbuilding company (SECN, Sociedad Española de Construccion Naval) managed by a British consortium consisting of Armstrong, John Brown and Vickers. Most Spanish warships were Vickers' designs, although Vickers' records also show destroyer and submarine designs developed (apparently independently) by 'la Sociedad'. In the late 1920s and early 1930s Vickers effectively owned the Spanish shipbuilding industry, but the Spanish Republic nationalised SECN. The earliest Spanish cruiser design listed in a Vickers 'Estimate Book' (papers to estimate ship cost) is dated 6 March 1913 (unnumbered; unfortunately the Estimate Book provides no details). It presumably developed into Design 706 (14 July 1914, described by Vickers as an improved *Birmingham*), which was built as *Reina Victoria Eugenia*, the first modern Spanish cruiser, built under the 1914 Navy Law (completed 1923). The next cycle comprised Designs 730 (10 March 1915) through 733, of which Design 731 became the two *Mendez Nuñez* class (1915 Navy Law, completed 1924–5). Design 731 displaced 4,500 tons (440ft pp, 455ft loa x 46ft x 25ft 4in x 4,650 tons x 14ft 6in mean) and would have been armed with six 6in/50, one HA gun and two broadside torpedo tubes. The design was modified in August 1916 to provide two twin deck torpedo tubes (later the ships were completed with four triple 21in, as in contemporary British light cruisers). The 6in guns were all in single mounts, two abreast at the break of the forecastle and two abreast superfiring over one gun aft. Unlike new British light cruisers, these ships burned both coal and oil. Unusually, they carried more oil than coal in the normal condition (175 tons and 75 tons respectively) than in the full load condition (500 tons and 718 tons respectively). Other contemporary Vickers cruiser designs were similar in this respect. Design 730/731 was intended to make 29kts on trial on 43,000shp. Designs 732/733 had slightly less power, but otherwise were similar. Two projected *Mendez Nuñez* class cruisers were cancelled in 1919, but the Spanish were still interested in further construction. Design 756 (July 1919) was a follow-on to Design 731, but had a new armament arrangement which Vickers favoured post-war: six 6in/50 in two twin mounts superfiring over two singles. Unlike the twin mount then being developed for British cruisers, Vickers used an open shielded mount. Other armament in the new design was four 4in HA and (like contemporary British light cruisers) twelve 21in torpedo tubes in triple deck mounts. The ship would have displaced 4,850 tons (450ft pp x 47 ft x 27ft depth x 14ft 6in) and would have made 29kts. A further Design 841 was offered in March 1920 (modified as 841A in April), with the same armament (5,250 [5,000 in 841A] tons, 450ft x 46ft 5in [47ft in 841A] x 27ft x 14ft 6in), 44,000shp. Three cruiser designs were offered for the 'third Spanish naval programme,' dated 1921 (in a notebook kept by George Thurston, Vickers' chief naval architect at the time): E (eight 7.5in/45, four 4in HA, four 4in/50 BL, two 3pdr, six fixed 21in deck torpedo tubes and two submerged tubes), F (eight 6in/50 [two single, three twin], four 4in HA and twelve torpedo tubes in triple revolving mounts), and G (six 6in/50 [two single, two twin], four 4in HA and twelve torpedo tubes in triple rotating mounts). Of these, F (dated 22 August 1921 in another Vickers book) was the basis of the *Principe Alfonso* ordered at about this time (545ft pp, 579ft 6in loa x 54ft 8½in x 30ft 6in x 16ft 6in; 7,850 tons. Side protection totalled 3in over the machinery (2in and 1½in ends) with 1in bulkheads closing off the ends of the magazines. It was designed by Sir Philip Watts, a former DNC, at that time chief naval architect for Armstrong. *Principe Alfonso* herself was renamed *Libertad* in 1929 and then *Galicia* (1939). The ships made 33kts on 82,000shp. This design seems not to have been, is often claimed, an evolved 'E' class cruiser. It was in line with many Vickers cruiser designs armed with twin 6in in 'B' and 'X' positions and singles in 'A' and 'Y'; in Design F, a fifth (twin) mount was inserted between the two funnels and the after controls. G was apparently Design 841. E was probably Vickers Design 894 (estimate sheet dated 22 December 1920: 580ft x 59ft [wl], 67ft 6in [over bulges] x 30ft 4in [to upper deck] 38ft 4in [to forecastle deck] x 17ft 6in, normal displacement 11,500 tons). Armament would have been four twin 7.5in/45, four 4in HA, four 4in QF, and two triple deck and two submerged 21in torpedo tubes. Twelve Yarrow boilers would produce 75,000shp for 31kts on the measured mile. Protection would have been similar to that of G, except that there was also 2in HT plating from the belt up to the forecastle deck. References to a forecastle show that this would have been a conventional rather than a flush-decked design.

Vickers' first 8in cruiser offered to Spain seems to have been Design 845 (estimate sheet dated 28 February 1924: 590ft x 61ft 6in x 33ft x 19ft [the figure is indistinct], 10,500 tons [normal], three triple 8in/50, four 4.7in HA, four triple 60cm torpedo tubes on deck, 120,000shp [eleven Yarrow boilers] for 35kts). The Thurston notebook shows designs 847 and 848 for 8in cruisers, apparently for Thurston's article on 'light cruisers' for the 1925 issue of *Brassey's Naval Annual*. As such they amounted to advertising for Vickers designs. They were probably variants of Design 845. Both displaced 10,500 tons in trial condition, hence were within the 10,000-ton Washington Treaty limit. Design 847 showed that such a ship (570ft x 66ft [over bulges] x 40ft 10in [depth] x 17ft 3in [trial]) could carry nine 8in/50 (triples), eight 4.7in HA (the new British heavy HA gun), twelve 24in torpedo tubes (rotating triple mounts on deck; the British were then experimenting with a 24in torpedo) and two depth-charge throwers, with machinery for 34kts (120,000shp, ten Yarrow boilers). Like a 'County,' she would have no side armour, relying mainly on a 1in–1½in deck over her machinery and a 1½in lower deck over her magazines (which would have 2in crowns and bulkheads). Design 848 was a modified version with four 4.7in HA guns but enough power (eleven boilers, 135,000shp) for 35.5kts in a longer hull (590ft x 62ft x 33ft x 18.5ft). The same lengthened hull could be driven at 35kts by 120,000shp. The lengthened hull had some side protection (2½–3–2½ –1½in) but no barbettes (1in in 847) and no deck over the machinery. Design 866 (cost estimate dated 6 June 1924) was the long hull (590ft x 65ft x 40ft 10in x 17ft) armed with a more conventional four twin 8in/50, plus six 4.7in HA and the twelve 24in tubes, capable of 34.5kts on 120,000shp (with the notation that this was on the measured mile, and that on an eight-hour trial the ship would make 33.5kts on 105,000shp). This design showed only deck armour (1½in over machinery and magazines), hence was comparable to a 'County.' Alternatives were the slightly shorter Design 867 (four rather than six 4.7in) and Design 868 (three twin 8in/50, 565ft x 66ft x 40ft 10in x 17ft 6in, 34kts for four hours on 120,000shp, 10,000 tons standard and 10,500 tons on trial, 3in rather than 2in magazine crowns). Roughly contemporary was Design 875 for Spain, a 14,000-ton cruiser (a sketch of her deck arrangement shows four funnels), 650ft pp, 684ft wl, 686ft oa x 74ft (over bulge) x 42ft (depth) x 19ft (mean) armed with eight 8in in twin turrets, eight 4.7in HA and twelve 60cm torpedo tubes. She was intended to make 35kts on 144,000shp. An alternative 876P (15,000 tons, 685ft [pp], 722ft 6in [oa] x 74ft x 42ft x 19ft, 138,000shp) showed four twin 10in/45. The sketched deck arrangement shows three funnels, but no other details. The calculation sheet mentions a drawing received from Barrow on 25 November 1924, and the associated cost estimate sheet was dated 2 January 1925. Another alternative (877) would have been armed with three twin 12in/50 guns (15,000 tons, 685ft pp, 720ft wl, 722ft 6in oa x 74ft x 19ft, 34kts on 120,000shp). Design 878 offered four 12in/50 on similar dimensions, with a speed of 35kts on 138,000shp. A slightly later (18 November 1924) Design 1143 would have been armed with four twin 10in/45, six 4.7in HA and four triple 24in torpedo tubes. This ship would have displaced 12,500 tons on trial (630ft pp, 665ft wl, 669ft loa x 67ft 6in over bulge, 63ft 6in at waterline x 38ft 9in to forecastle deck x 18ft). She would have made 35kts on 130,000shp. Design 1143A was a similarly-armed 15,000-tonner (670ft pp, 705ft lwl, 709ft loa x 70ft x 19ft) intended to make its 35kts on 135,000shp. Design 883 was described as a 'standard cruiser', hence was presumably the design offered to potential customers, including the Spanish. It reverted to earlier dimensions (590ft [pp] x 65ft [over bulges] x 40ft 10in x 17ft, 34.5kts on 120,000shp) and would have been armed with four twin 8in/50, four 4.7in HA, twelve 24in torpedo tubes and two depth-charge throwers. There was no side armour, but there was a 1in deck (1½in slope) and 4in over magazine crowns. Design 904 used the Design F powerplant (80,000shp, 32kts) but was armed with three twin 8in/50, six 4in HA and four triple 21in torpedo tubes. In modified form, as offered in March 1926, it displaced 8,650 tons in normal condition (550ft x 61ft x 40ft 6in x 17ft 3in)

A twelve-gun Design 910 was offered to Spain: six twin 8in/50, six 4.7in HA, four twin pompoms (a non-existent mounting), four triple 60cm torpedo tubes and two depth-charge throwers, all on 10,000 tons standard (10,500 tons on trial) in a 615ft x 65ft x 41ft 7in x 17ft hull. Protection (including a belt) was 1in thick. This design was probably the origin of claims that the *Canarias* class was originally to have had twelve 8in guns, but the fact that they were twins shows that this was a very different design. The ship would have made 35kts on 115,000shp (34.5kts on four-hour trial on 106,000shp). An alternative Design 911 had five twin mounts and eight 4.7in guns, other characteristics being much the same. The record is complicated because other Vickers documents do not show designs in numerical order. Another series of cruiser proposals apparently began with Design 1074 (18 December 1923: 590ft, 10,750 tons, nine 8in, four 4.7in HA, twelve 24in torpedo tubes; 1074C was an alternative with eight 8in). Vickers offered a version of 1074 (1074X: 10,750 tons, 590ft pp, 625ft lwl, 629ft loa x 64ft x 37ft x 17ft 6in, 115,000shp for 34.5kts) as an alternative to the *Kent* class Australia bought (dated 23 July 1924 in Vickers Design Book 2). Armament would have been three triple 8in/50, four 4.7in HA, four triple 24in torpedo tubes and two depth-charge throwers. There would have been ten boilers. Vickers also offered the Australians (on 26 November 1924) Design 1144, armed with four twin 8in, four 4in HA, two octuple pompoms and two quadruple torpedo tubes (it was probably simply a repeat *Kent*). The next offer (31 March 1926) seems to have been an 8,325-ton cruiser (Design 1242: 545ft pp, 579ft 6in oa x 56ft 6in ext x 16ft 6in, six 8in ['A', 'B', 'X' positions], four 4in HA guns).

Spain was soon offered a more powerful ship (Designs 923 and 924, 10,500 tons, eight 8in, six 4.7in, eight 2pdr or four 4in, and twelve torpedo tubes). A sketch of Design 1242 shows two twin 8in superfiring forward and the third on the forecastle deck aft, the break of the forecastle coming just abaft the bridge. Design 923 was dated 25 January 1927, and the dimensions were 600ft pp, 636ft oa x 64ft x 17ft 4in mean. Power would have been 112,000shp for 34.5kts; endurance would have been 7500nm at 15kts. The 2pdr anti-aircraft guns would have been in twin mountings, and the torpedo tubes were in rotating deck mountings, rather than in the fixed hull installation actually adopted. The attached sketch shows no catapult, but the ship has an aircraft perched on 'B' turret, with a crane forward of the bridge and a hangar in the bridge structure. Design 924 added a conventional catapult abaft the two funnels, but retained the aircraft perched forward. Designs 924/925 were clearly the basis for the *Canarias* design. These data are from Vickers design data books held by the Brass Foundry outstation of the National Maritime Museum. In Vickers papers they are clearly marked 'for Spain'.

Other early heavy cruisers in Vickers inter-war files are Design 835 for Brazil, Design 835X for Turkey (submitted 5 October 1923), Design 904 for Portugal, Design 1208 for Chile (see the caption to the HMS *York* drawing) and Design 1287 for Argentina (see the caption to the HMS *York* drawing). Design 835 was a 'modified *Raleigh*' (estimate sheet dated 20 January 1923) displacing 9875 tons normal (565ft x 57ft 6in [65ft over bulge] x 37ft 7in x 17ft 3in, 9,875 tons normal) and armed with three twin 8in/50, plus six 4.134in (105mm) HA guns, two Maxims and two triple 21in torpedo tubes. Quoted dimensions for 835X were slightly larger (567ft pp, 9,900 tons). Protection amounted to 3in over the machinery (to the upper deck, 2in to the forecastle deck) and 1½in at the ends, without any armoured deck, as in wartime British cruisers. The ship would have made 31kts. Design 904 was smaller (8,750 tons, 550ft x 61ft extreme x 40ft 6in x 17ft 3in, 32kts on 80,000shp, with eight boilers in three boiler rooms), armed with the same three twin 8in/50, plus ten 4in/45 HA, twelve 21in torpedo tubes and two depth-charge throwers. Like contemporary British heavy cruisers, this one would have had a 1in deck but no belt (she would have had 1in over her shell rooms and 2in over her 8in magazines). The 1930 London Naval Treaty banned further construction *in the United Kingdom* of heavy cruisers, but Vickers could keep building such ships as long as it retained its Spanish yards. It was therefore able to offer a heavy cruiser to Brazil in 1933 (it is illustrated separately). One other inter-war Vickers design qualifies as a more or less heavy cruiser. Although described as a protected cruiser, Design 819 (for Mexico) was more a fast gunboat (one twin 8in/45, two twin 6in, four 4.7in/45, two 4.134in HA, twelve 3pdr QF, two triple 21in torpedo tubes, 23kts [power is not given], 410ft x 47ft x 26ft x 15ft 6in, 4,850 tons, 3in side over machinery).

porated in the 1930 treaty, the escape clause: as proposed in 1927, the clause merely allowed any signatory which considered that another was using its tonnage in a manner inconsistent with the principles of the treaty to call a new conference. The idea was that signatories would be deterred from cheating or behaving badly, since a new conference might overturn any temporary advantage they had gained. Later the idea was to deter a non-signatory, like Italy, from outbuilding the signatories, by allowing any of them to announce that it felt compelled to renounce some limits (without renouncing the treaty structure altogether). The United Kingdom did exactly that with respect to the 1930 London Naval Treaty cruiser tonnage limit in 1936 – by which time it had already signed a treaty which eliminated such limits after 1936. The next treaty (1936 London Naval Treaty) contained an escalator clause intended to deter Japan from exceeding the battleship gun calibre limit (14in): if any non-signatory refused to abide by the limit, signatories could meet to set a new higher limit (which they did: 16in). For the Royal Navy, the main immediate effect of the rise and fall of the Geneva Conference was to lose a year of cruiser construction. However, the prospect of a conference led to re-analysis of the logic of cruiser designs.

In 1926 a 'reliable source' reported that the new Japanese heavy cruisers had eight rather than the previously reported twelve 8in guns, but twenty rather than eight torpedo tubes, all of 24in calibre. Two were submerged forward, the rest in twin deck mounts, and two torpedoes were carried for each tube. Speed was still the 35kts previously reported and they had 3in side armour.[51] It was too late to redesign the B Cruiser, the design of which had just been put up for approval, with many more torpedo tubes, but the 1927 cruisers might be given a more powerful torpedo battery. The Royal Navy still had many 'C' and 'D' class light cruisers with heavy torpedo batteries. Should torpedoes form a more significant part of the armament of the A and B Cruisers? Doubling their batteries would give them sixteen and twelve tubes respectively. It might also be possible to carry a spare torpedo for each tube. It seemed likelier that the larger A Cruisers would be given heavier torpedo batteries, because the B Cruisers were intended mainly for trade-route protection. However, it seemed that doubling the torpedo armament of the three A Cruisers planned for 1927–9 was hardly a sufficient answer to the Japanese. The two main fleet commanders (Atlantic and Mediterranean) and the Directors of the Tactical School and the Staff College were all asked to investigate the problem from the tactical point of view; DNC would do the same from the ship design point of view. Controller would report on the possibility of adding torpedo tubes to the 1926 A Cruisers. The Director of the Tactical School was asked to try such ships on his tactical (gaming) board. Large numbers of torpedoes had been fired in recent exercises: in one the First Cruiser Squadron fired sixteen torpedoes in double broadsides, representing a total of forty-two torpedoes, while in September 1926 the three ships of the Third Cruiser Squadron fired a total of twenty-four torpedoes in double broadsides

The 'reliable source' was muddled. The Japanese had never planned for twelve 8in (the British presumably thought they would be arranged like the six-gun *Furutaka*s, with twin mounts instead of singles). Although British naval constructors routinely checked reports of foreign ship characteristics for feasibility, they either had not reviewed the 'reliable' information or had failed to appreciate how impossible the figures were. However, the 'reliable' source certainly had discovered the Japanese fascination with large numbers of heavy torpedoes, but the British did not redesign their cruisers, and they did not begin carrying reloads in any numbers (the A Cruisers typically carried a single spare torpedo).

In July 1926 Controller asked ACNS to review cruiser armament; he feared that adopting the 8in gun had cost too much in protection, and that the new cruisers were too much like the battlecruisers of the First World War.[52] The alternatives were 7.5in and 7in guns. How much weight would be saved in an A or B Cruiser if they were substituted for 8in? How effective would they be against vertical and horizontal armour?[53] DGD evaluated British cruisers at medium (12,000yds) and long (20,000yds) ranges, at 70° inclination to the line of fire (shells hitting 20° off broadside). At medium range the magazine sides of all three classes could be penetrated, but not their crowns. *London*-class shell room sides and roofs could be penetrated, but the follow-on A and Bs had shell room armour as thick as that on their magazines. Machinery side armour could be penetrated, but not the 1½in deck overhead. At long range, the 4in magazine side was just penetrable, but the deck was immune. Sides and decks over machinery were probably penetrable. Destroyer fire could penetrate side armour over machinery, but not magazine armour and the deck over machinery. Overall, DGD considered British cruiser vertical armour insufficient despite its weight (deck armour seemed adequate). Because the gun positions were essentially unprotected, a single lucky hit could cost a ship a quarter or even a third of her gun power. Also, because British cruisers were slower than Japanese ships, they could not exercise what DGD called control of an area. Weight saved by switching to smaller-calibre guns could go entirely into armour, or into more guns plus extra armour. Substitution of 7.5in for 8in guns would save weight equivalent to almost a third of total armour weight, and 7in would allow about half again as much armour weight, probably enough to make the ship immune to 8in fire at all but very short ranges. However, the cost in offensive power would be considerable: if the eight 8in were rated by DNO at 800, eight 7in would be rated 472. That was particularly unfortunate if, as reported, the new Japanese cruisers were mounting twelve 8in guns, the new US *Pensacolas* ten, and new Spanish cruisers ten (which last was not true). On the available weight, if armour were not increased, a ship could carry five twin 7.5in or seven twin 7in guns. These batteries would have more offensive power than the 8in ship (820 for 7.5in, 827 for 7in).

Moreover, DNO had shown that the smaller-calibre guns offered greater deck penetration at long range. They could not penetrate vertical armour as well, but it appeared (correctly) that foreign cruisers had only limited vertical armour.[54] That should not have been a great surprise. Foreign cruisers generally had visible belt armour, and given the difficulty British designers had encountered trying to find weight for any protection at all, and the higher power of foreign cruisers, it is difficult to imagine how their designers could somehow have provided protection far beyond what the British had. The naval press generally characterised the new 10,000-ton cruisers in all navies as 'eggshells armed with hammers'. British naval constructors estimated that the Japanese hull was about a thousand tons lighter than the British (4,600 tons vs 5,540 tons in *London*), attributing the difference to the British requirement that ships be suitable for worldwide service, hence roomier above the waterline. That was the saving which, in their view, the Japanese translated into greater offensive power. However, DNC argued that in fact the Japanese had sacrificed strength and seakeeping, 'the performance of their new cruisers up to the present having, apparently, been unsatisfactory'.

If hull weight could not be reduced, what could be done to give British cruisers the fighting power of their foreign, particularly Japanese, rivals? DGD (Captain Brownrigg, Dreyer's successor) asked for a series of designs. In priority order, they were: a ship with ten 8in guns in twin turrets, the 'armament considered the minimum consistent with the reported progress of Foreign Powers'; failing that, a design with the armament bunched at one end in triple turrets (as in the *Nelson*s) to save enough weight to gain adequate protection; or a design with a larger number of smaller guns.[55] In the latter case 7.5in was surely the minimum calibre, with at least ten such guns. DGD regretted that British cruiser design seemed to be falling behind that of foreign powers; surely the aim should be twelve, not ten, guns. In 1924 DGD had commented that eight guns were wanted for commerce protection, but twelve for the fleet. It would be worthwhile to consider substituting deck for side armour (as proposed by Director of Scientific Research in connection with higher muzzle velocities for 8in guns). It would also be worth while to consider protecting the 8in guns against destroyer fire, as their ability to disable the guns would be a serious consideration for cruisers working with the fleet, i.e. as anti-destroyer ships. Of course much might depend on some revision of the Washington Treaty, but if the 8in gun were abolished altogether 'most cruiser actions will be indecisive since it should be comparatively easy for any nation to design an 8,000-ton cruiser which cannot be stopped with a 6in gun'.

Director of Plans (Captain G A Egerton) observed that although endurance did not figure in standard displacement, additional oil cost hull structure. The London class was now credited with 11,000nm, 3,000nm over the Staff Requirement. With the Singapore base now approved (the Labour government's cancellation was reversed by its successor), calculation should be based on the needs of the fleet proceeding East (ACNS, now Dreyer, forgot the calculation involving cruisers threatening Japanese trade routes). ACNS proposed fuel for 3,200 miles at 16kts with steam for 20kts (to get to Singapore) plus fuel for eighteen hours at 18kts with steam for full power (sortie from Singapore and return) plus fuel for eighteen hours at 28kts with steam for full power for the run-in to battle and the battle itself. For the Kent class, this amounted to about 2,000 tons rather than the 2,560 tons needed to meet the earlier requirement, a decided advantage.[56] It was impossible to make quick endurance calculations on this basis, so constructors continued to estimate on the basis of total mileage, and that was how Staff Requirements continued to be written. Even so, the new definition of endurance gives a good idea of what the Naval Staff thought cruisers, at least those working with the fleet, were likely to do in wartime.

By November British Naval Intelligence accepted (incorrectly) that the Japanese had adopted the eight-gun design with extra torpedo tubes, and it reported (also incorrectly) that the US Navy would arm its new cruisers with eight 8in guns, although it was mocking-up a triple 8in turret. ACNS concluded that foreign navies were finding it difficult to arm their ships with more than four twin turrets, so British firepower was not inferior. If weight could be spared, it should go into protection. 'For reasons of morale alone' he rejected any reversion to smaller calibre. However, having built triple turrets for the Nelsons, surely it was time for the Royal Navy to investigate a triple 8in turret, and to mock up gunhouse and magazine arrangements so that it could be adopted, if so desired. By February 1927, however, a new arms-control conference was in prospect, so Dreyer suggested deferring any work on the 8in triple. Once the Geneva conference collapsed, this discussion became the starting point for the next 10,000-ton cruiser design, which became the (abortive) Surrey class.

In the autumn of 1926 DNC sketched the 10,000-ton cruiser still planned for the 1927/8 programme.[57] The starting point was an 80,000shp ship with B Cruiser armament (890 tons rather than the 1,100 tons or more of a London). The main difference from York was a longer hull.[58] On this basis DNC gained 880 tons for improvements. That was enough for four twin 8in with 120 rather than 100 rounds per gun. Other armament was considerably different from that of earlier British 10,000-ton cruisers: four 4.7in (200 rounds per gun), four 3in (200 rounds per gun), four aircraft, two submerged tubes (24.5in calibre, as in the Nelsons), sixteen upper deck tubes, and thirty-six torpedoes (a mine load was proposed but dropped). This was the torpedo-heavy cruiser inspired by the Japanese reports. In a later version, the anti-aircraft battery was set at six 4in and the torpedo battery reduced to two triple 21in tubes.[59] Reduced hull weight also bought more protection (total 1,290 tons). The latter bought an extra inch of magazine protection (5in side, 4in deck, 4in ends). The improved ship also had a 90,000shp powerplant. Having considered a 100,000shp powerplant, Lillicrap chose 90,000shp, for about 33kts. This combination was feasible (weights totalled 9,960 tons). This design died due to the approaching 1927 Geneva Conference (no cruiser was ordered in the spring of 1927), but design work continued because cruiser construction would resume if the conference failed. By the spring of 1927 a new design was underway. It is described below.

Since the British objective at Geneva was to kill the 8in cruiser, in December DNC ordered Lillicrap to work out the characteristics of a cruiser armed with eight 6in guns, another a scaled-down 'County', in effect reversing the previous year's switch from 6in to 8in guns for Cruiser B (York). DNC suggested a somewhat longer hull and asked for shorter endurance (7,000nm, 1,900 tons of oil).[60] Armament weight fell from the 890 tons of Cruiser B to 680 tons. As in the larger cruisers, protection consisted of boxes over the magazines (4in sides, 2in crowns and screens; 1in over the magazine for 4in HA ammunition) and a belt (4in, with 5in glacis) and deck (1in) and bulkheads (2in) over machinery, plus 1½in over the steering gear, 1in over key control positions (as in recent large cruisers), 1in turret rings, and bullet-proof plating over bridges and controls aloft. Presumably side armour was emphasised because the ship was likely to fight at shorter ranges. Displacement was about 7,800 tons. In March 1927 Lillicrap sketched a somewhat smaller ship with 6in guns.[61] To gain a proposed thicker deck over the machinery, Lillicrap suggested eliminating box protection for the magazines and other spaces and instead extending the platform deck (which covered those spaces) out to the sides of the ship (and, presumably, fore and aft). Any shell had to pass through this deck before hitting the underwater spaces being protected.

Exeter

Because it was by no means clear whether the Geneva Conference would succeed, work began on a new 10,000-ton cruiser design in the spring of 1927. If no agreement was reached, it would be built under the 1928/9 programme (i.e. ordered in the spring of 1928). Initially the planned 1927/8 programme comprised one A Cruiser and two B Cruisers. In the aftermath of the failure of the conference, the new A Cruiser design was not ready. However, it was possible to order a repeat York. In November 1927 First Lord ordered the programme reduced to this single B Cruiser, which became HMS Exeter.

Initially the only change proposed was the greatly modified engine-room protection proposed for the 1927 A Cruiser (see the Surrey class, below). DNC pointed out that York had already grown from 8,000 tons to 8,400 tons, and he felt she would soon be much larger than had been imagined. He also offered the thicker magazine protection proposed for the new large cruiser.[62] Exeter had a modified Mk II* mounting which, unlike that in York, elevated to 50° rather than 70°. It is not clear when this change was made.

In December 1927 First Sea Lord approved the magazine armour but not the thicker machinery protection. The new design was submitted to the Board (as a slightly modified York) on 22 February 1928, receiving the Board Stamp on 1 March. That September it was decided to rearrange the funnels of HMS York, and Exeter followed suit, the forefunnel being trunked into the wide middle funnel. In February 1929 York's

BRITISH CRUISERS

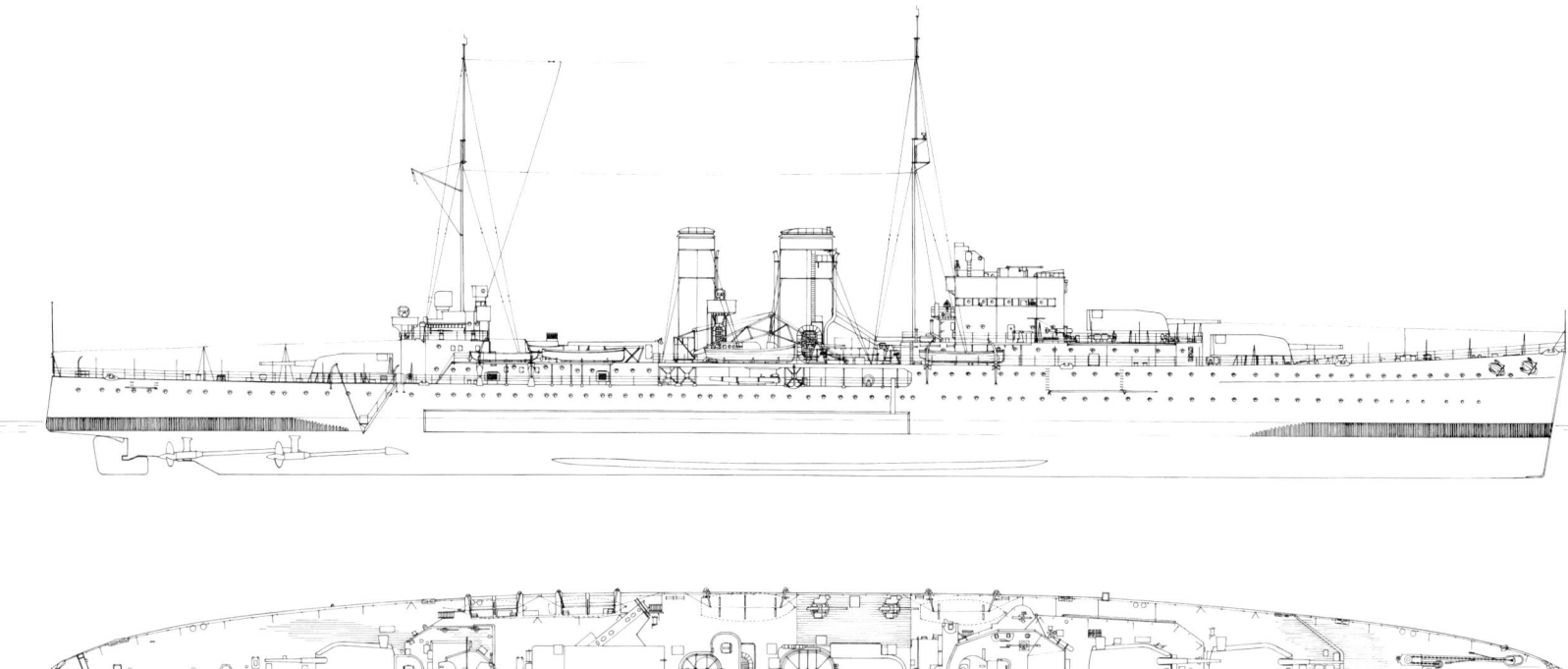

funnels were ordered raised 10ft (a decision on her near-sister *Exeter* was postponed pending trials results).

Initially changes to the new ship were intended only to balance additions made to *York* after she was designed. To regain stability, in December 1927 DNC suggested that beam be increased a foot to 58ft. To regain trim, the ship would either have her lines modified or her stowage changed. DNC suggested increasing allowed tonnage from 8,400 tons to 8,550 tons. He expected the greater beam and displacement to cost a quarter knot, so speed would be 32kts in standard condition and 31kts in deep condition. However, altering the ship's lines to correct trim by the bow would cost more speed, so it was desirable to solve the problem by changing stowage. Unfortunately the ship was already badly congested. DNC suggested stowing more oil aft. The spirit room would have to be raised partly above the waterline, an unusual step in a large ship but acceptable in smaller ones, and accepted in the past even in battleships and cruisers. The small-arms magazine was already above the waterline in HMS *York*. The design was extremely tight. In November 1927 DNO asked that a 4in handing room be provided on the same (protected) level as the 4in magazines, large enough to accommodate two endless-chain hoists serving the 4in anti-aircraft guns. DNC rejected the idea, as it would entail considerably more armour and possibly a longer hull.

In July 1929 DNC pointed out that although the approved design of *Exeter*, like that of *York*, included a catapult atop 'B' turret, no suitable catapult had yet been approved. As in *York*, the idea was dropped, but in this case early enough to allow for redesign. In January DNC had already offered a redesigned *Exeter* reflecting the experience with the new *Surrey* design (see below). It had the new fixed-catapult arrangement, vertical funnels and masts, a low bridge, and modified after controls. Approval was needed because the new arrangement would be more expensive. Other changes were also made: the lower bridge level was eliminated and the bridge lowered 8ft, deck height was reduced 3in (from 8ft to 7ft 9in), and bridge scantlings reduced. In July 1929, DNC observed that recent photographs of *York* showed that her widened fore-

Exeter as fitted, 7 November 1931. She shows the foundations for her unique crossed catapults, but not the catapults themselves. During her post-completion refit (1931–2) her side plating was extended aft to the second 4in HA gun on each side, and the catapults were installed. They were intended to accommodate two floatplanes, one on the catapult and one on the catapult support deckhouse. Two quadruple 0.5in machine guns replaced the single pompoms during her 1934–5 refit. During her 1940–1 refit the fixed catapults were replaced by a single trainable catapult sufficiently powerful to launch a Walrus. At the same time the single 4in guns were replaced by twins and two quadruple pompoms were fitted. Gun tubs were mounted atop 'B' and 'Y' turrets, ultimately for single 20mm cannon. A second HA director was added on the bridge. The ship was fitted with Type 279 air-warning radar and with Type 284 main battery fire-control radar. The bulletproof canopy over the bridge was apparently removed at this time. (A D Baker III)

funnel made her after funnel seem disproportionately small, and that it would be desirable therefore to widen the after funnel of *Exeter*, adding 4ft to its after side.

DNC pointed out that special catapults adapted to the new arrangement had not yet been developed, so *Exeter* had a pair of rotating-type E.II.H units without their turntables. No new catapult was developed because cruisers built immediately after *Exeter* were of a smaller type ill-suited to her arrangement. However, the fixed-catapult design led to the athwartships fixed type used in the *Southampton* and later classes. In January 1936 it was proposed that the two angled catapults be replaced by one athwartships D.I.H as installed in the *Southampton*s (this particular catapult had been ordered for the modernisation of HMS *Cornwall*). The catapult would be fitted at forecastle deck level, one deck level below the existing fixed catapults; one aircraft would be stowed on it, the other on a loading trolley at the port side. The change was not considered worthwhile, because to add a third aircraft it would have been necessary to trunk the after funnel into the forefunnel, which would have been too expensive.

When the design was approved, Legend displacement was 8,550 tons (draft 17ft). Later Controller directed that displacement be reduced

TREATIES AND HEAVY CRUISERS

From a design point of view, HMS *Exeter* was the bridge between the 'Counties' and the cruisers of the early 1930s. She introduced the new streamlined bridge, intended specifically to eliminate the drafts of the 'Counties', and vertical funnels and masts, to make it more difficult for an enemy to guess the ship's course and speed. The bow view was taken in January 1932, the stern view in 1936. Because she was assigned to the Western Hemisphere, *Exeter* was not modernised with the other British heavy cruisers (similarly, *York* was not modernised). After being damaged during the battle of the River Plate, she came home for a refit and modernisation at Devonport, from 14 February 1940 to 10 March 1941. Her single 4in guns were finally replaced by twins, and two octuple pompoms were added. She was fitted with Type 279 radar, and positions were arranged for two Oerlikons (the guns were not fitted at the time).

to 8,400 tons, corresponding to a draft of 16ft 9in; this figure was repeatedly used in public documents. In May 1931 DNC estimated that the ship would displace about 8,400 tons in standard condition (later he gave a corrected figure of 8,390 tons). Weight saved during construction more than made up for additions. On trials, *Exeter* made 32.01kts at 80,840shp at 8,700 tons. In deep condition (10,290 tons) she made 31.16kts on 81,463shp. These results were close to those predicted in the Legend (on 80,000shp, 32kts at 8,550 tons and 31kts at 10,700 tons). By way of comparison, *York* made 32.37kts at 8,440 tons on 79,900shp. Taking her greater horsepower and displacement into account, *Exeter* would have made 32.36kts under the same conditions as *York*, but the latter had run in shallow water and thus might have lost a quarter-knot. The result was impressive, since the EHP curve measured for *Exeter* at the test tank showed 2 per cent more resistance at 32kts.

The ship's officers were impressed by the lack of vibration; they said that it was difficult to realise that the ship was developing full power. Stiffening solved some localised vibration problems (forward 12ft rangefinders and after DCT, for example). There was no rough-water trial, but the ship did encounter a long swell on the beam while *en route* to Arran. Her period of roll matched that of *Devonshire* and *York*, her motion was described as easy and comfortable, and she was dry. Accommodation was somewhat cramped, and her CO criticised that for officers, but there was no easy remedy. DNC later commented that eliminating one turret had not much reduced the ship's complement and certainly had not added space for accommodation, and that cutting the forecastle down aft (to reduce topweight) had cost considerable volume (for, among other things, accommodation). Thus *Exeter* required 46 officers and 594 men, compared to 49 officers and 637 men in the eight-

BRITISH CRUISERS

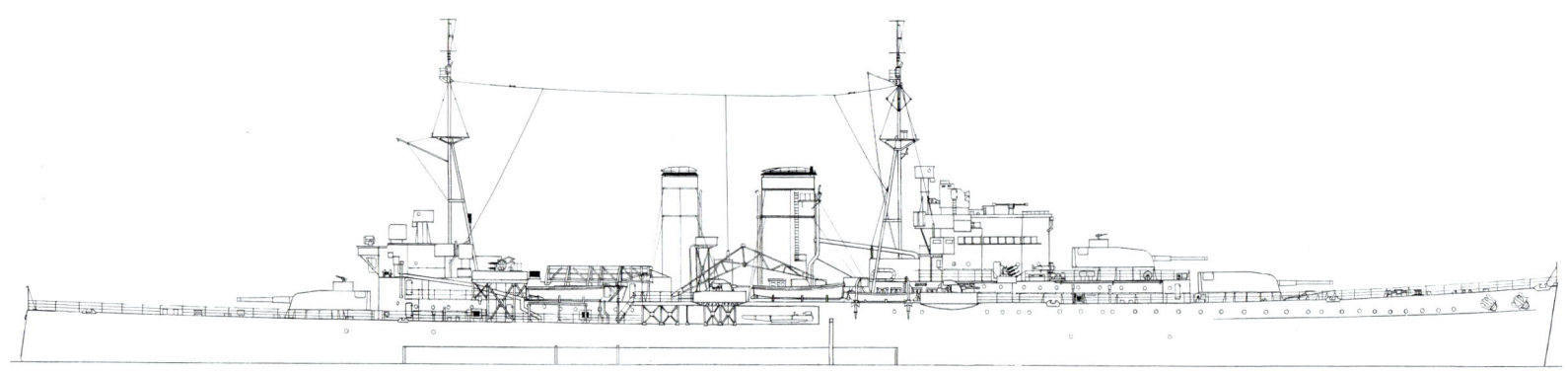

Above and below: HMS *Exeter* is shown at Balboa in the Canal Zone on 24 April 1934 (above), and in home waters in May 1941, as refitted (below). (1941 photograph courtesy of Paul Webb)

Above: HMS *Exeter* at the time of her loss, 1 March 1942. This drawing is based on as-fitted plans augmented by a Devonport Dockyard general arrangement drawing of the refit (unfortunately there appear to be no detailed as-fitted fly sheets showing details of the refit) and by photographs. The forecastle deck amidships was completely rearranged to suit the new heavy HA battery of twin 4in guns. The port and starboard after twin 4in were arranged asymmetrically because the crane was retained on the starboard side. The resting position of the crane arm was moved from the deck level to the top of the adjacent boiler room vent to provide working clearance for the after starboard 4in twin. The torpedo tubes were moved forward, to roughly abeam the fore funnel rather than roughly abeam the space between the funnels (i.e., abaft the fore-castle plating as it cut back, and forward of the plated-in area, presumably for an ammunition hoist, below the after twin 4in gun on the port side – there was no such plating on the starboard side). As refitted the ship retained her two quadruple 0.5in machine guns, but was fitted with tubs (zarebas) for single Oerlikons atop 'B' and 'X' turrets. They were off-centre, to the left side of the turret looking out over the guns. Initially each contained a pair of vertical stanchions for a temporary machine gun armament plus a tripod for the planned Oerlikon. A photograph taken in the Indian Ocean in October 1941, when the crew of the ship's crashed Walrus was buried at sea, shows two machine guns in place; they are either Vickers 0.303in or Lewis guns (British warships typically carried Lewis guns, but also typically placed them in standard twin mounts rather than atop single stanchions). A 1942 photograph appears to show two guns in the tub atop 'B' turret. It is not clear whether the ship ever received her Oerlikons. It appears that the shipment intended for Singapore (from which *Exeter* might have received guns) was diverted to Colombo. There is no evidence that the other two planned Oerlikon positions, abreast the main mast, were ever occupied. The ship carried one Walrus when she left the United Kingdom for the Far East, but it crashed at sea in October 1941. It may have been replaced, as reportedly she had one aircraft before the battle of the Java Sea. During that battle an 8in shell sprayed the aircraft position with fragments, and a photograph taken afterwards shows no aircraft at all. During the refit and repair in the United Kingdom the bridge platform was extended aft to allow fitting of a HA director. Flag-signalling arrangements were moved to the after end of this extension, so the flag-signalling platform at the rear end of the lower bridge deck (no longer needed) was reduced in size. At this level port and starboard were two submarine lookout shelters (little huts with eye slits); they are also visible on board *Fiji* class cruisers. (Alan Raven)

gun *Dorsetshire*. On the other hand, *Exeter* apparently had worse arrangements than *York*, which had the same complement and was slightly smaller. DNC suggested that the problem was that the dockyard building the ship did not realise early enough just how congested she would be, hence did not plan accommodation properly (it was responsible for detail design and for fittings).

The *Surrey* class

During the Geneva Conference, work proceeded on the next A Cruiser design, to be ordered if the conference failed to kill off such ships. The new design was shaped by the previous year's discussion of the limitations of existing British A Cruisers. In May 1927 DNC asked for a five-turret design with the existing powerplant, which sacrificed habitability and some protection.[63] However, what killed it was the problem of finding an appropriate position for the extra turret. Placed amidships, the turret would have restricted arcs (95° on either side) and would congest deck space. Placed on the forecastle (as in HMS *Nelson* and in Japanese heavy cruisers) the bridge would be affected by blast on many arcs. The alternative (requested in June) was a more heavily-protected ship, also sacrificing habitability. Cutting down the ship's hull bought another 370 tons for better machinery protection. Because the deck formed about 70 per cent of target area at expected fighting range, DNC proposed to make deck protection his priority; another inch on the deck would buy immunity out to 20,000yds range. The remaining weight could provide either better uniform protection over the whole machinery space, or could be concentrated on the central engine and boiler rooms (which could maintain a speed of 27kts). They could be protected out to 20,000yds (with 6½in side armour) if the end engine and boiler rooms were left essentially unprotected. This choice entailed providing the protected machinery with about 6½in side armour. These two alternatives were presented to Controller in July 1927. Meanwhile DGD circulated a new set of cruiser protection standards. The attempt to protect shell rooms was abandoned in favour of magazine protection out to 24,000yds, a new figure for cruisers.

The more heavily-protected alternative, with partial machinery protection, was favoured. The sketch design showed 2in bulkheads at the ends of the machinery spaces and ½in intermediate bulkheads, but that allowed a shell to penetrate the side and burst inside, its fragments passing through the ½in bulkheads into the protected machinery spaces. To solve that problem, the 2in bulkheads were moved to the protected machinery spaces and half-inch or thinner bulkheads left on the outer machinery spaces. That added some weight, because in the original design the end bulkheads had included the end bulkheads of the magazines. Also it was desirable to carry the 2in bulkheads lower, adding more weight. DNC favoured eliminating the box protection over the magazines, instead extending the deck armour over them to the ship's sides (the magazines themselves were below water). Armour boxes were retained because they offered protection against underwater hits both by torpedoes and by diving shells. First Sea Lord Admiral Beatty approved these choices on 21 July 1927. The new sketch design showed 1,620 tons of armour, compared to 1,100 tons for the most recent A Cruiser *Dorsetshire*. Magazines would be protected against 8in fire at all ranges below 20,000yds, and against 6in fire at all ranges. The protected part of the machinery would be protected against 8in fire between 7,000 and 20,000yds, and against 6in fire at all ranges up to 20,000yds.[64]

This cruiser was not built, but it became the basis for the 1928/9 design. Weight-saving measures included cutting down hull depth aft (as in HMS *York*) and moving the torpedo tubes down a deck (for reduced topweight). DNC offered two alternatives. DNC called the 1927 ship Design X. It was a quarter-knot slower than other A Cruisers because length was cut from 595ft to 580ft between perpendiculars. The other ship, Design Y, cut power to 60,000shp (six boilers, four shafts). Speed would fall to 30kts or perhaps 30.25kts. Length was reduced again, to 570ft. Hull weight was cut by reducing hull depth (hence less freeboard and reduced headroom in the living spaces), and the quarterdeck aft was cut down. Reduced freeboard aft and lower main turrets made it possible to reduce beam by 2ft. It was possible to reduce scantlings; the combination of a smaller hull and lighter scantlings saved about 500 tons. The weight saved in machinery and hull could provide protection for the entire machinery space. E-in-C observed that the weight per foot of length of machinery was about equal to the weight per foot of armour protecting it. On a fixed total weight of armour plus machinery, any reduction in the length of protected machinery allowed for an increased length (and power) of unprotected machinery. Compared to a conventional cruiser, with about 1,100 tons of protection, Design X offered 1,620 tons and Design Y offered 2,010 tons.[65] The reduction in power eliminated one funnel, clearing the deck for catapults and HA directors. Controller liked the added protection and the improved arrangement, but suspected that during the vital chasing period before action the slightly higher speed of Design X would be a major advantage. He also wondered whether the added protection might better be applied to the turrets. He asked ACNS to compare the two designs from a staff point of view.

ACNS (Dudley Pound) pointed out that it had already been decided that all cruisers should be capable of operating both with the fleet and on trade routes, and that it had been demonstrated that for a cruiser working with the fleet 30.5kts was a bare minimum. For trade protection, a cruiser had to be about as fast as potential adversaries. Design X was already better protected than foreign treaty cruisers. The slight reduction in speed was acceptable. ACNS pointed out that according to the Staff Requirements cruisers should be able to continue to function despite considerable punishment from enemy 8in cruisers, and also that they withstand 4.7in destroyer fire without materially losing fighting efficiency. Hence magazines should be protected against lucky hits which might destroy the ship, machinery should be protected well enough for the ship to retain mobility, and turrets should be protected to retain offensive power. The weakness of Design X was the lack of protection to turret trunks, which could be disabled by 4.7in destroyer fire. Ideally they should have 1in or better protection. Weight could be gained by reducing magazine and shell room roofs from 3in on ⅜in plating to 2½in on ½in, and the sides of the centre engine and boiler room to, say, 5in on ½in. A recent paper laid out cruiser aircraft requirements for trade protection and for fleet operations. Each entailed two aircraft on two catapults, but the trade protection ship needed two spare aircraft – which could be shipped in place of torpedo tubes when the cruiser was assigned to trade protection.

Designs X and Y were evaluated at a conference on 11 May 1928 attended by the First Sea Lord, Controller, DCNS, ACNS and DNC. DNC showed that he could provide the requested 1in to the 8in gun trunks; the conference preferred to reduce the bulkheads enclosing the machinery rather than magazine armour. Rethinking the value of speed in the new cruisers, the conference took up the idea that with fifteen fast 8in cruisers already under construction, it might be acceptable to build a few smaller ones with better protection. Moreover, all of these ships would probably outlast the Washington Treaty and its 10,000 ton limit; a new kind of heavier armoured cruiser might well materialise. The conference therefore recommended Design Y, which received the Board Stamp on 21 May 1928.

E-in-C tried to revive Design X, arguing that Y offered less complete protection than might be imagined. For example the forward pair of turbo-generators (out of a total of four) were not protected at all. Since the armament was electrically operated, their loss would be crippling. Also, the boiler room fans (absolutely essential while steaming) were entirely unprotected. Similarly, the engine room fans, 'without which the engine room staff could hardly live', were unprotected. Apparently no attention had been paid to protecting vital auxiliary machinery, e.g. by relocating fans below armour or protecting them separately. Controller rejected this argument (it was really too late to reconsider Design Y), and the Legend and drawings were approved by the Board on 22 November 1928.

This was far too late for the ordinary 1928/9 programme. However, by assigning both ships to Royal Dockyards (*Surrey* at Portsmouth and *Northumberland* at Devonport) Controller was able to order them in May 1928, before any design had been approved (he avoided the bidding process, which would have required designs). Although cancellation was surely not then being considered, this assignment limited the financial consequences when it happened.

Although the DCT was now being adopted for the latest heavy cruisers, the sketches of Designs X and Y still showed the earlier 'island' structure with a rangefinder-director on top. Besides its unfortunate aerodynamic effects, the tower blocked the view of the stern from the bridge, an unfortunate feature when ships entered or left harbour. After the trials of *Devonshire*, Controller (Rear Admiral Roger Backhouse) suggested that streamlining the bridge might solve the wind problem: 'We streamline submarine bridges and everything that goes fast is streamlined nowadays. Why not . . . the bridges and structures of our fast ships?' Alternative square and streamlined shapes were tried in a wind tunnel, as well as a bridge proposed for the new 6,500-ton cruiser (which became *Leander*), modelled on the angle-sided tower bridge of the battleship *Nelson*. DNE thought that a square-front bridge would be less drafty under all wind conditions (British destroyers designed at this time had square-front bridges). Tests showed that projections from the bridge structure helped cause the problem. In May 1929 DNC proposed a new bridge for the new A cruisers and for *Exeter*. It had angled sides in front and rear. To some extent the advent of the new kind of DCT reduced the problem by eliminating the need for the 'island' atop the bridge. Work on the bridge of HMS *Exeter* had been suspended pending redesign. The *Leander*s and their immediate successors had much the same bridge.[66]

Weight seemed to limit even large cruisers to a single HA director (HADT). In the earlier A Cruisers it was located aft, near the after 8in director; the same officers operated one or the other, it being assumed that the ship would not have to engage HA and LA targets simultaneously. DNO preferred a bridge position, which would eliminate smoke interference and distortion due to funnel gas between director and target. DNC objected that it was already difficult enough to isolate the main battery director from vibration, but in the end it proved feasible to accommodate a second HADT on the bridge (the after one remained). Note that the forward HADT appeared in *Exeter* and in *Leander*, both of which had the new bridge, but not in sketches of *Surrey*.

While the design was being developed, the director of the Senior Officers' Technical Course suggested via C-in-C Portsmouth that the masts and funnels of future cruisers (beginning with this one) be made vertical. Raked masts gave a good indication of target course at a distance, as did raked funnels under conditions of low visibility. By this time it was well known that estimating enemy course was a major gunnery problem. ACNS (Dudley Pound) agreed, adding that an estimate of course would also help an enemy submarine sighting the cruiser screen ahead of the battle fleet (but the value of such information would be much reduced if the enemy had reconnaissance aircraft). Masts and funnels should be made vertical, provided there was no tactical cost. DNC (Berry) pointed out that, in current cruisers, making the funnels vertical would move their smoke closer to the bridge. It would be difficult to move the funnel further back, owing to the position of the bridge abaft the twin turrets. However, in Design Y, the forefunnel of the earlier design was eliminated, so vertical funnels might be practicable. Controller directed that this idea be considered for Design Y, and it turned out that making the funnels and masts vertical provided more deck space for catapults and aircraft, and also for derrick operation. The new cruiser was accordingly modified. So was *Exeter*.

A new catapult arrangement was developed to suit the shorter length of Design Y: two fixed athwartships catapults (each at an angle to the centreline). Because the catapults were mounted on the deck, it was possible to carry two ready aircraft on them. DNC argued that the contemporary US arrangement of two rotating waist catapults could not accommodate ready aircraft, hence was ill-adapted to quick launches. Also, should one of the two catapults break down, the other could launch both aircraft. The new arrangement took up considerable space (but less length than a rotating catapult on the centreline) and it might block the run of the forecastle deck. *Exeter* received this new arrangement.

In May 1929 a general election brought a Labour plurality (though not a majority). Ramsay MacDonald, who had hoped to substitute arms control for naval construction in 1924, regained office. All ships of the 1928/9 programme not yet begun were suspended late in June 1929, and negotiations were immediately opened with the US government to call a new naval conference. This killed off the two cruisers.

The 1929 Cruiser

For the 1929/30 programme the Admiralty planned more 10,000-ton cruisers. The *Surrey* class design came out 100 tons under 10,000 tons, so the question was what additional protection could be provided on that weight.[67] Increases were better protection to the boiler room fan chambers (4in side, 3in ends, 2in deck rather than 2in side and 1in deck and ends; an earlier proposal showed 3in sides and 2in ends) and 2in rather than 1in deck over the transmitting station and the turret ring supports in way of the cable working gear. Bulkhead depth was increased by a foot. The new endurance requirement was applied; the ship needed 1,629 tons of oil with a clean bottom, or 1,906 tons when six months out of dock. Design capacity, unchanged compared to *Surrey*, was 2,200 tons, but the actual figure was 2,405 tons.

There was some hope that speed could be improved; calculations showed that with 72,000shp the ship could attain 30kts deep rather than the 29kts of *Surrey*. However, the Legend approved by the Board did not show the extra power. Early in July 1929 DNC asked Lillicrap to investigate restoring the full 80,000shp of the earlier heavy cruisers. On 80,000shp the ship would make 31.5kts at standard displacement and 30.5kts when deep, which would give the 1.5kts the Staff wanted. Some weight could be saved by using faster machinery (300 rpm). *Surrey* needed 1,450 tons for her 60,000shp powerplant; *London* would have needed 1,430 tons at 300 rpm (she needed about 1,750 tons at 280 rpm). That would be a 300-ton increase, and the longer machinery box would add some armour, for a total of perhaps 450 tons. It was possible to claw back 395 tons, not quite enough (but most ships were completed much further below their design weights).[68]

In response to an NID paper describing US cruisers, DNC asked for the effect on *Surrey* of substituting three triples for four twins (the effect of using triple mounts was also investigated for the new 6,800-ton 6in cruiser described in the next chapter). Although there would be little trouble fitting the two mounts forward, the wider handing room of a triple mount aft would foul the propeller shafts. Lillicrap wrote that 'it is very difficult even now to get satisfactory arrangements and at present I cannot see any possibility of doing as well a 6ft extra width in handing rooms'. Otherwise adding a gun and its ammunition entailed a small increase in weight. The triple 8in would have a wider roller path, 22ft rather than 18ft in diameter, and some oil capacity would be lost. Substituting triples for twins would save 144 tons on the 80,000shp ship, leaving 60 tons in hand. That would restore the deck over the machinery or add ½in to the belt armour over the machinery.

With the same protection as in the 1929 cruiser, a ship with three triple turrets would displace 10,450 tons using *London* class machinery. However, destroyer leader machinery could be installed. E-in-C weight for 120,000shp would weigh 1,975 tons (350 rpm); the equivalent for 80,000shp would be 1,375 tons. Allowing for reduced revs and including auxiliaries, the total would be about 1,500 tons, saving 250 tons compared to *London*, and only 50 tons over the 60,000shp plant in *Surrey*. The total excess would be 180 tons. Belt armour would not be reduced. This took no account of the effect of a shorter machinery box. About 250 tons of oil stowage would be lost, reducing endurance from about 9,000nm to about 8,000nm at 12kts. This study was completed on 29 July. The modified *Surrey* design received the Board Stamp on 11 July 1929, but the ships were never ordered. By this time preparations were under way for the 1930 London Conference, which the Admiralty

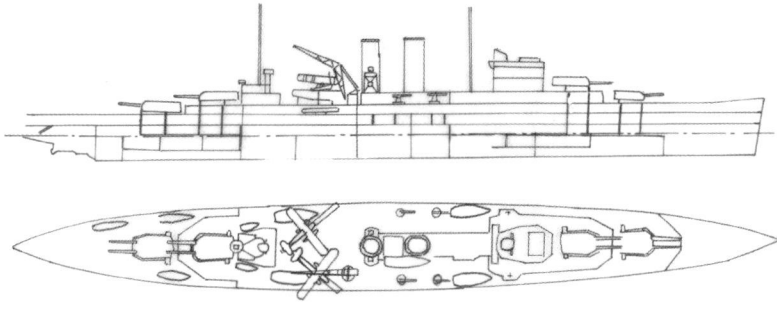

The A-type Cruiser of 1929, which was aborted by the 1930 London Naval Treaty, introduced important later features which were then incorporated in the modified design of HMS *Exeter*: the new streamlined bridge, perpendicular masts and funnels, and fixed catapults. A Legend dated 2 July 1928 showed a ship the same size as HMS *Surrey* (also cancelled: 570ft pp, 600ft loa x 64ft x 14ft fwd and 19ft aft), with a freeboard forward of 30ft 6in, 24ft amidships, and 14ft 6in aft. Her 60,000shp plant was expected to drive her at 30kts at Legend (standard: 10,000 tons) displacement and at 29kts at deep load. Armament was the usual four twin 8in (100 rounds per gun), four 4in HA (200 rounds each), two quadruple 0.5in machine guns, and two quadruple torpedo tubes. She would have 5½in armour on her ½in shell over her machinery, closed by bulkheads of similar thickness and covered by a 2½in deck on a 1½in deck. The most striking improvement over *Surrey* was in protection to the boiler room fans (necessary for speed): 4 rather than 2in sides, 3in rather than 1in ends, and 2in rather than 1in deck. Magazines and shell rooms were comparable in the two designs: 5½in on ½in sides with 3in on ⅜in decks and 3in on ½in bulkheads. (Norman Friedman)

(and the British Government) saw as another opportunity to cap the expensive class of 8in cruisers.

As a postscript, at the beginning of the 1935 London Conference, the Tactical Division laid out a Staff Requirement for a new 8in cruiser based on *Surrey*.[69] Presumably the object was to see what would happen if the forthcoming conference failed to sustain the ban on such ships. Replacing the relatively heavy machinery planned for *Surrey* with the lighter 64,000shp plant developed for the small cruiser *Arethusa* saved 235 tons on machinery, allowing a slight increase in auxiliaries such as generators. The unit machinery in the smaller cruiser occupied 23ft more length (which might be reduced to 20ft given the larger ship's greater beam). The Staff wanted fifty more rounds per 8in gun (70 tons), six twin 4in anti-aircraft guns (as in *Belfast*) rather than the four singles in *Surrey* (129 tons, with 250 rounds per gun), and two quadruple pompoms and two quadruple 0.5in machine guns (the current standard light anti-aircraft battery, weighing 53 tons). Aircraft should be as in the *Southampton* class (a fixed catapult and three TSRs). From this total of 252 tons added weight could be subtracted single pompoms and the original *Surrey* aircraft outfit. Enlarged 4in and other magazines needed extra armour (87 tons). The net increase would be 338 tons, against the saving of 235 tons on machinery. These were very rough figures. The ship would have to be longer (Lillicrap tried 16ft, so overall length would be 610ft). Welding might save about 6 per cent (say 100 tons) on the hull. The *Arethusa* plant used larger (taller) boilers, so the armour over them had to be on the upper deck, and current practice (see the next chapter) was to extend the belt armour up to that level. That would extend over about 140ft (210 tons). However, this added armour covered the fan chambers, which in *Surrey* absorbed 65 tons of armour. The net increase was thus 145 tons. A Legend showed a displacement of 10,090 tons. Lillicrap produced no sketch, but presumably the result would have resembled *Southampton* with twin rather than triple turrets, and with two more twin 4in guns. The ship would probably make 31.1kts on 64,000shp. To some extent this study showed that a large modern cruiser was feasible. It is not clear whether it had any connection with the *Belfast* design.

Comparison

In May 1929 the British constructors had their first opportunity to compare their new heavy cruisers with foreign practice when the French government traded detailed data on its *Duquesne* for data on the *Kent* class. The French ship developed 120,000shp and displaced 9,668 tons (*Kent* was 9,880 tons). Weight-calculating practices did not match in the two navies, but it appeared that *Duquesne* devoted 2,190 tons to machinery, compared to 1,800 tons in *Kent*, but only 328 tons to protection, compared to 960 tons in *Kent*. Armament weight was 1,340 tons, compared to 1,250 tons for *Kent*, for much the same main battery (eight 8in). The French figures were corrected because the French listed some British machinery weights under general equipment. The British estimated that a more comparable figure would have been 2,290 tons; on the same basis *Kent* would have needed about 2,500 tons to produce the same output.

In the Far East, the Royal Navy had the opportunity to see the US treaty cruisers, though it did not obtain their technical details. The six ships of the *Northampton* class impressed British observers with their roominess and simplicity. Although machinery weights were not calculated in the same way, US figures showed 1,945 tons for 107,000shp, compared to about 1,700 tons for 80,000shp in a 'County'. The US ship produced about a third more power on about 14 per cent greater machinery weight. Had E-in-C matched the US weights (assuming they were in consistent terms), a 'County' would have needed only 1,454 tons of machinery. Protection was much lighter, but the US ship came out well below the treaty limit, and a redesigned version (*Portland*) using the full treaty tonnage was much better protected. On the other hand, the British disliked such weight-savers as mounting the three 8in guns in each gunhouse very close together, on a single cradle (this practice was abandoned only in the *New Orleans* class). The US press reported that the ships were lightly built, that they vibrated badly aft (and had to be stiffened), and that they rolled heavily and quickly (US design practice favoured a stiff ship which could withstand considerable flooding). Speed somewhat exceeded that of the British cruisers, but given the length of such ships even a knot was very expensive in terms of power. A British naval visitor to USS *Chester* remarked that she was lightly built by British standards, and that her boiler rooms became far too hot at low speed in the summer (e.g., 140° in the Mediterranean summer).

Modernisation

Modernisation of the *Kent* class, by fitting additional protection, was first considered in 1932.[70] Given the standard displacement of the ships when completed, DNC thought 200 tons per ship was available. Cutting the ships down to main deck level abaft 'Y' turret would add another 50 tons, requiring some internal rearrangement. Further examination suggested about 250 tons in *Cumberland*, *Cornwall* and *Suffolk*, plus 50 tons if the stern were cut down plus another 50 tons to be gained by ransacking the ship, a potential total of 350 tons. *Suffolk* was 200 tons underweight, so for her the total was 300 tons. *Kent* was worst of all, only 150 tons under 10,000 tons, so the potential in her case was only 250 tons. These differences explain why different ships of the class received very different forms of modernisation.

DNC's proposal in 1932 was to fit a 4in belt over the machinery

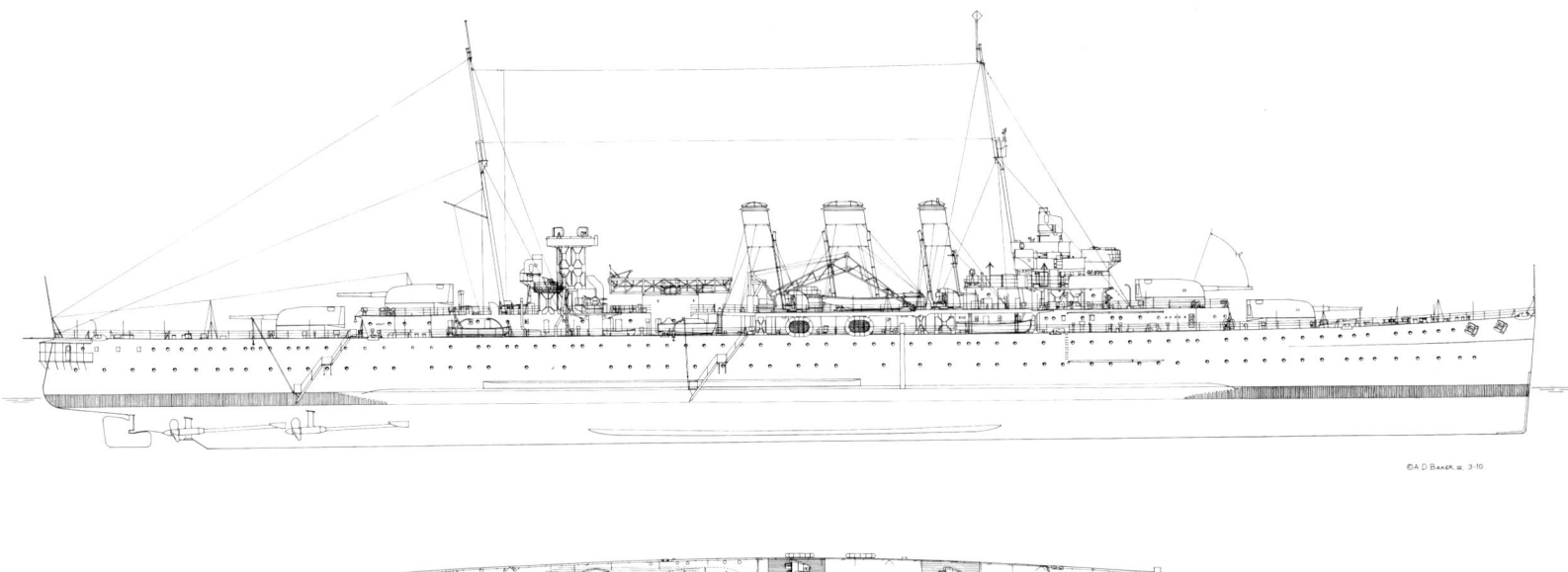

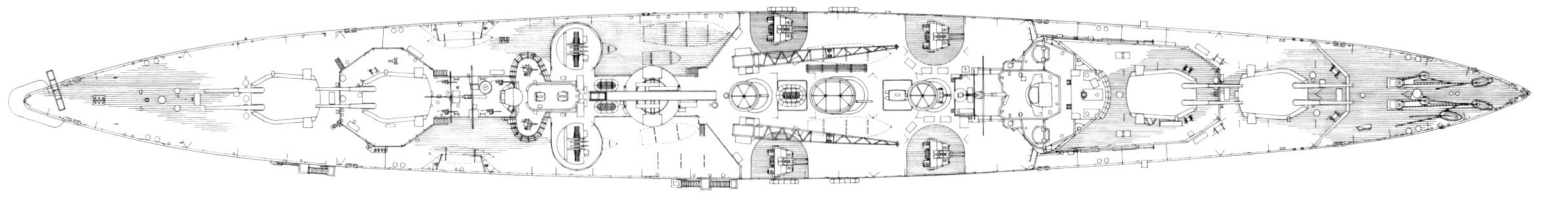

spaces (175 tons). No additional transverse bulkheads would be fitted. Controller asked for protection for the boiler room and engine room fans, now entirely unprotected. It would take about 100 tons to give these large spaces even 1in plating. However, the engine-room exhaust fans could be brought down into the after ends of the engine rooms, where they would be behind armour. Since there were alternative sources of intake air, this would provide the engine rooms (but not the boiler rooms) with protected ventilation. The boiler-room fans could be provided with fore and aft protection (in the absence of thick bulkheads to the machinery, there seemed to be little point in adding athwartships protection). This left 80 tons, which could be used for an extra inch of belt armour; to extend the 4in belt forward to cover the dynamo room, transmitting station, etc (but these spaces would have no deck protection); or to fit 3in transverse bulkheads to the machinery spaces (which would be quite difficult). In February 1934 DNC offered Controller schemes X, Y and Z. X was the 4in belt, the 4in extension over dynamos A and Y and the transmitting station, and 3in over the boiler room fans, a total of 250 tons. Scheme Y was a 5in belt with a 5in extension over the dynamos and the transmitting station, and 4in over the fans, a total of 320 tons. Scheme Z was a 5½in belt, extending over the dynamos and transmitting station, and 4in over the boiler room fans, a total of 352 tons. On 16 July 1934 Controller recommended Z for the three lightest ships, Y for *Suffolk* and X for *Kent*.

In February 1934 the Naval Air Directorate had proposed other changes: a hangar, a new cross-deck catapult and a second heavy crane, entailing a net increase of about 180 tons. In his Notebook, Lillicrap pointed out that this would preclude extra protection. Director of Tactical Duties (i.e. of Staff Requirements) did not like the larger silhouette the hangar would produce. The hangar might also block supply of ammunition to the 4in anti-aircraft guns, unless the 4in magazine was relocated aft. It might be difficult to find space for the aviation fuel (11,000 gallons rather than the present 1,000 gallons) needed by the three large aircraft (TSRs) envisaged to replace the single Osprey the

HMS *Kent* as modernised at Devonport, July 1938. Note the belt armour added during the reconstruction, its lower part contained within the anti-torpedo bulges. She lacked the large hangar erected amidships in other ships of the class, because she had a smaller reserve of weight against the 10,000-ton Treaty limit. Note that the twin 4in gun mounts were further apart than on other ships of this class; she never did receive 4in gun crew shelters with blast protection, as on others of the class. Her torpedo tubes were removed during the refit. Note that the upper deck wooden planking extended to the bow. During her first wartime repair/refit at Devonport (January–August 1941) the pole masts were replaced by tripods, a surface-search radar was mounted on the after end of the searchlight platform (the searchlights were relocated side by side). During a refit completed at Liverpool during November 1942, the 0.5in machine-gun mounts were replaced by pompom directors, six Oerlikons were added, a DF loop was added on a platform forward of the bridge, and the air-defence position deck was extended forward and provided with wind-shielding at its fore end. The catapult was landed at this time. During her September to October 1943 refit, six of the dozen single Oerlikons were replaced by twin power mounts. The after starboard anchor was removed during the Second World War. (A D Baker III)

ships currently carried.

In addition to the cut-down aft, torpedo tubes and other fittings were removed. In May 1936 DNC pointed out that the calculations evaded the reality that ships experienced natural growth during their lifetimes, so that *Cumberland* would displace much more than 9,750 tons when taken in hand, and would emerge at about 10,250 tons.[71] However, he argued that this procedure was legal under the Washington Treaty, because the ship could have been completed at 10,000 tons, and would still have grown during her lifetime. DNC did suggest that the distinction being drawn between *Cumberland* and *Suffolk* and the remaining ships was not valid; there could be no question of an additional 300 tons being available unless it was accepted that they would emerge from modernisation displacing 10,550 tons standard. *Cumberland* was the first taken in hand.

This was a delicate matter, because *Cumberland* was being completed

HMAS *Australia* in July 1947 after a refit in the United Kingdom. (Paul Webb)

before the Washington Treaty expired. A degree of 'tolerance' was accepted in the follow-on London Naval Treaty of 1936. Japan did not sign the treaty, but did have an observer present at the committee which developed the 300-ton tolerance rule. This was later described as a kind of 'gentlemen's agreement' not actually written into the 1936 treaty, to allow for 'natural growth'. Did that mean Japan had implicitly approved the change to the limits set in the Washington and first London treaties? If not, surely it had to be consulted – and informed of the additions to *Cumberland*. DNC argued against any need for notification, because the net addition to the ship's standard displacement was within the 250-ton difference between her original standard displacement and the 10,000-ton limit. The next ship, *Suffolk*, would not complete her modernisation until January 1937, hence would not come within the notification requirement. For the rest of the ships Director of Plans Captain T S O Philips (responsible for treaty compliance) proposed that modernisation should not increase the displacement of any 10,000-ton cruiser to more than 10,300 tons based on her original displacement. 'Any other basis, such as taking into account "natural growth", would result in measures having to be taken at frequent intervals to remove either portions of the ship or guns or torpedo tubes, etc., to ensure that the standard displacement not exceed 10,300 tons, and the same principle would apply to all other classes of ships.' Controller agreed. The Foreign Secretary Anthony Eden rejected the argument that Japan had tacitly accepted the 'tolerance' concept, but helpfully added that modernisation would generally be completed after the treaties the Japanese had signed had expired.

Modernisation added a 6ft-wide 4½in belt of C armour at the ships' waterline inside the bulge. The 'island' and its rangefinder-director were replaced by a DCT. A large hangar was installed abaft the catapult, with the existing after rangefinder-director (modified, hence called the after DCT, accommodating additional personnel, including the spotting officer and rate officer) atop it. HADTs were installed on the compass platform. Not all ships received the desired twin mounts immediately. For example, *Suffolk* was initially fitted with two twin and two single mountings. The ships received the two multiple pompoms (two quadruples in *Cumberland* and *Suffolk*, two octuples in the others) of the original design, plus two quadruple 0.5in machine guns installed earlier in the 1930s. A D.I.H cross-deck catapult was installed, with a hangar for three TSR aircraft; *Kent* had a revolving E.IV.H and no hangar, hence only one aircraft. *Suffolk* was unique in retaining her 'island' and rangefinder-director.

Reconstruction of the *London* class was discussed in 1938. The narrow belt added to the *Kent*s did not extend down to the light waterline, so wider thinner belts were considered. The 4½in belt could easily be penetrated by 8in guns, but belt armour made it easier for the cruiser to get within decisive range (about 10,000yds) of a 6in cruiser. Controller ended up favouring a 3½in belt of the same length as in the *Kent* class, but 2ft deeper, supplemented by a 1½in deck over the dynamo room. Controller pointed out that, because the *London*s did not have bulges, their new belts 'would be stuck on the ship's side in full view of everybody, whereas in the *Kent* class it is hidden inside the bulges'. Presumably that meant that the shallowness of the belt would be much more obvious. As weight compensation, the torpedo tubes would be landed, as in the *Kent* class. Modernisation coincided with a wider fleet anti-aircraft upgrade. In April 1936 it was ordered accelerated, presumably in response to the Abyssinian crisis and the increasing expectation of war. For heavy cruisers the immediate requirement was to add four more single 4in guns (total eight), to add a second set of HACS, and to bring short-range armament up to a total of two octuple 2pdr and directors and two quadruple 0.5in machine guns. The extra single mountings were considered an interim measure; at large repair the ships would receive twin instead of single 4in guns. The single mounts were expected to become available more quickly, as they were landed by capital ships and 6in cruisers.

As Controller, Backhouse had advocated better subdivision of cruiser boiler rooms. As First Sea Lord, in August 1938 he could try to apply

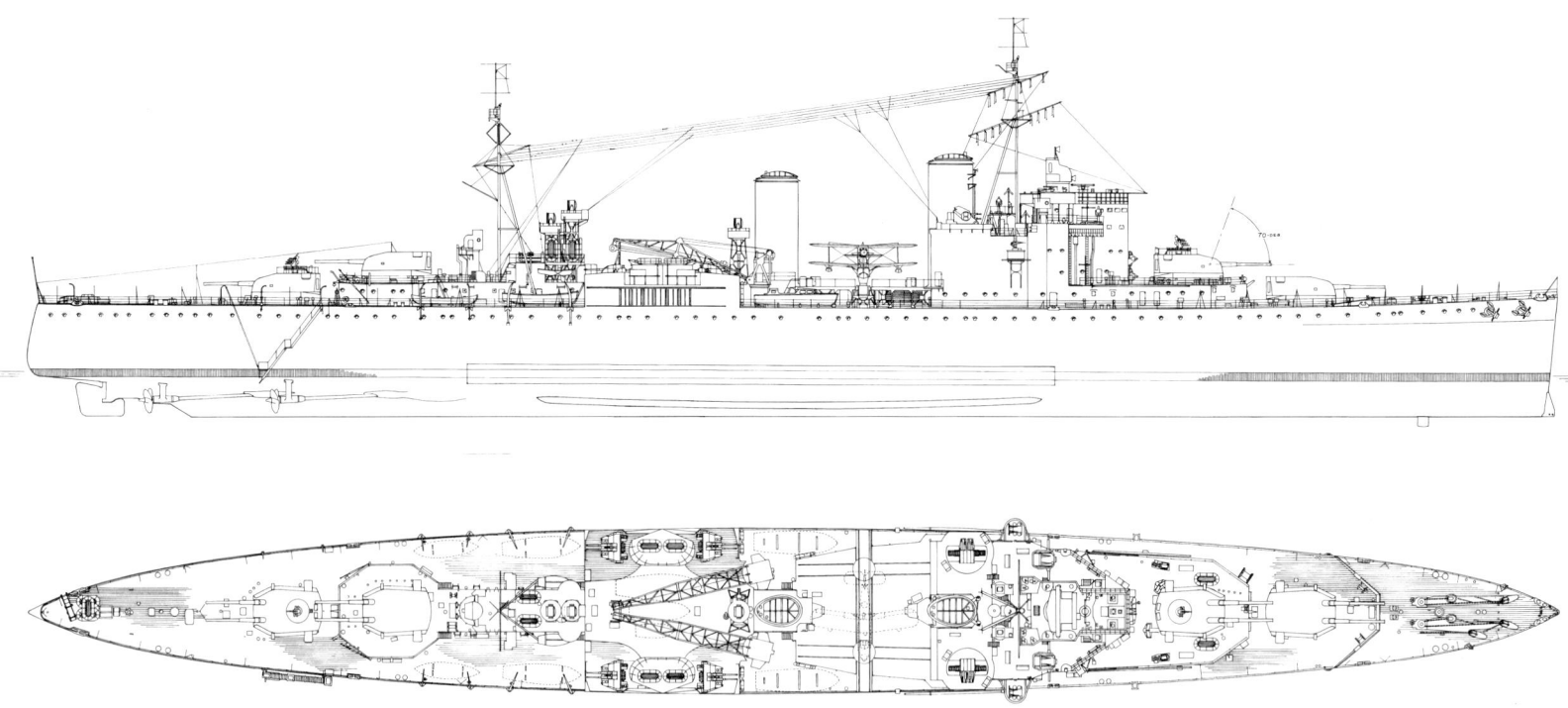

HMS *London* at the completion of her reconstruction, March 1941. Had war not intervened, her entire class would have been rebuilt in this way. With the various treaties no longer an obstacle, she could gain considerable weight: she emerged displacing 10,820 tons light, 11,015 tons standard, and 14,578 tons fully loaded. Note the 3½in cemented armour belt added over the original 1in armour over the machinery spaces. The ship had her original pair of quadruple 0.5in machine guns, which had been relocated to the roofs of 'B' and 'Y' turrets (that both were not relocated to the high turrets suggests topweight problems). The original lower-deck scuttles were reopened post-war. The after anchor hawse pipe was plated over during the Second World War. Note the DF coil retained on the mainmast, below the after Type 279 (not 281) antenna. She received her Type 273 surface-search radar during a repair and refit on the Tyne, 30 October 1941 – 25 January 1942. At that time her quadruple 0.5in machine guns were removed and eight single Oerlikons fitted. (A D Baker III)

his ideas to the reconstruction of the *London*s. He offered two possibilities: (a) six boilers split into three boiler rooms instead of eight in two; and (b) *Fiji* machinery. Either would have improved survivability, but both possibilities were very expensive. As completed, although *London* looked like a *Fiji*, under her plating she retained her old machinery, including all eight boilers.[72]

London was the only ship of her class rebuilt; she was taken in hand in 1939 for scheduled completion in 1942. Her upper superstructure was stripped; she emerged resembling an enlarged *Fiji*-class cruiser, with the same block bridge. She had a combined star shell and air-defence position on each side at the after end of her upper bridge. The fore end of the upper bridge carried an 8in DCT, the after part an HADT on each side. The existing Admiralty Fire Control Table was retained but some additional receiving and transmitting equipment added. Plans called for only the director towers on the bridge structure, but as completed *London* had a third HADT aft, just forward of 'X' turret. The single 4in guns were replaced with twin Mk XIX mounts (250 rounds per gun, plus 200 star shell per ship), and the allowance of 8in shells increased to an average of 172.5 rounds per gun, plus thirty-six LA and six HA practice shells. Had the war not intervened, the other ships of the class would have been rebuilt. *Sussex* would have followed *London* (expected, in 1939, to complete in August 1940) at Chatham (taken in hand in March 1940). *Devonshire* would have been rebuilt at Chatham between April 1941 and September 1942, *Shropshire* being taken in hand (presumably in the same drydock) in March 1942.

In April 1939 it was proposed to improve the deck protection of ships after HMS *London* (beginning with HMS *Sussex*) by adding a second layer of 1in D.1 plating (over the existing 1⅜in) over the machinery spaces (it was soon clear that it would be better to replace this deck altogether). In addition, splinter plating could be worked into the bulkheads between the boiler rooms and between the engine rooms, to localise the effect of hits in one boiler room or in one engine room. The extra deck protection would cost 200 tons, the bulkhead plating 18 tons. DNC pointed out that, since the deck was above the waterline, it could not prevent flooding. The extra bulkhead plating might keep out splinters, but would not strengthen the bulkheads. Controller saw the extra bulkhead plating as 'little more than a gesture' but was interested in enough extra deck protection to keep out 6in and 8in shells beyond 13,000 or 14,000yds, understanding that nothing could be done for HMS *London*. First Sea Lord (Backhouse) agreed, 'much as I regret doing so, as I think the poor subdivision of these important ships is most deplorable'.

Removal of torpedo tubes was a contentious issue. Experiments with a tactical table (reported in January 1939) showed that the *Kent* class would gain considerably if their torpedo tubes were replaced; opportunities to use torpedoes occurred frequently, particularly at night or in poor visibility (although it was pointed out that in such circumstances the cruiser would probably be seen first). Controller (Henderson) noted that the torpedo tubes had been removed to keep the cruisers within the weight limit. The First Sea Lord (Roger Backhouse) strongly favoured torpedoes for all cruisers, as they would be used as advanced forces, and would probably have to press in; at night or in bad visibility they might well encounter enemies at short notice and at short range, when torpedoes would be invaluable. He wanted tubes restored at least to ships not yet rebuilt (but without elaborate long-range controls) and also wanted the new *Fiji*s, which lacked weight for tubes, fitted with at least two tubes on each side, assuming that could be accommodated (he

TREATIES AND HEAVY CRUISERS

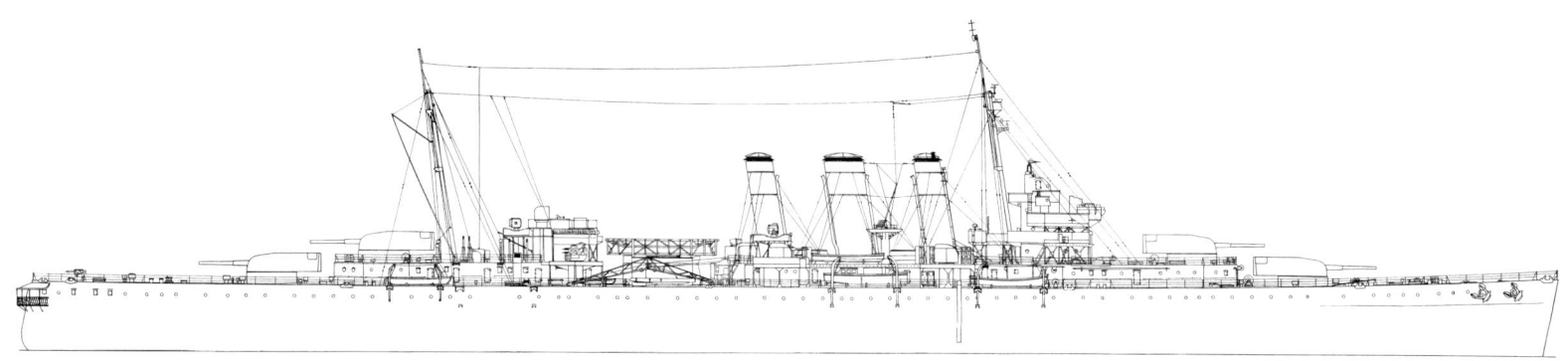

HMS *Dorsetshire* at the time of her loss (April 1942), showing limited improvements made in the 1930s. Radars were fitted during her 1941 refit: Type 290 on her foretop and Type 284 (with Type 285 array, as in several cruisers). She was fitted for but not with Type 285 for her Mk III HACS director. Fore and main topmasts were dropped 20ft, probably to reduce topweight. No Oerlikons had been fitted; she was due to receive four single Oerlikons, a batch of which had recently arrived in Colombo (probably diverted from Singapore). Under the threat of Japanese air attack on Colombo, she had to leave harbour (with HMS *Cornwall*, which was also sunk) before they could be mounted. The only additional light anti-aircraft weapon was the single pompom on the quarterdeck (Mark unknown). This drawing is based on as-fitted plans plus photographs, including one taken at Trincomalee the week before her loss, just after she was undocked. (Alan Raven)

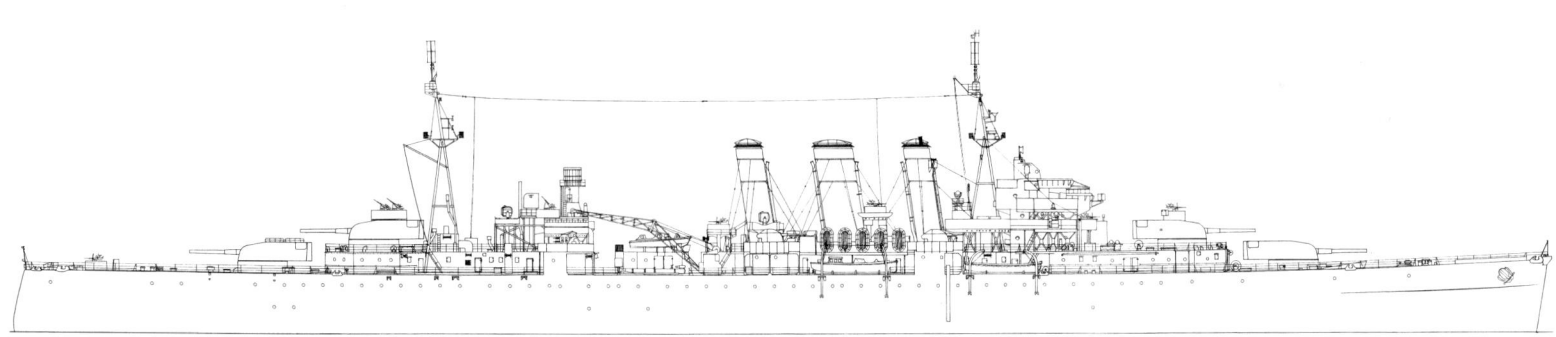

HMS *Norfolk* is shown as at the battle of North Cape, when the *Scharnhorst* was sunk (this is her appearance on completion of her mid-1943 refit). Note that she had two full sets of FV1 intercept/jamming antennas, one on the fore starfish and one on the main starfish. In addition to her pompoms, she had thirteen single Oerlikons on board. (Alan Raven)

preferred three or four). No new cruisers should be built without an adequate torpedo battery of at least four tubes on each side. It appeared relatively difficult to replace the tubes in the *Kent* class.

By this time *London* had grown substantially since completion, to the point where as rebuilt she would probably displace 10,650 tons rather than the 10,300 tons deemed acceptable. If that were accepted, the 44 tons needed to reinstall torpedo tubes in a rebuilt *London* was insignificant. Controller suggested that it would be entirely acceptable 'in the circumstances of today', i.e. in the shadow of a very likely war after Munich. Installing triple or twin tubes would save little weight. Backhouse concluded that the tubes should not be installed in peacetime (as removal had already been publicised) but that space and weight should be reserved so that a quadruple tube could be installed on each side in wartime. That proved entirely practicable. The special oxygen machinery would be removed, however, the ships receiving the simpler Mk IX torpedo.

The Kents were rebuilt, the outbreak of the Second World War precluding mass reconstruction of the later classes (except for HMS London). This amidships view of HMS Berwick shows the most important new features: the new DCT atop the bridge (the tower having been cut down), the two octuple pompoms, the twin 4in guns, and the narrow waterline belt. The old director was modified and moved to the top of the hangar. Not visible here is the fixed athwartships catapult. (Fahey Collection, US Naval Institute)

How much could be done for each ship depended on how much margin was left against the 10,000-ton Treaty limit (with a 300-ton allowance for growth). As the heaviest ship in the class, HMS Kent received the most limited modernisation; she did not get either a hangar or an athwartships catapult. She did receive a pair of HA directors alongside her bridge, in place of the earlier centreline unit aft. Note the vertical fore topmast for her DF antenna. (Cribb, Southsea)

Above and below: HMS *Suffolk* received the full modernisation. The HA director aft was replaced by a pair of directors alongside her bridge. Her stern was cut down to reduce weight (for Treaty reasons), not to save topweight. That proved unwise, since it also reduced reserve buoyancy. She almost sank after suffering bomb damage on 17 April 1940 while bombarding Sola air station at Stavanger, Norway. She reached Scapa Flow with her stern awash, and had to be beached to avoid sinking. *Suffolk* is shown in 1937. Changes through April 1940 were minimal, apart from installation of Type 79 radar antennas on both masts.

HMAS *Australia* was modernised broadly like the *Kent*s, between April 1938 and August 1939. There were important differences. Her twin 4in guns were mounted on her main deck instead of a boat deck. Like *Kent*, she received neither a hangar nor an athwart ships catapult. One of two octuple pompoms is visible against the forward part of 'X' turret. Her sister *Canberra* was never modernised, because the war intervened.

HMAS *Canberra* was only slightly modified. She is shown leaving Wellington, 22 July 1942, shortly before being sunk at Savo Island. Additional close-range armament amounted to Oerlikons on the tops of 'B' and 'X' turrets in zarebas, at the after end of the shelter deck (replacing the earlier quadruple 0.5in machine guns), and possibly on the quarterdeck. The ship appears to have been fitted with Type 271 radar at the after end of the platform supporting her 8in director. She still had her four single 4in guns, and she had never been given multiple pompoms. As of October 1940, the close-range armament was still the pre-war four single pompoms and two quadruple 0.5in machine guns. HMAS *Canberra* arrived in Sydney for refit on 17 February 1942. She was given positions for two quadruple pompoms, but the guns themselves were not installed, the positions being plated over.

TREATIES AND HEAVY CRUISERS

Above and below: Kent is shown in 1941 after a Devonport refit, 1 January – 20 September 1941. The main improvement was installation of radar (Types 281, 284 and 285), and she was given six Oerlikons. Surface-search radar (Type 271 and its derivatives) was still in too short supply for installation.

Kent received a further refit at Liverpool, 18 July – 7 November 1942. The most important change was installation of a surface-search radar aft, on the lattice tower visible here. It replaced the catapult. The two quadruple 0.5in machine guns were removed, and six more Oerlikons fitted. During a further refit at Chatham (22 September – 4 October 1943) six single Oerlikons were landed and three twin power Oerlikons fitted.

Later ships initially received only limited modernisation. *Shropshire* is shown at Scapa Flow, 28 August 1941, with four single 4in HA guns on each side. The most obvious improvement is installation of a second High-Angle Director Tower atop her forward fire control tower. She did not get her two octuple pompoms until a refit at Simonstown, 10 March 1941 – June 1941. Major improvements such as twin 4in guns did not come until a further refit at Chatham, 17 October 1941 – 16 February 1942.

TREATIES AND HEAVY CRUISERS

Above and below: Cumberland typified *Kent* class ships with the full modernisation package. She is shown on 21 February 1942, as yet without a surface-search radar. She had been refitted at Chatham, 1 July – 11 October 1941. Five single Oerlikons were installed; note that she retained her quadruple 0.5in machine guns. They were landed when she was repaired on the Clyde, 3 – 14 February 1943. One of the single Oerlikons was landed, and five twin power Oerlikons installed. When it was fitted, the 'lantern' of the surface-search radar was mounted atop the hangar, just forward of the after main battery director.

Above and below: Shropshire is shown as modernised in 1941–2 (these photographs were dated 27 April 1942). She was typical of her class, which retained the fire-control tower forward. At about the same time *Devonshire* and *Sussex* similarly had their surface-search radar forward of the bridge (the fourth ship of the class, *London*, was much more elaborately rebuilt). This position was not well-liked, because it restricted visibility from the compass platform. However, as long as ships retained their catapults, there was no other centreline position for the radar.

TREATIES AND HEAVY CRUISERS

A further refit (at Chatham, 26 November 1942 – 12 June 1943) eliminated the catapult, leaving space aft for the surface-search radar. Before this refit the ship had a total of ten Oerlikons and the two pre-war 0.5in machine guns; she emerged without the 0.5in guns but with six single and seven twin Oerlikons. *Shropshire* is shown in June 1943. She was transferred to the Royal Australian Navy on 25 June 1943, replacing the sunken *Canberra*.

Below: This photograph appears to show the blast shields of rocket launchers (UP projectors) atop the 'B' and 'X' turrets of HMS *Devonshire*. The US Office of Naval Intelligence dated it 11 March 1942, but it is probably earlier, as it shows air-warning radar but no fire-control radar, and the ship still has the single HA director aft. *Devonshire* was refitted at Liverpool between 19 February and 22 May 1941, two octuple pompoms and Type 281 radar being fitted. Two single Oerlikons were added that September. In this photograph the barge alongside is marked 'US Navy', but the photo is unlikely to show her Norfolk Navy Yard refit (24 January – 7 March 1942) because it included installation of a Type 273 surface-search radar, forward of the ship's bridge. This refit added six single Oerlikons. Only by late 1942 were the ship's eight single 4in anti-aircraft guns replaced by four twin mounts. The ship did not gain the usual pair of octuple pompoms until mid-1943. At that time six of her eight single Oerlikons were replaced by twelve twin Oerlikons.

Below: HMAS *Australia* was refitted several times during and after the war; she was the last 'County' to survive in commission as a cruiser. She is shown as photographed from USS *Wasp*, 31 August 1942. At the foretop is a radar which appears, when this photograph is enlarged, to be the New Zealand air-search set. However, it is not clear whether the 8in director on the bridge carries the corresponding fire-control antenna (part but not all of the necessary structure seems to be present). The ship retains her pre-war HF/DF coil, bracketed to the front of the bridge structure. 'B' and 'X' turrets carry single Oerlikons, but it is not clear that any others were mounted. The zarebas for these guns do not appear in a photograph of the ship at Suva, in the Fijis, taken in February 1942 (the guns themselves may have been in place by late 1941). (US Navy photo courtesy of Rick E Davis)

Above and left: Shropshire, having been transferred to the Royal Australian Navy, is shown in Sydney Harbor at or after the end of the Second World War and also in post-war livery (she was laid up in 1949). Note the single Bofors atop 'X' turret and forward of the mainmast. The gun atop 'B' turret was also a single Bofors. Further single Bofors are visible below the bridge, on the shelter deck (two each side), atop the box structure abaft the bridge, and on the upper deck just abaft the octuple pompom, which is alongside the after superstructure. An additional Bofors, right aft on the upper deck, seems to have been landed before either photograph was taken. In addition she had seven twin (power) and seven single Oerlikons. Although the ship's radars were modernised (she had Type 277 on a foremast bracket), she retained the 'lantern' of her earlier Type 273 surface-search set. The small objects on the starfish of the mainmast, just above the gaff, are the antennas of an FV1 radar intercept and jamming system. These photographs show the ship as refitted after Lingayen Gulf. Her torpedo tubes had been landed and their spaces used instead for boat stowage.

A somewhat later photograph of HMAS Australia in Sydney Harbor shows a British Type 286 air-search radar at her foretop and a Type 271 surface-search set aft. Later a US surface gunnery radar (Mk 3) was installed on the fore face of the forward DCT; it was still in place when the ship was hit by multiple kamikazes at Lingayen Gulf, 5–9 January 1945. By this time the ship had tripod masts. She had landed her torpedo tubes in 1942.

TREATIES AND HEAVY CRUISERS

Above and below: Australia was heavily refitted after Lingayen Gulf, her radars brought up to date and her light anti-aircraft battery replaced. She apparently underwent two refits in close succession. The first, probably after Kamikaze damage at Leyte Gulf (October 1944), provided some updated radars (Type 277 bracketed to her foremast and perhaps a US SG; she retained the Type 273 'lantern' aft). She was given eight single Bofors: one each atop 'B' and 'X' turrets, and three on either side at the shelter deck level at the after end of the forward superstructure and two apparently side by side in a separate box abaft that. All Oerlikons were landed. After Lingayen Gulf she had a larger refit. Her radars were modernised, the single-antenna Type 281B replacing the two-antenna Type 281, so that a target-indication radar (Type 293) could be fitted, its antenna on the foremast. The old Type 284 gunnery radar was replaced by Type 274. Barrage directors seem to have been fitted at this time, forward of the bridge and presumably aft as well (the may have been moved from positions alongside the bridge). The single Bofors guns on each side near the bridge were replaced by quadruple mounts. She was credited with three quadruple Bofors guns, six twin Bofors, two single power Bofors, and four quadruple pompoms, but the photographs suggest that she still had two octuple pompoms (aft), plus two quadruple Bofors (and possibly, for a time, a third in place of 'X' turret), and two single power Bofors (on the upper deck abaft amidships). The bow quarter view was taken on 30 March 1949; the view from aft is undated, but was presumably contemporary. *Australia* paid off for disposal on 31 August 1954. (Photos by Allan C Green via State Library of Victoria)

137

Above: HMS *Norfolk* was modernised on much the same lines as the *London* class. She is shown in December 1942 after a refit on the Clyde, 14–23 October 1942, when the Type 273 surface-search radar aft was fitted. In addition, three single Oerlikons were added to the four installed during a refit on the Tyne (4 July – 8 September 1941), when the UP mountings fitted on the Clyde (during repairs after bombing, 27 March – 14 June 1940) were removed. Her catapult was removed and another three Oerlikons added during a Portsmouth refit (16 March – 31 May 1943).

Below: London was the prototype for planned modernisation of the *London* and *Norfolk* classes, but the Second World War intervened before any further ships could be put in hand. She initially suffered some tank leakage, probably due to difficulties in the connections between the massive new superstructure and her hull, but they seem to have been overcome. The ship is shown in August 1942, after repairs on the Tyne (30 October 1941 – 25 January 1942). At this time her two quadruple 0.5in machine guns (one atop 'B' turret and, unusually, one atop 'Y' turret) were removed and eight single Oerlikons fitted. Note the Oerlikons atop 'B' and 'X' turrets. Type 273 radar was fitted, aft.

TREATIES AND HEAVY CRUISERS

Above and below: London is shown in May 1943, having been repaired and refitted on the Tyne, 21 December 1942 – 17 May 1943. Her catapult was removed and her hangars converted to other purposes, and seven single Oerlikons were added; three were removed and four twin Oerlikons added when she was docked at Rosyth, 27 December 1943 – 2 February 1944.

Above: London is shown at Sheerness in 1949; that year she was badly damaged during the *Amethyst* incident on the Yangtse, when the Chinese People's Liberation Army attempted to keep that sloop from running down-river. She shows some of the changes made during her final wartime refit, at Simonstown, 27 April – 2 July 1945. A Type 277 surface-search radar replaced the earlier Type 273, offering better low-altitude air cover, but the usual further change of replacing the two-antenna Type 281 with a single-antenna Type 281B, so that a Target Indication radar could be fitted on the foremast, was not carried out. She never received a full Action Information Organisation outfit. Four single hand-worked Bofors and four twin Oerlikons (for a total of eight mounts) were added, and eight single Oerlikons landed (leaving a total of four), presumably to improve her chances against Kamikazes. Despite their age, the big 'Counties' were valued after 1945 as foreign flagships. She and *Sussex* both served in the 5th Cruiser Squadron in the Far East in 1947–9. *Norfolk* served in the East Indies in 1945–9. Despite their popularity in the Far East, all were retired because their machinery was worn out, and they were not considered worth rebuilding. None of the other Royal Navy 'Counties' served post-war as cruisers.

Below: With the downfall of Armstrong after the First World War, Vickers was the only British shipbuilder which could realistically offer cruisers to foreign governments. Despite numerous design competitions, pickings were slim. Fortunately for Vickers, in the 1920s and early 1930s it effectively owned the Spanish shipbuilding industry, hence had a dominant position in whatever orders the Spanish did let (this position collapsed when the Spanish Republic nationalised the industry). The company's great success was two *Canarias* class heavy cruisers, the name ship being shown off Barcelona in February 1955, photographed from USS *Randolph*. Vickers' unique relationship with a *non-signatory* of the Washington and London treaties allowed it to offer ships which they banned, such as heavy cruisers (to Brazil) in about 1933. (US Naval Historical Center, from the C W Beilstein collection, 1980).

TREATIES AND HEAVY CRUISERS

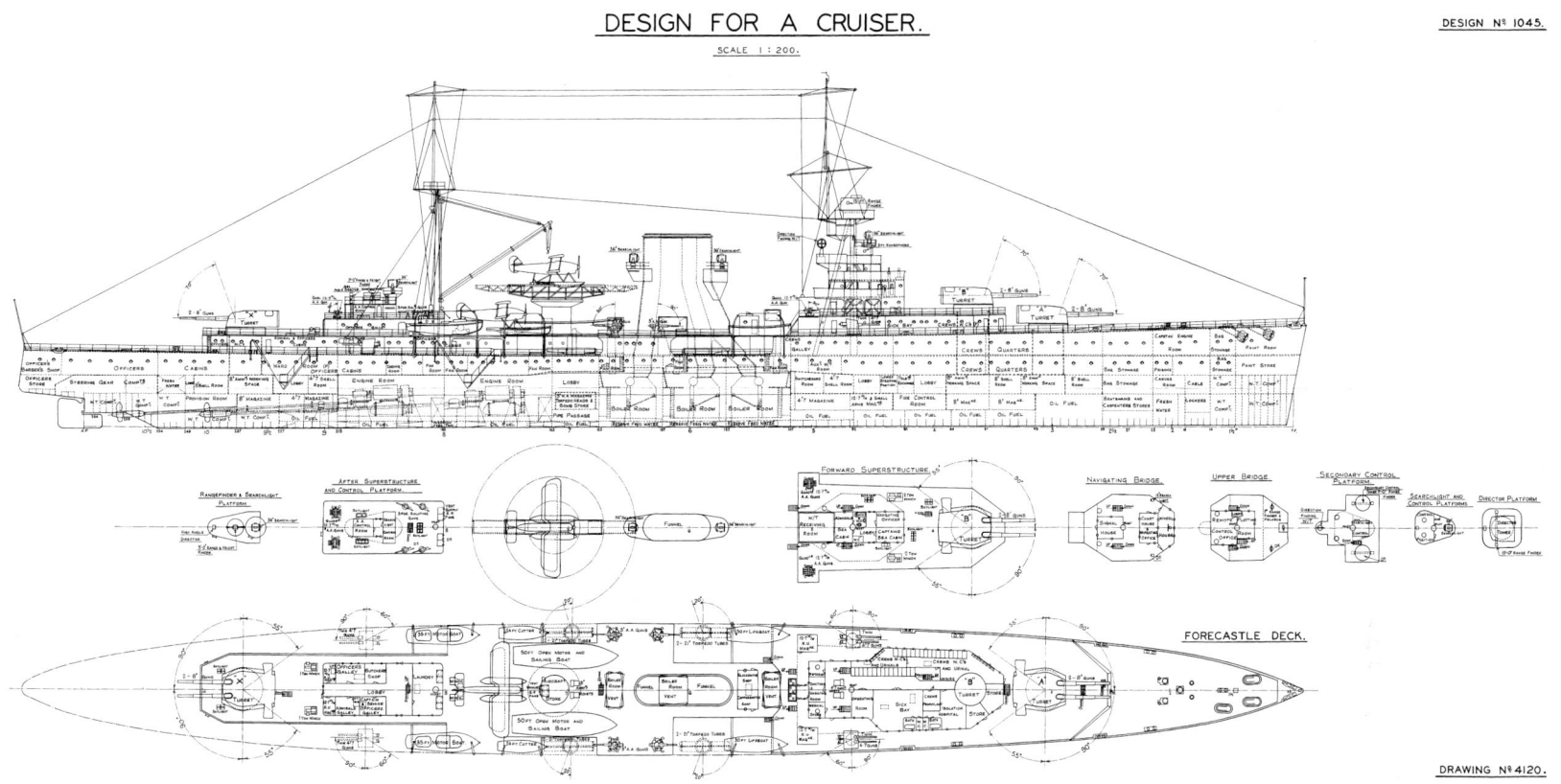

DESIGN FOR A CRUISER. Design No 1045. Drawing No 4120.

Given its Spanish yards, Vickers could continue to offer cruisers armed with 8in guns after the 1930 London Naval Treaty prohibited their construction in the United Kingdom. This is Vickers Design 1045, a 1933 design for a Brazilian cruiser armed with six 8in guns, presumably conceived as a reply to the two Italian-built *Almirante Brown* class Argentine cruisers armed with 7.5in guns. Vickers design records show a series of such designs, all very similar, beginning with Design 1045 and including Design 1052 and Design 1059 (similar but with six rather than four twin 4.7in guns). All had tripod foremasts and three twin 8in turrets. They had three boiler rooms, typically separated from the two engine rooms by a magazine for HA guns. Data for Design 1045, which was typical: standard displacement 9,100 tons (10,160 tons normal, 11,540 tons fully loaded); dimensions were 570ft pp, 600ft lwl, 606ft oa x 60ft 8in x 41ft to forecastle deck/32ft 6in to upper deck x 18ft. The powerplant would develop 80,000shp, and the ship was rated at 32kts. Fuel capacity was 2,200 tons. Armament was three twin 8in/50, four twin 4.7/50, four single 3in/55 HA, four twin torpedo tubes, eight depth charges (five on a rail), and two seaplanes with one catapult. Compared to a 'County', sacrificing an 8in turret bought a much heavier secondary battery and a 100mm (75mm on 25mm shell plating) belt. Complement was 660. This design was dated August 1933 in the Vickers design book. A note in the summary book of Vickers designs mentions a ship of this size displacing 8,500 tons standard, which may have been the first design in the series. Brazil had bought two light cruisers from Armstrong as part of the programme which produced its *Minas Gerais* class dreadnoughts. It sought new cruisers as it tried to buy more powerful dreadnoughts (*Rio de Janiero*, which became HMS *Agincourt*, and the abortive *Riachuelo*) during the run-up to the First World War. This business did not go to Vickers, but the company offered cruisers several times after the First World War. The earliest such design were apparently 1010 (10 March 1923) and 1013 (19 January 1923 – the dates may not be those on which the designs were actually offered). They had, respectively, three and two twin 6in guns, in each case plus one single, for totals of six and seven 6in/50. Both had the usual British light cruiser battery of two 4in HA and four triple 21in torpedo tubes. Design 1010 would have displaced 7,000 tons normal (8,300 tons full load): 520ft pp, 550ft lwl, 560ft loa x 52ft 8in x 37ft 8in [to forecastle], 29ft 6in [to upper deck] x 15ft 9in. She would have made 33kts on 76,000shp, with an endurance of 5,000nm at 15kts. The smaller (6,000/6,900-ton: 456ft pp, 481ft 6in lwl, 488ft 6in loa x 46ft 6in x 26ft x 14ft 8in) Design 1013 would have made 31kts on 50,000shp. In addition, Brazil was offered the modified *Raleigh* (Design 894A) also offered to Turkey at this time (the Brazilian version is dated 30 January 1923). Surviving Vickers records do not appear to list further designs until the 1933 heavy cruisers. By the late 1930s Brazil was again interested in buying cruisers. Surviving records include calculations dated 19 June 1939 for a ship whose hull form would have been based on the Argentine cruiser (presumably *La Argentina*, Vickers Design 1076) scaled up to a normal displacement of 9,000 tons (assumed length was 510ft pp and 535.4ft on the waterline). There are no other details.

CHAPTER 6
THE 1930 LONDON TREATY AND ITS CRUISERS

Cruiser Policy in the Wake of the Geneva Conference

With the collapse of the Geneva Conference, the British government pressed the Admiralty to develop a new policy which might be acceptable to the Americans. The Admiralty now proposed a split into categories A and B (8,000 tons), each capable of mounting 8in guns, of which the A type would still be limited.[1] C-in-C Mediterranean (Admiral Sir Frederick Field) and others argued that for fleet work 6in cruisers were preferable to 8in for pressing home torpedo attacks and for breaking up enemy torpedo attacks, due to their greater volume and rate of fire and their greater handiness. They were also better adapted to night screening and to shadowing, because they had much smaller silhouettes. He envisaged the A and B types as a way of meeting enemy 8in cruisers in a fleet action, as a striking force, and as a way of controlling the Fleet Area (in the First World War, the North Sea). The Staff College proposed a fleet of twenty A, fourteen B, and thirty-six G, of which the G cruiser would be armed with eight 6in guns, steam at 33.5kts (with 30kts written in pencil), and cost £1.2 million (First Sea Lord thought, correctly, that this was a lot to hope for on the tonnage and cost). For the Staff College, the large force of G cruisers was a way of holding down costs, justified by the claim that some could be used on the safer trade routes and to protect convoys against armed merchant ships. To First Sea Lord, this independent investigation demonstrated the wisdom of the British policy at Geneva. Madden also pointed out that new conditions would make surface raiding more difficult.

In October 1928 the Head of the Tactical Section (Captain Bruce Fraser) circulated a paper on the use of a small 8in cruiser, in effect a test of possible post-Geneva policy. A modern fire-control system needed at least four guns to produce adequate salvoes.[2] At one end of the scale was the minimum 8in cruiser with the facilities of an A or B Cruiser (Type I). At the lower end (Type II) was a ship equivalent to a small 6in cruiser but armed with four 8in guns, which were considered superior to six 6in. At this time the requirements of the small 6in cruiser (*Leander* class) were being developed (see below). They amounted to a minimum displacement of 6,000 tons, endurance 7,000nm at 16kts, speed 30.5kts deep and magazine protection against 6in fire. Type I would have the

THE 1930 LONDON TREATY AND ITS CRUISERS

HMAS *Sydney* in Melbourne, 9 November 1936, with her long catapult installed, and with four single 4in guns. (Photo by Allan C Green via State Library of Victoria)

BRITISH CRUISERS

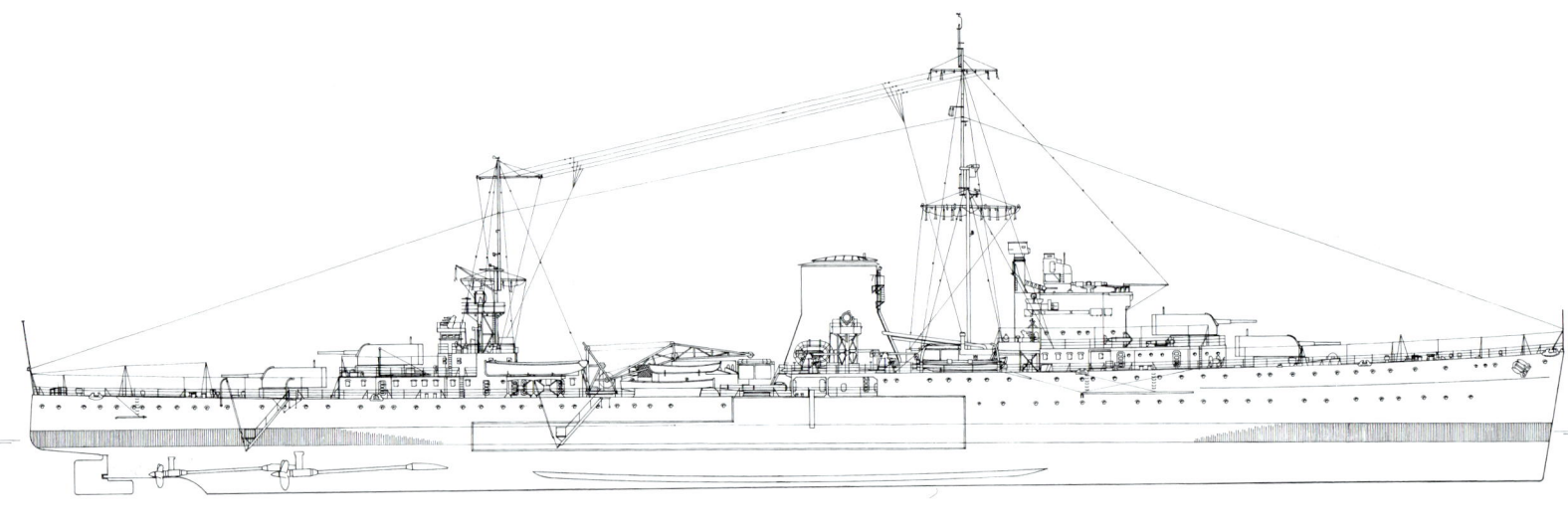

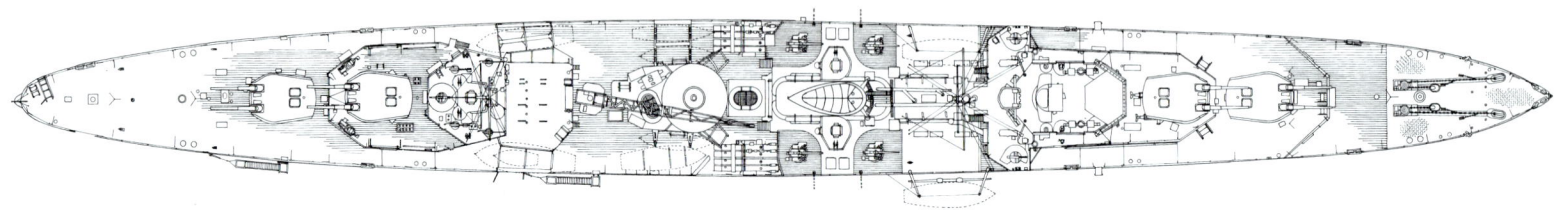

same role as A and B Cruisers, with the fleet and on trade routes, 'but naturally with a lower standard of performance'. Type II would operate with the fleet, but would be handicapped by the poorer performance of her 8in guns against enemy destroyers. On trade routes, her 8in guns would give her a considerable advantage, it seemed, over enemy 6in cruisers, despite her light protection. Director of Plans (Captain R M Bellairs) was less than enthusiastic: the point of the small cruiser was to be less conspicuous, handier and able to develop a rapid yet sufficient fire, but neither Type I nor II seemed to offer these qualities as well as a 6in cruiser. Bellairs concentrated on the role of the cruiser within the fleet; any virtue these ships might have in trade protection was secondary. For DGD, because a 6in cruiser would be considerably over-matched by a ship with four 8in guns, 'it is therefore essential to ensure by legislation or otherwise that we shall not find our 6in cruisers pitted against small 8in cruisers'. If that could be done, 6in cruisers would enjoy clear advantages because they could be built in greater numbers.

In theory, the design of the 6in gun *Leander* class left open the problem of trade protection, for which it seemed 8in guns were needed. The Admiralty reviewed the issue at length in 1928.[3] The big cruisers were clearly unaffordable. The Staff College suggested building less expensive 6in cruisers to protect trade in less dangerous areas, while Controller (Chatfield) objected that 6in cruisers should be used with the battle fleet, because they could fall back on supporting heavy ships and because their gun was the best anti-destroyer weapon. However, they would have little chance against the 8in cruisers an enemy would most likely use against British trade. He therefore favoured a convoy escort cruiser protected like the new Y Cruiser (*Surrey*) but with a speed of 21kts rather than 30kts. Chatfield thought the ship would cost £1.6 million, compared to £2.25 million for *Surrey* or £1.25 million for a 6in cruiser. First Sea Lord was interested enough to ask for details: displacement of about 7,500 tons, with six rather than eight 8in guns (as in

Ajax as fitted, April 1935. When the 46ft catapult was fitted, one Fairey Seafox was stowed atop it and a second athwartships (nose to port) on the after deckhouse between the boat skids. The ship was originally to have carried two Hawker Osprey floatplanes, spare floats for which could be stowed forward of the funnels. The two 27ft whaleboats were carried on light portable davits outboard of the 4in guns when the ship was in port. The forecastle deck plating was extended aft when the ship was rearmed with twin 4in guns in 1938, and the masts were later replaced by light tripods. (A D Baker III)

York), plus a good anti-aircraft battery, and one or two aircraft. First Sea Lord added in pen that cruiser units might be multiplied by attaching cruising destroyers or light cruisers of destroyer type to an 8in cruiser.

The slow cruiser idea collapsed because, in addition to escorts, there had to be fast cruisers to hunt down enemy raiders. ACNS (Pound) and DCNS (Fisher) asked whether convoy was the best protection against surface raiders, particularly since at best the Royal Navy would ultimately have no more than seventy cruisers, of which forty-five could be assigned to trade protection. With very few escorts, convoys would have to be large, hence tempting, targets. Convoy tactics against raiders were not the same as the anti-submarine campaign of the First World War, DCNS in particular pointing out that the slow cruiser could drive a raider off, but the raider could survive to attack again.

Leander

Formal work on a 6in cruiser began in 1928, although feasibility studies of such a ship had been going on for several years, sometimes in connection with attempts to reshape the treaty limit. The Royal Navy rather suddenly discovered that the policy of using the same cruisers with the fleet and for trade protection was impractical. The main fleets reported

THE 1930 LONDON TREATY AND ITS CRUISERS

Above and right: HMS *Leander* in her initial configuration, 1933. She represented an austere approach to cruiser design, with only a single main battery director (forward). Later ships of the class also had a much-simplified fire-control computer based on destroyer rather than battleship practice. Essential equipment was slow to materialise. *Leander* was completed without her HA director or her quadruple 0.5in machine guns, which were still missing in 1933. The aircraft is a Hawker Osprey.

that the big *Hawkins* class was proving less effective for major fleet roles (supporting and repelling destroyer attacks, night attacks and shadowing) than the 'C' and 'D' classes (the 'Counties' had not been in service with the Atlantic and Mediterranean Fleets, hence they did not figure in the reports). The Staff argued that the fleet cruiser should be armed with the most powerful gun which could be mounted on the displacement (elsewhere set at 6,000 tons), and the broadside should not be not fewer than six guns.[4] If six were the maximum they should be in single shields, but if four twin 6in mounts could be provided, that would be better. The guns should elevate sufficiently to range 18,000yds.

In December 1928 First Sea Lord asked for five sketch designs, differing mainly in armament: (1) five or six 6in guns in single mounts; (2) five or six 5.5in guns in single mounts, and endurance of 6,000nm at 15kts, with catapult; (3) four twin 6in, 6,000nm at 15kts, with catapult; (4) three twin 6in, 6,000nm at 15kts, with catapult; and (5) two twin and two single 6in, 6,000nm at 15kts, with catapult. The Staff later studied and rejected 5.5in rather than 6in guns for the ship. Open mounts were attractive because they were lighter, were easier to maintain, offered small individual targets and made firepower less vulnerable by being spread out. They could not sustain the rate of fire of power-operated mounts, but it seemed unlikely that sustained fire would be needed. As DNO had argued in 1920, centreline space limited the number of such mounts, and it seemed that the new twin mount could fire as fast as two open single mounts. If four twins could be mounted on the same hull as six or fewer singles, they were more attractive. Even without a catapult, a ship could not mount more than seven single centreline guns, if that. With twins, she could have eight, plus the increasingly important catapult. Controller recalled that wartime experience much favoured weatherproof mountings. Single open mounts were, moreover, impracticable if anything beyond 30° elevation was wanted. DNO used the short-trunk arrangement he had first described in 1920. The only problem was protection of the ammunition lobby.

As completed (1930), *Leander* shows one of two tactical rangefinders (one each side) atop her bridge and two of her four 4in HA guns, but not (as yet) their director.

DNO was content with the 1in already provided around the hoists, but this seemed so inadequate to DNC's representative that he wanted DNO's statement put on record. DNO's solution was to pass cordite up the hoists in metal cans, which were removed only at the gun.

The cruiser also needed a powerful anti-aircraft armament, because she might be used both to screen aircraft carriers and probably to operate ahead of the main fleet, and hence be in a position to attack enemy aircraft approaching the fleet. The Staff therefore suggested four 4in HA guns and two quadruple 0.5in machine guns (but no pompoms).[5] Although torpedoes were not the cruiser's primary armament, she might well find herself in position to attack when supporting destroyers, so the Staff therefore suggested one quintuple or quadruple destroyer-type torpedo tube on each side.[6] The Staff also pointed out that both the Japanese 5,500-ton cruisers and the American *Omaha*s, broadly equivalent to what was being planned, carried aircraft, and the British must do the same to remain equivalent.

The Staff Requirement demanded that shell rooms and magazines be immune from 6in shell (at 10,000 to 16,000yds, setting side and crown thicknesses, respectively, at 3in and 2in). They also had to be immune to destroyer fire (4.7in shell) beyond 7,000yds.[7] The minimum acceptable

145

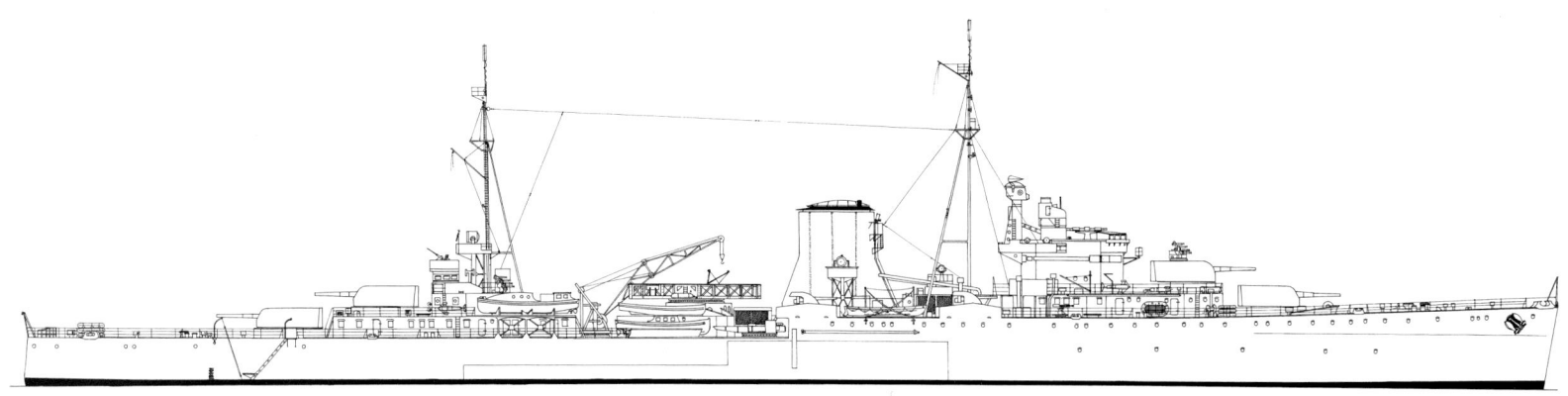

Above: Neptune as in 1941. Note quad 0.5in on 'B' turret roof. (John R Dominy)

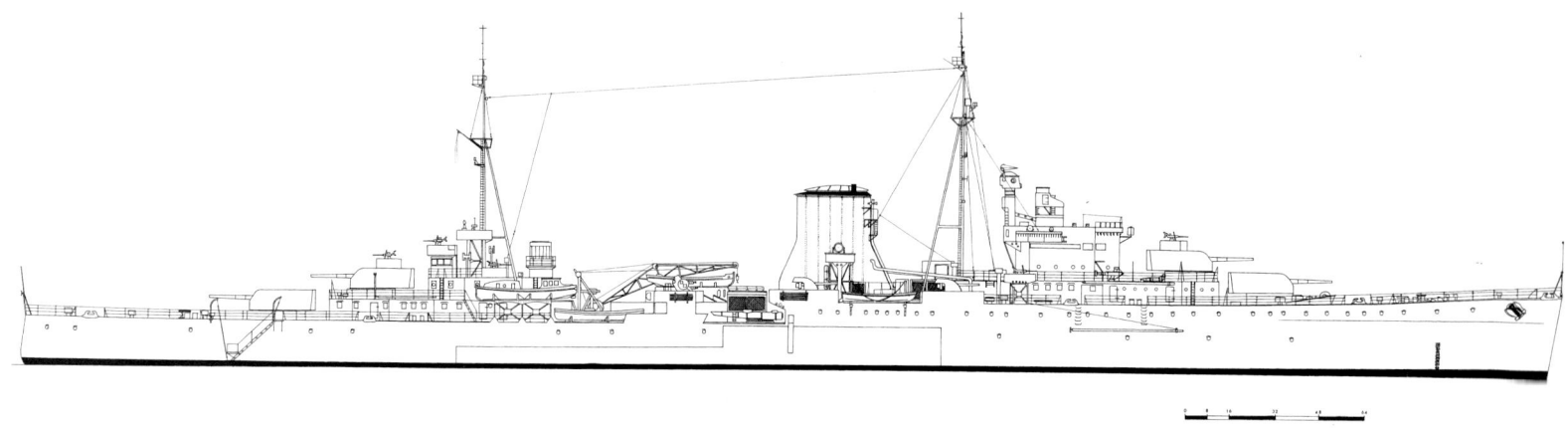

Below: Orion in 1943, after completion of outstanding 'As and As' in the UK after her US refit. (John R Dominy)

displacement (to hold down unit cost) was 6,000 tons. That choice was based on the relationship between displacement and sustained speed in a seaway: 'Visualise a seaway of 4 to 5, steady monsoon conditions, and a fleet proceeding at 17kts.' Cruisers needed enough additional speed to change position within that fleet. Experience with the *Arethusa*, *Birmingham* and *Durban* classes suggested that 6,000 tons was a minimum to provide adequate high-speed seaworthiness in average weather. Analysis suggested a minimum speed of 30.5kts (at deep load). The ship might gain from higher speed (to escape large enemy cruisers), but she would be working with larger cruisers, which would support her. Plans Division wanted an endurance of 7,000nm at 16kts because the fleet was to be capable of carrying out a complete operation in the Far East without refuelling (it was considered difficult to fuel cruisers at sea).[8]

DNC found that he had to provide four-shaft machinery (rather than lighter-weight three shafts) to provide centreline space aft for magazines and shell rooms under the turrets. The usual oil stowage spaces (below the machinery and below the lower deck forward and aft) would accommodate about 1,600 tons; any more oil would be stowed in some lower deck spaces aft normally occupied by provisions and storerooms, and the ship would therefore become somewhat congested. DNC expected endurance at 16kts to be 6,000nm with 1,600 tons of oil and 7,000nm with 2,000 tons, based on the recent steaming trials of HMS *Berwick*. These were unusually large amounts for so small a ship. The new cruiser had Asdic. Communications capabilities should approximate those of the large cruisers.

DNC submitted the five desired sketch designs, with their different armaments, on 23 January 1929. He estimated that 60,000shp machinery would provide 31.25kts at 6,000 tons and 29.75kts at deep displacement if the ship carried 2,000 tons of oil (30kts if she carried 1,600 tons). The ship's length was set by her desired speed, and DNC pointed out that the relatively long hull required stronger plating (1in) amidships over the machinery. This plating would also provide some protection, but unlike other protection it could not be traded off against other weights in the ship. Overall, the new cruiser had the same type of box plus belt and deck protection as the earlier 8in cruisers. In addition to the 1in of structural plating, she had 2in of armour over her machinery (total 3in, but because it was in two thicknesses, it was not as effective as the single 3in thickness over the magazine sides). From directly abeam (the worst case) the side was proof against 6in shell beyond 16,500yds and against 4.7in beyond 9,000yds; the deck would protect inside 13,000yds (against 6in) and would keep out destroyer shell at all ranges. In the twin mount designs turrets were close together, adjacent mounts sharing one ammunition lobby. A single hit below the turrets could knock out their entirely unprotected hoists and put two turrets out of action at once. The battleship design then being considered reduced this threat by covering the hoists with 1in protection, and the cruisers were given similar protection, despite its 100-ton price.

Designs 1 and 2 (five or six single guns) met the 6,000-ton requirement; Design 3 (four twins) would displace 6,400 tons standard, and Designs 4 and 5 (three twins or two singles and two twins), 6,200 tons. The extra displacement in Design 3 would cost 9in in draft and slightly more than 0.5kt in speed. The heavier ships could be brought down to

THE 1930 LONDON TREATY AND ITS CRUISERS

Above and below: As an early beneficiary of the anti-aircraft rearmament programme, in 1937 *Leander* showed four twin 4in guns and the planned quadruple 0.5in guns abaft the searchlight platform aft. The aircraft is a Supermarine Walrus, newly introduced into British service.

BRITISH CRUISERS

6,000 tons by cutting protection, but except for Design 5 there was not enough weight left for adequate magazine protection (without machinery protection beyond the 1in hull sides). In Design 3, for example, only 245 tons were available for protection, but 375 tons were needed for the magazines and other items beyond machinery protection. In this case magazine protection would have to be cut to 2in side and 1¼in crown.

The alternatives were discussed at a conference on 30 January 1929 chaired by First Sea Lord. The sketch designs convinced Controller that 6,000 tons was too small; better to add another 400 or 500 tons. E-in-C was already complaining that the proposed machinery arrangement

Below and right: Orion is shown at a southern US port in 1937. Note the windbreak, probably for the new air defence (control) position, and the new searchlight platform above the lower level of the bridge. At this time the ship had only two of her four twin 4in guns aboard. Of the five ships, only *Achilles*, assigned to the Royal New Zealand Navy, did not have twin 4in guns when war broke out.

Orion is shown escorting Convoy OA-196 in the North Atlantic, 10 August 1940. War modifications to date have been minimal: air-warning radar (vertical topmasts on both masts) atop new tripod masts and splinter shields for the bases of the twin 4in guns. Ships were soon fitted with Oerlikons, including single guns atop 'B' and 'X' turrets. Other standard positions were the signal deck (at shelter deck level alongside the bridge, in place of quadruple 0.5in machine guns), two abeam the mainmast, and one or two on the stern.

crammed the power of the recently-designed *Surrey* into a much smaller space, but DNC answered that he needed the full 60,000shp, and that some cramping was inescapable in the smaller hull. Given the difficulty of stowing the full 2,000 tons of fuel, the conference decided to have DNC consider stowing 200 tons rather than 400 tons in addition to the 1,600 tons he could easily stow, and therefore an endurance of 6,500nm at 16kts on 1,800 tons of fuel was tentatively adopted. The conference was unhappy with DNC's 30kts deeply loaded, but could not see any way to get more. To ACNS, rapid acceleration was more important than high speed for ships working with destroyers.

The Conference chose Design 3, with its four twin mounts. In May 1929 DNC submitted a sketch design for a 6,500-tonner (510ft between perpendiculars, 535ft on the waterline x 55ft); the Board approved it on 3 June 1929. Perhaps surprisingly, given the interest in anti-aircraft fire, it showed only a single HADT, aft on the centreline, with an LA DCT atop the bridge structure and an auxiliary LA director tower abaft the HADT. The bridge was the new angle-sided structure planned for *Exeter* and *Surrey*. As in those ships, masts and funnel were vertical. However, in this design the uptakes were all trunked into one massive streamlined structure intended particularly to limit objectionable smoke eddies abaft it. Gun armament was eight 6in Mk XXIII guns (200 rounds each) in Mk XXI turrets, four 4in HA Mk V or XIV (150 rounds each), and four quadruple 0.5in machine guns (2,500 rounds each). Tests of the new 6in gun showed that more armour was wanted to achieve the desired level of protection. As the design was developed, the magazines were enlarged (and hence also their protection), among other things to provide a separate handing room for each turret, instead of one for each pair of turrets, fore and aft. In 1932 the armament of these and later 6in cruisers was modified, the two quadruple 0.5in guns on the upper bridge being eliminated but one more added, for a total of three. Ships were also assigned a five-charge depth-charge rail (total stowage seven depth charges).[9]

DNC continued to consider alternatives. In July he assigned Lillicrap, who was estimating the effect of triple turrets on a heavy cruiser, to sketch a 6in cruiser with such weapons. The 6in mount had a much smaller below-decks footprint, hence did not interfere with shaft lines even in a smaller hull. As in the heavy cruiser, three triples would probably weigh about as much as four twin mounts, protection weights roughly balancing; additional ammunition (for one more gun) would weigh 15 tons. The 'tween decks working spaces would take up valuable mess deck space, to the point that Lillicrap doubted satisfactory arrangements could be made, and 50 tons of oil stowage would be lost. For this study he used the 80,000shp powerplant of the heavy cruiser *London*, giving 33.25kts at standard displacement and 31.75kts deep. With the same protection over machinery as in the larger cruiser, the ship would displace about 6,890 tons. If the ship had to be cut back to 6,500 tons, to retain magazine protection would require that all machinery protection be given up. To combine extra power and protection over the machinery he had to adopt destroyer-leader machinery, in which case displacement would be 6,640 tons. The overweight of 140 tons could be eliminated by cutting the belt and deck over the machinery by ½in. Lillicrap ended estimating that 100 tons of oil stowage would be lost, for a total of 1,700 tons rather than 1,800 tons, and an endurance of 6,100nm rather than 6,500nm at 16kts.

This study confirmed DNC's view that on 6,500 tons he could not go beyond four twin turrets and 60,000shp. On 26 July he or Lillicrap decided that the ship needed better subdivision. Instead of two boiler rooms (one with two and one with four boilers, 28ft and 44ft long) the ship should have three 28ft rooms, adding 12ft to machinery length (later E-in-C made the increase 19ft over the original 160ft). Some of the extra length could be gained by rearranging spaces at the ends of the ship, but the ship had to be lengthened by 13ft. Given added displacement, the beam had to be increased to maintain stability, and greater depth was needed to maintain hull strength. To maintain speed, another 3,000shp was needed (which seemed to present no great problems). As in the heavy cruisers, the machinery offered a kind of unit operation (originally the four boilers in the after boiler room and the two sets of turbines in the forward engine room formed a central unit).

A sketch dated 25 August 1929 showed dimensions of 547ft (waterline) x 54ft, and a displacement of 7,000 tons. Magazine protection was increased by another half-inch on the sides and ends. The main engine-space bulkheads were substantially thickened. Shell rooms and transmitting station were given 1in crowns and sides. The design was criticised: the side armour did not extend far enough below the waterline. In standard condition the lower edge was 2ft 6in below the waterline (5ft 9in below at full load). CNS wanted it simply lowered, arguing that the upper 3ft of armour was not very useful. Controller wanted the depth increased to 3ft in standard condition, so that in average condition it would be 4½–5ft below water. The extra armour added about 10 tons, which was acceptable. For the first time in a post-war British cruiser, the boiler room and engine room fans were protected. Without them, a cruiser probably could not exceed 10kts.

This design was submitted in October 1929, receiving the Board Stamp on 20 November 1929. DNC pointed out that although the new power-worked turrets needed many more men per gun than simple open single 6in mounts, the hull offered no more volume than that of HMS *Emerald*. More space was taken up by larger wireless offices, workshops and storerooms. Thus the standard of accommodation could not match that of the large cruisers, even though the ships were expected to operate in the tropics. The detailed design went to the Board in December 1929. Estimated speed was 31.5kts, and tank tests were being conducted to see whether a small bow bulb would help (they showed it would add about an eighth of a knot at 25–32kts, and would have no effect at all at cruising speed). DNC estimated that the 1,800 tons of oil fuel would give an endurance of 7,000nm at 16kts and 8,000nm at 12kts, an improvement over previous estimates. This design received the Board Stamp on 9 January 1930.

Meanwhile the Board laid out a 1929/30 programme. On 29 November 1928 it decided that if the US building programme (for fourteen cruisers) was approved, the Admiralty should recommend laying down three 10,000-ton cruisers (*Surrey* class) in 1929. However, by March 1929 it was clear that although the US programme had been authorised, it was possible that a new naval arms conference might be held before the last five US heavy cruisers were begun. The British already hoped to limit 8in gun cruisers to a total of 200,000 tons. On 7

Orion is shown on 6 March 1942, after a major post-damage refit at Mare Island Navy Yard (5 September 1941 – 15 February 1942). She had been damaged on 29 May 1941, then received temporary repairs at Simonstown (14 July 1941 – 5 August 1941). There her catapult was removed. Mare Island removed her quadruple 0.5in machine guns and installed two quadruple pompoms (one visible aft, abeam the boat on the centre-line) and seven single Oerlikons (including one each atop 'B' and 'X' turrets and three in zarebas on the after superstructure). The yard also installed new radars (Types 284 and 285 on her directors, and Type 279 on her masts [the mainmast aerial is not easily seen in this photograph, but others show it]; note that there was no antenna on the mainmast). The US yard lacked some of the desired material, so the ship went to Devonport (30 March – 24 April 1942) to complete outstanding approved alterations and to be fitted with a Type 273 surface-search radar. *Orion* was unique in having this radar aft, forward of the mainmast. In the other ships it replaced the HA director on the centreline; instead ships had two HA directors, one each side of the bridge. The only unit of the class which did not receive extensive improvement was HMS *Neptune*, sunk on 19 December 1941. She had only a single refit, at Chatham, 12 February – 1 May 1941, during which she received her allotted three quadruple 0.5in machine guns, three single pompoms, and radar (Types 281, 284 and 285).

THE 1930 LONDON TREATY AND ITS CRUISERS

Above: Ajax is shown after a refit at Chatham (27 May – 24 October 1942) during which she was fitted with two quadruple pompoms (in zarebas abeam her after superstructure). A single Type 272 surface-search radar replaced the centreline HA director, and to replace it two such directors were installed lower down on each side abeam the bridge. The Type 279 air-warning radar was fitted during an earlier Chatham refit (December 1939 – July 1940) to make good damage after the battle of the River Plate. At that time she received a larger catapult. It was, in effect, replaced by the pair of quadruple pompoms. Early in 1942 she had six single Oerlikons, and during the 1942 Chatham refit another three were added, the quadruple 0.5in machine guns being landed.

Right: Ajax is shown during a refit at New York Navy Yard (4 March – about 15 October 1943). The new object visible forward of the bridge was a barrage director with Type 283 radar. It was also installed on board *Achilles* and *Leander*. The two quadruple pompoms fitted at Chatham were replaced in New York by two quadruple Bofors guns, and four single Oerlikons were replaced by US-type twins (hand-worked). This refit counted, in the official list of British warships, as modernisation (no other *Leander* counted as modernised).

March 1929 the Board therefore proposed that the programme consist of one *Surrey* (called the Y Cruiser, of 10,000 tons) and two of the new 6,500-ton 6in cruisers (O Type). This programme would maintain the 5:3 ratio of heavy cruisers against Japan (assuming that the four *Hawkins* class were equivalent to two *Furutaka*s), maintain parity with the United States in 8in cruisers up to 1933, and begin replacement of the old 6in gun cruisers. It would also leave Britain free to resume building 8in cruisers if necessary. In fact the 1929/30 programme was cut to a single 6in cruiser, HMS *Leander*, as the 1930 London Naval Treaty finally stopped construction of 8in cruisers. Four sisters and three half-sisters (*Amphion* class) were later built.

During detailed design the new cruiser's power was boosted to 72,000shp to achieve 32.5kts. The two outer shafts carried 48,000shp, the two inner ones 24,000shp. The change in machinery arrangement had already cost 24 tons, but the cost of this further change was limited

Below and opposite: Photographs of HMNZS *Leander* taken from the destroyer *Nicholas* show the ship with a surface-search 'lantern' aft in addition to the New Zealand fire-control and air-warning radars. By this time she had an Oerlikon atop 'X' turret as well as 'B' turret, but she had not yet had a major refit to provide her with multiple pompoms. Probably four single Oerlikons in all were added at this time (the date of the refit is uncertain). Her catapult was removed during this refit. These photographs were officially dated 25 July 1943, but they were clearly taken earlier, since *Leander* was damaged by a Japanese torpedo on 13 July.

to 50 tons by increasing boiler pressure from 250 to 300psi, by increasing revolutions at maximum power from 280 to 300 per minute, and by using higher-quality materials. To extend range, in addition to the usual measures used to increase economy at cruising speed, the two forward sets of turbines could be disconnected from their shafts, the inner shafts alone providing enough power for up to 21kts. That was expected to reduce fuel consumption by 10 per cent at 15kts (20 per cent at 10kts). Also, skilled mechanics and ratings could be released from watchkeep-

THE 1930 LONDON TREATY AND ITS CRUISERS

ing duties and assigned instead to maintenance. These changes were apparently first proposed for the 1931 ships (Board Stamp 6 November 1930) and then retroactively applied to *Leander* (second Board Stamp, 4 June 1931).[10]

The revised *Leander* design approved in June 1931 was rated at 7,154 tons rather than 7,000 tons standard; the 1931 design approved in November 1930 was rated at 7,140 tons, and in revised form (on 4 June 1931) at 7,184 tons.[11] The 1930 London Naval Treaty made weight-

saving absolutely vital. Unfortunately the new twin 6in mounting was grossly overweight, much as the twin 8in had been.[12] *Leander* came out heavy, but DNC managed to cut weights on other ships. As inclined shortly before commissioning, *Ajax* (the last *Leander*) displaced only 6,837 tons standard. She was the lightest of the lot. Standard displacement deduced from weights was *Leander* 7,178 tons (7,140 tons reported); *Achilles* 7,018 tons (7,030 tons reported); *Neptune* 7,110 tons (7,030 tons reported) and *Orion* 7,128 tons (7,070 tons reported). For treaty purposes the ships were described as 7,000-tonners.[13]

In the three 1930 *Leander*s, the separate bomb room (forward in the hold between the 4in magazine and 'B' shell room) was eliminated. Bombs were stowed instead in the torpedo warhead magazine. That cleared space below the platform deck for a protected HA calculating position. This new space was placed between the transmitting station (6in calculating space) and the 4in HA magazine, the latter being moved forward. Reducing the plating over the new HA calculating position to the protection already provided to the transmitting station would save about 8 tons. Space formerly occupied by the HA calculating position would be used for a relocated auxiliary W/T office. This change was so attractive that it was applied retroactively to the two large

Above: Leander is shown on 16 October 1942, having been refitted in New Zealand, with New Zealand radars, late in 1941. This outfit comprised a fire-control set, its antenna atop the director, and an air-warning set whose antenna was mounted atop the lattice tower forward of the funnel. Five single Oerlikons were mounted (one atop 'B' turret – but, unusually, none atop 'X'), two on the forecastle abeam the bridge, and two right aft. These Oerlikons do not appear in a February 1942 photograph taken at Suva, Fiji, from USS *Curtiss*, but the two radars were already clearly in place. The centreline quadruple 0.5in machine gun was clearly present in the earlier photograph. She still retained her catapult at this time. *Leander* and the other New Zealand cruiser *Achilles* seem to have been unique at this time in retaining their pole masts.

Below: After her torpedo damage *Leander* went to Auckland for temporary repairs (in hand August 1943, completed December 1943), then to Boston for permanent repairs (3 January 1944 – 27 August 1945), and then to Rosyth to add additional work. She is shown at Boston on 24 August 1945. Note her new radars (single-antenna Type 281B, clearing foremast space for Type 293, and Type 277 on a stub mast replacing the earlier Type 273). 'X' turret was landed. At the end of the Boston refit she had two quadruple Bofors guns and three twin and four single Oerlikons. The additional refit at Rosyth replaced the quadruple Bofors with twin Mk V, and eliminated all single and one twin Oerlikons in favour of three single Bofors Mk III. *Achilles* also served with the RNZN. She was refitted at Portsmouth between 1 April 1943 and 20 May 1944. 'X' turret was landed (as in *Leander*), and the single 4in finally replaced by twin mounts. Her catapult was landed. She was fitted with four quadruple pompoms (two replacing 'X' turret), seven twin power Oerlikons, and four single Oerlikons. In October 1945 *Achilles* was credited with an additional five single Bofors (reduced to four by April 1946). One of the Bofors was apparently atop 'X' turret.

1931 cruisers. More weight was saved by adopting a light catapult and limiting the cruiser to lightweight (fighter-reconnaissance) aircraft. Given the limited total cruiser tonnage, every ton counted. This change was applied to the two large 1931 cruisers. All of these ships also had their forecastles extended, adding some more accommodation space.

Leander was generally well-liked at sea; her captain described her as 'an excellent ship with the handiness of a destroyer, but possessing a battleship bridge, *Kent* class stowage, and "C" class accommodation'. However, in his interim report (of December 1932) her Captain attacked her bridge for its poor look-out facilities. Earlier cruisers had wide bridges but limited depth, so personnel lined up along their fronts could easily look forward. The new type of bridge was narrow but deep. The captain and officer of the watch could see ahead over the narrow flat of the bridge, but the view for all others – admiral, signal officer, signalmen, lookouts – was poor. Looking forward through angled glass gave a distorted view. The windows had to be lowered if there was rain or spray, which would break so fiercely against them as to block the view – but which would also blow into the faces of anyone looking out without a window. This problem was experienced when the ship left harbour at night in rain. Also, the chart tables on the angled sides of the bridge cut out much of the available frontage where personnel could stand (matters would have been worse had one of the chart tables been replaced by a plotting table, as had been proposed). Finally, there was no sheltered position for lookouts. These were serious matters, as the same bridge design was being used in many other cruisers. For some reason the comments were not repeated in the formal report on the ship, so DNC did not have to answer them.

Cruiser Alternatives

With the *Leander* class design well underway, in July 1929 Controller proposed a study of a smaller ship. If the United States demanded parity at the next conference, the only way for the Royal Navy to get the numbers it wanted might be to reduce ship size. Numerous 'C' class ships of about 3,000 tons were approaching the end of their lives, and on a tonnage basis the Royal Navy would get only about one-and-one-quarter of the new ships for every two 'C' class cruisers it discarded. Cruiser size kept veering up and down; the 'Cs' were almost certainly too small, but the 4,800-ton 'Ds' were good ships, so a 4,500-ton design might be useful. First Sea Lord (Madden) asked ACNS to look into Controller's suggestion on the basis of two three-gun turrets, no vertical side armour and, say, 30kts. This idea in turn led to analysis by Tactical Division of the requirements which had led to the 6,000-tonner (which became the 6,500-tonner and then the *Leander*) in the first place. It took 6,000 tons to insure a sea speed of 26–27kts; a 4,500-tonner might be limited to 24kts, giving a margin of 7kts over the battle fleet at cruising speed. Any reduction in endurance would risk denuding the fleet of cruisers at a critical time, and hence was not acceptable. For deep speed, 30.5kts seemed to be a bare minimum. Tactical Division disliked mounting the whole 6in armament in two triple turrets; it preferred single mounts (it thought the ship might be too lively for a power-operated twin mount). Probably the ship would mount five such guns. The bare minimum HA battery was three guns and the bare minimum of protection was magazine cover against destroyer 4.7in guns.

Given these studies, at the end of September 1929 the Sea Lords asked for a range of alternatives, to be used as a basis for a British position at the naval conference due in January 1930. The most suitable might be something like the French super-destroyers of about 3,000 tons, which were then credited (incorrectly) with six (rather than five) 5.5in guns and a speed of 35kts. Japan had a similar ship in the 3,100-ton *Yubari*. Another possibility was something like the Italian *Bande Nere* (5,000 tons, eight 6in, 37kts, using destroyer machinery), while yet another was a ship with machinery and scantlings of destroyer-leader type, the speed of a British heavy cruiser, a 6in battery and magazine protection only. The fast ship would be a link between heavy cruiser and destroyer, which the slower *Leander* was not.

DNC assigned Lillicrap to develop the alternatives, beginning with a 4,000-tonner (5,500 tons deep), in effect a large, lightly-protected destroyer. She would be armed with five or six 6in guns.[14] Lillicrap soon settled on three twin 6in guns, two 4in HA, two single pompoms, and two triple torpedo tubes, but he found that even with destroyer machinery, 4,000 tons left nothing at all for protection. DNC had wanted at least a little protection for the magazines, a 1in crown, but by November he wanted a lot more: 2in box protection (sides, roof, ends) and 1in over the shell rooms and transmitting station and a 1in turtle-back over the steering gear. With bullet-proof bridge plating, that added up to 125 tons, hardly an insignificant amount in a small ship. Lillicrap estimated that he could work in about 750 tons of oil, enough for 5,000nm at 12kts. Instead of twin mounts, the final version of the ship had five single 6in guns, but it was also protected as desired. On the displacement, which by now was 4,200 tons, there was no possibility of protecting the machinery.[15]

The next study was a ship with three twin 6in and lower speed (32kts, but 33kts was desired). Lillicrap scaled her hull up from the 4,000-tonner and her armament down from the 7,000-tonner (*Leander*) and again destroyer machinery would be used to cut weight. As in the 4,000-tonner, protection was limited to the armament. The result displaced about 5,600 tons. Then Lillicrap produced a 6,000-tonner with four twin 6in, again with unprotected destroyer machinery but with the same magazine and other armament protection as the 5,600-tonner.[16]

At the other end of the scale, Lillicrap scaled down his 4,000-ton ship to 3,000 tons to produce a super-destroyer ('fleet scout'), broadly equivalent to contemporary French ships. He armed the ship with six single 5.5in guns (200 rounds each), two 3in HA guns, two pompoms and eight torpedo tubes, and sought a speed of at least 36kts. As in the 4,000-tonner, there was no space for a catapult or aircraft and no protection.[17]

These studies were reported to DNC on 22 January 1930. A set of small-scale sketches showed *Leander*-style single funnels and other features in the 6,000-ton and 5,600-ton cruisers. However, the 4,200-tonner had two raked funnels, a set of centreline torpedo tubes and five single gun mounts, one just abaft the second funnel. *Leander* was described as a 6,800-tonner. Estimated unit costs were: £1.6 million for the 6,800-tonner, £1.4 million for the 6,000-tonner, £1.3 million for the 5,600-tonner, £1 million for the 4,200-tonner and £750,000 for the fleet scout.

Controller (Rear Admiral Roger Backhouse) seems to have found the fleet scout particularly interesting. He was told that adding magazine protection (2in side, 1in crown) would cost about 75 tons and half a knot. He wanted to know whether some of the guns in the 4,200-tonner and 3,000-tonner could be paired. This was impossible in the 3,000-tonner, but on 4,500 tons the 4,200-tonner could have two twin 6in forward and two singles aft, which would make space for a catapult.[18] During the First World War Controller had commanded the light cruiser *Conquest* and he recalled the value of a waterline belt, which kept the waterline intact – 'we cannot hope to keep out all hits, but it is a great thing to keep out some'. He therefore asked for belt armour on the 4,500-tonner. That added 300 tons, including 80 tons for a 1in deck (only half of it for protection) and 40 tons for 1in end bulkheads.

The London Conference

The battleship-building 'holiday', the centrepiece of the Washington Treaty, was due to expire on 31 December 1931. By 1929 governments, at least in London and in Washington, had every reason to want to extend it, so they sought some sort of naval agreement. Mindful of the way that the 1927 conference had collapsed, both governments agreed not to allow naval officers to be delegates, although there were certainly naval advisors. The US Navy had reason not to resist, because a bill authorising fifteen cruisers, introduced in Congress in February 1928, passed in February 1929. This bill met the nominal US requirement for twenty-three 10,000-ton cruisers. Admiral Pratt, the US Navy's arms-control expert, was now Chief of Naval Operations. The feeling in the US Navy against any sort of treaty was so intense that some officers asked him to resign when he returned from the 1930 conference – despite his attempts to show that the United States had actually gained. For the British, there was increasing awareness that continuing to build large cruisers would be ruinous.[19] The financial crisis which began with the Wall Street Crash on 29 October 1929 helped convince both the US and British governments that it was time to make some sort of deal.

Much of the basis for an agreement between Britain and the United States seems to have been in place as early as July 1929. Meeting with American representatives, Prime Minister MacDonald and First Lord A V Alexander accepted the main US points – leaving the 5:5:3 ratio in place; parity should be enforced in overall combatant strength in each of five warship categories (cruisers, destroyers and submarines in addition to capital ships and aircraft carriers, with submarines to be abolished if possible); that the battleship-building 'holiday' of the Washington Treaty would be extended to 31 December 1936; and that relative levels of strength would be equalised by the same date. The United States offered to scrap sufficient destroyers and submarines to bring it down to the level of the Royal Navy. MacDonald repeated the British requirement for forty-five trade-protection cruisers, but not for the twenty-five to operate with the fleet. He felt compelled to offer to cut total British cruiser strength to fifty. The Admiralty considered this sacrifice acceptable only if it was temporary.[20] MacDonald also suggested freezing 8in cruiser construction at the fifteen already built for the Royal Navy and at eighteen for the United States, which could have another ten 10,000-ton cruisers armed with 6in guns (the Admiralty seems not to have realised this was a real possibility). There was some question as to whether the *Hawkins* class should be considered heavy cruisers or more akin to light ones of First World War design, and also of how the smaller 'C' and 'D' class should compare to the US *Omahas*.[21] The US government was less than enthusiastic, because the British proposal amounted to demanding a considerable superiority in cruiser strength, assuming that the US Navy built no further small cruisers (which it regarded as useless for Pacific warfare). The British saw their idea as a kind of parity, in which the US Navy would have superiority in large cruisers compared to a larger British number of smaller ones

In October 1929 MacDonald met US President Herbert Hoover in Virginia to settle as many of these issues as possible. On 7 October invitations were sent to all five major sea powers for a conference to convene in London in January 1930. In order to avoid problems, the organisers tried to exclude naval officers, although they had to accept admirals in the Italian and Japanese delegations. New Zealand sent her Governor-General, Admiral Jellicoe, who pointed out that to cut British cruiser strength to fifty would be to risk the security of the Empire. First Sea Lord similarly reiterated the requirement for seventy, which could be cut only if other powers drastically cut their own cruiser forces. The fifty-cruiser deal had already been made, however. For the Japanese, the main object was to raise their tonnage ratio, as they considered the one enforced at Washington humiliating. The ultimate Japanese objective was parity with the two Western sea powers, which would have made any Pacific naval campaign impossible for them individually.

The London Naval Treaty which emerged from the conference distinguished cruisers by main armament rather than by tonnage: heavy cruisers had guns of up to 8in calibre, and light cruisers guns of up to 6.1in calibre. The Royal Navy would build no more such ships; total Empire strength, including Australia, was fifteen. The US Navy was allowed a total of eighteen, despite its desire for twenty-three. The Japanese had to be satisfied with twelve. Total underage cruiser tonnage was also limited, the British being permitted a total of 192,200 tons of completed 6in cruisers as of 31 December 1936. Total replacement tonnage laid down after April 1930 and completed by 31 December 1936 was not to exceed 91,000 tons. Ships to be retained on 31 December 1936 could total 101,480 tons, which limited new construction to 90,720 tons. In effect the British government accepted an upper limit of fifty cruisers – which it did not have. To meet the fifty-cruiser goal by 1936 required a building programme of fourteen cruisers within the available tonnage, including HMS *Leander* of the 1929 programme. Since cruisers normally took three years to complete, the fourteen-cruiser programme spanned the four programme years through to 1933. The Admiralty planned three cruisers in each of 1930/1, 1931/2 and 1932/3, hoping that Australia would provide the fourteenth ship in the form of a replacement for HMAS *Brisbane*. Failing that (which happened) the Royal Navy would buy another ship in one of the years between 1931 and 1933.

Given the block obsolescence of British cruisers built during the First World War, the treaty set a sixteen-year age for ships laid down before 1 January 1920, a twenty-year age after that. As of 31 December 1936 the British Empire would have only eight underage 6in cruisers (as the situation stood in February 1932). By that time thirty-two light (6in) cruisers would be overage, and the four *Hawkins* class would have to be discarded. The Admiralty planned to retain fourteen of the seventeen ships which would be between sixteen and twenty years old in 1936, after which date a higher rate of construction would be needed. The Admiralty position was therefore that the 1930 treaty should be a short-term agreement, to be reviewed when it expired. The expectation seems to have been that the prospect of a renewed building race after 1936 would deter other sea powers from building too many cruisers during the life of the treaty.

With their large super-destroyers and small light cruisers, neither the French nor the Italians could accept separate limitation of destroyers and cruisers. Although both refused to ratify, they did agree to cease building 8in cruisers (in theory, both retained the right to resume doing so). As an incentive for them not to ignite a naval arms race, the treaty included an escalator clause: a signatory which considered itself endangered could announce that it was enlarging its fleet. That gained importance as the international situation darkened in the mid-1930s.

The Intermediate Cruiser: the *Arethusa* class

The fourteen new cruisers could be any combination of the designs on hand. Controller (Backhouse) considered the 3,000-tonner to be too specialised,[22] so he decided to mix the maximum number of 7,000-tonners with a new 5,000-ton design intermediate between a *Leander* and a fleet scout, which DNC could develop by scaling up from the 4,200-tonner previously offered. Once the 5,000-ton intermediate design was well in hand, in December 1930, Controller laid out a programme of ten *Leander*s

THE 1930 LONDON TREATY AND ITS CRUISERS

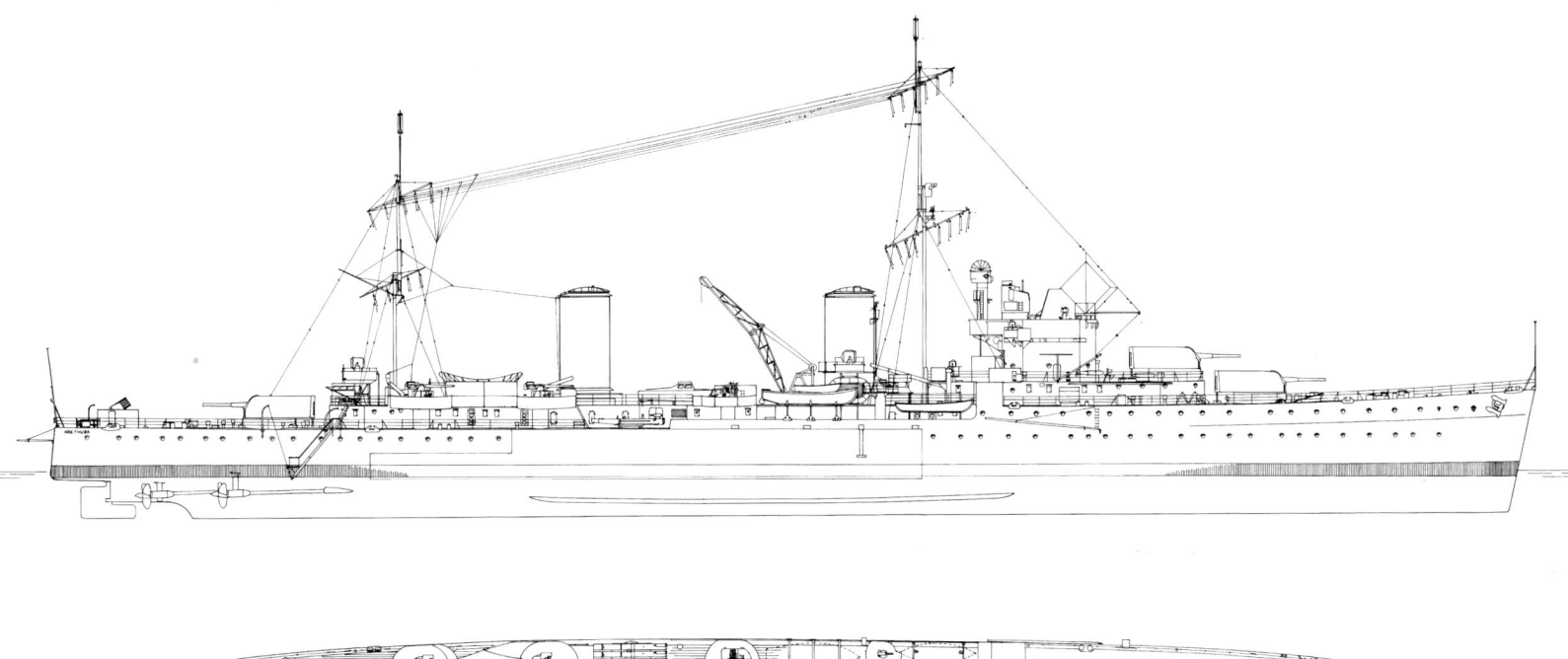

Arethusa as fitted, March 1941, with two quadruple pompoms added. Note the UP projector on the quarterdeck. She soon had a second atop 'B' turret (both UP projectors were apparently landed during a refit in 1942). The space above the torpedo tubes was later decked over. *Arethusa* was fitted with a Type 286 radar during a refit in July 1941, and had that set until Type 281 was fitted during a Chatham refit (6 February – 9 April 1942). At that time she received four twin 4in guns. (A D Baker III)

and four intermediates. Three repeat *Leander*s were included in the 1930/1 programme, during which the 5,000-tonner would be designed.

C-in-C Mediterranean particularly liked the intermediate as a fleet cruiser.[23] However, given sharply curtailed numbers, such specialisation was impossible; the 5,000-tonner seemed to be the smallest cruiser which could fulfill both fleet and trade-protection roles.[24] Controller saw the 4,200-tonner as a cross between an updated version of the well-liked 'D' class cruiser and the new French 2,600-ton super-destroyers armed with 5.5in guns. Later First Sea Lord described the new cruiser as an updated 'D'. Apparently it was easier to work up from the 4,200-tonner than down from the 5,600-tonner, the idea being to stay below 5,000 tons. Before that design was further developed, in June 1930 Rear Admiral Henderson, who headed the British Naval Commission to Romania, asked for a cruiser design. Lillicrap showed that if armour over the machinery was eliminated, a 5,700-ton ship could carry roughly a *Leander* armament.[25] That must have encouraged DNC to think that a viable six-gun ship could be designed on less than 5,000 tons. Work on a 4,200-ton design upgraded to three twin 6in guns began early in August 1930. It had the armament of the 5,600-tonner, except that there were two rather than four 4in guns (Controller wanted three), and, it was hoped, two triple rather than one quadruple torpedo tube. On the other hand, it had a seaplane and catapult.[26]

Controller asked for an endurance of 5,500nm at 16kts, for a good sea boat with a raised bow (but with height between decks minimised, to minimise the target). Magazines should be protected against 6in fire, with whatever machinery protection could be provided on the available displacement. This meant magazines with 3in sides, 2in crowns, and 2in ends where they met the machinery box, and by this time it was standard practice to provide an inch over the shell rooms and the transmitting station. A 1in splinter deck over the machinery was desirable, and structural strength demanded half-inch side and deck plating over the machinery. The desired endurance speed was higher than in the earlier design, which required extra length (10ft) for the same distance at 15kts rather than 12kts. Now the 940 tons of the earlier ship would have to increase to 1,140 tons (another 10ft longer than the original design).

It seemed impossible to hold down the size of the ship. The oil fuel requirement stretched her length to 500ft, which suggested a displacement of 5,200 tons, and size would spiral further because of the additional armament (compared to the 4,200-tonner) and protection. A 5,000-ton ship could be driven at 33kts on 60,000shp. Lillicrap provided machinery protection (3in side, 1½in deck). DNC understood that displacement had to be held down, so he suggested cutting power to 48,000shp in the hope that machinery weight could be dramatically reduced. A shorter machinery box required less protection. Lillicrap was to estimate what armament and protection could be had on the remaining tonnage. Not enough was left, but on about 4,500 tons Lillicrap could offer two twin and two single 6in guns on a 480-foot hull with nearly the same level of protection (machinery would be protected by 2in sides, a 1in deck, and 1½in end bulkheads).

Controller still wanted three twin mounts, and by mid-September a sketch design had been completed.[27] DNC became interested in alternating engine and boiler rooms to make the ship more survivable. Initially it appeared that the ship would need six boilers like those of the earlier heavy cruisers, and E-in-C suggested placing the forward engine room abaft the forward boiler room, with two more boiler rooms forward of the after engine room. Alternatively, four larger boilers could occupy two boiler rooms. They were taller than earlier ones, so the armoured deck

Above: *Aurora* as completed in 1938, as yet without her HA director forward. Note the deckhouse which replaced her catapult, as she was completed as RA(D) flagship.

over the machinery had to step up to cover them. The belt alongside had to extend higher to meet it. DNC put these ideas aside to develop a conventional 4,800-ton design with three boiler rooms forward of two engine rooms, as in a *Leander*, giving an endurance of 5,000nm at 15kts. Lillicrap provided DNC with rough particulars in mid-September 1930. He offered full magazine protection, but 2in over the machinery sides. On 60,000shp the ship would probably make 32kts. Only destroyer machinery would give more speed; E-in-C rejected it. Lillicrap calculated what would be needed to achieve 33 or 34kts.[28]

Lillicrap's alternatives were presented at a Sea Lords meeting on 23 October. The Staff wanted 33kts in standard condition, corresponding to a seagoing speed of 32kts, which Lillicrap's estimates suggested required 68,000shp in a 490ft hull. E-in-C wanted 8ft more of machinery box, which would also boost protection weight. The conference agreed that E-in-C could decide whether he could provide enough power to drive a 5,000-ton ship at 33kts. Displacement should not exceed 4,950 tons, Legend speed being 33kts and endurance 5,500nm at 15kts, with cruiser-type machinery with three boiler rooms. Controller felt strongly that effective subdivision of engine and boiler rooms had been neglected and should now be emphasised, as other navies were increasing the subdivision of their ships. DNC argued that extra subdivision made a ship much more complicated; with three boiler rooms instead of two, she had to be 8ft longer, with heavier machinery. The conference agreed that eliminating the mainmast was a virtual increase in the ship's protection, because it made the ship's course much more difficult to estimate. DNC's choice of triple tubes was accepted.

Machinery weight kept growing (it was 42lbs/shp rather than 40lbs/shp); Lillicrap doubted that everything the Sea Lords wanted could be provided within a 5,000-ton limit. Reductions in protection and

THE 1930 LONDON TREATY AND ITS CRUISERS

Above: Aurora shows some early-war modifications in a November 1940 photograph. She has no radars on board (the device at the foretop is a DF coil), but she has her quadruple pompoms and a UP projector on her quarterdeck (but no UP projector on a turret top). She had been refitted at Portsmouth, 30 May – 28 June 1940 after action damage. She was fitted with radars during a refit on the Tyne, 15 April – 7 May 1941 (Types 290 and 284).

armament would probably not be acceptable, so speed had to be cut. He suggested a 480ft ship developing 60,000shp to make 32kts, which could probably make the desired 5,500nm at 15kts. DNC accepted this idea, and Controller agreed. On 30 October DNC formally instructed Lillicrap to continue on this basis, with displacement absolutely not to exceed 5,000 tons. Initial feasibility studies were over, lines would be prepared and detailed calculations begun.[29] Protection was increased to roughly *Leander* levels: 3in side and 2in magazine crown, 3in side over machinery with a 1in deck. DNC produced a sketch design of the 480-footer in mid-January 1931, somewhat beamier than had been planned (49ft), displacing 5,000 tons. It had the three boiler rooms and single trunked funnel of a *Leander*. Two separate engine rooms were abaft the boilers. This cruiser had DCTs fore and aft and an HADT on a stalk abaft the forward DCT, atop the new-style bridge. E-in-C provided 64,000shp, despite the machinery arrangement having been a considerable problem. E-in-C and DNC had agreed on a four-shaft powerplant, as it had proven impossible to produce an adequate two-shaft arrangement. DNC expected a speed of 32.5kts (possibly slightly more) at standard displacement (31kts deeply loaded). There were three rather than the earlier two 4in anti-aircraft guns, one of them on the centreline.

Controller asked for some detail improvements, such as raising the after 4in HA gun on a platform so that it would interfere less with the two other guns on forward arcs. He wanted to reduce belt armour to 2¾in so as to protect 6in turrets, redoubts (rings above deck), and lobbies with uniform 1in, or at least ⅞in, plating. The bulkhead between the engine rooms should be strengthened to 1in because the spaces involved were so large (there were 1in bulkheads at the fore end of the boiler rooms, at the aft end of the boiler rooms, and at the after end of the engine room). The reduction in belt armour sufficed to strengthen protection over the armament to ¾in. DNC was able to thicken the bulkhead between the engine rooms to ⅞in ('a very stout bulkhead') by reducing the end bulkheads from 1½ to 1in. The after HA gun was raised on a 3ft platform. This revised design was submitted to DNC on 27 January.

The catapult was immediately abaft the funnel, its stowed length being quite limited. It was unlikely that a second aircraft could be stowed. To remain within the 5,000-ton limit, DNC proposed providing space for 200 rounds per 6in gun, but to include only 150 per gun in the weights. Standard displacement would be 4,980 tons.

ACNS pointed out that for trade protection the ship should have a heavy catapult sufficient to launch a spotter/reconnaissance aircraft.

Below: Penelope shows her full early-war radar outfit in this undated photograph taken in Malta. She was refitted on the Tyne (26 August 1940 – 2 July 1941) to make permanent repairs after grounding in Norway, 11 April 1940. She received her quadruple pompoms and Type 281, 284 and 285 radars, antennas for all of which are visible here. The ship's catapult was removed. Note the quadruple 0.5in machine gun visible just below the searchlight abeam the forefunnel, and the Oerlikons atop 'X' turret and on the quarterdeck. Penelope was mined off Tripoli on 19 December 1941, and then repaired in Malta, emerging with a total of four single Oerlikons were fitted (two of them replaced the quadruple machine guns).

Above and below: Arethusa is shown after a major refit at Chatham (6 February – 9 April 1942); she shows the standard appearance of this class after Type 272 surface-search radars were installed forward of the bridge. *Arethusa* was refitted at Chatham (17 August 1940 – 30 September 1940) and by August 1941 she had had her catapult removed so that two quadruple pompoms could be fitted abeam its former position, and she had Type 286 radar. By July 1941 she had UP projectors atop 'B' and 'X' turrets, and four Oerlikons. The UP projectors were removed at Chatham, and it was only then that she was fitted with twin rather than single 4in guns. She emerged from the refit with four more single Oerlikons (eight are visible here, including ones atop 'X' turret and on the quarterdeck), and three more were added by October 1942. She was torpedoed on 18 November 1942 while escorting a Malta convoy, and was refitted in the United States (Charleston Navy Yard, 30 March – 15 December 1943). Her quadruple pompoms were replaced by US-type quadruple Bofors guns, and she exchanged four twin US-type Oerlikons for three singles.

Controller suggested that replacing the centreline gun with one on each side (roughly abreast the aircraft crane) would free more centreline space – but it would add 40 tons, and the ship was already at the stipulated 5,000-ton limit. Controller then ruled that the standard displacement need include only enough ammunition for the three 4in guns previously planned (100 rather than 150 rounds per 4in gun). After further consultation with E-in-C, a six-boiler powerplant using three equal boiler rooms was adopted. The revised sketch design received the Board Stamp on 31 March 1931. Standard displacement was given as 5,000 tons.

The design was badly cramped, with fuel stowage squeezing magazine stowage in the hold, below the waterline. The transmitting station was protected there, but not the low power room necessary to transmit fire-control data, and the 4in handing room was on the platform deck. DNO was anxious to get both spaces under protection and to do that he suggested placing aircraft bombs in the 6in shell rooms, and 0.5in ammunition and the torpedo warhead stowage outside protection. Although the sketches showed DCTs fore and aft, it was soon decided not to provide the 5,000-ton cruiser with facilities for divided fire control (as were being provided in the 1930 *Leander*s). Fire-control semaphores were also omitted; these ships would not be master ships for concentration firing.

Controller submitted a detailed design on 30 September 1931. Given the financial crisis, tenders could not be invited for some months. Controller therefore asked DNC and E-in-C to reconsider the machinery arrangement, with its two large engine rooms adjacent to each other. Rearrangement seemed important given recent experience in shellfire trials against the battleship *Empress of India*. DNC proposed placing one of the three boiler rooms between the two engine rooms. He wanted the two engine rooms at least 40ft apart, so that explosion damage to the compartment between them would not disable them. So that the two outer shafts could pass through the after boiler room (between the engine rooms), it had its boilers in tandem rather than side by side, as in the two forward boiler rooms, hence was longer (44ft rather than 24ft). That further separated the two engine rooms. Longitudinal bulkheads enclosed the boilers, adding protection but also adding an off-centre flooding problem if the outer part of the room flooded. In action the machinery would operate in two units, the combination of forward boiler room and forward engine room being self-contained, as were the after boiler room and after engine room, cross-connections being provided. The forward engine room was shortened by moving its generators to the wing spaces abeam the after boiler room, making it less likely that both generators would be put out of action by one hit.

The change cost 20ft in length and 500 tons in standard and full load displacement; total power increased from 64,000shp to 66,000shp. It was claimed that there was no reasonable chance that the ship could be put completely out of action by one underwater hit, and that generators would be better protected. Accommodation would be improved, and DNC expected that the longer hull would be a better sea-keeper.[30] Against that, the ship would cost more (about an additional £80,000), it would be somewhat slower (by half a knot), and it could not accommodate the long catapult. Endurance would not be affected. In the previous design, the belt had to extend to the upper deck over the boiler rooms, because the big boilers were so tall. Now the two boiler rooms were no long contiguous, but it seemed unwise to drop the belt back down between them. The high part of the belt was 140ft rather than 76ft long, adding considerable weight. The second funnel crowded the centreline, so that the heavy 53ft catapult could not be accommodated. The 4in guns had to be relocated: now they were all grouped where the after pair had been. Greater length did alleviate overcrowding. DNC strongly advocated the change. So did Controller, who had asked for the redesign in the first place.

Unfortunately the change raised displacement to 5,500 tons. On 24 October DNC passed the word that Plans Division would not accept anything over 5,180 tons, based on total available cruiser tonnage. If the ship were squeezed down to 480ft, she might be cut to about 5,100 tons, which seemed to be about the maximum available. The situation was so bad that a member of the Naval Staff suggested giving up one boiler room, cutting power to 44,000shp and thus speed to 29.5–30kts (at about 5,000 tons). That inspired Controller to suggest using four more powerful boilers. The machinery box would be somewhat longer, but not so long as to break the tonnage limit. To do that, E-in-C accepted higher pressure and temperature than previously.[31] About this time the test tank (Haslar) estimated that the ship would need 63,000shp to reach 32kts. E-in-C offered 15,000shp and then 16,000shp boilers.

Lillicrap drafted a DNC memorandum for Controller: he now proposed a 5,180-ton design with the improved subdivision of the 5,500-tonner, reducing its 500ft length down to the 480ft of the original 5,000-tonner, using a four-boiler scheme proposed by E-in-C to Controller on 18 November. Estimated displacement was now the required 5,180 tons (480ft x 49ft 6in x 14ft 3in). The ship was badly cramped because the machinery box was so long. In the original 5,000-tonner it was 178ft long, which was bad enough (in a 'D' class cruiser of comparable size it was 146ft long). Moreover, the new ship was intended to have a larger complement than a 'D' (initially 550, compared to 470 in a 'D' used as a flagship, but now reduced to 500, or to 520 as a flagship). To cram all those men in, accommodation standards had to be reduced. In the 5,500-ton design, the machinery box was 188ft long, but that extra length was more than balanced by the 20ft greater length of the ship. Now the ship was squeezed back to her original length, but the machinery box was 185ft long, meaning even less crew space was available. Lillicrap told DNC that he could cut machinery box length to 185ft if he used six boilers, with four in the forward boiler room (and no bulkhead between the pairs). Perhaps they were trying to do too much on too small a tonnage. For a time it seemed the ship would revert to 500ft length and to six boilers (64,000shp). In the end the planners relented, and on 4 January 1932 Lillicrap was told to go ahead with a 5,500-tonner. Extra weight came from all sorts of sources as the design developed in detail. For example, each of the four boilers was so massive that it was pointless to keep it running in harbour. The ship therefore needed an auxiliary boiler, which was placed in the forward boiler room (this became a standard fixture in later British cruiser designs). One turbo-generator was placed in each engine room, and one diesel generator on each side of the after boiler room.

Subsequent British cruisers all adopted the new machinery arrangement alternating boiler and engine rooms, including the tandem boilers in the after boiler room, and the auxiliary boiler in the forward boiler room, but unfortunately it proved flawed. The spaces outside the two tandem boilers in the after boiler room had enough volume that, when an adjacent machinery space flooded, the ship could quickly capsize. This was entirely unsuspected, because the necessary calculation was too intricate for manual procedures.[32] As a result, several cruisers were lost during the Second World War to single torpedo hits. This problem helps explain why the final (unbuilt) British Second World War cruiser designs showed such widely-separated machinery units.

DNC and E-in-C wanted one ship with divided machinery spaces built for initial tests. The first ship therefore replaced one of the three repeat *Leander*s planned for the 1931/2 programme. The new design received the Board Stamp (as the 5,500-tonner) on 14 January 1932, the idea having been approved by First Sea Lord on 1 January. The Board formally approved the new sketch design on 18 February 1932. The building programme was provisionally cut to nine *Leander*s, with the understanding that the actual decision on programmes after 1932 would be subject to reconsideration by the Board at that time. HMS *Arethusa* was included in the 1931/2 programme alongside two *Leander*s. The Board approved the Legend and drawings on 23 June 1932.[33]

Arethusa was designed to displace 5,419 tons (standard), but was completed at 5,223 tons. Weight was saved by, among other things, welding, by omitting magazine cooling, and by substituting steel for iron cable. This saving made it easier to add weight after completion. By late 1941 approved additions amounted to 362 tons: splinter protection, protection to vital communications, bottle-rack stowage for 4in shells, 4in twins instead of singles, permanent degaussing, Oerlikons, Asdics, SA gear (self-protection against acoustic mines), radar (including tripod masts), pompoms in place of the catapult and aircraft, a second HACS Mk III, increased complement, etc. Another 375 tons were resisted or compensated for, including steam heating for the Arctic.

Roughly parallel to the intermediate cruiser study was a study of a new version of the slow trade-protection cruiser proposed in 1928, only this time armed with eight 6in guns. In January 1931 DNC asked Lillicrap to sketch a 21kt cruiser protected against 8in fire, to be based on the 5,000-tonner.[34] He expected to save so much length on machinery that the fourth 6in turret could be fitted in.[35] The ship might be shortened, but that would cost the catapult and considerable fuel oil. Lillicrap sketched 5,000-ton and 6,000-ton ships, both 480ft long (with 49 vs 52ft beam and 14 vs 16ft draft) with single funnels, short machinery boxes amidships, and a catapult between bridge and funnel; HA guns were abaft the funnel. Total magazine length (160ft) considerably exceeded machinery length (about 80ft). Lillicrap submitted his report on 14 January 1931. DNC asked about confining the belt to the machinery spaces but providing the magazines with belt rather than box protection, i.e. extending their protective decks out to the ship's sides. The belt and deck over the machinery weighed only 465 tons at this point, but that over the magazines would weigh another 930 tons, with another 300 tons for turret supports and another 150 tons for turret roofs, a total of 1,845 tons. To that it would probably be necessary to add protection to the magazine sides against plunging fire – say 100 tons for the lower sides of the magazines (3in thick) and more over the steering gear, and the bullet-proof plating now wanted for bridges – another 100 tons, for a total of 2,050 tons. As with the earlier slow trade-protection cruiser, nothing came of this idea.

Below: Arethusa at Malta, 3 October 1945. Although official records continued to show a Type 272 surface-search radar, by this time clearly it had been removed in favour of a small 'cheese' at the head of the foremast. The quadruple Bofors seem to be gone, the ship's close-range anti-aircraft armament reduced to Oerlikons, with twin power Oerlikons atop 'B' and probably 'X' turrets. Two single hand-worked Oerlikons are also visible. These numbers do not jibe with the officially-listed armament, which in October 1945 was two quadruple pompoms, four twin power Oerlikons, and three single Oerlikons. (Jack Blumfield, US Naval Institute Collection)

The Improved *Leander* (*Amphion*) class

On 2 March 1932 DNC assigned Lillicrap to investigate the desired Improved *Leander*, to be built under the 1932/3 programme.[36] He wanted to know the effect on a *Leander* of (a) making 'B' and 'X' triple turrets, and (b) of adopting three triple turrets, and also the effect on these versions of increasing belt armour to 4in, of improving machinery space subdivision, and of improving magazine protection. Existing sketch designs of triple mounts showed that they would weigh 50 per cent more than twins, so in weight terms, (a) was like adding another twin turret. However, there would be subtler increases in the size of ammunition lobbies, magazines, shell rooms and the turret ring on the deck. Lillicrap thought that these increases might be very large, and that it might be necessary to have separate ammunition lobbies for cordite and shell. He had already discovered as much during the 1929 studies of triple turrets. Increased complement, probably fifteen men per triple turret, would be a real problem, since the *Leander* was already badly crowded, and enlarged magazine and other spaces would consume yet more internal volume. Probably the ship would have to be lengthened, perhaps from the current 547ft to 555ft. With three triples, the added weight and volume would not be as large, so the hull did not have to be lengthened as much (Lillicrap suggested 3ft, to 550ft). Oil capacity would be reduced, because the tankage lost forward (to enlarged magazine spaces) would be less than what was gained aft (the forward tanks were much larger, the after ones being constricted by the propeller shafts). Lillicrap estimated that the four-turret ship would displace 7,500 tons (555ft x 56ft x 16ft 3in), or 7,600 tons with the extra belt armour. On the 72,000shp of a *Leander*, these ships would make 32.5kts and 32.25kts respectively. The three-turret ship would displace 7,325 tons (550ft x 55½ft x 16ft 3in), the same ship with more belt armour 7,425 tons. As with the four-turret ships, speeds would be 32.5kts and 32.25kts.

The improved machinery subdivision was the alternating arrangement already applied to the *Arethusa* class. To avoid gross overweight, Lillicrap suggested reducing power to the 64,000shp of the *Arethusa*, in which case the four-turret ten-gun ship would displace 7,520 tons, the three-turret ship 7,325 tons, and the up-armoured versions 7,640 and 7,445 tons. To accommodate the longer machinery box, the four-turret ship would have grown to 565ft, the three-turret ship to 560ft. Speed would fall below 31kts.

Leander offered magazine protection below 11,000yds and between 14,000 and 21,000yds (as set by the 2in magazine crown) to shells hitting right abeam; below 11,000yds the target would be immune because it was under water. They were altogether immune to 60° attack, and Lillicrap commented that this was 'very good indeed'. *Arethusa* was immune below 10,000yds and from 15,000 to 21,000yds. Lillicrap seems not to have estimated the effect of adding magazine protection. A ship with enough oil would probably displace 6,700 tons (545ft [wl] x 55ft x 15ft), which was not too much less than a *Leander*, or 6,800 tons with a 1¼in deck.

First Sea Lord also wanted to know what could be done with three turrets, with a twin in 'A' position, so Lillicrap scaled up the *Arethusa* to find out and also increased belt armour to 4in, as in the up-armoured Improved *Leander*s. He thought the combination would displace 6,200 tons (520ft x 53½ft x 14½ft) and would make 31.75kts. This data was passed to DNC on 17 March, who asked what would be required to

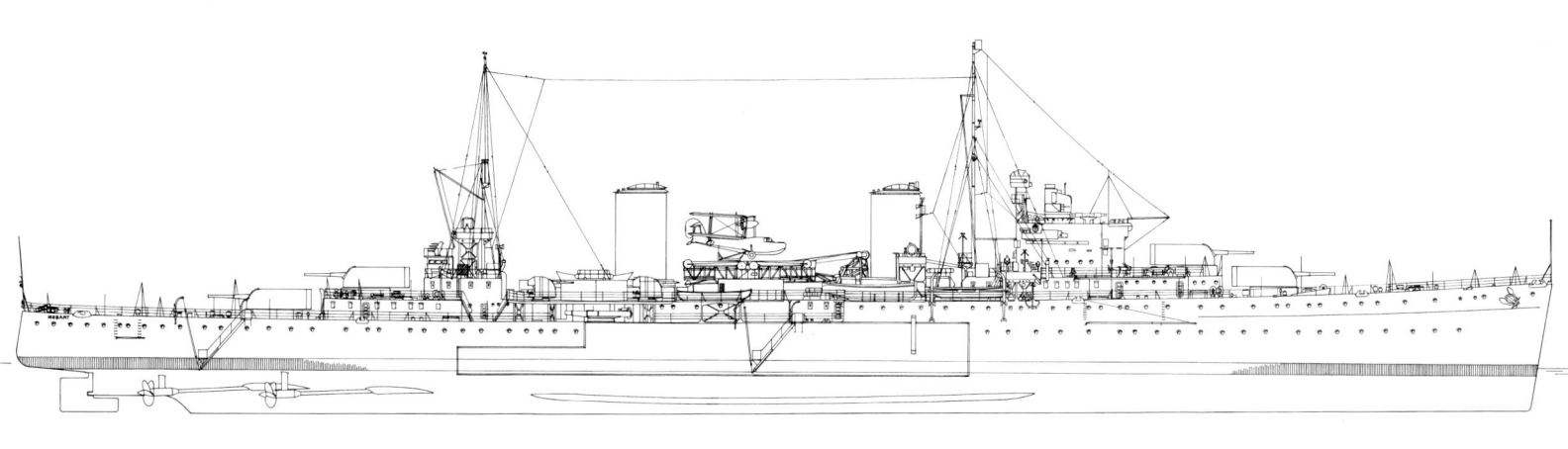

HMAS *Hobart* (ex-HMS *Apollo*) as fitted, 1939. She was completed with a short catapult fixed amidships for Ospreys and with single 4in HA guns. She was altered to the configuration shown in 1938 for transfer to Australia, at which time the formerly open boat-stowage skids had been decked over. The catapult was removed when she arrived in the Mediterranean in 1941 and replaced by a single quadruple pompom; two Italian Breda guns were mounted side by side on the quarterdeck and four single 20mm Oerlikons were reportedly added. (A D Baker III)

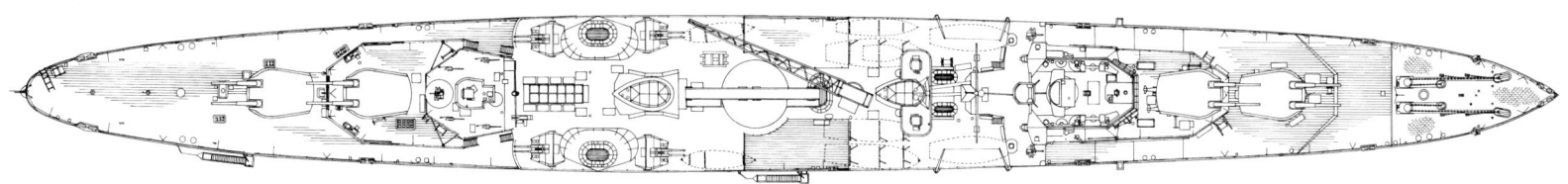

Below and left: HMS *Amphion* is shown before her transfer to Australia as HMAS *Perth*. She carries two aircraft, but the catapult had not yet been fitted. HMS *Phaeton* was completed in this form as HMAS *Sydney*, without a catapult. Her two sisters were given twin 4in guns before being transferred, and their original 46ft catapults were landed, for later replacement.

boost endurance to the 7,000nm (at 16kts) of a *Leander*. *Arethusa* managed 5,500nm at 15kts on 1,200 tons of oil (*Leander* carried 1,800 tons). It would take about 1,800 tons to drive *Arethusa* at 15kts for 7,000nm. Lillicrap estimated displacement at 6,750 tons. Later (in 1933) Controller wanted to know what could be done if only one of the three turrets was a triple, increasing *Arethusa* firepower to seven 6in guns.[37] The 7,100-ton figure given presumably included much more protection, as in the 1932 studies. Late in 1932 there was also apparently interest in gaining protection by reducing to *Arethusa* armament.[38]

The 1932 studies seem to have convinced the Sea Lords that little could be done immediately, apart from adopting the new machinery arrangement. To do that, E-in-C proposed using four 18,000shp boilers instead of the six 12,000shp of a *Leander*.[39] As in an *Arethusa*, the belt had to be raised to cover the higher boilers, and the upper part had to be lengthened to cover both boiler rooms and the space between them. The new machinery space was 9ft longer than in a *Leander*, but machinery was lighter (1,445 tons rather than 1,504 tons). The longer side protection, with its long upper strake, added topweight which had to be balanced by additional beam (there was apparently no hope of making the belt thicker), so the ship needed more beam. Unfortunately the *Leander* was already cramped, and a longer machinery box would worsen the situation, particularly since machinery occupied the middle part of the ship, which provided the greatest volume for personnel. The same problem had forced up the size of the smaller *Arethusa* class.

One of the methods used to ease crowding in *Leander* was to be abandoned. The *Leander* design allowed for only half the anti-aircraft guns and half the torpedo tubes to be manned in action. Providing crews for only two of the four anti-aircraft guns saved sixteen ratings; providing for one rather than two sets of torpedo tubes saved another five, and not providing ammunition-supply parties for the 0.5in guns saved another eight. In addition, 25 per cent of the stokers had been assigned to alternative roles (the rules allowed for 15 per cent), saving another seven ratings. The total of thirty-six ratings was more than 5 per cent of the planned number. As the new ship was being designed demands were heard that she be able to fight her whole battery with her designed complement. To shave the demand to a more practicable figure, DNO was willing to keep allowing the 25 per cent of stokers in alternative roles, and to use the Fleet Air Arm ground crew (about seven men) to supply ammunition to the machine guns. DNC had about the same deck space in the new design as in *Leander*. To squeeze in the additional twenty-three ratings eventually wanted he needed another 3 tons and took other measures: he relocated the auxiliary and second W/T offices and the central stores from the lower to the platform deck. Elimination of a separate bomb room had allowed him to move the HA calculating position from the platform deck (as in a *Leander*) to the hold, where it was protected.

The heavy catapult of the *Leander* class had to be sacrificed (this decision was later reversed). The Board had accepted growth to obtain both more survivable machinery and better accommodation, but it had also promised not to reduce the number of cruisers to be built under the

THE 1930 LONDON TREATY AND ITS CRUISERS

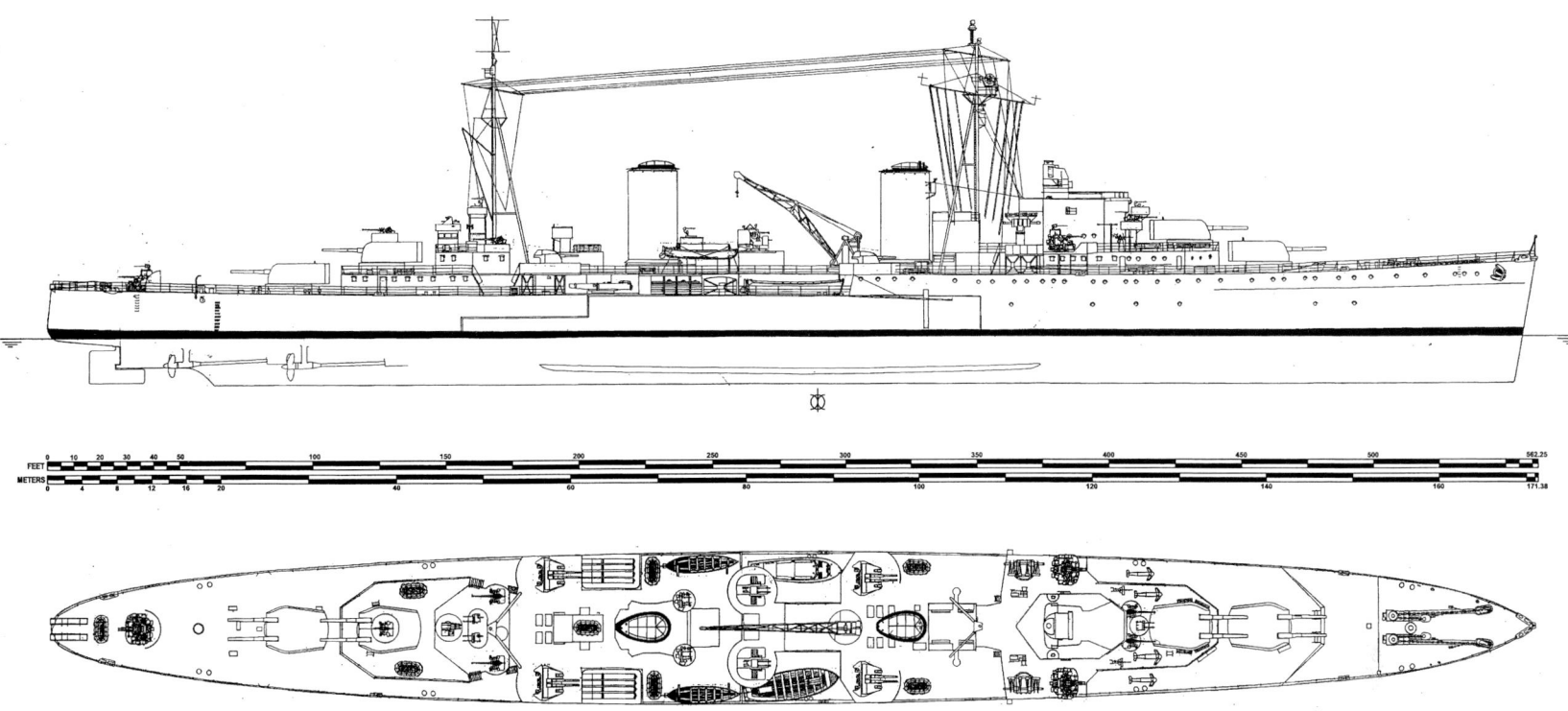

Above: HMAS *Hobart* in December 1944 as repaired following torpedo damage. As refitted, she had RPC 50 for her two quadruple pompoms. Reportedly they had survived the torpedoing, but they were relocated when the ship was refitted. When 'X' turret was removed in June 1946, its replacement by a quadruple Bofors was approved, but it is not clear whether this was done. Arrangement drawings for 1945 show a single or twin Bofors on the quarterdeck (the ship had three twin and five single Bofors). Radar at this stage was Type 281B air search, Type 277, Type 276, SG (US-supplied), Type 285 (two sets), Type 282 (two sets), Type 283 (three sets), IFF, Type 251M beacon, FV1 and a jammer. (Paul Webb)

Below: HMAS *Sydney* was apparently little modified during the Second World War, apart from the provision of splinter shields for her 4in and quadruple 0.5in guns. She was delivered to Australia before her single 4in guns could be replaced by twins. In Australia *Sydney* (and the other two ships) received a 53ft catapult.

London Naval Treaty. Moreover, space was very tight. The forward engine room (above which the catapult was placed) was 4ft shorter in an *Arethusa*. There was just no space for a heavy catapult, although the ship could accommodate two catapult aircraft, either one fighter/reconnaissance and one spotter/reconnaissance, or two fighter/reconnaissance, or two light reconnaissance seaplanes (the spare was stowed on the superstructure).

On 28 July 1932 the Board approved the modified *Leander*, accepting an increase to 7,350 tons (soon corrected to 7,250 tons). This was the *Amphion* class, all three of which were later bought by the Royal Australian Navy. Ships would be built under the 1932/3 and 1933/4 programmes, the latter being the last likely to be completed under the 1930 treaty. Total tonnage, including the 5,450-tonners, would be 91,300 tons, 580 tons more than the treaty allowed. Controller expected sufficient savings in actually building the ships to make up for this overage. At this point the planned 1932/3 programme comprised two 7,000-ton cruisers (one contract, one Royal Dockyard) and one 5,450-ton cruiser. On laying down HMS *Amphion* the British government notified foreign powers that she would displace 7,000 tons.

While detail design was proceeding, Controller and ACNS asked for further simplification, to cut both cost and complexity. DNO and DTSD were willing to cut what might be considered important gunnery equipment, accepting elimination of the after DCT, halving the number of director positions. They hoped to gain back some measure of alternative fire control by fitting a small fixed control position aft. With the after DCT went one of the ship's two duplex rangefinders as well as the ability to split the armament to engage two targets (this applied only to the 7,000-ton cruisers, as the smaller cruisers never had this capability). Costs could be cut further by substituting a destroyer-type fire control clock for the Admiralty Fire Control Table Mk V (a somewhat more complicated Admiralty Fire Control Table Mk VI was adopted). Automatic functions were reduced, increasing the chance of human error, and the fire-control personnel would have to be more highly trained. Recent tests of destroyer fire control clocks (in the shop and at the gunnery school, not at sea) suggested that the approximations

Above and below: HMAS Perth was practically unique in being fitted with First World War-style anti-rangefinder baffles on her funnels, though she did not have the spiral wires used in the First World War to break up the images of masts in coincidence rangefinders. She was refitted at Sydney, 31 March to 15 April 1940 (when the baffles were fitted). *Perth* arrived at Alexandria on 24 December to relieve *Sydney*, and was painted in a camouflage pattern. Perth was repaired at Alexandria between January 1941 and 22 February 1941, when the Type 286 radar on her mainmast was fitted. At that time her catapult was removed and replaced by the two light anti-aircraft guns seen in the camouflage photograph between the funnels. These 20mm Breda guns captured from the Italians (and much liked in the Mediterranean Fleet) were cross-decked from HMAS *Sydney* to *Perth* when *Sydney* left the Mediterranean on 12 January 1941. A third was mounted on the quarterdeck, at least during the Crete evacuation (June 1941). A quadruple pompom (ex HMS *Liverpool*) was mounted on the catapult deck, probably in May 1941 (presumably for Crete), but possibly during the early 1941 refit (note that it is not visible in the camouflage photograph). The Bredas and the pompom were cross-decked to *Hobart* when she relieved *Perth*. *Perth* left the Mediterranean (relieved by HMAS *Hobart*) on 18 July 1941, after a short refit at Port Said. At that time she was fitted with a new catapult (from HMS *Ajax*). When she arrived home, she was further refitted at Sydney (11 August – 30 October 1941). The photograph in gray paint shows the ship after this refit, with Oerlikons on the tops of 'B' and 'X' turrets. Oerlikons probably replaced the two quadruple 0.5in machine guns in the zarebas on the shelter deck near the bridge, and these guns were remounted right aft, on her quarterdeck, the original third gun remaining in its position abaft the searchlight platform just forward of the after two turrets. A fifth was mounted in a tub abaft the after funnel. At this time the ship was provided with a Walrus aircraft. She was sunk in this form on 1 March 1942. (RAN Historical Branch)

By the time HMS *Amphion* was delivered to the Royal Australian Navy as HMAS *Perth*, she had twin 4in guns. Note the absence of a catapult. She was photographed in New York City *en route* to Australia by Ted Silberstein. (US Naval Institute)

involved would not unduly reduce accuracy. CNS concurred with this. The proposal received the Board Stamp on 10 November 1932. These changes were applied to the *Leander* class cruisers already building (but not to *Leander* herself).

In the spring of 1933 DNO proposed to use long-trunk mountings, which would save personnel by eliminating the ammunition lobby halfway up from the magazine. In an *Arethusa* class they would save eight men per mounting, and also reduce mess deck congestion (by up to 300ft^2). Against that, the 6in magazines and shell rooms had to be lengthened, and magazine armour extended over the shell room. In an *Arethusa*, that would probably cost 50 tons in protection and about 20 tons of oil fuel stowage would be lost. The idea therefore died.

The detailed design received the Board Stamp on 10 November. E-in-C had shaved a foot from the after boiler room, so the ship was a foot shorter than the design previously submitted, but 8ft longer than a *Leander*, with 1ft more beam, 6in more depth, and drawing 3in more. Displacement had grown by 250 tons; DNC had managed 100 tons less than initially reported. The situation was actually better, because the standard displacement ultimately approved for the *Leander*s (Board Stamp 4 June 1931) was 7,154 tons. The modified ship was expected to be less than 100 tons heavier. Beardmore, the machinery contractor for the 1931/2 cruiser (*Amphion*) being built at Portsmouth, accepted the new machinery design so that this ship was built to the new design. The two large cruisers of the 1932/3 programme (*Apollo* and *Phaeton*) were built to the new design. This programme included the second *Arethusa*, HMS *Galatea*.

A New Look at Air Defence

While the *Arethusa* class was being designed, a new naval anti-aircraft gunnery committee was appointed in November 1931 and it reported in April 1932.[40] As in 1921, the committee had to rely largely on theory and on exercise experience.[41] The threat had certainly grown.[42] Compared to the situation in 1921, this Committee had to deal with faster and more numerous attackers, which might fly at higher altitudes. The 1932 report emphasised the threats of precision level bombing and torpedo bombing. It included a kind of dive bombing, but with dives beginning at low altitude (about 800ft) to achieve a good percentage of hits. The report also described (and discounted) US work on higher-altitude dive bombing, of the sort which was to prove so effective during the Second World War.[43] Gas attacks were also considered, as was the possibility that enemy aircraft might control unmanned explosive boats (the British actively considered such a weapon in the 1920s, but had abandoned the idea by 1931–2). Yet another possibility was the use of explosive gliders controlled by aircraft.[44]

Fleet experience suggested that individual aircraft would probably be spotted at a range of about 6–8nm; formations might be seen at slightly greater ranges. The longer the range of the anti-aircraft guns, the longer they could keep attackers under fire, and therefore the better the chance that they would achieve results. As aircraft speed increased, time under fire would decrease. The bursts of long-range AA shells might also be used to indicate to fleet fighters where enemy aircraft were. In theory, two 6in Mk XXII guns (as on new light cruisers) were equivalent to three or four 4.7in or five or six of the new experimental 4in Mk XIV anti-aircraft gun or to eight or nine of the older lower-velocity 4in Mk V. The larger shell reached further, retained its velocity longer, had a flatter trajectory, and had a greater effect when it burst. On the other hand, heavier guns fired more slowly and took up more space and weight. Taking rate of fire into account, one 6in gun was equivalent to 2.3 4in guns (one 4.7in was equivalent to two 4in). A new 5.1in gun being proposed as a dual-purpose battleship weapon was considered far better, equivalent to three 4in guns. The report advocated six 4.7in guns on each side as a minimum; 'An increase in the number of 4in guns in existing ships can, at best, only be considered a partial remedy.' The executive summary of the report simplified this conclusion: where the 1921 AA committee wanted four guns able to fire anywhere in the sky, this one wanted six.

The number of AA guns bearing on each broadside was insufficient, and ships were likely to be ineffective because they had only a single HA fire-control position, so could not cover multiple simultaneous attacks. Given available air room no more than eighteen torpedo bombers could attack the head of a line simultaneously. Since typical bases could support several times as many bombers, the fleet might well find itself attacked by several groups independently and in close succession. Succeeding waves had to be countered before the first wave attacked: a ship should be able to engage at least two air targets at the same time. To do that she needed two independent control systems, including director towers aloft and computing positions below decks.

The maximum usable gun elevation was 70°, because much above that the roll and pitch of the ship rendered effective control nearly impossible. Twin mountings were much preferred to single because they took up far less space and thus could have better arcs, were easier to keep clear of main battery gun blast, needed fewer personnel, and they were easier to control. Except for elaborate (between decks: BD) mountings, a twin did not weigh much more than two singles. However, BD mountings offered better protection to crews and their rate of fire seemed to be independent of gun elevation (because the gun could be served from a position well below the deck). They did weigh more and cost more, and they required more structure around them, because they cut substantial holes in the strength deck through which they projected.

HMAS *Hobart* was torpedoed by a Japanese submarine on 20 July 1943, returning to Sydney on 26 August for a refit which lasted into January 1945. She is shown at its completion. She gained quadruple pompoms with directors abeam the after funnel. The forward pair of twin 4in guns was moved forward. As in other British-designed cruisers, her centreline HA director was replaced by two directors, one to either side of her bridge, carrying standard British Type 285 radars. Note, however, that the face of the 6in director carries the antenna of a US Mk 3 surface fire-control radar. The Oerlikons forward of her bridge were replaced by single Bofors. A third is visible atop 'X' turret, and two more are visible at the foot of the forward leg of the tripod mainmast. On the shelter deck abeam the bridge and in the zareba fight aft on the quarterdeck are three Hazemeyer twin Bofors guns. Single Oerlikons are on the shelter deck near 'B' turret. Tubs on the face of the bridge and side by side on the mainmast carry Mk 3 barrage directors with their Type 283 radars. The foremast carries an unusual combination of radar antennas: a US SG surface-search antenna at the peak, with a Type 276 surface-search radar below it, two 'hayrake' IFF interrogators, and then the dish of a Type 277 (pointed skyward). The mainmast carries the single antenna of a Type 281B radar, with its interrogator above it. The starfish at the top of the tripod carries the four sensing/jamming antennas of the FV1 system. Barely visible forward of the mast is a structure supporting two canted 'hourglass' antennas, which serve the associated Type 91 jammer by providing accurate frequency data on the radar being jammed. *Hobart* was refitted again in 1946, her 'X' turret removed, before being placed in reserve in 1947. There may have been further improvements, since later editions of the Commonwealth ships' characteristics books showed the new Type 274 main battery fire-control radar and two Simple Tachymetric Directors (STDs, equivalent to the US Mk 51). By that time her close-range anti-aircraft armament had been slightly reduced, the five single Bofors and two single Oerlikons being replaced by two power and one hand-worked Bofors and two twin power Oerlikons. However, later editions of the ships' data book credited her with six STAAG Bofors and four single Bofors Mk 7. This may have been a planned armament. (Photo by Allan C Green via State Library of Victoria)

HMAS *Hobart* survived the war. She is shown off Wellington, New Zealand, on 22 July 1942, before the Guadalcanal operation. She operated in the Red Sea during the reconquest of British East Africa and Ethiopia, then refitted at Colombo, 19 October through 19 November 1940 *en route* home. During a further refit in Sydney (June 1941) her catapult and aircraft were landed, so that she could receive further anti-aircraft weapons during her coming Mediterranean deployment. Thus she received the cross-decked quadruple pompom and some Breda 20mm guns when she reached the Mediterranean to relieve HMAS *Perth* (she berthed alongside at Alexandria on 16 July 1941 for cross-decking). Two photographs taken at about this time (she has a dhow alongside, so they were taken somewhere like Alexandria) show the pompom atop the former catapult platform, and a platform for a light anti-aircraft gun on the fore side of her bridge structure. According to a file in the RAN Historical Branch, she and *Perth* were both given four single 0.303in Vickers guns late in 1941, two forward (side by side on the platform forward of the bridge) and two aft, all of which were to be replaced by Oerlikons, one for one (*Perth* was lost before receiving the four Oerlikons allocated to her). Other early war modifications were degaussing, protective plating (total about 35 tons) for the bridge and vital communications, and machine gun shields. Ships also had additional depth charge stowage. The ship still had quadruple 0.5in machine guns on the saluting gun deck roughly abeam her HA director forward, and on the centre-line abaft and below her after searchlight. At this time she had no radar, as her foretop was still occupied by her circular DF loop. She operated in the Mediterranean between July and December 1941. The catapult support structure itself survived until the major post-damage refit (August 1943 – December 1944). *Hobart* was ordered back to Australia when Japan entered the Second World War, arriving in Fremantle on 11 January 1942, having experienced a Japanese air raid the previous 3 January at Singapore. Her captain disobeyed an order to land the pompom, and she retained it as she steamed east. *Hobart* was refitted in Australia some time between early March and late April 1942, at which time a second quadruple pompom (shown in the July photograph here) was mounted on her quarterdeck. The origin of this weapon is unclear, since pompoms were in short supply at the time. Two were shipped from the UK to refit *Canberra*, but they arrived too late for that ship (one went to HMAS *Platypus*, and the Admiralty asked in 1943 that both be returned). *Canberra* was not lost until August 1942, but this July photograph seems to show both pompoms in place. On 22 April 1942 the admiral commanding Anzac Squadron wrote to the Australian Chief of Naval Board that 'it is considered that the 4-barrelled pompom fitted on the quarter-deck of *Hobart* together with the six Oerlikons will provide adequate close-range fire power' which can be read either to mean that the ship had only the pompom there or, more likely, that the *addition* of these weapons would finally provide adequate close-range firepower. This refit gave the ship two Oerlikons abreast on the platforms forward of her bridge and four more aft, near the quadruple 0.5in machine gun abaft the searchlight (the 0.5in guns were retained). The photograph shows three splinter-shielded positions around the base of the superstructure carrying her after searchlight and her after quadruple 0.5in machine gun; the middle and forward ones clearly show Oerlikons (what appears to be a third, after, one on the starboard side is apparently a control position). She was given 175 tons of pig iron ballast, increasing her estimated deep load displacement to 9,908 tons (compared to 9,130 tons as delivered). At this time she seems to have been fitted with a Yagi-type radar, presumably the New Zealand air-warning type, at her foretop. The ship suffered slight damage in a collision on 6 September 1942 with a US ship. She was repaired and again refitted, this time at Devonport (New Zealand), and after the completion of this refit late in October she spent two weeks in Sydney. According to a note dated 6 October 1942, proposed upgrades included a Type 271 (sic) surface-search radar and replacement of the three quadruple 0.5in machine guns by Oerlikons, for a total of nine. The two navigational rangefinders were removed (they were not needed once the ship had a good surface-search radar). *Hobart* was fitted with a Type 273 surface-warning set in a 'lantern' on the centerline midway between her 4in mounts (*Leander*, serving with the Royal New Zealand Navy, received a similar 'lantern' at about the same time). Either in Devonport or (more likely) in Sydney she was fitted with a US-supplied SC air-search radar on her foremast (using a square reflector, with a characteristic X-shaped brace behind it, rather than the rectangular reflector of the later SC-2 on board many US destroyers) and a US-supplied Mk 3 surface gunnery radar (antenna on the lower part of her DCT). Other Australian warships received these radars, which suggests that they were fitted in Sydney rather than in Devonport. For example, HMAS *Australia* had Mk 3. Many smaller Australian ships later had SC antennas, examples being HMAS *Warrego* and the 'River' class frigates. The SC antenna is visible in photographs taken after the ship was torpedoed in July 1943, and the Mk 3 is visible in silhouette in a photograph taken just before she was torpedoed. During this refit the Oerlikons were rearranged. Two of the six were moved from the after searchlight structure onto a zareba atop 'X' turret. Instead of being mounted on the saluting gun deck, as the forward 0.5in guns had been, it seems that two new Oerlikons were placed atop 'B' turret. Two unfortunately unclear 8 June 1943 photographs appear to show some kind of gun in the former saluting deck 0.5in positions, but the usual Oerlikon shield is not visible. Perhaps they were Bredas left over from the Mediterranean, and never carried on official lists. Atop both turrets the Oerlikons were placed side by side. A US Navy recognition silhouette produced at this time clearly shows the turret-top zarebas and the guns on the fore end of the bridge, but unfortunately it is blacked in, hence does not show the structure near the searchlight platform. Both pompoms survived the torpedoing, but during the big 1943–4 modernisation they were replaced by fresh RPC pompoms (with pompom directors) shipped from the United Kingdom.

Small-calibre anti-aircraft weapons presented greater problems, because as aircraft speed increased the available firing time was drastically reduced. The higher the bomber, the shorter the available time, and it was unlikely that level bombers would attack from below 4,000ft (9 seconds firing time). The range at which a torpedo bomber would likely launch its weapon determined how long it could be held under fire. Outside 1,200yds (17 seconds firing time) the target ship could evade (1,650yds for a 40kt torpedo). Future torpedoes might be dropped at 4,000yds, in which case small-calibre guns would be altogether useless. The Mk M pompom was not quite enough, as last-ditch defence envisaged required not its 720 rounds per minute, but at least 1,250. Because there was no associated predictor form of fire control, the gun was fired over simple sights and had to overcome range errors. Performance could be improved, and keep-out range increased (from the currently estimated 1,500yds) by increasing muzzle velocity (ideally, from 1,920ft/sec to 2,500ft/sec) and by streamlining the bullets. The Committee doubted that the quadruple cruiser pompom would meet requirements, preferring some smaller-calibre alternative. It might be time to begin a shift towards a smaller-calibre weapon which could be fired more rapidly. The result was an abortive attempt to develop a multiple 0.661in gun, which featured in many designs in the late 1930s, but never entered service.

Although the value of large-calibre (essentially LA, whatever their maximum elevation) guns for long-range anti-aircraft fire was questionable, the Committee pointed out that barrages by these weapons – firing at preset ranges – could usefully back up the short-range machine guns and machine cannon. Guns would fire at a spread of preset ranges. For example, a heavy cruiser would fire salvoes of at least four rounds. Cruisers armed with 6in guns presented more possibilities, because their shells could be man-handled, hence the guns could quickly switch between anti-ship and anti-aircraft shells. Cruisers prior to the *Caledon* class had 20° elevation; the later First World War cruisers had 30°, which already offered useful height performance. Only their age and limited space and weight made it unwise to provide them with a modern HACS to engage aircraft at long range. The *Leander*s had space reserved for a 6in HACP.[45] The 6in guns could also be used for close-range barrage fire.

In August 1952 *Hobart* was taken in hand for conversion to a training cruiser in Newcastle, New South Wales, but the project was cancelled and she was towed back to Sydney incomplete in 1955. She was sold for scrap in 1962; here she awaits the tow to Japan. The most visible modification was replacement of the tripod foremast with a lattice mast like that on board modernised British cruisers. Reportedly she would have been fitted with two sided Mk 6 directors in place of her British-type HA control towers (she would have been training men who would use those directors on board newer Australian ships). Reportedly the directors made for *Hobart* were installed instead on the Indian cruiser *Mysore*. (RAN Historical Branch)

Cruisers probably could not accommodate the Committee's favoured 5.1in gun, so for the future 10,000-ton cruiser the Committee proposed six twin 4.7in on each side (four 5.1in as an alternative). For smaller cruisers, there was no alternative to hand-worked 4in in the lightest possible mountings. Ideally HA mountings should be bunched together to limit complication in control. All 6in guns should be usable against aircraft at long range, and two HA/LA control positions placed on the centreline, plus one or two 4in HADT, depending on gun arrangement.[46] If size allowed, the ship should also have two Mk M pompoms and four quadruple 0.5in machine guns.

Like the 1921 committee, the 1931–2 committee devoted considerable effort to long-range anti-aircraft fire control. It proposed two alternative forms of rate-measuring (tachymetric) control. The simpler one measured target angular velocity, vertical and horizontal, directly, and fed those values into a computer. The more complex 'Flyplane' adopted after the Second World War translated data into the plane in which the target was flying (it had just been proposed by an assistant to DNO). Both methods were more complicated than the existing HACS, in which an officer estimated target speed directly as an input into the fire-control computer. Both systems entailed complex calculations, not least to translate between data taken relative to a rolling, pitching ship and the aircraft following a more or less consistent path through the sky. None of these systems could be put into service very quickly. In addition to proposing new methods of calculation, the committee pressed for introduction of longer-base anti-aircraft rangefinders (15ft rather than 12ft).

Firing at bombers flying level, ships whose guns could elevate even to 40° might contribute significantly to the fleet's defence. Thus the committee strongly recommended development of a dual-purpose fire-control system for destroyers and, by extension, for cruisers. It estimated that the combination of mechanical computing and a mechanically-set time fuse (set on the basis of computed aircraft motion) offered a threefold improvement in anti-aircraft effectiveness. Overall, the committee considered improved long-range defence more important than short-range, because long-range fire provided the fleet with collective defence. However, it accepted that a proportion of enemy aircraft would survive long-range fire, and that pompoms were an essential back-up. Smaller-calibre guns (quadruple 0.5in and Lewis guns) could not shoot down torpedo bombers before they attacked, but they could deal with strafers and with aircraft which had to get closer to bomb, such as those dropping the new 'B' bombs (buoyant bombs which could rise to explode under a ship's keel). They could also destroy aircraft and thus prevent them from reloading and reattacking. The *Southampton*s were the first ships affected by the new recommendations.

The Triple Turret

Once the London Naval Treaty had been ratified, the question was how to fit more guns into a ship of limited displacement. Reviewing attempts to develop a 4,500-ton design on 9 August 1930, Controller wrote that the greatest gain possible in the next few years would be a 7,000-ton ship (a follow-on *Leander*) with better protection. Triple turrets might provide that weight, so Controller requested a turret design.[47] The gunnery school (HMS *Excellent*) approved the idea of a triple 6in mount in 1932. Triple mounts featured in 1932 discussions of an Improved *Leander*, which became the *Amphion* class, but the technology was not yet mature enough to figure in the design adopted. Vickers-Armstrong drawings of a triple 6in turret with cordite hand-up (rather than full hoist) were circulated in April 1933. Like the twin, this was a short-trunk design with lobby or handing room not too far below the gun mounting. DNO estimated a revolving weight of 130 tons, compared to 93 tons for a twin mounting.

Calculations for a sketch design of an *Arethusa* with a twin turret in 'A' and triple mounts in 'B' and 'X' positions (dated 18 March 1932) are in the Constructor's Notebook for W G John (Vol 6). Belt armour was increased from 2½in to 4in. Revolving weight of the mounting was taken as 150 tons. John tried various lengths; standard displacement was about 6,200 to 6,500 tons. For example, a ship 520ft (pp) 540ft (wl) x 55ft x 30½ft x 15ft would displace 6,550 tons standard and 8,470 tons deep. A summary sheet dated 31 March 1932 shows two alternatives, both using the 64,000shp *Arethusa* class powerplant, but with endurance of 5,500nm at 15kts (as in *Arethusa*) or 7,000nm at 15kts; standard displacement would be, respectively, 6,200 tons and 6,550 tons (deep displacement 7,720 tons and 8,470 tons). Dimensions would be, respectively, 520ft (wl) x 53ft x 29ft 6in (depth) x 15ft and 540ft (520ft pp) x 55ft x 30ft 6in x 15ft. Speeds would have been 31.75kts at standard displacement and 30.25kts deep; and 32kts and 30.5kts for the larger ship, whose length would have more than compensated for her extra displacement.

Compared to foreign contemporaries, the one unusual feature of the Mk XXII triple mount was that the centre gun was set further back than the other two. This was described as a way of saving space.[48] Maximum elevation was reduced from the 60° of the twin mount to 45°, presumably indicating a loss of interest in anti-aircraft fire (no heavy-calibre anti-aircraft fire-control system had been introduced). The Mk XXIII, introduced in the *Belfast* class, was a long-trunk mounting, i.e. it eliminated the ammunition lobby (and its personnel). It equipped the *Fiji* class and its modified versions.

Surprises

Although British proposals at the London Conference included allowing the US Navy 10,000-ton cruisers with 6in guns, the Admiralty seems not to have taken such ships seriously, since *Leander* could stand up to any 6in gun cruiser, and she was a far more economical proposition. Surely foreign navies would see this logic. Unfortunately that was not the case. Both the US Navy and more ominously the Japanese Imperial Navy ordered large 6in gun cruisers, of the *Brooklyn* and *Mogami* classes.[49] Controller added that the newest French cruisers, most clearly comparable to the *Leander*s, displaced 7,500 tons, and that the additional 500 tons would have been most useful. It would be better to abandon hopes of getting fourteen ships within the Treaty limit in order not to accept gross inferiority. DTD favoured reducing the number to thirteen to ensure that ships more fully met requirements, but did not want to imitate the new foreign ships. Director of Plans pointed out that 6in cruisers were under the same 10,000-ton limit as 8in ships. It might therefore be possible to postpone building large cruisers until the 1934/5 programme in hopes that a competition in cruiser size might be avoided and some agreement reached at the next arms-control conference, expected in 1935. DCNS wanted to wait until late in 1932 to decide the 1933/4 programme, because an arms-control conference was beginning at Geneva in 1932. It might be possible to point out at that conference that cruiser size, hence cost, was being pushed up, and perhaps some action might be taken to cap cruiser size. ACNS agreed. Fourth Sea Lord wanted to concentrate on cruiser quality rather than quantity; numbers could be made up in some other way. Second Sea Lord considered British cruisers deficient in ahead fire, but it would be unwise to follow the foreign navies towards larger, more expensive ships.

A meeting of the Sea Lords and ACNS concluded that it was essential that British cruisers be able to stand up to foreign 6in fire. Protection had to be increased despite the cost in displacement, and gun power had to be increased also, as Second Sea Lord had pointed out. However, it was also important to avoid precipitate action either by an announcement at the Geneva conference or by altering British policy during the conference, which might cause foreign powers to continue building their large treaty-busting cruisers. The agreed policy was therefore to push at Geneva to limit cruisers to 7,000 tons, in the expectation that an 8,000-ton limit could be achieved (7,000 tons would be better from a British point of view). Controller should immediately begin sketch designs for a 7,500-ton improved *Leander* with improved protection and either ten guns in four mounts ('B' and 'Y' turrets triples) or nine guns in three triple mounts, two forward. Work should be pressed so that a ship of this type could be included in the 1932/3 programme. Controller should instruct DNO to proceed with the design of a triple 6in turret. First Lord added that if the Board decided to abandon the fourteen-cruiser plan it would have to provide the public with a convincing explanation.

All of this mattered because the international situation was worsening. Japan invaded Manchuria in September 1931 and attacked Shanghai early the next year. The Japanese made unpleasantly clear their determination to expel Westerners from Asia, which directly threatened the vital British economic interest in China. There were already rumblings in Europe, which a Board memorandum said reflected 'pre-war' thinking. On 15 July 1931, before the invasion of Manchuria, the Cabinet decided that the Ten Year Rule should be reviewed the next year. By March 1932, it was clear that the Geneva disarmament conference was deadlocked, and the Cabinet accepted Chiefs of Staff and Committee of Imperial Defence (CID) papers recommending that the rule be abandoned. The Ministerial Disarmament Committee (set up to develop a British position for the 1932 conference) was converted into the Defence Requirements Committee (DRC) which reported on what was most urgently needed to make up deficiencies largely traceable to the Ten Year Rule – which was scrapped without any formal statement to that effect. With Hitler openly rearming Germany, in 1934 Sir Robert Vansittart of the Foreign Office sought to reverse previous thinking by declaring 1939, five years away, 'the year of maximum danger'. To make matters worse, in August 1934 the Japanese government announced that it was renouncing the Washington Treaty, effective in two years. Although Japanese delegates attended the 1935 London conference intended to replace the 1930 treaty, their demands could not be met.

Britain already faced an increasingly aggressive Mussolini in the Mediterranean. That sea was an essential Empire line of communication to the East, and the Far Eastern War Plan assumed that the Mediterranean Fleet would be sent East upon the outbreak of war. Any threat in the Mediterranean was therefore to be taken very seriously. Italy attacked Abyssinia in 1935, and the British announced economic sanctions. For a time in 1935–6 it seemed the two countries would go to war. The crisis highlighted inadequate fleet anti-aircraft armament, and the need to counter the large number of Italian motor torpedo boats.

It became clear that Hitler was determined to build a navy grossly violating the terms of the Versailles Treaty. All the British could do was to negotiate a treaty which somewhat limited what the Germans might build. As this agreement was outside the scope of the 1930 treaty, it enraged the French, who felt limited by their own treaty obligations (under the Washington Treaty and under some agreements made after the 1930 treaty). The combination of Japanese withdrawal from the treaty system and the deal the British felt compelled to make with the Germans and also Italian aggressiveness in the Mediterranean all showed that the assumptions made in 1930 were no longer at all realistic.

On 15 July 1936 the British government invoked the escalator clause of the 1930 treaty in time to stop scrapping overage cruisers built during the First World War, and to provide additional tonnage for new construction. By this time the United Kingdom had already agreed to the new London Naval Treaty, which did not include any limits on overall cruiser tonnage, just qualitative limits. There was clearly little point in such scrapping. The British declaration was widely described simply as a decision to retain 150,000 tons of overage destroyers, but it covered 'C' and 'D' class cruisers as well.

The *Southampton* class

Given foreign construction of far more powerful 6in cruisers, Director of Plans (H R Moore) considered it very desirable to keep building Modified *Leanders* in the 1933/4 programme. The official statement of plans to buy 'probably one *Leander* and three *Arethusas*', was vague enough to allow for the preferable pair of Modified *Leanders* and two *Arethusas*. Unfortunately their total of 25,400 tons was 2,330 tons over the allowance (the alternative with three *Arethusas* would be only 530 tons over, which could be dealt with). There was no real hope of absorbing 2,330 tons. Any weight saved on the *Arethusas* would probably best go into improving belt protection on the 1933 ships. This programme was sent to the Cabinet.[50]

The Royal Navy could not continue to build small cruisers which would be outclassed by the new foreign ships. As First Sea Lord (Admiral Chatfield) wrote, 'We ought to build ships for the defence of our trade similar to those of Japan, but to build the number of ships we require of the size being adopted by other Powers would ... be financially ruinous.' Moreover, under the treaty the four *Hawkins* class, which were well-adapted to trade protection, had to be discarded. Given the large number of existing (if old) small cruisers suited to fleet work, Chatfield decided to defer such construction for a year while building the sort of large cruisers the Americans and, more importantly, the Japanese, were building.

Some time in mid-1933 DNC was asked to stretch the ten-gun ship to four triple turrets, trading speed (reduced to 30kts) for firepower.[51] Apparently a satisfactory combination could be achieved on 7,800 tons, which seems to have been the maximum available. It is not clear how that figure was reached, because the available tonnage would have supported an 8,600-tonner. First Sea Lord proposed to switch from the two *Leanders* and two *Arethusas* planned to three such ships. The 30kt speed recalled that chosen for the *Surrey* class; Chatfield had been Controller when those ships were designed.[52]

The Staff Requirement for the 1933 cruiser pointed out that the ship was conceived primarily for trade protection.[53] It called for twelve 6in guns in triple turrets with elevation for maximum LA range (40–45°) controlled by a DCT forward and by an after control position as in *Amphion*, with no provision for divided control. The long-range anti-aircraft battery would be twin 4in guns in hand-worked weather deck mountings (in the design stage) arranged so that two guns could fire on any bearing, and four guns on the widest possible arc, e.g. 25–165° each side. That could mean two sided mountings plus one on the centreline. All the HA mountings should be grouped together to avoid displacement errors and complications in ammunition supply and would be controlled by a single HACS forward. The close-range AA battery would be three quadruple 0.5in machine guns arranged as in *Leander*. For trade protection a torpedo armament was not essential; but the ships might have to work with the fleet, so torpedoes were desirable if they did not interfere with the main trade-protection function. If torpedo tubes were

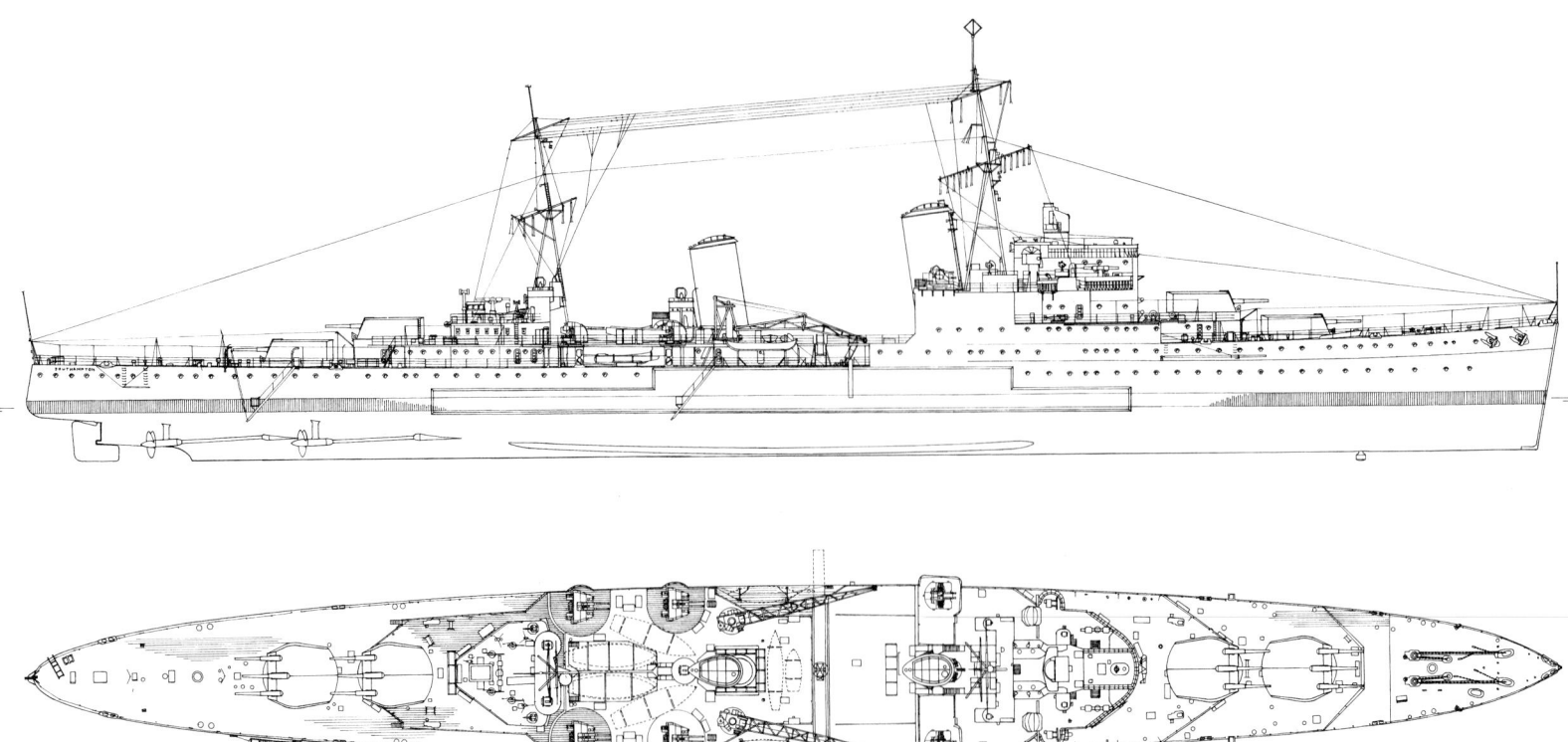

Southampton as fitted on completion, March 1937. Note the DF loop on her foremast and the Asdic dome (a feature of many British cruisers) under her bow. The plating at the sides of the forecastle deck was extended aft about 15ft after completion. Covered crew shelters with blast shielding were added to *Southampton* and *Newcastle* after completion; the others were completed with the structures. *Sheffield*, *Glasgow* and *Birmingham* all had a third HA director (Mk III) on the centreline forward of the after control station, and also more rounded bridge fronts. Dotted lines in the plan view indicate the extension of the telescoping fixed athwartships catapult; note that no deckrails for the launching trolley were fitted and that the trolley itself rotated instead of having two rotating sections of the catapult railing, as on the later *Fiji* class. The 36ft motor and sailing pinnace and the 35ft motor pinnace (fast type) were stowed on skids above the two 35ft motor boats, which were accessed through two large openings in the forecastle deck by the two aircraft and boat cranes. (A D Baker III)

not fitted to the ships as built, they should have provision for quick installation of one triple mounting on each side, firing 21in torpedoes capable of 10,500yds at 35kts or 14,000yds at 30kts. As in *Amphion*, the Staff Requirement called for sufficient complement to operate the entire armament simultaneously 'but a lesser standard of comfort must be accepted'.

The ship should have a heavy-type catapult capable of launching the heaviest aircraft (the TSR, Swordfish), of which it should stow and operate a minimum of five (two hangared). The TSR offered a cruiser operating on trade routes the ability to sink a raider at considerable range, the ship being supplied with a torpedo for each aircraft, as well as a considerable bomb load.[54] There was also still interest in the use of cruiser aircraft to strike and fix an elusive enemy. DNAD proposed a fixed athwartships catapult instead of the usual rotating catapult, whose upper surface would be about 8ft above the upper deck. The hangar floors would be at the same level, with a deck or platform at this level on either side between the after end of the hangar and the catapult. Aircraft could simply be wheeled onto the catapult. The additional 50 tons of structure could be compensated for by reducing 6in and 4in stowage in standard condition by fifty rounds per gun (space would suffice for the full 200 rounds). DNAD's new catapult had not yet been designed, however, and the dimensions and weight of the TSR had not yet been fixed.

The ships should be protected against the *Leander*-class 6in shells. Magazines and shell rooms should be immune at all ranges up to 21,000yds, and machinery at ranges up to 16,000yds. Turret trunks should be adequately protected. Other protection should follow that of *Leander*. Requirements for gas protection were laid out. Speed should be 30kts at standard displacement, and endurance 7,000nm at 16kts (as in *Leander*). Communications would be as in *Leander*.

DNC ordered alternative sketch designs KVIII through KX, which were presented to the Board in September 1933.[55] All had the previous combination of magazine box protection and belt and deck over their machinery. A comparison with *Leander* suggested that the armoured part of the ship should be considerably extended to preserve stability in riddled condition (magazine box protection remained). All of these designs had the new alternating boiler and engine room arrangement.[56] Length was set mainly by aircraft arrangements and the need for accommodation. KVIII was lengthened (600ft x 62ft x 16½ft, 8,625 tons). Given greater length, it might be possible to achieve higher speed. Thus KIX (600ft x 61ft x 16½ft, 8,740 tons) had 65,000shp rather than 50,000shp and could make 31.75kts.

KX (610ft x 62ft x 16½ft) was conceived to carry five aircraft and to make 32kts (on 70,000shp), and to displace if possible 9,600 tons. The belt armour was extended to cover the bases of the barbettes. It was 5in amidships, tapering to 3in at the ends, with 3in bulkheads angled around the end barbettes. The entire belt was covered by a deck (30lb armour over 20lb structure, total 1¼in). As before, fitting the complement into a relatively small hull was difficult, but this time the Permanent Complement Committee noted 'with satisfaction' that the complement of this ship (unlike certain former classes) was receiving due consideration as an integral part of the design. At this early stage, DNC expected less difficulty with officers' accommodation than with that of the men.[57] KIX (presented as Design D) was adopted. A four-

Above and below: The *Southampton*s marked a dramatic change in British cruiser design, to counter large foreign types such as the Japanese *Mogami*s. HMS *Gloucester* is shown, newly completed, in 1939. Earlier ships in the class could be distinguished by the absence of a third HA director, on the after superstructure just forward of the after 6in director. *Southampton* and *Sheffield*, for example, were completed without any directors aft.

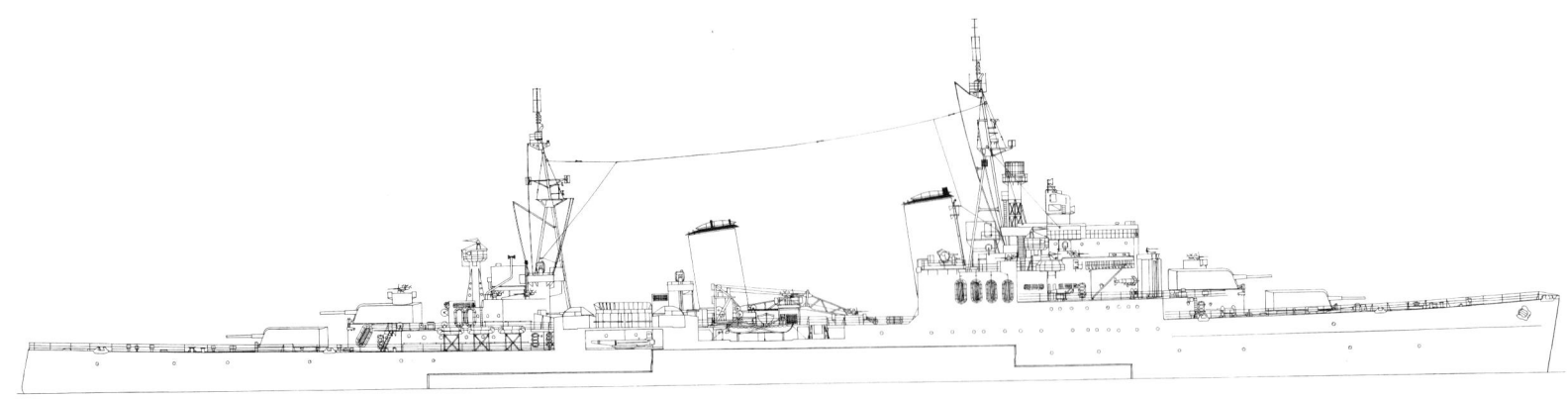

Below: Photographed in November 1942, *Glasgow* displays typical wartime modifications: zarebas atop 'B' and 'X' turrets for Oerlikons and Type 273 surface-search radar atop her bridge (installed in August 1942). During her first wartime refit (Liverpool, 14 May – 7 July 1940) *Glasgow* was fitted with UP mountings atop 'B' and 'X' turrets and with Type 286 radar. She was torpedoed at Suda Bay on 3 December 1940 and travelled nearly around the world for repairs: first to Alexandria for temporary repairs, which enabled her to reach Singapore for more temporary repairs (29 June – 29 August 1941, including removal of the UP launchers), then to New York Navy Yard for permanent repairs and a refit (6 May – 25 August 1942). Her Type 286 radar and her quadruple 0.5in machine guns were removed, and nine single Oerlikons, which are visible here, were fitted. She also received standard large-cruiser radars: Types 281, 282, 284, 285 and 273. Further work was done at Portsmouth (4 September – 12 October 1942) and on the Clyde (6–23 December 1942); during the latter availability she exchanged five single Oerlikons for eight twins. Two more single Oerlikons were added during a Devonport refit the following year (26 August – 6 October 1943). She was damaged by German shellfire when bombarding Cherbourg (25 June 1944), and was further refitted and repaired at Portsmouth (June 1944 – May 1945). By this time ships had no further reserve topweight to allow installation of more close-range anti-aircraft weapons, so 'X' turret and the ship's aircraft arrangements (catapult and one crane) were removed. Now there was space and weight for two more quadruple pompoms and for four single power-operated pompoms. Two twin and four single Oerlikons (including those on turret tops) were landed. The ship's radars were modernised: she was fitted with the single-antenna Type 281B, which cleared the foremast for the Type 293 target-indication radar. Her Type 284 fire-control radar was replaced by a centimetric Type 274.

Above: Sheffield is shown as she appeared from late July 1943 to the beginning of her early 1944 refit, i.e., as when she participated in the Battle of North Cape against the German *Scharnhorst*. Her aircraft arrangements had been removed, four single Oerlikons (two to port, two to starboard) being mounted on the former flight (catapult) deck. The rest positions of the cranes were moved to clear these guns (they were originally at the deck edge, and had to be moved inboard, almost touching on the centreline. That gave just enough clearance. She also had ten other Oerlikons. (Alan Raven)

shaft 72,000shp plant replaced the previous two-shaft one.

A Legend constructed in October 1933 showed a displacement of 8,835 tons (standard). Before submission to the Board on 25 October waterline length was reduced from 600ft to 584ft, so that the ship could dock more widely.[59] Because shaft horsepower was not increased, speed at standard displacement fell from 32kts to 31.75kts. The length of protected waterline was extended at the cost of reduced thickness.[59] Two of the three 0.5in machine guns were replaced by two quadruple pompoms (1,400 rounds per barrel), the remaining machine gun being mounted on the after control positions.[60]

Controller submitted this 9,000-tonner to the Board on 25 October. Controller wanted the pompoms relocated to better positions for dealing with dive bombers. The most obvious change in the design was a reduction from five to three aircraft, suggested by DNC because there was not enough space.[61] Controller doubted that the fixed catapult envisaged by DNAD was worth the extra 50 tons, but First Sea Lord

THE 1930 LONDON TREATY AND ITS CRUISERS

(Chatfield) overruled him; aircraft were very much what a cruiser needed. It was always possible, moreover, that larger amounts would be saved during construction. DNAD's fixed catapult characterised later British cruisers. The ship could carry three TSR, one on the catapult and one in each of the two hangars, with their three torpedoes. The Board approved the 9,050-ton sketch design on 9 November 1933. By this time the ships were called the 'M' class, the first ship being *Minotaur*. With their greater tonnage, there was no longer room for three of them in the 1933/4 programme, so an *Arethusa* (*Penelope*) was substituted for the third ship. Later HMS *Minotaur* was renamed *Southampton*, that becoming the class name.

DNC submitted the Legend and drawings on 28 February 1934.[62] To better support the heavier belt and to stiffen the hull, the ship reverted to 4ft rather than 6ft frame spacing over the main part of hull, at some expense in weight.[63] Turrets were lowered by sinking their roller paths below the weather deck, the ring bulkheads being structurally independent of decks near them. The trawler bow of the *Leander*s was retained. The HADT was on the bridge, immediately abaft the DCT, controlling three twin 4in between the after funnel and the mainmast (one on either side on forecastle deck level, one centreline on superstructure level). The pompoms were on either side atop the hangars, with directors on either side of the foremast. DTM's proposal for quadruple torpedo tubes was rejected. Space sufficed for 790 officers and men (as a wartime flagship). The hangars created a blind arc for the waist 4in guns on forward bearings at low angles, so they were sponsoned out. Waterline length remained at 584ft, but shaft horsepower increased to 75,000, to give 32kts at standard displacement and 30.5kts at deep load. Main machinery was arranged as in *Amphion* and *Arethusa*. Endurance was as in *Leander*, 7,000nm at 16kts. The Legend and drawings were approved on 8 March 1934. Further work showed that 200 tons more fuel could be carried at the cost of 5 tons of standard displacement, giving an endurance of 8,900nm at 16kts. The Board approved this change on 15 March 1934.

In September 1934 D of TD suggested withdrawing the 1932 orders to simplify cruiser fire control, as they were inappropriate for a ship the size of the new 9,000-ton cruiser. The 10,000-ton cruisers were already getting a second HACS each, so that they could engage aircraft on both sides simultaneously, so surely a 9,000-ton cruiser needed the same capability. The single centreline HADT was replaced by two, one on each side of the bridge atop the hangars, the bridge being narrowed to clear arcs for them.[64] Hold space was found for a second calculating position (computer space). *Leander* had experienced DCT interference from the cordite smoke from the fore turrets, so the DCT in the new ship was raised by raising the upper bridge so that its screens were just above the tops of the HADTs on either side, with a clear view all around except right aft, and overhead. The roof atop the compass platform was omitted. It proved possible to move the HADTs slightly outboard so that wind deflectors could be fitted around the entire compass platform. DNC proposed raising the forefunnel 10ft to minimise smoke interference with the raised bridge. The searchlights could move to positions

HMS *Glasgow* off New York Navy Yard on 11 August 1942.

alongside the funnel.

At the same time a fourth twin 4in was added. That had been considered impossible (hence the centreline mount), but space was freed by eliminating the torpedo parting (assembly) space. The magazine could not be enlarged, so only 150 rounds could be provided for each 4in gun, despite recent policy favouring more rather than fewer rounds per anti-aircraft gun. Adoption of bottle-rack stowage solved the problem. There was interest in replacing the 4in hoists with duplex hoists, which could more easily supply two pairs of guns on each side (the existing hoists provided thirty-six rounds per minute, sufficient for one pair of twin mounts). Adding the fourth twin 4in reduced boat stowage on the after superstructure deck, but some boats were stowed instead on the upper deck in a well in the superstructure deck.

D of TD wanted a second LA director or an HA/LA director aft (although there was not enough space for a third HA calculating position below decks), the single existing LA director being wooded on bearings from about 136 Red to about 135 to 156 Green. The two 1933 ships received the fixed control position of the previous classes, but 1934 ships had an HA/LA director. The centreline position for the one quadruple 0.5in machine gun was eliminated when the mainmast was moved to clear arcs for the new after director; it was replaced by two such guns, one on either side. The existing after 36in searchlight and two displaced from forward were grouped around the mainmast, one being immediately abaft the after funnel. The central part of the after searchlight platform became an alternative conning position for the ship.

A review of the building drawings suggested to DNC that the ships would come out at 8,947 tons rather than the planned 9,060 tons. His suggested use of the small difference for more armour was approved by the Board on 4 October 1934.[65] New drawings were approved on 9 May 1935, and new building drawings issued accordingly in July. Displacement increased to 9,060 tons. Slightly later the Board decided to rake the masts and funnels at a slope of 1:6.6.[66] This seems to have been considered preferable to moving the forefunnel abaft the bridge. The bridge front was curved to improve air flow around it.

Designed for 8,947 tons standard, *Southampton* was competed at 9,083 tons; approved additions between design and completion were small additions to armament and to protection, bridge modifications, additions to hangar structure, ring main modifications, etc. By late 1941 approved additions without compensation amounted to 324 tons. They included splinter protection, protection to vital communications, additional 4in bottle-rack stowage, permanent degaussing, Oerlikons, SA gear, radar, cutters as sea boats, hangar spraying, additional bridge stiffening etc. Proposals resisted or compensated for included Type 271 radar and steam heating for the Arctic.

Repeat *Southampton*s

Under the London Naval Treaty of 1930, additional cruisers could be laid down in 1934 to replace the fifteen ships which would become overage on 31 December 1937 (67,350 tons) and in 1935 the four which would become overage on 31 December 1938 (19,000 tons). No further tonnage would become available for replacement in 1936. However, the Royal Navy was reluctant to keep scrapping its overage cruisers. Thus the 1934/5 programme amounted to three more big cruisers (*Glasgow*, *Sheffield* and *Birmingham*) and the last *Arethusa*, totalling about 32,500 tons. The only important change was continuously flared bow lines in *Birmingham*, apparently for comparison with the trawler bows in the others.[67]

By May 1935 it had been decided to order three repeat *Southampton*s (*Gloucester* class) in the 1935/6 programme. Initially the only planned change was to accept E-in-C's proposed higher rate of forcing, giving 82,500shp at a cost of 70 tons. In November, DNO proposed additional protection to the turrets, to provide a 4in face, 2in roof and sides, and 1½in floor, at a cost of 35 tons per turret; previously the gunhouses had been 1in with ½in floors. The additional armour would help protect the 6in magazine from bombs. The magazine crown was open for the ammunition hoists directly under the turret. To compensate for the added topweight, the ships needed 8in more beam. DNO (Fraser) suggested that protection to the ammunition supply should also be increased. Controller (Henderson) approved 2in over the ammunition supply below the turret, pointing out that the hoists from magazine to handing room were not protected and that they too should receive 2in protection. The weight involved was small, and would not require more beam (which would have required that ships' lines be relofted).

DNO pointed out that in the past boilers were protected not only by the armour above them but also by the unarmoured decks higher in the ship. In these ships (and, for that matter, in the *Amphion*s and *Arethusa*s) the boilers were so tall that the decks over them were the upper decks. He suggested adding another quarter-inch to the 1¼in deck already present, balancing the extra weight by removing the torpedo tubes. ACNS rejected the idea. Torpedo tubes had been removed in the *Kent*s, and would probably be removed when the *London*s were modernised, but other cruisers should retain them; torpedoes were a potent weapon for night and low visibility. First Sea Lord agreed. The added weight (195 tons in all) and slight additional increase in beam were accepted. The modified design received the Board Stamp on 13 February 1936.

In service the thicker deck caused an unexpected problem. In rough weather with the Home Fleet in 1940, the abrupt change in thickness at the ends of the thick upper deck caused structural problems due to stress concentration in HMS *Manchester*. In the *Gloucester*s the change was abaft the forecastle rather than in a space covered by the forecastle (as in *Belfast*s and *Fiji*s), which was a strength deck. The cure, to be applied to future classes, was to make the transition between thicker and thinner decks smoother – at a significant cost in topweight. In 1940 the ships involved, the *Fiji*s, were already so badly overweight that DNC was fighting every proposed addition, threatening that the after two 4in mountings, catapult and aircraft would have to be surrendered if extensive splinter protection was added. Fortunately the *Dido* class structural design, similar to that of the *Arethusa*s, showed no such problems.

Belfast and *Edinburgh*

There was plenty of weight left for new cruisers, particularly after the United Kingdom invoked the escalator clause in the 1930 treaty. The 1936/7 programme included seven cruisers, two follow-ons to the *Southampton*s and five *Dido*s (see below). These were the last ships to be laid down before the 1936 London Naval Treaty, which restricted new cruisers to 8,000 tons, came into force on 31 December 1936. The Royal Navy was therefore free to use all 10,000 tons available in HMS *Belfast* and *Edinburgh*.

Initially the basis of the design was a new quadruple 6in turret. It was not expected to be much heavier than the triple, but additional protection would be needed. Later, when the quadruple design had been worked out, it was expected to weigh 188 tons, with 3in shield and 2in roof. The mount had a long trunk, i.e. no ammunition lobby to isolate magazines from turret (hence fewer personnel). On 8 October 1935 DNC asked for a cost estimate for a 10,000-ton cruiser with four such turrets, speed 32.5kts, and other features similar to those of the

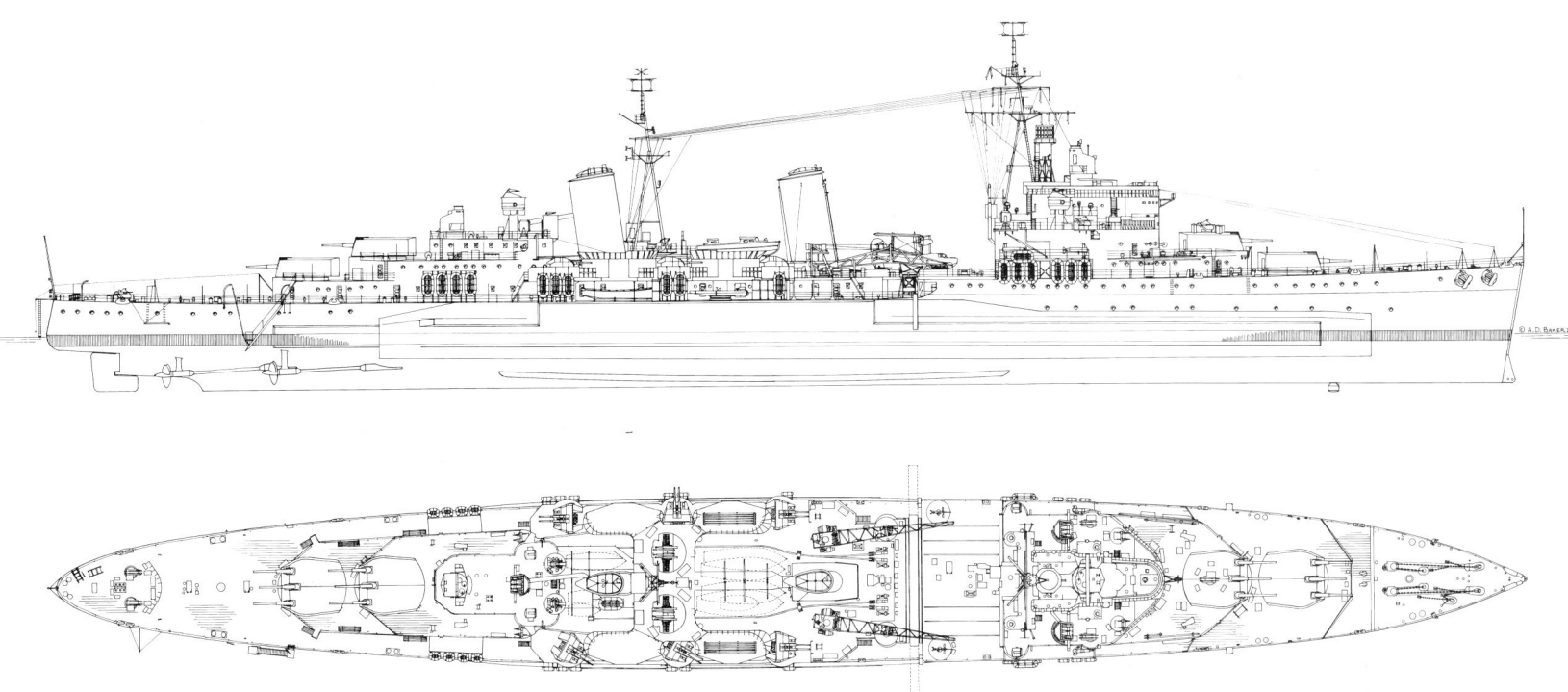

Belfast as fitted, November 1942, after her major reconstruction, when she was bulged. She had been mined in the Firth of Forth on 21 November 1939. The post-reconstruction beam reported in Admiralty documents varies between 66ft 4in and 69ft, the latter seeming more accurate when scaled off surviving drawings. Note the long railed system on the upper deck to supply ammunition from the magazine, forward of the machinery, to the 4in guns (this arrangement was adopted to minimise propeller shaft length, hence vulnerability to exactly the kind of under-the-bottom attack which put the ship out of action for three years). Aviation facilities were removed, but the catapult, which was inoperative, remained until the 1944–5 refit; one boat/aircraft crane was removed and the other relocated to the centreline just abaft the former aircraft hangars to handle the relocated boats. The 27ft whaleboats in the 'pockets' amidships were launched and recovered via an extended overhead rail system. The sliding doors that could cover the torpedo tube area are shown in their open position behind the vertically-stowed Carley life rafts. Ship's boats in November 1942 included one 36ft motor pinnace, three 35ft fast motor boats (one of Vosper design), two 32ft cutters, one 25ft fast motor boat, one 16ft motor dinghy, and two 14ft dinghies. *Belfast* was refitted between 4 August 1944 and May 1945, in part to prepare her for intense air attacks to be expected when she deployed to the Pacific. The two after twin 4in were removed and she emerged with two octuple, four quadruple, and two single power pompoms, three single hand-worked Bofors Mk III, two Boffins (single powered Bofors), two twin power Oerlikons and one single Oerlikon. Boat stowage was moved forward to the former aircraft-handling deck, but the 32ft oared cutters were still stowed beneath individual gantries on the main deck abreast the after funnel. *Belfast* was further refitted to prepare her for Korean War service, and then modernised in the late 1950s. At that time her tripods were replaced by lattice masts and her bridge superstructure completely rebuilt and enclosed. (A D Baker III)

Southampton class.[68] The hull had to be lengthened (by about 20ft) to allow wider turret spacing. Scaling the 9,100-ton *Southampton* to 10,000 tons gave a similar length, about 602ft (say 605ft) based on the 584ft of the smaller ship.

On 5 April 1936 Controller (Henderson) reported to the Board two studies of modified *Southampton*s: one with three turrets forward, arranged as in the battleship *Nelson*, and one with five triple turrets.[69] The ship with turrets forward had the same aircraft arrangements as in *Southampton*, but she was longer, so belt armour had to be trimmed (by not more than ½in) to stay within the 10,000-ton limit, and the *Southampton* machinery would drive her at a slightly slower speed. The five-turret design required about 450 tons more, but in that form it did not trim properly, and it would be difficult to fit in the required complement. Controller added that he expected to hold the *Southampton*s to 9,000 tons rather than 9,100 tons, so he could devote 10,000 tons to each of the 1936 ships without exceeding the planned total of 95,000 tons for the entire ten-ship *Southampton* class.[70] The 1936 ships became the *Belfast* class.

On 24 October DNC asked for two new sketch designs: KXIV and KXV.[71] KXIV would have three triple 6in turrets (if possible, all superimposed) forward and two aft. He hoped that space could be found for more twin 4in anti-aircraft guns. Power would be 82,500shp. All of this should be done within a 10,000-ton limit – which was not impossible, given that the US *Brooklyn* class had five triple 6in mounts on under 10,000 tons. KXV had four quadruple 6in turrets. Protection, aircraft arrangements and other features were as in the *Southampton* class. To add the extra turret, KXIV needed another 36ft between 'B' turret and the foremast (23ft magazine and 13ft shell room). This length would also provide accommodation for the sixty-seven extra personnel needed. Stability could probably be maintained by adding a foot of beam. That would give 620ft (wl) x 62ft 8in x 17ft, and a displacement of 9,720 tons based on the *Southampton* hull form (but not the necessary weights). If necessary, a bit more could be added for stability (say, a beam of 63ft 8in) and the bilge could be filled out. This was not bad; a preliminary check of weights based on the *Southampton* class expanded to the new dimensions, with the new 'C' turret, suggested a standard displacement of 10,005 tons, which could easily be shaved to the desired 10,000 tons. Estimated speed was 32.5kts. This design did not allow for any additional anti-aircraft armament, because it lacked both space and weight. Weight could be clawed back by shortening the ship slightly, but her topsides were already congested. However, the weight gained that way would buy a fifth twin 4in mount. As that would not offer a symmetrical layout; the torpedo tubes could be traded for a sixth 4in mount. The longer ship would not,

Above and below: HMS *Belfast* and *Edinburgh* were the largest pre-war cruisers. *Edinburgh* was lost in 1942, and *Belfast* was almost lost in 1939 when a magnetic mine broke her back. The blisters used to regain her hull girder strength also provided enough reserve buoyancy that, unlike contemporary British cruisers, she never had to lose one 6in turret (although two of her six 4in mounts were removed). HMS *Edinburgh* is shown newly completed.

THE 1930 LONDON TREATY AND ITS CRUISERS

Edinburgh is shown *en route* back to Scapa Flow, as seen from USS *Wasp*, April 1942. Her escort is a US destroyer. She received her Type 279 radar (antennas on both masts) during a refit at South Shields (20 March – 28 October 1940). Six single Oerlikons were fitted when she docked at Gibraltar in July 1941. When this photograph was taken, she had only recently emerged from a refit on the Tyne (17 January – 4 March 1942). She was fitted with a Type 273 surface-search radar atop her bridge plus gunnery radars (Types 284 and 285). Note that she retained her HF/DF coil on her mainmast, below the radar antenna. Her Oerlikons were unshielded: one is visible atop 'B' turret, one is on the shelter deck level below the bridge and abaft 'B' turret, and two more are visible on her quarterdeck; presumably the sixth was atop 'X' turret, hidden by the after director. She was sunk on 2 May 1942.

however, be able to use three graving docks open to the *Southampton* class: Hong Kong No. 1, Gibraltar No. 2, and Malta No. 5.

DNC asked for six twin 4in in KXV. The 4in gun deck had to be lengthened at least 28ft (which might be managed on a ship 21ft longer). The ship could be shortened by making the 6in magazines and shell rooms wider (18ft instead of 14ft) to make them 10ft shorter. All of that might be possible on a 605ft hull, which scaled up to 9,900 tons. Tentative weights showed another 92 tons for the quadruple mounts with heavier shields, a relatively small increase. It might make sense to abandon the earlier box protection for the magazines and shell rooms, relying on the external belt (180lbs, or 4½in) and deck armour, with 100lb (2½in) bulkheads.

In slightly modified form (614ft, six twin 4in) KXIV was presented to the Board as alternative A. KXV was presented as Design B. After a Sea Lords meeting on 7 November 1935, DNC ordered a third alternative (C, KXVI): four triple 6in turrets, six twin 4in anti-aircraft guns, two quadruple pompoms, and two 0.5in machine guns. Protection improvements were a 3in rather than a 2in magazine roof, 2in rather than 1¼in over machinery, turret roofs as in the projected quadruple mounts, and long trunks as in the quadruple mounts (with 3½in protection). The Staff wanted a speed of at least 31kts. The triple-mount designs had long trunks like the quadruple 6in. Eliminating their handing rooms made it possible to bunch turrets more closely, their centres 32ft rather than 36ft apart.

A quick estimate suggested that this KXVI would come out to 10,203 tons and would make 32.25kts on 70,000shp (31.5kts using *Arethusa* machinery). In view of the excess displacement, DNC ordered a fourth study, using the *Southampton* hull (584ft long) and the same power (82,500shp) with the same armament but with improved protection: 2in rather than 1¼in deck over machinery; 3in instead of 2in over magazines; turret shields thickened (but not quite as in *Gloucester*: 4½in face, 2in roof, 3in sides); and turret trunks and hoists covered by 3½in sides and 2in ends. The belt was the same 4½in thickness, but it was extended aft. Estimated total protection weight was 2,010 tons, compared to 1,478 tons for the 1935 version of the *Southampton* class. The increase of about 600 tons could be divided into 300 tons to improve turret protection against shellfire and 300 tons to improve deck protection against dive bombing. A 3in deck was considered proof against 500lb bombs dropped from 8,000ft or below, a 2in deck proof against 250lb bombs dropped from that altitude. Standard displacement would probably rise to at least 9,700 tons (increased protection weight would also increase hull weight, to avoid extra stresses). Speed would be about 32kts. Draft would be a foot more (18ft), so freeboard would be a foot less. The ship would lose a quarter-foot of metacentric height, making that about 2.9ft.

The next step (KXVII, Design D) was a *Southampton* with a new machinery layout, requested by Controller on 13 November. Shafting length was reduced to make the ship less vulnerable to shock (which could bend a propeller shaft) from 'B' bombing and to torpedoes (presumably particularly those with non-contact, i.e. magnetic, fuses). The alternating engine and boiler room arrangement adopted in the *Arethusa* class and continued in later cruises made for particularly long shafts from the forward engine room. Controller wanted revert to the earlier arrangement of two boiler rooms forward of two engine rooms, both boiler rooms being short, and with the 4in magazine moved from aft to forward. He envisaged a 10,000-tonner, but DNC decided to apply the idea to a *Southampton*. The existing alternating arrangement was 187ft long; the earlier type of machinery arrangement would be 168ft long, but would have an extra generator room on the centreline forward of the forward boiler room. The boilers in the after room would not be in

BRITISH CRUISERS

tandem. The older arrangement would entail only a single funnel, and 18 tons might be saved, plus another 18 tons due to the reduction of shaft length and the shift in position of the 4in magazine. The ship would not be shorter overall (due to the length of the generator compartment outside the engine and boiler rooms), but the side armour would be reduced, saving another 84 tons. The new arrangement would raise a few other problems, such as difficulty in arranging the wireless office aft under protection, and the fact that one or two more cranes would be needed (25 or 50 tons more) because the boats would be separated from the aircraft. Another study made at this time showed that eliminating the torpedo tubes would save about 40 tons. Design C was favoured, but it was too stiff, so in a revised version (K18) beam was reduced slightly to reduce metacentric height to the desired 3ft in standard condition.

On 17 February DNC asked what protection the big cruiser could have if speed was set at 32.5kts. The ship would revert to *Southampton* class protection except for 1¾in decks over machinery spaces and main W/T; 1½in deck over transmitting station etc; 3in decks over magazines and shell rooms; and 4½in turret face, with 2in sides and roof. This and designs since the *Amphion* had added considerable armour weight by having a belt of uniform height (up to the upper deck) over the forward engine room as well as the boiler rooms. How much could be saved if the belt and deck over that engine room were dropped to lower deck level? In that case a large part of the upper deck amidships would provide only strength; how thin could it be? It turned out that even if it were not considered protection, the upper deck had to be fairly thick (45 to 50lbs, i.e. 1⅛–1¼in instead of 2in, 80lbs).

The Board chose triple rather than quadruple turrets. This decision was later attributed to uncertainty about the necessary distance between turrets. The weight saved went into deck protection. Magazine spaces were shortened, so the ship was shortened by 10ft. To save weight, the belt and the deck over it were shortened at the fore ends, the 6in magazines and shell rooms receiving the earlier kind of box protection. That left all important control positions under the remaining extension of the belt and deck forward of the machinery spaces. Added weight made it possible to thicken the deck from 1¼in D to 2in NC armour. The 4in HA mountings were rearranged to minimise mutual blast interference, and the pompoms moved forward to be clear of the blast of the after 6in guns (they were still vulnerable to blast from the midships 4in guns, which was unavoidable). Design was slightly simplified because reduced clear 6in arcs were accepted, 140° each side rather than 145° in

Opposite: Belfast as refitted, 1942, with a very visible blister. The refit seems to have entailed about as much effort as building a new cruiser; but *Belfast* gained so much reserve stability that she had more capacity for improvement than any other modern British cruiser, and hence survived longer. She was mined on 21 November 1939, repaired temporarily at Rosyth (21 November 1939 – 27 June 1940), and then rebuilt at Devonport (3 July 1940 – 8 December 1942). While work was proceeding, requirements and technology (particularly radar) were changing, which must have complicated the project very considerably. Unlike smaller British cruisers, she had what was considered a nearly adequate close-range battery, so the main change was exchange of the previous quartet of quadruple 0.5in machine guns for five twin and four single Oerlikons. Note the twin power Oerlikon atop 'B' turret and the two right aft on the quarterdeck. Modern radars were fitted: Type 281 on the masts, Type 273 in a 'lantern' on the bridge, Type 284 for main battery control, Type 285 for heavy HA control, Type 282 for pompom control, and Type 283 for barrage directors. The FV1/Type 91 radar intercept/jamming combination was installed on her mainmast. Four single Oerlikons were added at Rosyth during a docking (18–29 June 1943). In this form she helped sink the German battlecruiser *Scharnhorst*. When docked afterwards at Rosyth (5 April – 8 May 1944) she had one twin Oerlikon removed and six more singles added.

Above: Belfast is shown after her last wartime UK refit, on the Tyne between 4 August 1944 and May 1945. Note the elimination of the twin power Oerlikon atop 'B' turret, but the retention of the two guns (with blast screen) on the quarterdeck. Two of the six 4in mountings and the aircraft arrangements were removed, as the ship was finally stability-critical. Four quadruple pompoms and four single pompoms were fitted, and eight single Oerlikons were landed. The ship's radars were modernised, her foremast cleared by replacing Type 281 with Type 281B. That left space for the Type 293 target-indication radar at the foretop and the Type 277 surface-search/height-finding radar (its dish is turned horizontally). A Type 268 navigational radar was installed, and the earlier Type 284 main battery set replaced by a Type 274. When the ship reached Sydney she was docked (August 1945) and additional close-range weapons added: three single hand-operated Bofors (one atop 'B' turret, one each side atop the former hangar); two Boffins (power Bofors on the mountings developed for twin Oerlikons) replaced two twin power Oerlikons.

BRITISH CRUISERS

Southampton (135° had been accepted in the quadruple-turret design). DNC was instructed to go ahead with detail design (on 13 May 1936). At this stage ACNS asked that octuple pompoms be substituted for the quadruples, at a likely cost of 50 tons (this was done). He also wanted better steering-gear protection. For greater propulsive efficiency, propeller revs were reduced from 350rpm to 300rpm, so on the same size and weight the machinery produced 80,000shp rather than 82,500shp. This revised sketch design received the Board Stamp on 29 May 1936.[72] Legend and drawings received the Board Stamp on 21 July 1936. Armour was again somewhat revised.[73]

As in previous cruisers the engine and boiler rooms alternated, but the machinery box was moved aft, presumably to shorten propeller shafts as a defence against 'B' bombing. That in turn squeezed the after 6in turrets, which were moved up a deck. The result was a somewhat odd appearance, the funnels being moved well aft of the bridge structure, leaving a conspicuous gap amidships. To make space for the machinery, the 4in magazine was moved forward, making for an awkward arrangement of ammunition supply along the deck. *Belfast*, one of two ships so redesigned, was the only British cruiser to fall victim to a non-contact explosion, when her back was broken by a German magnetic mine in November 1939.

The 6in turrets were the long-trunk type, served by DCTs fore and aft, and there were three HACS, one each side forward above the hangar, and one on the after superstructure. The 0.5in machine guns were placed

one each side of the hangar top forward, the pompoms being moved back to platforms aft platforms above and inboard of the middle 4in mounting, their directors on either side of the after superstructure.

The ship was now very close to 10,000 tons – on paper. On 3 October, in connection with a proposal to make the 2in deck continuous over belt and forward magazines, Controller decided to apply a margin <u>above</u> the treaty limit (in this case 10,000 tons) in the expectation that, as in the past, DNC would find it possible to shave weight during construction. In the new cruiser, that made it possible to provide a continuous 2in deck from forward to after magazines, at an estimated cost of 135 tons. The same policy was applied to other ships, such as the next class of large cruisers (the *Fiji*s). At one time British designs had all included a Board Margin allowing for growth up (not down) into their nominal displacements. Once the treaties were signed, such margins had been eliminated.

The two *Belfast*s were the largest British cruisers built since the 'Counties', with the heaviest anti-aircraft batteries of all. During construction *Edinburgh* had gained 302 tons in approved additions: bridge extension, pompom platforms, blast screens to 4in guns, protection to lower decks, bullet-proof screens, extra armament, increased complement, electric ammunition hoists, etc. By late 1941 another 145 tons had been added without compensation: strengthening for her upper and forecastle decks, additional stiffening and pillars, permanent degaussing, Type 279 radar, paravane clump and chains, hangar spray pump, side scuttles, SA gear; etc. DNC resisted or compensated for another 260 tons, including 75 tons of splinter protection successfully resisted: 42 tons for items including Types 271, 282, 284 and 285 radars, Oerlikons, steam heating, and increased complement for which compensation was demanded.

When *Belfast* broke her back after triggering a German magnetic mine in 1939, she was considered worth rebuilding despite the effort involved. She did not emerge until late in 1942, by which time her sister-ship *Edinburgh* had been sunk. The most prominent changes was blistering, to regain stability due to topweight growth.[74] Nearly all other British cruisers had 'X' turret removed during or immediately after the Second World War to regain stability so that radars and additional light anti-aircraft weapons could be added. Given her blisters, *Belfast* had sufficient stability, so she retained all four 6in turrets to the end. As inclined in October 1942 her light displacement was 11,400 tons. She had the full radar suite, Remote Power Control (RPC) for her two octuple pompoms, and ten Oerlikons plus ten twin Lewis guns in place of her two quadruple 0.5in machine guns. By September 1944 she had two twin power Oerlikons and fourteen single mounts. During a spring 1945 refit an Action Information Organisation (AIO) was fitted and aircraft facilities removed, close-range anti-aircraft armament being upgraded by adding four quadruple pompoms and four single power-worked pompoms, for a total of thirty-six pompom barrels. She now had four twin Oerlikons and six singles, a total of fourteen Oerlikons. As inclined on 2 May 1945 her light displacement was 11,919 tons (against an estimated 11,758 tons taking account of known additions). These additions included the Type 274 fire-control radar and fittings to fuel at sea. The unknown 161 tons was considered reasonable growth over 2½ years in a ship which had not been asked for compensation. She was refitted again at Sydney (completing in August 1945) for Pacific Fleet service. Two twin 4in were landed, two of the twin power Oerlikons were converted into Boffins (single powered Bofors guns) and three single Bofors were added, one atop 'B' turret in a zareba. Two of the single Oerlikons were also landed. By this time all the multiple pompoms had RPC.

When she fought in Korea, *Belfast* was in much the configuration she had when she emerged from the Sydney refit in 1945. She is shown on 5 August 1952. All Oerlikons had been eliminated. She had the wartime pompoms (two octuple and four quadruple), five single hand-operated Bofors guns, two Bofors Boffins, and two single power-worked Bofors of post-war design. By way of contrast, in April 1946 she had three single Bofors and two Boffins plus four twin power-worked Oerlikons and four singles. Her four single power-operated pompoms had already been landed. (USN photo courtesy of Rick E Davis)

The *Dido* class

In 1933 the Royal Navy was on the verge of building the big *Southamptons*, but many cruiser duties, such as supporting fleet destroyers, demanded much smaller ships. On 16 August 1933, at First Sea Lord's direction, Captain Tom Philips, Director of the Tactical Division, sent a letter to the two main fleet commanders asking them whether they were interested in a 4,000-ton cruiser, or whether the *Arethusa* was the minimum acceptable. The ship would replace the ageing 'C' and 'D' class cruisers. Fleet duties were reconnaissance, flotilla support, day and night screening, and detached operations.[75] Replies seem not to have been recorded, but they must have been positive. In August 1934 First Sea Lord (Chatfield) asked for designs for a small cruiser in hopes that something smaller and less expensive than an *Arethusa* could be produced, and also to counter the large leaders being built by the United States and Japan (the latter were the *Fubukis*, which the Japanese considered their new standard destroyers).

DNC produced Scout Cruiser sketch designs designated P through U, the tonnage limit having been relaxed to 4,500–5,000 tons.[76] As in the 1930 cruiser designs, DNC tried to cut tonnage by reverting to single 6in guns. Design P had six such guns plus the now-standard four 4in HA guns with *Southampton*-class protection. Speed was 30kts, but to increase that to a more desirable 33kts (Q), displacement was increased to 5,000 tons and one gun mount given up. To get 33kts on 4,500 tons (R), protection had to be cut drastically.[77] Alternatively, the level of protection could be maintained and the gun battery halved (Design S). Design T had two triple turrets and full protection (5,600 tons). Instead of scaling down from a cruiser, the ship could be scaled up from a destroyer leader. On this basis a 3,500-ton ship (Design U) could mount five single 6in guns and achieve 38kts, albeit with no magazine protection and limited machinery protection (1in side, ⅜in deck). She was the largest unprotected ship considered, the smallest of high speed which could mount 6in guns satisfactorily, and she offered good sea-keeping. Although she was largely unarmoured, her size in itself offered some protection. She seemed expensive for production in numbers, but destroyers or conventional leaders might not offer enough individual gun power. That led to the next step down, a real destroyer leader, called the 'V Leader'. She would cost about half as much as Design U. A new twin 4.7in gun offered her considerable firepower. The Naval Staff liked this ship, which did not impinge on available cruiser tonnage, and she became the 'Tribal' class. Design discussions of this ship show that she was envisaged very much as a substitute for cruisers in roles such as reconnaissance. The 'Tribals', rather than small cruisers, were included in the 1935/6 programme.

Of the cruisers, Design Q was best liked. The design worked only because it employed single open 6in mounts, but 'a strong body of opinion' held that turrets offered marked fighting advantages with regard to morale, arcs of fire (which would otherwise be limited by blast), ammunition supply and weather, and that open mounts occupied more space than turrets. Design U was the most powerful against light craft, but could not stand up to destroyer fire, was too expensive and offered too large a target. Against all of these alternatives, C-in-C Mediterranean (Admiral Sir William W Fisher) called for a different kind of 'small fleet cruiser' intended specifically to strengthen fleet anti-

THE 1930 LONDON TREATY AND ITS CRUISERS

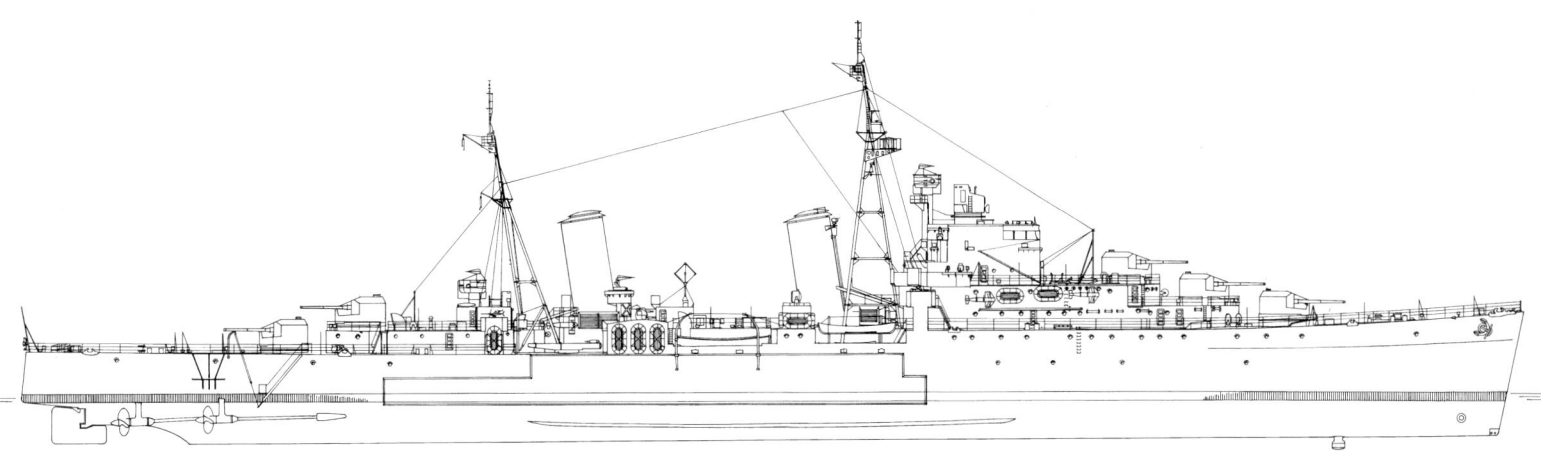

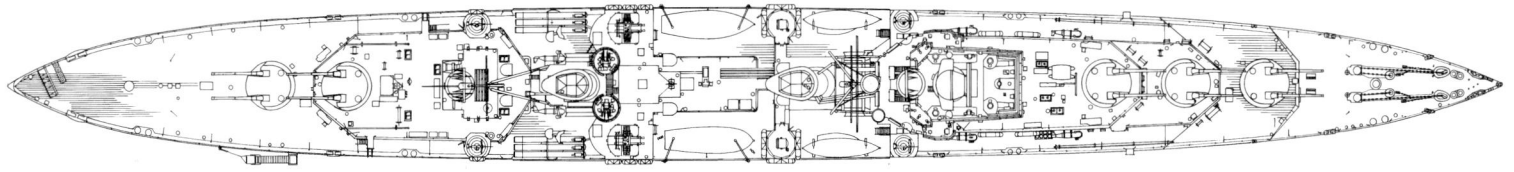

Above: Argonaut as fitted on completion, September 1942. In the elevation drawings, the yards are shown in direct profile. Six aerial wires ran from the upper yard to the after yard. The circular objects with a smaller internal circle in a rack parallel to the depth charge rack, at the fore end of the forecastle, and on the platform with the after pair of 20mm guns, are smoke floats. The main deck and forecastle deck abaft the breakwater were planked with 'Borneo White' hardwood; the other decks were covered with a 'latex' compound. Asdic was fitted, and six paravanes were carried for protection against mines. On return to the UK in June 1944 after battle damage repairs at Philadelphia Navy Yard (begun March 1943), 'Q' turret was replaced by a quadruple pompom and the close-range armament was further augmented by five twin power Oerlikons and four single Oerlikons. The depth-charge rack was removed. (A D Baker III)

Left: HMS *Euryalus* is shown in much her original configuration, probably in the Suez Canal in 1943. She did not yet have a surface-search radar, and she had a Type 285 on only her after HA director. That suggests a relatively low emphasis on anti-aircraft fire: these ships were by no means anti-aircraft cruisers (she did have a surface fire-control set, Type 284, on her LA director). She was also unusual in retaining the HF/DF coil on her mainmast. Her two quadruple 0.5in machine guns were removed about September 1941, not long after she was completed, and five single Oerlikons mounted; another two were added by September 1942, and two singles were exchanged for five twin Oerlikons by mid-1943. At that time, too, her Type 279 radar (visible here) was replaced by Type 281, and she received a Type 272 surface-search set in a small 'lantern' on her foremast. She also received Type 282 radar for her pompom directors.

aircraft defences; such a ship would be of great value for other cruiser duties. To gain anti-aircraft firepower he would accept much lower speed.[78] Much this idea had been raised and rejected in about 1925, the counter-argument being that it would be better to provide all ships in the fleet with better anti-aircraft armament. In 1934 Tactical Duties argued that the ship would be too specialised and too vulnerable to conventional cruisers. Moreover, measures were being taken to strengthen fleet anti-aircraft armament.

A sketch of the small fleet cruiser (3,500 tons) showed her forward guns in 'A' and 'B' positions, two guns aft in 'X' and 'Y', the fifth gun on the superstructure deck aft, forward of 'X' gun, and four sets of torpedo tubes. The sketch of the 'V Leader', however, showed a new kind of arrangement, three of her five twin 4.7in guns being arranged to superfire forward of the bridge, the third mount above 'B' mount, to achieve the sort of concentrated ahead fire which was wanted. The 4.7in guns, moreover, were considered a contribution to fleet anti-aircraft protection, despite their limited elevation (40°). They could open fire on a relatively distant enemy formation flying at medium altitude, hopefully breaking it up and making conventional bombing difficult. C-in-C Home Fleet approved the 'V Leader' idea, but wanted more close-range armament, while C-in-C Mediterranean thought the 'V Leader' was a bad destroyer and an even worse cruiser. He wanted a fast-firing dual-purpose gun with a calibre of about 4.7in, or the usual combination of a main armament of fast-firing guns (4.7in to 5.5in) and 4in anti-aircraft guns; cruisers would usually fire rapid bursts of short-range fire, for which 6in guns were not suitable. In a Japanese war, the main threats to a British fleet would be air and submarine attacks, plus a lesser threat from destroyers at night. To deal with such threats he liked the Design U cruiser, but wanted better protection to her vitals, perhaps at a cost in speed. Although battle fleet anti-aircraft weapons were being improved, any new cruiser should contribute to overall fleet anti-aircraft defence.[79]

BRITISH CRUISERS

Above: HMS *Argonaut*

Below: When the *Dido* class was being built, twin 5.25in dual-purpose guns and mounts were in short supply. HMS *Phoebe* was completed with a 4in QF starshell gun in 'Q' position (*Dido* was similar); *Bonaventure* had her 4in gun in 'X' position. Note the quadruple 0.5in machine gun on shelter deck level below the bridge, as in earlier cruisers. This photograph was censored to eliminate radars, but the antenna on the mainmast was missed.

Rear Admiral, Destroyers (RA(D)), the Mediterranean destroyer commander (Rear Admiral Andrew B Cunningham) considered it unwise to concentrate on large cruisers (i.e. *Southampton*s) because smaller ones were likely to be far more effective in support of destroyers, given their higher rates of fire. It was argued that aircraft could take over the reconnaissance and shadowing roles, but Cunningham noted that there were still too many occasions when weather would not allow a C-in-C to rely fully on aircraft. RA(D) also needed a new flagship to replace the small old cruiser he currently used.[80] Small cruisers were needed in general because the 'C' and 'D' class cruisers were approaching retirement age, and while the 'E' class and the new *Southampton*s were fast enough to work with destroyers, they were too big. RA(D) wanted something no larger than the 'C' class cruiser he had formerly used, but extra requirements might boost that to 4,500 tons.

RA(D)'s flagship was the rallying point for a torpedo attack. It required firepower mainly to enable him to exercise command and leadership freely, providing moral support and clearing up obscure or difficult situations personally. The ship therefore needed a reasonable ability to press forward despite enemy destroyer gunfire, i.e. the minimum required to menace destroyers (and protection against destroyer fire). It had to be about as fast as the new destroyers (the 'Cs' and 'Ds' certainly were not), with sufficient command spaces (including plotting spaces), with maximum handiness, and with a small silhouette, as they would probably lead night attacks. In 1935 it appeared that such fleet operations would be common in a Japanese war but not in any other, so the ship had to be adaptable to other roles. Independent missions, such as those of the old Harwich Force, would probably be the rule in anything but a Japanese war.

Late in January 1935 Director of Plans rejected D of TD's idea that the 1,830-ton destroyer (which became the 'Tribal') would suffice, because the RA(D) ship had to accommodate a rear admiral, seven

THE 1930 LONDON TREATY AND ITS CRUISERS

Above: Given the shortage of 5.25in gun mounts, two ships were completed instead with 4.5in anti-aircraft guns: they were the only *Dido*s which could be characterised as true anti-aircraft cruisers. *Scylla* is shown in June 1942. Her sister *Charybdis* lacked the pair of Oerlikons forward of the bridge.

officers and two warrant officers in addition to the usual ship's company. First Sea Lord sought to separate the small cruiser and large destroyer issues, approving the 'V Leader' but leaving the cruiser issue open. He suspected that C-in-C Mediterranean and RA(D) rejected the 'V Leader' only because they feared that approving it would let the Board out of building any small cruisers at all. Director of Plans suggested that the single remaining 1936 ship (as the programme was then understood) be the RA(D) flagship, because only one such ship was needed.

Director of Tactical Division laid out tentative Staff Requirements in a memo dated 11 April 1935. He suggested two alternative designs, one (A) with six guns, a second (B) with twin mounts of smaller calibre. Design A would be based on the Q design, but with five rather than six 6in single mounts, protection to be reduced to hold displacement within 4,500 tons, but as close as possible to that of Design Q. The ship would have maximum anti-aircraft armament consistent with these conditions. It was not certain that Design B would be steady enough to handle the large fixed ammunition of the 4.5in gun (55lb projectile) then planned as the dual-purpose secondary armament of the new battleships so, since HA fire was important, the choice might lie between the new 4.7in gun (62lb projectile) and the new 5.1in (80lb projectile), the latter 'showing promise in loading trials'.[81] As neither the new 4.7in nor the new 5.1in would be developed for a time, the initial B design should employ the twin 4.7in gun planned for the 'V Leader', plus two quadruple pompoms. Design B should be protected against the new 4.7in shell at 5,000yds. The flagship's contribution to flotilla torpedo fire would be so small that torpedoes were not necessary.[82]

The ship required good plotting arrangements, so that RA(D) could disentangle a complicated night battle situation, 'more room being required in the plotting house than normally fitted in a cruiser' and good communications facilities were essential. Wireless would be as in *Arethusa* plus a set for use in the Attacking Force frequency. Ships would

Above: HMS *Argonaut* is shown after her major refit at Philadelphia Navy Yard, March 1943 – November 1944, after she had been torpedoed on 14 February 1943 in the Western Mediterranean. The big refit, which eliminated 'Q' turret, was roughly analogous to contemporary attempts to gain topweight in larger cruisers by landing 'X' turret. A quadruple pompom and its director replaced 'Q' turret, four single Oerlikons were landed, and five twin Oerlikons installed. The ship's radars were modernised: adoption of the single-antenna Type 281B cleared the foremast for installation of a Type 293 target-indication set, and she was given a Type 277 surface-search set (its dish is horizontal in this picture). When the ship returned to the United Kingdom, another six single Oerlikons were mounted. By August 1945 the ship had been refitted to better enable her to beat off Kamikazes. She had been given five Boffins (power Bofors on twin power Oerlikon mountings, hence replacing twin Oerlikons) and two single hand-worked Bofors Mk III. In October 1945 she had three quadruple pompoms (the two original mounts plus the one which replaced 'Q' turret), five Boffins, four single Bofors, and four single Oerlikons. *Cleopatra*, *Euryalus* and *Phoebe* had all been reduced to four twin 5.25in guns, but *Dido* and *Sirius* retained all five mounts. *Euryalus* was rebuilt on the Clyde, 20 October 1942 – 19 July 1944, emerging with six twin power Oerlikons and five singles, plus her pompoms. *Phoebe* was rebuilt at New York Navy Yard, 15 January – 14 June 1943, having been torpedoed off Pointe Noire on 23 October 1942. She was given three quadruple Bofors. At the end of the war she also had six power twin Oerlikons and four singles. She had already been torpedoed on 27 August 1941 and repaired at New York Navy Yard between 21 November 1941 and 21 April 1942; at this time her 4in star shell gun was replaced by 'Q' twin 5.25in turret.

make 33kts. For each design, DNC should show the effect of fitting the new type of subdivided cruiser powerplant, and also the effect on displacement and protection of reducing speed to 31.5kts. ACNS approved the Staff Requirement on 16 May.

Submitting sketch designs in September, DNC pointed out that B had a heavier armament and slightly heavier protection, hence displaced 200 tons more. He tried the requested variations in the 6in ship; their effects on Design B would be similar.[83] Controller saw no point in spending 4,700 tons to mount five twin 4.7in on a cruiser hull, when a 'V Leader' offered the same battery on 1,830 tons. He was far more interested in five twin 5.1in guns, with their much more powerful shells. Their 80lb shells represented the upper limit for man-handling in a lively ship. This weapon would not be available for a 1936 ship, but it was worth asking whether the delay in building such a ship would be worthwhile. The 5.1in gun morphed into a 5.25in dual-purpose gun. ACNS agreed with Controller that the 5.25in dual-purpose armament was best, because it covered both HA and LA requirements.[84] In November 1935 the Sea Lords decided to defer construction until the gun was ready.

About January 1936 it was decided that the RA(D) flagship should become the basis of a more general-purpose small fleet cruiser. Although the first of class, RA(D)'s flagship, need have no torpedo tubes, follow-on general-purpose light cruisers should have adequate torpedo armament, which might mean two quadruple tubes. ACNS rejected Staff arguments demanding an aircraft, because that would make the ship too large. By February 1936 plans called for three such cruisers.[85]

DNC produced two alternative designs, each with five twin 5.25in, with the old and new machinery arrangements; in February 1936 ACNS preferred the new subdivided machinery spaces.[86] At a meeting on 8 April 1936 First Sea Lord, Controller and ACNS decided to proceed with the ship, which was expected to displace 5,100 tons, be armed with five twin 5.25in guns and cost about £1.3 million. Of the guns, the second and third pairs would not normally be used for HA fire due to the blast of the foremost pair of guns, which at this point were envisaged as open upper-deck mountings. Machinery spaces would alternate, and speed was held at the 32kts of the earlier design studies. Magazines were given 3in crowns and sides, and the machinery a 3in belt and 1in deck. A BD version of the twin 5.25in was in development which was preferable to the open upper-deck mounting because mountings would not suffer from the blast of adjacent guns. On 8 May Controller noted that the design of the BD mounting was much more advanced than had been imagined; as long as ships were not laid down before February 1937 they could be armed with it. On 13 May First Sea Lord (Chatfield) chose the alternating-machinery option.

It was immediately obvious that the ships would be badly cramped, but there was apparently no interest in making them much larger. For example, to hold down displacement when the BD mountings were adopted, ammunition per gun was cut from 300 rounds to 250. ACNS later pointed out that the ships would have an unusually high proportion of HE shell, due to their dual-purpose role, and that would make them more vulnerable than usual to explosions near their shell rooms. He therefore wanted the heavier magazine (cordite) protection extended to the shell rooms, at a cost of about 35 tons. The design was so tight that Controller rejected even this addition, the improvement to be explored after tenders had been invited. Protection as described in the final Legend for Board approval was somewhat lighter than in the earlier

THE 1930 LONDON TREATY AND ITS CRUISERS

design studies.[87] This was the *Dido* class. The Legend and drawings received the Board Stamp on 14 December 1936. Five were ultimately included in the 1936/7 programme alongside the two *Belfast*s.

In February 1937 heavier long-trunk 5.25in mountings were adopted.[88] The magazines could be pushed under water, where they would not need side protection (but would retain their ends, since shells could pass through the structure of the ship). The ammunition lobbies of earlier designs were eliminated. That made it possible to bring the mounts closer together, so 'Q' mounting was 6ft further from the bridge (and less likely to cause blast problems at high angles) and the quarter-deck became 22ft longer. Fewer men were needed to serve the guns (complement was cut from 558 to about 485), so some of the congestion was relieved (similar long trunks were adopted in larger cruisers for similar reasons). Now 360 rounds per gun could be stowed for the forward mountings, 320 rounds for 'X' mount, and 300 rounds for 'Y'. Magazine crowns were reduced to 2in, but the same armour could cover the shell rooms (the ammunition spaces were larger than before). Displacement increased to 5,450 tons. Boilers were re-rated to 62,000shp under temperate conditions (which bought a quarter-knot) by increasing their forcing rate (output in the tropics continued to be 58,000shp; much the same re-rating was being done in the contemporary *Fiji* class). That added 20 tons, which DNC considered acceptable, but at the very limit the ship could accept. At the same time tripod masts were substituted for poles. A memo indicated that beam had to be increased half a foot, but the ships ended up with the same 50ft 6in beam chosen earlier. Not surprisingly, estimated cost rose throughout the design process.[89]

The ships had an unavoidably high silhouette due to the three superimposed 5.25in guns forward. Attempts to cut it down, both by sinking the turrets into the deck, and by eliminating the wheelhouse under the bridge, failed.[90] The bridge had to be high enough to provide clear vision over the upper guns.

Given their ancestry, these were hardly single-purpose anti-aircraft cruisers. Low-angle fire was far too important in their conception. The 5.25in calibre was chosen not because it made for good anti-aircraft fire, but because it was good for stopping destroyers and lightly-armoured cruisers. The Royal Navy considered the 4.5in gun, with its lighter and handier shell, a more ideal heavy anti-aircraft weapon – which is why it armed British aircraft carriers. The ships had a LA DCT atop the bridge, with a HADT abaft and above it, and a dual-purpose (H/LA) DCT aft.[91] There was a single combined transmitting station (for surface fire) and HA control position (HA computer room) below decks, containing an Admiralty Fire Control Table Mk VI* and a High-Angle Calculating Table Mk IV*. At the outset, DNO pointed out that without dedicated anti-aircraft guns, the ships had no way to fire star shell (the principal means of night illumination) other than their main batteries. The Royal Navy was increasingly interested in night combat, so DNO proposed a separate 4in star shell gun, which would not be an anti-aircraft weapon (development of a separate star shell thrower was considered and rejected). The idea died because too much – perhaps both pompoms – would have been sacrificed. The design allowed for two of the new quadruple 0.661in machine guns in place of the usual quadruple 0.5in guns, but the new weapons never entered service.

Three repeat *Didos* were included in the 1937/8 programme, and two more repeat ships in the 1938/9 programme. No *Didos* were included in

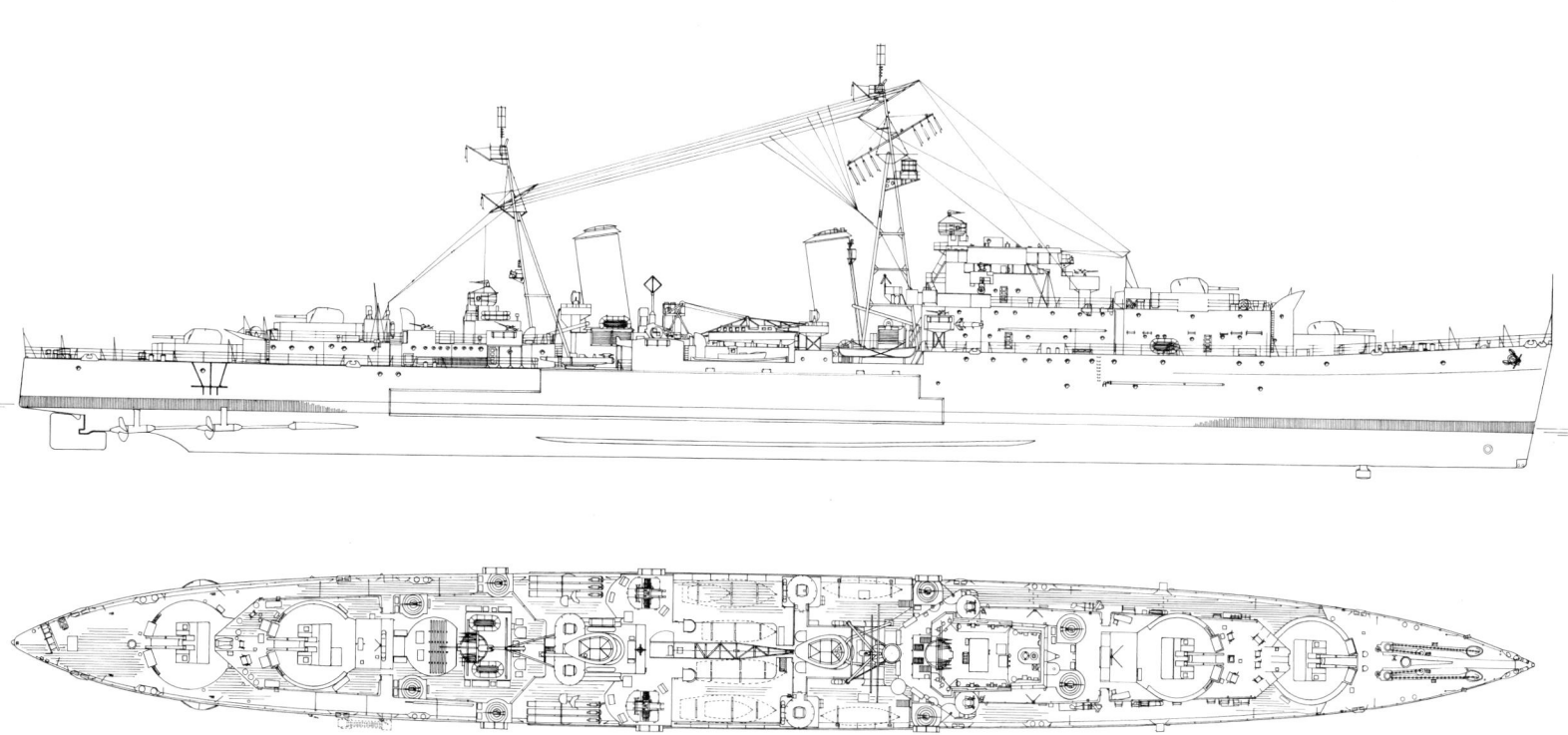

Scylla as fitted on completion, June 1942. Although the plan view shows provision for a 4in gun to fire star shell mounted between the aftermost pair of single Oerlikons, no such gun was ever fitted, and the surrounding splinter shielding and ready-service ammunition lockers seem to have been removed before commissioning. Shortly after completion, two more Carley floats were added to each side beneath the forward pair of 44in searchlights. In 1944 the single Oerlikon mountings were replaced by three twin power mountings. Note that the 20mm mounts did not have on-mount shields for the gunners as of June 1942. (A D Baker III)

BRITISH CRUISERS

Cleopatra (shown on 3 April 1945) was modified similar to *Argonaut*, but had US-type quadruple Bofors instead of pompoms. She was modified at Philadelphia Navy Yard, 24 November 1943 – November 1944, having been torpedoed off Sicily on 16 July 1943, and having been given temporary repairs at Malta. In her case other close-range armament comprised, by the end of the war, six twin power Oerlikons and four singles.

the regular 1939/40 programme. However, by mid-1939 the projected Emergency War Plan included four ships. At this time (see below) the 'D' class cruisers were to be rearmed with standard twin 4.5in mounts, but this rearmament was to be cancelled in the event war broke out before it could begin (as was the case). The projected emergency programme was set partly by the expected production of twin 5.25in guns. DNC pointed out that if two cruisers could be armed instead with 4.5in mounts freed by cancellation of the 'D' class reconstruction, it might be possible to build five rather than four emergency *Dido*s. That was done: in September 1939 *Charybdis* and *Scylla* were ordered armed with four twin 4.5in mountings each.[92] They were fitted as RA(D) flagships.

Dido, *Bonaventure* and *Phoebe* were each completed with four rather than five twin 5.25in guns, with a single 4in LA gun in 'C' position. Only *Dido* eventually had her proper 'C' mount fitted. Like their contemporaries the *Fijis*, these ships came out badly overweight. Designed for 5,450 tons, *Naiad* was completed at 5,677 tons. The 220 tons of approved additions included extra light armament, heavier main machinery, electrical increases, and larger complement. By the end of 1941 another 72 tons had been added without compensation: radar, RPC for pompoms, internal degaussing, SA gear, Asdic, etc. Another 92 tons of proposed additions, including a UP ('Unrotated Projectile' – anti-aircraft rockets) mounting, had been rejected or, in some cases, compensated for.

The original close-range battery was the standard two quadruple pompoms and two quadruple 0.5in, but by January 1942 all 0.5in guns had been removed from *Dido*, *Hermione*, *Naiad* and *Euryalus* and replaced by five Oerlikons. *Cleopatra* had three Oerlikons. *Argonaut* was completed with two single power-driven 2pdr Mk VII and four single Oerlikons (Oerlikons temporarily replaced the single 2pdrs). By 1944 her 'C' mount had been replaced by a third quadruple pompom, and in addition she had six twin power and five single Oerlikons. *Sirius* was similarly armed, but by 1944 had seven Oerlikons. In July 1942 *Cleopatra* had two single 2pdr Mk VII, three single Oerlikons and two quadruple 0.5in machine guns, the latter to be replaced by Oerlikons or by 2pdr Mk XVI. With less topweight, in January 1942 *Phoebe* had eleven Oerlikons. When she was refitted in the United States, the 4in gun was replaced by a US-type quad Bofors, two replacing her pompoms, plus six twin power and four single Oerlikons. *Cleopatra* was similarly rearmed. *Euryalus* had 'C' mounting replaced by a third quadruple pompom, and by 1944 she had six twin power and four single Oerlikons. The Emergency War Programme ships completed to a modified design as the *Black Prince* class are described in a later chapter.

Cruiser Designs for Export

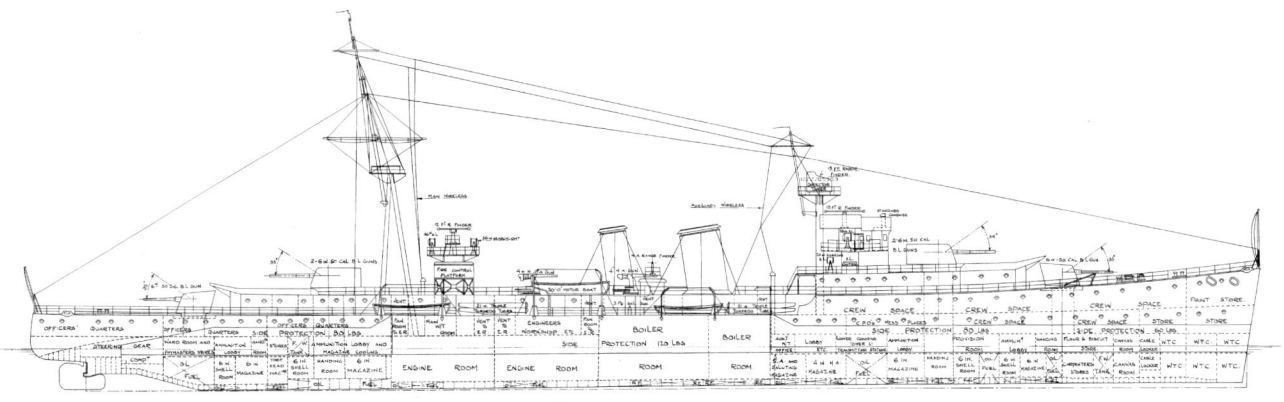

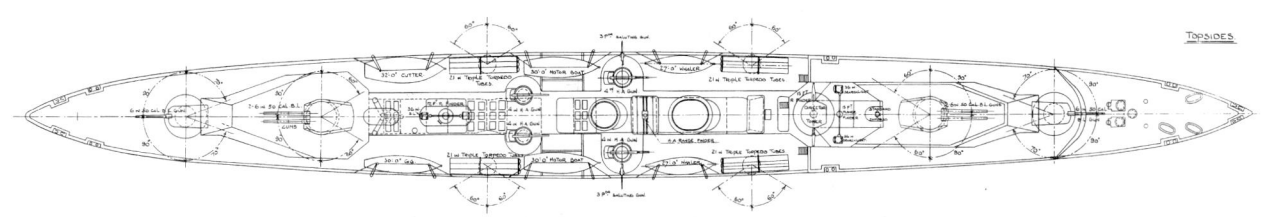

Vickers-Armstrong's light cruiser for Portugal (June 1930, Design 1005) would have been armed with six 6in guns, four 4in HA, one machine gun, four triple torpedo tubes and two depth-charge throwers. The latter appeared on nearly all Vickers cruiser designs of the inter-war period. This ship would have displaced 5,000 tons (450ft pp, 476ft loa x 46ft 6in x 27ft x 14ft 6in). Power would have been 40,000shp (29kts), and endurance would have been 5,000nm at 12kts. This design seems to have been offered to Romania in August 1930. The Portuguese Navy suffered from ambitious but grossly-underfunded plans, and Vickers offered designs to match. The Navy Law of 1907–8 called for six 5,000-ton protected cruisers (which come before the period covered by this book). The monarchy was dissolved in 1910, and a new 1912 Navy Law called for three battleships and three 2,500-ton scout cruisers, among other ships. There was insufficient money, so in 1913 new plans called for a smaller programme including two small cruisers (2,500 tons, 20kts, two 6in and six 4in guns). The cruiser contract was provisionally awarded to a British consortium (John Brown, Cammell Laird, Fairfield, Palmer's, Thornycroft), but there was no money. No new programme was announced until 1930, and it included no cruisers. However, it is clear from Vickers' record that numerous designs were offered. The Vickers version of the slow protected cruiser was Design 614 (20kts, 305ft x 37ft 6½in x 22ft 6in) of 7 January 1913. During the First World War, when Portugal was a British ally, Vickers offered an enlarged *Cassandra* design (Design 817A of 22 August 1916, described as having 12 per cent more power: 450ft pp, 480ft oa x 44ft x 25ft x 14ft 9in, 4,820 tons, armed with six 6in/50, two 3in, two 3in HA and two twin 21in torpedo tubes). Slightly later it offered slightly larger Designs 750 and 751 (460ft x 44in x 25ft 4in x 14ft 6in, 4,930 tons) with the same armament. There was also a related 'fast cruiser' (Design 819 of 18 September 1916). These seem to have been the last export cruiser designs Vickers offered before the First World War ended. Note that Vickers' design numbers were apparently anything but consecutive. The *Cassandra* design itself was later offered to Uruguay (about 1921) and to Romania (as Design 805). Vickers continued to offer the country warship designs, including several aircraft carriers. Portuguese desires (and, presumably, resources) waxed and waned. In November 1921 Vickers offered a 2,000-ton, 310ft scout cruiser (Design 933) armed with four 6in guns in single upper deck mounts, two 3in HA and two twin 21in torpedo tubes. The Vickers design book gives no other dimensions, so these data may indicate what the Portuguese wanted rather than what Vickers could provide in a realistic design. Vickers later offered a 3,300-ton (normal displacement), 400ft cruiser armed with six 6in guns, two 3in HA and two triple torpedo tubes, which may have been an outgrowth of this project. In August 1925 Vickers offered Portugal Designs 1010 and 1013, which had previously (1923) been developed for Brazil (7,000 tons, seven 6in guns, and 6,000 tons, six 6in guns). At the same time Portugal was offered Design 1187, in which the eight 6in guns were all in twin mounts. There were also four 4in HA and four triple 21in torpedo tubes. The design showed a 3in belt over its machinery. It was described as similar to the Spanish F design (*Principe Alfonso*), but lengthened by 10ft, displacing 8,000 tons (555ft pp, 585ft loa x 54ft 6in x 30ft 6in x 16ft 6in). Estimated power was 80,000shp for 33kts. There was no conning tower. The Design F powerplant (82,000shp) would drive the ship at 33kts. In effect this was much the design to which two of the Spanish ships were later rebuilt. By February 1926 Portugal was clearly shopping for something more impressive. Vickers offered Design 1228, 8,250 tons (555ft pp, 585ft lwl, 589ft loa x 56ft x 37ft 6in x 16ft 6in), designed for 33kts on 80,000shp and for a radius of action of 5,000nm at 15kts. Armament was three twin 8in/50, four single 4in HA and two triple torpedo tubes. Configuration was very unusual, all three turrets being forward, arranged as in the later USS *Brooklyn*, leaving the after part of the ship open. All armament was on the forecastle, the triple torpedo tubes being forward of the anti-aircraft guns. Vickers records also show an un-numbered design for Portugal dated 12 July 1926, for a 5,000-ton cruiser (450ft x 46ft 6in x 27ft x 14ft 6in) armed with six 6in/50, four 4in AA and four triple torpedo tubes. She would have made 29kts on 40,000shp, and she had wartime-type side protection. No drawing has survived, but she seems to have been a precursor to Design 1005. The Portuguese soon abandoned these projects, and in 1927 they asked for a gunboat, although they called it a cruiser: Vickers offered Design 918 (3 March 1927), displacing 2,500 tons on trial (330ft x 39ft 6in x 28ft 9in x 13ft trial) powered by 3,000ihp reciprocating engines (17kts). Armament would have been two 6in, two 4in HA and four pompoms. This ship might be considered the ancestor of the big gunboats Portugal ordered a few years later. In 1930 the country moved back to something more impressive, in the form of Design 1005. Somewhat later Vickers offered Design 1070. (National Maritime Museum)

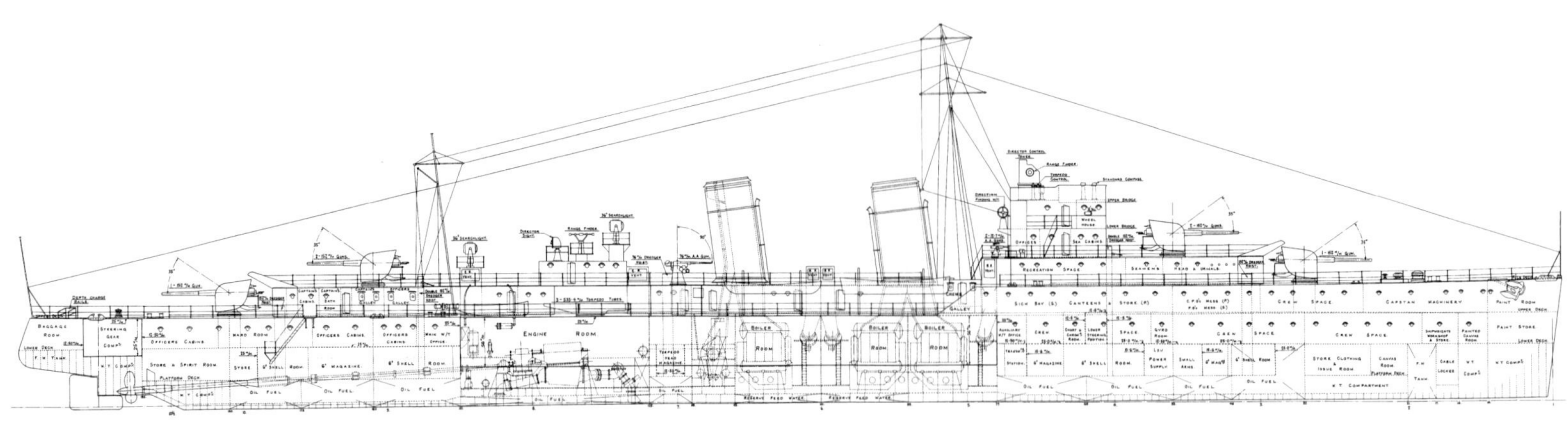

Above: Vickers' Design 1070 for Portugal was armed with six 6in/50 in two twin mounts superfiring over two singles. It is undated, but it is included in a list of designs of the early 1930s. Note the resemblance to contemporary British destroyers. (National Maritime Museum)

Below: Design 1071 was the larger alternative to Design 1070. It would have displaced 4,550 tons rather than 3,350 tons, and would have been 450ft rather than 400ft long. Note the gun arrangement, in which one twin mount is forward and two are aft. (National Maritime Museum)

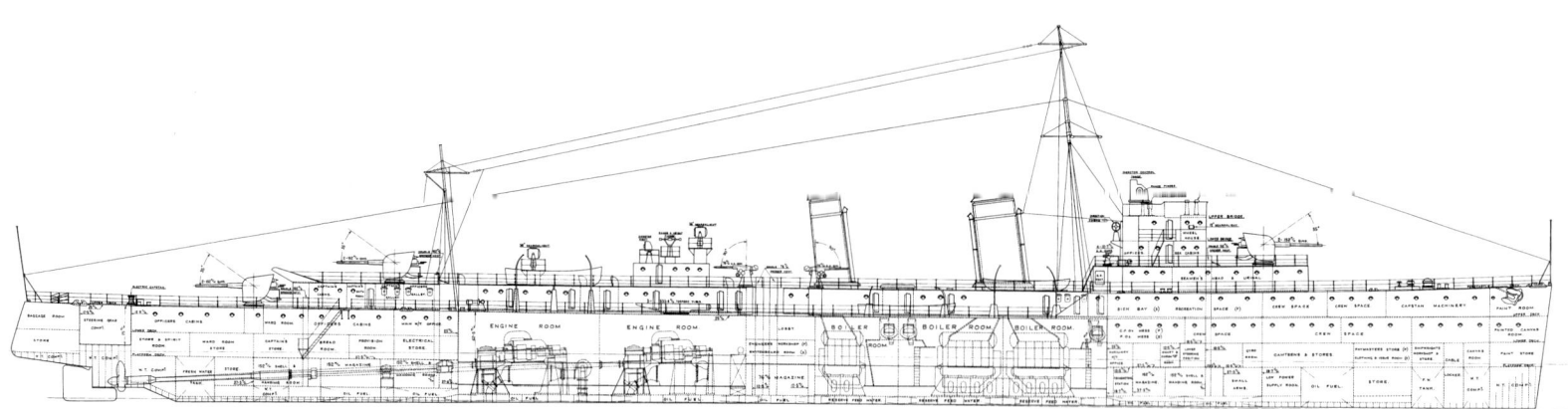

THE 1930 LONDON TREATY AND ITS CRUISERS

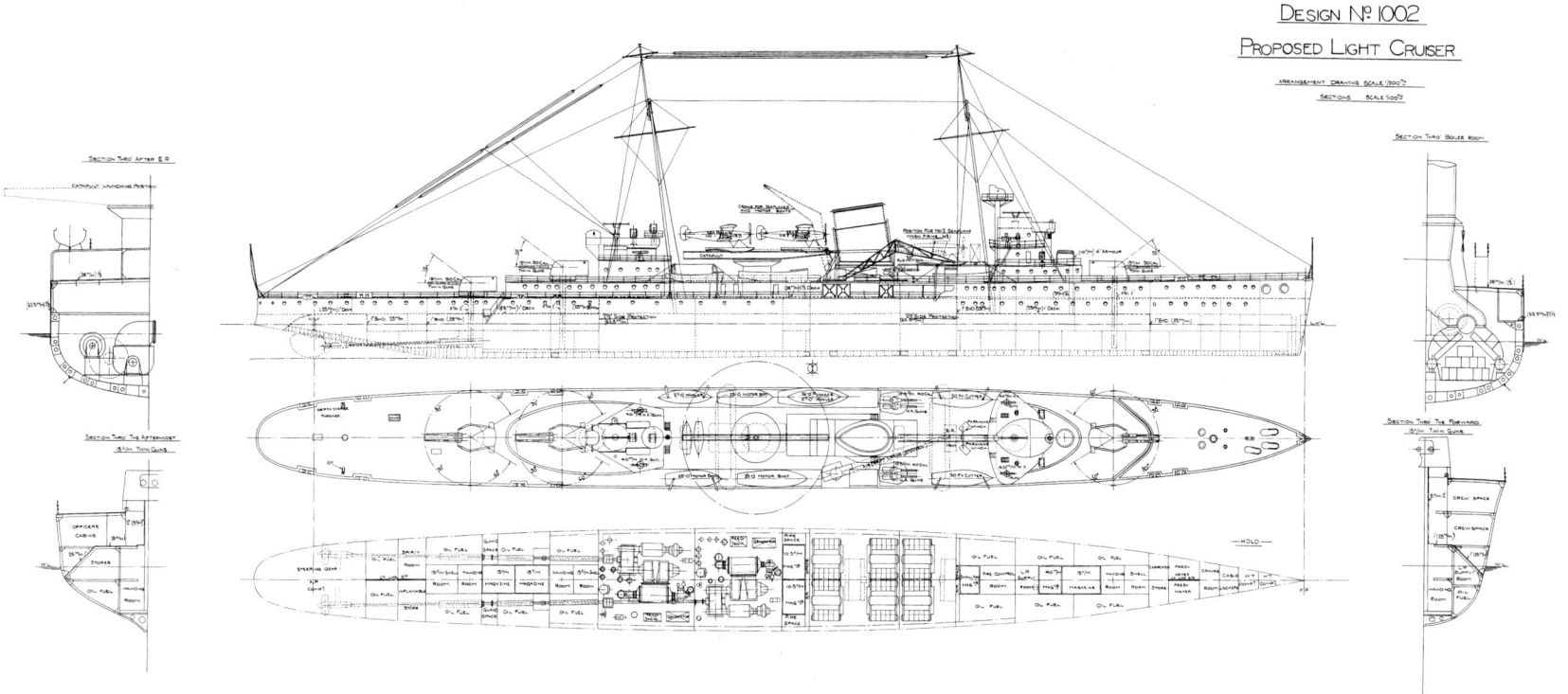

Design 1002 was a light cruiser for the Royal Netherlands Navy. It appears to have been Vickers-Armstrong's bid for what became the cruiser *De Ruyter*, as a paper in the file, dated October 1932, is a sketch of the latter (which was now to be built) forwarded by Vickers' Dutch agent. Vickers notes on Design 1002 were dated 1 May 1930. The Vickers plans archive at the National Maritime Museum includes Design 1002A, which has just the sort of tower bridge/foremast the Dutch adopted; perhaps it was what the Dutch built. There were, after all, few firms capable of producing such a design at the time. Armament was three twin 15cm (one forward, two superfiring aft), four 10.5cm anti-aircraft guns, four 40mm AA and two depth-charge throwers. Unusually, there were no torpedo tubes. The ship would have displaced 6,300 tons normal and 6,900 tons fully loaded (500ft pp, 530ft lwl, 534ft loa x 50ft 5in ext x 35ft 6in to forecastle deck, 28ft to upper deck x 15ft 6in). She was expected to make 32kts on 60,000shp (six Yarrow water-tube boilers), with a radius of action of 5,000nm at 12kts. Total complement was 450. She would have had one catapult and two seaplanes; the Royal Netherlands Navy placed considerable stock on the use of such aircraft in the East Indies, and even carried them on board its destroyers. From the Dutch point of view, this project was an attempt to revive an expansion programme rejected by the Dutch parliament in the early 1920s. In September 1920 Vickers offered the Dutch its Design 767: five twin 6in/50, two 4in HA and four triple tubes on 5,150 tons (455ft x 47ft 6in x 27ft x 14ft 6in, 29kts on 40,000shp [all-oil]). Protection followed wartime British practice (side but not deck). The ship's power and armament suggest that she was an enlarged version of the Vickers-built *Cassandra* with twin 6in guns replacing the single mounts on board that ship. The design developed for the Netherlands was later offered to Romania (as Design 808) and to Turkey (in May 1923, as Design 836X). Design 834X (sent to Turkey 5 October 1923) was larger than the earlier Dutch design but had the same five twin 6in/50 (plus four 105mm/50 HA, two 3pdr QF, two single pompoms and four triple torpedo tubes) on 7,000 tons (520ft x 53ft x 30ft 3in x 16ft, 31kts on 60,000shp using six Yarrow oil-fired boilers). Protection followed British wartime practice (side but not deck). Vickers continued to court Portugal, Romania, and Turkey between wars. Despite evident interest in cruisers, the first post-First World War Romanian naval programme (1927) amounted to four destroyers (two of which were bought). The next programme (1937) called for a cruiser and lesser warships. Unfortunately Vickers' design files for this period are too incomplete to show whether the company offered anything to Romania at that time. Before that, other designs offered to Romania included Design 1290 of February 1927, with eight 6in/50 in three twin and two single ('A' and 'Y') mountings; two 3in HA, four twin torpedo tubes and two depth-charge throwers. The arrangement was as in the Spanish *Principe Alfonso*, with the amidships twin mount between the two funnels and the after controls. The bridge resembled that of the British 'County' class with its tower surmounted by a director. A second director was abaft the pole mainmast. Displacement was 6,100 tons (normal), 7,105 tons (deep) (500ft pp, 527ft lwl x 50ft 5in x 28ft 6in x 15ft 3in). The ship would have made 33kts on 70,000shp, and endurance would have been 3,500nm at 15kts. Design 927 (18 March 1927) was similar, and the difference in designation may indicate mainly that it was developed by a different designer, possibly George Thurston. The ship would have been faster (34kts, 87,000shp) and slightly smaller (6,000 tons, 500ft x 48ft x 28ft 6in x 15ft 8in). In August 1930 Romania was offered Design 1008, 5,750 tons normal and 6,600 tons fully loaded (480ft pp, 500ft wl x 49ft 7in x 27ft 6in x 15ft), armed with four twin 6in/50, two 105mm/40 HA, two twin 40mm HA, two quadruple torpedo tubes and two depth-charge throwers. She would have made 30kts on 50,000shp, and endurance would have been 7,000nm at 15kts. Complement would have been twenty officers and 480 ratings. Vickers offered a minelaying cruiser to Turkey (see the Appendix). It also offered the Design 881 light cruiser (estimate dated 20 January 1925) displacing 4,100 tons (430ft x 43ft x 24ft 3in x 13ft 9in) armed with six 6in/50, two 105mm HA, two pompoms and two triple torpedo tubes. Machinery would have been similar to that of HMS *Curlew* (40,000shp, 29kts). It is not clear to what extent this design answered a formal request for proposals connected with a programme; the largest surface ships Turkey actually bought between wars were destroyers. In 1938 the Turkish government sought large cruisers from Germany and from the United Kingdom, but it is not clear whether any designs were prepared (Vickers' material, which is very incomplete this period, does not mention any). Vickers' records also show a cruiser offered to Cuba (18 December 1929, not numbered): 4,878 tons (455ft x 46ft x 26ft 1in x 14ft 3in), six 6in, four 3in AA, two triple torpedo tubes and two depth-charge throwers; 40,000shp, 29kts. (National Maritime Museum)

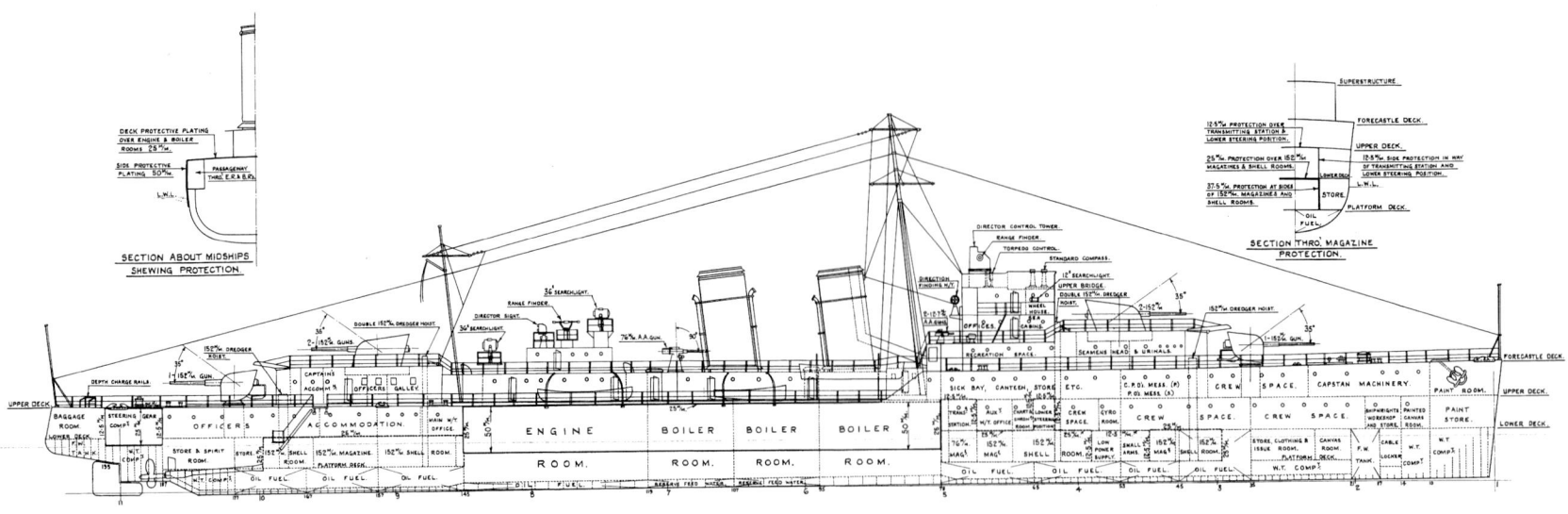
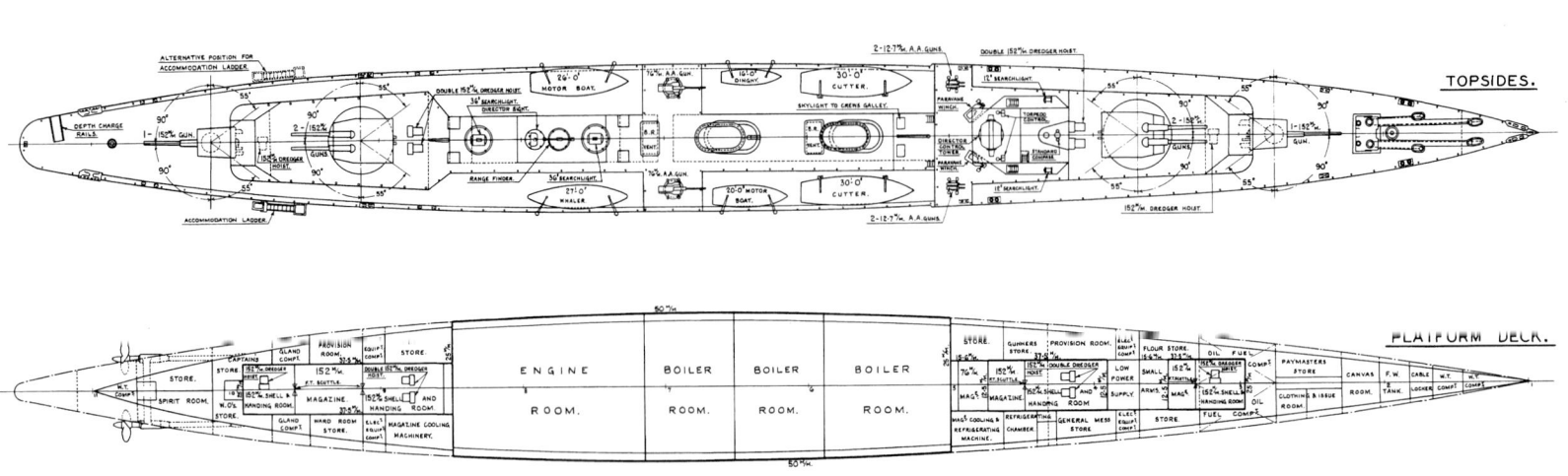

Vickers offered this Design 1054 to Norway in the early 1930s. The ship would have displaced 3,000 tons in trial condition (384ft pp, 407ft lwl x 38ft 6in x 22ft 6in x 12ft 6in). She would have made 30kts on 34,000shp. These data were undated, but the adjacent page (in a Vickers design notebook) of design data for an Estonian coast defence ship is marked 'to Thornycrofts 13-10-33'. The cruiser was part of a programme which included coast-defence ships (Design 1053, data dated 16 September 1933). Vickers archives also show an un-numbered small cruiser offered to Estonia, probably in the late 1920s, to be based on Design 925 (which, unfortunately, does not appear in surviving Vickers material): 3,250 tons (405ft pp, 429ft loa x 40ft x 12.5ft), 27kts (endurance 4,500nm at 15kts). Armament would have been four 6in, two 4in HA, four single pompoms and two triple torpedo tubes. The comparable Design 887 was a light cruiser for Yugoslavia (offered in conjunction with 888, a flotilla leader): two twin 6in/50, two 4in/45 HA, two single pompoms and two triple torpedo tubes on 3,000 tons (390ft x 37ft 6in mld x 22ft 6in x 12ft 6in, 27kts on 22,500shp with six Yarrow oil-burning boilers). Vickers' most interesting inter-war small-cruiser design was probably Design 776, for Japan (estimate dated 7 February 1921). The competition involved seems to have been used to obtain British ideas for what became the small Yubari. The central requirement was high speed on a limited displacement, so Design 776 was 4,080 tons (440ft x 44ft 6in x 25ft 6in [hull depth] x 14ft), designed to make 33kts on 59,000shp (six boilers burning only oil, three burning coal and oil). She would have been armed with Japanese weapons: four 14cm/50, one 8cm/40 HA, two single pompoms and four twin 53cm (21in) torpedo tubes. Protection would have followed wartime British practice (2½in side over machinery, 1in to the upper deck, and 1in bulkheads at the ends of the machinery spaces). The smallest alternative in the series was 776C (3,700 tons, 430ft x 43ft x 25ft 6in x 14ft, 31kts on 45,000shp with four all-oil and three oil and coal boilers). In May 1921 Vickers sketched an (unnumbered) cruiser armed with only two 6in/50, one 3in HA and four twin torpedo tubes: 4,100 tons (425ft pp x 42ft 3in x 24ft 7in x 14ft), with 40,000shp engines (six Yarrow boilers) for 29kts. The estimate sheet does not explain why the ship was so lightly armed. (National Maritime Museum)

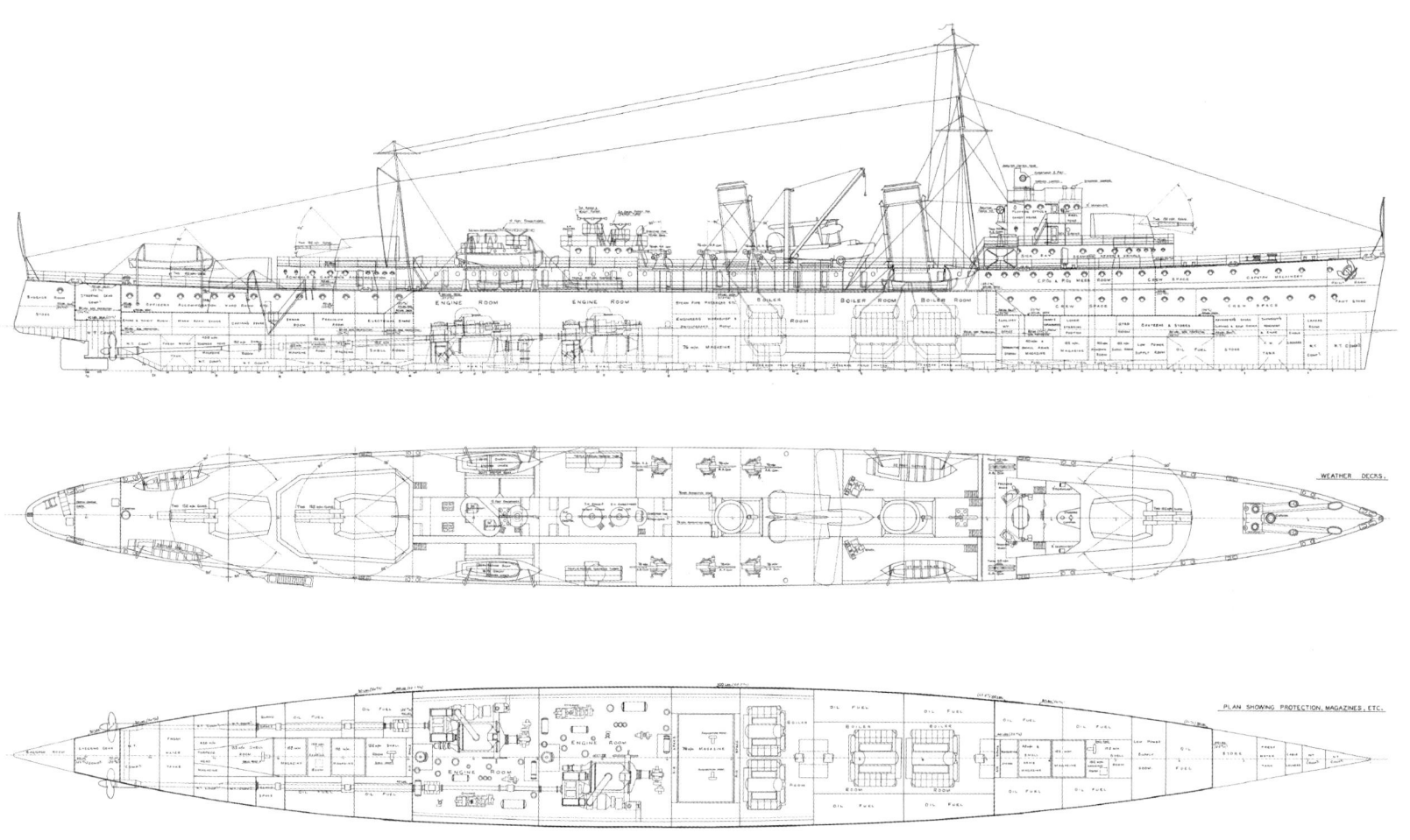

Design 1087 was offered to Siam; it presumably competed unsuccessfully with an Italian design, some time in the late 1930s (the Walrus amphibian also suggests this period). The ship would have displaced about 4,000 tons and would have made 30kts (no horsepower is given in Vickers papers). Note that she would have had fully-enclosed turrets for her six 150mm/50 guns. Other armament was six 75mm/50 HA, four twin 40mm, and two sets of triple 45cm (18in) torpedo tubes. Protection was 70–75mm amidships, 40–50mm fore and aft, and 35–40mm on deck. Dimensions: 450ft pp, 475ft wl, 478ft oa x 47ft 6in x 26ft 3in x 14ft 3in, 4,550 tons standard displacement. This seems to have been one of a series offered to the Siamese government, the others being a nominal 2,000-tonner (350ft pp x 37ft 6in x 21ft x 10ft 9in, 2,320 tons fully loaded) and a 3,000-tonner (3,400 tons: 400ft pp, 420ft wl x 42ft x 12ft 6in). Armstrong having won a Thai contract for the coast-defence ship *Ratanakosindra* in 1924, Vickers-Armstrong sold another such ship to Siam in 1928, but Kawasaki built the follow-on *Sri Ayuthia* class in 1937–8. Other Vickers papers identify the 2,000-tonner as Design 1056, armed with two 4.7in guns, two 3in HA, and two 21in torpedo tubes (2,160 tons). No armament details of the 3,000-tonner seem to have survived. Vickers also negotiated with China. A 6,000-ton (500ft x 52ft x 28ft 6in x 14ft 6in) light cruiser (Design 970, 7 September 1929) would have been armed with two twin 8in, four 4in HA, four twin pompoms, and two depth-charge throwers. She would have made 30kts on 46,000shp. The two considerably more modest Japanese-designed *Ning Hai* class seem to have been bought instead. (National Maritime Museum)

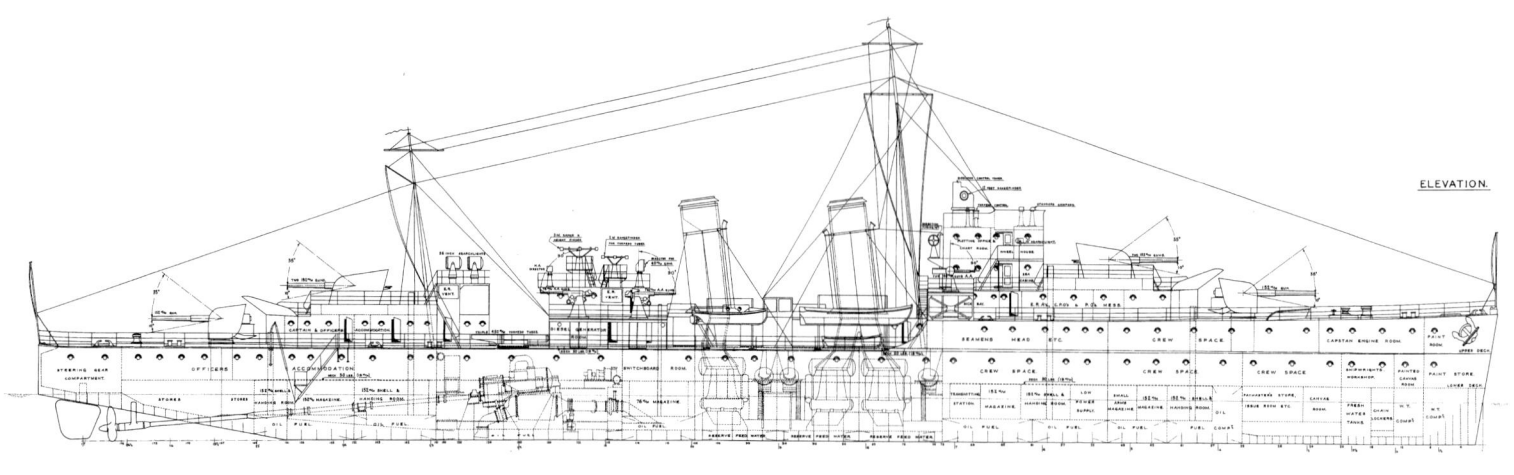
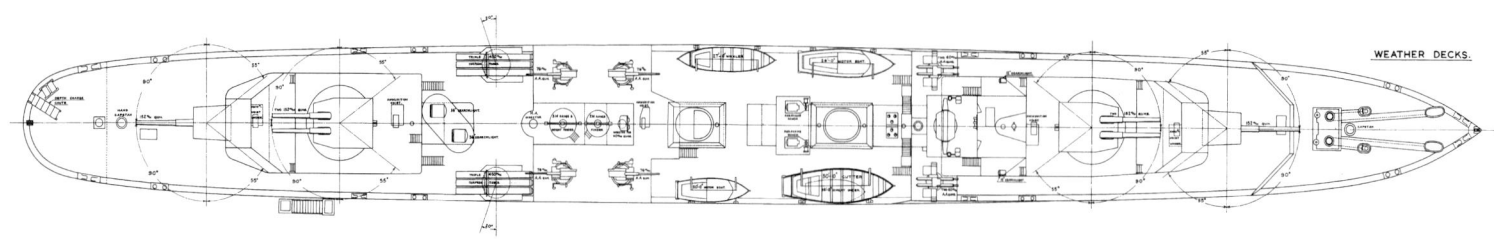
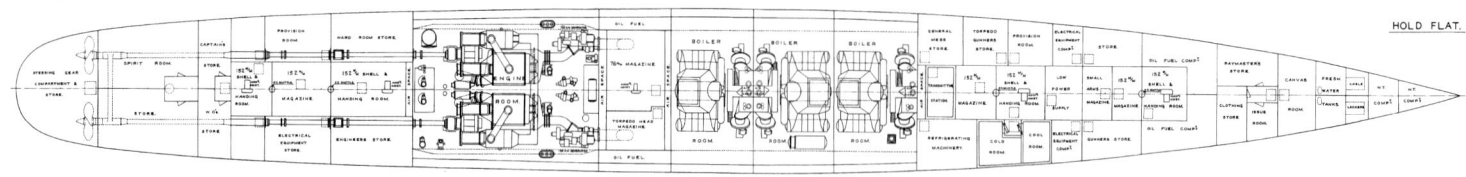

Vickers-Armstrong's Design 1089 was the late-1930s version of its small cruiser with six 6in guns, two twin and two single. Secondary armament was two twin 40mm (Bofors guns) at the break of the forecastle and four 76mm HA guns abaft the second funnel. Unfortunately Vickers-Armstrong records of designs for foreign customers in this period seem not to have survived. (National Maritime Museum)

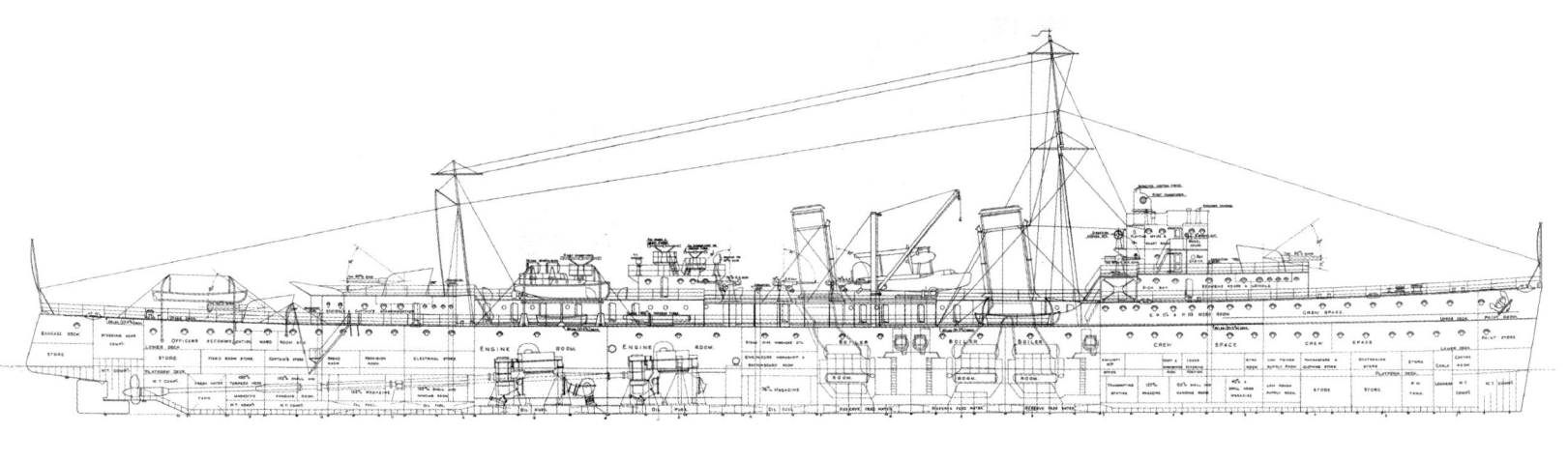
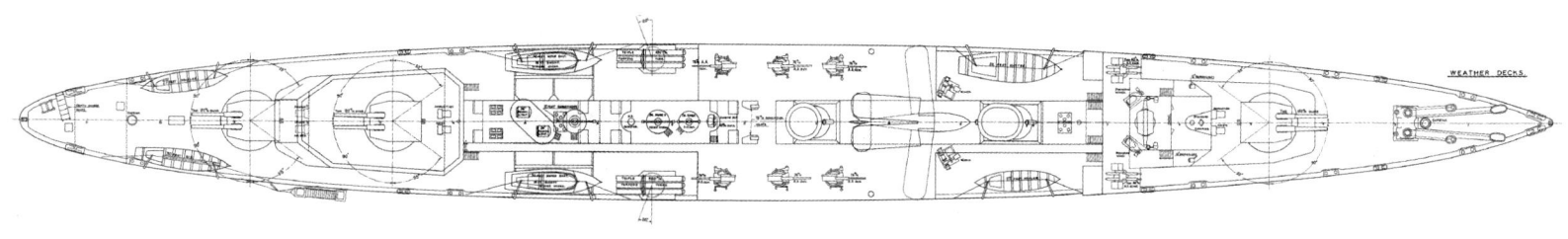
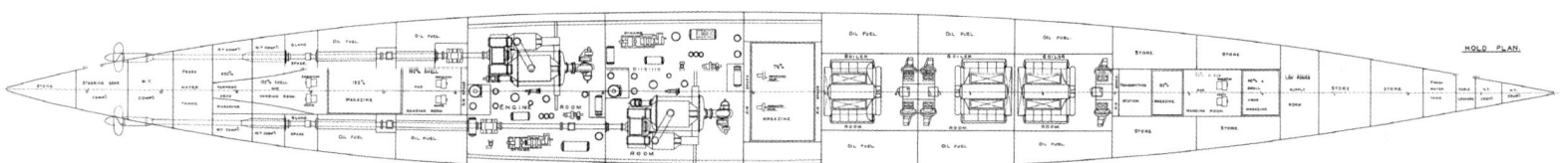

Design 1094 was a larger-cruiser counterpart to Design 1089, with three twin 6in mounts and a Walrus aircraft amidships (served by a crane, without a catapult). The Walrus was a Vickers Supermarine product, hence worth advertising as part of the design. It could operate without a catapult because it was designed to survive in ocean swells. Abreast the torpedo tubes are the rangefinder and height-finder (abaft it) for the AA battery, with a larger rangefinder (for the 6in battery) further aft. The 6in in this ship are still in open twin mounts; presumably Vickers also offered larger designs with enclosed mounts. Design 1094 or something similar may have been offered to Chile, which chose an Italian Ansaldo design instead in about 1937. The Chileans then decided to have something larger, touching off a diplomatic crisis as the British tried to ensure that they could not buy a treaty-busting 8in cruiser. When the crisis subsided, it turned out that Chile could not have afforded anything so expensive, and just before the outbreak of war Vickers-Armstrong was offering Design 1111, armed with four twin 5.25in guns. This design had alternating engine and boiler rooms (all four boilers in tandem, with oil fuel outboard). One of several unusual features was that the HA director forward was stacked atop the LA director, with a rangefinder forward of both. The identity of Design 1111 is clear because it figured in 1939 discussions between Vickers-Armstrong and the Admiralty.

CHAPTER 7
THE SLIDE TOWARD WAR

The 1936 Treaty

The seventy-cruiser policy collided with the demands of the 1930 treaty that many older cruisers be scrapped by 31 December 1936. On 6 October 1934 First Sea Lord Chatfield signed a Board memo advocating retention of many of these ships. He estimated that on 31 December 1936 the British Commonwealth cruiser force would include thirty-seven underage ships built and building plus twenty-two overage ships eligible for scrapping and replacement, including three ships earmarked for scrapping in 1935 (*Brisbane*, *Constance* and *Castor*). Ten of the latter group, including the four *Hawkins* class, would be scrapped under treaty terms. Given the grave world situation, Chatfield considered the shortage of cruisers a critical problem. The government had now accepted that seventy was the minimum acceptable, but there seemed to be no hope of paying for more than three or four a year. Chatfield hoped to gain more tonnage at the coming London Conference (1935) and also to change the scrapping programme. He also needed more personnel, having only enough for forty-four ships, Australia contributing another four and New Zealand two. The United Kingdom would have to provide the remaining twenty ships, of which eleven would be in full commission and nine in reserve. He estimated that he could meet manning requirements by 1942 without opening a new Boys' Training Establishment, so he therefore made 1942 the target year for filling out the seventy ships. Instead of

THE SLIDE TOWARD WAR

Curacoa typified the production version of the anti-aircraft conversion, armed with twin 4in guns and one quadruple (rather than octuple) pompom in place of her original five 6in guns. Two quadruple 0.5in machine guns replaced the two 3in anti-aircraft guns originally mounted abeam the forefunnel; by the time of this photograph they had been replaced by Oerlikons (*Curacoa* had a total of five, plus two single pompoms, which are probably the objects abaft No. 3 twin 4in gun). Conversion originally entailed erection of a pole mainmast carrying a DF coil, but when ships received air-warning radar (Type 279 in this case, during conversion, in January 1940) the DF coil was moved to between the funnels. The Type 271 surface-search radar was installed some time before September 1942 (the ship was lost that October). Type 285 HA control radar was installed in June 1941.

All of these factors suggest that this photograph was taken shortly before the ship was lost in a collision with the liner *Queen Mary*.

scrapping the four *Hawkins* (which exceeded the limit on cruisers with guns of more than 6.1in calibre), he planned to rearm them with 6in guns, on the grounds that their high endurance and good sea-keeping made them more valuable than 'C' or 'D' class cruisers.

The Naval Staff hoped to 'isolate' the big new 6in gun cruisers at the coming conference, in which case the *Hawkins* class would be among the limited number of such ships the United Kingdom could retain, although they would clearly be inferior to newer foreign counterparts. If no such isolation proved possible, they should be rearmed and retained, 'C' class cruisers being scrapped to make up for any tonnage ceiling. *Frobisher* was already used for cadets' sea training, and no other suitable cruiser was available, so if the class was retained, she would continue in this role. If not, she would be demilitarised and retained. In the end, Chatfield got both 'isolation' of the biggest cruisers and, when the British government invoked the escalator clause of the 1930 treaty in July 1936, the right to retain all of the ships which would otherwise have been scrapped, the latter.

The capital ship building 'holiday' of the Washington Conference and the London Naval Treaty of 1930 expired on 31 December 1936. Despite a dramatically worsening international situation, neither the British nor the US government could easily abandon naval arms control, because in both countries much of the population saw arms control as a guarantee that nothing as brutal as the First World War could happen again. Few wanted to acknowledge that several major governments seemed to welcome another war; surely no one could be that insane, particularly in Europe. Rearmament was therefore a reluctant process. During preparatory talks for the next naval conference, in March 1934, it became clear that neither the United States nor the United Kingdom would accept the Japanese demand for parity, although both hoped that some deal could be arranged.

The Admiralty Board revived its demand for seventy cruisers: sixty modern and ten overage ships (twenty overage if necessary). First Sea Lord Chatfield said that he would prefer no agreement at all on cruisers to abandoning the seventy-cruiser goal. The British also wanted to do away with the new large 6in cruisers, Chatfield proposing a 7,000-ton limit on future cruisers. The US Navy was willing to accept a limitation on the size of future cruisers, once it had completed the nine large *Brooklyn*s and it had its eighteen 8in cruisers. US negotiators pointed out that the American public expected arms reductions, not the increase envisaged by the British; in June 1934 President Franklin D Roosevelt called for a ten-year treaty which would, over that period, cut strength by 20 per cent. The British answered that they now faced both European and Far Eastern threats, and that they could not depend on the United States to help them against the Japanese, and they therefore demanded substantial modification to the 1930 London Naval Treaty. Complicating the situation was escalating rivalry between the two Washington Treaty powers which had not signed the 1930 treaty, France and Italy. Given the Japanese insistence on parity, which was utterly against British and US interests, by October 1934 Chatfield doubted that any treaty could be negotiated. British attempts to negotiate with Japan were fruitless, but it proved relatively easy to work with the Americans. Through late 1934 and early 1935 British also talked with the French, and with the Germans, who wanted an agreement allowing them a fixed percentage of British naval strength.

By July 1935 the British favoured qualitative limitation, such as a cap on cruiser tonnage (at this stage, 7,600 tons). Overall tonnage limits were not as important, though still desirable. The British came to value a requirement that each country announce its building programme in advance. Any power deciding to accelerate construction would risk other powers reacting before its planned ships could be completed. US reactions were mixed; the US Navy's General Board correctly predicted that the British would support qualitative limits and the announcement provision. Despite considerable pessimism, on 23 October 1935 the Cabinet decided to issue invitations to a naval conference. President Roosevelt told his delegates to continue to press for limits on overall tonnage, but admitted that once the Japanese demands were rejected that would become unlikely. He therefore hoped for a tripartite (with Japan) qualitative limit backed by an escape clause which would deter non-contracting parties. This was the evolving British position. The British and American delegates met informally in London the day before the conference opened, agreeing to resist the Japanese demand for a common upper tonnage limit (i.e. parity). In January 1936 the Japanese withdrew from the conference altogether, having been as disruptive as possible up to that point.

Only Britain, France and the United States ratified the resulting treaty, which eliminated all overall tonnage limits. It incorporated an Italian proposal that programmes had to be published annually within four months of 1 January, that at least four months had to elapse between announcement and keel-laying, and that programmes could not then be changed (Italy did not ratify the treaty). An 8,000-ton limit was placed on cruiser tonnage, and the ban on new 8in cruisers repeated. Without any limit on total battleship tonnage, any power might build super-cruisers it called battleships. To make that difficult, the treaty placed a lower limit on battleships, 17,000 tons and 10in guns (35,000 tons and 14in guns was the upper limit). To encourage Japan not to exceed that limit, the treaty contained an escalator clause allowing for increases in the event a non-signatory refused to accept these limits; the signatories were to consult together to minimise their departure from the agreed terms. On this basis battleship gun calibre was soon increased to 16in and two years later maximum battleship tonnage was increased to 45,000 tons. A special clause for cruisers allowed any party to announce that it was building cruisers of up to 10,000 tons, which could be armed with guns of greater than 6.1in calibre. This would release other signatories from their obligation not to build such ships. British constructors' notes suggest that on two occasions the British government considered activating the escalator clause for cruisers (but they did not do so). The treaty was signed on 25 March 1936, coming in effect on 31 December 1936. It was intended to run until 31 December 1942.

Anti-Aircraft Rearmament

Rearmament funds made it possible to carry out some of what the 1932 air defence committee had suggested. The multiple HACS (and pompom directors) envisaged by the 1932 committee made it possible to split anti-aircraft fire in various ways, so the Royal Navy devised a centralised system controlled by an Air Defence Officer (ADO) in an Air Defence Position (ADP), communicating both with the captain and other control officers, and with the guns.[1] Ideally the ADP was a central position high enough in the ship to be clear of funnel haze and cordite smoke, with an all-round overhead view, close to, but independent of, the compass platform. In 8in cruisers that meant a position at the front of the DCT barbette; in light cruisers, the roof of the captain's shelter. The ADO maintained an air plot which, in theory, allowed him to allocate his ship's anti-aircraft resources. That plot in turn was adapted to radar inputs during the Second World War. ADO was linked both to air-defence lookouts (who sent target bearings mechanically from their sights [target bearing transmitters in US parlance]) and to the weapons via their directors. Short-range weapons were grouped (one Mk M pompom, which had a director, and one quadruple 0.5in machine gun). Particularly when aircraft were attacking at short range, voice control

THE SLIDE TOWARD WAR

Above and below: Clearly connected with the anti-aircraft rearmament programme was a pair of prototype anti-aircraft cruisers, HMS *Coventry* and *Curlew*. HMS *Coventry* is shown as completed. She had two items in short supply at the time, a pair of HA directors (with their associated computers) and an octuple pompom in 'B' position. The main battery was ten 4in guns; conversion was simplified by using the existing structure: 'A' and 'B' positions for 4in and pompom, the positions for two 3in anti-aircraft guns abeam the forefunnel, the platform for No. 3 6in gun abaft the after funnel for two more 4in guns, the after 6in positions for 4in guns, the after 3in anti-aircraft position for the pompom, and an additional position on the quarterdeck. The after pompom was removed from both ships in 1938–9 and replaced by a pair of quadruple 0.5in guns. About 1939 the stub mainmast was replaced by a pole carrying a DF loop. (James Fahey collection of US Naval Institute).

was impossible due to the noise, so mechanical connections between ADP and other elements of the system were essential.

When the air threat was re-evaluated in January 1937, the emphasis was still on long-range barrage fire to break up attacking formations, including high barrages by guns without specific anti-aircraft control. However, the dive-bombing threat was fully accepted (and was expected to increase in importance). Future aircraft would fly faster at higher altitudes, so fire would have to be opened at greater ranges (8,000–12,000yds) to sustain fire long enough to inflict damage, and to keep enemy aircraft out of ranges from which they could make dive-bombing attacks. Attention was (and had been) paid not only to the needs of the fleet at sea, but also to fleet air defence at anchor, because the fleet would often have to operate from unprepared anchorages, even places like Alexandria, the main Eastern Mediterranean fleet anchorage.

A key point entirely missed by the Royal Navy and by all other navies at this time was that anti-aircraft fire tended to ruin pilots' aim rather than actually shoot aircraft down. Thus the fact that light weapons fired tracer was essential, because pilots saw the tracer flashing past them, but for the 1937 report, tracer was valuable only to the extent that it showed large errors and indicated (to those on the ship) which target was being engaged.

The upgrades, class by class, proposed by the 1932 anti-aircraft committee in effect formed the basis of the fleet rearmament programme. Each ship should have two HA control systems, at the least offering simultaneous control on both sides. Given foreign improvements, 'it is clear that an increase in the strength of AA armaments throughout the British Navy is overdue' (this passage in the report was underlined). Due to lack of space, the maximum battery possible in

Above: Coventry shows early-war modifications in a 29 June 1940 photograph, including the addition of splinter shields around her 4in anti-aircraft guns. She had been fitted with topmasts (to take air-warning radars), so her forward HA director was relocated to the fore end of her foretop. The two 4in guns abeam her forefunnel had been landed, presumably to reduce topweight. The ship did not receive Oerlikons until a refit at Alexandria in May 1942. She was sunk on 14 September 1942.

cruisers and capital ships was four guns on a side in twin mounts. Unfortunately that was not enough (given existing standards of control) to bring down one aircraft in a group of three.

In the committee's view the *Leander*s had the maximum battery they could accommodate, and they could not take any Mk M pompoms. Given the HA capability of the 6in guns, it was essential that they be given some form of fire control. There was no space for a second HADT. The positions selected for the four 0.5in machine guns were the best possible.

In *York* and *Exeter* a second HADT could be bracketed to the bridge structure, above and abaft the main DCT (the sided positions suggested for larger 8in cruisers were not practicable). The 4in single mounts could be replaced by weather-deck twins. A Mk M could replace the existing single pompoms, but the position was close to 'B' turret, and there was too little space. The single pompoms could be replaced by quadruple 0.5in machine guns, but these positions were untenable when the 8in guns fired. The alternative was to place them atop the roofs of 'B' and 'X' turrets. Ideally two could go on the turret tops and two in place of single pompoms.

Norfolk and *Dorsetshire* had their single 4in guns amidships, on a boat deck surrounding their funnels, rather than aft, as in earlier large cruisers. This position apparently made it possible to replace each single mount with a BD twin, as there was considerable space under the deck. Two Mk M pompoms could be mounted inboard of the 4in guns, between Nos. 1 and 2 funnels. Four quadruple 0.5in machine guns could be mounted on platforms built out from the deck supporting 'B' turret. HADTs could be placed alongside and below the DCT, and directors for the Mk M pompoms on the after corners of the upper bridge.

The *Kent* and *London* classes had their single 4in guns on a boat deck somewhat further aft, alongside the second and third rather than the first and second funnels. This platform was built above the torpedo tubes; there was no surplus space underneath for a BD mounting. Instead, the Committee proposed twin mounts for the 4in guns, and positions for the Mk M pompoms alongside the forefunnel. BD mounts could be installed if the torpedo tubes were moved (one in place of an existing 4in gun, one in place of single pompoms), but then the Mk M pompom would be subject to 4in blast. The ships had high bridge towers carrying their DCTs; the Committee envisaged placing the two HADTs alongside.

The two *Emerald*s already had an awkward anti-aircraft battery arrangement, with two guns on the sides and one at the after end of the after gun deck, abaft the superimposed 6in gun. The Committee suggested replacing all three with twins. The ships already had HADTs between their after control and after funnel, a position considered preferable to the original AA control position on the after control, to keep it clear of smoke. However, the ships were scheduled to receive new tripod masts just forward of their after funnels, which would block after arcs. No forward position was available, so the Committee proposed placing the director in the original AA control position atop the after control, accepting the smoke problem but expecting it to be reduced when the funnels were raised, as planned, 15ft (when the new mast was stepped). No position was available for a second HADT, or for pompoms, but

THE SLIDE TOWARD WAR

Above and inset: The anti-aircraft cruisers proved so valuable that two more ships, HMS *Caledon* and *Colombo*, were converted in 1943. HMS *Caledon* is shown on 1 February 1944. Improvements included more close-in (self-defence) weapons, on the theory that the ships would be valuable targets in their own right. Instead of pompoms, they had two twin Hazemeyer Bofors guns, visible amidships. Other close-in weapons were three single Bofors (not in *Colombo*), six twin power Oerlikons and two single Oerlikons.

there was space for two quadruple 0.5in machine guns. The Committee made no recommendations either for the 'C' and 'D' class cruisers or for the *Hawkins* class.

The Board understood the importance of anti-aircraft modernisation, and planned to upgrade ships when they came in for large repairs. Thus it began to consider AA modernisation of the *Kent* class heavy cruisers, in 1932, and definite plans were completed in September 1934.[2] At this stage the HA improvement was to have been limited to replacing the four single 4in guns with two twin and two single 4in, and there was no proposal to install Mk M pompoms. In May 1935 it was decided to replace the one twin 4in mount on each side in *Penelope* and *Aurora* (*Arethusa* class) with two, and to add a second HACS, which brought the ships into line with the *Kents* (but the 1936 programme did not include a second HACS for anything but heavy cruisers). Again, there was no mention of Mk M pompoms. At this stage the Naval Deficiency Programme included capital ship but not cruiser modernisation, and it did not include new anti-aircraft weapons for the cruisers.

Emerald and *Enterprise* were included in the list of prospective improvements, but nothing could be done before the outbreak of war, and then the planned 1940 refits were abandoned. However, both ships were modernised in 1943.[3]

The crisis between Britain and Italy over the invasion of Abyssinia nearly brought the two countries to war. Italy had a large air force, and the threat of war dramatised the need for heavier anti-aircraft armament

205

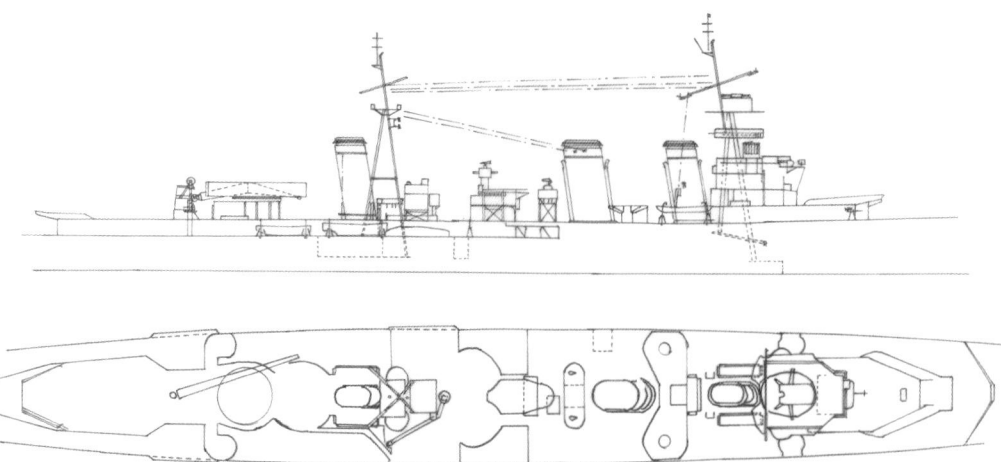

Emerald as refitted in 1943, from Portsmouth plans dated 24 March 1943. The dashed box at the foot of the foremast, below decks, was the Second (and Auxiliary) Wireless Office; the main office was the dashed box under the mainmast. The yard shown on the foremast carried the small crosses of Type 86 tactical radio at its ends (one is indicated). The multiple strands of the ship's 'flat-top' antenna are indicated, the line leading down above the forefunnel being the receiving antenna (one on each side). Below it, the ship's foretop carried a rangefinder. The 'lantern' atop the bridge carried the ship's Type 273 surface-search radar. The masthead carried the receiving antenna of her Type 281 air-warning radar, with the transmitter on the mainmast. Above the forward Type 281 antenna was the 'hayrake' of the associated Type 243 interrogator (for IFF). The small rectangles on the main starfish are the FV1 countermeasures antennas. The plan associates the two 'hourglass' omni antennas below with the ship's Type 91 jammer, one for receiving and the one below for transmitting (but the contemporary FV1/Type 91 manual claims that the FV1 antennas were generally used to transmit, which would make sense because they were somewhat directional; the omni may have been an alternative). The only wireless antenna shown aft (of several) is the emergency receiver shown against the after funnel. The quad pompom on this side is not shown, but note its platform, with the pompom director (with Type 282 radar, and Type 282 radar office below it) abaft the second funnel. Abaft it is the after control structure, with its HA director topped by a Type 285 radar antenna (with Type 285 office below). The light gun under 'B' mount is an Oerlikon, the only one included in this elevation drawing. Boats shown are a 27ft whaleboat abeam the forefunnel and, aft, a 30ft motor boat forward of a 25ft motor boat. On the opposite side of these boats was a single 36-footer. The object on deck forward of the after funnel is the main (wireless) aerial trunk. Not visible in this drawing, but evident in the original, were 25ft booms on each side under the slight overhang aft, and two booms rigged from each side forward, from the superstructure just forward of where the bridge met the shelter deck level: two sounding booms abaft two lower booms. Note the absence of a main battery fire control radar (Type 284). (Norman Friedman)

in British battleships and cruisers. *Coventry* and *Curlew* were rearmed on an emergency basis during the crisis, Mediterranean Fleet anti-aircraft armament being considered too weak. They were presumably conceived as exactly the sort of fleet anti-aircraft ships envisaged by the 1921 anti-aircraft committee. The 1935 provisional armament statements for both ships described them as 'special duties' ships. Armament was ten single 4in guns (250 rounds each) and two octuple pompoms (later one: 1,800 rounds per barrel), and an HACS was installed, its HADT occupying the top of the foremast. Torpedo tubes were landed. Both ships had two single 4in removed in 1939–40; *Coventry* later landed a third gun.

In the wake of the crisis, a two-part fleet anti-aircraft upgrade was proposed in 1936. Class I modernised the main fleet, while Class II provided trade-protection anti-aircraft ships.[4] The Treasury approved Class I spending in April 1936, and an AA Rearmament Committee was formed. Plans called for adding four 4in guns (making a total of eight) to all 8in cruisers, as well as a second HACS, and bringing short-range firepower up to two octuple pompoms with directors and two quadruple 0.5in. Ships would receive twin 4in guns at large repair, eight single guns being an acceptable interim battery, the guns becoming available as battleships and 6in cruisers were rearmed with twins. The *Kent*s were already being rebuilt to this standard.

In 6in cruisers (*Leander*s, *Amphion*s [including *Sydney*], and *Arethusa* and *Galatea*) single 4in guns would be replaced by twins, but there was no mention of pompoms. Ships would not receive a second HACS. *Achilles* was not refitted because she was in New Zealand waters (the guns were replaced only in 1943–4), and HMAS *Sydney* (ex-*Phaeton*) was not refitted because she was already in Australian waters (she was sunk carrying her single 4in guns). *Arethusa* and *Galatea* only had their upgrades completed in December 1940 and September 1941.[5]

Including capital ships and carriers, the programme required fifty twin 4in guns (Mk XIX twins), twenty-five sets of HACS, twenty-two octuple 2pdr with their directors and ammunition, and ten quadruple 0.5in guns. Equipment on hand would be used in 1936–7; deliveries of new twin 4in guns (two per month, rising to four in July 1937) could begin in February 1937. HACS deliveries (one or two per month) could begin in June 1937. The only really new item was the proposal to fit octuple rather than quadruple pompoms in the 8in cruisers. The programme would complete AA upgrades in two to three years; otherwise the programme of large repairs would not have been completed until 1942 or later. Eight cruisers would be rearmed in 1937 and *Apollo* (HMAS *Hobart*) in 1938.

Class II included rearming cruisers, 'V' and 'W' class destroyers, and sloops as convoy AA escorts. Initial funds bought weapons for six 'C' class cruisers and HMS *Whitley*. The programme responded to a shift towards the possibility of a European war during which British shipping would be threatened by shore-based aircraft (that was far less likely in a Far Eastern war). Escort rearmament was initially given priority over smaller fleet cruisers for which definite dates had not yet been settled: the rest of the *Leander* and Improved *Leander* classes, *Arethusa*, *Galatea*, *York* and *Exeter*, and replacement of singles by twins in *Cumberland* and *Suffolk*.

In December 1936 the British government assured the Japanese and US governments, the other signatories of the 1930 London Treaty, that five of the 'C' class cruisers it was retaining would be converted into anti-aircraft ships, and that they would be scrapped by 1941. As of March 1937 plans called for taking four cruisers in hand during 1938

THE SLIDE TOWARD WAR

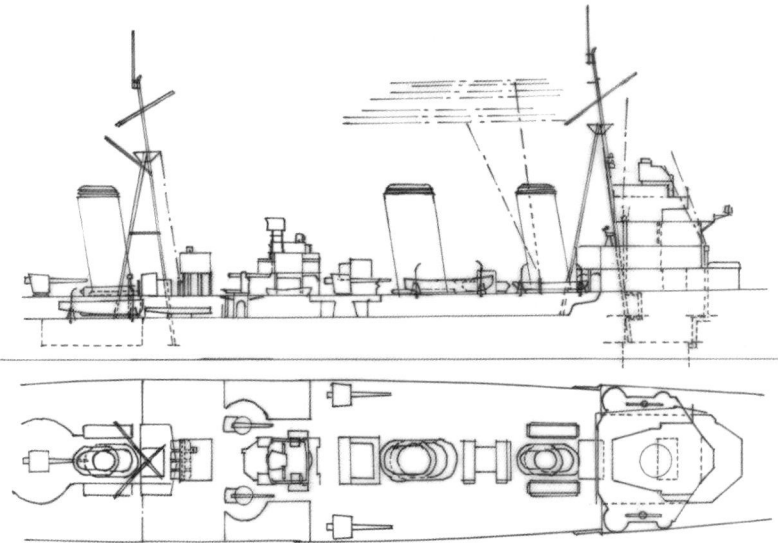

Enterprise as refitted 1943, based on a 17 February 1942 drawing of her planned rig, done by Devonport Dockyard. The dashed lines below her bridge structure are cable passages leading down to her Second Wireless Office, below decks, the trunk for its main aerial being visible as a vertical pipe forward of her bridge structure. Both of the more or less vertical lines near the bridge indicated antennas for the Second Wireless Office, terminating on the 32ft signal and wireless yard shown. Below it were two shorter yards (not shown) with fixed manoeuvring lights at their ends. The trunk higher up in her superstructure, nearer her mast, served a secondary Type 30 wireless. The ship's LF and MF wireless sets used 'flat top' antennas, for which wires were rigged between the two masts, with vertical elements connected to them. These wires are shown in part as dashed lines. They served a Type 36S wireless. The two slanting dashed lines abreast the fore funnel were receiving aerials. The raked dashed line is simply the foot of the fore leg of her tripod foremast. Topmasts were the receiver (foremast) and the transmitter (mainmast) of her Type 281 air-warning radar. Not shown is an additional line leading up to an insulator on the fore side of the forefunnel. It is described as part of the Type 281 receiving antenna. The large dashed space below decks aft is the main wireless office. The slanting vertical dashed line leading up to the mast is one of two receiving antennas for this office. The corresponding transmitter emerged from a trunk at the fore side of the after funnel. An outrigger on the side of the after funnel (not shown) carried the upper end of a short-range wireless antenna, which was a nearly vertical wire. The object which appears to be carried by stanchions is the Type 281 transmitting office, serving the transmitter atop the mast. The Type 281 receiving office occupied the after extension of the shelter deck, atop stanchions. This mast carried a 24ft yard. The box abaft the second funnel was the Type 285 office. The Type 284 office occupied the after part of the second bridge level, the one at whose forward end an MF DF coil was bracketed. In the plan view, the positions just abaft the open (12ft) rangefinders in the bridge structure are for pompom directors (the pompom positions were not shown). The two long rectangles flanking the forefunnel were boiler room vents, as were the two rectangular objects athwartships between the first and second funnels. Two more boiler room vents flank the after funnel. Note the two searchlights sided just forward of the HA director on the control structure between the second funnel and the Type 281 transmitting office. No others are shown. The object atop the Type 281 transmitter office is an insulator group. The object below the fore starfish is an electric steaming light above an oil-burning steaming light. Boats (omitted from the plan view) were a 27ft whaler on each side near the break of the forecastle. Abaft this boat on the each side was a 32ft cutter. The boat abeam the after funnel was a 30ft motor boat. (Norman Friedman)

for completion by that December. The programme applied to six 'C' class cruisers and eight 'D' class cruisers, in addition to the two earlier conversions. Selected merchant ships would receive the same set of weapons and fire controls as a converted 'C' class cruiser.[6]

Basic requirements for the 'C' class conversion were laid down in around April 1937. The conversion design was adapted to the later 'C' class cruisers with superfiring guns in 'B' position; converting any of the earlier ships would require a separate design, and hence take too long. Ten such ships had been built, of which *Coventry* and *Curlew* had already been converted into anti-aircraft cruisers. The five ships initially earmarked were *Cardiff, Ceres, Caledon, Calypso* and *Caradoc*, but *Caledon* and *Caradoc* were removed from the list because they were of the earlier non-superfiring design. By July 1937 plans called for taking five ships in hand during 1938: *Calcutta, Cardiff, Cairo, Ceres* and *Colombo*. The remaining six 'C' class cruisers would follow during 1939: *Capetown, Carlisle, Curacoa, Caledon* and *Calypso* (in December 1937 three *Caledons* [non-superfiring] were omitted from the programme). Given delays in the programme, in June 1938 the Admiralty proposed that instead of modifying five ships soon to be scrapped, they or an equal number of 'C' class cruisers of equal or greater tonnage should be scrapped before 1941, and an equal number of ships with longer remaining lives (to 1943 or 1944) substituted as anti-aircraft cruisers. The five named ships would be refitted for continued service as 6in cruisers. Other candidates for scrapping by 1941 might be a 'D' class cruiser. The other warships envisaged for AA conversion at this time were thirty-six 'V' and 'W' class destroyers, two *Stork* class, six *Grimsby* class and sixteen *Milford* class sloops, and one other escort. The list was considered provisional, e.g. more destroyers could be converted at the expense of 'C' class cruisers.

Sometime in 1937 Class I received first priority, Class II cruiser and destroyer rearmament moving to the end of the programme, beginning in 1939 (although initially a 1938 start was planned). Main fleet ships (capital ships, carriers and cruisers) were rearmed first, followed by modern escorts (sloops and minesweepers), and then by the AA escort programme. A parallel programme provided existing destroyers with modern anti-submarine armament (Asdic and increased depth charge batteries).[7]

As of February 1939, two cruisers (*Cairo* and *Calcutta*) were in hand at Portsmouth and Chatham respectively. They were completed on 3 June and 6 March 1939 respectively. *Calcutta* was followed by *Curacoa* (to be taken in hand June 1939, completed 24 January 1940). The first

Because she was so much larger than the 'C' and 'D' class, *Emerald* (and her sister, *Enterprise*) was much more extensively modified during the Second World War. She was refitted at Portsmouth between August 1942 and 6 April 1943. These photographs were taken on 14 March 1943. Her bridge was rebuilt. The old pair of single pompoms and the pair of quadruple 0.5in machine guns were landed. Instead she was given two quadruple pompoms and six twin 20mm guns. Two of the four quadruple torpedo tubes were landed. She retained her three single 4in anti-aircraft guns. Radars installed at this time were Types 281, 273, 285 and 282 (the latter for the pompom directors). The catapult was removed during her 31 March – 5 April 1944 Rosyth refit.

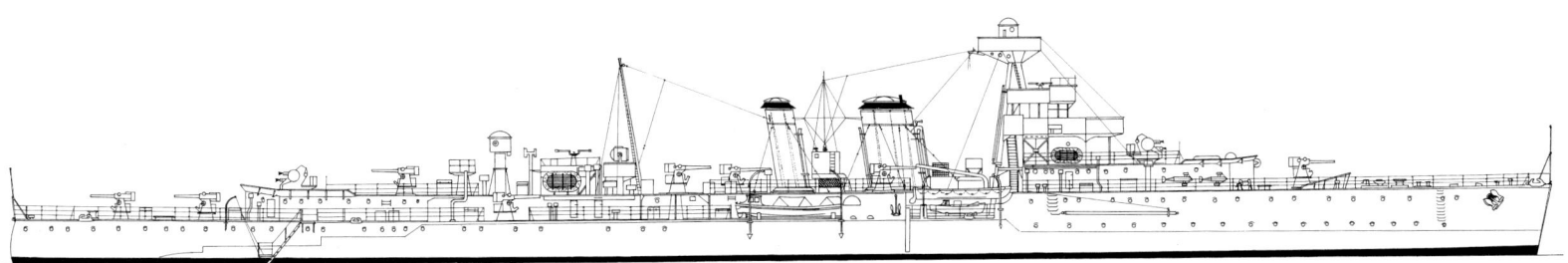

Above: *Coventry* as converted into an anti-aircraft cruiser in 1936. (John R Dominy)

Below: *Coventry* as an anti-aircraft cruiser in 1940. She shows topmasts and a tall mainmast for her Type 279 radar, plus splinter bulkheads. (Henry R Dominy)

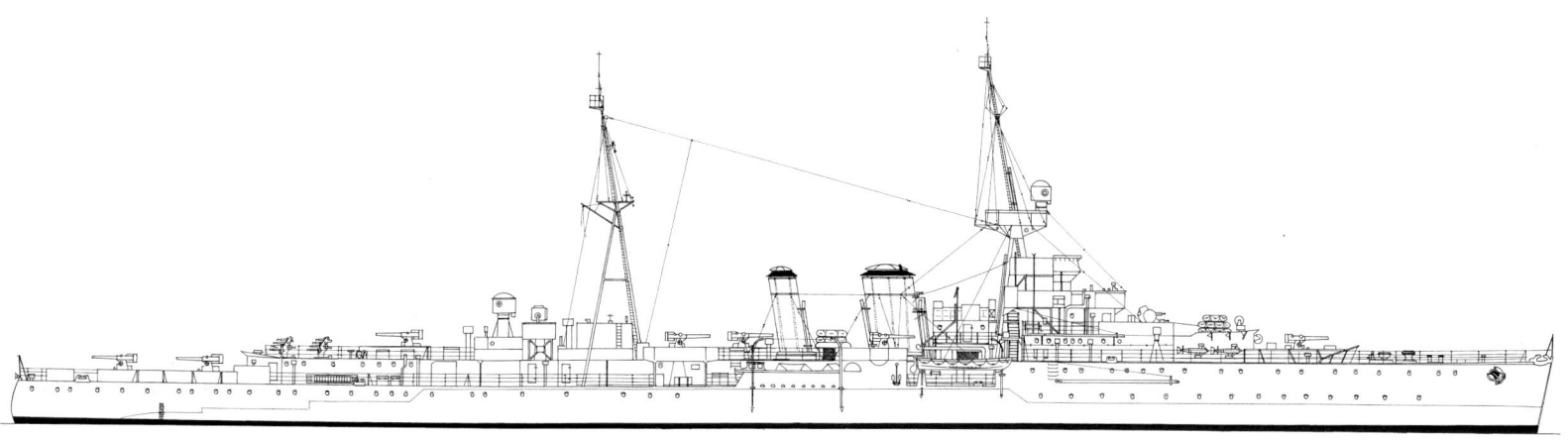

ship converted at the third yard, Devonport, was *Carlisle* (taken in hand April 1939, completed 20 January 1940), to be followed by *Colombo* (to start September 1939 for completion by the end of January 1940) and *Capetown* (to be taken in hand in November 1939). The outbreak of war precluded conversion of *Capetown* and *Colombo*, so only four ships were completed.

Initial plans called for all existing armament to be replaced by four twin 4in HA guns, two quadruple pompoms, and two quadruple 0.5in machine guns.[8] The HADT was bracketed on the tripod foremast, on the ADO platform. DNO preferred a fifth twin 4in gun to the pompoms, self-defence weapons being of little or no value for area convoy defence. The two 0.5in machine guns would offer some defence against point-blank attacks 'which an AA cruiser, if properly fought, was more likely to encounter than close range air attack', i.e. dive bombing. If required, a second HADT could be mounted on the boiler casing aft, as there was space aft for a second HA computer. However, the second HADT could not be delivered until 1942–3, by which time (as seen in 1937) many of the ships involved would already have been scrapped. It would not be capable of LA fire control. It turned out that weight precluded substitution of an octuple for a quadruple pompom. Meanwhile DNC was producing a sketch design. It turned out that topweight precluded mounting a twin 4in gun in 'B' position, so the sketch showed a pompom there, with a twin 4in abaft the funnels and two in 'X' and 'Y' positions. Formal Staff Requirements were developed at a meeting on 26 August. To the basic role of helping to protect a convoy against persistent attacks by land-based aircraft, it added as secondary roles the protection of advanced naval bases and, 'in emergency, supplementary protection to United Kingdom seaports'.

The Staff did want the second HACS, because it wanted to be able to split the guns into two groups of four guns each. For self-defence the ship should have one quadruple pompom, plus the usual pair of 0.5in machine guns. Space would be left for a second HACS, but no wiring would be run. However, the second HACS Mk III featured in a summary of the ships' fire-control systems produced in August 1940. The fire-control computer space forward was a combined transmitting station (for surface fire) and HA computing position; the analogous space aft contained only an HA computer. A provisional armament statement (1940) for the first four ships (*Cairo*, *Calcutta*, *Carlisle* and *Colombo*) showed eight twin 4in (270 rounds per gun), one quadruple pompom (1,800 rounds per barrel), and two quadruple 0.5in machine guns, plus six Lewis guns. In common with other British cruisers, they carried depth charges: two chutes with six First World War Type D depth charges.

Interest in conversions revived in 1942, a Staff Requirement being issued that April. Now it was recognised that the ships would be important targets for enemy aircraft, hence that they needed more close-range anti-aircraft weapons, which could be fitted only at the expense of one of the four twin 4in. Plans thus called for three twin 4in ('A', 'B' and 'Y' positions, all with RPC), two twin (Hazemeyer) Bofors in 'Q' position and two more on each side in 'X' position; and eight Oerlikons (four on the bridge structure, with good forward arcs, two abreast the forefunnel and two aft). They would have one HACS director above an ADP, directly abaft the compass platform, the corresponding computer (table), and a barrage director aft. They would be fitted for fighter control, with the necessary automatic plot. *Colombo* and *Caledon* were

Above and below: HMS *Dauntless* shows typical mid-war modifications in these 1 May 1942 photographs. At this time she had three 4in anti-aircraft guns, two single pompoms (the weapons at shelter deck level abeam the bridge), and eight Oerlikons. She had a Type 280 air search radar on her foremast and the 'lantern' of a Type 271 aft. The radars and the guns were probably fitted during a Portsmouth refit (23 February – 8 March 1942). *Durban* showed similar features in September 1942. Some ships were never really modified: as late as April 1943 *Diomede* and *Durban* had the pre-war battery of two single pompoms and two quadruple 0.5in machine guns. They also had eight single Oerlikons. *Despatch* lacked the machine guns, and *Durban* had no 2pdrs. *Diomede* was converted into a training ship at Rosyth between July 1942 and September 1943.

Above: HMS *Danae* received an elaborate 'D' class upgrade, as shown in these August 1943 photographs. She was the prototype for a projected modernisation of the 'C' and 'D' classes. Her No. 3 6in gun was replaced by a pair of quadruple pompoms, each with its radar-equipped director (using Type 282 radar). All the single 4in guns were landed, but a twin 4in was mounted to superfire above the aftermost 6in gun. No separate 4in director was fitted. All torpedo tubes were landed. The ship was unusual in that she had the FV1/Type 91 intercept/jamming system on her foremast, the FV1 antennas being distributed around the spotting top. *Danae* was given to the Polish Navy in September 1944 as ORP *Conrad*, replacing ORP *Dragon*, which had been damaged by a Marder human torpedo at Normandy and then expended as a blockship. *Dragon* had been loaned to the Polish Navy in 1943. Modernisation of *Danae* was so extensive (and expensive) that it led to reconsideration of the future of the 'C' and 'D' classes. In mid-1943 plans called for modernising the next three cruisers, *Dauntless*, *Capetown* and *Ceres*, beginning in June, August and September 1943. Not modernising them would release valuable shipyard labour and would also release ships for subsidiary tasks such as aircraft targets and training. On 27 April 1943 DOD(H) (Director of Home Dockyards) therefore proposed cancelling the planned modernisations. For his part, C-in-C Eastern Fleet valued the old cruisers as convoy escorts for the Indian Ocean. However, it appeared that he actually preferred armed merchant cruisers for this work. The situation was further complicated because existing armed merchant cruisers were being converted to troop transports, depot ships and the like. An additional factor was the appearance of modern French cruisers released when deadlocks over the French fleet in Alexandria were resolved. *Caradoc* had already been earmarked by Home Fleet as a gunnery training ship because her full speed was too low for fleet operations. On 27 May 1943 Deputy Controller circulated a decision: ships would be refitted but not modernised. Alterations to gun armament would be limited to substitution of twin Oerlikons for singles and also for single pompoms. 'D' class cruisers would have their after (centreline) 4in guns removed and as many Oerlikons as possible substituted. The two sided 3in in the 'C' class and 4in in the 'D' class would be retained to fire star shells. *Dragon*, which was a Polish ship, was modernised similarly to *Danae* during a Chatham refit (April – 16 May 1944). She had the same twin 4in gun aft, the two quadruple pompoms replacing No. 3 6in gun, the torpedo tubes removed, and eight twin power Oerlikons. Prior to this refit she had no pompoms at all, but No. 5 gun had already been landed, and in April 1944 she was credited with five twin power Oerlikons and four single Oerlikons. She had been turned over to the Poles the previous year after a refit at Cammell Laird. Initially the intention was to use her in the Eastern Fleet, which would have justified a limited anti-aircraft battery. When they found that she would serve instead in home waters, the Poles asked for the increased anti-aircraft battery. Initially they were offered a single quadruple pompom, but the Polish naval commander pressed for two. First Sea Lord personally ordered two pompoms to be provided even though they were in short supply (they were earmarked for new destroyers), in view of the valuable service the Poles were providing. *Durban* also apparently lost a 6in gun (but that is not reflected in the April 1944 armament list), and by early 1944 only *Despatch* and *Durban* retained torpedo tubes. *Dauntless* was refitted at Chatham for training. It seems unlikely that she ever received any additional close-range armament. In April 1943 she and *Dragon* were both credited with two quadruple pompoms (she was also credited with eight Oerlikons), but that probably reflected plans rather than reality, particularly since neither ship was listed as having surrendered a 6in gun to provide the necessary space and weight. *Despatch* lost all of her 6in guns (she retained one 4in anti-aircraft gun) and gained sixteen single Bofors in an October 1943 – 6 May 1944 Portsmouth refit. This was apparently to prepare her for service as a gunnery school firing ship, possibly in combination with *Durban* (as was being considered by C-in-C Portsmouth in February 1944). However, it was soon evident that her hull was in too poor a condition for such employment, and in March 1944 C-in-C Portsmouth asked that she and *Durban* be made available as accommodation ships. She was recommissioned on 5 April 1944, and her radars were made operational. Off Normandy she was used as a headquarters for the northern Mulberry harbour, berthed in Gooseberry III, and then as a shipping control vessel until relieved by HMS *Waveney*. It was pointed out in August that full value was not being gained from the heavy close-range armament because it lacked fire controls. No such equipment could be made available, and by late August 1944 the ship had been relieved from her 'Neptune' (Normandy invasion) duties for reduction to reserve. *Durban* was expended as part of the Arromanches breakwater.

BRITISH CRUISERS

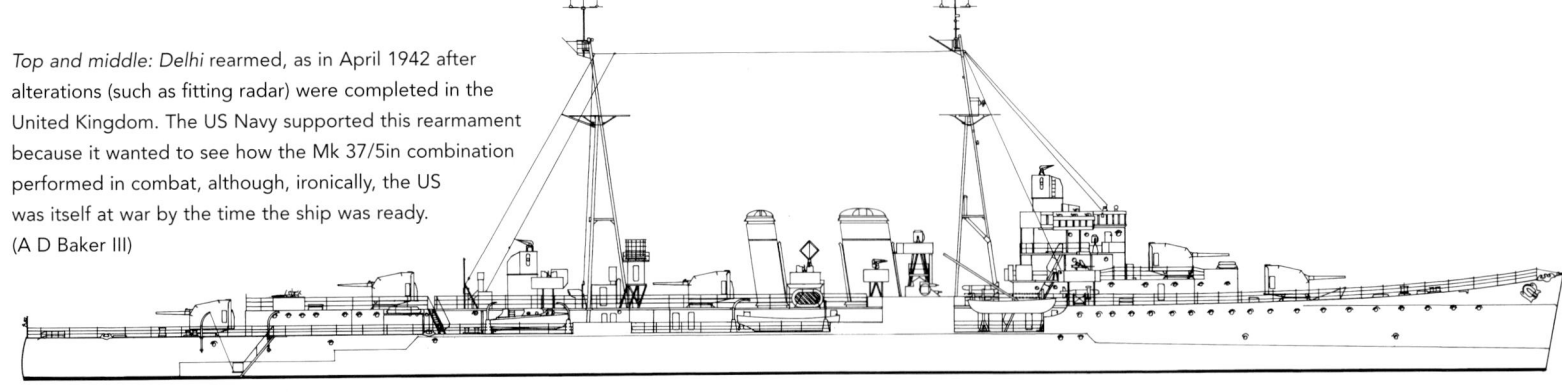

Top and middle: Delhi rearmed, as in April 1942 after alterations (such as fitting radar) were completed in the United Kingdom. The US Navy supported this rearmament because it wanted to see how the Mk 37/5in combination performed in combat, although, ironically, the US was itself at war by the time the ship was ready. (A D Baker III)

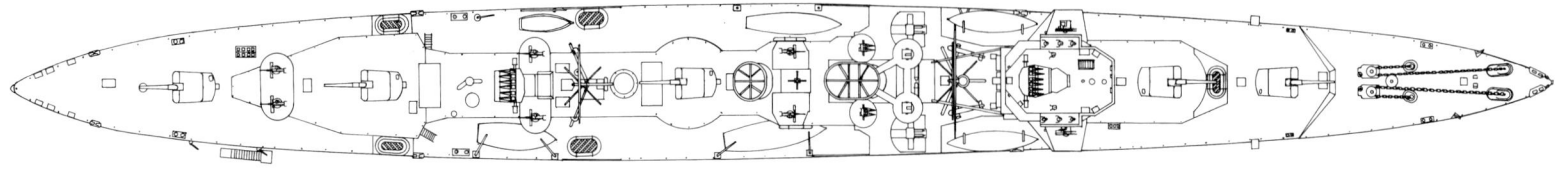

Below: Delhi rearmed (inboard profile). (A D Baker III)

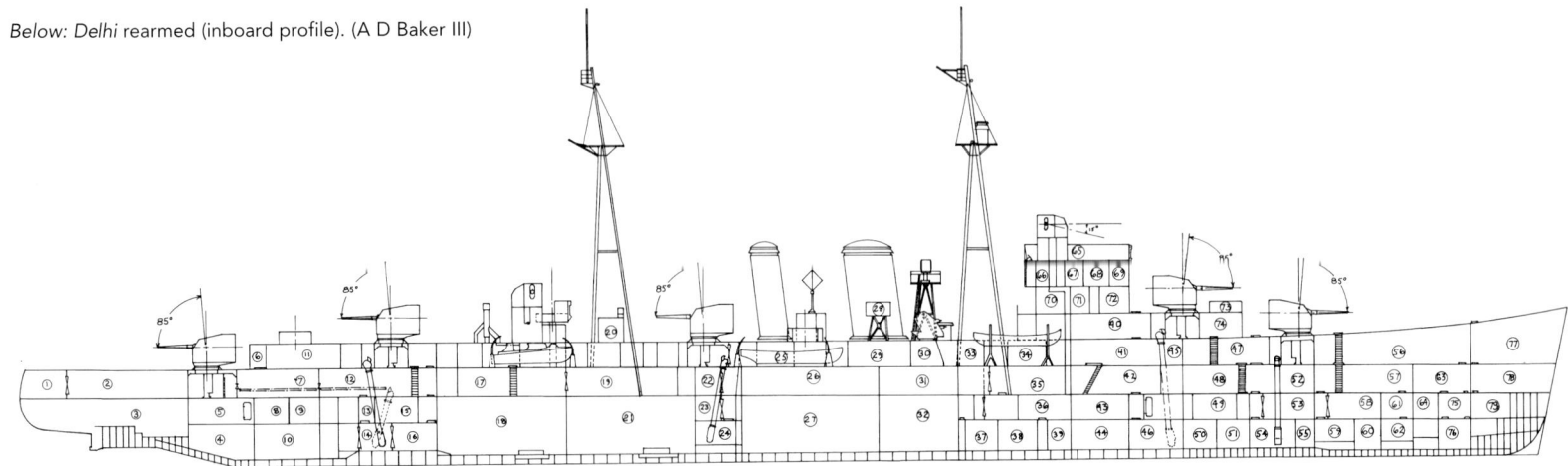

1. Officers' heads (bathrooms)
2. Passageway with officers' cabins port and starboard (P&S)
3. Steering machinery room
4. Fuel oil tankage
5. Wardroom stores (P); paymaster's and marines' stores (S)
6. Gun crew shelter
7. Centreline passageway with 5in ammunition passing trays; wardroom to port, officers' cabins and offices to starboard
8. Provision room (P); captain's stores (S)
9. Warrant officers' stores
10. Fuel oil tankage
11. Captain's day (in port) and dining cabin
12. Passageway with officers' cabins (P) and warrant officers' mess and pantry, captains' pay office and engineer's office (S)
13. Trunk
14. 5in projectile stowage and handling room
15. 20mm ammunition stowage
16. 5in powder magazine (centreline); fuel oil tankage (P&S)
17. Main radio office (S); gunnery office and mechanics and chief stokers' messes (P)
18. No. 2 steam turbine engine room
19. Centreline passageway with engineer's workshop (P) and miscellaneous workshop (S)
20. Vegetable locker
21. No. 1 steam turbine engine room
22. Passageway (P&S); 2pdr pompom equipment (P): magazine cooling and ice-making machinery (S)
23. 2pdr pompom magazine
24. Fuel oil tankage (P&S); 5in projectile magazine and handling room (centreline); 5in powder room (P)
25. After funnel trunk
26. No. 2 boiler room (passageways P&S at 1 platform level)
27. Lower portion of No. 2 boiler room
28. 2pdr pompom director (P&S)
29. Forward funnel trunk
30. Boiler room vent space
31. No. 1 boiler room (passageways P&S at 1 platform level)
32. Lower portion of No. 1 boiler room
33. Type 281 radar receiving office
34. Crew's galley
35. Degaussing machinery room centreline, flanked by coal and wood stowage
36. Lower gunnery plotting room
37. Fuel oil tankage
38. Fuel oil tankage
39. Fuel oil tankage (pitometer trunk at centreline on after bulkhead)
40. Passageway and crew's recreation space
41. Passageway centreline; chief petty officers' mess (P); sickbay (S)
42. Passageway centreline; stoker petty officers' mess and dry canteen (P); petty officers' mess and issue room (S)
43. Lower conning tower (P); canteen stores (S); flanked by gunner's stores (P) and central storeroom (S)
44. Fuel oil tankage
45. 5in ammunition hoist
46. Fuel oil tankage (P&S) flanking 5in projectile stowage and handling room
47. Crew berthing (with ammunition hoist space centreline at forward end
48. Crew berthing with artificers' mess to starboard
49. Cold and cool provisions storerooms centreline; refrigeration machinery room (P); dry stores room (S)
50. 5in powder room (flanked by fuel oil tankage)
51. 5in powder room (flanked by fuel oil tankage)
52. Crew berthing space (mess)
53. Central stores, with 5in ammunition hoist centreline
54. 5in projectile stowage and handling room centreline, flanked by fuel oil tankage
55. Fuel oil tankage
56. Crew berthing space (mess)
57. Crew berthing space (mess)
58. Flour and biscuit stores
59. Stores
60. Fresh water tanks (P&S)
61. Canvas stowage
62. Inflammable stores
63. Chain locker
64. No. 1 storeroom
65. Upper bridge with lookout positions (P&S)
66. Remote 5in gunnery control room
67. Plotting office
68. Chartroom
69. Wheelhouse (pilothouse)
70. Radio direction finder (RDF) office (P); gun crew shelter (S)
71. Radio signal (message) distribution office (P); head (S)
72. Captain's (sea) cabin (P); navigator's cabin (S)
73. Vent shield
74. Passageway, with mail office (P)
75. No. 1 storeroom
76. Lower chain locker
77. Paint room (boatswain's stores)
78. Watertight compartment
79. Watertight compartment

converted to meet this requirement. A provisional armament statement showed four twin and two single Oerlikons, with the note that twin Oerlikons could replace twin Bofors if the latter were not yet available. Neither ship was complete until early in 1944.

Plans called for refitting the eight 'D' class cruisers with four twin 4.5in mounts, one quadruple pompom and two quadruple 0.5in. Orders for the 4.5in mounts were let in 1938 to keep gun and mounting production running at a steady pace, even before the full Class II programme had been approved, since they could also be used in new ships and in the Class I programme. Large repairs and conversion of a 'D' class cruiser were expected to take twelve months. Three 'D' class would be taken in hand on 1 January 1940 for completion by 1 January 1941, one was to complete by 1 May 1941, and the others were to be rearmed beginning in 1941. By mid-1939, however, the British war plan included cancelling the 'D' class conversions. Their 4.5in guns became available for other purposes, including arming the *Dido*-class cruisers *Scylla* and *Charybdis*.

In the autumn of 1940 the Admiralty became interested in buying US 5in/38 guns to arm new destroyers, and there was naturally parallel interest in using these guns for a new anti-aircraft cruiser conversion, which, given the size of the weapons, had to be one of the 'D' class. A Staff Requirement envisaged two directors and either four twin or six single mounts. On 9 November DNC wrote to Controller that he had little information on the weapons, but that it might be necessary to accept fewer mounts; by mid-November he had settled on five singles and two fire-control systems. The next day the British mission in Ottawa provided full details. By late December Controller wanted a feasibility study (arrangements to buy the guns and fire controls were still being settled), and in January DNC calculated that the ship could carry five single 5in/38 and two Mk 37 fire-control systems, which at 113 tons would weigh slightly less than the four twin 4.5in and two HACS (130 tons) planned for the 'D' class conversion. Another alternative for a 'D' class cruiser was apparently five twin 4in with two HACS (82 tons).

A design conference met in Bath on 8 January; this was an urgent project. The five guns would occupy positions formerly occupied by 6in guns, so that they could use the same magazines. On 10 January DNC drafted a letter to the senior British naval officer (Admiral Evans) in Ottawa, explaining that the British yards were too overloaded for the job, so 'the best, in fact the only, way' to get the conversion done quickly would be send the ship to North America. Ideally the whole job would be done in an American yard, but 'apart from the political aspect' American yards were undoubtedly full, 'and they might not like to convert a British ship that is to be returned to the British service'. The next best, and likeliest, possibility was conversion in a Canadian yard under American supervision. Conversion was on standard British anti-aircraft cruiser lines, with two pompoms (with directors) and four quadruple 0.5in machine guns. Because the US guns used AC electric power, whereas the British used DC (and had limited generating capacity), the ship had to be, in effect, rewired.[9]

The passage of the Lend-Lease Act (11 March 1941) transformed the situation, since the ship could be converted in a US yard using US equipment. HMS *Delhi* was chosen (she had been about to join the East Indies squadron at Singapore). Like the smaller anti-aircraft cruisers, she became a two-target ship (a US destroyer armed with the same five guns had only a single fire-control system). The two Mk 1 computers of the Mk 37 systems were installed in a single plotting room. British radars were of course installed: Type 285 for the 5in directors, Type 282 for the pompom directors, and one Type 281 for air warning. A D/F office and coil were installed between the funnels. Like other modern British

HMS *Delhi* was rearmed at Norfolk Navy Yard as a possible prototype for further 'D' class conversions, but with the entry of the United States into the war so elaborate a conversion was no longer practicable. *Delhi* had the same main battery as many US destroyers, but she had two (rather than one) Mk 37 fire-control systems, hence could engage two targets at the same time. Her British radars were installed only after she returned to the UK. She is shown in April 1943.

cruisers, *Delhi* was fitted with Asdic on conversion (in her case, Type 128A). While *Delhi* was being rearmed, plans were made to provide her with Oerlikon guns.

In May 1941 word was received that the US authorities might be willing to rearm a second 'D' class cruiser along the same lines. Although he had only observers' reports of the efficiency of the US long-range anti-aircraft guns, DTSD urged further conversions, asking Operations Division to release either a second 'D' or even an unconverted 'C' class cruiser. A sketch design was drawn up for 'C' class conversion.[10] If that proved impossible, he suggested asking instead for a merchant ship conversion similar to the merchant ships already converted into anti-aircraft escorts. Director of Plans welcomed the idea 'since it seems almost axiomatic that a 5in HA/LA equipment already in supply to the US Navy and with tachymetric control must be more effective than the British 4in HA/LA controlled by HACS. In addition the heavier calibre renders it just possible to retain the re-armed ship within the cruiser category instead of declassing her to AA duties only.' Availability depended entirely on the refit programme. All the 'D' class cruisers other than *Dauntless* (which completed a large repair at Portsmouth in July 1939) were due for large repairs when war broke out, but no such refits had been scheduled, because ships were so badly needed in service. *Dragon* was to follow *Delhi* for a US refit, but not for rearmament (2–3 months rather than 5–6 months' work). Despite boiler defects which reduced her speed, *Dunedin* was being kept in service. Due to her defects, she was the best candidate for conversion, but cruiser strength could not be reduced before the autumn of 1941, when new ships would enter service. She and *Dragon* were currently on the South Atlantic station.

The position of the 'C' class was similar. Ships not scheduled for AA conversion were to have been refitted to extend their lives to December 1941. As Gunnery Firing Cruiser, *Cardiff* was not available, while *Capetown* (which had been earmarked for AA conversion) was being repaired at Bombay following torpedo damage. *Caledon* and *Ceres* were to be refitted at Colombo. *Caradoc* was to have her turbines rebladed.

In July *Dunedin* was approved as the second ship, but only subject to successful trials of *Delhi*, experience with US-made 14in guns in the First World War having been less than happy. By the time *Delhi* had been completed, the United States was at war, and the opportunity for a second conversion was gone.

The *Fiji* class

With the 1936 London Naval Treaty the British finally stopped construction of big cruisers. What could DNC design within the new 8,000-ton limit? Apparently the first attempt was to replace the four triple turrets with three quadruple ones; it came out to 8,500 tons (this design may have been K20).[11] A successor K21 (Design P) was developed on 23 January 1936 to answer DNC's question about a ship with ten guns (twins in 'A' and 'Y' positions) with anti-aircraft armament, speed and protection as in the 1935 *Southampton*s (the *Gloucester* class).[12] An initial estimate showed that 15ft could be saved on length (total 569ft), and that standard displacement would be about 8,900 tons. That assumed that the turrets were the short-trunk type as in *Southampton* and *Leander*. K22 (Design Q) was a nine-gun ship with three triple turrets. In that case 36ft would be saved, and displacement might be about 8,500 tons. K23 (Design R) showed how much protection Design Q could retain if she were limited to 8,000 tons. All three designs retained *Southampton*-class machinery producing 82,500shp. For Design R length was based on that of the three-turret ship, beam and draft being adjusted to give the desired displacement. Subtracting the estimated hull, machinery, armament and equipment weights (adjusted

for the saving in personnel) left about 1,210 tons for protection. The protection provided for HMS *Liverpool*, a *Gloucester*-class cruiser, appropriately scaled down for size and number of guns, required 1,415 tons. Reverting to the reduced protection of the *Southampton* class would save 155 tons, leaving another 55 tons above the desired tonnage. Taking ½in off the belt would save another 75 tons. Alternatively, the belt could be cut to 2½in rather than 4in and turret protection be reduced (from that of *Liverpool*) to a 2in roof, 1½in sides and a 3½in face.

Now that DNC knew what he could have on 8,000 tons, the question was whether he could get more by making further reductions. Design S (K21 modified) was the ten-gun ship with reduced protection. Length could be reduced to 566ft. She could make 32.5kts on 77,000shp (330rpm machinery). However, at 300rpm it would take only 74,000shp to drive her at the desired speed, so it might be possible to use *Amphion* class machinery (slightly pressed). Machinery weight would be reduced from 1,545 tons to 1,475 tons, saving 70 tons. The ship would have to sacrifice two of the three aircraft of the *Southampton* class, reverting to *Leander*-class arrangements. She would have a 3½in belt, but otherwise would be protected like a *Southampton* (armour weight would be 1,083 tons, compared to 1,529 tons in *Liverpool*). Speed would be the 32.5kts of a *Southampton*.

All of this was encouraging, so on 3 February DNC asked what displacement the 8,500-ton ship (with protection like *Liverpool*) with three quadruple mounts would have if speed were cut back to 32 or 31.5kts. At 32kts the ship would probably displace 8,330 tons (77,000shp); she would need 71,000shp for 31.5kts (8,170 tons) and 66,000shp for 31kts (8,000 tons). This was presumably design K20A. Controller then returned to versions of the three-turret K23. One possibility (K23A) was to cut back side protection to that of the *Leander* (3½in rather than 4½in), but to leave 4½in magazine sides. Transmitting and other control positions had the belt protection instead of the 1in in a *Leander*. The alternative (K23B) was to trade off speed and length (530ft rather than 550ft, both on the waterline) rather than protection for the three-turret main battery (30kts). Both designs would clearly be badly cramped. K26 was a further development of K23B in which the catapult and hangars were eliminated, to open some midships space (a light reconnaissance aircraft could be stowed on deck and lowered over the side to be launched in calm weather, but its weight was not included in the ship's standard displacement). The space opened up could accommodate two more twin 4in guns (total of six, as in *Belfast*), and it also allowed for enough additional power to boost the ship to 30.7kts. To improve 4in ammunition supply, the magazine was placed amidships, between the forward engine room and the after boiler room (no such magazine could be fitted into K23A or K23B). Trading the weights of the two extra twin 4in for machinery would bring speed up to 31kts, making K26 fairly attractive. The three designs with three turrets were reported on 9 July.

Controller saw the 5.25in gun as an alternative, so on 11 February he asked how many he could get into an 8,000-ton ship with the same other requirements (speed, protection, aircraft) as a *Southampton*.[13] A first cut was discouraging: if the *Southampton* hull were simply scaled down to the smaller displacement (562ft long) and protection and other weights were left virtually unchanged, very little weight would be available. This was probably K24, and it was so discouraging that it was

Below and opposite: The *Fiji* class was an attempt to package the armament of a 'Town' class cruiser in a hull about a thousand tons smaller. HMS *Kenya*, newly completed, is shown in 1941. *Kenya* had the usual air-warning and gunnery radars, but not a surface-search set. The view from aft shows one of the methods used to compress the ships. Instead of an after DCT, they were given a fixed rangefinder and director atop 'X' turret. In the stern quarter view, note the surviving quadruple 0.5in machine guns and also the Oerlikon mounted right aft on the quarterdeck.

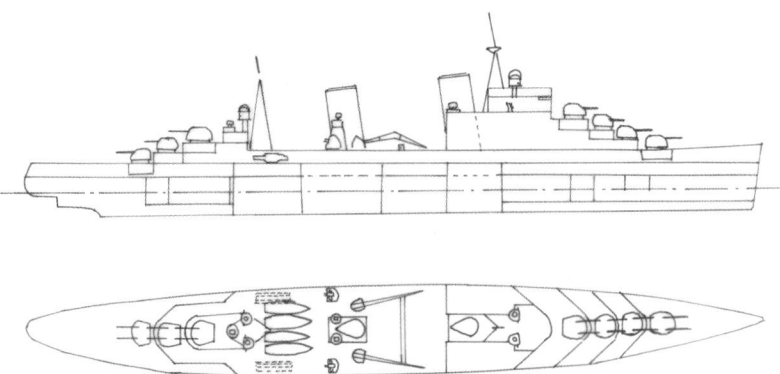

Above: The K25 design for an 8,000-ton cruiser armed with 5.25in guns. (Norman Friedman)

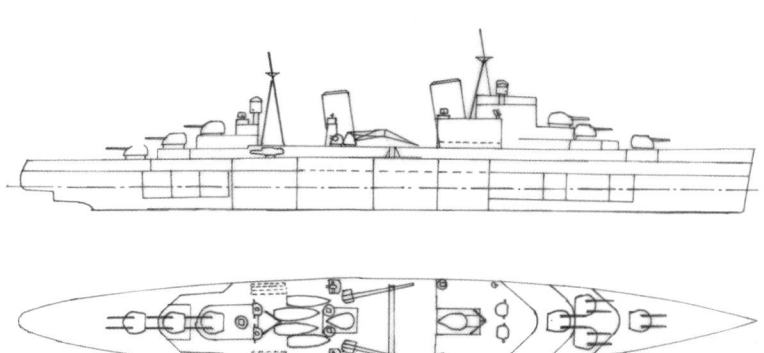

Above: The K25A-F design for an 8,000-ton cruiser armed with 5.25in guns. All had the same configuration, but they differed in speed and protection (the highest was K25A, 32.5kts). Protection in a typical version, K25E (31kts), amounted to a 3½in belt over machinery and magazines and shell rooms with a 2in deck atop it. The 5.25in turrets had 3in faces and 1in sides and crowns. Handling spaces under the turrets had 1in protection. Controller, Rear Admiral Reginald H H Henderson, much favoured the 5.25in cruiser, but the Board rejected his views. Henderson was responsible for several innovative ships, including the 'Tribals,' the *Dido*s, the armoured carriers, and HMS *Unicorn*, as well as for the *King George V* class. (Norman Friedman)

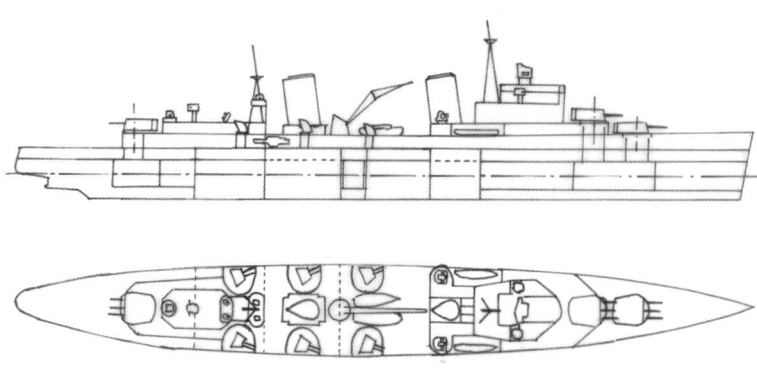

Above: The K26 design for an 8,000-ton cruiser armed with three triple 6in turrets and six twin 4in, as in *Belfast*. Speed 30.7kts, length 530ft. There was no catapult, but there was deck space for a single long-range seaplane. Machinery protection would have been 4½in side and 2in deck, with similar magazine and shell room protection (belt and deck rather than box). K23A was similar but had four rather than six 4in mountings. She would have been 550ft long and would have made 32.5kts. Machinery would have been protected by a 3in belt and a 1¼in deck; magazines and shell rooms would have had box protection (4½in side, 2in deck). (Norman Friedman)

not reported. Controller went back to DNC to ask what could be gained by reducing protection. This was the K25 series. DNC found that with machinery for 32.5kts and with minimum protection (3in belt and 1¼in deck over machinery, 3½in side and 2in deck over magazines, and 1in side and deck over shell rooms), there was enough weight for seven twin 5.25in with 300 rounds per gun (but space for 400). Four turrets were superimposed forward (but with two turrets side by side on 'B' level, so that there were three levels forward) and three aft. Shell rooms had to be placed above magazines, because there was not enough space for them to be side by side, as in a *Dido*. Unlike a *Dido*, these ships had protected turrets, with 3in faces and 1in sides and roofs. K25A was the basic version. In K25B through K25F speed was reduced to 32kts and the weight saved used in various ways: to substitute a complete 3in belt for the belt/box arrangement (K25B); to increase deck protection (in K25A) to 2in over the machinery (K25C); to increase machinery side protection to 3½in and give control spaces 1in box protection (K25D). In K25E, speed was cut to 31kts and a complete 3½in belt and uniform 2in deck provided over both machinery and magazines, 3½in thick. K25F added an eighth twin 5.25in turret to K25A at the cost of reducing speed to 31.7kts. This design showed four turrets superimposed both forward and aft. K25H showed that the cost of providing a 4½in belt over machinery and magazines (and 4in turret faces and 2in roofs and sides) was a reduction to 29kts, which was unacceptable. These alternatives were reported on 25 July 1936.

On 16 July DNC offered two more alternatives, K27 and K28, which reverted to 6in batteries. K27 (545ft) had three quadruple turrets and the lighter protection (3½in side, 2in deck); speed would probably be limited to 30kts (figures were preliminary because the size and weight of the quadruple turret had not been settled). The ship would be very cramped, which was what probably killed the design. K28 (575ft) was another attempt at a ten-gun ship, two twins superfiring over two triples. With the same protection as K27, the hull had to be longer and heavier, but not so much longer that the ship benefitted hydrodynamically from the longer hull. Speed would drop to 29.5kts, an even more depressing result. Again there were uncertainties, because no long-trunk twin 6in turret had been designed. Two more 6in designs were apparently prepared in July; they must have been K29 and K30 but details have not survived. The Sea Lords chose a 6in design, probably with twelve guns, because when they approved the final design in October 1936, it was described as an elaboration of a design approved in July. It seems clear that Controller was unhappy with this choice.

In mid-September Controller wrote to ask for a Sea Lords' conference on the 8,000-ton cruiser, as a decision was urgently needed. He offered what he considered the three best design alternatives, K27, K23A* and K25G. An entirely new quadruple turret design would impose at least a six-month delay, which he considered sufficient grounds to reject the idea. Moreover, recent tests had shown that guns had to be at least 6ft 6in apart not to interfere with each other. That sort of spacing would make a four-gun turret impractical. Spacing could be tighter if the turret used a delay coil (0.01 sec) to fire alternating pairs of guns together, but if the ship were rolling heavily, the delay would throw off her salvoes, and it would have to be cut out. The ship would become, in effect, a six-gun cruiser. Henderson considered the nine-gun ship underarmed for her size (she should have ten guns). To bring her speed up to that of the 'Towns', ½in had been taken off her belt. For enemy inclinations of 60° and 45°, the 3in belt would keep out shells at 10,500yds and 8,600yds respectively. The 2in deck was chosen to keep out bombs, and more than sufficed to keep out 6in shell. However, Henderson was inclined to consider the eight 4in guns of both K27 and K23A* insufficient. He much preferred the all-5.25in K25G, which was armoured against 6in shellfire. K25G showed no fewer than four superfiring mounts forward and three aft. Henderson liked the 5.25in design because it offered much more anti-aircraft firepower, all of it protected against bombs (and their splinters), and because at cruising stations and in harbour half the anti-aircraft battery could be kept manned. Training would be simplified. He was also 'inclined to think that to strike out on a "novel" design would be good generally for our prestige as leaders of design'.

HMS *Mauritius* is shown during the September 1943 Italian landings, about to shell German batteries near Naples. She displays the standard initial surface-search radar arrangement in this class, a Type 273 forward of the bridge – an unpopular arrangement, because it blocked the view from the bridge and the helmsman's position.

Bermuda as completed, September 1942. The ship was completed with large Vokes sound-reduction filters over the fixed air intakes for the forward boiler spaces; these had vertical intake openings and a semi-circular plan form, and they were soon removed. Only one aircraft/boat crane was fitted (aircraft facilities were removed during an April to May 1944 refit). The big Type 273 'lantern' forward of the bridge was replaced by Type 272 (in the foremast) about April 1943. As with all Royal Navy cruisers that served after the Second World War, *Bermuda* had additional portholes added to her hull sides to improve internal light and ventilation. As built, she appears to have had a 'non-latex' covering over the exposed decks abaft the spray deflector, but post-war at least the quarterdeck had wooden planking. An additional six single Oerlikons were added during September 1943; in April 1944 she had eight twin 20mm. During a major refit (June 1944 to April 1945) 'X' turret was removed and replaced by two quadruple pompoms. A third was mounted on her centreline forward of her mainmast and four single pompoms were added. During August 1945, in preparation for planned Pacific deployment, the close-range armament was further altered, with two twin and two single 20mm removed and replaced by two single 40mm Mk III (hand-worked) and two Boffins. At this time the two-antenna Type 281 was replaced by a single-antenna Type 281B to clear the foremast head for a Type 293 target-indication radar antenna; the surface-search 'lantern' was replaced by the dish of Type 277. Post-war the 40mm and 20mm mounts were removed but the pompoms survived. In the late 1950s the pompoms were replaced by seven twin Bofors and fire controls were updated, the two forward 4in mounts being given on-mount radars of the US-supplied Mk 63 system. After 1954 the ship was given an enclosed conning position at the fore end of the bridge and Type 274 replaced Type 284 on the 6in director. She never received new lattice masts or barrage directors. After 1955 the ship carried NATO pennant number C52. (A D Baker III)

On 10 October, with a decision even more urgently required, Controller (Rear Admiral Henderson) had two more designs, K31 and K25G*, prepared, both 550ft long. K31 had four triple 6in turrets, K25G* seven twin 5.25in.[14] Both had 3½in side and 2in deck over both machinery and magazines. Unlike a *Dido*, K25G* had serious turret protection, a 3in face and 1in sides and roof, compared to a 3½in face, 2in side and 1in roof in K31. There was no question that the 6in armament of K31 was superior to the 5.25in battery of K25G*, but K31 had too little HA armament (but the same four twin 4in, two quadruple 2pdr and two quadruple 0.5in as the *Southampton*s) and it was 0.75kt slower than K25G*. Henderson also noted that the problem of blast interference among the 4in guns had not been solved. On balance, he considered the 5.25in ship a better choice, at least for the first five 8,000-ton cruisers. However, the Board approved K31, the Minute describing it as a worked-out version of the sketch design approved the previous July.

Overhanging the choice was the fact that the Japanese had not signed the 1936 treaty; no one knew what sort of cruisers they were building. Henderson argued that nothing could affect the British requirement for numbers of cruisers, hence for a relatively small affordable ship. 'When we know what the Japanese design is we can build to meet it.' In October Vickers offered its own 8,000-tonner armed with 5.25in guns.[15]

K31 traded off power rather than armour for armament, with 66,000shp rather than 74,000shp. On a thousand tons less than a *Southampton*, it offered the same armament: four triple 6in (but in long- rather than short-trunk mountings, with the usual 200 rounds per gun),

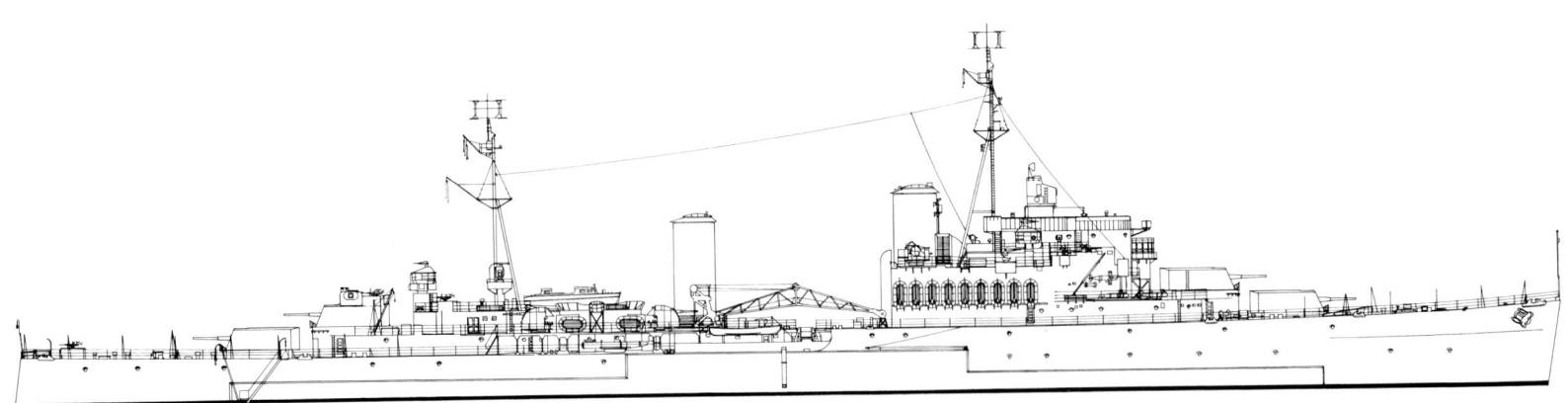

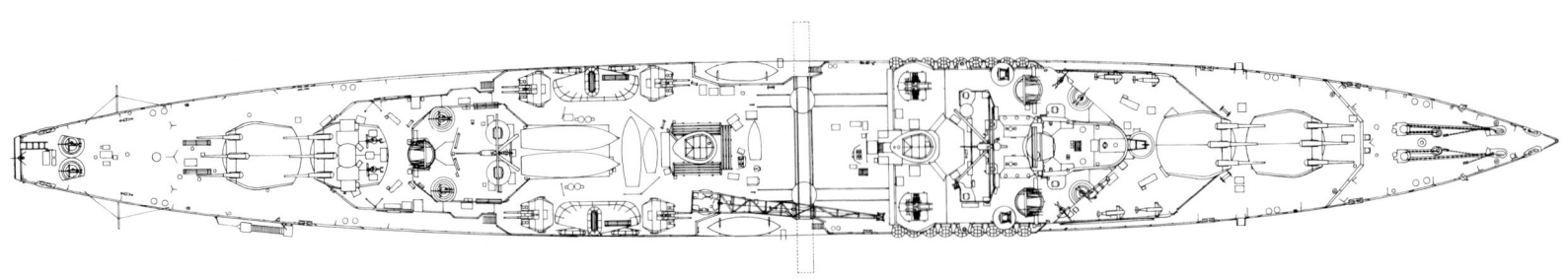

THE SLIDE TOWARD WAR

four twin 4in HA (250 rounds per gun, and 200 star shells per ship), two quadruple pompoms and two quadruple 0.5in machine guns, two triple torpedo tubes, and ten Lewis guns in the usual twin portable mounts, plus depth charges (fifteen on ships with Asdic, six on those without). K31 also offered the same aircraft arrangements, with two hangars and the athwartships catapult. That must have seemed miraculous, but it came at a high price: the design was extremely tight, just when naval technology was changing with the advent of radar and the demand (in three years) for much more anti-aircraft firepower. Obviously the ship would be much more crowded than a *Southampton*, so complement had to be held to a minimum. For example, the 4in magazines were placed immediately below the guns, saving the four men who, in a *Southampton*, passed it along the deck under the guns. War complement was cut (as of late 1936) from the 46 officers and 734 ratings of a *Southampton* to 48 officers (two more for additional fire control) and 662 ratings (the original pair of 4in hoists was replaced by four hoists, one per mounting, in January 1938).

The endurance requirement (7,000nm at 16kts when six months out of dock) conflicted with the demand for at least 30kts at deep (full-oil) displacement; at the required tonnage, K31 could not make 30kts. The solution was twofold. First, E-in-C was asked to provide for a 10 per cent overload. Expected overload speeds were 31.6kts in standard condition and 30.24kts deep. W G John, the constructor responsible, suggested an alternative. The ship's nominal deep fuel load could be limited (she could make 30kts at 9,770 tons).[16] It would not be enough to make the required endurance, so additional reserve oil tanks would be provided. Alternatively, the ship could be required to make her speed after she had burned off enough oil. Four- and two-shaft machinery arrangements were offered, the former (with alternating boiler and engine rooms) being chosen.[17] The 4in magazine were placed between the after boiler room and the after engine room.

The design was on the edge of being feasible within the 8,000-ton treaty limit. In November 1936 estimated standard displacement was 8,360 tons. To get down even to this figure, the ring bulkheads and turret deck rings (i.e. barbettes) had to be reduced from the 2in and 1in of earlier designs to 1in. Further weight might be saved by taking into account the shorter trunks of the 6in triples, as compared to those in *Belfast*. On the other hand, recent figures for the 6in triple were 20 tons

Left and below: Photographed in 1943, HMS *Jamaica* shows mid-war adaptation of the *Fiji* class. The view from the bridge was cleared when a lightweight surface-search set (Type 272) suitable for mounting on the tripod foremast, became available. The heavy director/rangefinder aft was eliminated in favour of more light anti-aircraft guns (Oerlikons, in this case) atop 'X' turret, to help deal with aircraft attacking from aft, along the centreline of the ship. The view from aft was taken at Portsmouth, 18 September 1943; the view from ahead was taken at Scapa Flow in 1943.

Clearing the front of the bridge made space for the two barrage directors, superfiring, visible in this November 1943 photograph of HMS *Gambia*.

greater than had been estimated, a total weight growth of 80 tons. Estimates suggested inadequate stability, and the preliminary designer suggested 6in more beam – which would add a few more tons. It seemed that standard displacement might be 8,412 tons – 5 per cent over the limit. British builders had managed to shave considerable weight from earlier cruisers, but that was a lot to expect. DNC asked that displacement be held to 8,300 tons, and that metacentric height be held to 2.5ft.[18] By this time beam was 61ft 3in or, later, 60ft 6in (it had been 59ft 6in). Possible ways of holding displacement down were omission of torpedoes and shortening the 4in HA magazine. It might seem that another way to save weight was to revert to box magazine protection outside the belt, but that actually added weight (about 20 tons). The 8,000-ton limit was artificial, but the stability problem suggested that any additional topside weight would cause unacceptable loss of stability – which was the problem throughout the Second World War.

To bring down displacement enough for submission to the Board in December 1936, DNC had to reduce armour: the 3½in NC turret faces of K31 were cut to 2in NC, and the trunks were reduced to a uniform 1in instead of 2in tapering to 1in. The crowns of the magazines and shell rooms had to be cut from 3in NC (which would keep out a 500lb dive bomb) to 2in.

Policy was to provide a 10 per cent margin in boiler power to meet tropical conditions, so that the ship would develop her trial power anywhere in the world. The nominal performance of the ship could be improved by taking the 10 per cent into account in calculating trial performance, which would be valid for temperate waters.[19] Thus the 66,000shp of the new cruiser was 72,500shp in temperate waters, offering 31.75kts at 8,000 tons and 31.5kts at 8,300 tons, rather than 31kts at 8,300 tons.

Meanwhile (late in November) an entirely different arrangement was suggested. Three of the four turrets could be placed forward (No. 3 slightly lower than No. 2), and the hangar placed where No. 4 turret normally was, the new No. 4 superfiring over it.[20] The arrangement moved the 4in HA battery clear of 6in blast, and it was well-adapted to autogyros, helicopter-like aircraft which were considered very promising. The ship would have been 570ft long (waterline). The hull would weigh more; as in the new 8,000-ton cruiser, the machinery would be reduced (to about 50,000shp) to make up for that. However, speed in standard condition would fall to about 29.9kts, which was too low. An alternative was to revert to a ten-gun battery (two twin, two triple), saving about 137 tons on armament, which could go into more power (57,400shp, for about 30.8kts).

When the Italian heavy cruiser *Gorizia* docked at Gibraltar in September 1936, analysis suggested that she was about 10 per cent heavier than her stated 10,000-ton displacement (actually it was 20 per cent). In December DNC asked what 10 per cent overage would buy for the 8,000-ton cruiser: half a knot (32kts in standard condition) and an inch more belt armour; the ship would also be lengthened for higher speed.[21] This idea was not followed up. However, on 10 October Controller applied the same inverted margin policy he had used for the *Belfast*s. As submitted to the Board, the design included a 200-ton margin above the 8,000-ton limit.

DNC submitted a sketch design dated 22 December 1936. Shell rooms were above the magazines, to save length and to provide additional protection. As in the earlier design, the two pompoms were mounted above the hangars; the two quadruple 0.5in machine guns were on either side of the forward superstructure. As in a later *Southampton*, there were two DCTs (fore and aft), three HADTs (one each side of the bridge, one aft), and two pompom directors (on the bridge wings). To hold down standard displacement, ammunition capacity in that condition was held to 150 rounds per gun of 6in and 4in, and 1,200 per 2pdr barrel, but there was sufficient space for the usual 200 rounds per 6in and 4in, and 1,800 per 2pdr barrel. The ship had the two triple torpedo tubes of the *Southampton* design, and the same aircraft facilities. However, instead of the three strike aircraft (TSR) of a *Southampton* or a *Belfast*, the Legend showed three Walrus flying-boats better adapted to trade protection. In his covering memo to the Board, DNC mentioned that alternative aircraft arrangements, including a catapult aft, had been considered. That was probably a reference to the design with three turrets forward.

The greater temperate climate performance was made explicit, as was the special deep load condition excluding 500 tons of reserve fuel oil. Thus at deep load the ship would carry 1,200 tons, but at extra deep load she would carry her full 1,700 tons, speed being 0.5kt less that the rated deep load figure. Speed in a temperate climate was 31.5kts at standard displacement and 30.5kts at a modified deep load, without fuel in reserve tanks. Corresponding figures for all climates were 31kts and 30kts.

The 3½in belt (1in thinner than that of *Southampton*) covered not only the machinery but also the magazines and shell-rooms, as well as the fire-control computer spaces (transmitting stations and HACPs), switchboard room, LP switch rooms (for data transmission), and wireless transmitting and receiving offices, thus covering more of the waterline than in a *Southampton*. The belt was closed by 2in bulkheads and a 2in deck, the latter superior to that of *Southampton* over machinery and control spaces but not as good over magazines and shell-rooms. Turret and steering-gear protection matched that of *Southampton*. A request to protect magazine sides to the level in *Belfast* could not, apparently, be met.

Reviewing the drawings, ACNS argued that masts and funnels should be vertical instead of raked, as in a *Southampton*. Controller agreed that raked masts and funnels made it easier to judge a ship's course, but they also made it more difficult to take her range (using a British coincidence rangefinder, not the stereo type other navies were adopting). Appearance could not be ignored, however; raked masts and funnels were more attractive. But First Sea Lord found upright funnels more modern-looking, and rejected the argument that sloping kept smoke away from the bridge (it was effective only if the funnel were radically sloped, as in Japanese cruisers). Nearly all foreign navies had adopted either straight or very slightly sloped masts and funnels, and a very slight slope gave no advantage at all. The design was therefore modified with vertical masts and funnels. The issue of clearing arcs of anti-aircraft fire, which has often been used to explain the switch to vertical masts and funnels, seems not to have arisen.

First Sea Lord saw only one problem: accommodation. Something had to be done, particularly for ratings, to overcome the trend towards more and more crowded ships. When the Board approved the design on 14 January 1937, it asked Controller to look into improving ratings' accommodation.

Building drawings were circulated to departments on 28 February, but the story was not yet over. In March DNC observed that the bottom of the armour belt might be uncovered by the ship's wave pattern when she steamed at high speed. That required 1ft 6in of extra belt depth. The ship might also be vulnerable to a diving shell hitting short; that seemed to require 3ft below the bottom of the belt when at standard condition and not at high speed. The Japanese were interested in exactly such shells, and intelligence to that effect may have become available in 1936.[22] He suggested adding 40lb (1in) D plating below the belt, or perhaps 60lb plating over a smaller area, enclosed by 60lb bulkheads. W G John, in charge of the design, pointed out that the waterline in action condition was 2ft above the waterline at standard displacement. On this basis the protective strips in way of the forward and after magazine blocks could be omitted, but a 2ft 6in belt should be worked below the main belt in way of the 4in magazines, because they were uncovered by the hollow in the ship's wave pattern. John pointed out that this did not take into account extra exposure as the ship rolled; there was an even chance of a hit against the protected or unprotected area. It is not clear that this protection was provided, as DNC did not mention it when submitting the modified design to the Board a few months later.

Just how tight the design was became obvious when, in April 1937 Controller (Admiral Henderson) asked how much boiler rooms had to be shortened to save 70 tons (having learned that a foot less boiler room saved 3.1 tons on armour). Nothing could be done, because the space between 'B' and 'X' turrets was set by the demands of topside arrangement: bridge, hangars, catapult, 4in HA guns and boats. No matter how much speed was reduced, the ship could not be materially shorter. Controller then asked what speed the ship could make with 30 per cent *more* power. The power-speed curve was so steep that only about 2kts would be gained, and that would require a new design (which would take six months to produce).

Controller wanted to know what he could get by deleting the after DCT (14 tons) and perhaps an after 6in turret. Other items which could be deleted to save weight were the aircraft (143 tons) and the torpedoes (40 tons). Every ⅛in off the upper or lower deck (except over magazines) would save 35 tons. A ½in off the belt abreast the magazines would save 14 tons, and ½in off the belt abreast machinery and control positions would save 30 tons. More powerful machinery typically added something over 200 tons.

A series of designs was produced in which extra power was obtained at the expense of armour and, in many cases, aircraft and torpedo tubes.[23] Boiler design kept improving, so that it appeared that on much the same weight 83,000 to 86,000shp might be achieved (total machinery weight would increase about 180 tons, the machinery box being 2ft longer).[24] DNC offered two series of sketch designs, A and B, the latter lengthened from 526ft to 538ft (pp). In modified versions (A' and B') boosted horsepower was cut to 80,000shp, which gave a knot higher speed but reduced the necessary sacrifice to ⅛in of side armour, ½in of deck armour over the forward engine and boiler rooms, and two triple torpedo tubes. It might be possible to save enough weight to install the torpedo tubes upon completion; at the least, space could be provided so that they could be installed on the outbreak of any war. Controller chose Design B'. The delay in tendering due to redesign (to October 1937) was acceptable because it would not delay the programme; the crucial items were guns and mountings. Laying-down dates could be put back and the second armoured carrier (HMS *Victorious*) brought forward. First Sea Lord approved the idea on 28 May, but DCNS wanted a staff meeting called to decide whether greater speed was really worth the sacrifice of protection and the torpedo tubes. The Staff was particularly unhappy with the steady loss of torpedo power in the fleet. The British fleet considered that it enjoyed a particular advantage in night battle, in which torpedoes would be important. The meeting ruled in favour of retaining the original level of protection except that the belt abreast the machinery was reduced to 3⅛in, the torpedoes were sacrificed, and a speed of 32.25kts (standard) accepted.[25] New towing-tank results showed that speed had been over-estimated by a quarter-knot. Controller decided that the side abreast the machinery would be 3⅛in and the side over magazines 3½in. All decks would be 2in thick. E-in-C would calculate machinery box length for 1,440-ton machinery (rather than 1,390 tons in the most recent B2 design).[26]

About this time DNC adopted a transom stern, in effect a longer

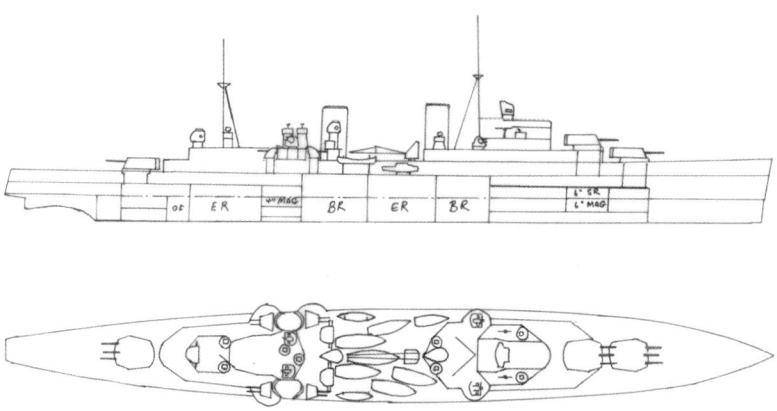

K34, the 1939 alternative to continuing *Fiji* class construction, had the main battery reduced to three triple turrets but with four quadruple pompoms. Note also the sided HA directors abeam the second funnel. (Norman Friedman)

stern chopped off, which offered the wave pattern, and hence the resistance, of a longer hull. The idea had been tested, somewhat unsuccessfully, in the minelayer *Adventure*.[27] The towing tank (Haslar) estimated that effective horsepower (hence shaft horsepower) would be reduced about 4 per cent at 32kts and 3 per cent at 25kts. DNC expected about a quarter-knot more at full power, so that with 80,000shp he expected 32.5kts at standard displacement and 31.25kts at deep load (1,200 tons of oil).

The Board approved the amended design on 8 July 1937. The after DCT was omitted and no weight was allowed for the torpedo tubes but space was provided for their installation. Elimination of the after DCT was balanced to some extent by placing a control position atop 'X' turret.[28] All three HACS were retained, with two (not three) computer positions below deck. Adopting an 80,000shp plant (at the new high forcing rate) made it possible to reduce boiler size slightly and thus to reduce citadel length slightly. However, if the earlier forcing rate were accepted, power would be only 72,500shp. The higher maximum power required somewhat heavier turbines. DNC thought the added 90 tons well worth while, but in February 1940 E-in-C refused to sanction operation at full 80,000shp power because destroyers and fast minelayers had not yet forced their boilers even to the rate corresponding to 72,500shp. DNC was impatient with E-in-C's proposal for full-power trials at both 72,500shp and 80,000shp: he had paid for the power, and he meant to get it. Trials reports show that ships typically developed something very close to the full 80,000shp. *Bermuda* made 31.091kts on 79,340shp at 10,900 tons

Shorter machinery spaces freed somewhat more space for accommodation, and DNC freed more by squeezing store-rooms. Space was also freed by using only turbo-generators, all in the main engine rooms (there were no separate diesel generator rooms) and by enlarging the superstructure somewhat. Beam was increased slightly, to 62ft (61ft 5½in exclusive of armour).

The Legend and drawings received the Board Stamp on 4 November 1937. Calculated standard displacement was 8,170 tons, within the margin Controller had offered a year earlier. Five ships were included in the 1937/8 programme, ordered in December 1937: *Fiji, Kenya, Mauritius, Nigeria* and *Trinidad*. Another four were included in the 1938/9 programme (ordered March 1939): *Ceylon, Jamaica, Gambia* and *Uganda*.

Pressure grew to install torpedo tubes; First Sea Lord was willing to have two on each side if he could not have three, and ACNS considered the torpedo 'the decisive weapon at night or in low visibility'. It would be most desirable to have a three-torpedo salvo to deal with a modern fast target. The triple tubes earmarked for the ships were already being manufactured against the possibility that they would be installed.

British shipbuilders and designers had already wrung all they could from the design. Weight grew as minor improvements were ordered, such as a D.IV.H catapult instead of the earlier D.I.H. and a revised generator arrangement. In January 1939 the ship's calculated standard displacement was 8,268 tons (8,298 tons if torpedoes were carried). At her launch on 31 May 1939 it became obvious that *Fiji* was badly overweight: she completed at 8,631 tons (standard). Somewhat later DNC credited the ship with a design displacement of 8,250 tons. Approved additions before completion added up to 319 tons, although more was clearly involved: growth in electrical equipment (70 tons), turret overweight (52 tons), upper deck stiffening (52 tons), added generating power (45 tons), the new catapult (35 tons), torpedo armament (33 tons), degaussing gear (10 tons), heavier cranes (10 tons), Type 281 radar (7 tons), and RPC for pompoms (5 tons). Another 33 tons was partly compensated for (further radars and Oerlikon guns). About 1941 DNC wrote that he had successfully resisted (or compensated for) another 432 tons, including 85 tons for jibs for the cranes and Mk V rather than Mk IV HADTs. The ships kept gaining weight, however. Inclined on 2 January 1942, *Gambia* displaced 8,846 tons in standard condition (average action condition was 10,167 tons).

By this time there was some question as to just how good a bargain the *Fiji*s were. First Sea Lord wanted something less expensive (requiring a smaller crew), so Controller had DNC design a nine-gun ship, K.34. The initial version had 3½in side armour, but the Staff wanted more (4½in, as in a *Southampton*).[29] On 19 May 1939 the Sea Lords met to decide whether the cruisers of the 1939 programme should be *Fiji*s or K.34s. They were uncomfortably aware that foreign navies might disregard the 8,000-ton limit altogether (First Sea Lord referred to a 15,000-ton 'monstrosity' that Germany might build). As long as Britain continued to obey the treaty, at the least she should build to the treaty limit. Something midway between *Fiji* and *Dido* (presumably the original intent with K.34) was rejected. Second Sea Lord bitterly remembered the previous attempt to inspire foreign powers to limit themselves, the *Leander*s. 'We had had to reply by increasing gradually up to the limit and the pendulum had swung back and forth. While we had wobbled, other powers were building up to the limit.'

The question was armour vs guns. Fourth Sea Lord and ACNS argued that a nine-gun salvo would be good and perhaps preferable to twelve guns and weaker armour (in the run-up to the *Belfast* class, twelve rather than fifteen guns had been chosen on gunnery grounds). However, ACNS also said that he saw no point in talking about quarter-inches of armour: particularly after a talk with DNE, he favoured twelve guns. DNC pointed to the delay inherent in any new design, while the Staff pointed out that the nine-gun ship would cost little less than a *Fiji*. The decisive argument was the basic British policy of closing and defeating the enemy with gunfire.

The Sea Lords refused to give up either aircraft or torpedo tubes (First Sea Lord wanted no more cruisers built without torpedo tubes). When DCNS said that the displacement limit could be disregarded, DNC said that was not the problem: stability was. The two *Belfast*s had both lost significant stability due to changes during construction. *Fiji* lacked their margin against such weight growth.

The Sea Lords decided to continue building *Fiji*s. DNC thought it worth recording that Second Sea Lord (Admiral Sir Charles Little) said that in the past the Board had been criticised for switching from design to design 'as though the Board did not know their minds'. The *Fiji*s were the result of careful consideration, with the understanding that their powerful armament was paid for with protection; 'on the grounds

Above and below: Like all other British cruisers of their day, the *Fiji*s were very tightly designed, hence could not easily accommodate all the additions required during the Second World War. Removal of 'X' turret was part of the solution; only HMS *Nigeria* retained her 'X' turret beyond 1945. HMS *Jamaica* is shown on 12 June 1945. Her radars have been modernised. The stub mast forward of her after funnel carries a US YE aircraft homing beacon, important for fighter control (cruisers so equipped were sometimes called 'pylons', i.e., reference points for fighters). The Royal Navy was much more interested than the US Navy of that era in using surface ships to control fighters, both carrier- and shore-based. At this time the ship's close range anti aircraft battery amounted to no fewer than five quadruple pompoms, four single power pompoms, four single Bofors, two twin power Oerlikons, and six single Oerlikons. *Mauritius*, *Kenya* and *Nigeria* all still had four triple turrets, but all but *Nigeria* were refitting.

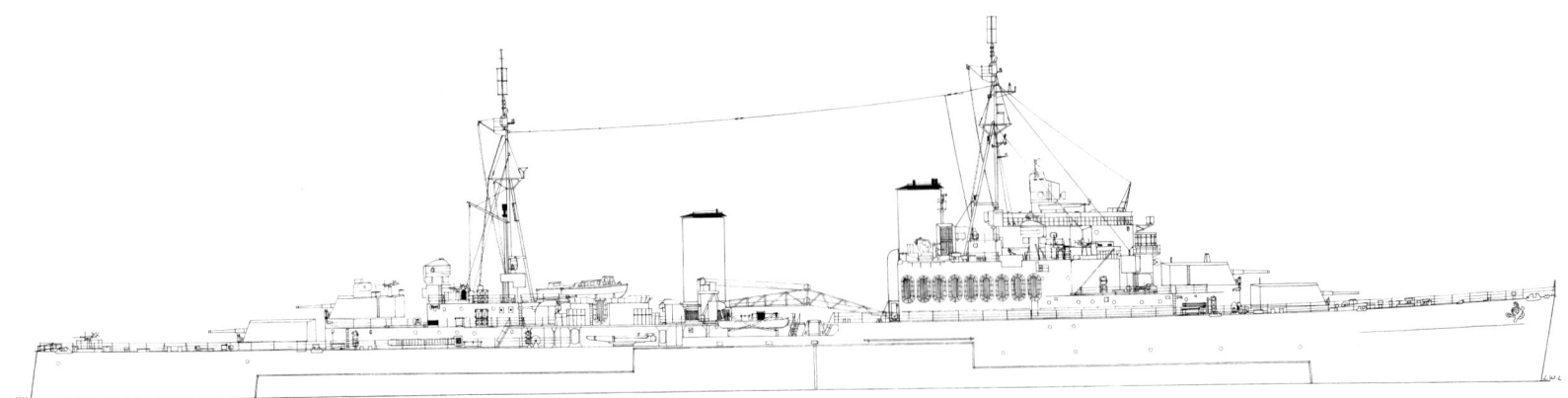

of consistency alone there was an argument in favour of perpetuating the *Fiji* class.' The 1939/40 programme, the last before war broke out, was held to four cruisers (all *Fiji*s) because overall defence spending was reaching the limit of affordability, but once war broke out six *Dido*s were substituted for two of the *Fiji*s because they could be completed sooner. That left *Bermuda* and *Newfoundland*, making a total of eleven ships. At this time it appeared that British industry could produce enough armament for ten cruisers a year (four *Fiji*s, six *Dido*s).[30]

After the ships had been in service for two years, 10th Cruiser Squadron answered a standard questionnaire. The ships were seaworthy, but very wet at moderate and high speed, and speed did not fall off excessively in sea and wind – 20kts could be maintained in rougher weather than might have been expected, actual loss of speed depending largely on the length of swell. Once the ship started to bump her speed dropped quickly. Steaming into a full gale with a moderate swell, the ship not bumping badly, she made about 9.5kts on revs for 12kts. Ships turned well, but could be much affected by the wind, given their large sail area. Thus a wind on the beam made them turn faster to windward and slower to leeward. Trying to turn while at rest into a strong wind caused the ship to drift bodily to leeward. This behavior was considered more marked than might have been expected. The bridge was well-arranged but cramped. Habitability seemed satisfactory given the ships' size, but they were badly overcrowded, and became very uncomfortable when lower deck accommodations were closed up at night. The suggested solution was 'restaurant messing', also referred to as the American system. Boats were inadequate and inaccessible.

After war broke out in September 1939, the torpedo tubes were ordered installed despite overweight, but stability was so critical that weight had to be surrendered.[31] More was wanted: in May 1940 the captain of the newly-completed *Fiji* complained that he had too little ahead anti-aircraft fire. He wanted UP (rocket) projectors, which the Royal Navy was then installing on board some ships, and additional light anti-aircraft guns at the fore end of the hangar-room deck (outboard of the navigational rangefinders). The Admiralty approved replacing the navigational rangefinders with two additional quadruple 0.5in machine guns. Typically a quadruple 0.5in could be replaced by a single 20mm Oerlikon machine cannon.

In September 1941, painfully aware that there was no margin for growth, the captain of HMS *Nigeria* suggested simply keeping some oil in his tanks at all time (he misinterpreted DNC's comment that stability was satisfactory with 150 tons of oil on board).[32] *Nigeria* already had two approved additional light weapons (Oerlikons) and had added two more (she had retained her original pair of 0.5in machine guns). DGD wanted the four-Oerlikon battery but DNC was unenthusiastic. 'Stability owing to additions already accepted is below what it should be in a fighting

Jamaica as she appeared from completion (June 1942) to at least the end of Operation 'Torch', during which she suffered blast damage (from her own guns, firing at low elevation) to the Type 273 'lantern' forward of the bridge. Note the unique prominent aerial outrigger at the fore end of the compass platform, on the starboard side. Close-range armament at this time was two quadruple pompoms and ten single Oerlikons. She was docked on returning to the United Kingdom from the Mediterranean, and she may have been fitted at this time with the lightweight 'lantern' of a Type 272 atop her forward starfish. The big Type 273 'lantern' remained, but probably was not repaired, and thus probably was no longer operating. Certainly a photograph dated 24 April 1943 shows both 'lanterns'. However, note that an account of the use of radar in the Barents Sea battle (31 December 1942), in which *Jamaica* and *Sheffield* proved very successful, refers only to Type 273. It makes special mention of the excellent plotting work done on board HMS *Sheffield*, which may mean that *Jamaica* had inferior surface radar. The account of radar in the battle of the Barents Sea circulated by Director of the Signal Division (DSD) specially commended the plotting party on board HMS *Sheffield*, but warned against using the surface-search radar (at that time, Type 273) for rangefinding, despite its excellent ranging panel. DSD wrote that 'Type 273 enabled the Cruiser Force to detect and plot two enemy units, probably *Lutzow* and *Hipper*, and make a surprise attack on one of them. At this time the enemy were undoubtedly concentrating their attention on the targets to the southward (the destroyer escorts [i.e., the close escort of the convoy]) and were thus caught unaware. The Cruiser Force was similarly surprised by an enemy destroyer later on.' That referred to the use, by *Sheffield*, of her Type 273 for rangefinding, which made it possible for her plotters to miss the approaching destroyer. (Alan Raven)

ship intended to stand up to severe damage.' Free surface made liquid ballast dangerous: 'stability cannot be left to chance as, except very rarely, a ship can only be sunk once.' Too many additions had already been made without compensation.

In February 1942 it was decided that all wood decking would be removed from the forecastle and upper decks, including the quarterdeck, as compensation for fire-control radars (Types 284, 285 and 282). Compensation for the Type 273 surface-search set was to remove two searchlights and their platforms from abreast the after funnel, the two searchlights with the best arcs being retained. A second pair of quadruple 0.5in machine guns was added at the cost of the tactical rangefinder and the balloon-filling equipment. A pair of Oerlikons on the quarterdeck cost the wood there. DNC suggested removing the centre gun of 'X' turret to compensate for two more Oerlikons on the quarterdeck, while also removing the cradle, shell loading tray, recoil cylinders, run-out and elevating gear would buy a third Oerlikon on the quarterdeck or a total of two on the hangar roof. By also removing the shell and cordite hoists to the gun, the ship could have four Oerlikons on the quarterdeck and two on her hangar roof. Single power 2pdrs could replace the hangar-roof Oerlikons. This desperate step was never taken, but it shows just how tight the *Fiji*s were.

Wartime building practices ended the use of light alloys, which added more weight. Additions included RPC for 4in guns, two barrage directors for 6in guns, splinter protection for the HA armament (the largest new item), Arcticisation, spare 4in barrels, another two Oerlikons, and close-range predictors Mk I instead of pompom directors Mk IV.[33] Only partial compensation had been obtained for the addition of radar and a pair of Oerlikons (net addition 33 tons). The only remaining candidates for compensation were removal of the port aircraft crane and boat stowage lowered between the 4in guns. Aircraft could then be recovered only from the starboard side. By February 1942 this step had already been accepted, reluctantly, by DNAD for HMS *Southampton*. Free access between the port and starboard sides of the 4in gun deck would be lost (the intention had been that either side could be supplied from either set of hoists). Boats would be free of blast, although they would need protection from ejected 4in cases. One searchlight could be carried just abaft the after funnel, giving a total of three.

One way out was to remove 'X' turret, as in the *Uganda*s (see below) and work on drawings began in August 1941. Depending on the availability of materials, it would probably take six to eight months after working drawings had been prepared. Alternatively, centre guns could be removed from all four turrets, but in April 1942 DGD refused to sacrifice any more 6in guns because the Japanese had twelve- and fifteen-gun ships (the *Tone* and *Mogami* classes); he did not know that all of these ships had had their triple 6.1in turrets replaced by twin 8in turrets. It was considered undesirable to land the catapult and aircraft until more carriers were at sea. Alternatively, some desired items could be foregone. In order of priority, in the spring of 1942 these were spare 4in barrels, splinter protection (including over 2pdr ready-use stowage) and the two extra Oerlikons, which would be traded for 6in barrage directors and 4in RPC.

The required compensation was so large not because the desired items were all very heavy, but because they were so high up in the ship. In topweight terms the two proposed barrage directors were equivalent to all three guns (and their gear) in 'Y' turret, the lowest in the ship. DTSD, in charge of Staff Requirements, considered it outrageous that two directors for an important but auxiliary function should require as much topweight as the armament of a whole turret. He argued for a split approach. Cruisers working with the fleet needed guns but not aircraft, because surely carriers would be present, while cruisers on the trade routes needed aircraft but could afford less armament. The crisis eased somewhat when the call for new pompom predictors was dropped, and when DNC found that the after HA/LA director could substitute for a barrage director. It proved possible to trade the crane and lowered boats for a barrage director forward and 4in RPC. Ships also added two Oerlikons on the roof of 'X' turret and two on the after superstructure.

Problems returned as improved radars were demanded. In September 1942 Type 272 (in place of the lighter Type 273) and barrage directors were requested for *Kenya*, a choice strongly supported by DSD.[34] The attempt to rewire the after director as a barrage director failed (except for

Photographed on 24 November 1948, and little changed since 1945, *Jamaica* clearly shows her two quadruple pompoms and two single Bofors aft. The fifth pompom was on the centreline forward of the mainmast.

ships coming in for their rare large refits), so two barrage directors had to be installed aft. There was very little left to remove: either the third searchlight and the torpedo tubes or the aircraft and catapult. It was agreed that the latter choice gave a better-balanced ship. Ships involved were four-turret *Fiji*s and repeat *Fiji*s: *Nigeria*, *Kenya*, *Jamaica*, *Gambia* and *Mauritius*, of which *Kenya* was already in hand (hence required an immediate decision) and *Nigeria* was about to begin large repairs (there was enough time to convert her after director for barrage direction). DTSD approved this solution, and Deputy First Sea Lord agreed that there was no real alternative. First Sea Lord concurred on 2 October 1942.

Removing aircraft freed weight and space, both in short supply. In *Kenya* one hangar was fitted with a flat at middle height and used for accommodation, the other being left as a chapel or cinema. Six twin Oerlikons could be added (four on the flight deck and two on the quarterdeck). There was sufficient weight for two 80kW diesel generators. Similar modifications were made to *Jamaica* and *Nigeria*. If both hangars were given flats, two twin Oerlikons could be mounted on the flight deck, while if both hangars were used for accommodation at flight deck level (without flats), eight twin Oerlikons could be added (six on the flight deck and two on the quarterdeck). HMS *Bermuda* proposed a more elaborate conversion, the remaining crane being moved to the centreline and the waist closed in with plating (other ships just moved the crane to the centreline). Proposals to use the hangars as workshops were rejected, presumably because of the topweight involved.

Both Cammell Laird and Vickers promoted new-construction *Fiji*s for post-war export. Cammell Laird offered the design to Argentina in 1947, while Vickers offered a new-construction *Fiji* or *Swiftsure*, with suitable changes in equipment and electronics, to Chile in 1946, together with *Dido*. Vickers offered a somewhat modified *Dido* to Venezuela in 1948, but none of these deals succeeded.

The Heavy Cruiser Problem Revived

Through the late 1930s the Royal Navy faced a terrible possibility. The *Fiji*s symbolised the sacrifice it had made specifically to stop construction of heavy cruisers, but some countries capable of building such ships had signed neither the 1930 nor the 1936 treaty. They or some customer might build a cruiser armed with 8in guns, reviving the idea and making the treaties, and the sacrifices, irrelevant. In 1937 Chile presented exactly that possibility. Chile and Argentina were rivals, and in the 1920s Italy had built two cruisers armed with 7.5in guns for Argentina. Although they were often described as overarmed failures, the other two major South American powers, Brazil and Chile, naturally wanted equivalents and in 1933 Vickers designed an 8in cruiser for Brazil. Even though no British yard would have been allowed to build it, Vickers could have supplied the ship via its Spanish shipyards. For some reason the prospect did not alarm the Admiralty or the Foreign Office, perhaps because in 1933 there was no sense of headlong rearmament around the world. Also, the project may not have aroused much concern because it had little chance of coming to fruition, given the state of Brazilian finances at the time.

But in 1938 the situation was different. The three major South American sea powers all had ambitious building plans which seemed to have a fair chance of materialising. A ship built for any one country might easily be sold to or requisitioned by a European power, as South American battleships building in 1914 had been by the Royal Navy. In 1937 Chile was buying aircraft from Italy and Germany, hence might be moving towards them (the Chilean Navy was equipped with British-built ships). In July 1937 a Chilean naval representative came to the United Kingdom with a firm offer to order two 8,500-ton cruisers armed with six 8in guns in two triple turrets (the Argentine ships had

three twin 7.5in turrets).[35] Neither the displacement nor the armament was permissible. At first it seemed that only Italy and Japan could build the ships, but the Chileans later also contacted Finland, Germany, the Netherlands and Sweden. DCNS pointed out that the construction of such a ship would give any signatory a perfect excuse to invoke the escalator clause of the treaty in such a way as to kill the limit on 8in ships.[36] The Chileans approached three British builders: Vickers, John Brown and Scott's. Vickers offered two *Southampton*s (which could not be built under the treaty). In September 1937 the Chileans increased the pressure, offering a guarantee that 'if we [the British] are in trouble' they would turn the ships over, as in 1914. The Chilean Ambassador said it was his President's 'ardent wish' to have the two cruisers delivered before he left office.

Given the sheer size of the British rearmament programme, there was little hope of quickly building two cruisers for Chile. Most of the ships under construction or on order were destroyers, the main exceptions being the Argentine cruiser *La Argentina* and a Soviet order for cruiser/destroyer machinery (which would figure in early British wartime thinking). To Controller the great bottleneck was fire-control equipment, some new British cruisers (presumably *Arethusa*s) having only one instead of two HACS. Equipment had been diverted for the Argentine cruiser, and Vickers was making predictors (AA computers) for the army. The prospective Brazilian destroyer order would absorb capacity freed when the Argentine cruiser was completed. Controller wanted British yards to promote foreign fire-control systems for their export warships (new Greek destroyers had German systems) or have them made abroad.

By early 1938 Japan was outside the treaty system, and DNI cautioned that she might quickly revert to building 8in cruisers (no one knew that the 6.1in-gun *Mogami*s had been designed for wartime re-armament with 8in guns). When that did not happen, the British began to hope that nothing would happen – unless the Chileans found a builder for the 8in cruiser they wanted. The Chileans asked the British whether invocation of the escalator clause for battleships had freed the country to build 8in cruisers. It had not. The British thought (incorrectly) that it was working effectively to deter the Japanese from breaking out in their cruisers. In September 1938 Vickers was told that it (in combination with John Brown) could offer Chile two repeat *Fiji*s, to ensure speed of delivery. Details of the design, sufficient to allow bidding, were released to the Chileans. Rumours included what turned out to be a false claim that Bofors had already begun work on 8in guns for Chile. Krupp was reportedly willing to supply guns for hulls to be built in the Netherlands. As a backup, in case they were unable to buy 8in cruisers, the Chileans asked Great Britain, the United States, France, Italy, Sweden, Finland, Germany and the Netherlands whether they would be willing to build two 8,000-ton cruisers armed with 6in and 4.7in guns, with a speed of 32kts and an endurance of 8,000nm at 14kts.

The British felt compelled to go to each potential builder to secure an agreement that no 8in cruisers would be built. Negotiations between Chile and Finland broke down. The Germans rejected a British claim that the Anglo-German Naval Agreement of 1935 barred them from building such ships, but they were not yet ready to break with the British, so in September 1938 they agreed not to allow Krupp to build 8in guns for Chile (the '10,000-ton' *Admiral Hipper* class cruisers they

Opposite: HMS *Bermuda* at Melbourne, 9 September 1946: 'X' turret was replaced by two quadruple pompoms; two more are on the nearby after superstructure. She also had four single power pompoms, two single Bofors, two Boffins and two single Oerlikons. (Allan C Green via State Library of Victoria)

Below: HMS *Nigeria* (laid up) was the last *Fiji* to retain 'X' turret.

were building were still to be armed with 5.9in guns, the switch to 8in not yet having been announced). The Dutch abandoned negotiations because they were relying on Krupp 8in guns, although they were still interested in building 6in cruisers. The most extraordinary aspect of the affair was that the British offered to recognise the Italian conquest of Abyssinia (which had almost brought the two countries to war) in exchange for an agreement not to build 8in cruisers. In January 1939 the issue seemed that important.[37] The Chileans were not amused, and at a dinner in November the British Ambassador noted the bitterness of junior Chilean officers.

By January 1939 the Chileans had largely but not completely accepted they could buy only 6in cruisers. It was rumoured that the Dutch were building two cruisers 'more or less on the assumption that Chile will take them over, if and when she can pay for them'.[38] The Chileans continued to support the project despite the enormous cost imposed by a severe earthquake, the money having been set aside for that purpose. The Foreign Office was particularly concerned to maintain the British position with the Chilean Navy, and pressed for export credits. In May 1939 the Chileans abandoned their search for large cruisers in favour of two light cruisers of about 3,000 tons with 5.5in or 6in guns and a speed of 36kts (they described the ships as something like the French *Mogador* class super-destroyers), and Vickers tendered for the contract. The whole cycle began to repeat in October 1939, with reports of fresh Chilean approaches in Europe and in the United States, and an Italian offer of an 8in cruiser, but the outbreak of war stopped further developments. This episode shows just how badly the British government wanted to kill off 8in cruisers in the run-up to the Second World War.

Larger Cruisers

Although prohibited by the 1936 treaty, larger cruisers remained interesting, not least because there were continual rumours that the Japanese were building 20,000-ton super-cruisers, and because it was considered entirely possible that the Germans would do much the same thing. In January 1938 DNC asked his preliminary cruiser designer, W G John, to sketch a variety of heavy cruisers, beginning with a 20,000-ton ship (700ft x 70ft x 25ft), to make 33kts at standard displacement, with high endurance. Armament would be three quadruple 9.2in turrets, six twin 4.5in and four octuple pompoms, but there would be neither torpedo tubes nor machine guns (0.661in). The ship would have two aircraft and a deck catapult. Armour would be an 8in belt covered by a 4in deck, with underwater protection against a 750lb charge, as in a capital ship. The designer was to try to 'scrape down' to not more than 15,000 tons; cost should be about £5 million. Admiralty Board minutes do not suggest any interest in building such ships, but in January 1938 it was just becoming clear that Japan could not be enticed to abide by the new treaty restrictions (the British thought the Japanese were designing ships only slightly above the allowable tonnage), and the design may indicate British interest in a cruiser escalator agreement. The design was developed in some detail.[39]

The ship would be about the size of a carrier; DNC suggested starting with 132,000shp. DNO provided data on a quadruple 9.2in turret. The belt armour would resist 10in shells 40° from normal at 16,000yds; the deck was expected to resist 10in shells and also a 1,000lb AP bomb dropped from 8,000ft. DNC suggested a short forecastle design with a hangar and the usual alternating engine and boiler rooms, the machinery box being pushed well aft (as in *Belfast*). A cross-section showed a 16ft 6in deep belt extending 11ft 6in above the waterline, with an internal deck below the armour deck but well above water, and a side protective system inboard of the belt. Hull depth would have been 35ft 6in. The belt would have been closed by 6in bulkheads. As in recent much smaller cruisers, the belt armour, rather than armour boxes, would have protected the magazines. The 20,000-ton hull actually proved somewhat small, so John tried a 23,700-tonner as well.[40] To get back down, John tried triple 9.2in guns, but found he did not have enough machinery weight to make 33kts. He had to reduce armour.[41] By early February 1938 this design was being called an armoured cruiser. In mid-February John reported to DNC, who asked about the effect of substituting triple for quadruple 9.2in turrets. The docket and drawings were sent to Controller on 17 February 1938. As reported, the ship with three quadruple turrets would displace 20,750 tons and would cost £5.2 million, while the alternative with three triple turrets would displace 23,500 tons and cost £6 million.

DNC next asked what could be done with nine 8in guns (Design A), or with nine 9.2in (Design B). Speed was set at 33kts and endurance at 10,000nm at 15kts. Other armament was six twin 4.5in, four octuple and two single pompoms, and twelve torpedo tubes. Armour should defeat the ship's own guns at 90° inclination (worst case) between 8,000 and 25,000yds (with an inch less over machinery). The deck over the magazines should defeat a 500lb SAP bomb dropped from 10,000ft (over machinery, a 500lb dive bomb); and underwater protection should, as before, defeat a 750lb charge. For the 9.2in gun, this was more than had been asked for before: 10½in C over magazines and 9½in over machinery, with a 3½in deck over magazines and 2½in over machinery.[42] For 8in shellfire it was quite heavy, too: 9in (8in) belt and 3½in (2½in deck). Deck armour was the same in both cases because it was determined by the bombs, not the shells. This was the sort of protection the US Navy later provided in the *Alaska* class, not even approached in any conventional cruiser. Cutting the number of guns in Design B had little effect; the ship would still displace about 20,500 tons and would require 182,000shp to make the desired speed, probably using eight 25,000shp boilers. A sketch showed a ship with a waterline beam of 84ft, bulged underwater to 88ft, so that the battleship-style underwater (side) protective system could be 10ft deep. DNC received the report of this pair of designs on 28 February and in turn reported to Controller on 4 March. The ship with nine 8in guns was expected to displace 21,750 tons and to cost £5.5 million.

Based on this work, in December 1938 DNC asked for details of an *Alaska*-like super-cruiser, at 20,000 tons, armed with six 12in guns, with 7in belt and 3in deck, which was what the Japanese were (incorrectly) reported to be building.[43] Unlike the 8in and 9.2in cruisers, it was legal under the 1936 treaty, because it came in above the prohibited cruiser zone (8,000–17,000 tons, 6.1in to 10in guns). The estimate was based on the 9.2in ship reported the previous February (700ft x 84ft x 23ft), the twin 12in turret being comparable to the triple 9.2in. Work on such ships resumed in February 1939, with analysis of an 18,000-tonner armed with six 10in guns and twelve 4in HA/LA guns, protected against 8in fire, with a speed of 32/34kts and a cost of about £5 million. Protection was that calculated the previous year, 9in side and 3in deck, offering immunity (against 8in fire) between 8,000 and 25,000yds. The deck could resist 1,000lb AP bombs dropped from 4,000ft and 500lb SAP from 7,000ft. On 10 February John reported to Lillicrap that a ship with six 10in guns, otherwise armed as *Belfast* and protected against 8in shell, with a speed of 33kts, could be built on 18,000 tons (weights added up to 18,550 tons). The ship had 9in belt armour over her magazines and 8in over her machinery, covered by 3½in and 2½in decks respectively. References in the Notebook suggest that there was still interest in 9.2in main armament. Presumably 10in guns, which no major navy used, was interesting because it was the smallest calibre permissible for a capital ship (i.e. not a prohibited heavy cruiser) under the 1936 treaty. Estimated shell weight was 500lbs, roughly twice that of the 8in cruiser gun.

In June, DNC asked for studies of a 10,000-ton cruiser armed with 8in guns, in effect a 'County' using current technology.[44] Presumably Controller (and the Board) envisaged a partial but not complete breakdown of the treaty, so that 10,000-ton 8in cruisers could once more be built under the specific treaty clause allowing a breakout without mutual agreement as to new limits. Hence, as calculations were made, small excess displacements over 10,000 tons were rejected as unacceptable. DNC envisaged a flush-deck ship armed with three triple 8in guns, four twin 4in, four multiple pompoms, and two triple torpedo tubes, plus the usual aircraft and catapult, capable of 32kts, with a 2in deck. Could such a ship have 6in side armour? Variants should be (a) protected against 11in fire (i.e. German 'pocket battleship' fire) at fine inclinations; (b) with armament forward; (c) well protected against 8in fire; and (d) without 4in guns and pompoms. The ship would have new heavy 8in shells (290lbs). On DNO advice it was assumed that the 8in turrets would elevate to 60–70° and that the guns would be as far apart as in the twin mounts of the 'County' class – which guaranteed that turrets would be large. Analysis suggested that a 32kt ship would displace just over 10,000 tons. Data were worked up for both a 32kt (80,000shp) ship (Design A) and a 30kt (58,000shp) ship (Design B), both with 5in belt and 2in deck, Design B coming closer to 10,000 tons, but by surprisingly little.[45] As an alternative, John considered a ship with four twin 8in mounts. That lengthened the citadel and thus required more armour (1,920 tons vs 1,850 tons). The estimated weight of an 8in triple was 328 tons, compared to 218 tons for the twin, so in this analysis adopting twin mounts saved considerable weight (984 tons for triples, 856 tons for twins). Designs C and D were 32kt and 30kt alternatives with the four twin mounts (10,250 tons and 9,995 tons), Design D having ½in more belt armour. The next possibility was eight guns distributed as two triples and one twin: 10,165 tons. All versions used the same hull, presumably to make comparison easier. Different levels of protection were also evaluated.[46]

DNC preferred the ship with three triple 8in guns, and E-in-C was asked about machinery. It would be the 80,000shp *Fiji* plant, but possibly with the 4in magazine forward of the machinery instead of between the after boiler and engine rooms, pushing the machinery box aft. By July, DNO was working up a turret design. As sketched on 29 June 1939 the ship looked like a cross between *Belfast* and *Fiji*, with a substantial gap between her bridge structure and a hangar built around her forefunnel. Like a *Fiji*, the bridge carried a DCT on its centreline and HADTs on either side. The forward 4in guns were in this gap. The ship was flush-decked, but with cut-outs in the forecastle deck so that the after 4in guns (between the two funnels) and the torpedo tubes were a deck lower. An after superstructure just forward of 'X' turret carried a third HADT on its centreline, with the after DCT below and abaft it. This structure also carried the ship's pompoms. This version superseded a more *Fiji*-like 'first attempt'.[47] John ended up with a 580ft x 69ft 3in x 17ft ship (10,576 tons).[48]

Once war broke out, there was no longer any point in a 10,000-ton limit, and Winston Churchill, back as First Lord of the Admiralty, pushed for a powerful ship. John was assigned to develop a new 14,000–15,000-ton cruiser armed with 9.2in guns and protected against 8in shellfire, with a good radius of action, higher speed than the German 8in cruisers (say 33/34kts), six twin 4.5in anti-aircraft guns, and four quadruple pompoms; she would have no torpedo tubes. The ship would carry the usual two aircraft and one catapult, and she would be protected against aircraft torpedoes.[49] Given the earlier studies, John chose a 7in belt and 2in deck (3in if possible), considering a 9in belt excessive. Cruisers would generally fight at something other than 90° inclination. At 90° it took 9in to keep out 8in fire at 8,000yds, but 7in would be enough at 8,800yds at a reasonable 60° inclination. The end of treaty restrictions shows in DNC's instruction: 'Let the displacement come to what it will.'[50] The ship quickly grew back to what John had been sketching early in 1938, about 21,500 tons and 700ft long.[51] The design was reported on 8 September, and it formed the basis for a further armoured cruiser project described in Chapter 9. As an alternative, John was asked for a ship armed with four triple 8in, but otherwise as the 9.2in ship.[52] Both the ship armed with nine 9.2in guns and the one with four triple 8in were expected to displace 21,500 tons and to cost £5.5 million.

These designs suggested that the earlier 10,000-ton cruisers had been a mismatch of guns and ships, and that the ideal was something far larger – and far less affordable. Director of Plans concluded that it would be much better to build as many fast capital ships as possible. Big cruisers were clearly poorly suited to work with the fleet: they would rather fight enemy cruisers in minor actions and on the trade routes. Under such circumstances much would depend on effective range, because it might take hours for a cruiser to close from the range at which an enemy was spotted to firing range. The larger gun offered longer range, both absolutely and because its splashes were easier to spot. Against that, although a design for an 8in turret was proceeding, nothing was being done about the 9.2in. The pits used to build 6in turrets could probably be adapted to 8in, but not to 9.2in, so adopting the larger gun would make for further delays. If any ships were to be built, perhaps they should be faster 8in cruisers more likely to close their targets before nightfall. This reasoning led to the decision to design a new fast battleship using four existing twin 15in turrets, HMS *Vanguard*.[53] However, the 8in (but not the 9.2in) heavy cruiser survived in British building plans.

Director of Plans decided that sacrificing speed might be better than sacrificing armour, so John estimated the effect of holding speed to 27 or 29.5kts. By December 1939, John was working on a cruiser with three triple 8in turrets, other characteristics being as in the earlier 21,500-ton ship. This design is best described as part of the wartime design series, in Chapter 9.

The 8in and 9.2in cruisers were conceived in response to reports that the Japanese were building four or five 'pocket battleship' equivalents armed with 12in guns. The choice in 1939 (for the 1940 programme) was two such ships or a 40,000-ton fast battleship. The latter was chosen; it became HMS *Vanguard*.

In the months before war broke out, the Admiralty developed a new fleet plan including 100 (rather than the earlier seventy) cruisers, because it had to take into account three potential enemies rather than one: Germany, Italy and Japan. The 100-cruiser force would have comprised twenty heavy, forty medium, and forty small cruisers. The Royal Navy had fifteen heavy, thirty-six medium (including thirteen building) and thirty-five small (including ten building) cruisers, a total of eighty-six, but four *Hawkins* and twenty-one 'C' and 'D' class were either overage or soon would be. Pre-war planning was complicated by a treaty requirement to inform the Germans of near-term building plans, which the Germans in turn might use to justify breaking out of the limits imposed on them by the Anglo-German Naval Agreement of 1935.

HMS *Devonshire* is shown on 29 June 1944, her catapult having been removed and her surface-search radar (the 'lantern') moved aft. She had been refitted on the Tyne (18 May 1943 – 20 March 1944), her 'X' turret being removed, together with her catapult. Another two octuple pompoms were added. Her two-antenna Type 281 was replaced by a single-antenna Type 281B. In October 1945, *Devonshire* was credited with six quadruple pompoms, seventeen twin power Oerlikons, and six single Oerlikons. Of the surviving heavy cruisers, only she, *Norfolk*, and *Sussex* had had their 'X' turrets removed.

CHAPTER 8
WAR

The two most obvious wartime changes to existing cruisers were radar and increased close-range anti-aircraft weapons. Not so apparent was the introduction of degaussing coils to deal with the new threat of magnetic mines. All of these presented problems because existing cruisers were such tight designs.

Radar

British work on naval radar (RDF) began in 1935, work on a prototype warning radar (Type 79X) operating at metric wavelength (43 MHz) beginning in September 1937. The first operational set, Type 79Y, was installed in the cruiser *Sheffield* in August 1938, while the first production set (Type 79Z) was installed on board the anti-aircraft cruiser *Curlew* in August 1939. About 100 such sets were made. Work on a higher-frequency air warning (WA) set (Type 281, 90 MHz) began in December 1939. It was designed to range on against surface as well as air targets, exploiting its surface wave. The Staff Requirement to provide accurate surface ranging to 10nm required a masthead height of at least 110ft. DSD proposed that the first go into a *Dido*, because 'for the present war the development of an accurate wireless rangefinder for cruisers is of more importance than for battleships'. Higher frequency made for a smaller, lighter antenna, which was first tested using an army searchlight control set, as Type 280 tested on board the anti-aircraft

cruiser *Carlisle*. It proved possible to use a single antenna with a transmit/receive (T/R) switch, so ships with Type 281 were thus able to make do with a somewhat heavier antenna on their mainmasts, the foremast becoming available for a target-indication radar (Type 293).

These radars were primarily for broad-beam air warning, but given their low frequency they produced surface waves which gave them some surface-warning capability. RAF Coastal Command used much the same technology in its early surface-search (air to surface vessel, ASV) sets, operating at a higher frequency (214 MHz, about 3m wavelength). The surface-ship version of ASV Mk I, using a fixed antenna, was Type 286. Improved versions, including Type 290, had rotating antennas. Although designed (in theory) for destroyers, it was also installed on board cruisers.[1] Installation of such radars presented both topweight and electric power problems, as in 1939 many ships had barely enough generator capacity to cover their action loads. Initially radar required separate (but identical) transmitting and receiving aerials, mounted on the two masts. The displays showed only range, the antennas being rotated to look out on each bearing. Given this type of operation, WA required a broad beam and could not be used to track targets. Later radars were given T/R switches which made it possible to use a single antenna for both transmission and reception.

A parallel series of gunnery radars, also with separate transmitting and receiving elements, worked at 600 MHz (50cm). Work began in February 1938, sea trials following (aboard HMS *Sardonyx*) in June and October 1939. The first set developed was a pompom rangefinder (Type 282) tested in 1940. In 1939 a Staff Requirement was issued for a ranging set which could control main battery fire against aircraft. This Type 283 was installed in a barrage director, the ship's LA guns being

As completed in 1940, HMS *Curacoa* shows the two antennas of a Type 279 radar on her topmasts. There was as yet no T/R (transmit/receive) switch, so the radar had to have separate transmitting and receiving antennas (Outfit ATD). The single-antenna Type 79B (for carriers) appeared in February 1941. Type 279 was the production version of the prototype air-warning radar, Type 79 (Type 79Z was installed on board HMS *Curlew* and on board HMS *Suffolk* in August 1939; HMS *Sheffield* had the earlier Type 79Y in October 1938). Nominal maximum range, based on the scale in the display, was 120nm (accuracy 5nm), but actual ranges were shorter. The radar produced both a ground wave clinging to the sea (nominal range 2 to 6nm) and an air wave, and initially it was seen as a dual-purpose air- and surface-warning or ranging set. Thus it had a short-pulse (3 rather than 8 to 30 microsecond) operating mode for gunnery, and it had a special ranging panel (ranging scale 2,000 to 14,000yds, with transmission in 50yd steps). It reverted to air warning only with the advent of the gunnery radars, Types 282 through 285. The antenna proper consisted of four horizontal dipoles, each with a reflecting dipole a fifth of a wavelength behind it. This produced a broad (hence low-gain) beam (84°) which gave good warning but little idea of the direction from which an aircraft was coming. The frequency was low (39 to 42 MHz, depending on version, equivalent to a wavelength of about 7.5m) because it was the best that British radar engineers could do when Type 79/279 was designed. Peak power (Type 279) was 70kW in warning mode. Initially it seemed that the shorter-wavelength Type 281 was much superior. However, in March 1944 the US Naval Attaché in London reported that the Admiralty had a subtler view: there were trade-offs between the two sets. Type 279 offered better cover above 20,000ft and was less liable to be saturated by land echoes. Its performance also varied less on a day-to-day basis. However, Type 281 offered superior medium-height cover and its echoes could be displayed on a PPI, the basis for fighter control and shipboard air defence. Among cruisers, HMS *Euryalus* seems to have been unique in retaining her Type 279, and in having had it converted to a single-antenna Type 279B in July 1944.

In 1939 the United Kingdom had by far the largest merchant fleet in the world, and the Admiralty had long seen it as both a vital national resource and as a pool from which emergency warships could be drawn. Auxiliary anti-aircraft ships such as *Alynbank*, shown in 1941, were the mobilisation counterparts of the converted anti-aircraft cruisers, with much the same armament, including a scarce HA director (with the corresponding computer below decks) atop her bridge. She had the same four twin 4in guns, as two (sided) quadruple pompoms. The radar antennas served Type 280, an army set (GL1) developed to serve anti-aircraft batteries ashore (it was first installed on board the anti-aircraft cruiser *Carlisle*). The antenna developed for Type 280 was later used in the widely-installed Type 281 air-warning radar, whose wavelength was half that of the original Type 79/279. Type 280 operated at 3.6m wavelength (82 MHz) with a peak power of 25kW. This radar was limited to *Carlisle* and to the converted merchant ships.

aimed to produce barrages at one of two fixed ranges.[2] They were triggered when the range-only Type 283 found that the target was at the right firing range. The first barrage directors were tested on board HMS *Charybdis* in 1942, and the first four Type 283 went to HMS *Berwick*. Ideally one barrage director would be fitted for each main battery turret, but that was not always possible. The director was body-trained and handwheel-elevated as in the Mk I pompom director, the eye-shooting operator estimating deflection using a 300kt cartwheel sight. A second operator could train the director for blind-fire. The Type 283 radar office contained an Auto-Barrage Unit (ABU); the turret could be fired either from the director or from this ABU. The director transmitted future bearing and elevation to the turret(s) it controlled.

In May 1940 work began on low- and high-angle sets (Types 284 and 285 respectively) using the same transmitter and receiver but different antenna arrays. Type 284X was tested on board the battleship *Nelson* in June 1940, and Type 285 in December 1940. Staff Requirements for longer range and more accurate bearing (by beam-switching) were issued in July 1941. Antennas were sets of directional Yagis ('fishbones') or arrays of dipoles along a half-cylinder ('pig-troughs'). Different ships had different Type 284 arrays on their DCTs, and in some cases Type 285 arrays were used for Type 284.[3] Unlike Type 284, the anti-aircraft sets all used arrays of 'fishbones'. As long as WA procedure was to point the antenna in one direction and range on whatever was there, the gunnery radars could also be used for searching in much the same way.

Designs for ship installation were ordered in February 1939 by Director of Plans, for *Dido*s and *Fiji*s and for the eight projected 'D' class anti-aircraft cruisers, but sets would actually be fitted to two *Dido*s per squadron and to all the 'D' class conversions. On 30 May 1939 radar installations were approved in principle, the cruisers involved being *Suffolk* (August–September 1939); *Curlew* (as soon as possible); and *Curacoa*, *Capetown* and *Colombo* by the end of 1939. At the end of June 1939 WA installations were planned for *Suffolk*, *Curlew*, *Curacoa*, *Capetown*, *Colombo* and certain ships (not yet selected) of the *Fiji* and *Dido* classes. It was undesirable to stop work on all the *Dido*s, so DNC proposed that the two being built in dockyard, *Euryalus* and *Sirius*, be selected. About this time a Naval RDF Panel was formed to decide which ships to fit. Purchase of the first thirty WA sets was approved about this time. The first twenty-four WA sets were allocated at its second meeting on 27 July 1939, the cruisers being six of the first eight *Dido*s, three of the first four *Fiji*s (in both cases not the flagships), *Cairo*, *Calcutta*, *Coventry*, three 'D' class AA ships, *Canberra* and *London* (if not flagships), *Exeter* and the first ships of the America and West Indies Squadron (not flagships) coming home for refit. By this time RDF was considered so important that it should always be installed when available. For 'C' class cruisers converted to anti-aircraft ships, Plans Division was willing to surrender one or two guns as compensation. Initially it seemed that installation entailed a considerable effort, but by mid-October 1939 it was understood that time in dockyard could be reduced to that required to ship new tripod masts (a week or even less), so that ships in service could be refitted.

As of 18 December 1939, among the first thirty-one WA sets, cruisers were allocated nineteen: four of the five handmade sets (range panels to be supplied later): *Curacoa*, *Cairo*, *Calcutta* and *Coventry*; four of the five initial production models (range panels to be supplied when available): *Glasgow*, *Fiji*, *Edinburgh* and *Belfast*; three of four production models with range panels: *Naiad*, *Phoebe* and *Nigeria*; and eight of ten later production sets: two *Fiji*s and six *Dido*s. In February 1940 blanket permission was given for installations, so that cruisers could be fitted with radar whenever they became available. DSD sought approval to fit RDF in all anti-aircraft cruisers, and to all other cruisers except the non-

Although the early radars were conceived for air warning, they could certainly detect surface ships. HMS *Suffolk* is shown on 8 August 1941, two months after she successfully shadowed the German battleship *Bismarck*. She had been fitted with Type 79Z some time before September 1939, and then with Type 279 in August 1940 (it was replaced by Type 281 in July 1942). Damaged by bombing off Norway on 17 April 1940, she was refitted on the Clyde, 24 May 1940 – 12 February 1941. Improvements included the installation of Type 279 and Type 284 (gunnery) radars; her two quadruple 0.5in machine guns were landed and four single Oerlikons fitted. The gunnery radar is not visible here, but this photograph was presumably censored to eliminate it. In December 1941 a US officer attached to HMS *Renown* described this operation as the best example of radar shadowing: for thirty-six hours *Suffolk* managed to keep *Bismarck* in range at 14,000 to 26,000yds. It seems clear from this report that, until centimetric sets (e.g., Type 273) entered service Type 284 was frequently used for surface search/warning. For search operation, the director trainer (in constant communication with the radar operator below decks by telephone and handset) trained the DCT and thus the Type 284 mounted on it over the arc to be searched. The radar operator had an indicator of DCT train angle, hence could tell the bearing of any echo he saw on his A-scope (showing only range vs. echo strength). If he saw an echo, he reported it by voice pipe to both compass platform (i.e., the command) and to the plot below it (typically visible to the command via a window in the deck). At the same time he turned a switch in the radar office to 'office controlling' position. That in turn rang a gong in the DCT and lit up an open-faced indicator. Now the radar operator could control the DCT by turning his bearing indicator (the director trainer matched pointers on the open-faced indicator). A US officer wrote in December 1941 that 'this system demands great co-operation between DCT and RDF [radar] office and has not been satisfactory' (in HMS *Renown*). The radar could be switched between 'director controlling' and (radar) 'office controlling' modes, the latter being used for surface search. Typically it was switched to office control after an echo from another ship had been obtained by the director office, the Type 284 being kept lined up on the target by its operator below decks. He in turn sent ranges and bearings to the ship's plotting office as frequently as possible. The plotters used this data to produce a plot of target course and speed. According to the US report, 'to get bearing accurately requires a lot of experience and co-operation between Director Control Tower and 284. The DCT will have to sweep two or three times for each bearing cut so that the 284 operator can determine the bearing at which the echo was a maximum.' That is, the operator used direction-finder techniques.

anti-aircraft 'C' and 'D' classes. That was given in April. In March 1940 cruisers under refit (hence to be fitted with radar) were *Belfast* and *Adventure*. *Ajax* and *Exeter* were due from the South Atlantic. *Perth* was going to Australia instead of returning to the UK, hence could not be refitted as had been planned. Drawings were being prepared for installations on board all cruisers, so that work could go ahead as quickly as possible for ships unexpectedly coming to hand. Priority cruisers were *Glasgow*, *Edinburgh*, *Belfast*, *Ajax*, *Exeter* and *Adventure*.

The long-wave radars suffered from reflection off the sea. The shorter the wavelength, the less reflection interfered with searching for objects on or near the surface. In 1940 researchers at Birmingham University found that they could generate useful amounts of power at a much shorter wavelength (10cm), which was well-suited to surface search. The shipboard version of this radar was Type 271 (warning surface, or WS), its two stacked antennas (transmission stacked above reception) housed in a 'lantern' (cylindrical radome) immediately above its office (to

HMS *Mauritius* shows standard radars (1942), including the important Type 273 surface-search set, in the sixteen-sided 'lantern' (replaced by a perspex cylinder beginning in November 1942, to eliminate echoes from the frames of the 'lantern') forward of her bridge. On her director is the early version of the Type 284 main battery gunnery radar, with separate transmitting (bottom) and receiving (top) arrays, both of which are 'pig-troughs' consisting of a row of vertical dipoles in a long cylindrical reflector. The array at the masthead is for a Type 279 air-warning radar, fitted soon after the ship was completed. Alongside the bridge is a HA director carrying a Type 285 radar to control the ship's 4in guns. IFF antennas are not visible. Type 273 was the large-ship equivalent to the Type 271 used aboard many smaller units. It was the first British naval microwave radar (operating at about 3,000 MHz, a wavelength of about 10cm), and it offered unprecedented surface-detection capability, particularly of small objects. The catch was that the radar office had to be directly below the radar, to minimise losses in the waveguides connecting it to the radar. The 'lantern' was tall because it accommodated two separate antennas, one for transmission and the other for reception. The development of a more powerful radar power source made it possible to move the array away from the radar office, producing Type 272. Type 271 used a pair of 'cheese' antennas, but Type 272 used a pair of side-by-side 3ft dishes (paraboloids) offering higher gain. They produced a 10° wide beam. The operator turned a crank, the antenna being connected by cable to the office equipment, so the radar could not swing in a complete circle, but only through an arc of about 200° to either side of dead ahead. The same limit applied to the later Type 272. Thus it was impossible to connect Type 273 to a PPI (map-like) display, since it could not rotate continuously. The microwave radars (Types 271, 272 and 273) were given fire-control data transmission capability, since they offered more precise surface ranging than longer-wave (50cm) gunnery sets. They were also considered much better than the long-wave sets against low-flying aircraft. Type 273Q was a complete redesign to accommodate the new strapped-cavity magnetron radar tube, which offered an output of 70kW rather than 5kW. It did not enter service until mid-1943. It is not certain which if any cruisers were given it. Type 284 was a 50cm (600 kHz) gunnery radar, one of a series which included the pompom control radar (Type 282), the barrage director radar (Type 283) and the HA control radar (Type 285); it was the only one of the group not (except in the earliest installations) to use an array of 'fishbone' (Yagi) antennas. The long receiving element offered a narrower receiving beam, and later versions produced pairs of beams, which could be switched to find the bearing of the target more precisely. Of this series, the first in service was Type 282, because without radar there was little hope of rapidly providing the range to a pompom director. Type 284 was an adaptation of Type 282, not the other way around, using the same transmitter (initially 25kW, ultimately 150kW) and the same receiver. As of late 1941 (as reported by HMS *Renown*), the radar could be used either for surface warning (range scale to 48,000yds) or for ranging (24,000yds); it could detect a 'Town' class cruiser at 18,000yds and an aircraft (below 15,000ft) at 15,000yds. Bearing accuracy was 1°. Work on fire-control radar began in February 1938, and trials aboard HMS *Sardonyx* began in June 1939. The Staff Requirement for the radar for a barrage director to counter dive bombers (Type 283) was issued in 1939. Type 282 was first tested in 1940, 200 sets being ordered in April. The prototype Type 284X was tested on board HMS *Nelson* in June 1940, while work was already underway on common arrays for transmission and reception. The first production Type 284 was installed on board HMS *King George V* in December 1940, the

first Type 285 on board HMS *Southampton* at the same time. The standard 'pig trough' was 21ft x 2ft 6in, but at least at first there were numerous variations. The M and P versions introduced common antennas for transmission and reception, and the M3 and P3 versions introduced beam-switching for greater precision. These suffixes applied to Type 285 as well as 284. Types 282 and 283 used a pair of Yagis, initially one each for transmission and reception, but later both for both functions..

minimise signal loss in waveguides between antenna and display). The position of the radar depended both on topweight and on topside space available for the radar office. Ideally the radar was atop or adjacent to the compass platform, so that radar information was immediately available to the command. Thus *Fiji* class cruisers generally had the radar and its office immediately forward of the bridge structure, the DCT occupying the only available space atop that structure. With a larger bridge, a *Southampton* or *Belfast* typically had the lantern and office abaft the DCT, looking out over it. Type 271 was initially allocated to convoy escorts. Because it offered better surface performance than the longer-wave Type 284, Type 271 (and later equivalents) were sometimes used for gunnery ranging. Like the WA sets, the early WS sets had only range displays, and were manually rotated. When the set was being used for gunnery, the ship lost situational awareness. That actually happened at the battle of the Barents Sea (in December 1942), when a German destroyer unexpectedly appeared near HMS *Sheffield* – which promptly sank her. The upshot was that the use of surface-search radars for gunnery ranging was discouraged, although versions of these radars continued to have automatic range transmission to the ship's fire-control system (suffix M before 1943, PR afterwards). After 1943 the modified Types 271Q and 273Q had automatic rotation and PPI (map-like) displays.

As microwave radar developed, it became possible to locate the

HMS *Enterprise* shows standard British mid-war radars in a March 1943 photograph. The large array near the masthead served a Type 281 air-warning radar (the mainmast carried a similar array); it could be distinguished from a Type 279 array by its smaller size, because it operated at about half the wavelength. In fact the antenna (Outfit ATE) was a pair of scaled-down Type 279 antennas side by side, so it had eight rather than four radiating dipoles. Above it is an IFF interrogator for this radar, turning with it. The interrogator below the Type 281 antenna, bracketed to the mast, works with the Type 272 radar in the short 'lantern' on the mast, below the crow's-nest. The main battery director carries the 'pig-trough' antenna of the standard Type 284 main battery radar, in this case the version with a common transmitting and receiving antenna (earlier versions had separate antennas on the face of the director). Visible on the face of the bridge are two power twin Oerlikon mounts. Compared to Type 279, Type 281 produced far more power (a short [1.7-microsecond] pulse at 1,000kW and a long [15-microsecond] pulse at 350kW) and had a narrower beam (27°) with higher antenna gain. The original Type 281 could switch beams to find target direction more precisely. The interrogator, Type 243 (IFF Mk III), used a modified ASV Mk II (i.e., Type 286) radar and operated at 179 MHz (with an alternative frequency of 171 MHz if a ship had two interrogators). It superseded a slightly higher-frequency Type 241. There was an associated transponder, to identify the ship to other ships and to aircraft. Type 941 was a Type 243 modified to work with a radar using a PPI (map-like) display, such as the Type 281B fitted to many cruisers. Type 242 was a smaller-ship alternative to Type 241/242 mainly for centimetric surface-search radars. Corresponding transponders were designated in a 250 series, e.g. Type 253 to work with Type 243. The first ship with Type 281 was HMS *Dido* (1941). The single-antenna Type 281B was introduced late in 1941, but was not aboard many cruisers even at the end of the war and beyond. Type 281BQ (continuous rotation, PPI display) was introduced in 1945, and was an important post-war modernisation item. The arrays were rotated by Selsyn (a form of magnetic motor which followed a remote command) at 2 or 4 rpm, the direction reversing after each complete rotation. Alternatively the operator could point the radar antennas (transmitting and receiving) to a desired bearing by hand. Nominal range was the same 120nm as in Type 279, but performance was much better. Originally, like Type 279, Type 281 had a gunnery ranging function, but it was soon abandoned. In 1941, HMS *Renown* circulated a radar pamphlet claiming that its Type 281 radar could detect an aircraft flying at 20,000ft at 75nm (35nm for one at 5,000ft) and a cruiser at 15,000yds. Bearing accuracy was given as half a degree. Type 272 used the same superimposed pair of 'cheeses' as Type 271, but it had a more powerful transmitter, hence could be up to 40ft (i.e., up a mast) from the radar office. Like Type 273, Type 272 could be used as a gunnery ranging set. That could prove embarrassing. During the battle of the Barents Sea (December 1942), HMS *Sheffield* used her Type 273 for that

purpose, the radar being kept trained on a German cruiser. Because it was not being used to search (manually), the ship's command missed the approach of a German destroyer – which was very fortunately spotted visually, and destroyed. This practice, apparently common, was then strongly discouraged.

office 40ft from the antenna (Type 272). Thus in modified *Fiji*s Type 272 could be placed in front of the DCT barbette, the DCT being raised to clear it. Like Type 271, Type 273 had its antenna in a lantern below which the operator sat. Antennas of the next generation (Types 277 and 293) could be located well away from the radar office. Type 277 was a tiltable dish which could be used either for surface search (with a narrow beam) or tilted for limited height-finding. Type 293 was the first British target-indication radar. Some link was needed between the broad beam of the WA radar and the narrow beam of a gunnery radar, the target-indication radar in effect searching the broad beam. In 1943 the British became interested in Target Indication Units (TIUs) which functioned as track-while-scan memories. A TIU could, in effect, maintain multiple tracks which could be assigned either to different guns or to one gun in sequence as it dealt with nearby targets. After the war the TIUs were associated with Gun Direction Systems (GDS), which were rated in terms of how many targets they could handle. TIU III with its Type 992 fast-scanning radar was a staple of post-war designs. Microwave technology was also applied to gunnery radar, the most important wartime example being Type 274 for main battery fire control.

Backing up (and working with) radars were intercept devices and jammers. The first shipboard electronic sensors used by the Royal Navy were radio direction-finders. Work on shipboard HF direction-finding began in 1930, and by the late 1930s many cruisers had remote-control rotatable loops at their fore mastheads. These were not the instantaneous HF/DF devices used by convoy escorts during the Second World War. The British classed interception of enemy signals for exploitation without code-breaking as 'Y'; initially that meant exploiting enemy communications, but from 1940 on it included exploiting enemy radar emissions ('noises'). The first effort was directed against German coastal radars in the Pas de Calais which supported attacks on British coastal convoys; later efforts expanded considerably. Beginning about 1941 the FV1 VHF/DF was developed specially for ships engaged in 'Y' (it was first tested in the monitor *Erebus*). In 1942 all cruisers (and many other ships) were ordered fitted with it and with the Type 91 tunable spot jammer operating at 200–600 MHz, the German radar frequencies (Type 91M extended its range down to 90 MHz and provided noise modulation).[4] Type 91 was successfully used against German naval radar up to January 1945. Some cruisers assigned to bombardment on D-Day received additional jammers (AC III and ATDV). AC III was probably Airborne Carpet III. A second threat, which appeared in 1943, was radio-controlled missiles. The first jammer produced was Type 650,

The Type 285 HA gunnery radar is atop this HA director, shown on board HMS *Ajax* at New York Navy Yard, 16 October 1943. Type 285 began as a range-only radar, and it was developed with beam-switching so that it could track a target in bearing – but unlike its US contemporaries it could not track in elevation, hence could not be used for blind fire (the array did elevate with the sights). It was designed to provide fuse settings, as the associated shells were time-fused. Large HA directors like this one used an array of six Yagis, initially three for transmission and three for reception. Gain doubled when the entire array was used for both purposes, so range increased considerably. The aloft director was part of a larger High Angle Control System (HACS), which employed a below-decks computer called the High Angle Control Table (HACT). The director fed the computer with target range (optical and later radar), elevation (taken from the optical rangefinder), and bearing. There was no attempt to measure target motion using gyros, as in US systems such as Mk 37. The initial HACS I was fitted to all ships up to 1935, after which it was replaced in battleships by HACS III, the sets so released being installed as second sets on board heavy cruisers. HACS I was given improved fuse prediction gear for target speeds of up to 250kts, and ultimately all were to be converted to HACS I*** with maximum enemy speed of 350kts and also with gyro roll correctors (GRUB). The *Leander* class was fitted with HACS II, which was similarly modified. Compared to HACS I, the director was reshaped to accommodate junction boxes and the rangefinder. Later cruisers prior to HMS *Birmingham* were given HACS III, which was considerably improved mechanically, including a better means of data transmission. It was reshaped to suit a new 15ft rangefinder and had a slightly thicker shield. Mk III* had an additional rangetaker's position (III** was completely round and had no additional rangetaker's position). From *Birmingham* on, cruisers were given HACS IV, which had provision for roll compensation, cross-levelling (against pitch), and a new means of data transmission (Magslip). In earlier versions each director tower was associated with one calculating table, but in HACS IV any director could be connected to any table. Ultimately the number of targets the ship could engage depended not on how many directors she had but on how many calculating tables (i.e., how much internal volume) she had. Like Mk III**, Mk IV had a completely round shield, but it accommodated the second rangetaker. Mk IV* was designed to control 5.25in guns on *Dido* class cruisers. Mk V was a high/low angle director of a different type, used on board battleships and carriers. The early post-war standard types were Mks 4 and 4*. The successor director, carrying the Type 275 blind-fire radar, was Mk 6. Directors generally accommodated a control officer, layer, trainer, telephone operator and range taker. Later towers also had provision for a LA rate officer (for when the tower was used for LA control) and space for a second rangetaker, who was needed when duplex height-finders were introduced. The control officer had glasses which moved in step with the two director telescopes; the right-hand glass incorporated a graticule which could be rotated to indicate the apparent course of the target. In later directors the other glass could be used for spotting. Later directors also had a separate telescope (forward area sight, later High Angle Direct Eyeshooting Side [HADES], which was superseded post-war) with a spider's-web graticule for the control officer, to be used to indicate deflection. Gyro roll correctors stabilised the director telescopes. A contemporary HACS manual advised that, since it took time for the gyros to settle down, they should be kept running when air attack was considered probable.

which was limited by its narrow frequency range. It was soon followed by the broader-band Type 651. It was fitted on board most important warships by the end of the war.

A new generation of radar direction-finders was developed in 1944–5 for the Pacific: four RU series DF sets were being developed to cover the full radar frequency range, RU4 (2–6 GHz) being the most advanced.[5] The war ended before it could enter service, and none of the series survived post-war.

Beyond radars themselves, the great wartime development was the Action Information Organisation (AIO). The Royal Navy had invented and developed plotting as a way of maintaining situational awareness, and in effect the AIO extended the plotting idea to radar. The AIO was not quite equivalent to the US Combat Information Center, because the Royal Navy split it into a number of related spaces, typically (by 1945) a main tactical plot (Operations Room), a Radar Display Room, an Aircraft Direction Room (ADR), a Gunnery Direction Room (GDR) or Target Indication Room (TIR), descended from the earlier ADP, and often a larger-scale bridge or flag plot. This arrangement could be traced partly to pre-war practice, and partly to the accident that British air-warning radars could not be used directly for fighter or gun control, as the narrower-beam US sets could. The British arrangement entailed problems, because there was no way of making sure that the pictures in all the related parts of the AIO matched. Post-war this problem led directly to the development of the Comprehensive Display System, the first (albeit analog) integrated tactical data system. The British form of AIO was installed on board most ships of the *Southampton*, *Belfast*, *Fiji* (not in the four-turret *Nigeria*) and later classes. Due to the scale of work involved, it was installed on board only two heavy cruisers (*Norfolk* and HMAS *Australia*). *Sussex* and smaller cruisers had only a partial installation.

Below left: Achilles shows New Zealand-developed naval radars in this February 1942 photo taken from USS *Curtiss* at Suva in the Fijis. In February 1939 the British government invited the governments of the technically-advanced Dominions (Australia, Canada, New Zealand, and South Africa) to come to England for briefings on radar. All four produced radars, but only Canada and New Zealand produced naval sets, and the Canadian SW-1C was used only aboard destroyers and corvettes. New Zealand was represented at the British briefings by Ernest Marsden, Secretary of the Department of Scientific and Industrial Research (DSIR). After several months in England, he brought home considerable material, including parts of two television sets (as the basis for receivers) and parts of an ASV Mk I airborne surface-search radar, a predecessor of the British Type 286 naval radar. New Zealand set up a naval radar development organisation, the Radio Development Laboratory (RDL), which in turn set up a group at Canterbury University College in Christchurch (an air and ground radar group was set up in the New Zealand Post Office [responsible for radio] in Wellington). The first experimental gunnery set, probably designated SS-1 (66cm/450 kHz, peak power 5kW) was installed on board HMNZS *Achilles* at the end of May 1940 (some reports have it installed on board the armed merchant cruiser *Monowai* in July 1940). Performance was encouraging enough to warrant work on an improved set, which was installed on board *Achilles* in February – March 1941. It was accurate to within 50yds. From it were developed both a new gunnery set (SWG) and a warning set (SW), both of which were installed on board *Achilles* in August 1941. They proved successful during a daylight exercise with the liner *Aquitania*, the radar much outperforming an optical rangefinder at 15,000 to 20,000yds. The Chief of the New Zealand Naval Staff, Commodore Parry, considered the sets valuable enough to warrant production, and he called for a radar capable of ranging on an aircraft at 12,000yds. *Leander* was given pre-production SWG and SW sets during her October – November 1941 Auckland refit. Meanwhile, in September, the New Zealand Radio Development Board made the development of ship and aircraft sets its highest priority, and wanted three of each type provided for the ships on the New Zealand station. New Zealand offered radars to the Admiralty for the Eastern Fleet, to make up for shortages in shipments from the United Kingdom. Plans called for producing at least five of each type per month, and in November 1941 an officer arrived from Singapore to ask for thirty of each type (and also to ask the Australians for forty SW sets, mainly for destroyers). The New Zealanders decided to make the SWG set their priority, and by November 1942 fifteen complete SWG had been shipped to Australia. Another nine were sent to Ceylon, one of which was lost to the Japanese when the merchant ship *Haranuki* was captured. By then sufficient supplies of superior sets were available from the United Kingdom, and eventually the SWG sets went to New Zealand naval batteries ashore. At least eight sets of radar were produced for the Royal New Zealand Navy. SWG operated at slightly lower frequency than SS-1 (73cm/430 kHz) and had an effective range of 7nm (accuracy 50yds). It had two displays, one to give range (using a 30,000yd base with 1,000yd markers) and one to assist in training the director. The antenna was two fourteen-element Yagis, side by side, one for transmission and one for reception, backed by a semi-cylindrical reflector. A later version, on board *Leander* in 1943, used three Yagis, each backed by its own mesh reflector, presumably one for transmission and two for reception. SW was a 1.5m radar using two Yagis (side by side) atop a lattice pylon, visible between the cruiser's funnel and her bridge. Some of these sets may have been installed on board Australian ships (HMAS *Canberra* is sometimes suggested as a candidate). Note that a US report (November 1941) had *Achilles* fitted, *Leander* being fitted, and *Monowai* about to be fitted with an SW set. SWG was described as a 73cm set which might be redesigned to operate at 50cm. According to one account of the radars, in 1941 an improved SW antenna, using four dipoles and a mesh reflector, was proposed, but it appears it was never installed.

This 15 September 1943 enlargement of the mainmast of HMS *Birmingham* shows the four semi-directional arrays of the FV1 radar direction-finder, the rectangles in which are dark vertical elements indicating the framework and the vertical dipole in each. On the lower right are two hourglass-shaped omni-directional antennas, each canted to the vertical. The four FV1 antennas fed a radar intercept receiver which registered the approximate frequency of the target signal. The two omnis were used to measure its frequency more precisely, so that the associated Type 91 jammer could be tuned properly. Its jamming signal was sent out via FV1 antennas. The process was anything but automatic. During the *Scharnhorst* engagement jamming failed because it was tuned to the image frequency of the monitor receiver, rather than to the frequency used by the enemy battleship. Fortunately a cruiser shell soon destroyed the German's radar. There were also successes. In April 1944 HMS *Black Prince* successfully jammed enemy radar during a battle against three German torpedo boats. Type 91 jammed German shore fire-control radars successfully during the invasions of Normandy and of Southern France. The system was used for the last time in January 1945 (Operation 'Spellbound'), when HMS *Bellona* successfully jammed a German Giant Würzberg radar that was being used against her surface strike force. Denied radar coverage, the German convoy lost two merchant ships and an escorting destroyer. In 1942 FV1/Type 91 was ordered installed on board all British battleships and cruisers, but that seems not to have applied to the old 'C' and 'D' classes except for the modernised *Danae* and perhaps ORP *Dragon*. Unfortunately no official register of fittings has come to light. The following list was compiled by Alan Raven based on photography: 'County' class: *Australia* (by late in war), *Berwick* (on completion of August 1942 refit), *Cumberland* (not fitted at 1942 refit, not certain about later), *Cornwall* (not fitted when sunk), *Devonshire* (fitted by March 1944, maybe earlier), *Dorsetshire* (not fitted when sunk), *Kent* (no evidence of fitting as of early 1944), *London* (on completion of May 1943 refit, on foremast), *Norfolk* (fitted on completion of refit June 1943, uniquely fore and aft), *Shropshire* (fitted on each side of the bridge on completion of April 1942 refit, on mainmast as refitted October 1943), *Suffolk* (on completion of May 1943 refit), *Sussex* (by April 1945). *York* and *Exeter* were sunk before they could be fitted. 'Town' class: *Birmingham* (on completion of late 1943 refit), *Glasgow* (fitted by July 1943), *Liverpool* (on fore leg of mainmast at height of after funnel top on completion of large refit, August 1945), *Newcastle* (not certain), *Sheffield* (on foremast mid-1942, mainmast array added 1943); *Belfast* (fitted by 1944), *Edinburgh* (during large refit January – March 1942). *Manchester* was lost before she could be fitted. *Fiji* class: *Bermuda* (not fitted until August 1945 refit), *Gambia* (fitted after 1943 refit), *Jamaica* (upon completion of 1945 refit, not before), *Kenya* (as completed), *Mauritius* (not certain), *Nigeria* (by end of 1943 refit in United States, possibly earlier). *Fiji* was lost before she could be fitted. *Trinidad* also probably was not fitted. Improved *Fiji*: *Ceylon* (as completed), *Newfoundland* (as completed), *Uganda* (apparently not as completed, but had the system as refitted by the end of 1944), *Swiftsure* (as completed). *Dido* class: *Argonaut* (fitted during US refit, November 1943), *Cleopatra* (by end of US refit November 1943, by end of war aerials moved to higher position just below Type 281 aerial), *Dido* (not fitted), *Euryalus* (fitted by 1945, date not known), *Charybdis* (definitely not fitted as of early 1943), *Hermione* (not fitted), *Phoebe* (fitted during US refit June 1943), *Sirius* (no evidence of fitting as of 1944, may have been fitted later). Improved *Dido*: as completed. Anti-Aircraft Cruisers: *Caledon* (upon completion of 1944 refit), *Carlisle* (upon completion of November 1942 refit), *Colombo* (upon completion of April 1943 refit). *Delhi* was almost certainly not fitted, nor were the other ships. *Emerald* and *Enterprise* were both fitted when they were refitted in 1943. *Ajax* was fitted during her 1943 New York refit. *Achilles* was fitted from the completion of her May 1944 refit. There is no evidence of fitting in any other *Leander*, but *Orion* may have been fitted. *Hobart* was fitted during her big late-war refit, but her sisters were lost before this became an issue. There is no evidence that *Aurora* or *Penelope* was fitted, but *Penelope* may have been. The earliest installations, in *Sheffield* and *Shropshire*, used two antennas each, so the ship had to be swung to obtain a bearing. Both ships later received the full four-antenna arrangement.

Above: As modernised, *Danae* had FV1 antennas on the starfish of her foremast. She had the Type 291 air-warning radar typically installed on board smaller ships, descended from the Type 286 shipboard version of the Coastal Command air to surface vessel (ASV) radar. Type 286 actually used an ASV Mk I set; about 200 were placed in service beginning in mid-1940 (see HMAS *Perth* for an illustration). An improved Type 286M appeared in January 1941. Both Type 286 and 286M used a fixed antenna array (Outfit ATQ), in which two dipoles pointed along the ship's centreline transmitted, and two receiving dipoles were canted outward on each side. They were connected to the receiver for alternate pulses, the radar in effect beam-switching. The display was a vertical line, the right- and left-hand echoes being shown to right and left. When the ship was pointed at the target, the echoes on each set of dipoles were the same. The radar could detect targets 50° to either side of dead ahead. Type 286P and 286PQ (and Type 290 and 291) employed a scaled-down Type 281 array with common transmission and reception; Type 291(as here) had a PPI (map-like) display. Type 291 entered service in late 1942; it used a new 100kW transmitter. That represented impressive growth from the original 7kW of Type 286 and its M and P versions and the 50kW of Type 290 (fitted beginning in mid-1941). The rotating-antenna version of Type 286 entered service in February 1941.

Above: HMAS *Australia* shows the standard radars installed when British cruisers were modernised late in the Second World War. The tilted 'cheese' of the Type 293 target-indication radar occupies the masthead. Below it is the tiltable dish of a Type 277 surface-search radar, which could also be used for low-altitude air search or for limited height-finding. The director carries a Type 274 fire-control radar, the centimetric replacement for Type 284, a double 'cheese'. To the side of the bridge is a Type 285 gun-control radar on a HA director: the intended successor, Type 275, required an entirely new (and heavy) director. Below the compass platform is a twin power Oerlikon on a circular platform. Types 277 and 293 were part of a family of new 10cm radars using a 500kW magnetron and waveguides to carry the signals between radar office and antenna. All were to have PPI displays. Type 277 was intended to replace Type 271 and another member of the family, Type 276, was intended to replace Type 272 (it did so in a few ships, such as HMAS *Hobart*). Type 293 offered much better air cover, so it largely displaced Type 276 (the latter was reinstated late in 1944 as a temporary substitute when the original Type 293 antenna proved inadequate). Type 277 (Outfit AUK, the 'great auk') was stabilised in elevation and could elevate to 40° (277P could elevate to 70). The post-war Type 278, in 'County' class guided-missile destroyers, was a modified version remotely controlled in elevation by the ship's combat direction system.

WAR

Anti-Aircraft Improvements

Probably the most important single development was Remote Power Control (RPC), which greatly improved the precision with which a director could control light and heavy anti-aircraft guns. Before the Second World War, directors transmitted their orders to dials at a gun, and the gun crew matched dials or pointers to train and elevate their weapon. But the time lags inherent in such operation became less and less tolerable as aircraft performance improved. The Royal Navy did have a kind of RPC for heavy guns, as used in ships like HMS *Nelson*, but it was heavy and ill-adapted to light weapons. Magslip, which was much lighter, could move a dial in response to a remote movement, but it sent only low-powered signals. The question was how to amplify those signals sufficiently to drive a gun mount. When the Admiralty Research Laboratory (ARL) produced its first magslip in 1928, it immediately went on to produce a magslip-controlled hydraulic servo, which was tested on board the light cruiser HMS *Champion*. In effect this servo was the beginning of RPC, although no servo gun mount appeared for a decade. For a naval gun, RPC would both control the gun (in response to fire control calculations) and stabilise it. In 1937–8 ARL produced a hydraulically-driven twin 4in gun.[6]

Several alternatives were developed: RP 10, RP 40 and RP 50 series (RP 20 and RP 30 were presumably abortive alternatives). Generally the controlling signal (from a magslip) was initially amplified electrically, then either hydraulically or electrically as a final stage. RP 10 and RP 40 were hydraulic; RP 50 was electric.

In 1942 it seemed that the future for close-range control belonged to a new Close Range Predictor, which would supersede the pompom director. The entire predictor, carrying its trainer's and layer's sights, was stabilised, and it carried the same Type 282 radar as the pompom director. This director appears in several of the designs described below. It died when Type 282 was superseded by the much more massive Type 262 centimetric radar.

The Royal Navy entered the Second World War with the most powerful close-range anti-aircraft batteries in the world, but unfortunately cruiser designs were so tight that the increases demanded by war experience were difficult to absorb. By 1943 the only major weights left to remove were ships' boats, aircraft and 'X' turret ('Q' turret in *Dido*s). British officers noticed that the US fleet eliminated nearly all ships' boats in favour of boat pools at its bases. For example, by landing her boats a 'County' class cruiser could gain one or two quadruple pompoms

Above: Type 274, the double 'cheese' visible atop the director of HMS *Kenya* (photographed at Yokosuka on 19 May 1951, during the Korean War), produced a single narrow beam. Note the separate transmitting and receiving 'cheeses.' Late in October 1943 the US Navy officer responsible for fire-control radar development, temporarily attached to the Naval Attaché's office in London, reported on British naval fire control sets. He regarded Type 284 as roughly equivalent to the US Mk 3, which was then being superseded by the centimetric Mk 8. He felt that Type 284 offered only two advantages over the US set. First, it provided the trainer with a full view of all the targets in the radar beam, hence did not require co-operation from the range operator in setting the range gate before the trainer could see any signals. He was therefore protected from missing a signal by mis-setting the gate. Second, it provided a direct measure of range rate via a range aiding mechanism, which gave the correct rate when the target pip remained in the range notch even though the knob was not adjusted. Any change in range rate was immediately obvious. The US officer felt that there was no US counterpart to Type 274; the closest was the US Mk 27, just entering production. Both were S-band (10cm) sets using lobe-switching and precision ranging, but the US set could also be used independently, and offered a PPI display. He saw Type 274 as a much-refined Type 284 with greater gain than the US set. It had a separate spotting display, and a separate range marker to indicate where a salvo should be expected to fall. However, its narrow beam did not permit spotting in deflection. The contemporary US main battery radar (Mk 8) scanned its beam back and forth for that reason (but therefore could not be as precise as Type 274). Type 274 was 'probably the most powerful and precise lobe-switching type of fire-control radar in production today. It is undoubtedly superior to our Mk 8 in almost every respect except the two vital ones of presentation and of deflection spotting. The Type B presentation of the Mk 8 with its ability to present a bird's eye view of any given area, complete with all targets and shell splashes simultaneously, far outweighs all the advantages which the 274 has.' He reported that the British were working on a rapid-scan X-band successor roughly equivalent to the US Mk 13, to appear early in 1945, but it never materialised, and no Type number seems to have been assigned. Instead, the Royal Navy developed a broader-beam splash-spotter, initially an adapted army set (Type 930) and then the Type 931 visible under the right side of the Type 274 array. Type 931 was developed in Canada, as part of the wartime integrated Commonwealth weapons development programme, and twelve were made. Post-war the Admiralty 'anglicised' it for production as Type 932 (to limit the loss of foreign exchange); it is not clear how many were made (it was tested on board the cruiser *Sheffield* in 1952).

(with directors) and possibly four twin Oerlikons, plus the desired target-indication radar (Type 293), a 'Town' could gain two quadruple pompoms and a three-turret *Fiji* a quadruple pompom with director and six twin Oerlikons. The 6in cruisers could gain the target-indication set with a compromise aerial (providing surface search) if they surrendered their Type 272 or 273 surface-search sets. All of these figures assumed that the ship had already had her aircraft removed, and that she retained two boats. This February 1943 proposal was rejected because the Royal Navy considered it vital that the cruiser be a self-contained unit; it was unacceptable to lose the strategic mobility which (according to a submission to the Future Building Committee) 'has always been the pride of the British Navy; that is, the ability to go anywhere and do anything at the shortest notice … [which] more than any other, distinguishes us from the other Services and indeed from any other Navy'.[7] Boat cranes were also used for storing, embarking provisions, etc., hence could not lightly be eliminated (they were given up in the *Dido*s as compensation for Type 79Z radar). Moreover, to set up boat pools at specific ports required that which ports the fleet would use, and with what intensity, should be forecast, but the centre of gravity of the war could move.

In May 1943 DNC was asked to find compensation so that ships could be fitted with their 'ideal' close-range armament.[8] He suggested considering fitting pompoms and Bofors right aft on the quarterdeck, cutting away the after cabin (under that deck) to bring the weight lower and to clear main battery arcs. He was impressed with the new power-operated twin Oerlikons ('Beehives') which saved considerable weight. The use of main-battery barrage fire had to be taken into account, because previously main-battery blast on close-range anti-aircraft crews had not been taken into account. Ships should be divided into those with adequate, reasonable and poor anti-aircraft armament, and shown as such in the Pink Lists of operational units. As a typical 'County', *Kent* had two octuple pompoms and seven twin and two single Bofors; ideally she should have twice as many pompoms and thirty-two Oerlikons. *Ajax* had two quadruple pompoms and four twin and two single Oerlikons; ideally she should have twelve pompom barrels and twenty-four Oerlikons, twice as many. *Birmingham* had two quadruple pompoms (ideal sixteen barrels) and eight twin and two single Oerlikons (ideal twenty-eight barrels). *Kenya* had two quadruple pompoms (ideal sixteen barrels) and eight twin Oerlikons (ideal twenty-six barrels). *Dido* had two quadruple pompoms (ideal twelve barrels) and four twin and two single Oerlikons (ideal twenty barrels).

On 6 October the Board approved removal of 'X' turrets from the *Ajax*, *Birmingham*, *Fiji* and *Devonshire* classes and 'Q' turret from five-turret *Dido*s as compensation for increased anti-aircraft armament. By this time it was looking ahead to the war against Japan; while the Japanese might seldom have superior surface gunpower, they would always be able to deliver heavy air attacks against individual ships. Once 'X' turret had been removed from a four-turret *Fiji*, enough weight was available to use the hangar for an AIC. Cabins previously used for that function could be recovered. Removal of a turret was a major shipyard operation, and it was not done in all ships before the end of the war. In 1946, of the heavy cruisers, only *Norfolk*, *Devonshire*, *Australia* and *Shropshire* had had 'X' turret removed. Of the *Dido*s, only *Dido* and *Sirius* retained 'Q' turret. Of the *Fiji*s, only *Gambia* and *Nigeria* retained 'X' turret. Of the 'Towns', *Newcastle* retained 'X' turret. Only *Orion* of the three surviving *Leander*s retained 'X' turret.

The first new weapon was the Swiss 20mm Oerlikon, twenty of which the Admiralty bought in 1939. They could be distinguished from guns made in Britain and in the United States by their lack of a shield and by their muzzle collars. Up to about 1942 the main improvements to cruiser close-range batteries were the replacement of quadruple 0.5in guns by Oerlikons. It turned out that a twin unpowered Oerlikon did not weigh very much more than a single. The Royal Navy, but not the US Navy, invested in power-driven twin Oerlikons, which it introduced in 1942. Initially they were visually aimed, but a tachymetric sight was introduced in 1944.

Below: HMS *Sussex* is shown in modernised form, 4 April 1945. She had just been refitted at Sheerness between June 1944 and March 1945, her 'X' turret and her torpedo tubes being removed. She emerged with six octuple pompoms (four added) and four twin and six single Oerlikons. Her radars were modernised, her foremast being cleared by replacing the two-antenna Type 281 with the single-antenna Type 281B. That left space for the Type 293 target-indication radar and the Type 277 surface-search radar (the dish). Her previous major refit had been on the Clyde, November 1940 to August 1942, after having been bombed on 18 September 1940. At that time she was given twin 4in guns in place of her earlier eight single mounts, two octuple pompoms, ten Oerlikons, and radars, including surface search.

After the fall of France, the Dutch gunboat *Willem van der Zaan* demonstrated both to the Admiralty and to the US Navy her twin Bofors guns on a stabilised Dutch Hazemeyer (later Signaal) mounting, incorporating its own fire-control system. Both navies were very impressed. The US Navy concentrated on the gun, placing it on a simple twin (and later quadruple) mounting with an external director, while the Royal Navy liked the integrated combination the Dutch had developed, perhaps partly because it was better adapted to a ship with limited internal space. A later British account claimed that the Hazemeyer, whose design began in 1936, was a decade ahead of its time. However, it was also very difficult to maintain. The British placed their standard light anti-aircraft rangefinding radar (Type 282) on board the Hazemeyer. The Hazemeyer was placed in production, but it entered service, as the Bofors Mk IV, only in limited numbers, beginning with HMS *Whimbrel* in November 1942. HMAS *Hobart* and the anti-aircraft cruisers *Caledon* and *Colombo* were armed with this weapon, as were the fast minelayers *Apollo* and *Ariadne*. The British equivalent was the STAAG (Stabilised Tachymetric Anti-Aircraft Gun), which the single Mk I was a prototype only, but the twin Mk II entered production after the war to become a standard weapon. The British became interested in an even more self-contained mount carrying its own diesel generator, but this 'Buster' (Mk VIII) weighed too much (20 tons) and never entered service. The British also developed a six-barrel power-worked Bofors gun, Mk VII, which entered service post-war. After the war the Bofors company developed a 70-calibre follow-on to their wartime 60-calibre gun, which the Royal Navy planned to make its primary close-range anti-aircraft weapon but it was superseded by the Seacat missile (the L70 did enter British army service). A few British ships were fitted with US quadruple Bofors when refitted in the United States. Single Bofors guns were mounted in adapted Oerlikon power twin Mk V and VC mountings as 'Boffins' and were in widespread service by 1945. By that time some British ships, including cruisers, had single US-type Bofors (unpowered 'army' mounts). Post-war, many ships had twin Mk 5 and single Mk 7 power-worked Bofors, the Mk 5s in conjunction with off-mount Simple Tachymetric Directors (STDs). The pre-war single pompom was little used during the war, but a power-worked version appeared in some numbers.

In 1940 some ships were given 20-tube 7in rocket launchers (for 'unrotated projectiles' [UP]) on turret tops. Each rocket carried aloft a line, at the end of which was an aerial mine suspended from a parachute. In theory an approaching bomber fouled the line and brought the mine down on itself.[9]

Electric Power

Throughout the 1930s and the Second World War, a hidden pressure on British cruisers was the need for more electric power. Wartime modifications dramatically increased power loads, particularly the installation of radar and additional close-range anti-aircraft guns. Required generator (dynamo) power was based on the night action load and on the average maximum harbour loads. The standard requirement was to provide the action load with one generator out of action. Cruisers typically had four generators (dynamos), any three of which should be able to carry the action load continuously. Since one of the three might go down, the remaining two had to carry the full action load for a short time, giving time to reduce that load. In 1938 standard generators were designed for a 50 per cent overload for eighteen minutes. Unfortunately the turbines driving them were designed for a 10 per cent overload (for two hours). DEE wanted half the generators to be able to carry the full load for fifteen minutes. Both prime mover and dynamo should be able to run at 25 per cent overload for two hours. E-in-C much preferred turbo-generators to diesels because they were easier to overload. There was no interest in US-style emergency diesel generators. In 1941 the total connected load in an Improved *Fiji* class cruiser was 3,280kW, against an action load of 1,240kW. That was 40 to 60 per cent of the comparable US battle load. The ship's four 450kW generators added up to only 1,800kW. The US Navy required half the ship's generators to carry the full battle load, so it would have demanded a total of at least 2,480kW, of which half the generators should produce at least 1,240kW. That would have required six 450kW units, 2,600kW.

Electric loads began to rise when the Royal Navy adopted power-operated turrets. The worst before the *Southampton*s were the 'Counties' (980kW night action load); that fell to 760kW in a *Leander* and to 660kW in an *Arethusa*. Day action loads were smaller, because they did not include searchlights: 890kW for *Southampton* (but 930kW for *Manchester*), 537kW for *Arethusa*. Corresponding maximum harbour loads (under General Quarters conditions) were 526kW for *Kent* (420kW in *Norfolk*), 670kW for *Leander*, and 560kW for *Arethusa*. In 1933 the estimated night action load of the *Southampton*s was 950kW and harbour load was 770kW because the new ships had more extensive power installations, including their catapults, cranes and even torpedo heaters. The Counties had four 300kW turbo-generators, the *Leander*s four 225kW, so even three generators did not quite cover the night action load.

The situation was further complicated in July 1938 when DEE proposed a new and more survivable electric-power arrangement (ring main) in major warships. It would run behind armour to the extent possible (entirely behind armour in battleships and carriers). It should be split into independent elements, each with its own generator, the port and starboard sides in different main compartments. A generator should be adjacent to each engine room/boiler room unit. DEE wanted half the generators to be diesels, because they could operate independently of the ship's boilers. However, E-in-C considered diesels less reliable than steam. They were also bulkier, demanding more maintenance and staff. DNC wanted diesels kept to a minimum – one advance claimed for the *Fiji* design was their elimination. DNC also pointed out that a ship with unitised machinery had only short steam pipes to the turbo-generators in each unit, which limited their vulnerability. Eliminating diesel generators made it impossible for a cruiser to associate each turbo-generator with a separate machinery unit. The new requirement was applied to both *Fiji* and *Dido* classes. Their action loads were 994kW and 967kW respectively. DNC protested that the 50 per cent fifteen-minute standard would require 1,500kW for fifteen minutes in a *Fiji*, which would involve 10 per cent more weight. The 50 per cent margin seemed to be based on the earlier choice that Admiralty generators should have this sort of overload, not on any particular logic. *Fiji*s (except for the first five) got four 350kW turbo-generators; but the 1938 ships were credited with an 1,100kW action load.

The 1941 *Minotaur*s were originally given the same installation, supplemented by one 150kW diesel generator, but the turbo-generators were replaced by 450kW units (10 per cent overload for two hours). The late 1941 ships (*Tiger*, *Defence* and *Superb*) were upgraded yet again, with four 500kW units (10 per cent overload for two hours) and one 150kW diesel generator. By this time the ring main was normally split fore and aft to overcome action damage, so DEE wanted each half to carry the full action load, and to do so with one of its two generators out of action. The action load in the *Minotaur*s was estimated at 1,300kW, but the largest available British turbo-generator was rated at 500kW. His solution, four 300kW turbo-generators and four 300kW diesel generators, could not be accommodated, and neither could an alternative of six 450kW generators, with diesels moved to the ends of the ship for better survivability. Instead the ships were given four 500kW turbo-generators and two 150kW diesels, one forward of and one abaft the machinery box. The diesels were emergency generators, as in the US Navy. *Newfoundland* already had two such generators, and in February 1942 British policy favoured that practice. *Minotaur* and *Swiftsure* had enough weight margin to accept the two diesels, but not *Bellerophon*, for which heavier boilers had been ordered.

In March 1943 DEE proposed upgrades to cruisers already in service, a plan already having been approved for *Liverpool*, *Sheffield*, *Birmingham*, *Newcastle* and *Belfast*. He wanted a separate diesel generator in *Dido*s and *Fiji*s, and the 300kW units in the first *Fiji*s would be upgraded to 350kW. By June 1944 DEE wanted a fifth turbo-generator (200kW) and a diesel generator. Of modern cruisers only the small *Arethusa* and *Aurora* lacked space for an additional 250kW diesel generator outside their machinery spaces. A planned *Dido* upgrade had their 300kW turbo-generators replaced by 350kW units, but in June 1944 it was reported that only three of the four could be replaced in *Euryalus*, *Sirius*, *Scylla* and *Cleopatra*. In all the others existing turbo-generators either had been or were being upgraded. By July 1944 350kW turbo-generators were on order to replace 300kW units in the 'County' and *Southampton* classes. By October the surviving *Southampton*s (*Sheffield*, *Birmingham*, *Glasgow* and *Newcastle*) were to get two additional 200kW turbo-generators in their machinery spaces. *Hobart* was having her 225kW units replaced by 250kW turbo-generators, plus an additional 200kW unit. *Emerald* was adding two 75kW diesel generators (*Enterprise* was already up to standard). Further recommendations were: to fit a 300kW turbo-generator in place of the auxiliary boiler in *Sirius*, *Scylla*, *Cleopatra*, *Dido*, *Phoebe* and *Argonaut* (plus uprating existing 300kW units to 330kW in the last three); to add a 300kW turbo-generator in place of the auxiliary boiler in *Bellona*, *Black Prince*, *Diadem* and *Royalist*; to fit a 350kW turbo-generator in place of the auxiliary boiler in *Bermuda*, *Ceylon*, *Jamaica*, *Gambia*, *Uganda* and *Newfoundland*; and upgrade existing 300kW units to 330kW and add a 300kW turbo-generator in place of the auxiliary boiler in *Kenya*, *Mauritius* and *Nigeria*.

Suspension and Redesign

In the wake of the Norwegian campaign, on 29 April 1940 DNC asked the constructor who had been with the fleet what the future criterion for cruiser protection should be. Most German bombs seemed equivalent to the British 500lb SAP, and they were typically dropped in a dive from 3,000ft, equivalent to a bomb dropped in level flight from 5,000ft: therefore 2½in in one thickness; decks should be 3in thick to provide a margin of safety (to give an equivalent thickness in two layers, another 80lbs – 2in – would have to be laid on top of the existing 80lbs). Ships might have to be bulged or, if the state of work permitted, have their sides moved bodily outwards to maintain stability. *Dido*s nearing completion could be given extra upper deck armour in exchange for two turrets. Those in the war programme could be redesigned altogether, with 100lb NC added over machinery spaces and 80lb fore and aft. *Fiji*s nearing completion could surrender 'Y' turret to gain about 200 tons, which might give 80lbs (2in) on the lower deck from the stern to 'Y' turret, plus splinter protection over AA guns and torpedo tubes. DNC ordered calculations to show whether a more radically redesigned *Fiji* could get the sort of protection he was considering for the war programme *Dido*s.

Controller soon asked whether ships could be protected against near misses from bombs, whose splinters could riddle a ship's waterline: *Southampton* reported about 240 such holes, *Suffolk* also a large number. DNC was more impressed by internal damage: in *Norfolk*, for example, splinters passed through bulkhead and frame plating (total thickness nearly 2in) to flood spaces 40ft from the explosion. DNC proposed that bulkheads be strengthened so that splinters from a bomb bursting in the middle of a large compartment, e.g. a 48ft engine room, should not pierce either bulkhead at its ends. Experience of bomb damage suggested that 30lbs of mild steel should suffice 27ft from an explosion, so bulkheads should be 1in (40lbs); DNC noted that pre-war experiments (Job 74) suggested that 1½in NC armour was needed. Blast would ruin the watertightness of any bulkhead near an explosion, but bulkheads about 15ft away should stand up well. DNC proposed to take up these issues with builders of *Dido*s and *Fiji*s whose ships were not too far advanced. These ships had insufficient deck armour to keep out 500lb bombs, but the effects of hits could be isolated by strengthening internal bulkheads, both longitudinal and transverse. Surrendering 'Q' and 'Y' turrets in a *Dido* would release enough weight for the stronger bulkheads (210 tons). The holes would be covered with 2in NC plates, 'Y' mounting being replaced by a 4in star shell gun (100 rounds). Surrendering 'Y' turret provide a *Fiji* with ¾in bulkheads (500 tons). Nothing could be done for cruisers already in service or nearly complete.

The constructor assigned to consider extra deck armour in the *Fiji*s assumed that the armour deck should be increased to 3in equivalent. There was already a 50lb (1¼in) strength deck, of which ½in counted as protection, so another 2½in was needed on the upper deck at least from the forward boiler room back to the after end of 'Y' magazine, with 2in abaft that on the lower deck. Protection over the engine rooms could be 60lbs (1½in) instead of 2in. Forward of the forward engine room the forecastle deck should be 2½in thick to the fore end of 'A' magazine. Side armour could be left unchanged. The effect was more on stability than on weight, since most of the weight involved was already present in the armour deck carried lower in the ship. The higher weight would cost 2ft in metacentric height, and on completion *Fiji* had only 2.27ft altogether. A properly protected ship might displace about 9,500 tons, with 65ft beam. If heavier bulkheads and thicker decks were taken into account, the ship needed about 1,360 tons more armour, and about 12ft more beam to preserve stability.

These calculations convinced DNC that a new cruiser design was needed. In August 1940 he proposed a new design for a 10,000-ton cruiser armed with three triple 6in guns, four twin 4in and four (rather than two) quadruple pompoms, plus torpedoes and aircraft. It would be protected as a *Fiji* except for a thick upper deck (2in to 1½in NC) worked structurally; magazines might have an additional 1in on decks and ¾in on sides. Estimates showed a standard displacement of 10,400 tons; the 80,000shp *Fiji* powerplant would drive such a ship at 31.25kts in standard condition, or 30.25kts deeply loaded.

As a parallel idea, on 30 July DNC asked for a rough estimate of the displacement and speed of a ship with four 5.25in guns and *Dido* machinery, with protection amounting to 60lb (1½in) side, 2½in on the structural deck, and boxes over the magazines. He apparently had in mind a new design using the machinery already ordered for the suspended *Dido*s.

Late in May 1940, with invasion threatening, ships which could not be completed within the year were suspended, including five *Dido*s and three *Fiji*s, thus giving DNC the opportunity to redesign them. On 8 August he reported to Controller that work on the suspended ships, particularly the *Fiji*s, had gone so far that scrapping them would be uneconomical. The three *Fiji*s could be given better deck and splinter protection if they were given *Dido* armament (which would be available if three of the suspended *Dido*s were armed like *Scylla*, with 4.5in guns).[10] Of the *Dido*s, three could have improved splinter protection, while the other two could have more radically improved protection through total redesign. In the event, simpler options were chosen: the *Dido*s and *Fiji*s lost one turret each.

In August 1940 the Chiefs of Staff recommended resuming work on the eight suspended ships, and in October 1940 the Admiralty ordered this. In its view the greatest deficiency in cruiser strength was ships suited to trade protection against surface raiders, which at this time was considered very much a cruiser role. The *Fiji* design was not altogether

Left: Norfolk is shown in December 1942 after a refit on the Clyde, 14–23 October 1942. Damage to the ship from near-miss bombs led the DNC to reconsider splinter protection in cruisers.

HMS *Bellona* was typical of the *Dido* class cruisers completed after having been suspended in 1940. She is shown, newly completed, in November 1943. At this time her close-range armament was three quadruple pompoms, three twin power Oerlikons and four single Oerlikons (only in *Bellona*). By the end of the war the close-range batteries of these ships had grown: *Black Prince* had two single Bofors, she and *Bellona* had six twin power Oerlikons (the others had eight), and she had eight single Oerlikons (six in *Diadem*, four in *Royalist*). After the war she and *Black Prince* were both lent to the Royal New Zealand Navy.

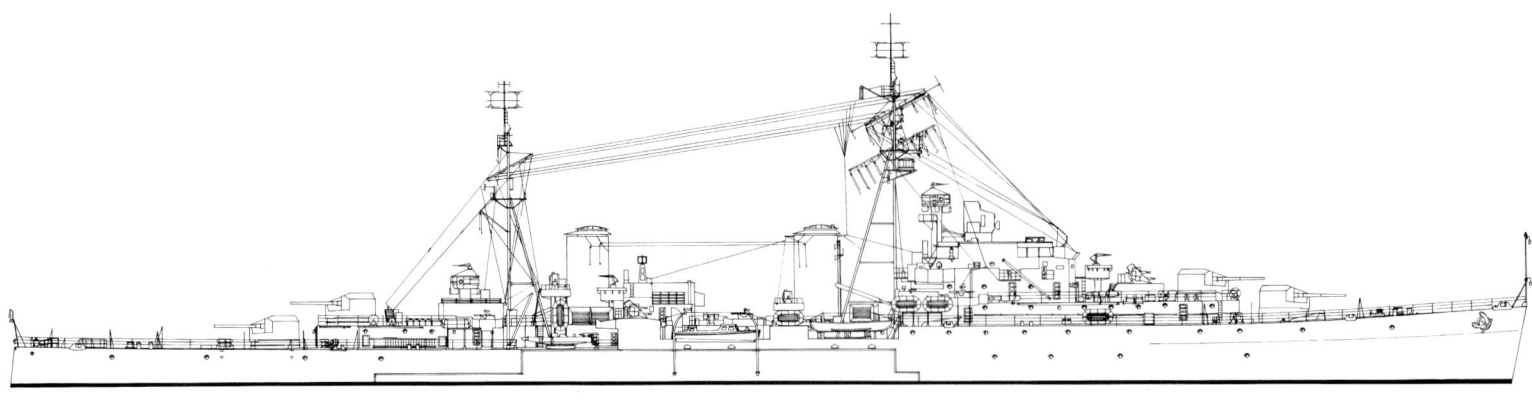

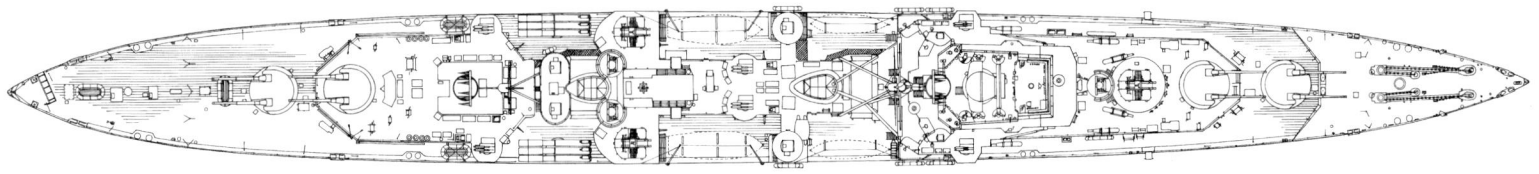

adequate, not least because it had insufficient endurance: it was noted, for example, that a *Fiji* could not escort a convoy from Great Britain to Freetown without fuelling at Gibraltar.

The Redesigned *Dido* class (*Black Prince* class)

By mid-August it had been decided to eliminate 'Q' mounting from the *Dido*s but to retain its magazine as a reserve, i.e. to avoid redesigning the hull, and to increase splinter protection.[11] Masts and funnels would be vertical rather than raked. That made it possible to lower the bridge one deck and move it slightly forward. 'Q' mounting would be replaced by a 4in star shell gun with ready-use ammunition only.[12] As in other British warships, cabins and messing accommodations were rearranged to bring officers and men closer to their action stations. Controller approved this redesign early in September 1940, a Legend for the revised design having been produced on 31 August. Estimated standard displacement was 5,770 tons. The new design incorporated radars. Work was resumed in February 1941.

Internal (rather than external) degaussing, bridge protection and reinforced support for 'A' turret were added. In January 1941 a 4in HA/LA gun replaced the 4in star shell gun in 'Q' position (it was not ultimately fitted, but that came later). The pompoms were given RPC and shielded (1 ton per mounting). Asdic was added. The ship was also somewhat rearranged internally. In July 1941 the 4in gun was replaced by a third quadruple pompom and an admiral's bridge added. Three single Oerlikons were added, and two power-operated single 2pdrs replaced the earlier pair of quadruple 0.5in machine guns. The ship was further stiffened. In December 1941 ships were ordered fitted for Arctic service. In April 1942 the 5.25in guns were ordered fitted with RPC. A 50kW diesel generator was added. Seven twin Oerlikons replaced the three single Oerlikons and the two single 2pdrs. The aftermost twin Oerlikon was removed and two blast cabs were ordered fitted to two twin Oerlikons: the ships were completed with three quadruple 2pdr, six twin Mk V Oerlikons, and three double ship mountings for Lewis guns. Estimated standard displacement was 5,956 tons.

Above: Spartan as fitted, December 1943. Her underwater hull has been omitted because it was identical to that of HMS *Scylla*. She had a retractable Asdic dome. Note that no boat crane was carried and that the number of boats was reduced from the complement of the *Dido* class. Some radio antenna rigging has been omitted for clarity. Her sister *Bellona* was completed with four additional single Oerlikons. HMS *Spartan* was sunk off Anzio by a German guided missile (glider bomb). (A D Baker III)

The Modified *Fiji*s

The three suspended *Fiji*s were *Ceylon*, *Uganda* and *Newfoundland* (two 1938 ships and one war programme ship). Of the war programme, *Bermuda* had not been suspended; it turned out that modifications applied to the other three could not be applied to her. The ships were called the Modified *Fiji* class, the three-turret *Fiji*s (until turrets were removed from the earlier ships), or the *Uganda* class.

On 19 August DNC, having offered alternative modifications to Controller, asked cruiser designer W G John to calculate what could be done if 'X' turret were removed and its main battery control function moved to 'Y' turret.[13] The alternatives of interest were to add two additional twin 4in and two quadruple pompoms (what additional protection could be added?); to add only the two additional twin 4in guns; or to add no further guns, only more protection. Priorities for added protection were splinter protection to the HA armament and ready-use pompom magazines; splinter protection to torpedo tubes; a deck over the small gap between the steering gear and 'Y' shell room; and splinter plating at the side of this gap (splinter plating was taken as 20lb [½in] KN). John found that the first option entailed adding twenty men to an already badly-crowded ship, and another two quadruple pompoms would be difficult to place clear of 6in and 4in blast while also keeping the after HACS in a satisfactory position. Controller chose to add the two twin 4in guns, in tandem on the centreline, one superfiring over the other, using the magazine originally provided for 'X' turret.

DNO pointed out that with only two HA computers the six-mount battery could not be used to best advantage. War experience showed that a 'four-cornered' system was desirable; provide a 'three-cornered ship'

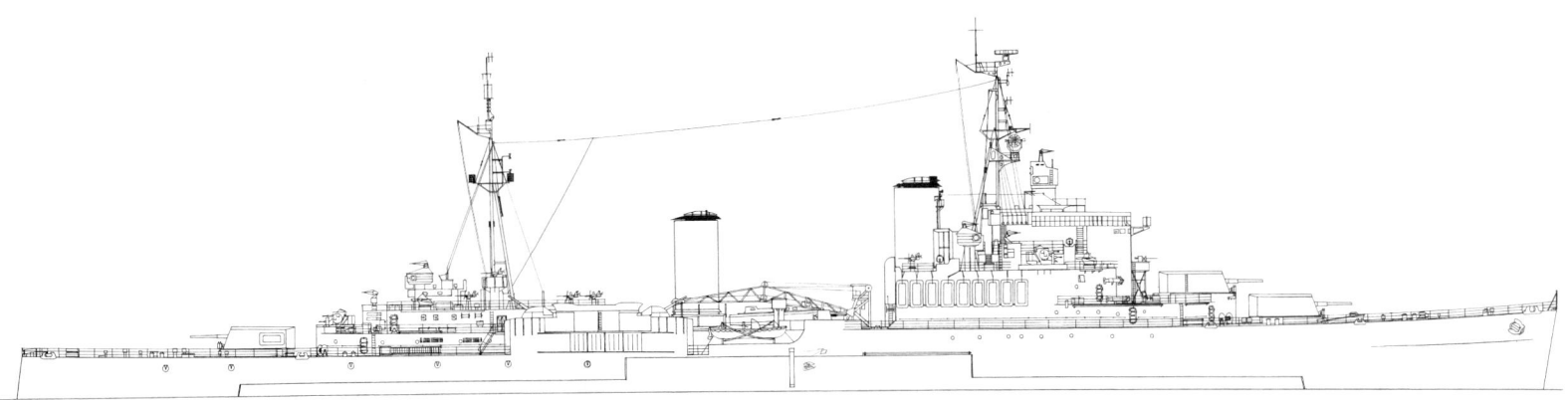

Above: Uganda after her Charleston Navy Yard refit (October 1943 – 14 October 1944) following missile damage at Salerno, 13 September 1943. This was her configuration up to the end of her Pacific service in May 1945. Her port crane was removed. A barrage director was installed in front of her bridge. Close-range armament amounted to three quadruple pompoms (one in place of 'X' turret), two quadruple Bofors (US type), and four twin power and eight single Oerlikons. The ship's radar outfit was modernisd. (Alan Raven)

Below: Newfoundland was one of three *Fijis* which, like five *Didos*, were suspended in 1940 and completed to modified designs, exchanging additional close-range weapons for one turret. She was photographed by a blimp of squadron ZP-11 on 19 April 1944.

using a third AA computer would be almost as good. The mismatch between three AA directors and two computers was particularly bad when radar (Type 285) was fitted to the directors – ideally each HA computer had a dedicated director and radar. Without 'X' turret, there was little point in having an elaborate means of divided main-battery control, using the Admiralty Fire Control Clock in the LA computer room (Transmitting Station). Eliminating that auxiliary computer would help make space for the desired third HA computer space. The third HACP would add thirteen ratings, and another twenty-six were needed for radar.

Weight would be available for splinter protection to the 4in guns,

Uganda in October 1944. She was transferred to the Royal Canadian Navy and later renamed *Quebec*. (Canadian Navy Historical Office)

torpedo tube positions, HACS supports, pompoms and their ready-use magazine. Estimated standard displacement was 8,640 tons, not too far from *Fiji* as completed. DNC convinced Controller that no formal Board approval was needed for this modification to an approved design, and Controller approved it on 26 September.

By this time war experience had shown that accommodation had to be rearranged so that officers and men could be closer to their action stations; the suspension of the three ships made that possible. DEE wanted the ring main electrical system modified to provide for radar (this for *Bermuda* as well). Assistant Controller wanted the bridge rearranged, the wheelhouse (below the compass platform) eliminated in favour of a new lower steering position. An alternative conning position should be provided below the existing wheelhouse (itself on the level below the bridge), for use if the bridge were damaged. The front of the bridge should be made square to provide more space.[14] However, the square bridge was not included in tracings circulated in December 1940.

When revised drawings were circulated in October 1940 the question of extra light anti-aircraft weapons was raised. ADNO suggested that Oerlikons or additional 0.5in machine guns could be added at the expense of splinter protection. He also wanted RPC for the 4in guns and pompoms, but DNC doubted that there was enough space for RPC if a third HACP was installed. An order of priorities was set: (1) Radar (Types 279, 284, 285), (2) RPC for pompoms, (3) a third HACP, (4) RPC for 4in guns; and (5) Type 282 radar for pompom control. In December, DNO felt compelled to switch the last two items because to take full advantage of pompom RPC the director had to be as accurate as possible, which required the radar. As weight compensation for RPC, DNO suggested eliminating the aftermost of the extra two twin 4in guns (neither mount could fire across the stern, because they were blocked by the remaining after turret). That would save twenty-two ratings (sixteen gun crew and six ammunition-supply party). This compromise was generally acceptable; it was approved by Controller (Rear Admiral Bruce Fraser) on 13 January 1941. All HA weapons would have RPC. The ships also had better splinter protection.[15] Two more quadruple 0.5in machine guns were added, for a total of four.

By October 1941 plans for the bridge structure included a Fighter Direction Office (FDO). Ships were to be fitted as flagships, with an admiral's bridge level below the compass platform. Requirements included ice protection as well as wind and blast protection, and experience of air attacks made it clear that the captain (and others) needed a clear view of the whole sky. The admiral's bridge level included the charthouse, plotting office, FDO and Radar Control Office (RCO), the last three adjacent to each other, the plotting office being near the Admiral's bridge so that the Admiral could keep track of the tactical situation. The RCO had to be on the deck immediately below the compass platform, and it also served as a wireless receiving office.

By the autumn of 1941 the fleet favoured the pompom over the 4in anti-aircraft gun, on the grounds that the main threat to individual ships was clearly the dive bomber. For example, there was serious interest in a destroyer armed with 2pdrs instead of any larger-calibre guns. The extra twin 4in planned was superseded by one quadruple pompom, which weighed about as much. However, the ships retained the three HADTs.

Like the original *Fiji*s, these *Uganda*s were subject to excessive weight growth. Based on the completed weight of HMS *Fiji* and calculated deductions and additions, in December 1941 estimated standard displacement was 8,691 tons and estimated deep load was 10,829 tons (half-load, taken as average action condition, was 9,983 tons). At that time armament included two single power-operated 2pdr and four Oerlikons (two on the hangars and two on the after superstructure). The *Fiji* design was so tight that the *Uganda*s could not sustain the 25 tons of extra weight involved in the main proposed upgrades (including Type 272 radar on the foremast). Space for the close-range directors was also a serious problem. The delay involved for ships expected to complete during 1942 was not acceptable. In February 1942 some further additions were wanted for both the *Uganda*s and the follow-on *Minotaur*: barrage directors, Arcticisation, spare 4in barrels, Mk VI HA directors instead of Mk IV, and two more Oerlikons. Elimination of the port aircraft crane and lowering the boats, as in the *Fiji*s, would make possible three of these five items (the Mk VI director proved far too heavy, and it was not ready until 1945). The situation was further simplified when it turned out that one of the two desired barrage directors could be eliminated in favour of modifying the HA/LA DCT aft.

Given the decision to remove aircraft and catapult from the *Fiji*s, the same change was ordered for the three *Uganda*s before they were delivered, so the catapult already installed on board *Ceylon* was removed during construction. One hangar had a flat installed, to function as an AIO on one level and accommodation on the other (as ordered in October 1943), while the other became a cinema or chapel, as in the *Fiji*s.

In April 1947 the long-term close-range armament planned for *Newfoundland*, and presumably for the others, was two STAAG, three twin Bofors with STD (Simple Tachymetric Directors, comparable to the US Mk 51), and a number of single Bofors to be approved. Substitution of Mk VI HADTs for the existing Mk IVs was being investigated.

Improved versions of the *Fiji* class remained in production because to shift to any alternative design would have imposed unacceptable delays. In 1940 there was annual capacity for ten cruisers, four with 6in guns and six *Dido*s. However, a proposed 1940 Supplemental programme included four 8in cruisers. That year no cruisers were ordered, because ASW and mine countermeasures craft had priority. For 1941, the Board considered further *Dido*s unnecessary, so it traded them for additional large cruisers. That left seven such ships, the four 8in cruisers (deferred to autumn 1941 and then to March 1942) and three Improved *Fiji*s: *Swiftsure*, *Bellerophon* and *Minotaur* (the Staff originally wanted all seven ships to be heavy cruisers). These ships had RPC anti-aircraft armament and added fuel-oil capacity forward, so endurance increased from 5,130nm to 5,550nm at 20kts and from 6,459nm to 6,900nm at 16kts.

A 1941 Supplemental programme developed in the autumn of 1941 replaced three of the four 1941 heavy cruisers (weapons for only one of which had been ordered) with three more *Fiji*s: *Defence*, *Superb* and *Tiger* (however, all four heavy cruisers remained in the notional building programme). The three Supplemental ships were built to a further modified design (*Tiger* class).

All seven 1942 cruisers were modified *Fiji*s. Only two were ever named (*Blake* and *Hawke*, the latter cancelled).[16] Due to her suspension, *Bellerophon* was completed to this design (as *Tiger*), which was heavily re-worked after the war and therefore is described in a later chapter. Of the other two, *Minotaur* was transferred to Canada as HMCS *Ontario*. *Tiger* was reordered (having been renamed *Blake*, she was renamed *Bellerophon*) as a

Below: Swiftsure is shown as in August 1945, with emergency close-range anti-aircraft additions (to deal with Kamikazes) in the form of eight Boffins and five single Mk 3 Bofors. All the Oerlikons were apparently removed at this time. In April 1945 the ship had four quadruple pompoms, eight twin power Oerlikons, and six single Oerlikons. She was refitted at Sydney, 29 June–August 1945, the twin power Oerlikons being replaced by Boffins (essentially single Bofors on the same power mounting). The single Oerlikons gave way to single hand-worked Bofors, at some gain in weight. (Alan Raven)

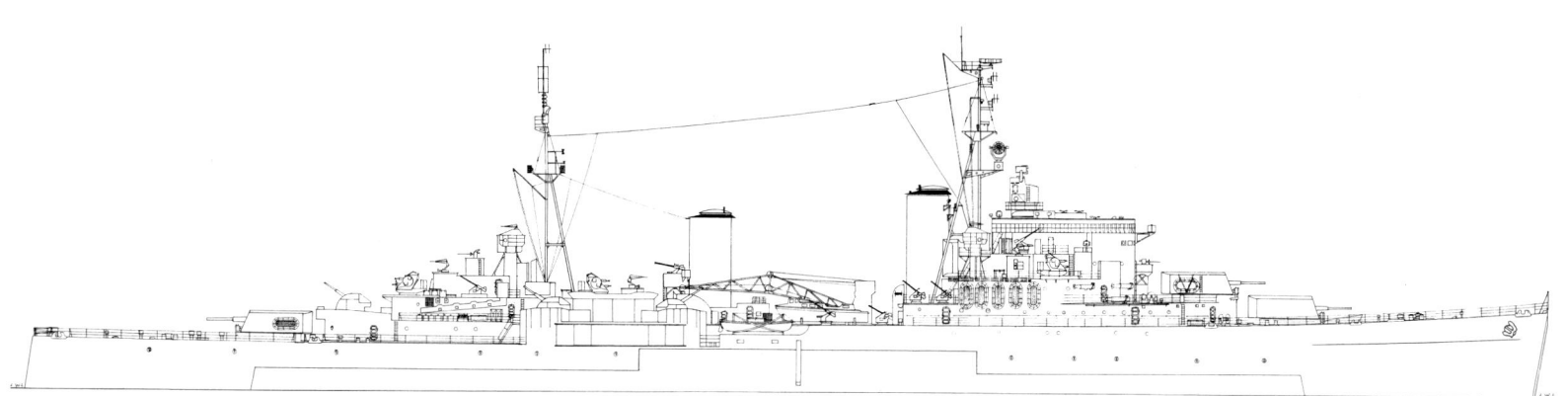

unit of the 1944 *Neptune* class described in the next chapter. The remaining three ships were suspended in 1945 and completed to the new design, *Bellerophon* being renamed *Tiger* and *Defence* being renamed *Lion*.

Staff Requirements were issued in September 1941, before the after twin 4in of the *Ugandas* had been replaced by a pompom. DGD wanted a pair of octuple or quadruple pompoms in tandem instead, with the new close-range predictors (instead of pompom directors). Forward, sided octuple or quadruple pompoms would replace the two HADTs. Close-range predictors would be superimposed immediately abaft the pompoms, and an HADT with Type 285 radar would be superimposed above each close-range predictor (in the position currently occupied by the quadruple pompom). Type 271 radar would be mounted on the foremast clear of the DCT. Oerlikons would be fitted as practicable. The need for better forward AA fire was well understood, and Staff Requirements for a cruiser of this size called for six quadruple or four octuple pompoms.

DNC rejected eliminating the after HADT because it was needed to provide secondary control of the 6in guns, and also because a centreline HA director could be essential in conditions of low readiness and surprise attack (as experience with HMS *Warspite* had shown). He could offer one octuple or two quadruple pompoms, preferring the former to avoid blast and congestion. It was decided to trade the extra twin 4in aft for two quadruple pompoms (total four, two in tandem aft) with close-range predictors on the after superstructure. Similar predictor/radar combinations would replace the pompom directors on the forward superstructure. The stabilised predictors needed gyro transmitting units in or adjacent to existing gyro rooms. Space was so short that magazine stowage was limited to that for two pompoms, but the ready-use stowage at each mounting would be equivalent to one-third of the usual full pompom outfit. All 0.5in machine guns would be landed, two of them replaced by Oerlikons. The control cabinet would be deleted from 'Y' turret, but a fire-control clock could be placed in an enlarged after HACP. The after HADT would be rearranged. Late in 1941 a 150kW diesel-generator was added to the 1941 ships in place of the auxiliary boiler (and 350kW turbo-generators were replaced by 450kW ones). To restore stability, the ships were given a foot more beam (increased to 63ft).[17] Requirements for clear lines of sight fore and aft made it nearly impossible to provide the usual galley funnels, so the ships used electric galleys for officers but retained oil-fired galleys for the crew. The planned increase in electric generating capacity more than covered this addition.

In October 1941 DNC compared the new version of a *Fiji* with the roomier *Southampton*. The two ships were not too far apart in tonnage (8,650 tons vs 9,100 tons). DNC estimated that to bring the latter fully up to date would require 12,000 tons and would cost a knot. *Fiji* was handier, and the *Southampton* armament was now outdated. Her thicker side armour covered a smaller area, and the deck over machinery was thinner. War experience (oil fires in *Southampton* and *Sussex*) had shown the box magazine protection of the earlier design was unsatisfactory. Due to her greater oil capacity, *Southampton* had greater endurance (7,130nm vs 6,450nm at 16kts). Both ships were badly congested (complement of 860 in *Southampton*, 800 in *Fiji*), but *Southampton* was somewhat more habitable.

The ships ordered in the autumn of 1941 (*Tiger* class) had another foot of beam added (64ft in all). The most visible improvement was the Mk VI HA director with its Type 275 radar, which was adapted for blind fire (the earlier Type 285 radar was essentially range-only and could not track a target sufficiently well). The modified design incorporated new target-indication facilities adapted to engaging unseen targets. As in the *Fijis* and *Ugandas*, catapult and aircraft were eliminated in the *Minotaur* and *Tiger* classes (this was decided on 15 March 1943). Rearrangement was considered urgent for ships expected to complete beginning in December 1944 (*Swiftsure*, *Minotaur* and *Superb*). Space originally allocated to the hangar and catapult was rearranged and close-range armament increased. The hangar was lowered by 2ft at the fore end and 4ft at the aft end, its top now level with the sea cabin flat. Flats were fitted at mid-height in the former hangar space, the upper part accommodating eight petty officers and fifty-four seamen, the lower part providing eleven cabins as well as recreation space and a chapel. The crane was retained and repositioned on the centreline, as in *Fijis*. The forecastle was extended over the former flight deck, closing in the waist of the ship. As of October 1943 they were expected to displace 8,864 tons standard (GM 3.05ft); deep load was 11,222 tons (4.8ft). Oil fuel capacity was 1,902 tons.

Anti-aircraft armament could now approach or exceed a cruiser specification set out on 18 February 1943: four twin 4in with four directors, four multiple 40mm and ten twin 20mm. The ships would gain one twin 4in, one quadruple pompom and director, and four twin Oerlikons, for a total of five twin 4in guns (three directors), four quadruple pompoms with directors and twelve twin Oerlikons. Two twin Oerlikons were moved forward from the quarterdeck. There was no space and weight for any more pompoms. The only ship completed to the original *Tiger* design, *Superb*, had four quadruple 2pdrs, eight single Bofors, two single 2pdrs and six Oerlikons. A 1947 report from the ship pointed out serious deficiencies, which would have been evident had the ship fought in the Pacific. The 6in battery had only rudimentary control for anti-aircraft fire, the 4in battery still relied on the pre-war anti-aircraft computer (the Flyplane later associated with the Mk VI director had not been installed, not yet being ready), and close-range weapons were inadequate both in quantity and in quality. *Superb* compared unfavourably with the slightly larger USS *Cleveland*.

At the end of the war the design was reviewed for updating, as they were badly crowded. *Superb* was too far along to change, leaving a *Tiger* class of four ships: *Tiger*, *Defence*, *Blake* and *Hawke*. The main new features were replacement of Mk XXIII by Mk XXIV 6in mounts (with RPC), fitting of 'ring main' main service and main suction systems, and making certain after bulkheads solid to the upper deck. The ships had a new ventilation system, an automatic telephone exchange and counter-flooding arrangements. Changes described as modernisation during the building process comprised armament and fire-control upgrades, radar and wireless upgrades, and habitability upgrades such as centralised messing.

At a meeting on 20 September the Staff pressed for quadruple rather than triple torpedo tubes, as in the 1944 cruiser, with power training and remote control (approved January 1946). New self-contained 40mm guns would replace the old multiple pompoms. Eight single power-worked Bofors (Boffins) would replace the twin Oerlikons.[18] Existing close-range fire-control systems would be replaced by three of the new MRS to control 40mm mountings.[19] The 1947 armament statement for these ships showed three triple Mk XXIV 6in turrets (200 rounds per gun), five twin 4in (250 rounds per gun plus 400 star shell for the ship), four twin STAAG (1,440 rounds per gun), and eight single Mk VII 40mm (1,440 rounds per gun). All guns would be fitted for blind fire. The AIO should be moved below armour and enlarged or moved to the centreline and enlarged, in either case to protect it. Radar offices should be moved below armour. All of this would have been difficult, given the ships' limited internal volume. New radars should be installed.[20] Estimated standard displacement was 9,420 tons, compared to 9,240 tons for *Superb*. All available weight having been expended, VCNS wondered whether some of what was wanted, such as RPC torpedo tubes and Type 268 stand-by radar, was really needed. In February 1946 at attempt was made to prune back requirements. For example, no place

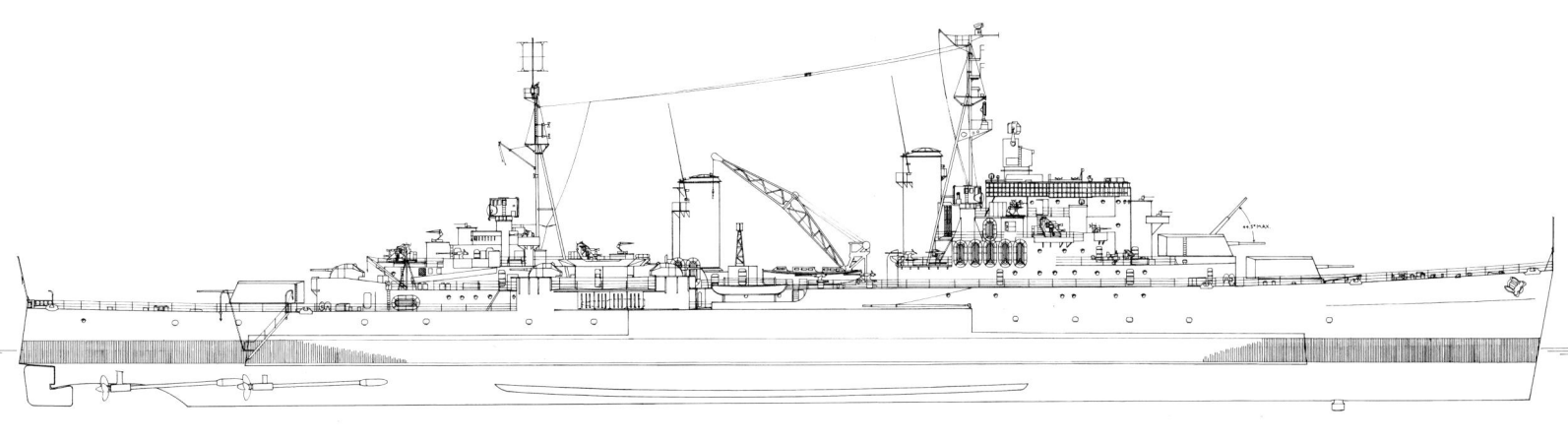
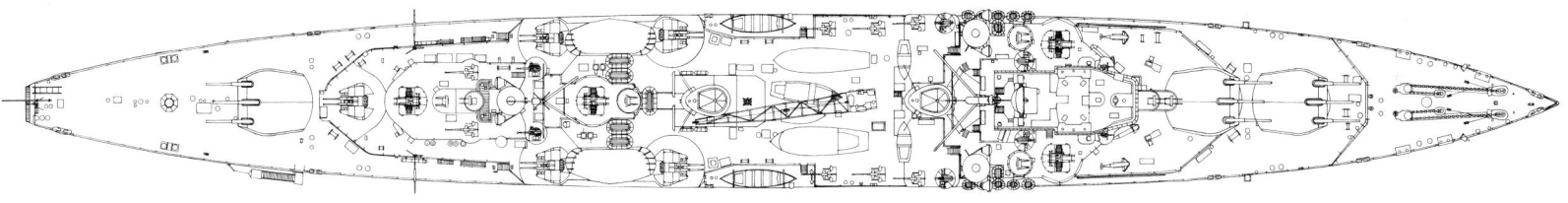

could be found for the YE beacon useful in fighter control.

In the end, none of this happened. *Hawke* was cancelled, and the other three were badly delayed as a shortage of electrical draftsmen made it impossible to supply armament drawings on time. In September 1947 the carriers *Eagle* and *Centaur* were given higher priority, and work on the three cruisers was suspended and the ships laid up without armament or fire controls. As DGD saw it in September 1947, the ships might still be worth completing. They could be ready in 1953, whereas the next-generation weapon, which became the Sea Slug missile, would not be ready until 1958 (which was optimistic). If a missile ship were given an anti-aircraft rating of 100, a *Tiger* would have 40 – but the best wartime ships, *Ontario* and *Superb*, would be rated at 20, and older cruisers at no more than 5. All but *Dido*s had 'extremely low value' as fighting ships mainly because their anti-aircraft fire-control systems were so abysmal.[21] Although plans were being drawn up to modernise the cruiser force, perhaps bringing ships up to a rating of 80, DGD doubted that much would be done. It would probably be less expensive to complete the *Tiger*s with new weapons (to reach a similar rating) because, unlike existing cruisers, they had new hulls and engines. Even as designed they were much better than any other British cruiser.

A similar estimate of AIO efficiency rated *Tiger* at 90, *Superb* at 70, an 8in cruiser partly fitted with AIO at 35, and a 6in cruiser partly fitted at 30. Communications ratings were 65 for a *Tiger*, 55 for *Superb*, and 40 for existing cruisers. The *Tiger*s would have a considerable advantage over other ships in that they would have the anti-torpedo system which was being developed (however this did not, in fact, ever enter service).

Preliminary studies by DNC suggested that one missile system

Superb as fitted upon completion, January 1946. She was the only *Tiger* class cruiser completed to the original design. As such she had both the new 4in director (Mk 6, with Type 275 radar) and the new radar-only pompom director, one per pompom (still equipped with the wartime Type 282 radar, incapable of blind fire). The Mk 6 directors were removed in 1955 (the port-side director remained briefly, cocooned). She had eight single Bofors Mk 3 (hand-worked), four quadruple pompoms, two single power pompoms, four twin power Oerlikons and two single Oerlikons. She could easily be distinguished from *Swiftsure* by her two (rather than one) barrage directors forward of her bridge. The single Oerlikons were removed shortly after completion. In 1949 the single 2pdrs were replaced by single Bofors guns and the four single Bofors on the boat deck were removed to allow more boat stowage. About 1952 the two twin Oerlikons abreast the LA director on the bridge structure and abreast the fore funnel were replaced by four single Bofors. By that time the depth-charge rack had been removed. (A D Baker III)

could be installed in a *Tiger* hull. The trials cruiser *Cumberland* had already been allocated for Guided Air Projectile (GAP) trials; she was considered far more suitable for that purpose than a *Tiger* hull (in fact Sea Slug was tested on board the converted merchant ship *Girdle Ness*).[22] DGD saw little point in wasting a new cruiser on weapon trials when ships of negligible value were available from reserve. Director of Airfields and Carrier Requirements suggested converting the ships into small carriers (DNC had recently studied conversion of 8in cruisers).

DNC favoured competing the three ships as cruisers, as they would be superior to the ships they would replace; on 22 January 1948 First Lord approved: they should be completed as soon as economic conditions made that possible. However, work would be suspended until armament requirements were better known. The outcome is described in a later chapter.

CHAPTER 9
WARTIME CRUISER DESIGN

The outbreak of war ended the treaty limits, but the initial war programme, which was actually an extension of the planned 1939/40 programme, included no new designs. Anything new would come in the 1940/1 or later programmes. Throughout early 1942 work proceeded on designs for large heavy cruisers. These have often been attributed to Churchill's personal interest, expressed once he returned to the Admiralty in September 1939, but from Constructors' Notebooks it is now clear (see the earlier chapter) that the idea considerably predated his arrival, and that the really large British cruiser designs seem to parallel the US *Alaska* project. The main cruiser design reflecting Churchill's personal ideas was an abortive torpedo cruiser (recalling his enthusiasm for a torpedo cruiser in 1913–15), which did not survive his move from the Admiralty to 10 Downing Street in May 1940.[1]

The 'Russian' Cruiser

In 1939 eight sets of cruiser machinery were being built in Britain under Soviet contract, two already being complete. Since machinery was a bottleneck in warship construction, naturally there was interest in incorporating it in new ships, as the Germans had done with Russian cruiser machinery in 1914. They were rated at 48,000shp on two shafts, running at 430rpm (overload rating was 54,000shp with 449rpm). Overall arrangements were similar to those in an 'L' class destroyer rather than in a larger cruiser, although the machinery required 16ft more length than in the 'L' class; presumably the machinery was intended for Soviet super-destroyers. The proposal was to use them in anti-aircraft cruisers with heavy (3in) armoured decks.[2] Such ships were

WARTIME CRUISER DESIGN

Diadem is shown at about the end of the Second World War. The short vertical dipoles at the yardarms forward and on one side of the light yard on the mainmast may be for Type 650/651 missile jammers.

badly needed, so in May 1940 Controller ordered the project to go ahead, even though it had not yet received Board approval. A Legend (Design C) showed a standard displacement of 5,800 tons (470ft [pp] x 54ft x 17ft 3in deep) and a speed of 29/28kts on 48,000shp, which was hardly impressive. Armament was four twin upper-deck 4.5in (as in *Scylla* and *Charybdis*), two rocket projectors (UPs), two octuple pompoms and two triple torpedo tubes. Dimensions were determined largely by the need for stability.

The project to use the Soviet machinery died because the sets did not include boilers, hence using them would not make for quicker construction. Moreover, it was British policy not to antagonise the Soviet Union, so on 22 May 1940 Metropolitan Vickers was told to deliver the machinery. However, Controller continued to be interested in an anti-aircraft cruiser using 'L' class machinery, and in June 1940 he asked for a design for a fast minelayer using 'L' class machinery. The Legend prepared has, however, been lost.

The Large 1941 Cruiser

For the 1940 programme the only large cruiser whose design was ready was *Belfast*, so the design was revised. Beam at the waterline was increased by 2ft (soon increased again to 2ft 6in), the shape of the upper deck being unchanged. 'X' and 'Y' turrets would be lowered ('X' to the level of 'A' turret, and 'Y' below that), presumably to make it possible to add new topweight. The machinery had to be moved forward, and propeller shafts lengthened (adding 29 tons). More extensive splinter-proof plating would be provided for the 4in gun crews and other topside personnel.[3] The forecastle deck from the catapult back to abaft the 4in guns would be lowered 6in and new (heavier) Mk V HADTs would replace the Mk IVs. Structural design would be improved by making the break from 60lb armour more gradual. Aluminium alloys used in pre-war ships were now unavailable, which added weight. Estimated standard displacement with all these changes was 10,866 tons.[4]

In December 1939 work was also proceeding on a large cruiser armed with three triple 8in guns, descended from the studies begun that June, in an alternative to the repeat *Belfast*. Because no cruisers were included in the 1940/1 programme, enough time elapsed for this design to become mature enough to be offered in the spring of 1941 for the 1941/2 programme. On 8 January 1940 DNC asked for a three-turret 8in ship based on the earlier 21,500-ton design, although he hoped to reduce tonnage to about 15,000 tons by cutting protection to that effective against 6in fire, on the grounds that it would resist 8in fire at the large inclination angles at which cruisers normally fought. Thus it would have a 6in belt covered by a 2½in deck (3in over magazines), and the turrets would have 6in face, 3in roof and 2in sides. DNC hoped the machinery could be more extensively subdivided than in earlier ships.[5] The ship should also have six twin 4in and two octuple pompoms (as in *Belfast*), as well as a fixed catapult and two aircraft. No torpedo tubes were specified, but provision should be made for them. Speed should be 33kts standard, 31.5kts deeply loaded. Boilers should all be below the lower deck, where they would be fully protected. The ship's GM should be 4 to 6ft. An alternative ship, which DNC hoped might displace about 10,000 tons, would differ in having four twin 4in guns, a 5in belt, and a speed of 32.5kts at standard displacement and 31kts deeply loaded. GM would be 3 to 6ft.

To keep boilers below the lower deck (for better protection), lower output per boiler had to be accepted. For the kind of power needed (96,000shp or 112,000shp or more) a ship needed more boilers, so W G John sketched eight-boiler arrangements based on the four-boiler layout in existing cruisers, ultimately with boilers paired side by side, their uptakes trunked together to form a funnel from each machinery unit. It might be possible truly to isolate the two machinery units by inserting an air space between them (i.e. abaft the forward engine room and thus forward of the after boiler rooms). Beam was so great that there was no point in putting the after boilers in tandem. E-in-C offered up to 25,000shp per boiler (John thought they could be pushed higher), but the individual boilers would be too tall, so the eight-boiler plant (14,000shp boilers) was preferable.

Visiting the designers at Bath on 8 January 1940, DNC clearly wanted the 8in ships to be built. DNO was to be asked when triple 8in mounts would be ready for the ship. If that was (as apparently expected) June 1944, the ships would be laid down in June 1941, under the 1941/2 programme. The desired pair of Legends were dated 22 January 1940. The 15,000-tonner (670ft x 77ft 6in x 20ft, with 18ft 6in freeboard) required 125,000shp and was credited with 9,000nm endurance at 16kts. The 10,000-tonner (610ft x 72ft x 18in 6in with 17ft freeboard) required 96,000shp and was expected to have the same endurance. These sketches became the basis for further design development. DNC had earlier asked whether the 10,000-tonner would actually displace about 12,000 tons. He could hold down standard displacement by limiting standard ammunition stowage to 100 rounds per gun, providing space for 150. Tonnage was also being held down by adopting a higher forcing rate. The result was a paper speed equal to that of the Germans, 'but experience shows that they don't over-state their speed so that actually would be slower than the Germans'.[6]

Early in August 1940 DNC became interested in the better-protected nine-gun 6in cruiser he preferred to a modified *Fiji*, with *Fiji* side armour but the upper deck increased to 2in – 1½in armour throughout, worked structurally, rather than upper deck armour over only the boilers and the engine room between them. If possible another inch would be added to the deck, and ¼in to splinter protection. *Fiji* machinery would be used and the ship would be armed with four twin 4in and four (rather than two) quadruple pompoms. After an initial design was completed DNC ordered it modified with a 3in upper deck over boiler and engine rooms, the 6in turrets to have *Belfast* protection (by making barbettes 120/60lb instead of 80/40lb), and HA armament increased to six twin 4in and four quadruple pompoms. Splinter protection would be increased. DNC also wanted a twelve-gun version prepared.[7] The nine-gun ship was expected to displace 10,790 tons,

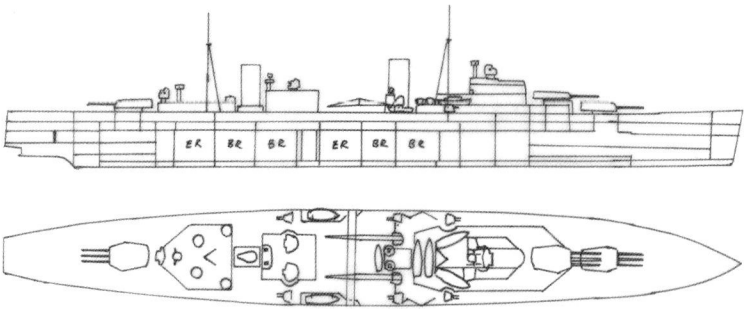

The 15,500-ton (670ft x 77ft x 20ft mean) heavy cruiser as sketched 20 January 1940; a 12,500-ton version was broadly similar. Armament was three triple 8in, six twin 4in HA and four octuple pompoms. Note the two HA directors sided aft, just forward of the after 8in director. There were no torpedo tubes, although later versions of the heavy cruiser had them. Machinery spaces would have been covered by a 2¾in deck, increased to 3in over the magazines and the 4in HA magazine and then the control space forward of the forward boiler room. (Norman Friedman)

although another estimate gave 11,016 tons for an updated *Belfast* with nine guns. Additions included the more powerful D.IV.H catapult of a *Fiji* (30 tons), additional splinter protection (65 tons), weight due to the non-availability of light alloys (100 tons), radar (15 tons), UP (rocket) mountings (15 tons), new Mk 4 HADTs (6 tons), and modified machinery (30 tons). Beam would have been increased by 2ft 6in to maintain stability despite considerably more topweight. A nine-gun updated *Belfast*, again with increased beam, would displace about 10,510 tons. A Legend was prepared for a nine-gun ship with *Fiji* side armour (3¼in and 3½in) but with 3in decks and *Fiji* machinery (10,500 tons). E-in-C disliked forcing boilers to the extent resorted to in the *Fiji* class, so an 80,000shp plant in any new ship needed larger, heavier boilers, as in *Belfast*. The desired ship had *Belfast* machinery and side armour (4½in). At the beginning of September 1940 DNC was instructed to proceed with the nine-gun improved *Belfast* with increased protection, and to work out the tonnage of a comparable ship armed with twelve 6in guns. Legends were prepared in October 1940: the nine-gun ship would displace 11,450 tons, the twelve-gun ship 12,980 tons.[8]

Once estimates had been completed, the two 8in designs were updated to reflect war experience. E-in-C no longer wanted to accept the rate of forcing which had provided the *Fiji*s with their nominal 80,000shp, so machinery would have to be heavier. More electric power was needed, and requirements for reserve power were becoming more stringent. Fire control, including radar, was becoming more and more complex, demanding more power rooms and more men. Late in 1940 DNO wanted RPC for 4in and pompom mountings, which added more weight and internal space. The structural problems revealed by the *Fiji*s demanded more armour in way of previously abrupt changes in thickness and more splinter protection was wanted. On this basis the nine-gun ship was expected to displace 11,945 tons rather than the 10,500 tons estimated in October 1939, and to make 30.75kts rather than 32kts in standard condition. The eight-gun version displaced 12,500 tons.[9]

Attention then shifted back to 6in cruisers, estimates being ordered for nine-gun *Fiji*s with more deck armour: either (a) 3in over magazines and 2in over machinery or (b) 3in over magazines and 60lbs with 2in armour on top, on the upper deck. On 16 October DNC asked for quick studies of 'Mr. Lillicrap's proposal', i.e. twelve 6in, *Fiji* speed, 2in rather than 3in over magazines and 1in bulkheads, and a ship with *Fiji* speed (or less, if that could not be achieved on four boilers) with *Belfast* protection (4½in side, 2in deck over magazines, 1in over DCT, bridges, and wiring, and 1in bulkheads to machinery spaces). DNC became unhappy with even the deck protection he advocated, which would suffice against a 500lb bomb dropped from 5,000ft. A re-estimate showed that a dive-bombing drop from 3,000ft at 300 mph would be nearly equivalent to a level-bombing drop from 11,000ft, against which the ship should have 3½–4in deck armour.

Four quick studies were produced: a modified *Fiji* with twelve guns (Design X); Design X with nine guns; a twelve-gun ship with *Belfast* protection (Design Y); and a nine-gun version of Y. As might be imagined, these were not small ships; Design Y would have displaced 14,050 tons.[10] On 16 November Controller asked for a further twelve-gun design (Z) modified from Design Y, with a 4in rather than 4½in belt and 2in deck (3in rather than 4in over magazines), to save, respectively, 110 tons and 230 tons. This ship had a speed of 30.5kts in deep condition. Required machinery and magazine spaces did not allow much reduction in length; the designers cut 10ft (to 625ft) to save 200 tons on the hull (6,400 tons rather than 6,600 tons). Standard displacement would be about 13,500 tons, and the ship would need about 97,000shp rather than 100,000shp. Controller also wanted to know what thickness was needed to resist 112lb 6in shell at 45° inclination at 8,000–9,000yds (the answer was 3½in NC).[11] Design Z was carried forward, DNC asking in mid-January for the effect of carrying belt armour up to the upper deck throughout the citadel, rather than stepping it down at the end of the machinery space. That added about 256 tons; the ship would probably need a foot more of beam to maintain stability, and that would add another 20 tons for the extra strip of deck. Probably the ship would grow by 400 tons, counting extra hull weight, to 13,600 tons, and she would lose a quarter-knot, to make 31.75kts standard and 30kts deep.

Controller held a meeting on future cruiser designs on 5 December 1940. Opinion favoured the 6in cruiser, largely for fear that a post-war

In 1952, in New Zealand service, *Bellona* had no single Oerlikons, but she had six single Bofors guns. Her sister *Black Prince* had four single Bofors, four Boffins (power-driven single Bofors on twin Oerlikon mountings), and only three single Oerlikons. Radar modernisation of *Bellona* was limited to replacement of the old Type 272 by a Type 277 dish.

Above: Photographed post-war in Melbourne, HMNZS *Black Prince* shows her heavy post-war 40mm battery. Her radar suit had been modernised to the point of converting her Type 281 air-search set to use a single antenna, and replacing her Type 272 with a Type 277 dish (here horizontal, hence not very evident). (Photo by Allan C Green via State Library of Victoria)

naval limitation conference would find the British with four 8in cruisers under construction and the prospect of a new naval building race – i.e. with the problem the *Hawkins* class had posed in 1922. Any hope to kill off the 8in gun cruiser depended on an agreement with the United States. Unfortunately the US Naval Attaché said unofficially that the US programme included fourteen heavy cruisers (four of them with 12in guns [the *Alaska* class] and ten with nine 8in each) plus thirty-four cruisers armed with 5in or 6in guns. Corrected figures supplied in January 1941 showed six super-cruisers and eight 8in ships. A few days later Controller decided that the Staff must take cruiser design up again with VCNS, his view being that the Royal Navy could not go beyond 10,000 tons and twelve 6in.

In mid-January 1941 First Sea Lord asked the two main fleet commanders (Home and Mediterranean) to comment on three cruisers which might be laid down in the near future: (a) a modified *Fiji* (nine 6in, 8,650 tons), (b) a heavy 6in cruiser with twelve guns (14,000 tons, 635ft long), and (c) an 8in cruiser (15,000 tons, 650ft long). They could be completed in March 1944, March 1945 and December 1945 respectively. Both of the larger cruisers had the same protection: 4½in belt, 2in deck (4in over magazines); the modified *Fiji* had a 3½–3¼in side and the same 2in deck over machinery and magazines. It was faster at deep load (clean), 31kts rather than 30.5kts, but endurance at 15kts was 8,000nm rather than the 11,000nm of (b) and the 12,000nm of (c).

Lillicrap, in charge of cruiser design, envisaged a flush-deck design

Below: Bellona in April 1947. (Photo by Allan C Green via State Library of Victoria)

Above and below: Uganda in October 1944, newly transferred to the Royal Canadian Navy. She and her sisters gained only one quadruple pompom in return for 'X' turret, a bad bargain given what *Fiji*s often gained. However, she also had a pair of US-type quadruple Bofors, their shields visible behind the bulwark just forward of the break of the forecastle. She also had eight twin power Oerlikons and eight single Oerlikons. She had been damaged by a German guided missile at Salerno on 13 September 1944, and repaired at Charleston Navy Yard (October 1943 – 1 October 1944). In the process her port crane and six twin Oerlikons were landed, and the two quadruple Bofors, eight single Oerlikons, and the barrage director, an important factor in the ship's redesign (the object forward of the bridge), were installed. Visible atop the former hangar are her HA director and a tower supporting one of her three pompom directors. Her sisters retained their original close-range battery of three quad pompoms, eight twin power Oerlikons (ten in *Ceylon*) and six single Oerlikons (eight in *Ceylon*). *Uganda* was later renamed *Quebec*. (Canadian Navy Historical Office)

if the large cruiser was chosen. The forecastle deck would be the strength deck, and the deck below it (the upper deck in British parlance) should be the thick armoured deck, over the whole citadel (he cited the loss of HMS *Southampton* as justification for armour as high as possible in a ship). The upper (armour) deck should also be considered a strength deck, and stresses calculated assuming that the forecastle deck and side plating above the upper deck had been destroyed. Lillicrap added a new idea. Strength should also be calculated with the ship heeled 30°, which would increase stresses at the deck edge.

The Board decided to reintroduce 8in cruisers because the US Navy was doing so, because radar made long gun range more usable, and because it was willing to accept the compromise 6in protection ('8in guns on a 6in hull') to hold down size. There was some unhappiness that at the battle of the River Plate (December 1939) British 6in shells had not penetrated the belt of the German 'pocket battleship' *Admiral Graf Spee*; the Japanese were rumoured to be building four or five such ships. Also, even the improved 6in cruiser would displace 13,000–14,000 tons.[12] Ships would have cruiser speed (32.5kts, corresponding to 29kts in half-oil condition when six months out of dock in the tropics) and the most up-to-date scale of anti-aircraft firepower. Endurance would be 8,000nm 'deep and dirty' (six months out of dock) at 16kts or 12,000nm at 16kts in trial condition.[13] In October 1941 an endurance speed of 24kts was suggested (this would also be adopted for a new 6in cruiser). Endurance at 24kts would be 6,150nm. Thus the 1941 programme included four 8in cruisers, to be laid down as soon as possible. They would have been given 'Admiral' names.[14] This was the 'large cruiser' which the Future Building Committee (formed in August 1942) rejected in comparison with a *Dido* follow-on.

Controller's instructions in March 1941 led to preparation of four alternative designs, all armed with three triple 8in guns.[15] All had a 4½in belt and 2in deck (4in over magazines), as well as the same speed, aircraft arrangements, close-range armament and torpedo armament.[16] They differed in long-range anti-aircraft armament: (i) four 4.5in BD mountings, (ii) eight 4in twin, (iii) six 4in twin, and (iv) four 4in twin (forward pair omitted). Controller had emphasised the need for a small turning circle, a difficult problem in so massive a ship. Special efforts

would be made to improve seakeeping at speed in bad weather.[17] DNC observed that reinforcing the strength deck around the 5ft openings for BD mounts entailed an appreciable weight gain. The Staff wanted new higher-velocity close-range weapons (described as 50–60mm twins, i.e. not Bofors guns), but DNC had no information about them. He therefore provided four quadruple pompoms.[18] As in the earlier ships, he reduced standard displacement by including 100 rounds per gun of 8in ammunition, but provided space for the usual 150, and 200 rounds per 4in or 4.5in gun (with space for 400). The belt would keep out 6in shells at 5,500yds (60° inclination) and 8in at 11,000yds (40° inclination). The armour deck was the upper deck, offering advantages both in structure and in protection, but heavier than decks in previous cruisers. It might prevent fires started by bombs, which had burned out some British cruisers. The deck would keep out 8in shells below 21,000yds, and 500lb SAP bombs from 2,500ft (level bombing at 200mph). The 4in over the magazines would keep out 500lb SAP bombs from 10,500ft and 1,000lb SAP bombs from 6,000ft, both for level bombing at 200mph. DNC proposed placing thick bulkheads where the 2in and 4in

Swiftsure was a further development of the modified *Fiji* class. She is shown on 31 January 1951, little modified. Immediately after the Second World War she had four quadruple pompoms (note the two aft, on the centreline, before and abaft her after superstructure), five single Bofors, and eight Boffins. By April 1950 two of the single Bofors had been replaced by single power-worked guns of a new type, and there were only four Boffins. Not long after this photograph was taken, she was refitted, the pompoms being replaced by twin Bofors. She was left with nine single power-worked Bofors of a new design, and four Boffins. As modernised she retained her old-type directors. In 1953 plans called for modernising the ship to near-*Tiger* standards, with 3in guns. She collided with the destroyer *Diamond* south of Iceland on 16 September 1953 and was repaired and then laid up in the Nore through 1955. She was taken in hand at Chatham in February 1956, for completion in December 1959. In 1958 her planned secondary armament comprised four (rather than the original five) twin 4in and two twin Bofors Mk 5, with a total of six MRS 8 fire-control systems (two for 4in guns only, two for dual-control of 4in and 40mm guns, and two for 40mm guns). By this time it was probably obvious that the projected 3in guns could not be accommodated. Costs escalated, and work was stopped in August 1959; at that point she had two lattice masts and a new bridge broadly resembling that of a *Daring* class destroyer.

decks met. DNC particularly cautioned that a margin should be provided against the sorts of additions he had faced during less than two years of war. If 100 tons at forecastle level were to be provided, displacements would increase by 250 tons.

Legends were dated 25 March 1941.[19] The design as approved was for 15,000 tons (635ft). With deck armour all on one deck, and with eight 4in guns, it would displace 15,500 tons (640ft); with eight 4.5in, 16,100 tons (650ft); with twelve 4.5in, 17,000 tons (670ft); and with sixteen 4in, 16,600 tons (670ft). DTSD liked the 4in gun but wanted it in a BD mounting; he would go to 4.5in if only that calibre could be so mounted. He also thought that with three directors, eight long-range AA guns were enough (as was then being decided for the modified *Fiji*s).

A sketch design ('XY') for a 670ft ship dated 30 April 1941 showed eight twin 4in guns and four multiple pompoms, two on the hangar roofs and two on a deck forward of the bridge. There were two 4in magazines, one roughly below the bridge under four forward 4in twins, one further aft, just forward of the after boiler room, under the after group of 4in twins. This design had three 3in internal bulkheads, which were soon cut to 2in to save weight. The belt was closed by 4in bulkheads (extending 3ft below the bottom of the belt, and 2in thick below that), to keep out shells entering the side plating forward of the main belt. Protection rings were 4in, and ring bulkheads varied from 4in to 2in. In September DEE estimated that the ship needed six 400kW generators, half of which should be diesel and half steam, a considerable jump up from the four 400kW units envisaged in January.

Design work continued through 1941, a new draft Staff Requirement being framed that October (the March 1941 papers had gone astray). Now the ship was expected to displace about 17,500 tons and to be 670ft long, about the size of the US *Des Moines* class conceived two years later, but without the latter's automated 8in guns, and with a lighter anti-aircraft battery. By this time she was to have had five octuple pompoms (rather than four quadruples, a considerable jump, which brought displacement to 17,500 tons), as many Oerlikons as could be fitted in, and two quadruple torpedo tubes rather than triples. Protection and speed were unchanged. A formal Staff Requirement issued in December 1941 added RPC for the 4in guns, which would be controlled by four directors, two sided forward and two aft, capable of bearing across bow and stern to handle crossing targets. All versions of the Staff Requirement included Asdic, which British cruisers used to evade torpedoes and submarines.

In October 1941, Controller set priorities for new designs: destroyers, a new armoured carrier, 8in cruisers, and then 6in cruisers. DNC thought that with these priorities new 8in cruisers could be laid down by the end of 1942. By November 1941, 8in mountings for the first ship were on order. Design work continued through early February 1942, as indicated by a request that the model tank (Haslar) provide a power estimate for an 18,500-ton ship (690ft x 82ft x 21ft) to make 32.5kts. Controller stopped most design work about 27 February 1942, although some minor work continued, and the four heavy cruisers remained in the official list of projected ships.

This was a very large ship. In October 1941 DNC sketched an alternative 10,000-ton 8in cruiser with endurance about as in the 'County' class, with an 80,000shp powerplant. In mid-January 1942 DGD asked for a quick estimate of a ship with two twin 8in turrets, one forward and one aft. Speed and endurance would be as in *Southampton*, and protection as in a *Fiji*. Anti-aircraft armament should be on the lines of a *Fiji*. DNC offered an 11,140-ton ship (615ft x 70ft) using *Fiji* machinery and lighter armament (four twin 4in, four quadruple pompoms [two superimposed aft], ten Oerlikons, and two triple torpedo tubes). Beam was scaled from the 64ft of the 1941 Supplementary *Fiji*s (*Tiger* class).

The 1943 Cruiser

Staff Requirements for a 1943 6in cruiser were formulated in mid-1942 as an alternative to more *Fiji*s.[20] ACNS(W) began the process in May by asking whether separate LA and HA armaments were required, and also whether a 5.25in gun would be acceptable for LA fire. Could weight be saved by making guns only partly HA, as in the 55° destroyer mounts then being proposed? Could fire control be simplified to save topweight? Could aircraft be omitted? Displacement should not exceed 10,000 tons, and ten LA guns might be desirable. DTSD thought that aircraft could be omitted, given the future availability of carriers not only with the fleet but on the trade routes (a paper written not long afterwards for the Future Building Committee envisaged trade protection units consisting of a cruiser plus a carrier). DTSD wanted at least four torpedo tubes on each side, although ACNS(W) considered torpedoes 'a bit of a luxury'. The key issue turned out to be endurance, now that the war had spread to the Far East. Even the improved *Fiji* lacked it; DTSD wanted something more like a 'County', rated at 4,140nm at 25kts (7,600nm at 15kts), compared to about 3,550nm for the improved *Fiji*.

Proximity fuses having been developed, large-calibre anti-aircraft guns were once more attractive. DGD considered anything smaller than a 6in gun insufficient LA armament for a 10,000-ton cruiser, and wanted a fully dual-purpose main armament. That would simplify fire control. It would be further simplified if the self-contained close-range guns ('Busters') under development proved satisfactory (it was not yet apparent how heavy they would be). It might be impossible to mount more than three triple turrets in a 10,000-ton ship, particularly as DGD wanted them well separated from the superstructure for maximum firing arcs. The four-cornered close-range armament should be well clear of heavy gun blast. To DNO, the rationale for separate LA and HA armaments was the pre-war idea that ships engaging other ships might have to engage aircraft at the same time, but that had never happened. A unified main battery was a risk worth taking, particularly if the ship could use divided control to engage a surface and an air target at the same time.

Any dual-purpose weapon had to use cartridge cases rather than bagged charges.[21] DNO doubted that a 6in cartridge case could stand up to ramming at high angles under a 112lb shell; perhaps 55° elevation might be acceptable. In any case, it was unlikely that a dual-purpose 6in gun would be available in time for the 1943 cruiser. DNC pointed out that the shipyards could not lay down such a ship until 1944, for completion in 1946, so the Staff chose nine 6in guns in three triple turrets, with 80° elevation. It was hopeless to attempt to provide deck protection against heavy bombs.

Director of Plans concluded that the cruiser should be able to neutralise ships of similar class and to destroy inferior ships (destroyers, raiders) and also to provide anti-aircraft cover on the most ample possible scale. She should therefore be armed with 6in guns, and should have carrier endurance and slightly greater speed. That was taken as 32kts and 6,000nm at 24kts. Endurance forced displacement up to 12,000 tons despite a 10,000-ton limit laid down by ACNS(W).

The resulting Staff Requirement called for nine 6in HA/LA guns in three triple turrets (400 rounds per gun), ten twin Hazemeyer Bofors, twelve twin power-worked Oerlikons, and two quadruple torpedo tubes. Speed should be 31.75kts in standard condition, 30kts deep. Endurance should be 6,000nm at 24kts when six months out of dock in home waters. Protection should be 3½in side abreast magazines and 3¼in abreast machinery, with 2in decks and 1in main bulkheads bounding the machinery spaces. DNC had already produced the corresponding Legend and sketch design.[22]

DGD was unhappy that this was *Fiji* protection and pretty much a *Fiji* armament on 3,000–4,000 more tons, but DTSD considered the displacement reasonable given the increased close-range armament, fully HA main battery and much-improved endurance. The Staff Requirement defined Design A in a new series. It was still being developed as late as October 1942, but by that time alternatives armed with 5.25in guns were also under consideration. The main new development was a proposed triple 5.25in mount.[23] In September 1942 the Staff Requirement was passed to the new Future Building Committee.

The Future Building Committee

In August 1942 a Future Building Committee was formed to develop building programmes in view of the desired balance of the future fleet. Deputy First Sea Lord was chairman, ACNS(W) acting as Deputy Chairman. Controller was represented by his Naval Assistant, and membership included representatives of DTSD, DOD (Director of Dockyards), DNC and DNAD. Although it considered all types of ships, the Committee seems to have concentrated on the carrier programme and on carrier aircraft; presumably it was formed largely to change the orientation of the Royal Navy towards a more carrier-based force.[24]

The Committee discussed the 1943 programme at its meeting on 2 November 1942. Naval Assistant to Controller wanted three cruisers so that design work could be divided among three firms. However, to order them in 1943 would perpetuate the existing *Tiger* design, which the Committee did not want to do, so it decided to defer any new cruisers to 1944. Meeting on 16 November 1942, the Committee compared two alternative 5.25in guns cruiser designs, one with three twin and one with three triple mountings, Designs K and L in the ongoing series.[25] Design L was a step down from some four-turret designs. The 6in HA/LA design was moot, the only available truly dual-purpose cruiser gun being the 5.25in.[26] The six-gun design was approved for further development, on the grounds that it was a better carrier escort, and that in front-line cruiser fighting it was desirable because it could be built in greater numbers. DNC developed designs with four twin 5.25in as alternatives.

DTSD demanded that any future cruiser be designed to survive two torpedo hits; *Dido*s had sunk too easily from one hit in the machinery spaces.[27] That would spread out the ship's machinery spaces. Against increased machinery volume, DTSD proposed trading off speed to get better protection. If, as the Future Building Committee asserted, the carrier was the core of the future fleet, then a cruiser, primarily a carrier escort, need not be much faster than a cruising carrier (the carrier needed higher speed for air operations) – as slow as 26kts. A *Dido* used nearly half her power to go from 28kts to 32kts. The Sea Lords agreed, and in December they asked DNC to sketch 28kt cruisers armed with three or four twin 5.25in mounts. Endurance should be 6,000nm at 18kts, a considerable step down from the 24kt endurance speed of the big 6in cruiser and even the 20kt endurance speed of the 5.25in ship proposed a month earlier.

DAWT and Chief Advisor on Operations Research (CAOR) set the case of the small cruiser armed with six or eight 5.25in guns and numerous close-range weapons (7,000–8,000 tons) against the large cruiser (a 15,000-ton ship armed with eight 8in guns and 4in HA guns). Nearly twice as many small cruisers could be built in place of large ones. The large cruiser would enjoy greater endurance and slightly better seakeeping, and it would be somewhat better protected, but the smaller cruiser was a smaller target, and was more manoeuvrable (hence better able to evade bombs and torpedoes). Large cruisers were nearly always intended to counter the threat of large foreign cruisers – if a future enemy built large cruisers, would the best counter be large British cruisers or carriers? Looking forward eight or ten years, 'the only war we can reasonably foresee is an extension of the war against Japan, let us say after a pause while each side has recovered from the effects of the present world war. Any such war will be a mixture of oceanic operations in the Pacific and Eastern Indian Oceans, and Combined Operations in the Indonesian Archipelago'. Cruisers would generally operate with carriers, whose aircraft would be the primary anti-ship weapons. There was therefore no point in building larger cruisers.

DTSD and Director of Plans (D of P) disagreed, however. The future carrier screening role would be filled by 5.25in and 6in cruisers, but a larger and more heavily-armed type was needed for the future balanced fleet. In their view a small cruiser design became unbalanced when required to make more than 28kts – hence the 5.25in cruisers in the design stage were given that speed. However, future aircraft would require higher carrier speed. The 33kt *Ark Royal* would have a 5kt margin of speed over her escorts. It was undesirable for the carrier to have to conform to the movements of her supporting force, since the strategic mobility of carrier forces would likely become 'one of the most potent factors' of a future war; at times high transit speed might materially alter the entire situation. At night or in bad visibility, the carrier escort had to be able to deal with a chance encounter by enemy heavy cruisers and a small 5.25in cruiser would not be enough. Something more powerful than 6in guns was needed. ACNS(W) came down on the size of the small cruiser, agreeing that in the past the Royal Navy had generally preferred smaller ships (he imagined that the *Hawkins* class, the exception, had been an answer to foreign designs). Cruisers were built to work with the battle fleet and to protect commerce. The former now meant screening carriers; and with the growth in size and seakeeping capability of destroyers, it seemed less vital. Carriers were now taking over the commerce-protection role, although cruisers were still vital for this role in some waters. Balanced cruiser design had never been very practical, and now large cruisers were almost as vulnerable to bombs as smaller ones. After reviewing historical experience, DTSD was compelled to admit that the carrier escort justification for a heavy cruiser was questionable.

On 8 March 1943 the Committee reviewed DNC's designs (K4, M1, N1 and P1), of which K4, M1 and P1 were three-turret ships.[28] K4 offered the greatest speed and endurance (30kts, 7,000nm), and M1 was the smallest (7,150 tons standard, compared to 7,600 tons and 7,250 tons for the others). P1 offered the best subdivision. N1 was the four-turret ship. The Sea Lords chose the four-turret ship because it offered four-corner anti-aircraft fire and increased LA fire. DGD argued that each of three turrets could have its own HA control, but each of four could not. Although it was impractical to demand sufficient speed (36kts) to keep up with a 33kt carrier, there was interest in higher speed, and DNC was asked to consider an increase to 30kts, with after boilers arranged as in P1 (which suggests that in its case they were not in tandem), as well as increased endurance and the effect of bringing the funnels closer together to give better director arcs. This design had somewhat heavier close-range armament. The revised N2 design (still not capable of 30kts) was discussed at the meeting on 12 April.[29]

At a meeting in July 1943 the Sea Lords agreed to drop the 8in cruiser. They also authorised preliminary work on a 6in dual-purpose gun mount, while recognising that little could be done for some time, given the press of other work. That left the 5.25in gun as the only available dual-purpose weapon for the next cruiser. The N2 design received the Board Stamp on 16 July 1943, a Staff Requirement being written around it. DNC was later asked to provide more torpedoes, up to eight on each side, the recommended broadside against a fast modern ship. The ship should be able to divide LA fire against two targets. Machinery should be in well-separated units, with two turbo-generators in each unit. Each turbo-generator would be paired with a diesel generator of equal power, as far as possible from the main machinery spaces, and as high as possible (consistent with reasonable splinter protection).

First Sea Lord Admiral Sir Dudley Pound resigned as of 9 October 1943, dying shortly thereafter. He was succeeded by Admiral Sir Andrew B Cunningham, who had been Mediterranean Fleet commander. At its meeting on 17 January 1944, the Future Building Committee was told that First Sea Lord disliked the 5.25in cruiser, preferring a 6in battery, presumably reflecting Cunningham's experience with *Dido* class cruisers in the Mediterranean. It was suggested that the 5.25in cruiser be dropped in favour of an improved *Belfast*, but a paper was written comparing that ship, in nine- and twelve-gun versions, with the 5.25in cruiser the Committee still favoured. The 6in gun was clearly superior (on a shell by shell basis) in surface (LA) fire, since it could penetrate 3in side armour (at 30° to the normal) at 12,500yds, compared to the 5.25in shell at 9,500yds. It could penetrate a 2in deck (which was proof against the 5.25in) at 22,000yds, or a 1in deck at 14,500yds (above 16,000yds for the 5.25in). Both guns had much the same maximum effective range (i.e. the range above which a target could

HMS *Superb* was the ultimate development of the *Fiji* design. The most obvious change was substitution of the heavy Mk 6 director, carrying the two antennas of a Type 275 blind-fire radar, for the earlier HA director with its Type 285. However, she was not fitted with the Flyplane computer associated with this combination in later ships; instead she had a derivative of the Fuse-Keeping Clock (FKC) which often equipped destroyers. She was also distinguishable by the two barrage directors forward of her bridge. In this post-war photograph they have been mothballed (their Yagi antennas are not visible) and the pompom directors are missing from the side of the bridge. As completed she had only three quadruple pompoms, three single Bofors guns, and eight Boffins.

successfully evade fire), but 6in splashes could be spotted at longer ranges (about 20,000yds vs 17,000yds). The 6in burster was somewhat heavier (3.75lbs vs 3.25lbs), so the effect of hits was in the ratio of 8 to 7. On the other hand, the 5.25in fired faster, twelve rounds per gun per minute compared to five for the new 6in Mk 24 turret. Given radar control, both guns would obtain about the same hitting rate per shell fired, so if nine 6in were taken as a rate of 1.0, the eight 5.25in offered 1.9 and the twelve 6in, 1.5. Overall, therefore, the 5.25in ship offered better LA fire. To achieve equivalent long-range HA fire, the improved *Belfast* would need six twin 4.5in guns of the new type then under development. It seemed to follow that the 5.25in ship was better until a new rapid-fire 6in mounting became available. DGD pointed out that no such mounting could become available for any 1944 ships, so the comparison was between a fully up-to-date 5.25in mounting and a fifteen-year-old 6in mounting. In June, DGD circulated a comparison of possible alternative 6in gun mounts for next-generation cruisers, in effect beginning work on the dual-purpose mounts which eventually armed the *Tiger* class.[30]

A Sea Lords' meeting on 6 February favoured the twelve-gun improved *Belfast*, with a secondary armament of six twin 4.5in guns

(explicitly mirroring the US *Cleveland* class). Five such cruisers were included in the 1944 programme. The comment on equivalent HA fire may have been reflected in a decision to provide these ships with six twin 4.5in guns. The close-range armament proposed for all three options was eight 'Busters' (self-contained twin Bofors). An Outline Staff Requirement dated 29 February 1944 called for nine 'Busters,' as many 20mm as possible, and eight torpedo tubes on each side, with protection similar to that in *Belfast*. Low speed was abandoned: the ship should make 32.5kts fully loaded and 29.5kts when six months out of dock (presumably in the tropics). Endurance should be 4,500nm at 20kts when six months out of dock. Continuing the series of cruiser designs, this one was designated R.

Wartime cruiser construction had been deliberately deferred in favour of light aircraft carriers. Now new cruiser construction seemed urgent, as the existing fleet was ageing, and it was no longer possible to add much to existing cruisers of the 'County', *Leander* and *Arethusa* classes. Including the five *Tiger*s, the Royal Navy had fifty cruisers (other than First World War types), whose average age in 1950 would be 12.5 years (fifteen would be at least fifteen years old).

There was still some interest in smaller cruisers. In March 1944 ACNS(W) asked whether they would be wanted in the post-war fleet. There were three choices: the four-turret 5.25in cruiser already sketched, a small 6in cruiser with 4in HA armament comparable to the new Japanese *Agano*, and a 4.5in gun cruiser comparable to the US *San Diego*. That August DTSD and Director of Plans framed the case for a small future cruiser, which they envisaged as a 6,000–7,000-tonner armed with three twin 5.25in (two forward) and two quintuple torpedo tubes, plus close-range weapons, capable of 32kts, with an endurance of 5,000nm at 20kts, good protection, and a low silhouette. This was much the previous year's argument, an additional point being that the small cruiser could operate in mineable waters where a larger and more valuable ship would not be risked, e.g., the British East Coast in 1940. First Sea Lord raised this question in August 1944. At the meeting on 8 August 1944 DTSD noted a tendency for a cruiser – necessarily a small one – to be assigned to lead groups of destroyers. Compared to emerging large destroyers, such a ship would offer better sea-keeping, better armament and better equipment (including, though not mentioned, better command and control). In the past this had been the argument for small destroyer-leading cruisers such as the pre-1914 *Arethusa*s. Others in the Committee noted that such a ship need not have great endurance; it would be used for escort and inshore work. Such ships (presumably mainly the *Dido*s) had proved their value in the Mediterranean and off the French coast. It also might be very difficult to gain political approval of many of the very large cruisers now contemplated. This ship would displace 6,000–7,000 tons, have a speed of at least 30kts, and endurance of about 5,000nm at 20kts. DNC should prepare sketches of such a ship armed with two triple 6in and 4.5in anti-aircraft guns for comparison with the two earlier 5.25in gun designs. It was soon clear that any ship armed with 6in guns would displace at least 8,500 tons, and with secondary armament that would rise to at least 9,000 tons; 6in guns were 'out of the question' below 7,000 tons.[31] On the other hand, the small cruiser would be needed as leader of a destroyer striking force. This was becoming more important given the complexity of radar and fighter direction. On this basis it was agreed that if a small cruiser was eventually approved displacement should be limited to 7,000 tons and main armament to three twin 5.25in with as many torpedo tubes and Bofors as possible. Protection should be by subdivision rather than armour, as armour offered no real protection against air attack. Endurance should be that of the 'Battle' class destroyers, with amenities on a similar scale. Speed should be 30/32kts. In effect the committee was accepting that the separate categories of small cruiser and large destroyer were merging.

The 1944 Cruiser

Work on a new 6in cruiser began with a note from DNC on 9 February 1944: the 5.25in cruiser was dead. The preferred alternative was a 6in ship (improved *Belfast*) with four triple Mk 24 mountings, displacing about 12,000 tons. A flush deck hull (or a forecastle extending two-thirds the length of the ship) would avoid weakness at the break of the forecastle. Bridge height should be a minimum, and splinter protection should be reduced. Secondary armament should be six (at least four) twin 4.5in guns. Since the twin 4.5in Mk VI was comparable to the US 5in/38, these requirements were about those of the US *Cleveland* class. By April 1944 DNO was pressing for a dual-purpose 6in mounting heavier than the Mk 24, but not as well protected as the guns in *Belfast* (a required level of protection). DGD considered the high rate of fire the only way to exploit fully the latest radar and fire-control developments; the US Navy already had a fast-firing 6in turret. The Mk 24 was a further development of the existing turret, now 'decidedly old fashioned', limited to six to eight rounds per gun per minute. DGD would 'feel the utmost concern' if another programme slipped by without some of its ships being armed with a modern turret. If some could not have the new gun, then two designs should be prepared, one with Mk 24s and one with new ones (which became the Mk 25). Because a modern cruiser should contribute to the air defence of a heavy unit, such as a carrier, which she might be supporting, he fully agreed with the battery of six twin 4.5in guns.

The first of the new design series was Design Q (March 1944), superseded in May by the very similar Design R, with four Mk 24 mounts. The ship was not large enough to take DNO's projected new mounts.[32] As specified, the secondary battery was six twin 4.5in Mk VI guns, as in the new *Daring* class destroyers, with two Mk VI directors sided forward and one aft. Close-range armament was ten 'Busters' and twenty-eight Oerlikons (powered and hand-operated twin mounts in roughly equal proportions). The ship would also have four quadruple torpedo tubes. For anti-submarine self-defence she would have Asdic and fifteen depth charges, the usual cruiser load. Protection was similar to that of *Belfast*, but the gap between the main belt and the steering gear was filled. The belt would be proof against 6in fire outside 8,000yds or against German 5.9in fire outside 11,000yds (and a Japanese 5.5in shell outside 7,500yds). Deck protection was proof against 6in fire inside 22,000yds, against German 5.9in inside 27,000yds, and against Japanese 5.5in inside 22,500yds.[33] Unlike earlier British cruisers, this one had both of her two pairs of boilers abreast. The two units were separated by a block of compartments for magazines and for oil fuel. Boilers were at the after end of the forward unit and at the forward end of the after unit, to reduce total machinery length and to put the funnels closer together, for better gun and director arcs. DNC provided eight 500kW generators, twice the power in the *Tiger* class. Ships would maintain themselves in half-oil condition by flooding oil tanks after half the fuel had been burned, a settling and sullage tank system ensuring against contamination by sea water. Maintaining this liquid load reduced the risk that the ship would list badly after underwater damage. Design R was chosen in preference to Design Q (of the same size) developed in March 1944.

The 4.5in gun was expected to fire so rapidly that it could add to close-range fire, so DGD asked that the two waist mounts each have their own local director (an MRS), at the cost of Oerlikons (which turned out to be two twin mounts). He also had the DCT atop the compass platform moved back to provide more space for the compass platform, battle bridge and ADP. He also wanted a small closed bridge at the fore end of the structure. All of this could be done, but some clear arcs might have to be reduced. Director of Radio Equipment (DRE) was unhappy that the mainmast was only 8ft abaft the after funnel; funnel gas was already causing considerable problems. He suggested that mast and funnel be interchanged, or that a 'Weapon' class solution be adopted, a lattice mast being built around the after uptake. None of this was a problem, as long as shorter radio aerials (strung between the masts) were acceptable.

Alternatives V and W were developed to scale down Design R mainly by changing the secondary battery.[34] Apparently these changes were not considered worthwhile. Deputy First Sea Lord asked DGD to review the type of 6in gun mount the ship would carry, for consideration by the Future Building Committee. That practically guaranteed adoption of the dual-purpose mount. Design Y was effectively Design R with Mk 25 dual-purpose triple 6in mounts instead of the earlier Mk 24s.[35] That cost about a thousand more tons. There would be two separate long-range LA control systems and three separate HA systems. The design was formally submitted on 12 September 1944. The most obvious difference from the previous design was that the engines were designed to develop 108,000shp so that they could be relied upon for 100,000shp in the tropics. Close-range armament was ten twin Bofors and thirty-two Oerlikons in twin powered and hand-worked mountings.

Writing late in October 1944, ACNS was not altogether pleased by Design Y, because on a much larger displacement it carried what seemed to be little more than the armament of a *Cleveland* class cruiser, with a total of twenty-one directors. If, as DNO offered, the 6in mounts could have 80° elevation, why bother with the 4.5in? The US Navy had done as much in its J design (the *Worcester* class). He did not really like a 6in HA gun, and suspected that much of the growth of the ship could be traced directly to the high angle of elevation. Cutting that to 40–45° might make for a smaller 6in turret and thus for a smaller ship. Armour, in this case comparable to that in *Belfast*, also contributed to sheer size. Should the Royal Navy continue to armour against shells of the sort the ship fired, or should it concentrate on bombs, or should it cease to armour against shells and instead strengthen bulkheads to localise damage?

DNC took this occasion to point out why he had preferred the LA Mk 24 in the first place: the growth of Design Y over Design R could be attributed completely to the heavy new gun. For the rest, it was for the Staff to rethink its requirements. He pointed out that from the start he had predicted that carrying the heavy secondary armament would make for a big ship, and had proposed that the ship be armed only with dual-purpose 6in guns and with close-range weapons – which was just what the US Navy was doing. Now ACNS said that he disliked the dual-purpose 6in gun. 'I am constrained to ask "what *does* the Staff want?".'

In mid-November the Staff met to discuss cruiser armour. No affordable cruiser armour could keep out large bombs; the only defence against them (and torpedoes) was subdivision by strong bulkheads. Underwater weapons already did the worst damage, and future weapons would probably exploit that fact. Even so, a cruiser needed some protection against shellfire 'as the occasions on which a ship is subjected to shell fire are just those in which it may be important for her to be able to continue in action and perform [her] functions'. Perhaps a good compromise would be to reduce the standard of such protection from 6in to 4.5in fire, i.e. against destroyer fire. The minimum range for such protection should be that at which the destroyer torpedo became a greater threat than the destroyer gun, tentatively 5,000yds at a bearing of about 60° to bow and stern. It was agreed that a 2in deck offered no real advantage over a 1½in one, since neither would protect against a 500lb bomb. The loss of protection against 6in fire at long range was accepted. Even so, it would be vital to continue to protect magazines. Early in December, therefore, DNC had his cruiser designer look at modified protection. The belt would be limited to the machinery spaces, its thickness reduced to what was needed to keep out 4.5in shells (4in C armour). As in earlier cruisers, the magazines would revert to box protection against bomb, shell and torpedo attack, their deck area minimised. The number and thickness of protective bulkheads would increase.

As a first cut, Scheme 1 provided 4in C on belt and bulkheads (reduced to 100lb NC below certain level) and 60lb NC on deck. The belt was closed by 4in bulkheads. The magazines had 4in sides and bulkheads and 6in crowns. Turret roofs were 6in NC, to cover the openings in the magazines through which ammunition was passed up to the turrets. An alternative Scheme 2 was the original armour scheme with thicknesses reduced to keep out 4.5in shell. It was rejected as impracticable, because the ship would have been lengthened about 25ft to carry the increased weight (due to the greater area protected). To cut weights, DNC decided that deck protection against 4.5in shellfire could be reduced (where not needed for strength) from 60lb NC to 25lb DW (on the lower deck over the machinery). On the other hand, provision had to be made against diving shells. Thus 4in armour might have to extend for the full depth of the 6in magazines at the ends, and also over the sides of the 4.5in magazines amidships. The ends of the midships 4.5in magazines should have 100lb (2½in) NC to protect against splinters. Some reduction in steering gear protection was possible. These changes gave a standard displacement of 15,694 tons.

A third possibility was to minimise the magazines' deck area by working them on two levels (hold and platform deck), with shell rooms in the space saved. However, the upper part of the magazine would be above water, and the usual flash-tight arrangements could not be fitted. To protect the engine rooms against plunging 4.5in fire between lower and upper deck, the 4in belt could be extended up between those decks over the length of the forward engine room.

A fourth possibility was to put the magazines down on the inner bottom, working two magazine flats, thus shortening the length of the magazines. This idea took advantage of the considerable space below the magazines devoted to watertight compartments (this space amounted to 7.5 per cent of the displacement). The ship could be shortened about 15 per cent and perhaps the beam reduced as well.

Overall, Design Y had 2,344 tons of armour, about 15 per cent of her standard displacement (as rearmed, *Belfast* devoted 17 per cent of her displacement to protection). That included a 984-ton belt (4½in) and 862 tons of deck armour. The 4.5in standard proposed by the Staff could be met by a 4in belt (saving 176 tons) and making the deck thinner would save another 312 tons. Even more could be saved by reverting to box protection for the magazines (468 tons on the belt, 779 tons on decks). The weight saved could improve protection with priority (a) underwater, (b) bomb, and (c) shell.

DNC pointed out that tighter subdivision would not improve survivability, because what counted was the length over which a given weapon would tear up a ship's hull. The longest compartment in the ship was 36ft long, but a modern torpedo would disrupt a ship over a length of about 70ft. Maximum floodable length would be 110–140ft. Something could be achieved by increasing the number of thick bulkheads from four to six, and increasing their thickness to 60lbs below the platform deck, and to 40lbs between platform and lower decks. The weight involved would be modest (68 tons). The magazines were not adequately protected against diving 4.5in shells or from such shells penetrating below the side armour at a 10° heel. To solve that problem would take 370 tons: 4in C on magazine sides and 100lb NC on longitudinal bulkheads. That would consume almost all the saving achieved by reducing armour thickness. Reducing the extent of armour and boxing the magazines could make another 800 tons available. To make the magazines safe, DNC wanted 6in on the deck over them and 4in and

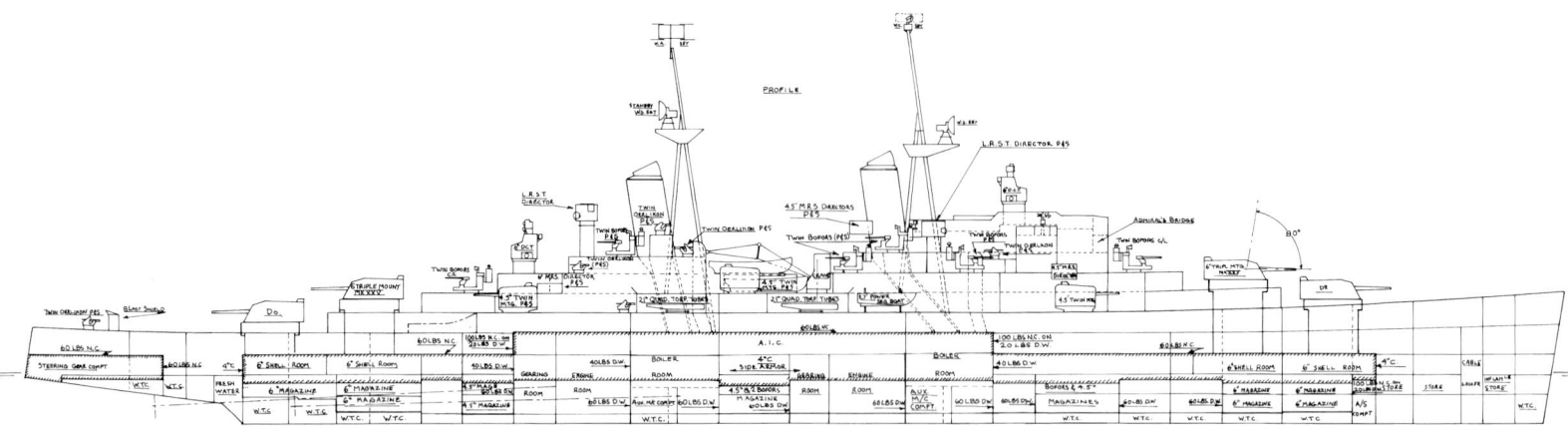

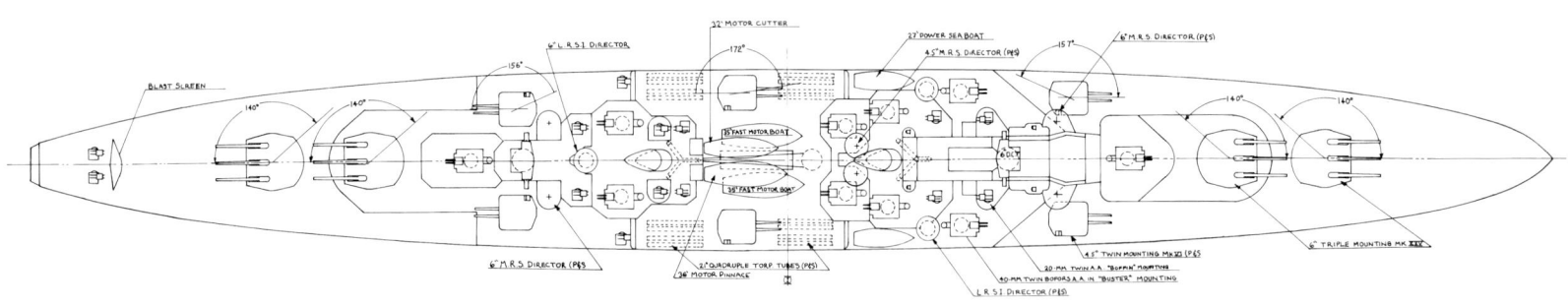

100lbs on their sides and ends, and also to increase turret roofs to 6in Including 60lb on the steering compartment, that came to 1,493 tons, well beyond the saving envisaged, adding 250 tons to the ship. ACNS also wanted magazine vulnerability reduced by reducing their deck area. For example the magazines could be enclosed entirely within turret ring bulkheads extending down into the ship's hold, on two levels. The circular shape offered problems, however. DNC offered an alternative which preserved the rectangular shapes of the magazines. The combination involved added 1,372 tons of armour.

DNC preferred the scheme using the space in the hold but did not like reducing general splinter protection to communications and important services to buy 6in NC over magazines. He decided to keep protection over the area in the original design but to reduce thicknesses to keep out 4.5in shell: 4½in C side reduced to 4in, 80lb NC on upper and lower decks reduced to 60lb and 100lb NC on transverse bulkheads. Weight saved would be devoted to thickening bulkheads and to fitting additional side protection below the main belt in way of magazines. Side armour would be continued to a depth of 1ft below the waterline at 10° heel in half-oil condition. The lower edge of the side armour below the waterline in way of the various magazines was then 4ft 9in for the forward block, 5ft 9in amidships, and 4ft 3in aft. Underwater protection was improved by increasing the number of thick bulkheads and increasing their thickness from 40lb to 60lb DW (cost 70 tons). Weight freed by reducing magazine length provided additional splinter protection to magazine boundaries: 60lb DW sides and ends, 40lb DW roofs and flats. To preserve trim, the citadel was moved about 6ft aft.

Overall, the hope that rearranging protection would cut the size, hence the cost, of the ship was not fulfilled. Ship size was governed mainly by the weight and space required for the main and secondary armament. In this ship armament weight was about 50 per cent more than in *Belfast* as rearmed. The length of the citadel was governed by the

Design Y was to have been built as the *Neptune* class. Note the unit machinery, separated by a 4.5in magazine. Displacement would have been 15,350 tons standard (18,740 tons fully loaded); dimensions would have been 655ft (waterline) 662ft (overall) x 76ft x 24ft 9in (full load). Four Admiralty boilers would have produced 108,000shp for 33kts. Endurance would have been 7,500nm at 20kts (2,850 tons fuel oil). As a flagship she would have accommodated 1,351 men. Armament was twelve 6in Mk 25 using two DCT for surface fire; six twin 4.5in dual-purpose guns, ten 'Busters' (self-contained twin Bofors), and fourteen twin power Oerlikons, plus four quadruple torpedo tubes. Anti-aircraft control for both 6in and 4.5in guns would have been exerted by three LRS I directors with a new radar (apparently Type 901 in its initial incarnation), plus six MRS (two for 6in, four for 4.5in). 'Buster' mountings had on-mount radars. Armour would have been a 4in belt amidships with 4in bulkheads and 1.5in extensions fore and aft and 2.5in on 1in in machinery spaces. The upper deck would have been 1.5in over machinery and steering gear, and 1in elsewhere. The lower deck would have been 1in. Turrets would have had 4in faces and 2in sides, rear, and crowns. Note the effort to separate the forward and after machinery units. (A D Baker III)

layout of the upper works, which was already cramped. There was no hope that the ship could displace less than about 15,000 tons.

The design was modestly revised in January 1945, its displacement increased slightly.[36] In February 1945 DNC assembled building drawings and specifications for Board approval. At the end of December 1944 Director of Plans pointed out that, as the end of the war approached, national shipbuilding capacity should be shifted towards replacing the merchant fleet. The naval programme would be cut back, ships expected to complete after 1946 being reduced to slow building rate. The suspended *Tiger* class cruiser of the 1941 supplemental programme, which had been suspended, should be built to the new design, making a total of six such ships. They should be completed no later than 1950.[37] By this time the 1944 Cruiser was being called the *Neptune* class.

Apparently it took time for DNC's comment about 70ft damage length to sink in, because only in mid-June 1945 did he point out that the boiler rooms were not far enough apart. A single properly-placed hit would knock out all power. He could increase separation from the original 52ft to 62ft, by placing the forward boiler room on the fore side of the forward engine and gearing room, reducing the midships block to that required for the 4.5in magazine. To straighten uptake lines and to keep the foot of the foremast from passing through the fore uptake, he had to rake that funnel and mast, and therefore the after pair as well. Thick bulkheads would be rearranged. E-in-C wanted the boilers at the fore end of the forward boiler room (to simplify the run of main steam pipes), so the uptakes had to be brought even further forward. DNC observed that earlier arguments for and against raking masts etc, because of range-finding and inclinometry were surely no longer valid with the advent of radar. At about the same time, detailed design of the Mk 25 mounting showed that the original roller path diameter was inadequate.[38]

The *Neptune* project survived the end of the war. In November 1945 First Lord decided that two ships should be laid down as soon as possible (Controller suggested waiving tender rules so that information could be forwarded to firms by April 1946). The required crew grew, and improved habitability standards (soon to be '25 per cent increased amenities') demanded a larger ship, perhaps with a flush deck. DGD now proposed for future cruiser armament a combination of dual-purpose 6in and 3in guns as in the US cruiser *Worcester*. His proposed 6in mount became the Mk 26 of the post-war *Tiger* class.

At a Deputy First Sea Lord's meeting on 29 January 1946 DNC reported that Design Y was well advanced; he was ordered to redesign it with the desired improved habitability. DGD's new design should also be considered (as Design Z). Proposed armament was five twin dual-purpose 6in guns and eight or ten twin 3in. The ship should displace about 12,000 tons, with a complement of 800–900. Elimination of the 4.5in battery would, it was hoped, greatly reduce the number of directors needed. DNC's instructions of 8 February were to retain the flush upper deck of the earlier design, but try to reduce its length by

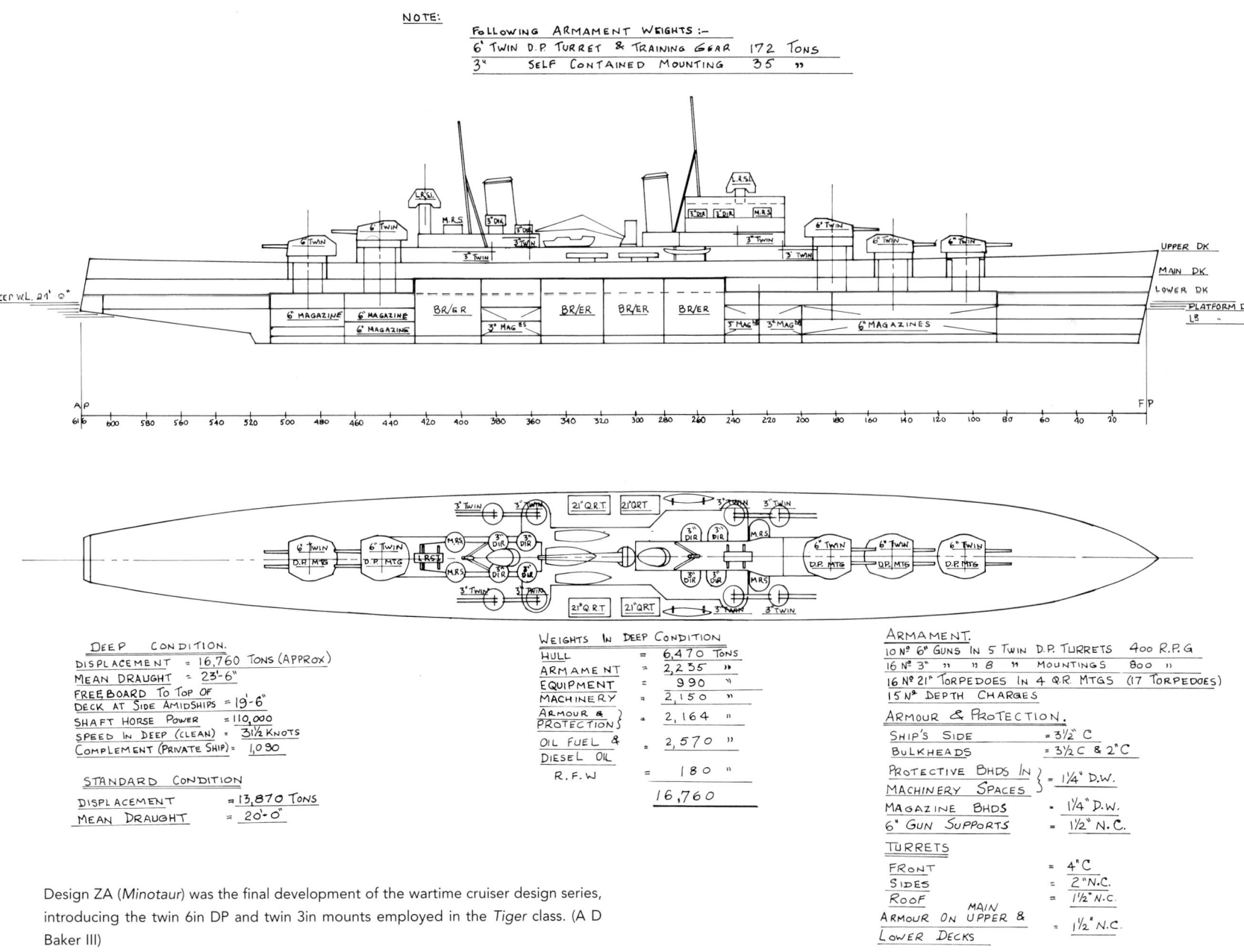

Design ZA (*Minotaur*) was the final development of the wartime cruiser design series, introducing the twin 6in DP and twin 3in mounts employed in the *Tiger* class. (A D Baker III)

about 30ft. Alternative arrangements should include three turrets forward and two aft, and two forward, one amidships, and two aft. To reduce weight, side protection should be cut from 4in to 3½in if that sufficed against 4.5in shells at 8,000yds range. If possible the 60lb deck of the earlier design should be retained, but bulkhead protection could be reduced from 60lbs to 50lbs. The separation between machinery units would be retained. DNC doubted that the complement goal could be met. Eliminating the 4.5in secondary battery could help, but only if the 3in mounts were self-contained, like Bofors guns, not if they were on the lines of 4.5in guns and mounts. DNO thought the twin 6in would weigh about 160 tons, but when Vickers supplied three designs in March, he considered only the largest and heaviest (184 tons) plausible.[39]

DNC pointed out that for LA fire the heaviest shell was best, whereas for high angles the smallest shell which could accommodate a proximity fuse and enough explosive to be effective at the range of the fuse (20yds) was best, meaning a 3in shell. The lower limit for a LA shell was 5.25in. The US Navy was beginning work on a twin 3in/70, which might be equivalent to the existing British 4.5in.[40] DNO eventually decided to adopt the same calibre, apparently on the theory that in wartime the Royal Navy would have to depend on the United States for supplies of spare barrels and ammunition. DNO's initial estimates were based on the sextuple Bofors guns being installed on board the battleship *Vanguard*.[41]

As an initial estimate, total armament weight (five twin 6in, ten twin 3in and four quadruple torpedo tubes) was 2,193 tons. Based on past designs, that suggested a standard displacement of 15,960 tons rather than the 12,000 tons imagined in January. DNC developed a series of designs with three, four or five Mk 26 mounts, all designated as different versions of Design Z.[42]

A Sea Lords' meeting on 5 June 1946 compared *Neptune* with three versions of Design Z: B, with five twin 6in, C with four, and D with five and a revised machinery arrangement for better survivability. First Sea Lord considered only D acceptable. It was armed with five twin Mk 26 and with eight twin 3in (not yet 3in/70) plus four quadruple torpedo tubes. Three superfiring twin 6in mounts were forward (as in a *Dido*, for example) and two aft.[43] The project was designated *Minotaur*, to distinguish it from the series of *Neptune* designs. It replaced *Neptune* in the building programme, but only limited design work was ever done. *Minotaur* would displace 15,280 tons in standard condition.[44]

DNO was scrambling to produce higher and higher rates of fire, because fast new aircraft (and missiles even more so) had to be destroyed so quickly. Thus in 1947 he proposed a quadruple 3in gun, and previously a sixteen-barrel Bofors had been considered as an alternative to the 3in. All of these weapons were massive, and they pushed up ship size. DNC produced a new series of designs, P, Q, R and S, with five, four, or three twin 6in, with varying numbers of 3in, and with and without torpedoes.[45] The designs suggested that two quadruple 3in were about equal to one twin 6in in both weight and the volume required below decks. Given a series of unpalatable alternatives, DNC suggested that a future cruiser should be able to engage two surface targets, hence should have either two or four twin 6in, plus six or four medium range anti-aircraft (twin 3in), plus the desired torpedoes. That led him to consider a *Fiji*-size cruiser with a twin 6in at either end. An attempt to design such a ship with three quadruple 3in and with the widely-separated unit machinery developed for the recent designs produced an unacceptable 15,500-tonner, 570ft long. DNC was just about to work out plans to modernise the fleet using twin 6in guns and twin (not quadruple) 3in, the only fruit of which was the *Fiji*-sized *Tiger*. She displaced about 12,000 tons deeply loaded. The key difference with the 1947 study was probably not so much the use of twin 3in (which were a good deal heavier than expected) but the omission of the unitised machinery.

In 1947 DNC compared the *Minotaur* design with USS *Worcester*, details of which had just become available. Despite having an extra 6in turret, *Worcester*'s armament weighed only 74 tons more. The US twin turret was 36 tons heavier than the British Mk 26 due to its heavier armour, but less ammunition was carried. Mk 26 also allowed greater depression (15° rather than 5°). Over 30° arcs to each side of the centreline the US Nos 2 and 5 turrets had to elevate to at least 15° to clear the adjacent main deck turret – reducing silhouette and top hamper. The board margin in *Minotaur* was equivalent to one 6in turret in *Worcester*. The open US single and twin 3in/50s were not comparable to the British twin 3in gun envisaged for *Minotaur*, which offered twice the rate of fire and completely mechanised loading to cut the reloading period to a minimum. Magazines had to be in the same fore and aft position as the guns they served. The estimated weight of the British mounting was 35 tons, compared with 14 tons for the US twin. On a weight for weight basis, the lack of torpedoes in *Worcester* was comparable to the lack of aircraft in *Minotaur*.

Both ships devoted about the same amount of weight to armour and to hull. The lower structural weight of the US ship was consistent with the higher stresses accepted in the design. Some weight was also saved by welding, which was more extensive in the US ship, which also devoted only about half as much weight to general fittings.

DNC noted that the total space above the lower deck in *Minotaur* was about the same as in *Worcester*, although the latter had a larger complement. Space taken up by boilers in *Minotaur* had to be compensated for in the superstructure, adding weight and reducing stability. The British design could be made substantially smaller by reducing space requirements, but the smaller *Neptune* had been rejected precisely because she was too small to meet the space requirements.

Although *Worcester* had far higher power (120,000shp plus provision for 10 per cent overload), her powerplant was only 50 tons heavier than *Minotaur*'s. DNC attributed this disparity to fewer electric generators and to the omission of propeller disconnecting couplings. He noted that the more compact American boilers lay entirely below the lower deck, whereas in *Minotaur*, as in previous British cruisers, the boilers extended up to the main deck. The *Worcester* arrangement reduced armour extent and weight, which freed valuable lower deck space. However, the *Worcester* machinery spaces were much longer, making her more vulnerable to torpedo hits. One major hit between the after boiler rooms could flood both engine rooms and immobilise the ship; *Minotaur* would do better. Both ships used the same type of machinery, but *Worcester* was rated at 8.0 rather than 8.8 tons per hour at 20kts.

The comparison file includes a drawing of alternative Z4C, modified as in July 1947, with combined engine and boiler rooms. DNC developed two smaller alternatives to *Minotaur*, designs ZA and ZB, which offered the same armament, speed, etc. but not the estimated required space (ZA did not provide a Board margin). In effect these sketches showed the effect of wartime space demands. ZA was the smaller version, ZB being enlarged to include the 2 per cent Board margin of standard displacement.

By 1947 Royal Navy planning was based on a ten-year period before war might once again be imminent.[46] Construction of six *Minotaurs* was justified because in 1956/7 so many cruisers would be overage. Thus the proposal called for laying down two ships in each of 1951, 1952 and 1953, for completion in 1954, 1955 and 1956 respectively. This plan collapsed both because it was utterly unaffordable, and because priority went to carriers and to anti-submarine craft.

Cruiser Designs for Export

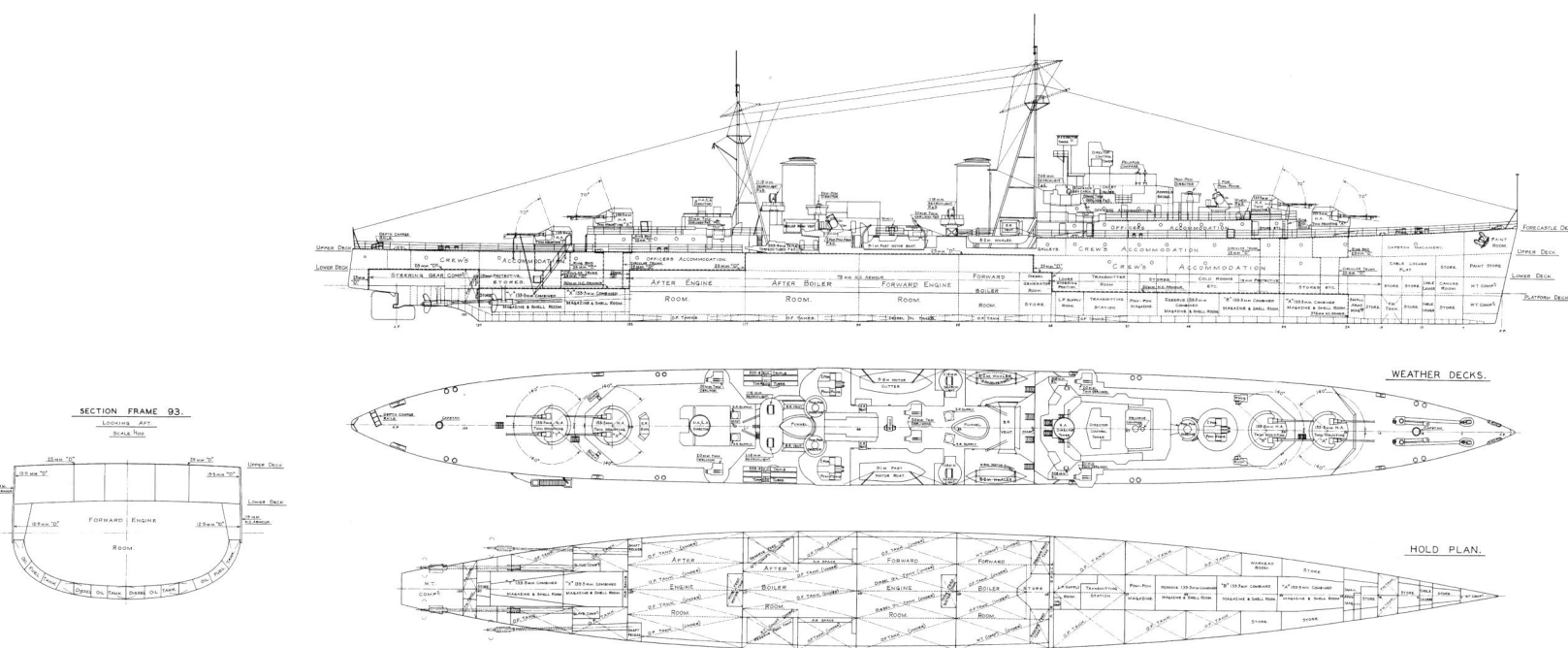

Above: Vickers Design 1122 was the wartime Admiralty *Black Prince* design offered for export post-war (Design 1123 was HMS *Swiftsure*). Neither sold, and Vickers developed a new Design 1124 embodying systems it hoped to manufacture. The plan of the hold shows the characteristic machinery spaces of British wartime and late pre-war cruisers, with the unfortunate longitudinal bulkheads surrounding the after boiler room, the boilers being in tandem. (National Maritime Museum)

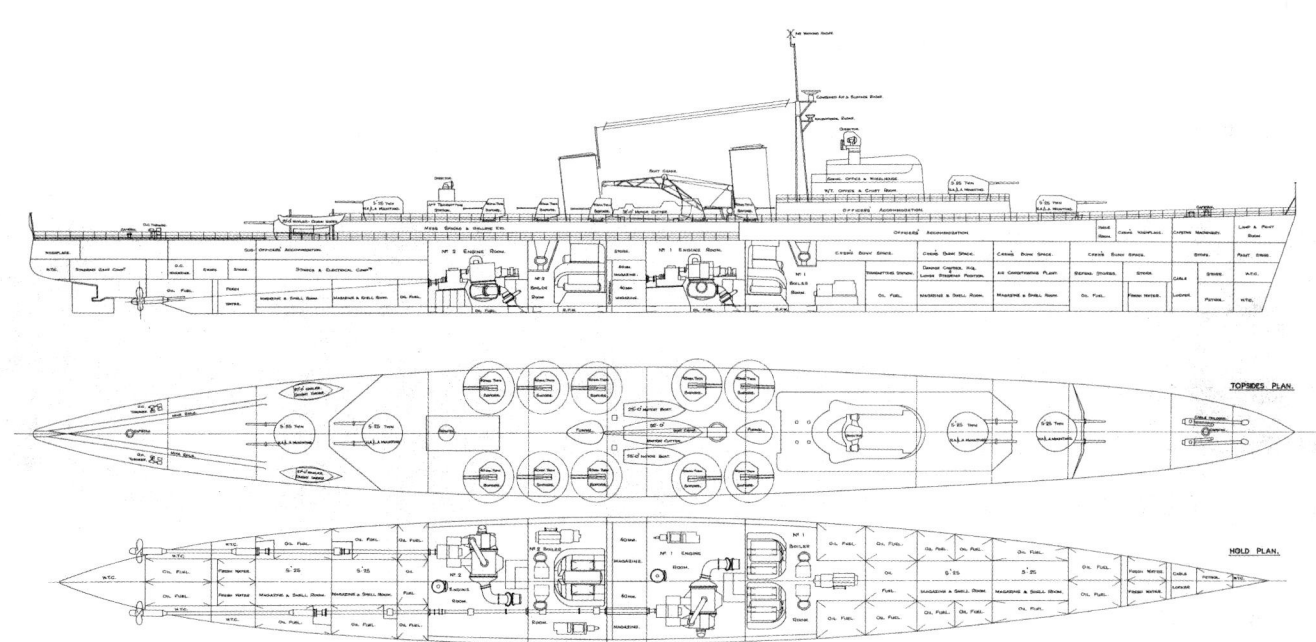

Above: Vickers developed the *Dido* into this small cruiser, which it offered to post-war customers. Only the South American governments were likely to be able to buy such ships, and the likely target customer was Venezuela, which in the late 1940s tried to buy a *Dido*. Note that the earlier boiler arrangement (two side by side, two in tandem) was retained. The twin mounts are 5.25in, the smaller weapons are twin Bofors guns. There was also a three-turret version with twelve single Bofors, six of which occupied roughly the same space as 'X' turret here. This version also had a seaplane amidships, but not a catapult, where the four-turret version carried no aircraft. Unfortunately no details of the design have survived, apart from a sketch of planned protection. (National Maritime Museum)

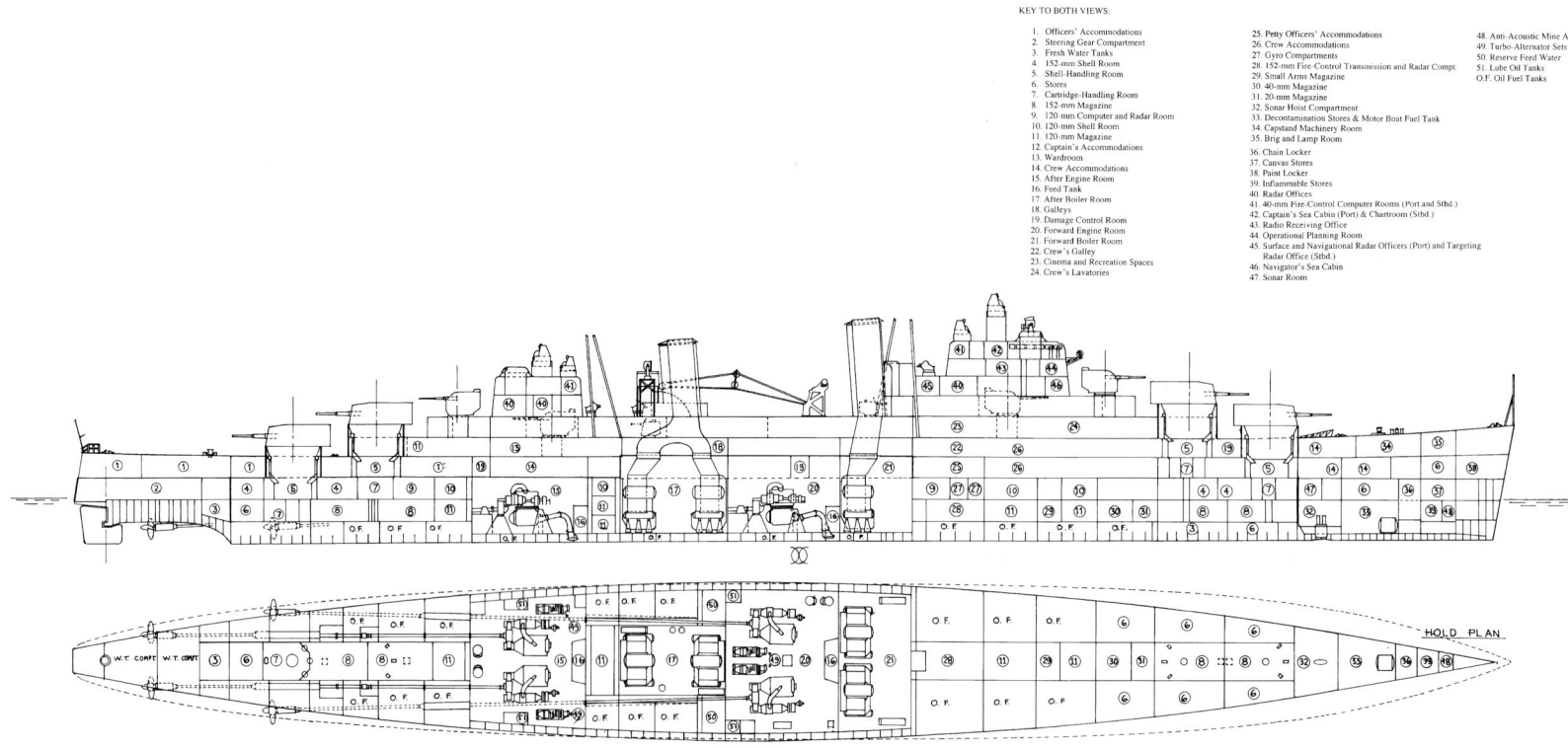

Above and below: Vickers designed a new large cruiser as Design 1124A. In this sketch the heavy guns are triple 6in, and the smaller turrets are 120mm (4.7in, presumably Bofors twin) anti-aircraft guns. The search radars are British, but the directors are not. For example, the director immediately atop the bridge, for the main battery, seems to resemble the US Mk 34. The 120 mm director seems to use the British Type 275 radar, but it is shaped more like a US Mk 37 (it certainly is not the usual British Mk 6). The absence of helicopter facilities dates the design to the late 1940s or early 1950s. This design was also described as the Vickers 'stock cruiser', i.e., as the basic design it offered potential clients before developing follow-ons to suit their tastes. It had an 81mm belt (over 12.5mm side plating) raised to cover the boilers, extending between the outer parts of the end barbettes. It was covered by a 50mm deck and closed by 50mm bulkheads (including bulkheads closing off the raised section). The end bulkheads below the belt were 37.5mm thick. Barbette armour thickness varied around each barbette, from 19mm at the ends to 25mm at the sides. The steering gear was protected by 37.5mm sides and 31mm deck and inboard end. A cross-section at the after boiler room shows the same longitudinal bulkheads that had caused trouble in the Second World War, but with voids between them and the oil tanks outboard, to limit flooding in the event of underwater damage. A list of Vickers drawings identified Design 1124 as an Argentine cruiser. Hopes that British builders could sell to the three large South American navies must have been damaged when the US government gave each two *Brooklyn* class cruisers. Even so, in 1952 (after the US deal had been arranged) the Brazilian naval attaché approached the British to build his country a small fleet, including two cruisers and one or two escort carriers. The British refused because the programme would have disrupted their ongoing rearmament – and because Brazil planned to pay in sterling that it had accumulated during the Second World War, rather than in the dollars the British wanted. Overall, in the wake of the American deal Venezuela, oil-rich and ambitious, seems to have been the likeliest customer for new cruisers. (A D Baker III)

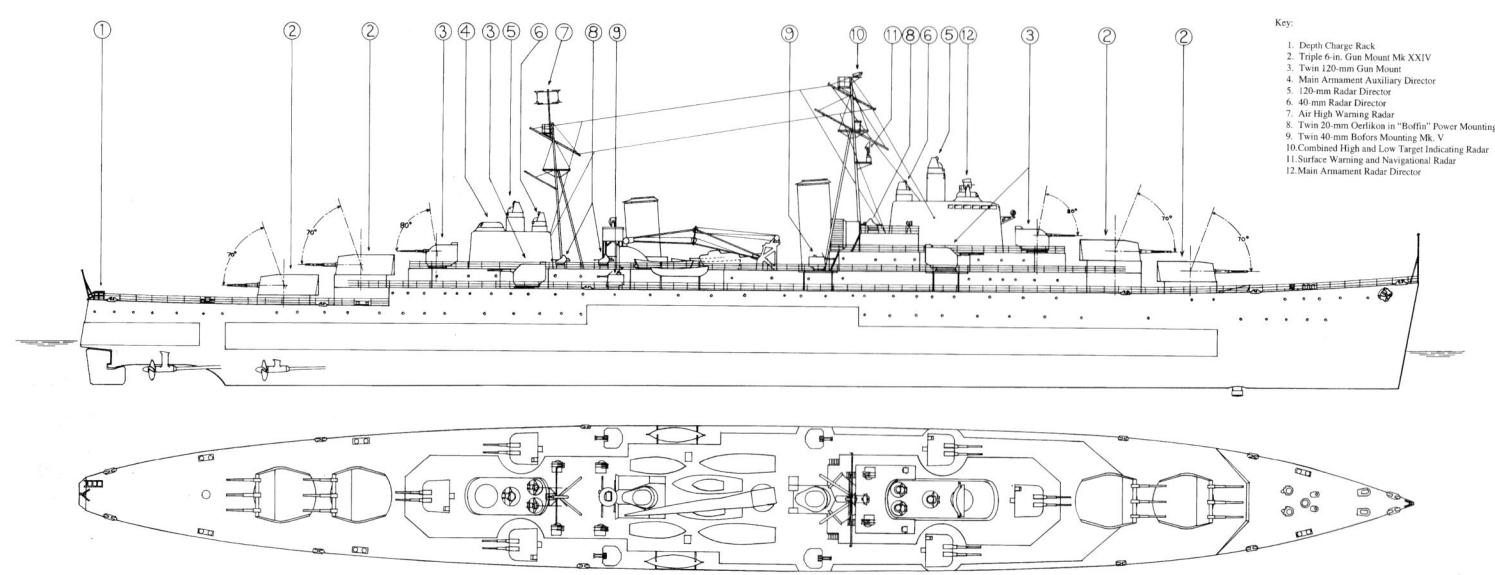

WARTIME CRUISER DESIGN

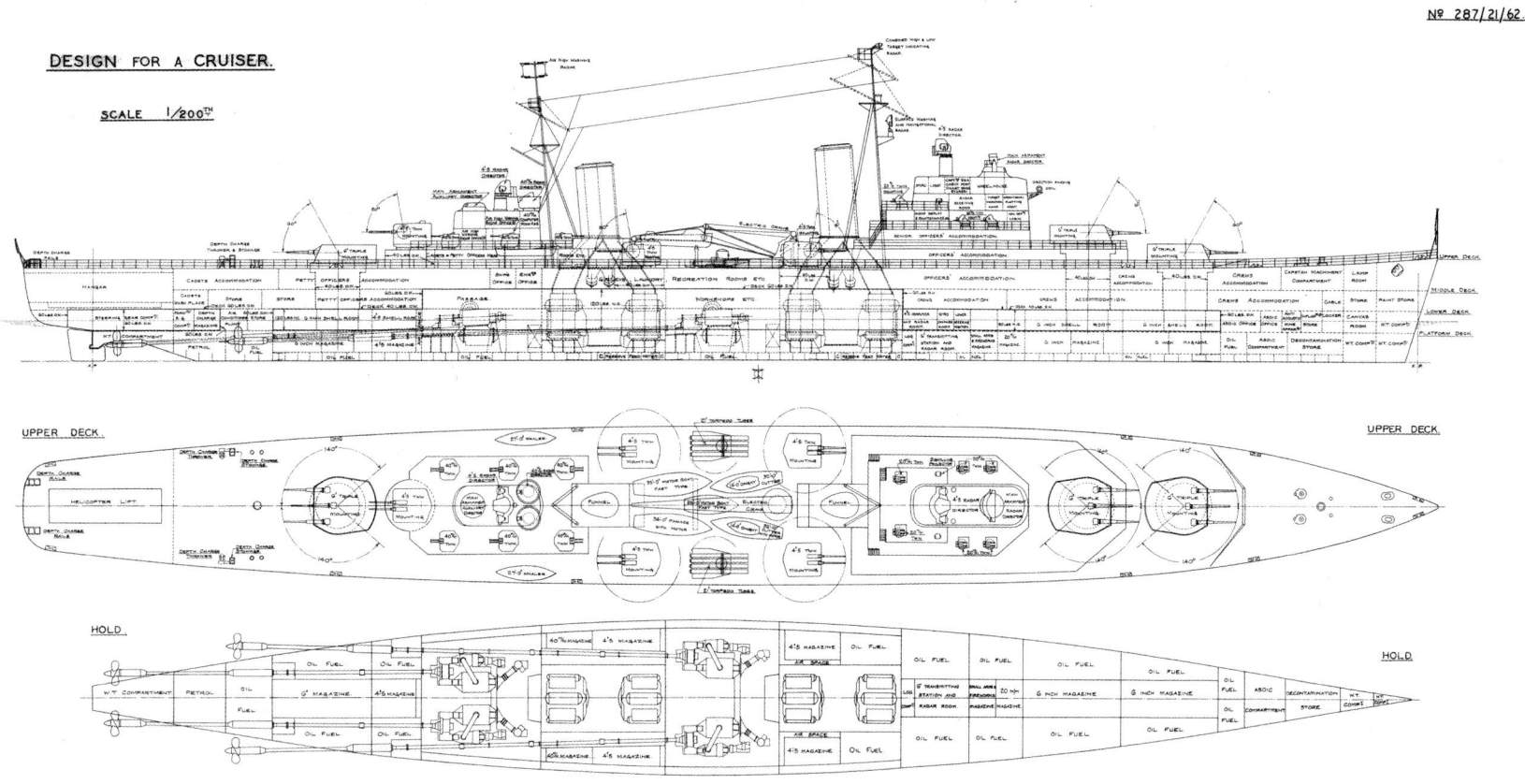

A follow-on large Vickers cruiser might be compared to the Admiralty's *Neptune*, with 4.5in secondary guns plus twin Bofors and twin power Oerlikons. This drawing is undated, but the helicopter hangar suggests it was drawn in the mid-1950s, when Vickers hoped to sell a cruiser to Venezuela in the face of US official opposition. (National Maritime Museum)

The modernised *Sheffield*, seen in July 1957.

CHAPTER 10
POST-WAR CRUISERS

In March 1944 Churchill ordered the Royal Navy to begin post-war planning, on the theory that the rationale for its force structure should be laid down before any world organisation (like the United Nations) could be created, with its implication that smaller forces would suffice. As in the past, the fleet would operate station units abroad and fleet units which could quickly reinforce them. The rise of carriers had changed the fleet, however. Because large numbers of carriers could not operate effectively together, a carrier-centred navy would naturally operate several separate fleets. The fleet unit envisaged in 1944–5 was a maximum of four carriers, each with a light cruiser supporting her, plus two supporting battleships, each with a heavy cruiser in company, plus a destroyer screen. As in the past, there would be two main fleets, one in home waters and one in the Mediterranean, ready to swing to the Far East. There would also be Home and Far East station forces, with two carriers, each with its light cruiser: the cruiser force would total twenty-two large and twelve light cruisers. A projected reserve fleet would include eight large and eight light cruisers, for a total of thirty large and thirty-four smaller cruisers. Nine cruisers were programmed: three *Tiger*s and six *Neptune*s. In 1945 another forty-five more or less modern ships were on hand: ten heavy cruisers, five *Leander*s, two *Arethusa*s, six 'Towns,' eleven *Dido*s, and eleven *Fiji*s and modified *Fiji*s. But in fact this plan was unaffordable, among other things because there was insufficient manpower: the Admiralty needed 220,000 men (30,000 from the Commonwealth) compared to a pre-war 138,000, but after the war the

RAF was much larger and the British army found itself badly stretched, particularly after manpower from India was no longer available. The first step down was to accept that there would be only two fleets, each containing, among other things, a heavy squadron modelled on the wartime Force H and a light squadron modelled on Home Fleet destroyer forces which had escorted Russian convoys. The cruiser scouting role (for bad weather) remained. The Admiralty also argued that the pre-war station forces, which had shown the flag and (it argued) had promoted British trade, should be revived: the South Atlantic, East Indies and Pacific stations, each with four cruisers, and an Africa station, with two. Again, this was not affordable. The situation was so bad in fact that in 1948 ships had to be immobilised for lack of fuel. At that time the operational fleet included sixteen cruisers, and two more cruisers were engaged in trials and training duties.

With the end of the Second World War, the immediate British problem was to restore order in the Empire, particularly those places which had been occupied by the Japanese, and also to restore order elsewhere in the wake of war. In this context a cruiser offered a valuable combination of firepower, potential landing parties and command and control. The big 'Counties' were particularly attractive for their internal volume, so sadly lacking in later ships.

Looming beyond this was the very different role of the Royal Navy against the most likely future threat, the Soviet Union. If war came, it would likely be a land invasion of Western Europe by the Soviet army, backed by a large-scale submarine offensive in the Atlantic. Before the Second World War the Soviets had operated the largest submarine force in the world. British sailors who had faced land-based German aircraft both in the North and in the Mediterranean, and marauding surface ships mainly in the North were well aware of the threats to Allied shipping posed by a large Soviet land-based naval air arm and by the modern Soviet cruiser force. Given much the same strategic reality, the US Navy chose to emphasise its ability to strike at the Soviet land mass and thus prevent the Soviets from concentrating on a land offensive. The Royal Navy was precluded from similar emphasis by the monopoly the RAF guarded on land attacks.

Some time in 1947 the British Joint Intelligence Committee concluded that, however hostile the Soviets might be, they were unlikely to attempt a major war before 1957, which they called the 'year of maximum danger', just as 1939 had been designated in 1934.[1] Therefore, 1957 became the target year for defence modernisation to incorporate the new technology developed during the War, and to face the new threats. A 'Nine Year Rule' formulated in 1948 was not unlike the 'Ten Year Rule' as the Royal Navy – but not the government – had understood it (and without any self-renewing feature). This time it was obvious that the only way to buy what would be needed in 1957 would be to limit any spending until about 1954, to give the economy time to recover and, as importantly, to provide time to develop the weapons and systems needed for a 1957 war. The fleet envisaged under the February 1948 Nine-Year Plan included twenty cruisers (eight *Dido*s and the twelve best 6in). Another four 6in cruisers were to be retained pending a decision as to their disposal.

Even this would be expensive. Confronted by the high estimated costs of future cruisers and destroyers, in 1949 ACNS (Rear Admiral Edwards) proposed merging the two categories to build fifty cruiser-destroyers (thirty of which would be operational in peacetime) instead of ten large and four smaller cruisers and thirty-four destroyers. The attempt to design the resulting ship is described in an earlier volume on British destroyers. First Sea Lord Admiral Bruce Fraser did not wholly accept the idea. He planned a cruiser-destroyer in the 1955/6 programme, but a large cruiser (13,000 to 14,000 tons) the following year.

In 1949 it was decided to retain eighteen cruisers, three of the existing ones being scrapped when the *Tiger*s were completed. The three surviving 'Counties' not being used for trials and training were to be discarded at once, and three more cruisers would be discarded when large refits were necessary. Under this plan new construction would be deferred to 1957/8, when one large cruiser and four cruiser-destroyers might be ordered. A mobilisation plan envisaged building four cruisers, among other ships. In 1950, fifteen cruisers were active and another twelve were in reserve; the target fleet for 1957 included eighteen cruisers.

British membership of NATO tended to stabilise the Royal Navy, because after 1952 the British government had to declare forces available ninety days after mobilisation; NATO strategy relied on a nuclear deterrent backed by large reserves. In 1952 the declared force for 1955 included fourteen cruisers. Further reserves were justified by the idea that an initial nuclear battle would leave both sides fighting a drawn-out 'broken-backed' war in which second-line units might make a decisive difference. In particular, the survival of what was left of the United Kingdom would depend on the defence of shipping against what was left of the Soviet fleet.

Contrary to the idea of the 'year of maximum danger', while design work was proceeding, the Korean War broke out in June 1950. The British government decided to mobilise, despite its financial problems. A £4,700 million three-year defence programme proved over-ambitious, so that by the end of 1951 cuts had already been ordered in the 1952/3 programme. This plan included two small cruisers (cruiser-destroyers) in its 1954/5 programme, plus another two in 1955/6 (later all four were pushed to 1955/6), i.e. it brought forward the 1957/8 small cruisers. Rearmament was stretched to 1958 under a £1,610 million programme formulated in January 1953.[2] It too proved unattainable, and a Radical Review was ordered to reconcile resources with goals. In July 1953 the four cruiser-destroyers were notionally replaced by a large cruiser, the large 1957/8 cruiser already being in the long-term programme. The 5in gun crucial to the cruiser-destroyer was cancelled in September 1953. As of May 1954 plans extended up to the 1959/60 programme.

By this time the Sea Slug anti-aircraft missile was well advanced. What sort of ship should be armed with it? Until about 1952 the answer was obvious: the missile was needed to protect convoys, so the first missiles should go aboard a slow ship well suited to that task.[3] In mid-1954 the long-term fleet plan (for the period beyond 1957) called for four fleet missile cruisers and ten convoy missile ships. The convoy prototype would be laid down in 1957. In May 1955, however, plans called for building one new missile cruiser per year from 1955/6 onwards. A missile cruiser (plus complementary missile destroyers) was included in the 1955/6 Estimates, but it was soon cancelled.

Mobilisation showed that the sort of effort envisaged was unaffordable until the British economy recovered. Ideas of building major new ships in 1952 (a carrier was also envisaged) evaporated. The unhappy experience of Korean War mobilisation led to a new approach to British sea power and to a new kind of cruiser described in the next chapter. No cruisers were recommissioned to fight in Korea, but two fast minelayers were brought back into service.

In the wake of the Korean War, the British Chiefs of Staff, among others, began to wonder why that war had been limited. They concluded that, particularly with the advent of thermonuclear weapons, deterrence might become effective.[4] But that would not end all conflict. Instead of World War III, East and West would fight on the Eurasian periphery, in places like Malaysia and the Middle East (the US Navy drew similar conclusions). The intense emphasis on anti-submarine warfare might give way to an emphasis on mobile naval forces capable of dealing with insurgencies and local aggression – for the Royal Navy, culminating in the Confrontation with Indonesia in 1964. The British began writing about 'Warm War' intermediate between the Cold War and a possible Hot War (World War III), which was a lot like what the Royal Navy had done in the immediate aftermath of the Second World War. It probably explains the survival of HMS *Sheffield* and HMS *Belfast*, as they could supply what was often needed: guns and landing parties to back them up. Modern air defence was not nearly so important.

Throughout the 1950s the British government enforced further cuts. As of 1 July 1952, thirteen cruisers were active, another (*Cumberland*) was assigned to trials, two were refitting for service, and ten were in reserve (three in Category A, one in B, and six in C, the lowest level). The 1953/4 Estimates cut two cruiser modernisations.[5] In July 1954 it was decided to cut the lowest reserve category ships, including the *Dido*s. At this point the planned operational fleet included a single cruiser squadron, divided between the Home and Mediterranean Fleets, plus another five operational cruisers spread as necessary around the world. As of May 1955 plans called for reducing the number of active surface combatants of cruiser size and above (the only non-cruiser being the battleship *Vanguard*) from thirteen in 1958/9 to six in 1965. This idea caused considerable debate. Meanwhile the suspended *Tiger*s were completed with many of the new weapons and other systems, to become the last British gun cruisers. They were the closest approach to a general cruiser fleet modernisation proposed in 1948, presumably as part of the planned nine-year fleet upgrade.

A further plan submitted to the Board on 8 March 1956 envisaged a minimum force of eight cruisers. It proposed modernising the cruiser *Superb*, the only one of the *Tiger*s completed to the original design, as a fourth *Tiger*. Although such expensive modernisation of an older ship might seem wasteful, it would create an effective class of four ships with the best conventional armament. If it began in mid-1958, it would take 3½ to 4 years. *Superb* was, moreover, already of *Tiger* configuration. On the other hand, she would have to be stripped before she could be rebuilt. There was some feeling that the public would react badly to spending so much money on a gun cruiser in the early 1960s, when missiles were either already in service or were clearly coming very soon.

Three missile cruisers and sixteen missile destroyers should be built. This force was unaffordable, not least in terms of manpower. In February 1957 an '80 Fleet' (i.e. 80,000 personnel) was proposed; it was essentially the force which emerged from the large-scale Defence Review managed by Duncan Sandys. The Royal Navy would maintain three fleets: Home, Mediterranean and Indian Ocean. Each would be built around a carrier battle group comprising a carrier, a cruiser, four destroyers and four fast frigates. In addition, a Far East Fleet would comprise a cruiser, two destroyers and two frigates. Further ships were needed to make up for refits, so the '80 fleet' was more or less equivalent to five carrier task groups plus lesser units such as cruisers. This last study justified a new type of ship, the Commando Carrier.

The new strategic vision kept the British carrier force alive after the 1956–7 Defence Review. The most important future role was support of British Commonwealth allies East of Suez, where Malaya was approaching independence as Malaysia. The area was of great economic importance to Britain, whether the countries involved were colonies or independent. This was the sort of 'informal Empire' consideration which had made big cruisers important in China in the 1920s and 1930s. In this case Indonesia (which at the time seemed to be a Chinese proxy) posed a particular threat, as it demonstrated during the Confrontation of 1964–6. It seemed, from the late 1950s on, that the Soviets were willing to provide allies like the Indonesians with modern anti-ship weapons, such as submarines, boat-fired missiles and bomber-fired missiles. The Royal Navy became much more interested in classic anti-ship and strike operations based on carrier battle groups. In this context a new kind of cruiser emerged, albeit largely on paper.

Victory in the Confrontation was the high point of the general-purpose force created by the Royal Navy. It coincided with the attempt to replace the existing carriers, which had been laid down during the Second World War. That proved unaffordable, and two successive British governments abandoned both the carrier replacement programme and, largely, the 'East of Suez' role which had justified a carrier-cruiser force. The official emphasis turned towards NATO and the North Atlantic, although the Royal Navy continued to point out that Britain remained a global power economically, dependent on events far beyond Europe and the North Atlantic. As the next chapter explains, a new kind of cruiser proved relevant to the new kind of seapower evolved after the end of 'East of Suez'.

New Weapons and Systems

Modernised ships would take advantage of the radars and combat direction systems under development in 1945. The main new radar was Type 960, an alternative to Type 281 with a similar antenna but greater range. For air interception, there was a combination of two long-range precision radars (Types 982 and 983), one with a vertical and the other with a horizontal fan beam. The bearing and height of a target could be determined by matching up their displays in an AIO. Although they had impact on cruiser thinking, they were never installed on board a cruiser. Wartime development also produced the Target Indication Unit (TIU) Mk III with its associated Type 992 radar, capable of tracking multiple targets on a track-while-scan basis. One TIU could feed initial target information to several directors, or it could provide a director with a new target as soon as one had been destroyed. By 1947 the Admiralty Signal Research Establishment was working on something far better, in effect a TIU capable of handling dozens of tracks fed to it by a massive three-dimensional radar. The analog memory was the core of a Combat Direction System (CDS), one virtue of which was that it showed the same radar picture to any summary display in a ship. That alone made for the faster decision-making which modern aircraft demanded. The associated Type 984 radar occupied a big nacelle. Since its single beam scanned in both azimuth and elevation, there was no need for operators to be certain that two separate radars (Types 982 and 983) were looking at the same target.

The last of the earlier generation, the Mk VI director with Type 275 precision radar, entered service aboard HMS *Superb*. On that ship it was associated with earlier types of HA computers, but later ships had the Flyplane computer, first suggested as long ago as 1932, which was finally approaching service. In theory Flyplane offered much longer-range performance than MRS 3, but it was also considerably more massive and more complex.[6] The late-war fire-control programme envisaged a series of new systems using electric calculators and more sophisticated operating concepts: LRS 1 for surface fire (long-range); medium-range MRS 1, 2 and 3; and a Simple Tachymetric Director (STD) for small-calibre guns, equivalent to the US wartime Mk 51. MRS 1 and 2 were abandoned. MRS 3 was an Anglicised version of the late-war US Mk 56, which could control both 5in and 3in guns.[7] MRS 5 was an abortive all-British alternative, and MRS 6 combined the late-war Mk VI director with a new electric computer, for 4in control. MRS 7 was a related 4.5in gun control system. MRS 8 was a short-range 40mm director. Unlike earlier British systems, LRS and MRS were integrated, a particular computer being associated with a particular type of

Above: From about 1948 on British naval planners thought in terms of a new generation of weapons and, if possible, wholesale modernisation of the cruiser force. The old heavy cruiser *Cumberland* was test ship (it proved impractical to use her to test British guided missiles, but she did test guns and even an above-water torpedo tube for frigates). She is shown on 5 May 1956, with a prototype twin 6in mount forward and a prototype twin 3in mount aft. The 6in gun was the first of its size in Royal Navy service to rely on cartridge cases, and during tests the gun repeatedly jammed due to poor quality control in their manufacture. It is not clear whether it performed better in service (it had a bad reputation for jamming). The British 3in/70 twin mount seems to have performed far better than its US equivalent, and it was also adopted by the Royal Canadian Navy. The identity of the object on the stern is unknown. The guns seem to have been brought aboard in 1955–6 (photographs taken in 1954 show no guns at all). Although the ship retained her engines, authorised full power was limited to 31,000 shp (24kts), with permission to make up to 26kts only either for special trials or in an emergency.

director above decks. After 1950 the United States supplied the Mk 63 fire-control system, which it used for 3in/50 control, under the Mutual Defence Assistance Programme (MDAP); it was installed on board several cruisers, to control 4in guns (they had radar dishes placed on their shields as part of the system).

The two new cruiser weapons under development immediately after the war were the automated twin 6in Mk 26 and the twin 3in/70. As of 1949, the new guns were expected to be available about 1954. The Mk 26 was a fully dual-purpose mounting intended to achieve a high rate of fire by using cased charges rather than the bagged ones of earlier guns. The twin 3in/70 was intended to use US guns and ammunition, but it was an entirely British design, intended to achieve the very high rate of fire of ninety rounds per gun per minute. By June 1948 DNO had adopted the US idea that the gun should less mechanised (even so, the US design at least proved far too ambitious). A further new gun, the

5in/70, was expected to appear in 1957; in effect it could replace both 6in Mk 26 and 3in/70. It had only a limited impact on the cruisers, because it was bound up with a follow-on cruiser-destroyer programme which does not figure in this book.[8] Unlike the 6in Mk 26 and the 3in/70, it never materialised. The new guns required AC electrical power, but existing cruisers had DC generators, and in 1949 no turbo-alternators (AC generators) were on order.

Fleet Modernisation

By October 1947, plans called for fleet modernisation by 1957, and they included a Staff Requirement that as many cruisers as possible have dual-purpose armament by that time. It appeared that six 6in cruisers, including the three *Tiger*s, could be modernised by 1957.[9] The only available dual-purpose cruiser gun was the twin 6in Mk 26.[10] Once plans for the large new cruiser had been abandoned, the question was whether existing ships could accommodate their weapon. A quick weight and volume analysis by DNC department for DNO (October 1947) suggested that only the *Southampton*s could have all three of their turrets replaced by Mk 26. The *Fiji*s and their successors had their forward turrets too close together for the necessary magazines below decks, although they could, however, have the new turrets in place of 'A' and 'X' mountings, and perhaps a smaller-calibre HA mounting (e.g., 4in) in place of 'B' turret. The *Minotaur*s had accommodated 400 rounds per gun. A *Southampton* would be limited to 300 rounds per gun, but a *Tiger* could have 400. Further investigation showed that a *Tiger* could take two twin mounts and two fire-control systems (Flyplane plus clocks for 6in control, Flyplane or MRS 3 for 4in control). Although a *Fiji* could take the two 6in mounts, its hull was more congested, and it might accommodate less ammunition, and it would have only a single 6in fire-control system (to engage one rather than two targets at a time).[11] Although a *Tiger* could take the new 3in/70, a *Fiji* could not. The hull could not accommodate Flyplane for 4in control, although MRS 3 could be fitted. Similar further investigation limited a *Southampton* to two 6in mounts, but additional space would have to be found (i.e. modernisation would be a larger job). The 4in battery would have to make do with MRS 3, as there was no space for Flyplane. The dramatic reduction in surface firepower was considered acceptable, as it had a much lower priority than anti-aircraft fire. In effect this reversed the priority in the previous dual-purpose cruiser design, the *Dido* class. The new turret needed a great deal more electric power, a problem in tightly-designed ships.[12]

The only existing cruisers with dual-purpose batteries were the ten surviving *Dido*s (not counting *Scylla*, with 4.5in guns). Their HA computers could be replaced with Flyplanes (presumably the existing directors would have been replaced by Mk 6). For DGD 'this represents a far bigger return than that provided by the same effort to any other class of cruiser at the present time'. However, their turrets would be obsolete by 1956, and no substitute would be available for a few years beyond that. The ships lacked enough internal volume to take enough ammunition for the 3in/70 which might replace their 5.25in turrets (they might have enough for only a minute and a half of fire).

In November 1947 it was assumed that modernisation would be conducted at the two Royal Dockyards (Portsmouth and Devonport), and that it would take two years per 6in ship: eighteen yard-years in the nine-year programme, ten of which would be consumed by the one-year modernisations of the *Dido*s, leaving four ships. More 6in cruisers could be modernised at the cost of some of the *Dido*s. For example, if only five *Dido*s were modernised, the total available for 6in cruisers would grow to eleven yard-years, perhaps enough for six ships. DGD was more pessimistic (or realistic); at a meeting in November 1947 he suggested that money would not be available for more than two 6in cruisers, and that they would be *Tiger*s. A parallel study of the shape of the desired 1957 fleet showed eighteen war-worthy cruisers (another showed a total of twenty-three cruisers).

A Staff Requirement for fleet modernisation was drawn up. In May 1948 the Ship Design Policy Committee was given schemes for broader cruiser fleet modernisation:

• *Fiji*s (Scheme A): two twin 6in Mk 26 (one Mk VI director with Flyplane, one MRS 3), three twin 3in/70 with three MRS 3 and one TIU 3 (Type 992 radar); estimate based on *Kenya*. Although weight was not a problem, internal space certainly was. Space was needed for TIU 3/992, MRS 3 computing positions and new electric generators and power rooms. Requirements were difficult to estimate, but based on the Flyplane installation proposed for *Tiger* about 1,000–1,500ft^2 would be needed, at the expense of accommodation, store rooms and possibly offices, shops or oil fuel tanks. About 60 tons would be added and about 0.3ft of GM lost. The ships were already in critical condition, with any increased stress or loss of GM unacceptable. The heavy Type 992 aerial would probably need a new mast.

• Modified *Fiji*s including *Swiftsure* (Scheme B): same but with two Mk VI/Flyplane directors for main battery; estimate based on *Newfoundland*. The situation was similar to that for the *Fiji*s except that stability was not quite so critical. A note to the Staff Requirement for *Swiftsure* modernisation suggested that the new 4.5in Mk VI, 5.25in, guided missile, or fighter-direction facilities might replace some or all 6in turrets. It might be difficult to provide adequate 3in ammunition stowage, so the ship might be limited to a half outfit.

• *Dido*s (Scheme C): two Mk VI/Flyplane directors (fore and aft) for 5.25in battery, one TIU 3 (Type 992 radar); estimate based on *Euryalus*. The scheme appeared practicable but ships had only limited space and weight margins left. If any attempt were made to upgrade anti-aircraft firepower, the 5.25in guns should be replaced by 4.5in rather than 3in/70 guns, because they had only enough space for three minutes' worth of 3in/70 fire.

The Staff Requirement also envisaged rearmament of the 'Towns': two twin 6in (one Flyplane, one MRS 3) with 4in guns retained but controlled by four MRS 3 systems, and installation of TIU 3/Type 992.

Cruisers would retain their torpedo tubes wherever possible, so that they could fire the improved anti-ship torpedoes under development. Missiles might later be installed on board some of these ships. Anti-torpedo systems under development should be installed, but in the event no satisfactory system emerged (nor has one been deployed in the sixty years since).

The new threat of nuclear attack required further measures, suggested by reports by British naval observers at the US tests at Bikini Atoll in 1946. Bridges should be enclosed: the Royal Navy had to abandon the open compass platform it had found so important during the Second World War. Funnel tops had to be reshaped to protect boilers from the air blast of a nuclear explosion, and machinery and its operators had to be protected from the radioactive air the boilers would suck in. In 1951 E-in-C proposed a new boxed boiler whose intake air would not circulate into the ship. Ships had to be able to washdown, or pre-wet, to clear nuclear fallout from their decks, and ideally ships should be shaped so that radioactive water would not cling to them. The superstructure had to be fully enclosed, and ideally placed under positive internal air pressure. Later it became clear that many ships risked having their superstructures simply blown off because the structure linking them with the hulls was not sufficiently continuous, but that was not evident for some years.

In March 1949 the Ship Design Policy Committee recommended completing the three Tigers as soon as possible (by 1954) and modernising five Didos as soon as possible starting with HMS Argonaut, modernising two Southamptons as anti-aircraft cruisers, and maintaining the remaining seven Southamptons and the other Didos by making large repairs (i.e. extended refits). That would provide the desired eighteen war-worthy cruisers for the target year of 1957. Director of Plans preferred the 'Towns' to the Fijis and their ilk. He also assumed that a Dido modernisation would take more time, hence that the total available would be smaller. In 1949, also, the available time had been cut to the eight years between the 1950/1 and 1957/8 programmes. He therefore proposed doing only three Southamptons and two Didos. Of the 'Towns', Belfast offered the greatest potential thanks to her size and improved stability (due to her wartime reconstruction after having been mined). Liverpool was next best, for stability and for the latest completion date. The third ship chosen was Glasgow. The Didos chosen were Royalist and Diadem. Ceylon and Newfoundland, the Improved Fijis, were better than the earlier Fijis. VCNS had a different view, because he saw cruisers as valuable surface escorts against Soviet cruiser attack. Superb already had a very effective 6in fire-control system, and could probably deal with anything the Soviets could produce over the next five or six years.[13] It

Below: After the Second World War the Royal Navy gradually rearmed its existing cruisers. It replaced quadruple pompoms with twin Mk 5 Bofors using off-mount directors rather than with the elaborate self-contained STAAG developed during the Second World War, and it eliminated Oerlikons in favour of additional single Bofors guns (Mk 7s). These guns are shown aboard HMS *Sheffield* on 30 January 1954.

would be inexpensive to bring other cruisers up to the same standard, in order of merit Swiftsure, three Fijis (Kenya, Ceylon and Nigeria), and three 'Towns' (Sheffield, Newcastle and Birmingham); Belfast might similarly be worthwhile.

By September 1949, plans called for beginning cruiser modernisation in 1950. Kenya and Nigeria would be scrapped when the three Tigers were completed. Swiftsure, Superb and two Didos would receive large repairs. Large repairs to Birmingham, Newcastle, and Newfoundland would begin at once. DNC suggested scrapping the Didos, and it was agreed that Dido, Phoebe and Argonaut would be surveyed.

Alternative levels of anti-aircraft modernisation could be rated. DGD rated a modernised Tiger at 10. VCNS laid out lesser levels:

X: Four or five single 3in/70, two or three MRS 3: 7 or 8.
Y: Four or five twin 4in, two or three MRS 3: 6 or 7.

He recognised that single 3in guns would be in great demand for smaller ships. Despite its low rate of fire, when combined with MRS 3 the 4in gun retained some limited value. Surely three cruisers could be modernised to X standard and three others to Y. At least four Didos could become anti-aircraft ships. By this time the cruiser-destroyer was being discussed: VCNS thought that such ships could enter production in 1955, and that work on a new 6in cruiser could begin in 1956. First Sea Lord favoured a 6in cruiser displacing 13,000 to 14,000 tons, something like DNC's 'medium cruiser of 1960' described below.

Fleet planning made evaluations of different existing classes clear. Director of Plans considered the roomy 'Counties' particularly well suited to hot climates, and they were in demand as flagships for foreign stations and they had good endurance. However, they were at least eight years older than all other cruisers, and hard wartime service had reduced their remaining lifetimes. Machinery in particular would be expected to give constant trouble. This applied even to the heavily rebuilt *London*, whose machinery had not been replaced or modernised. Next best for flagship service because of their roominess and habitability and endurance were the Southamptons, but they were ten years old. On the other hand, they had better (longer-lasting) finish than later cruisers because they had been built before the war. Having been designed under less stringent treaty limits than the Fijis, they were larger and more heavily built. The newer Fijis were too cramped and would make poor flagships, and Swiftsure was just a more modern Fiji. The Didos were even more cramped and had poor endurance as well; their only positive points were handiness and low fuel consumption, and they were considered modern. Full AIO facilities either had been installed, or had been authorised, in all classes from the Southamptons on.

The outbreak of war in Korea emphasised the need to modernise cruisers. In March 1951, DGD proposed modernisation plans for three 'Town' class cruisers, but the Ship Design Policy Committee decided instead to give the two best 8,000-ton cruisers, Swiftsure and Superb, a more ambitious armament improvement than the 'Towns' when they were taken in hand in June and September 1957. They were deficient mainly in anti-aircraft fire control. Superb had Mk 6 directors with Type 275 radar, but they were served by the inadequate combination of a destroyer-type FKC computer and the SEDC. The committee suggested a simple upgrade, fitting MRS 7, which had been adopted as interim for 'C' class and 'Battle' class destroyers. Destroyers had it fitted during a three-month refit. MRS 7 would be available in mid-1952.

Swiftsure had the earlier HACS Mk IV with Type 285 radar, which was obsolete. One possibility was to install two Flyplane systems (FPS 5); the other was to replace the three HADTs with three Close Range Blind Fire (CRBF) systems. FPS 5 entailed considerable expensive work.

Above and left: Manpower was always a serious problem, and the Royal Navy operated many of its ships with parts of their armament mothballed. HMS *Bermuda* is shown in October 1954 with her barrage directors and the pompoms atop her former hangar all cocooned for preservation without use. She also had some weapons removed. The two pompoms on her after superstructure, forward of the former 'X' turret position, have been eliminated in favour of single Bofors guns, plus a centreline pompom between and above her after twin 4in guns. Her pompom directors are gone, and it is not clear that her 4in directors are operational.

BRITISH CRUISERS

POST-WAR CRUISERS

It required the heavier Mk 6M director; only two directors could be fitted instead of the existing three. This arrangement had already been accepted for 'Town' class modernisation. No provision had been made for such expensive work in *Swiftsure*, so DGD preferred a simpler solution. A CRBF could directly replace each HADT, its radar room being placed nearby. That this was better than the existing system showed just how awful that system was: CRBF was intended to control pompoms or 40mm Bofors, not longer-range 4in guns. Of course, it would eventually have to be replaced by something more appropriate and adequate, like FPS 5. FPS 5 would be available in 1955 and CRBF in mid-1954.

The best air-warning radar (Type 960) gave 20 per cent more range than the 'fair' radar (Type 281BQ), which was 20 per cent better than a 'moderate' radar (Type 281B or Type 281).[14] Ships without PPI displays and with poor bearing accuracy were considered 'poor', although they might enjoy ranges like those of 'moderate' ships. DND, who was responsible for fighter control, argued in March 1949 that nothing short of a combination of Type 960 and the Type 277Q surface-search/height-finding radar would be of much use against jet aircraft – but production of both was so slow that they would probably be available only for new ships. In 1948 no cruisers were in the best category. *Kenya*, *Belfast* and *Mauritius* were in the 'fair' category, and *Argonaut*, *Cleopatra*, *Bermuda*, *Jamaica* and *Newfoundland* in the 'moderate' one. Most cruisers were slightly worse: *Diadem*, *Dido*, *Phoebe*, *Royalist*, *Sirius*, *Ceylon*, *Gambia* and *Swiftsure*. On the 'poor' list were *Euryalus*, *Nigeria*, *Superb* and *Liverpool*. For interception and height-finding capabilities 'moderate' meant that a ship had Type 277 radar (good meant instead the developmental pair of intercept radars, Types 982/983). A 1948 list showed the best cruisers in the moderate category (*Argonaut*, *Cleopatra*, *Euryalus*, *Belfast*, *Bermuda*, *Jamaica*, *Kenya*, *Liverpool*, *Mauritius*, *Newfoundland*, *Superb* and *Swiftsure*) and the rest credited with nil capability.

The fleet badly needed radar air defence ships, which the British called FADES, but they were unlikely to materialise.[15] The alternative was a slightly modified cruiser; there was some feeling that the *Leander*s would have been particularly well suited to such a role (but they had been discarded before any serious planning could be done). Cruisers converted for air defence should have Type 960P search radars and a pair of intercept radar combinations (Types 982 and 983 for precision horizontal and vertical target tracking). Other ships would make do with Type 960P, the Type 277Q surface-search/height-finding radar, and the Type 992 target-indication radar, with a form of Type 275 for fire control. Installation of one Type 982/983 pair would be possible only if some armament were surrendered, and adding a second pair aft would require major reconstruction. The 'Town', *Fiji*, *Dido* and even *Swiftsure* classes required a 200 per cent increase in tactical (V-UHF) radio sets (for *Dido*s the figure was 200 to 250 per cent). MF and HF radio also had

Left and inset: There were several attempts to modernise the cruiser fleet. Plans eventually called for large refits for the 'Towns'. HMS *Birmingham* is shown on 6 June 1952. Her bridge was streamlined but not enclosed (as was later done for other cruisers). She was given a pair of Mk 6 directors with Type 275 radars (there was insufficient weight margin to add a third aft) and a lattice foremast for new radars. They were part of the MRS 6 fire-control system. Note also the stub mast for the YE aircraft beacon, abaft her mainmast. Her pompoms were replaced by six twin Bofors, each controlled by a Simple Tachymetric Director (STD) Mk 6. There were no lighter guns. The wartime air-warning radar was replaced by a new Type 960 with a similar antenna but much better performance (for a time this radar was to have had a narrower-beam mattress antenna, but that project died at the end of the Second World War). There were three barrage directors. HMS *Newcastle* was similarly modernised and *Sheffield* underwent a somewhat similar modernisation in 1955.

Sheffield did receive a simplified close-range battery. She is shown on 7 July 1952 at Balboa in the Canal Zone. By this time her pompoms had been replaced by four twin Bofors (Mk 5) and she also had ten single Mk 7 Bofors – and two single power pompoms, a wartime type of gun nearly extinct by that time. This armament change was probably seen as a way station to full modernisation, since she had the six STDs of the fully-modernised ships.

Above: For *Sheffield*, modernisation included enclosing her bridge, presumably to help protect her command from nuclear fallout; the view above, so important during the Second World War anti-air actions, was no longer necessary. She is shown in July 1957. The ship had eight twin Bofors (Mk 5), each with its STD, and two single Bofors (Mk 7), plus the two MRS 6 HA fire-control systems of the modernised ships, but the official book of ship data did not show her as having been modernised like her two sisters. MRS 6 employed a new electric computer rather than the Flyplane system usually associated with the Mk 6 director. The other two ships in the class, *Glasgow* and *Liverpool*, retained their old 4in directors. Their pompoms, presumably no longer maintainable, were gone, but they had mixed close-range batteries. *Glasgow* had six single power pompoms and eight single Bofors guns (Mk 7s), with six STDs. *Liverpool* had an even less capable close-range battery: six single power 2pdrs, two single 2pdrs (probably hand-worked), six hand-worked Bofors Mk 3, and two single Mk 7s, with the six STDs.

Above and below: As the largest surviving British cruiser, and the ship with the greatest reserve of stability, *Belfast* was well worth modernising even after modernisation plans had been abandoned for other cruisers. Her 1959 refit was intended to extend her hull and machinery life by a decade (parallel refits were planned for the two newest large cruisers, *Superb* and *Swiftsure*, but neither was completed). *Belfast* lasted longer than projected. Note that she was not fitted with Mk 6 directors for her 4in guns; instead she had eight of the new MRS 8 systems, controlling both her 4in and her 40mm guns. MRS 8 was essentially an expanded Bofors director, using the Type 262 radar originally developed for the integrated STAAG radar-gun combination. *Belfast* is shown in June 1959, newly refitted.

Above and left: Early post-war modernisation projects included the *Dido*s, because they alone offered dual-purpose heavy guns. In the end only *Royalist* was modernised. By the time she was complete, she was no longer wanted by the Royal Navy, so she went to New Zealand in exchange for *Bellona*. She is shown in British waters in 1956 (the aerial view is dated 11 July 1956). She was given two of the new Mk 6 directors (with their Type 275 radar) and, unlike the larger modernised British cruisers, she had the Flyplane computer (FPS 5 system). In effect she had two independent 5.25in batteries, because she had two GDS 2* direction systems, one associated with each director/computer combination. Close-in weapons were two self-contained STAAG Bofors guns (twin guns with on-mount Type 262 radar) and two 40mm Mk 9 guns.

to be modernised, but not to anything like the same extent. Modern radio warfare (EW) equipment had to be installed.

In September 1950 Director of Plans pointed out that the two key cruiser roles were (a) anti-aircraft protection of convoys and of carrier task groups, and (b) close cover of North Atlantic convoys. The planned peacetime fleet consisted of six 'Towns', *Swiftsure*, *Superb* and *Newfoundland*, and six *Dido*s. Another two *Fiji*s and a *Dido* were to be retained until the three *Tiger*s were completed. Another five *Fiji*s and a *Dido* would be discarded, although not for some time. Only the *Dido*s could fulfill the most important task, (a) above. Director of Plans therefore proposed emergency rearmament of other cruisers. The Mk 24 triple 6in turret, despite its limited anti-aircraft potential, was attractive because it was suited to RPC. Unfinished Mk 24 turrets left over from the original *Tiger* programme could be made available more quickly than the new Mk 26. DGD asked for emergency Staff Requirements for a *Tiger* thus completed. It might have two such mounts in 'A' and 'B' positions (and four twin 4in ['X' and 'Y' positions and two waist positions]) instead of the planned combination of two twin Mk 26 and three twin 3in/70. Alternatives were two triple 6in and four (later two) twin 4.5in, each with its own Mk 6M director and Flyplane computer below decks. The 4.5in mountings would go in 'X' and 'Y' positions, the 6in turrets in 'A' and 'B'. Directors would be sided abreast the foremast. The close-range battery would be the three STAAGs then planned for the full conversion. Another possibility was a secondary battery of six single destroyer-type 4.5in guns: one each in 'X' and 'Y' positions and four in the waist, with the same two directors.[16] Since nothing had yet been done with her hull, a *Tiger* could also be completed as planned, with three Mk 24 mounts. DNO rejected all the emergency secondary armaments as inferior to that of HMS *Superb*. Since it was highly desirable to mount the entire 6in battery forward in view of the predominantly surface role, he feared the ship would trim excessively by the bow (which turned out not to be the case). DGD pointed out that the emergency plan would advance completion no more than a year or eighteen months, to produce a cruiser little better than existing ones. The fleet would not suffer any great shortage of cruisers over the next four or five years, so the project was hardly urgent. The project died late in 1950.

A plan in August 1951 called for fully modernising *Liverpool*, *Royalist*, *Sirius* and *Phoebe*, but in December planned modernisation extended to five *Dido*s: *Royalist* (prototype) to begin in January 1953, *Phoebe* in August 1953, *Sirius* in April 1954, *Diadem* in June 1955 and *Cleopatra* in November 1955. They would get a new main battery fire-control system and a new bridge with an enclosed compass platform and space for a modern AIO, plus improved radars and communication systems. The two 'Town' class cruisers to be modernised were *Liverpool* (prototype) and *Glasgow*. In 1952 the beginning of *Liverpool*'s modernisation was moved from February 1955 to April 1954 and instead of beginning with *Royalist*, the *Dido* programme would begin with *Phoebe*.

POST-WAR CRUISERS

Above and below: Nearly a decade after the end of the Second World War, HMS *Jamaica* was still in much her late-war appearance, albeit with a reduced close-range battery. She is shown on 13 May 1954. At this time no ship of the class had been modernised. Close-range armament was five quadruple pompoms, two Bofors Mk 7 (power-worked single mounts), and six hand-worked Bofors Mk 3.

BRITISH CRUISERS

Left and below: HMS *Gambia* shows the effect of rearmament and other improvements in these 14 November 1955 photographs. Note the superimposed barrage directors (covered) on the fore side of her bridge. She had an all-Bofors close-range battery: each pompom was replaced by a twin Bofors Mk 5 controlled by an STD. In addition she had two single Mk 7s. Later the single mounts were replaced by two more twins, and her 4in directors were replaced by six US-supplied Mk 63s. No British ship of this class was fully modernised.

Above: HMS *Bermuda* is shown on 4 September 1960 in the Mediterranean (this photograph was taken from USS *Des Moines*). She was given a covered bridge, and her 4in guns were controlled by US-supplied Mk 63 gun control systems whose directors are the small canvas-covered objects at the after end of the former hangar and on the after superstructure. She had an all-Bofors close-range battery: seven twin Mk 5 mounts controlled by seven STDs. Although the old 4in director still seems to be present, it was no longer operational. The Mk 63s were supplied in the early 1950s under the Mutual Defence Assistance Programme (MDAP), which was also the origin of the US-type 3in/50 guns on board HMS *Victorious*. *Gambia* (1957) and *Kenya* (1955) had similar extended refits. *Gambia* emerged with the same armament as Bermuda. *Kenya* had only two Mk 63s, and she had five Mk 5 and six (single) Mk 7 Bofors, with five STDs.

The last two *Dido*s would be reboilered (for nuclear protection). *Belfast* was later added (as of June 1954, she would be taken in hand in August 1955), and then *Swiftsure* and *Superb* (as of June 1954 they would be taken in hand in November 1956 and February 1958 respectively).

Unfortunately the British had been right to think in terms of 1957 as their target year for modernisation. Korean War mobilisation was expensive because, in common with other Western governments, the British government saw Korea as the beginning of a new world war (there is some evidence that Stalin had something like that in mind). Cruiser modernisation was cut because there were much more urgent requirements, like building escort ships. On 20 November 1953 Director of Plans announced that *Belfast*, *Swiftsure* and *Superb* would be modernised as planned. *Sheffield* would receive the MRS 6 fire-control system in a six-month large refit rather than modernisation. Other cruisers would receive the US Mk 63 fire-control system in short refits. Work on *Liverpool* and *Glasgow* and modernisation of the three *Dido*s beyond *Royalist* (already being modernised) was cancelled. *Nigeria* (which went to India) and the other seven *Dido*s (not counting two in the Royal New Zealand Navy) would be discarded. Although the Royal Navy lost interest in them, they attracted foreign attention. In 1953 Turkey asked for *Dido* or *Sirius*. The British suspected that the Turks did not realise how expensive these ships were to operate, and upon discovering what was involved, in January 1954 they withdrew. The Spanish government expressed similar interest, and then shifted to an (unsuccessful) attempt to obtain an *Atlanta*-class light cruiser from the United States. As modified, *Royalist* was lent to New Zealand to replace *Bellona*, while *Diadem* was transferred to Pakistan to balance the transfer of *Nigeria* to India.

Director of Plans classified ships by standards of anti-aircraft capability and showed planned modifications through 1959.

1. *Tiger* standard (also upgraded *Swiftsure* and *Superb*): self- and force defence under all conditions (GDS 3, MRS 3, L70 Bofors). Ships required AC electric power systems.
2. Modernised *Belfast* standard (as for 1, but with 4in instead of 3in/70 guns): *Belfast* only.
3. *Newcastle* standard: self-defence under all conditions, limited force defence in visual conditions (three MRS 6): ultimately *Newcastle*, *Birmingham*, *Newfoundland* and *Sheffield*.
4. *Kenya* standard: self-defence in visual conditions and to a limited extent in blind conditions (US Mk 63 fire control in place of HACS). This would ultimately apply to *Kenya*, *Liverpool*, *Ceylon*, *Jamaica*, *Bermuda*, *Gambia*, *Mauritius* and *Glasgow*.
5. Present standard: hardly able to defend themselves against air attack.

The shock of the Korean War mobilisation failure continued to affect British naval spending, as new systems, such as new aircraft and submarines, consumed money. In 1955 only three cruisers remained in the programme: *Belfast*, *Swiftsure* and *Superb*. *Belfast* was already in hand at Devonport. In January 1956 Deputy Controller proposed that *Belfast* not be modernised at all, instead brought up to *Ceylon*/*Kenya* standards. She was downgraded partly because her modernisation interfered with work on the carriers *Centaur* and *Eagle*. Instead she would be refitted to *Birmingham* standards. *Swiftsure* too would not be refitted, but would be brought up to *Kenya* standard. She would get a bridge similar to that of *Birmingham*, new radars (Types 277Q, 293Q and 960), and the GDS 2* gun-direction system plus four MRS 8 fire-control systems, for her 4in and smaller guns. *Superb*, however, would receive the full *Tiger* modernisation.

Jamaica never had a full refit, but she was modernised piecemeal, her bridge covered. She was probably the last British cruiser to retain the old type of 4in director in service. She was given the new radars (Types 960, 277Q and 293Q, and the Type 974 navigational radar), though not a lattice foremast. Close-range armament was five twin Mk 5 Bofors (each with an STD), two single power Bofors (Mk 7), and six single hand-worked Bofors Mk 3.

Before being transferred to India as INS *Mysore*, HMS *Nigeria* was completely refitted to the standard which would have been met had the *Fiji*s been modernised in the 1950s. She is shown on 5 October 1964.

POST-WAR CRUISERS

In February 1955 Controller called a meeting to see what would have to be done to keep *Sheffield* and the four *Fiji* and Improved *Fiji* class cruisers going until they were replaced by the three *Tigers*. They were expected to finish their lives about 1962; if the *Tigers* and the modernisations went to plan, all would be placed in reserve about 1959. They could not yet be discarded because all had been declared to NATO as part of the British wartime contribution. Their existing HA control systems were considered useless, so plans called for fitting US Mk 63 systems. Given their limited remaining lifetimes, upgrades were urgent. Once a cruiser was in good material condition but did not yet have improved anti-aircraft fire control she could be held in Class II reserve. *Ceylon*, which was in poor condition, was refitted. On this basis *Jamaica* would be taken in hand for an eight-week refit in August 1955. The main planned improvements were an enclosed bridge (as in *Ceylon*), replacement of Type 281 radar by Type 960 (but without additional displays), installation of UHF radio, modernised wireless and improved accommodation. *Gambia* was scheduled to begin her refit in March 1956. Her bridge would be enclosed, she would receive the Mk 63 fire-control system, her AIO layout would be improved, and she would receive improved communications as in *Jamaica*. *Bermuda* was ordered brought up to the same standard as *Gambia*, beginning in November 1955. A planned refit at Portsmouth for *Sheffield* was put back to August 1956 due to the press of other work at that yard. The major planned improvements were MRS 6 fire control and improved communications and Type 960 in place of Type 281.

In April 1955, *Superb* retained her wartime control systems and her close-range battery of four quadruple pompoms, plus two single power-worked Mk 7 Bofors and eight hand-worked Mk 3s. Her half-sister *Swiftsure* was in reserve. Her only remaining close-range weapons were two twin Bofors Mk 5s. Of the two remaining *Ugandas*, *Ceylon* had the US Mk 63 controlling her eight single power-worked Mk 7 and two hand-worked wartime Mk 3. *Newfoundland* had the same close-range battery but MRS 6 instead of Mk 63. *Bermuda* was refitting, and would emerge with Mk 63s controlling her 4in guns, plus seven twin Bofors Mk 5, each with an STD. *Jamaica* had five quadruple pompoms and two power-worked (Mk 7) and six hand-worked Bofors. *Gambia* was being fitted with Mk 63s, and would emerge with seven twin Mk 5 Bofors, each with an STD. In reserve, *Mauritius* lacked Mk 63s and had the five quadruple pompoms and eight hand-worked Bofors. *Kenya* had the Mk 63s and five twin Bofors Mk 5 (each with an STD) and six power-worked Mk 7 singles. *Belfast* was in reserve awaiting modernisation. She had an extended refit instead (she received MRS 8 for each 4in mount, plus six twin Mk 5 Bofors and another four MRS 8).[17] Of the 'Towns', *Liverpool* was in terminal reserve, her modernisation having been cancelled; she had six quadruple pompoms, two single power-worked pompoms and two powered and six hand-worked Bofors. *Birmingham* and *Newcastle* both had had a large refit in 1952, which gave them two MRS 6 for 4in control; they also had six

Below: HMS *Ceylon* is shown on 4 December 1956, soon after her extended refit. Her bridge was covered and she received an all-Bofors close-range battery: five twin Mk 5 and eight single Mk 7s, the Mk 5s being controlled by five STDs. Her 4in fire-control system was replaced by four US-supplied Mk 63s. Mk 63 used an on-mount radar dish (in the US Navy it controlled 3in/50 guns, and these dishes were very conspicuous). Although it is not obvious, the refit included replacement of her Type 281B radar by a Type 960; she also had the extended antenna of a Type 277Q radar instead of the roughly circular dish of Type 277. Her masthead target-indication radar was Type 293Q. The array barely visible up her fore topmast is a collection of microwave horns for the UA-1 system, a series of twelve horns, four for each frequency band (S, C, X). In effect UA-1 was the direct successor to the wartime FV1, in that it was a wide-open system well adapted to transient signals.

twin Bofors (each with an STD) and six power-worked single Bofors. *Glasgow* still had her wartime fire controls, and she had six quadruple pompoms (each with an STD) and eight hand-worked single Bofors. *Sheffield* was refitting, and would emerge with two MRS 6 and eight twin Bofors Mk 5 (each with an STD) and two powered single Bofors.

By 1956 the active cruiser force planned for 1958 was eight ships: two in the Far East, two in the Mediterranean Fleet, one on the East Indies station, two in the Home Fleet, and one on the South Atlantic station (though the manpower shortage precluded any permanent presence there). By the autumn of 1957 *Jamaica*, *Swiftsure* and *Mauritius* were all in reserve, and *Belfast* was refitting. The 1956–7 Defence Review required further economies, so the expensive modernisation of HMS *Superb* was cancelled and she was discarded. Early in 1958, a list of near-term cruiser employment covered ten ships including HMS *Tiger*, but not her two sister-ships. *Belfast* was due to commission for foreign service in March 1959. Having gone to the Mediterranean in January 1958, *Sheffield* was due to return in January 1959 to go into Operational Reserve, which meant that she would probably be further employed.

Birmingham was scheduled to go to the Mediterranean. *Newcastle*, the remaining ship of the class, was expected to return to the UK in August 1958 to go into Extended Reserve, from which she did not emerge. Of the four *Fiji*s and *Uganda*s, *Bermuda* was operating in the Mediterranean, scheduled to return in April 1959. *Ceylon* was scheduled to go to the Far East in June 1958. *Gambia* was available for the 1959 Home Fleet spring cruise, after which she would go to the Mediterranean. *Kenya* was expected to begin reducing to Extended Reserve in September 1958. *Newfoundland* was in the Far East.

Of the 'Towns', only *Belfast* survived, to become a museum ship in London. *Sheffield* reduced to the disposal list in 1964 (she was scrapped in 1967), while *Newfoundland* and *Ceylon* went to Peru.

HMS *Newfoundland* received a large refit or modernisation in 1953, hence used the Mk 6 directors (MRS 6 system) available at the time. She is shown coming alongside the US heavy cruiser *Bremerton*, in the Sea of Japan on 8 September 1954. Modernisation had included an all-Bofors close-range battery: five twin Mk 5 (each with an STD) and two Mk 7. Note that her bridge was not closed in.

Above: HMCS *Quebec* (ex-*Uganda*) was reduced to training duties before her sister-ship. She is shown in Guantánamo Bay on 10 April 1954. At that time she retained all four twin 4in guns and all three wartime directors (plus one barrage director, before her bridge), but her close-range battery had been reduced to one twin US-type Bofors (aft) and four Boffins. The reduction was probably carried out in 1953; in April 1952 she was credited with two quadruple Bofors, four twin power Oerlikons and six single Oerlikons, which were soon changed to two quadruple Bofors, one twin (US type) Bofors and eight Boffins.

Below: HMCS *Ontario* was the sister of the British cruiser *Swiftsure*. The Royal Canadian Navy used a mixture of US and British equipment. Unlike *Swiftsure*, she was completed with three Mk 6 directors (like *Superb*). In 1954 her close-range battery was four quadruple US-type Bofors (two abeam the bridge, one between mainmast and funnel, and one superfiring over the aftermost 4in mount) and nine Boffins. The following year she was a training ship. She retained all her 6in guns but had only the after Mk 6 director and the after 4in gun and quadruple 40mm, and one Boffin. It appears that her close-range battery was upgraded some time in 1952–3 from the mix of pompoms and Oerlikons installed in the UK to four US-type quadruple Bofors and nine Boffins. This undated photograph shows the ship in Melbourne. (Photo by Allan C Green via State Library of Victoria)

Below and opposite: HMS *Superb* was photographed by US Navy squadron VU-7 on 11 July 1955. Note that one of her Mk 6 directors (starboard side of the bridge) has been removed, presumably for repairs, while the opposite director on the port side has been cocooned. The ship still retained her four original quadruple pompoms. She also had two single power-operated Bofors Mk 7 and eight hand-worked Mk 3s. *Superb* was refitted in 1955–6 but was laid up in 1957. At that time she had six rather than eight hand-worked Bofors, but little else had been done (she did have a Type 960 radar, but that may have been installed earlier). There was interest in modernising her to be, in effect, a fourth *Tiger*, but that was not done. During 1956 the Admiralty Board considered partial modernisation, to be carried out in 1957 or early 1958. The 4in battery would have been reduced to four mounts, as in *Swiftsure* (but some proposals showed five), each controlled by an MRS 8 system, and the short-range battery replaced by six twin 40/70 Bofors (a type then of great interest, but not bought; the Royal Navy adopted the Seacat missile instead). These guns would have been controlled by another two MRS 8 (the original 1956 proposal envisaged the more elaborate MRS 3). The torpedo tubes would have been landed. The bridge would have been covered and the AIO modernised to the extent possible without radical redesign. The ship would not have the Comprehensive Display System (CDS) then being installed on board HMS *Victorious*, but she would have the associated digital data link and a means of displaying the tactical picture it carried (a note in the modernisation Cover pointed out that *Belfast* and *Swiftsure* were being given completely new bridges, largely to accommodate modernised AIOs with the data link). DND and DTSD wanted her at least arranged to be fitted with the new WAIR radar (Type 965), if sufficient information was available. In February 1957 DTSD decided to ask for both Type 965 and the data link. There was also interest in fitting the new-generation Type 184 scanning sonar, which offered better torpedo detection (for evasion) capability than the planned searchlight type. This was the prototype Long Range Self-Protection Asdic for cruisers and carriers, hence should be tested as soon as possible. It would, however, add considerable weight and cost. A reduction of speed to 29.5kts deep and dirty was considered acceptable. A hull-machinery refit would have made it possible for her to serve for another decade. Work would have been similar to that then planned for *Belfast* and *Swiftsure*, both of which were described as extended refits. A 1956 note in the Cover mentions a proposed missile-ship conversion with the US Terrier aft. By December 1956 *Superb* and *Swiftsure* were being considered together for modernisation (*Superb* would have been taken in hand early in 1958). Estimated costs were rising rapidly. The extended refit had been proposed because a complete modernisation like that of HMS *Birmingham* would have cost £4 million and have taken two years. Initially the more limited refit envisaged was expected to cost £2.5 million, but in August 1956 that was increased to £3 million. The original estimate was lower than that for *Swiftsure* because *Superb* was newer and had not suffered collision damage. *Superb* was dropped from the refit programme in April 1957.

The *Tiger* class

Of the 1948 plan for cruiser fleet modernisation, only the rearmament of the three suspended *Tiger*s survived, and that took much longer than expected. Ships would be stabilised to make fire control simpler. Equipment was chosen based on an estimated completion date of 1954. In March 1948 DTSD provided a rationale for the anti-aircraft cruisers. The new *Hermes* class light fleet carriers had sacrificed their entire medium-calibre (4.5in) armament for a second catapult, hence could not survive without surface escorts with medium-range anti-aircraft guns. A sketch Staff Requirement in July 1948 made the primary role of the ships 'to act in conjunction with aircraft carriers as escort against aircraft and surface attack'. The other most important role was to support destroyer striking forces. Traditional cruiser roles were also listed: attack on and defence of trade, support of combined (i.e. amphibious) operations, reconnaissance, and independent operations.[18] DNC considered it necessary to warn ACNS that his hope that the ships might be given useful aircraft-direction facilities was misplaced. He emphasised that the main hull structures of the ships were practically complete, limiting any modernisation. They were nearly as tight as the original *Fiji*s, which had given so much trouble in the past.

In December 1947 the Ship Design Policy Committee asked how many twin 3in/70 could be fitted to a *Tiger* with or without any 6in mounts, although the 3in/70 was only in the early design stage.[19] The mixed-calibre design envisaged a twin 3in/70 in 'B' position and two more in the waist. In the all-3in version, a pair of mounts would be in 'A', 'B', and 'X' and 'Y' positions, with the same two in the waist. Given the two 6in mounts, a *Tiger* could add three twin 3in/70, each with its own medium-range control system (MRS IV or V). Without any 6in mounts, she could accommodate six twin 3in/70, each with its own MRS. Using existing magazines, the ship could accommodate 800 rounds per 3in gun forward and 500 per gun for the two planned amidships mountings. In the all-3in configuration, the ship would have 700 to 800 rounds per gun for the two forward and after mounts, and 500 per gun for the two amidships mounts. DGD pointed out that, given its high rate of fire (ninety rounds per gun per minute), the 3in/70 had only six minutes' worth of ammunition. The Staff rejected the all-3in armament for a *Tiger* on the grounds that the ship would be so vul-

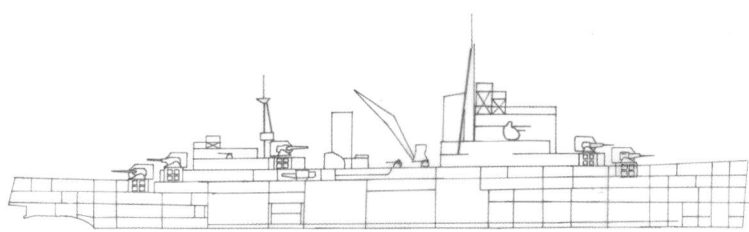

Above: The March 1948 proposal to complete the *Tiger*s with a main battery of twin 3in/70. This was Scheme II; Scheme I (the one chosen) envisaged two twin 6in Mk 26 and three twin 3in (described as MR/CR, or Medium/Close Range) plus two STAAGs (self-contained twin Bofors) and six single Bofors, with two quadruple torpedo tubes. Standard displacement was given as 9,550 tons and deep load as 11,700 tons. Scheme II, shown here, would have had six twin 3in plus the same Bofors and torpedo batteries. Also, each 3in gun would have had 600 rather than 550 rounds of ammunition, using space freed by eliminating the 6in guns (400 rounds each in Scheme I). Each 3in was served by a double-deck magazine. It took a large ship to carry enough ammunition to make the 3in/70 worthwhile (a *Dido* conversion was rejected because so little ammunition could have been carried). Each 3in mount had its own director: two of them superfiring aft, two alongside the tripod mainmast, in tubs, two superfiring atop the bridge. In each case armour was unchanged over that of the original design. Standard displacement would have been 9,190 tons, and full load, 11,350 tons. The ship would have been a bit faster: 30.9kts rather than 30.75kts deep and clean, 29.4kts rather than 29.25kts deep and dirty (six months out of dock). The drawing itself was dated 3 February 1948. (Norman Friedman)

BRITISH CRUISERS

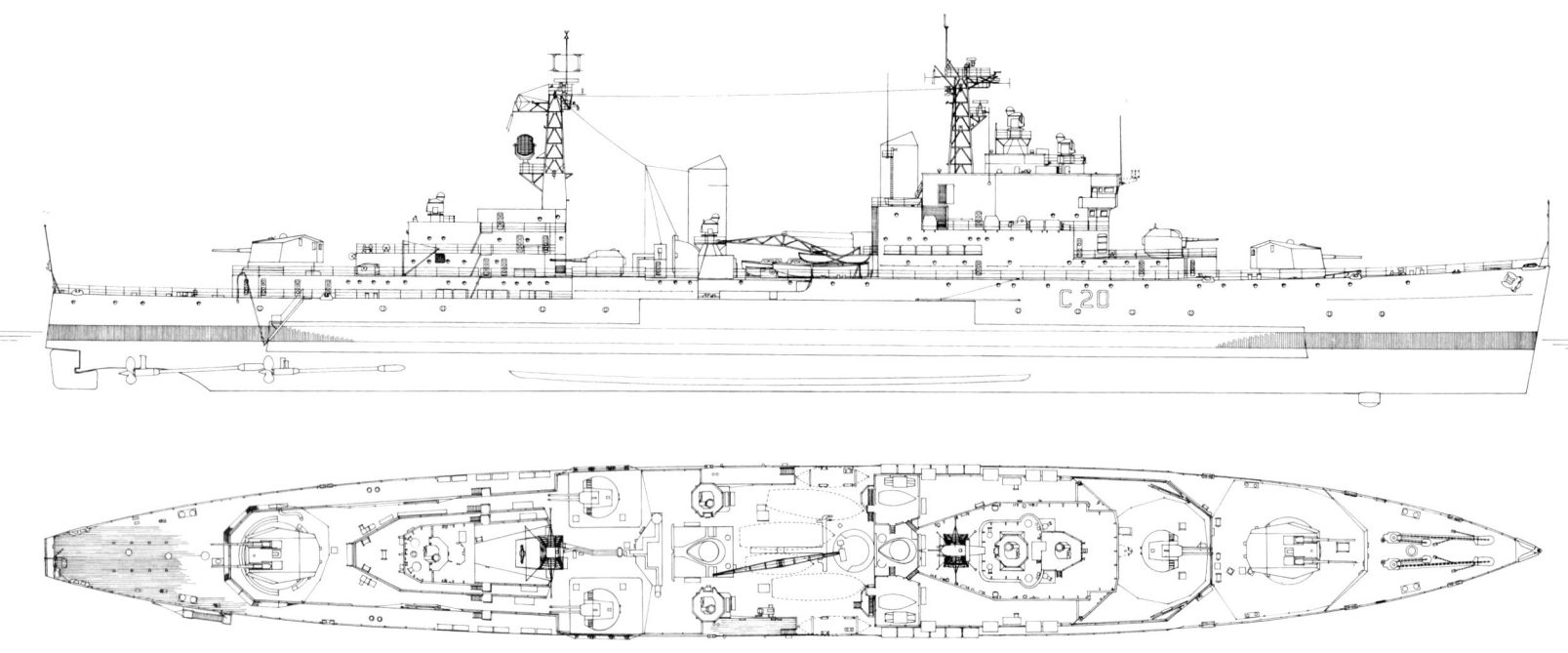

Above: Tiger as completed, March 1959. *Lion* and *Blake* had a raised circular platform at the end of the highest deckhouse on the after superstructure for the magnetic compass; they also had larger bridge wings to the admiral's (lower) bridge, extensions to the forecastle, and trunks on either side of the stacks leading from vent fans atop the boiler space casings. The hemispheric domes atop the five gun directors were later replaced by flat-panelled covers. The MFDF loop and platform shown in dotted lines forward of the bridge were added in 1961. (A D Baker III)

Below and opposite: HMS *Lion*, one of the three *Tigers*, is shown on 17 May 1961 as completed.

POST-WAR CRUISERS

nerable to any ship with heavier guns, such as an armed merchant cruiser. DNC provided Controller with Legends for the two alternative *Tiger* modernisations, including the 3in/70 mounts, in February 1948.[20] DGD proposed placing the four twin 3in/70 in a diamond arrangement ('A' and 'X' positions plus two in the waist), but that was found to be impractical.[21]

The proposals for new gun armament did not immediately affect the 1945 decision that these ships should have quadruple rather than triple torpedo tubes. The tubes were, moreover, a considerable advance on previous practice, as they had RPC. In March 1948 Director of Underwater Weapons (DUW) pointed to a 1946 Board decision approving an eight-torpedo broadside for future cruisers, and suggested that the *Tiger*s were a good place to start (the abortive 1944 design had had this broadside). But if cruisers were mainly anti-aircraft ships, there was little point in so powerful a torpedo broadside. The resulting deliberations by the Ship Design Policy Committee led to a series of policy papers and then to the design of the '1960 cruiser' described below. In June the Board decided to omit torpedo armament from the *Tiger*s altogether, because it would interfere with their new anti-aircraft battery.

The Ship Design Policy Committee asked whether the ships could be fitted with a third Mk 26 mount and with 4.5in rather than 3in/70 guns. There was no question of adding the third mount, but (with difficulty) two twin 4.5in could replace the three twin 3in/70. This must have been disappointing on about two-thirds the displacement of the recently-abandoned *Neptune* design.[22] It appeared that both new mounts might be available by 1953; if given special priority the 6in might be available earlier, as it was the more orthodox of the two. At this stage the modernised *Tiger* was also to carry two 40mm STAAG mountings. The desired four such mounts had to be cut to three.

Given space limitations, the considerable power demands of the two Mk 26 mountings were difficult to fill. Each needed 500kW. DEE proposed adopting the AC power required for the guns, for the entire ship substituting four 750kW AC generators for the four 500kW DC generators of the original *Tiger* design. In that case all electrical equipment on board the ship would have to be adapted to AC power. DEE was already making the case for fleet-wide adoption of AC power, which was being tested on board four of the eight *Daring*-class destroyers.[23] The formal decision that the ships would have AC electrical plants was taken in May 1949. The ships had therefore to be redesigned.

The new mountings would train at considerable speed (38°/sec for the 6in, 60°/sec for the 3in), hence would build up considerable momentum. To stop clear of structure, they would have slow down well before reaching that point. Overall performance would suffer badly. DNO suggested that they be placed so that they could train 360°, i.e. could not collide with superstructure or with each other. With so few gun mounts, each was much more important, hence should have clearer arcs. This idea was reflected in the '1960 cruiser' studies produced at about this time. The requirement presented some problems, because the 6in turrets were located above the magazines originally provided for the single-purpose 6in guns of the earlier design. However, the forward 3in mount could be moved aft. Its after arcs would be more constricted. In theory the bridge structure, foremast, funnels, etc could be moved about 12ft aft, to clear its after arcs, but that would require considerable effort, including removal of structure already in place (DNO and DGD considered

this effort perfectly justified). Another possibility was to transpose the forward 6in and 3in mounts. That would add appreciably to topweight (the 6in mount would be in 'B' position), and the forward and after 6in mountings would no longer be interchangeable, hence would require more design and production work. The 3in mounting would be more exposed to the sea, and also to the blast of the 6in guns firing over it. DNC considered this solution the worst of all. It would be difficult but not entirely impossible to provide the waist guns with all-round training because the centreline deck structures there could not fit into either the superstructure or the crowded hull. DTSD pointed to serious congestion and to the loss of scarce accommodation space, particularly if the superstructure were moved aft simply to clear the arcs of a single 3in mount. He preferred clearing the arcs of both 6in guns and of the waist 3in guns. That was approved.

Something new was emerging in the shape of the Comprehensive Display System (CDS), the first, albeit analog, automated combat direction system. DND wanted both CDS (sixteen-track version) and the associated data link (DPT), but neither would be ready until 1958, and there was also some question as to whether space and weight were available. By September 1954 the Board had approved in principle installation of sixteen-track CDS and DPT on board cruisers.[24] However, CDS did not figure in the final Staff Requirement dated January 1958, which in effect described the ships as completed.

In June 1948 DNO sought authority to order the 6in mountings. If he received it (and the necessary priority) in the 1949/50 budget, the first three mountings could be delivered between September and November 1953, followed by the other three between November 1953 and January 1954. This schedule depended on splitting the order between the two plants (Barrow and Elswick), and thus on buying two sets of tooling. Money was not available, and in 1949 the size and shape of the fleet were reviewed as part of a larger defence review. Given a possible cut in naval spending, completion of the ships had to be deferred, and with it the order for gun mounts. Action was deferred to June 1949, when the future of the ships would be clearer.

The *Tiger* redesign clashed with British mobilisation for Korea, as the limited design capacity had to cover more urgent projects. Moreover, some key items, such as the gun mounts, were not yet completely designed, so the ship design could not be ready until 1951 for Board approval. In March 1951 it seemed that work could resume in the summer of 1952. However, the sheer size of the mobilisation programme made savings essential. One of them was to defer all work, other than on the gun mountings, for a year, from the 1952/3 to the 1953/4 programme.[25] Meanwhile the design matured. In 1952 a new proposal was approved to replace the STAAGs with twin L70 Bofors guns (Mk 11) controlled by MRS 3 (one would be sacrificed if an anti-torpedo weapon materialised).[26] There was also a proposal to restore torpedo armament, this time using the only available space, on the quarterdeck. There was also interest in mounting a single 4.5in star shell gun in 'X' position, in place of a twin Bofors. These new early 1952 proposals somewhat delayed completion of the design. The multiple spaces of an AIO were to be replaced by a combined Bridge Operations Room (BOR) and Aircraft Direction Room (ADR) (about January 1954). The three ships were not completed until 1959–61 and CDS was never fitted to them or to any other British cruiser.

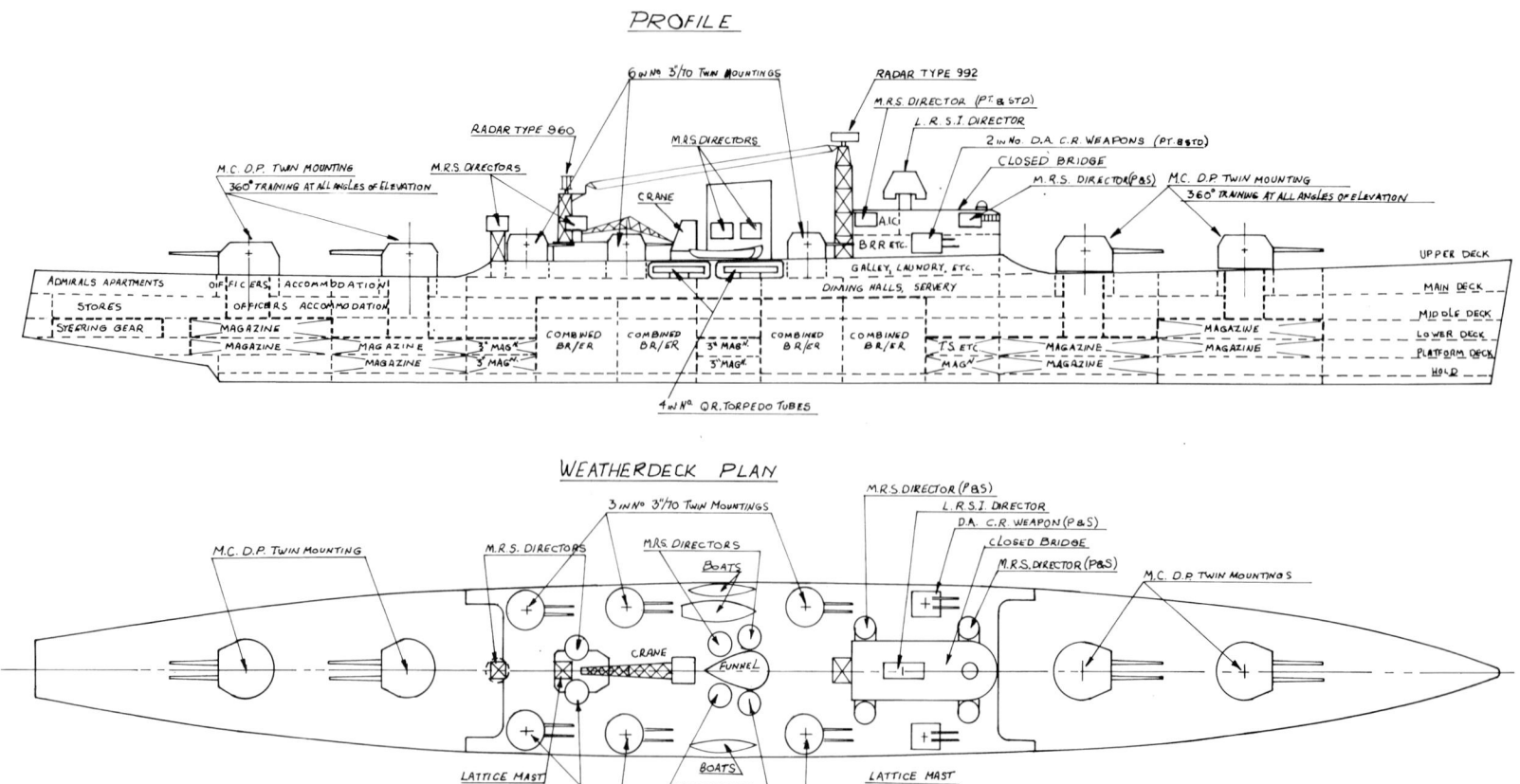

POST-WAR CRUISERS

The '1960 Cruiser'

Given questions raised in the *Tiger* class modernisation, in 1948 the Ship Design Policy Committee reviewed the role of the cruiser. Late in 1948 DNC developed three sketches to show what sort of ship could incorporate the new weapons and meet modern requirements.[27] He saw them as concrete reflections of the Ship Design Policy Committee discussions. In DNC's view the fleet required three kinds of ships: large fast well-protected surface ships with good communications and headquarters accommodation; carriers; and smaller ships including submarines. The country could not afford sufficient numbers of capital ships, so large cruisers had to take over the first role, as small ships alone could not fulfill all duties. DTSD had already argued that cruisers had to be able to protect trade against cruisers, large destroyers and armed merchant ships with speeds of up to 25kts. They had to screen other ships and to protect themselves against air attack. They also had to be self-supporting for long periods. The first requirement reflected British experience,

Opposite: In 1948 DNC sketched a series of cruisers embodying the new weapons and sensors and the new powerplants then being developed. He called them '1960 cruisers' because the full range of new equipment would become available then, and also because it seemed unlikely that new cruisers would be laid down before 1957, for completion that year. This is the large gun-armed cruiser (Sketch II). (A D Baker III)

Below: By 1960 the new naval surface-to-air missile, just named Sea Slug, would surely be ready. This version of the big cruiser, armed with the new missile, can be compared to the full missile cruiser design in the next chapter. (A D Baker III)

particularly in convoys to Russia which faced German surface attackers. The second reflected experience of air attacks in all theatres of war, from the Mediterranean (e.g., the Pedestal convoy to Malta) to the Pacific. The third in effect defined a cruiser. DNC added an offensive function against enemy trade, and noted that at some point ASW weapons might demand cruiser-sized ships. He mentioned that the US Navy was already designing the ASW cruiser *Norfolk*. In her case cruiser size was required so that the ship could operate at high speed in all weathers.

As a gauge of what a medium-term ship might look at, DNC listed equipment with its projected ship-fitting dates (which turned out to be rather optimistic):

6in Mk 26 main battery weapon	End 1953
Medium-calibre DP weapon (perhaps 5in/70)	1957
3in/70 for secondary battery	1953
New DA close-range weapon	1957
LRS I long-range director and TS equipment	1953
MRS III director and predictor	1953
MRS IV director and predictor	1957
TIU Mk III	1954
Type 960 radar	Existing
Type 992 radar	1954
Sea Slug missile	1958

There was no near-term alternative to steam turbines, so DNC assumed that the future cruiser would be powered by the machinery envisaged for the recent *Minotaur* design (an improved *Daring* power-

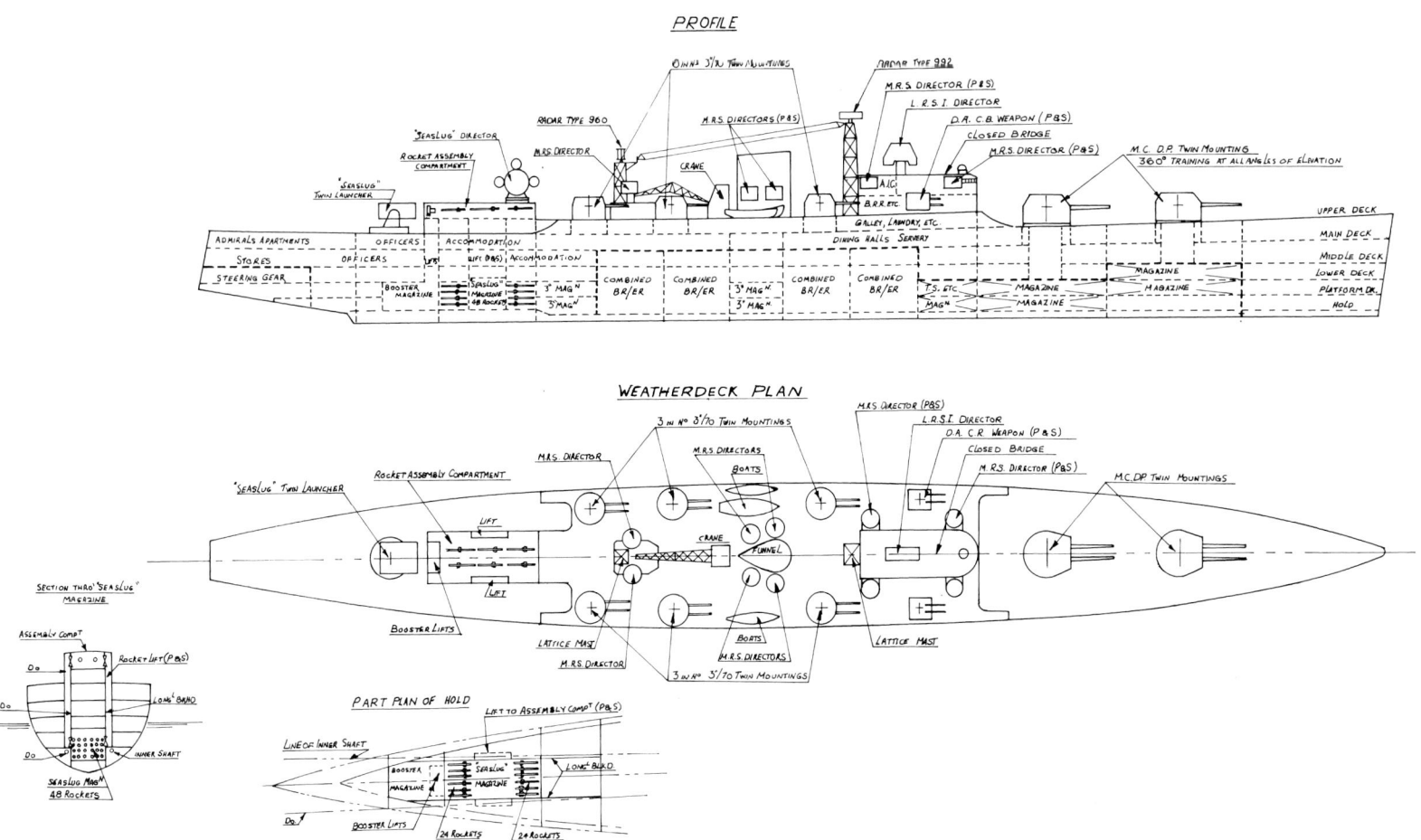

BRITISH CRUISERS

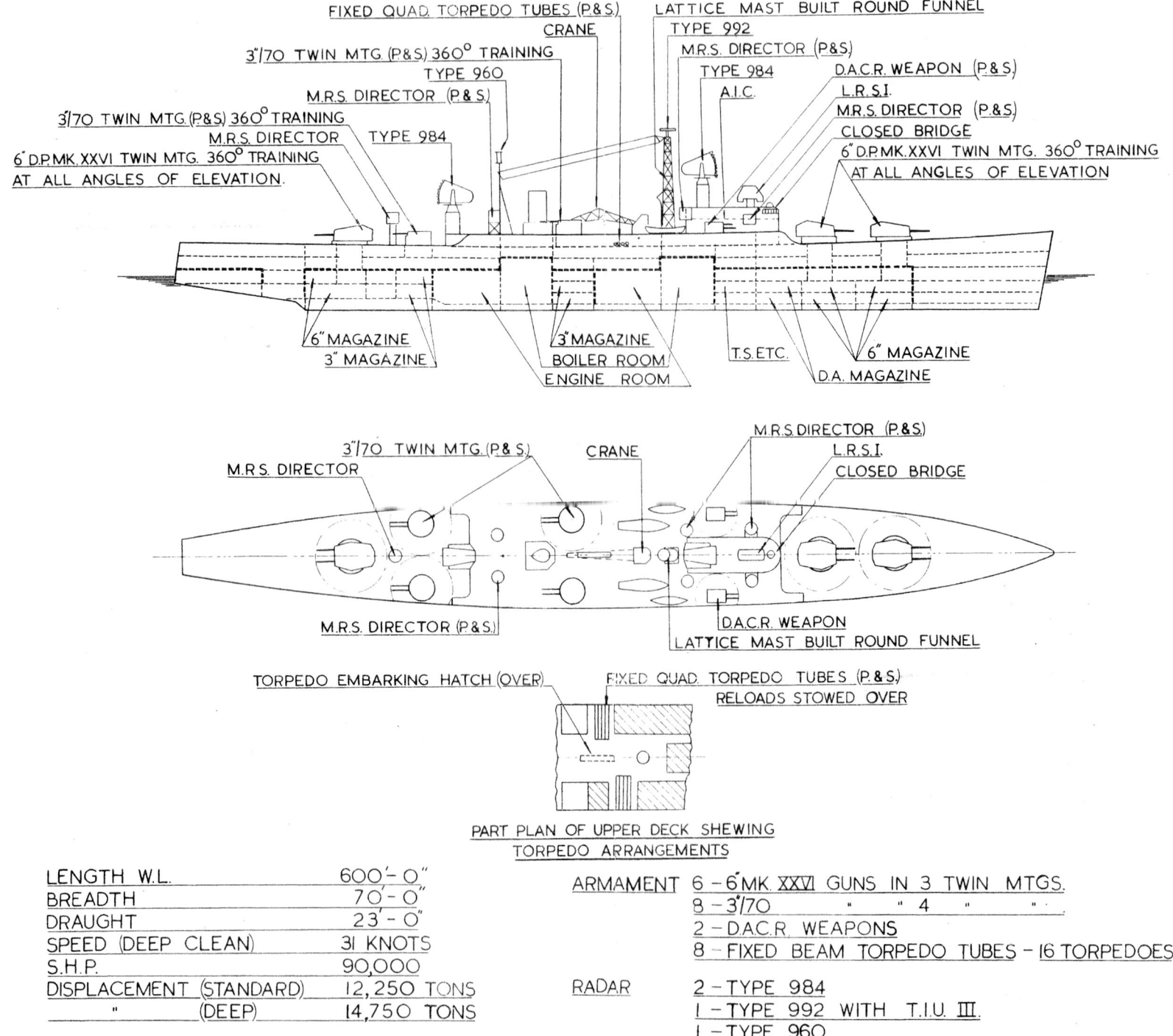

The medium version of the 1960 cruiser with a four-shaft powerplant, Sketch IV. (National Maritime Museum)

POST-WAR CRUISERS

THE MEDIUM CRUISER OF 1960 (3 SHAFTS).   SKETCH V.

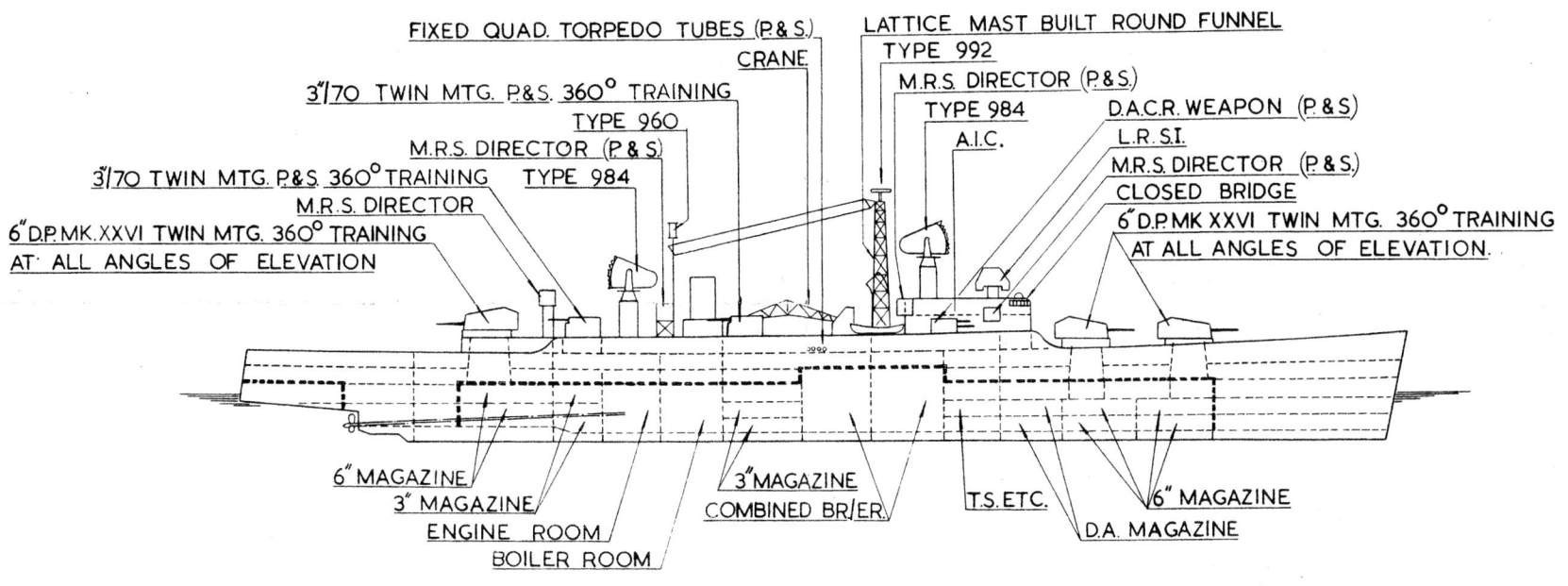

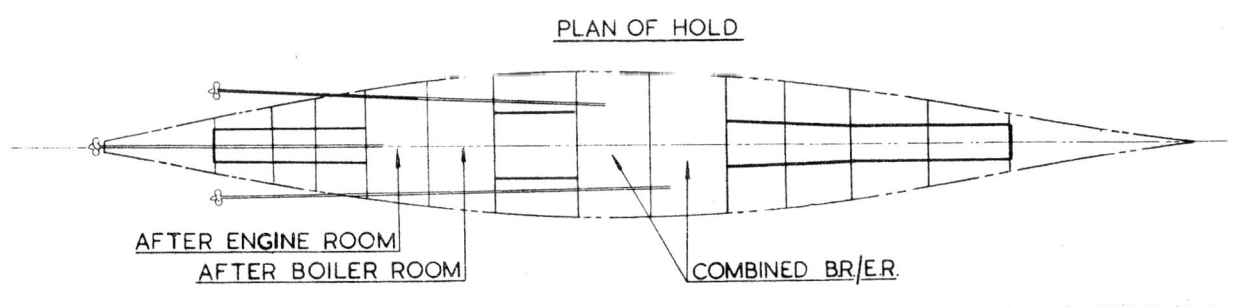

LENGTH W.L.	580'-0"	ARMAMENT	6 - 6" MK. XXVI GUNS IN 3 TWIN MTGS.
BREADTH	70'-0"		8 - 3"/70 " " 4 "
DRAUGHT	23'-0"		2 - D.A.C.R. WEAPONS.
SPEED (DEEP CLEAN)	31 KNOTS		8 - FIXED BEAM TORPEDO TUBES 16 TORPEDOE
S.H.P.	90,000	RADAR	2 - TYPE 984
DISPLACEMENT (STANDARD)	12,000 TONS		1 - TYPE 992 WITH T.I.U. III
" (DEEP)	14,250 TONS		1 - TYPE 960

The medium version of the 1960 cruiser with a three-shaft powerplant, Sketch V. (National Maritime Museum)

The small version of the 1960 cruiser. (A D Baker III)

plant). Funnels could be reduced to one if machinery were concentrated sufficiently.

On this basis DNC offered three alternatives, none including all these features, as they would have made any ship too large. Sketch I was a large cruiser, Sketch II a missile version of Sketch I, and Sketch III a small cruiser.[28] The large cruiser was armed with four twin 6in, six twin 3in/70, two DA (direct attack, also called DACR because they were for close range) weapons, and four quadruple torpedo tubes. DNC arranged his main-battery turrets as far apart as possible and at about the same deck level, to give each all-round training. He considered such a design necessary given the high training speeds envisaged (so the guns could engage fast-moving air targets). This design requirement in turn demanded that superstructure be cut to an absolute minimum, so that the hull had to accommodate much more than in earlier cruiser designs (much the same happens in a modern stealthy warship). DNC estimated that a flush-deck hull could accommodate about 650 men to modern standards. Probably a larger number would be needed, closer to 900 or 950, in which case a superstructure deck would have to be built up amidships and 'B' and 'X' turrets raised, together with secondary and close-range armament. The superstructure deck would also provide space for torpedo tubes on the upper deck. This ship would probably displace 14,500 tons standard (17,500 tons deep).

Wartime experience of multiple targets showed in the sheer number of directors envisaged. Enough directors were provided to give separate control of each mounting on any bearing, as some directors would always be blocked by superstructure. For the main battery, DNC provided one LRS I forward and an MRS IV aft, plus four sided MRS directors (two forward, two aft) for separate control. For the secondary battery, he provided two MRS IV and four MRS III, again allowing the ship to engage numerous multiple targets simultaneously. Gun direction was by TIU III with its Type 992 radar, and the air-search radar was the new Type 960. The ship would have a 3½in belt plus box protection for machinery, magazines, transmitting stations, steering gear, etc. The bridge would probably have to be closed, and adjacent to the AIC.

On 95,000shp the ship should make 31kts 'deep and clean', enough to give 30kts under all conditions. Endurance would be 7,500nm at 20kts 'deep and clean'. DNC placed his machinery in two combined engine/boiler units, based on those in the *Daring* class destroyers, well separated to guard against damage. It might be necessary to reduce to one funnel so as to provide guns with sufficient arcs, at the cost of providing long horizontal uptakes. DNC recognised that future ships would have great electrical loads (at this stage each medium-calibre dual-purpose turret [5in/70, for example] was expected to require 600kW). He therefore expected the ship to require 6,000–8,000kW, compared to a typical wartime load of about 1,200kW.

The Sketch III cruiser followed the same logic as Sketch I but was reduced roughly to *Tiger* armament (two twin 6in Mk 26, three twin 3in/70, four light anti-aircraft guns and four quadruple torpedo tubes). The DA weapons were shown as twin mounts. There would be insufficient beam for the combined turbine/boiler units, and machinery might be arranged as in a *Fiji*, except that the forward and after units would be well separated. DNC described it a an improved version of the modernised *Tiger* (11,700 tons deep at this stage) with a better machinery layout, speed 31kts, heavier close-range armament, and a 2 per cent Board margin. Estimated displacement was 10,500 tons (standard) and 13,000 tons deep.

At its meeting on 27 January 1949 the Ship Design Policy Committee took up the characteristics of future cruisers. Director of

Plans hoped that two ships intermediate between DNC's alternatives I and III could be laid down in the next eight years. The 5in gun then being developed probably would not be available until 1957, so any ship laid down in 1954 would be armed with 6in guns. A 1954 cruiser could not be much more than an improved *Tiger*. DNC wanted better machinery protection, given wartime experience of oil fuel fires. After HMS *Southampton* was lost it was agreed that a belt at the armoured deck level was essential. Director of Plans had hoped to omit side protection to simplify design. Controller thought there should be two types, large and small, as during the war. The Committee invited DNC to sketch a cruiser of intermediate type on the lines suggested by Director of Plans, armed with three twin 6in (Mk 26), four twin 3in/70, DA or Bofors short-range guns and four fixed torpedo tubes on either side, using the new Type 984 three-dimensional radar.

In April 1949 DNC presented a fourth sketch to the Ship Design Policy Committee, meeting the requirements Director of Plans proposed. The most striking new feature was a Type 984 radar nacelle above the bridge, Type 992 being raised on a lattice mast. It had four twin 3in/70 and two DA weapons. It had the other features shown in his earlier Sketch I, but due to the smaller beam it might not be able to adopt the combined engine/boiler room arrangement of the earlier design. Waterline length would be 600ft, and deep displacement just under 15,000 tons, which might make sense since it had 50 per cent more main battery than the 10,000-ton *Tiger*.

E-in-C wanted to standardise on the 30,000shp units used in the *Daring* class, so the ship could have three or four shafts, each with that power. DNC offered both 120,000shp and 90,000shp versions of his ship. In a three-shaft ship, he could combine engine and boiler rooms for the two outer shafts. As in earlier attempts at three-shaft cruisers, the magazine for one twin 6in mount would present problems, and it might have to be raised.

Director of Plans wanted protection for magazines, machinery, steering gear, etc but no belt (to alleviate design problems). DNC made the argument against such armour, that few cruisers suffered shell damage, because the enemy (except in the Pacific) had only weak surface forces. Only three British cruisers were sunk by shellfire (two by the Japanese), nine were more of less badly damaged, and twenty-two were lightly damaged. Of seventy-five known hits, only five were on side armour, which stopped all of them. Had the shells penetrated, they would have caused serious damage (*Berwick*'s belt protected her machinery from an 8in hit). One other shell was stopped by a 4in vertical magazine bulkhead. Six US cruisers took a total of 209 hits, only four of which were on belt armour (US belts were lower, because US boilers were not as tall as British ones), none of which penetrated. Thus it might be thought that belt armour was not worth the weight involved, but DNC cautioned that it would be unwise to leave a ship as expensive as a large cruiser vulnerable to the smallest-calibre projectiles, including aircraft rockets, or to splinters from near-miss bombs. Belt armour protected not only a ship's vitals but also her stability and buoyancy, and it formed the sides of the citadel covering the vitals. He therefore wanted at least 2in of side armour. If the belt were this thin, magazines should have box protection to bring their total side protection to at least 3in. Deck armour should be of similar thickness, though it might have to be thinner in places to limit the weight involved. All offensive capability, including AIO and important cable leads, should have splinter protection at least. Like the earlier large and small cruiser studies, this medium cruiser study envisaged protection on the scale of the 1946 *Minotaur* design, including a 3½in belt. The weight involved could now be redistributed, but it would not change very much.

The Emergency Cruiser

The British mobilisation plan included four repeat (actually modified) *Dido*s, because the 5in cruiser-destroyer design was not yet ready, nor was its gun and its powerplant.[29] As of March 1951, it was understood that in the event of a larger war the ships would be laid down between April 1952 and April 1953 (i.e. during the 1952/3 financial year), to complete in 1954–5 (another note said that they would be laid down only if there were an emergency before April 1953). The natural choice for fire control was the new MRS 3, but it would not be ready in time. DNO therefore proposed the best existing system, using Mk 6M directors and the FPS 5 (Flyplane) computer, with the Admiralty Fire Control Box Mk 10 to provide surface fire-control. The directors would replace the DCT forward and the HA director aft. In addition, for individual fire control, the ships would have two CRBF directors, one replacing the HA director forward and the other abaft the mainmast on the centreline. Guns would be switchable in pairs between the Mk 6M/FPS 5 system and the CRBF. DNO considered the two CRBFs well worthwhile for their blind fire capability and because they gave the ship four-cornered anti-aircraft capacity. Three STAAGs would replace the three pompoms of the wartime ships: one forward and one each side amidships.

To make up for stability lost to wartime additions, DNC wanted another foot of beam (which he said would not delay construction). The ships could then accommodate their wartime battery of four twin 5.25in. They could accommodate four twin 4.5in Mk VI (as in *Daring* class destroyers) without any change in hull design. There was originally hope that they could have five twin 4.5in, but DNC said that was impossible without complete redesign, which would take far too long. The 5.25in gun and mounting were long out of production, so any mounts would have to come from existing ships, without replacement: from the *King George V* class battleships recently laid up or from existing *Dido*s (in which case some redesign would be needed). It seemed pointless to restart production of an obsolete weapon. DGD considered the 4.5in a better anti-aircraft weapon, as the 5.25in was limited to eight rounds per minute in practice (though it might be modified, and then might perhaps achieve twelve). The 5.25in had such poor armour penetration that policy was to withdraw its AP shells and substitute 100 per cent HE, to damage an enemy's upperworks. This was still better than the 4.5in, but the difference was likely to be less than the ratio of shell weights (80lbs to 55lbs). Against a surface target, the 5.25in did enjoy a slight range advantage, but the 4.5in fired considerably faster. DTSD plumped for the 4.5in gun. Nothing more was done at this time. DNC was not to go any further without specific approval.

Below and overleaf: HMS *Blake* is shown as newly converted (from aft) and off Gibraltar in May 1975. Initially it seemed impossible for these ships to operate Sea King helicopters (as in the view off Gibraltar), because there was not enough length aft to accommodate a long enough hangar and a long enough flight deck, as long as the two waist 3in guns were retained. The solution was to extend the hangar forward and to replace the guns with Seacat point-defence missile launchers.

CHAPTER 11
THE MISSILE AGE

Gun Cruisers

Design work on a large cruiser began in about the spring of 1951, based on the large cruiser sketched in 1948–9.[1] Work was loosely based on the *Minotaur* design, but the assumed armament was four twin 5in and six twin 3in/70 (discussion of the relative merits of the 6in and 5in guns was, however, continuing). Armour weight was recalculated on the basis that instead of jogging up to cover the boilers, the deck armour would all be at the same height, the belt being raised to match. Armament of this Sketch I design would have been four twin Medium Calibre Dual Purpose (MCDP) (later 5in), six 3in/70, two DA guns (as yet undefined), controlled by one LRS 1 system, five MRS directors for the 5in guns, and six for the 3in/70s, with TIU III/Type 992 radar and Type 960 for air search.[2] Like *Minotaur* and *Neptune*, the ship would have four quadruple torpedo tubes. Deep load displacement would have been 17,350 tons (14,850 tons without 2,500 tons of oil fuel and reserve feed water).

By this time E-in-C was developing standardised high-temperature high-pressure (hence lightweight) machinery: for a large cruiser he offered 30,000shp YEAD(1) units (it would take disproportionate effort to go to 25,000shp or to 40,000shp). Each unit would combine a boiler and a turbine. Except for the aftermost unit, the boiler could be kept below the lower deck by placing it between the shafts. Each unit could also accommodate a 1,000kVA AC turbo-generator. Given the nuclear threat, the favoured arrangement was a closed-front boiler which took its

air in through a downtake surrounding the uptake (which would have a flap to filter outside air). The machinery space would be arranged to circulate its air so as to avoid contamination, and to be completely independent of boiler air. E-in-C thought he could provide 120,000shp on 1,900 tons, against which the wartime design of the large US cruiser *Worcester* required 2,491 tons for the same power.

In May 1951 work began on a 'New Armoured Cruiser' (CR 2, CR 1 presumably being Sketch I).[3] DTSD had recently issued a Staff Requirement for a ship with four twin 6in and four twin 3in/70 (sided). Given added armament weight, the ship would probably displace 19,000 tons deep. That was very close to *Neptune* (18,740 tons deep), so much the same dimensions could be used. Estimated dimensions were 650ft x 70ft x 26ft, with a hull depth of 41ft (as in Perry's 5in cruiser – whose depth had been chosen to suit the 5in mounting as initially understood). Available weight would provide *Neptune* side and deck armour (4in belt, 1½in deck), with 600 tons left over to protect the AIO and bridge (say, 2in over AIO and 1½in over the bridge). As an indication of how important command and control was, the box to protect the AIO would have been 60ft x 100ft x 18ft (two decks high). Skeptical of E-in-C, Paffett took 2,100 tons as his machinery weight, and sought an endurance of 9,000nm at 20kts.

The Staff considered CR 1 and CR 2 too big; it wanted nothing above 15,000 tons. At a meeting on 30 June 1951 further designs were ordered. The Staff wanted each to have either two of the big new three-dimensional Type 984 radar or one Type 984 and one Type 960 air-search set. The project was now called the New Design Armoured Cruiser (NDAC), and the new versions were CR 3 and CR 4. CR 3 would have four twin 5in but no 3in at all, with four of the new DA close-range weapons and four torpedo tubes (instead of eight) on each side. CR 4 would be larger: three twin 6in, three twin 3in/70, four close-range weapons, and four torpedo tubes on each side.[4] Estimated dimensions were 625ft x 70ft (no dimensions were calculated for CR 3). In addition to the weapons, CR 3 would have one Type 984, one Type 960 air-search radar, the GDS 3 gun-direction system (including TIU III/992), and four MRS 3 for main battery control. Estimated deep displacement was 13,500 tons. If the only difference between CR 4 and CR 3 was armament, the latter would probably displace about 16,000 tons deep. These were not really surprising figures, because it took 11,700 tons (deep) for a *Tiger* to accommodate two twin 6in and three twin 3in. Worse, at 16,000 tons CR 4 would be very inadequately protected, because only 1,790 tons was likely to be available for that purpose. Because the design was being worked on the basis of deep displacement, to some extent fuel could be traded for protection; minimum endurance was 5,000–6,000nm 'deep and dirty' (the 1,790-ton figure was associated with an endurance of 9,000nm clean and 6,820nm 'deep and dirty'). It turned out that a shorter ship could devote more tonnage to protection (2,300 tons for a 600ft ship), but only 1,550 tons for a 675-footer, hull weight increasing with length.[5] Maximum speed (for the fixed power) rose with length, too. The question was which was most important.

In an attempt to cut weight further, in July H S Pengelly, who was in charge of the project, added three more versions, displacement not to exceed 10,000 tons and speed to be 30kts deep: CR 5 (four single 5in), CR 6 (two twin 6in, two twin 3in), and CR 7 (three twin 5in). CR 7 would have been 536ft x 65ft, CR 8, 600ft x 65ft. Each would also have four torpedo tubes on each side plus reloads (a new feature), and the usual quartet of close-range DA weapons. Pengelly guessed well; it seemed that all three could be accommodated within 10,000 tons, CR 5 possibly offering embarrassingly little armament on that displacement (contemporary cruiser-destroyer designs offered three 5in on about 4,750 tons). All could probably make 30kts on three machinery units, but a three-shaft design made for awkward magazine arrangements (as had been made obvious in the 1930s). In fact E-in-C had some further units in mind; he planned to develop a 45,000shp unit for a large carrier (in which case the ships could have twin screws; but no work had yet been done). He also had a 15,000shp Y100 unit for frigates, and a ship could have two 30,000shp units and two 15,000shp. E-in-C preferred using three 30,000shp units, which were also planned for the cruiser-destroyer. He would refuse to design any non-standard unit. There was also some good news. The weight of the YEAD(1) unit had been grossly overstated, the correct figure being 25lbs/shp, or 335 tons per unit and 1,440 tons, rather than 1,900 tons, for a four-shaft installation. As a measure of progress, a 'County' class cruiser needed 1,800 tons for 80,000shp and a *Fiji* about 1,500 tons. On this basis it would have taken 1,690 tons (rather than 1,005 tons) for 90,000shp and 2,250 tons for 120,000shp (the designer allowed 400 tons per unit, to be safe). CR 6 was interesting because the armament lent itself to a short hull, even if the two twin 3in were placed on the centreline. A version with a 525ft x 65ft hull was tried, but it offered very little weight for protection (the 'long' version, 560ft, offered even less).

In a letter dated 31 October DTSD stated that the Naval Staff highly valued speed, and was prepared to pay for 34kts 'deep and clean'. A new CR 8 was therefore prepared based on CR 6, to make 34kts on four shafts, with an endurance of 4,500nm at 20kts. Could this be done on 11,000 tons deep? That machinery came in discrete units forced up the ship's length, since she would obviously need four units. She could not be much shorter than CR 4, despite the much lighter armament.[6] On the desired displacement, she would have virtually no armour. The next attempt was CR 10, the same ship on 12,000 tons (i.e. also 600ft x 65ft), her speed therefore reduced to 33.6kts 'deep and clean'. Given the additional displacement, the ship needed more fuel oil to make her required endurance, so the increase in armour weight was considerably less than the thousand tons added. This was not a particularly good bargain. The next step (CR 12) was 13,000 tons (still on a 600ft x 65ft hull). It finally offered much more armour weight (1,558 tons instead of 883 tons in the 12,000-ton version and only 212 tons in the 11,000-ton version).[7]

Late in November 1951 DTSD decided to prepare a table of possible cruiser types for Board action, comparing heavy and medium cruisers. The 16,000-ton CR 4 (three twin 6in, three twin 3in) was therefore revived. In this version it had sixteen torpedo tubes but no Type 984 radar. Armour had to be reduced to a 2½in side, 1½in deck, and 1in bulkheads (total protection weight was 1,730 tons). The long-hull version of CR 6 was re-evaluated as the medium cruiser.

Orders were given late in December 1951 to investigate CR 11 (three twin 5in) 'to the extent of getting a plan of weather deck and a plan of hold. The Staff will probably come down in favour of the 5in ship.' This ship would not have any 3in armament, just the four DA weapons and two quadruple torpedo tubes, with an MRS 3 for each twin 5in gun. Data were derived from those for CR 10: 600ft, 12,000 tons.

Further attempts (in January 1952) were CR 14 and 15, 6in and 5in ships displacing 12,000 tons but with three machinery units. In effect CR 14 was CR 6 with displacement increased from 10,000 tons to 12,000 tons. On 560ft length she would make 31.4kts 'deep and clean'. Most of the extra tonnage would go into armour: 1,802 tons instead of 352 tons. By way of comparison, the 12,000-ton CR 10 offered 883 tons of armour. On this tonnage, CR 10 could not even provide a 1⅔in deck once allowance had been made for steering-gear protection, and she could barely manage a 1in side and 1in deck. CR 14 could have a 2½in belt and 1½in deck.

At this point NID supplied data on the new Soviet cruiser *Sverdlov* (assumed deep displacement 18,000 tons), and the British cruiser designers tried to reverse-engineer to estimate her armour. They got a 4in belt and 2½in deck. That was not too bad; the belt was actually 3.9in (100mm) but the deck was only 1.9in (50mm). The sheer size of the *Sverdlov* probably made a big British cruiser seem more acceptable.

The CR series might be seen as high-end cruisers. However, DTSD was also interested in a much smaller light cruiser armed with two twin 5in, in effect an enlarged version of the cruiser-destroyer (three single 5in) then being designed.[8] Initially DNC ordered the ship designed to destroyer standards, but in April 1953 Controller decided that it should be developed instead to cruiser standards. The result was something like a *Dido*. Displacement rose from the 4,750 tons of the earlier cruiser-destroyer (in effect a super-destroyer) to something like 7,000 tons.[9] It made sense to scale up to three twin 5in mountings and three machinery units (575ft x 65ft, 9,900 tons). Later displacement was given as 11,500 tons. With four guns and the four-unit machinery, displacement rose to 16,300 tons.[10] In effect these calculations confirmed some of the conclusions of the 1951–2 studies.

By June 1953 DTSD wanted a range of data for decisions as to the future building programme. One was a cruiser armed with four twin 6in and four twin 3in, with two quadruple torpedo tubes, a sped of 30kts, and an endurance of 4,000nm. An alternative cruiser would be armed with three twin 6in and three twin 3in/70, with the same torpedo tubes, speed and endurance.[11] The four-mount ship was likely to be 675ft long and displace 18,000 tons; the three-mount ship, 660ft and 16,950 tons. In both cases there was enough weight for reasonable protection (about 2,200 tons and 2,100 tons respectively, enough for 2½in side armour and 1½in deck armour).

These and other studies were collected about mid-1953 into a table of 'Existing and Possible Cruisers, Destroyers, and Fleet Escorts' as guidance for the Sea Lords as they prepared the future construction programme. The table took no note of the cancellation of the 5in gun. Seventeen different types were included, beginning with a 570ft, 30kt, 15,000-ton missile ship armed with one twin Sea Slug launcher, two twin 3in/70 and six twin Bofors. It would cost about £10 million. The larger of the two heavy cruisers (four twin 6in, four twin 3in, 18,200 tons) would cost about £13 million, the smaller version (three twin 6in, 16,850 tons) £12 million. The next step down was a pair of 660ft, 16,000-ton ships: one with four twin 5in (and six sextuple Bofors) was another step down (16,000 tons, 31kts, £11 million), and one with two twin 6in and four twin 3in/70 (and four twin Bofors: 16,000 tons, 31kts, £11.5 million). An alternative ship with four twin 5in and six sextuple Bofors, but with reduced speed (29kts rather than 31kts) could be built on a length of 620ft (13,000 tons: £9.5 million). A new *Tiger*, 550ft long (12,000 tons, 29kts at deep load) would cost about £9 million. The 450ft, 7,900-ton cruiser version of the cruiser-destroyer with two twin 5in and four twin Bofors would make 28kts, and would have an endurance of 3,500nm rather than the 4,000nm of the larger cruisers; it would cost £6.5 million. The more destroyer-like version of the same ship would have somewhat less endurance (3,000nm) on 4,750 tons (435ft) but would be somewhat faster (29.5kts deeply loaded); it would cost £5.5 million. All of the cruisers, but not two cruiser-destroyers, had the Camrose anti-torpedo weapon.

British Cruisers

Cruiser Designs for Export

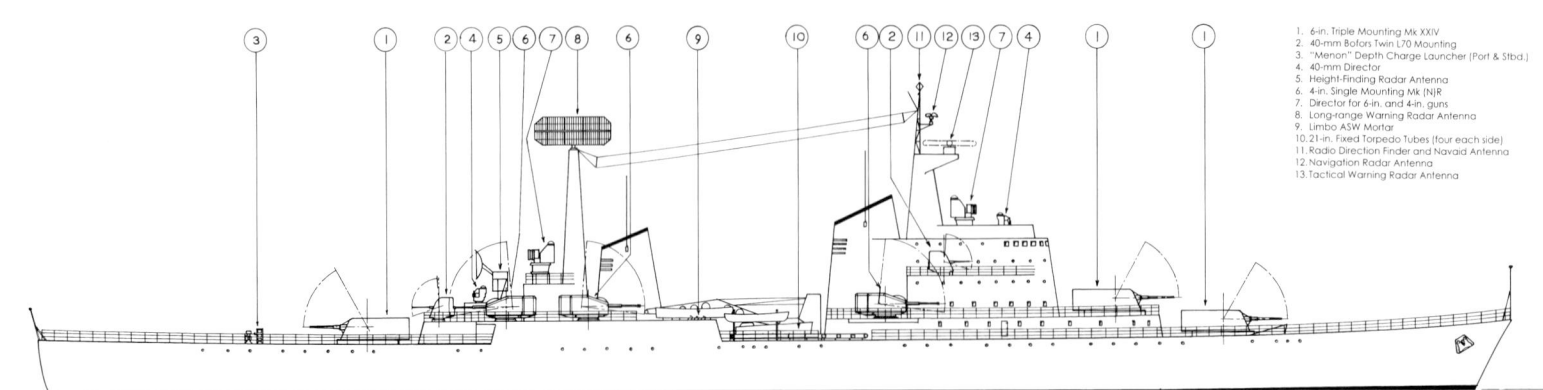

Above: This design marked the beginning of serious attempts by Vickers-Armstrong to sell new cruisers to Venezuela as a second phase of the naval expansion programme begun with the *Zulia* class destroyers. From about 1948 onwards Venezuela was interested in buying or building a *Dido* class cruiser. The cruiser programme was the personal project of dictator Perez Jiminez, whom the British Foreign Office alternately disliked and courted. By about 1955 he was personally negotiating with Vickers-Armstrong for a carrier and two cruisers, at an estimated cost of £40 million. He also approached firms in France and Italy. The US government opposed the Venezuelan programme, on the ground that heavy expenditures on such showy warships would weaken the local economies unduly (it may have seen transfers of older cruisers to the 'ABC' powers as a way of dissuading them from spending heavily to buy new ships). After the United States gave two cruisers to each of the 'ABC' powers, it seems that the Venezuelans wanted something even more impressive, in particular with guns more powerful than the US-supplied 6in/47. Vickers wanted to offer the new twin Mk 26 dual-purpose gun which armed the rebuilt *Tigers*, but export was rejected (it was not scheduled for declassification until 1962), not least because it would have slowed the British programme. The alternative offered to Vickers-Armstrong was the Mk 24 of the original *Tiger* class, twelve of which were stored, partly complete, at Rosyth (one 80 per cent complete, seven 75 per cent, one 50 per cent, and three 20 per cent). This sketch shows the ship so armed, with the new Vickers 4in gun (Mk N) designed for the Chilean *Almirante Williams* class destroyer as secondary weapon. The Bofors guns are the new 70-calibre version which the Royal Navy planned to adopt (but which was dropped in favour of the Seacat missile). Directors are unidentified, but may be proprietary Vickers types. Similarly, the big search antenna is unidentified, but its shape suggests a non-British type. The other radars may be the types used on board the Chilean destroyers, which were equipped entirely by Vickers. It is not clear why the ship was to have had the Italian Menon depth-charge mortar or its British equivalent, Limbo, but before the Second World War Vickers generally provided depth-charge throwers in its cruisers. The ASW weapons may have been an extension of the standard British practice of providing cruisers with small depth-charge tracks as a means of driving off submarines. The new design had a 120lb (3in) belt covered by 40lb (1in) decks and closed by 60lb (1½in) bulkheads. Barbettes were 40lb (1in). Steering gear was protected by 60lb deck, side, and ends. The Venezuelans rejected the design with Mk 24 turrets about September 1955. The new Venezuelan designs could be distinguished visually from Vickers' 'stock' cruiser (Design 1124) by their more sharply raked bows and their cruiser (rather than transom) sterns. Nominal displacement was 8000 tons; unfortunately surviving records seem not to provide further data. (A D Baker III)

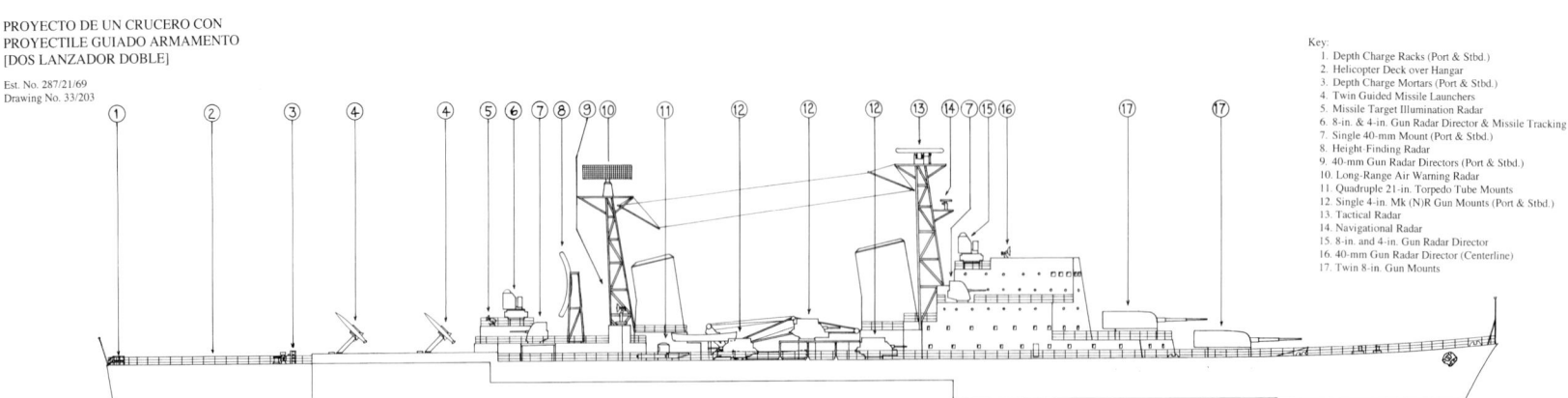

Above: Vickers-Armstrong had something more exciting to offer: the first missile cruiser in Latin America, armed with the Swiss Contraves-Oerlikon anti-aircraft missile, the first to be offered commercially. The missile cruiser was adapted from the heavy cruiser, with the same type of protection. Its belt stepped up to cover the boilers and then stepped up again to meet the upper deck abreast the missile launchers. The MoD (all-service) Arms Export Working Party agreed in May 1956 that the company could proceed with the design. The three ships were expected to cost £40 million. The 1957 papers of the working party show that the British government considered that it could not approve the carrier without consultation with the US government, but that it approved the cruiser plan in December 1957. The working party decided to proceed, on the grounds that Venezuela could afford the programme given its oil income. It did think that the Admiralty should be consulted concerning the strategic implications of the programme, and the Commonwealth Office might be concerned because of possible implications for South African security. Two weeks after the project was provisionally approved, Jiminez fell in a coup, and his personal big-navy project collapsed. (A D Baker III)

THE MISSILE AGE

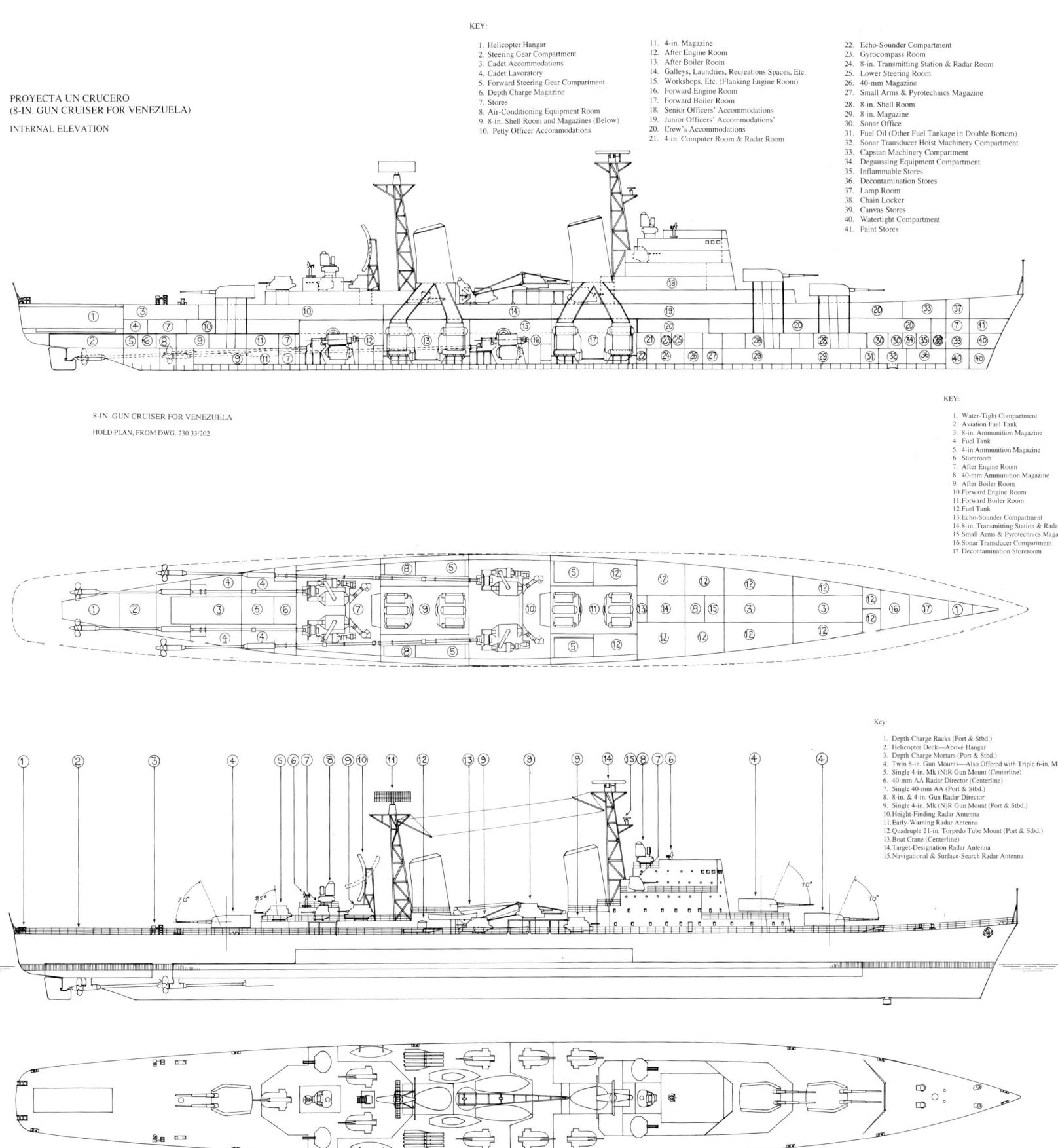

After the Venezuelans rejected the ship with Mk 24 turrets, Vickers offered the twin 8in/50 sold to Spain for the *Canarias* class. When it became available, the Mk 26 could replace these weapons. It reported this initiative to the British Naval Arms Export Working Party in October 1955. (A D Baker III)

KEY:

1. Helicopter Hangar
2. Steering Gear Compartment
3. Depth Charge Magazine
4. Aviation Fuel
5. Missile Checkout Room
6. Galleys, Laundries, Recreation Spaces, Etc.
7. After Engine Room
8. Electric-powered Missile Transport Cart

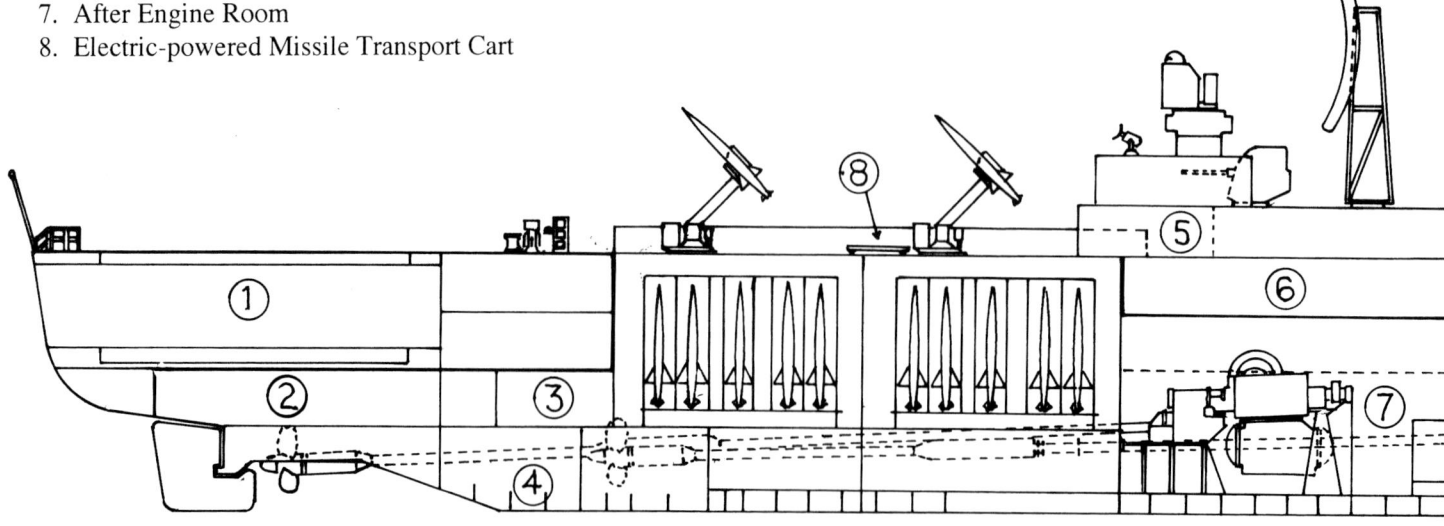

Above and below: The Vickers-Armstrong missile cruiser would have been armed with a Contraves-Oerlikon surface-to-air missile. The missile was never installed on board a warship. Work on this weapon began in 1947, and development was completed in 1950 (tested in France) under the designation RSC-50; it was the first anti-aircraft missile to be commercially available (the slightly improved RSC-51 was evaluated by the US Air Force in 1952 [programme MX-1868]). Later versions (RSC-54, -56, -57 and -58) were bought by Switzerland, Italy, Sweden and Japan, but were used for training. All were liquid-fuelled. At least the later versions had tail fins (RSC-50/51 had four mid-wings but no tail fins; the wings stabilised the missile, which was steered by deflection vanes in the exhaust). RSC-51 was 5m long and 40cm in diameter (weight 250kg, with 20kg warhead) and had a range of 20km (ceiling 20,000m) and a speed of 750m/sec (2,460ft/sec). Beginning in 1959 Contraves worked on a solid-fuelled version, MiCON (Missile Contraves), but it enjoyed no commercial success. (A D Baker III)

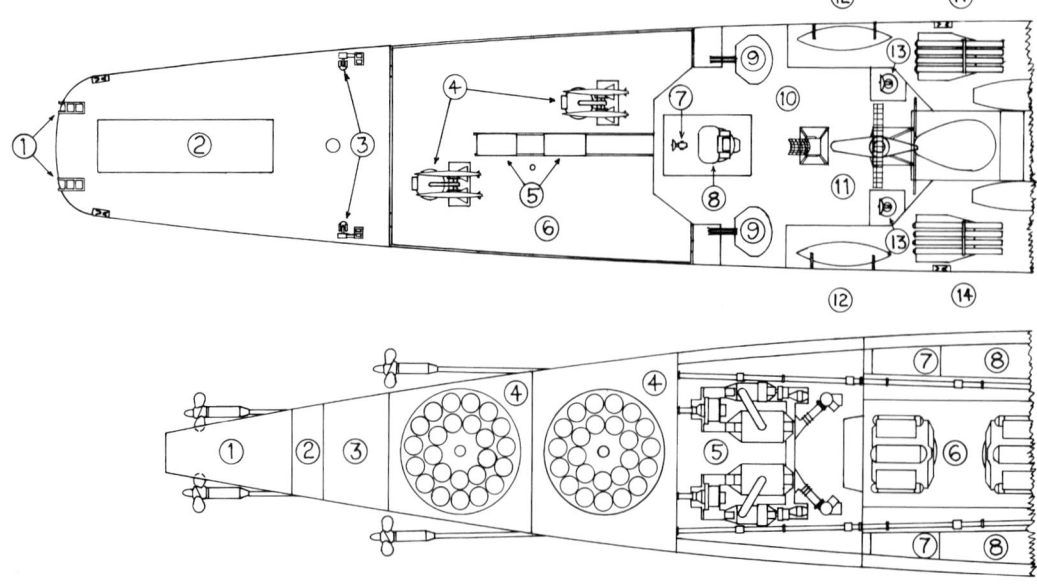

PROYECTO DE UN CRUCERO CON PROYECTILE GUIADO ARMAMENTO
[DOS LANZADOR DOBLE]

Main Deck and Platform Deck Plan Views

Key:
Main Deck:
1. Depth Charge Racks
2. Helicopter Elevator
3. Depth Charge Mortars
4. Twin Guided Missile Launchers
5. Missile Trolleys (Rails to Checkout Room on Main Deck)
6. 4-ft. Blast Protection Bulwark around Missile Launch Area
7. Missile Target Illumination Radar
8. 8-in. and 4-in. Fire Control and Missile Target Tracking Radar
9. Twin 40-mm Gun Mounts
10. Height-Finding Radar
11. Air Early Warning Radar
12. 26-ft. Motor Whaleboats (Not Shown in Elevations)
13. 40-mm Gun Fire Control Directors
14. Quadruple 21-in. Torpedo Tube Mounts

Platform Deck:
1. Steering Gear Compartment
2. Forward Steering Gear Compartment
3. Depth Charge Magazine
4. Revolving Stowage for 24 Missiles
5. After Engine Room
6. After Boiler Room
7. 40-mm Magazine
8. 4-in. Magazine

Overall, detailed design work on gun cruisers lapsed in 1952–4 as effort shifted to the 18kt missile ship.[12] However, in April 1954 DNO told a senior constructor, W G John, that although the first guided weapon ship would probably commission in 1961, it would be 1970 before missiles were in full use. Until then the most effective anti-aircraft gun would be the 3in/70. For a light cruiser, DNO considered the twin 5in best. The new twin mount designs then under consideration would be less complicated and lighter than the designs under consideration up to mid-1953. Favoured armament for a light cruiser would be three twin 5in and four twin Bofors Mk 11 (with two MRS 3 to control the 5in guns and two CRBF) or three twin 5in, two twin Bofors, and one sextuple Bofors (with two MRS 3 for the 5in guns, two Tachymetric One-Man (TOM) directors for the Bofors Mk 11, and MRS for the sextuple Bofors Mk 12). The ship should be fin-stabilised. DNC wanted to know what such a ship would be like. This sort of armament should fit a *Dido* hull. John guessed that the new light-weight twin mount would weigh 150 tons with a ¾in shield. He estimated that the ship would be 550ft long, displacing 8,000 tons (it had to be lengthened to accommodate the estimated complement); with two of the new Y102 Combined Steam and Gas (COSAG) units (30,000shp each) it would make about 29.25kts deep. By paring down protection (to 1in belt and deck over magazines, and ½in belt and deck over machinery) the ship could be cut down to 525ft and 6,750 tons, a dead minimum, which John reported to DNC. The 8,000-ton design was reported to the Sea Lords, as part of a spectrum of possible cruisers, in November 1954.[13]

The existence of the *Sverdlov*s must have been chilling to a Royal Navy which had fought German surface combatants under conditions in which aircraft were grounded a decade earlier. In November 1954 Director of Plans argued that new gun cruisers should be included in the long-range programme specifically to deal with the Soviet cruisers. The only suitable gun was the 6in. The *Sverdlov* problem presumably killed the twin 5in, and made the big cruisers with twin 6in guns attractive.

Detailed work on gun cruisers resumed in November 1954, with the emphasis on high-end designs. The CR series seemed inadequate because machinery weight was too optimistic (unless gas turbines were adopted) and because equipment weight seemed inadequate at 5 to 6 per cent (a more typical figure would be 7 to 8 per cent of deep tonnage). The 18,000-ton *Minotaur* (1947) was still the only recent cruiser design worked out in enough detail to form the basis of a new design. It was therefore designated CR 16. Although as yet there was no formal requirement for an all-gun ship, the constructor assigned to the project wrote that 'we suspect that there may soon be'.

It was worthwhile to re-think the last of the CR series. CR 17 (three twin 6in, four twin Bofors) would have a two-shaft COSAG powerplant: one 30,000shp YEAD(1) and two 7,500shp gas turbines on each shaft. The goal for protection would be 3in side and 2in deck, with a 2in deck over the AIO spaces. This time the ship was to have the big Type 984 radar, with CDS and its DPT data link, plus GDS 3 with Type 992.[14] The constructor planned to start with a 560ft x 70ft hull. On this basis the ship would displace about 12,000 tons, and would make 31kts 'deep and clean' (30.25kts 'deep and dirty'). As in the CR series, this was not enough to provide the desired level of protection. CR 17 ended up as a 14,000-ton ship (600ft x 74ft x 20ft, hull depth 42.5ft).

Missile Cruisers

By this time there was some interest in a combination cruiser and missile ship.[15] It was assumed that two twin 6in and two double missile

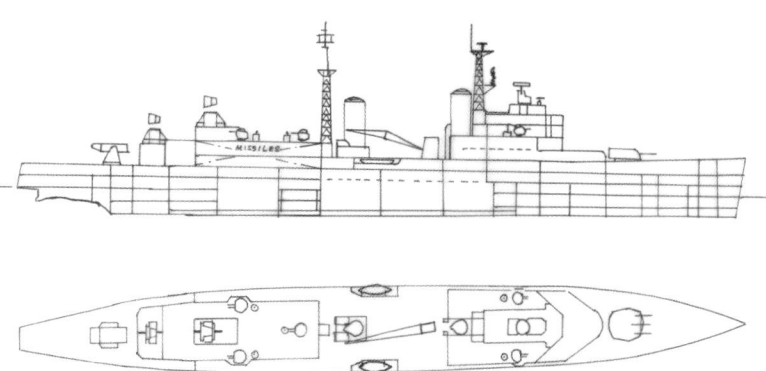

A *Fiji* class cruiser converted to a missile ship, as sketched in October 1954. This design can be compared to roughly contemporary US conversions of *Cleveland* class cruisers. The *Fiji* would have retained 'A' turret. She would also have had three twin Mk 5 Bofors, the type then replacing pompoms on board such ships, and twenty-four Sea Slugs. Deep displacement would have been 10,950 tons (draft 20ft 3in), and her 80,000shp powerplant would have driven her at 30.5kts deep and clean (29.5kts deep and dirty, i.e., six months out of dock, but in temperate waters). Operational endurance would have been 4,900nm at 20kts. She would have had two missile-guidance radars (Type 901), but no big three-dimensional Type 984, leaving her somewhat inadequately equipped. Search radars would have matched those on board a 'County' class destroyer: Type 960 for long-range air search, Type 992 for target indication, Type 277Q for limited height-finding, and Type 974 for navigation. The 6in turret would have retained its existing DCT, with its Type 274 radar; the Bofors guns would have had local directors (STDs). Protection would have been that existing before the refit (it is not clear how much extra protection was planned for the missiles). Maximum accommodation at existing standards (not current ones) would be 790. The file does not include any modified designs showing conversion of *Swiftsure* or *Superb* or a *Tiger*. (Norman Friedman)

launchers could be accommodated on about 20,000 tons (which proved insufficient). This was not, at least initially, a variation on the CR series of gun cruiser designs. The missile outfit included the Type 984 radar and Type 901 or 902 fire-control radar, the Type 984 being necessary because the missile control system needed target height.

Another possibility of immediate interest was a *Fiji* (not a *Tiger*) class conversion, the missile system (with forty-eight missiles) replacing all after armament. The ship would be fitted with a Type 984 radar and full fighter-direction facilities. The ship would retain only her forward triple 6in turret and would have four twin L70 Bofors guns. As in contemporary US plans to convert *Cleveland* class light cruisers, the entire missile magazine would be built atop the hull as a superstructure. Unfortunately the result was not stable, so the missile magazine had to rethought, the number of missiles halved, and the massive Type 984 replaced by the far less capable Type 982/983 combination and the Type 960 air-search radar.[16] Unfortunately, as long as any protection was provided, the rethought missile stowage saved too little weight, and it represented too much weight too high in the ship. The next step down was to abandon fighter control altogether, replacing the Type 982/983 combination with an even less capable Type 277Q height-finder, the radar eventually installed on board 'County' class missile destroyers. Further steps down included a new bridge (because the ship had grossly inadequate AIO facilities), and the replacement of the L70 Bofors by wartime type twin L60s. None of this was particularly attractive, and although stability was recovered, in the end the design showed an unacceptable 3ft 2in trim by the stern.

A series of 'GW' designs for missile ships began in March 1953, but did not begin to include cruisers until September 1954. To provide a C-in-Cs' meeting with an idea of what was possible, DNC assigned a constructor to produce four sketch designs, GW 24 to GW 27, representing the full range of possibilities. The first was a small missile ship (called a missile destroyer, but not related to the later 'County' class), designated GW 24: one launcher, twelve missiles, one guidance radar with Type 992 short-range search radar (no height-finder), and gun armament limited to two twin Bofors or one or two twin 3in/70 or a combination. Like the cruiser, it would have fleet speed (30kts 'deep and clean'). Endurance should be the maximum possible, say 4,500nm. GW 24 was imagined not as a cruiser, but as a scaled-up *Daring* class destroyer. It and its series of successors were eventually superseded by the 'County' class, which offered both a similar missile battery and two twin 4.5in guns (guns must have seemed more useful as the Royal Navy became more interested in limited war). Target displacement was 3,500 tons.

The trouble was that missile stowage was a matter of volume much more than of weight. GW 24 used 'tube' stowage, the missiles being carried in two sets of side-by-side tubes, three missiles deep, each leading to a launcher rail, each tube being three missiles long (but with spaces left so that missiles could be checked out, so tubes were actually two missiles long). It was impossible to adopt the denser 'coke machine' or revolving-drum methods being adopted by the US Navy for its contemporary Terrier, because Sea Slug was a relatively fat missile, with its boosters wrapped around it.

The next study (GW 25) was a full missile cruiser, armed with two twin 6in, two twin 3in/70, four twin L70 Bofors, and eighty-four Sea Slugs, with the Type 984 radar. Starting characteristics were 645ft x 79ft x 22.5ft, 17,725 tons, 120,000shp, and 32.5kts. Because the missiles themselves were not too heavy (1.645 tons each), it seemed that the full complement would weigh less than a twin 6in turret (138 tons vs 195 tons). On 17,500 tons a Legend suggested adequate weight for protection (2,842 tons), and it seemed a reasonable basis for further development.

The GW 25 cruiser was further developed, its bridge raised a deck (which implied considerably more topweight, since the big Type 984 array (and fire controls) were all raised a deck. This change was connected to a decision to grow the ship by 275 tons, without changing dimensions. Another possibility investigated at this point was to replace one or both twin 6in with missiles forward: GW25A had one 6in (in 'A' position) and one launcher forward, with one missile control radar (Type 901); GW25B had one launcher forward and two radars. It is not clear how a missile launcher and a twin 6in turret could possibly have been combined at the fore end.

Somewhat later DNC realised that the ship could not accommodate so many missiles, so in GW25C missile capacity was cut from eighty-nine to forty-eight. Various weights, including the hull weight itself, had been mis-estimated, and had to be revised. Dimensions were held constant at 645ft x 79ft, with a depth amidships of 43½ft.

At a Ship Design Policy Committee meeting on 20 December 1954, Controller said that the Sea Lords had broadly approved GW 25C. Sketch Staff Requirements (TSD 2291/54) based on it were issued on 30 December. However, a Sea Lords meeting on 28 January 1955 decided that no further preparatory work on the new cruiser should be undertaken for the time being. Meanwhile, in December, DNO proposed laying down a class of 8,000-ton cruisers (deep displacement) armed with two twin 5in forward and twenty Sea Slug aft. This proposal led to the GW 35 sketch design, which had insufficient speed (see below). On 17 December 1954 the Sea Lords decided to send back the A/A Cruiser (small gun/missile cruiser) for further study.

A big all-missile ship could be conceived as such from the beginning (GW 26, October 1954): one twin launcher (sixty-six missiles rather than the original sixty) with two guidance radars and a limited gun battery (two twin 3in/70, four twin L70 Bofors). Radars would be limited to Type 992 for short-range (but relatively precise) detection and to Type 982/983 for three-dimensional tracking to feed data to the missile fire-control system. It seemed that the armament and the radars could fit into a 10,000-ton hull, with two or three machinery units. In calculating endurance the designer reverted to the much earlier idea of splitting the ship's mission into segments. He imagined a 225-hour mission comprising 180 hours at 20kts, thirty-four hours at 80 per cent of full power and eleven hours at full power. By way of comparison, 4,500nm at 20kts equated to 225 hours at that speed.

Higher authority considered sixty-six missiles insufficient, so the design was recast with ninety. Tube storage was no longer practical, so a new scheme was devised. The problem was that missiles had to be lined up. The more missiles, the more of the centreline of the ship they occupied, hence the worse their interference with the uptakes. This was not too different from the problem of a minelayer with a mining deck, but the analogy seems not to have been seen. The initial scheme showed two side-by-side lines of eleven missiles each, a total of twenty-two on each level. At the fore end were two more lines, outboard, of six missiles each. Abaft all the lines of missiles was a lift/traverser leading to the loader and thence to the launcher. If missiles were stowed in lines four deep, this arrangement offered ninety-four missiles, including one level of extra missiles on the sides (it was rated at ninety missiles). The whole magazine was raised so that it could vent into the open in the event of an explosion. One consequence of raising the stowage section was that it made the hull girder deeper and hence stronger for given scantlings; clearly this structure would run much of the ship's length. It turned out that there was enough weight to provide 1½in armour over magazine (including missile magazine) sides, with 2½in–3in sides and 2in deck and bulkheads over the machinery. A ship thus arranged would displace 10,800 tons (530ft [w]) x 70ft x 18ft 6in).

GW 26 was not entirely satisfactory, so a new design (GW 29) was developed, with YEAD(1) steam machinery instead of the combined steam and gas turbine plant, because it was thought that the senior officers would prefer all-steam, and because with the more compact steam plant it was possible to separate the two units by about 42ft, for better survivability. The 1½in over the missile magazine was increased to 2in, on the basis of 'decisions from above' that a bit more magazine protection was wanted. That entailed adding 125 tons relatively high in the ship, quite unlike earlier gun magazine protection. The ship was expected to grow slightly, to 10,950 tons, drawing half a foot more.

Another possibility (GW 28) was the same sort of ship, but without fleet speed – the convoy escort. It would be powered by two frigate machinery units (15,000shp each), which were expected to give 24kts on about the same dimensions as GW 24 (520ft x 70ft, 10,000 to 10,300 tons). More refined estimates showed 25.2kts 'deep and clean', and 23.7kts 'deep and dirty' at 10,000 tons.

Finally, GW 27 was a 7,000-ton missile ship using the sketch Staff Requirement for a missile ship (TSD 2282/54) 'as a guide to the Naval mind'. The limited number of missiles desired (thirty) was seen as a welcome reduction from the ninety of GWS 26; GWS 27 would carry thirty to forty. Other armament would be one twin 3in/70 and four twin L70 Bofors; radar would be limited to Types 992, 960 and 974 (surface search). Endurance would be 4,500nm, broken down by proportions: 80 per cent at 20kts, 15 per cent at 80 per cent of full power, and 5 per cent at full power. Desired protection was 1in over AIO and missiles. The missile magazine was a box 172ft long, 18ft wide and 16ft high, missiles being stowed in two lines, two high. The AIO was a 60ft x 30ft x 16ft (two-deck height) box. No speed was specified, only the powerplant: two YEAD(1), 60,000shp, or two COSAG Y102s (also 60,000shp), the latter version being GW 27A. This was not too different from the light cruiser with two twin 5in guns, and much the same dimensions were offered: 450ft (wl) x 63ft x 16ft, with a light displacement of about 5,900 tons (7,200 tons fully loaded). On that basis 60,000shp would provide 28.8kts 'deep and clean' (28.25kts 'deep and dirty'), which was considered satisfactory. DTSD unofficially requested a version of this design, GW 31, in which ASW weapons replaced the 3in/70s, on the theory that such a convoy escort would have to fight both air and underwater threats. Endurance speed was 15kts rather than the fleet endurance speed of 20kts. GW 32 was a further entirely unprotected version, displacement falling to 4,800 tons (450ft x 53ft x 12¼ft), with an estimated speed of 27.25kts 'deep and clean', or 25.5kts 'deep and dirty'.

After the Sea Lords met, they asked DNC for a further missile cruiser, presumably based on the light cruiser with three twin 5in guns: GW 35, in which the after twin 5in mount was replaced by a twin launcher with twenty missiles.[17] Other armament would be four Bofors. Radars were limited to Types 992 and 277Q, and a single Type 901 guidance set. The ship was to have fleet speed (interpreted as 28kts 'deep and dirty') and an endurance of 4,500nm at 20kts. Missiles would be stowed in two lines, each three missiles high and four long, to minimise overall length. The magazine thus occupied a box 144ft long, 17ft wide and 28ft deep, set into the top of the machinery box. Capacity was twenty-four missiles, but each feeding line had two blank spaces so that missiles could be withdrawn for checking. In all, this was not far from GW 27. Machinery would be a pair of YEAD(1) units, 30,000shp each. A quick estimate gave an 8,500-ton ship (480ft x 65ft x 17¼ft). Estimated speed was 28.75kts 'deep and clean', 28.25kts 'deep and dirty'.

A design (GW 36) was drawn up for a ship with two 5.25in turrets (plus four Bofors) and twenty-four missiles, conceived with ultimate conversion to an all-missile ship in mind. In this sketch design the missile magazine seems to have been stretched lengthwise by reducing its height. A three-shaft (three YEAD(1)) powerplant was proposed, as greater speed was desired (estimates were 31kts 'deep and clean', and 30.25kts 'deep and dirty'). Because the missile magazine was high in the ship, there was no difficulty with a centreline shaft. This was a 10,600-ton ship (530ft x 63ft x 18ft, hull depth 45ft).

Almost all the designs had unrealistically austere radar systems. To be effective, Sea Slug really needed something more like the massive Type 984. In mid-January 1955 the sketch designer, T J O'Neill, was asked to find the effect of fitting Type 984 to GW 36 (as GW 37). It had both a direct effect (weight and volume) and an indirect one (complement, increased hull weight). Simply to give the radar nacelle clear arcs, the ship had to be made 20ft longer. This seems to have been one of the first times that the extra deck area required was calculated and used to find how much the hull had to grow.[18] That not too much more ship was needed suggested that it was reasonable to demand Type 984 in any missile cruiser. Given a reasonably satisfactory ship armed with 5.25in guns, a follow-on GW 38 was requested, armed with twin 6in guns, but they were so much more massive than the 5.25in that the designer went back to work from scratch. To get suitable arcs of fire for the guns, he lengthened the ship to 580ft and increased beam to 72ft, but kept the same hull depth (45ft); he expected the ship to displace 12,200 tons (draft 18½ft). This was a considerable contrast with the 18,000-ton ships previously offered with two twin 6in guns. It appeared that the 90,000shp plant could drive the ship at 31.5kts 'deep and clean' or 30.5kts 'deep and dirty'. GW 39 was a step in the opposite direction, the two twin 5.25in being replaced by two twin 5in. It would displace 10,900 tons.

A Sea Lords meeting on 28 January 1955 decided that DNC should prepare a sketch design for a missile cruiser on the lines of GW 36/36A but with twin 5in guns, which could later be replaced by Blue Slug, the projected surface-to-surface version of Sea Slug. Thus GW 39 inspired a Staff Requirement (TSD 2295/55) issued in mid-February, for a ship with two twin 5in guns, forty-eight missiles, and a speed of 31kts 'deep and dirty' (GW 41). The higher speed required more power. Because E-in-C offered plants only in unit amounts. E-in-C offered 105,000shp – three 30,000shp units and two 7,500shp gas turbines. Although all necessary items could be accommodated within a 530ft length, that would be difficult to drive at high speed, so minimum length was pushed to 550ft. Desired metacentric height (GM) set beam at about 74ft. Protection was revised: instead of 1in over AIO and missiles and the rest over the machinery, the ship had 1½in over the AIO, missiles, 5in magazines, deck (1in) and side over machinery, all worked into the ship's structure. Heavy bulkheads were built into the machinery space. So much length was needed for the missile magazine that it extended under the AIO and the bridge. It was a 210ft-long box, 18ft (three missiles) high, extending beyond the machinery spaces, which were only 162ft long. In contrast to earlier cruisers, the box for the 5in magazines forward did not have to be contiguous with the machinery. It turned out that the ship would displace 12,600 tons (50ft x 74ft x 18ft 8in, with a hull depth kept to the 45ft of the earlier studies). The great problem in these designs was that the heavy part of the ship was forward, where the ship was less buoyant. It was a struggle to keep the centre of gravity far enough abaft amidships to keep the ship from trimming by the bow. In connection with a later design, the constructor in charge of the series of GW ships commented that whether or not that could be done would determine the success of the design. Part of the problem was the way the missiles were stowed, their weight spread along the amidships part of the ship. A more concentrated form of stowage would have simplified overall design considerably. E-in-C was dubious about gas turbines, so he offered a 35,000shp steam plant, presumably based on the 30,000shp YEAD(1). On this basis GW 41 was redesigned as GW 42. Displacement increased slightly, to 12,500 tons.

The next step up was to return to the earlier pair of twin 6in and pair of twin 3in/70 (plus two twin Mk 11 Bofors), with forty-eight missiles and fleet speed (over 30kts), as GW 44.[19] Compared to GW 42, such a ship had her length between perpendiculars pushed 50ft forward, the amidships section being pushed 25ft forward. 'B' mounting would be in the same position as 'B' 5in mounting, but 'A' 6in mounting would be forward of 'A' 5in mounting. Type 984 radar, AIO protection and fire controls (two MRS 3 for the 6in guns, two for the 3in) would be pushed up a deck, to clear the higher 6in turrets. One MRS 3 would be added. This cruiser would use the 105,000shp plant planned for GW 42. A sketch of protected spaces showed a 234ft x 18ft 6in missile magazine, for two lines of missiles, each three missiles high. The checkout section was 25ft long, with two 26ft-wide bays. Abaft it was the loader, also protected. The AIO structure formed a separate box, 26ft wide, 16ft deep (two decks), and 74ft long, well separated from the machinery and immediately below the big Type 984 radar. A separate 6in magazine box forward was 86ft long, 16ft wide and 18ft deep. This ship would have displaced 14,500 tons (600ft x 77ft x 20ft, 45ft hull depth). GW 45 (mid-March 1955) was identical except that it was armed with single automatic 6in guns (which did not exist). That saved 140 tons on armament and 20 tons more on personnel; displacement was given as 14,340 tons. In GW 46, the two single 6in were replaced by one twin mount ('B' mount eliminated), the result being a lower bridge structure. The ship could also be shortened: 580ft x 76ft x 19½ft, 13,550 tons.

These GW studies provided the Sea Lords and the Staff with useful guidance, but all were defective, many being inefficient in their use of space for magazines. Missile layouts had changed considerably, as it became clear that Sea Slug with its wrap-around boosters could not be broken down further for stowage. That made 'tube' stowage, in which missiles sat in tandem in lattice-work tubes, preferable at least for twelve to forty-eight missiles, with multiple tubes providing stowage for larger numbers. Attempts at deep stowage had not succeeded.[20] Furthermore, machinery weights (supplied by E-in-C) proved unrealistically low. Staff requirements for speed kept rising, from 12kts to 22kts for the convoy escorts. The only studies with the preferred scheme of protection were those from GW 41 onwards. Many, and probably all, of the studies had space and accommodation problems. The first design to reflect the full Staff Requirements was GW 47.[21] It reverted to two twin 5in guns, and was interesting enough to make a general arrangement worth sketching. However, it was not entirely satisfactory: it could not accommodate its full complement, and it had one rather than two fire-control systems for its two gun mounts (GW 50 was a better version).

Only at this point were some basic decisions taken. Formerly Sea Slug had been conceived as a liquid-fuelled rocket, at least in its initial form; now it would be solid-fuel only. Missile characteristics were set by the manufacturer, Armstrong Whitworth, only in April 1955: weight 4,200lbs, overall length 20ft, maximum width 63in. At about the same time DGD finally proposed killing the twin 5in gun, as it would be unprofitable to start a new major gun project when full efforts would have to be devoted to guided weapons.

Given the work already done, in April 1955 Controller asked for a new series of designs based broadly on GW 42, each of which would have two main guns forward: 4.5in, 5in or 6in, with separate directors for the 5in and 6in versions. The 4.5in was considered the smallest gun likely to be effective against enemy small surface craft (at this time Sea Slug had no surface-attack capability). The secondary battery would be either Bofors Mk 11 or 3in Mk 6, with separate directors. Radar would include Type 984, which was finally understood to be essential. Protection would amount to a 1in deck over machinery, and 1½in side over machinery. Smaller ships would have 1in over AIO and missiles, larger ones, 1½in. Speed 'deep and clean/dirty' should be about 31/30kts. Endurance should be 4,500nm 'deep and dirty'. Ships should have two or three 35,000shp shafts.

This series became GW 48 to 52, GW 48 being the smallest. GW 48 had two machinery blocks separated by 38ft for survivability; laying out its components, it had to be at least 520ft long. Hull depth was set by machinery depth, 24ft, plus a double bottom (3ft) plus the missile magazine (18ft), as in GW 41 to 47 (45ft). An initial sketch showed thirty-six missiles in a magazine 190ft long; that was later increased to the desired forty-eight, in a lengthened structure. The ship was expected to displace 10,300 tons (520ft x 72ft x 17½ft). Unfortunately, two machinery units were not quite enough: on 70,000shp she would make only 29.5kts 'deep and clean' (28.5kts 'deep and dirty'). GW 49 was a similar ship with twin 5in guns. While the design was being worked out, the 5in twin mount began to grow to the size of the twin 6in, which forced the bridge up one level to clear the two gun mounts. The resulting ship displaced 11,100 tons (550ft x 73ft x 18ft). No speed calculation seems to have been made, perhaps because GW 48 was so disappointing. GW 50 was a more satisfactory version of GW 47 (two twin 5in), with three machinery units to regain the desired speed. The missile magazine was 226ft long, sitting atop a 179ft machinery box. Not surprisingly, this was a considerably larger ship, displacing 13,000 tons (570ft x 76ft x 19½ft). The extra power provided sufficient speed: 31.5/30.75kts. GW 51 was the 6in version of GW 50, the bridge being raised accordingly. The ship would displace 13,900 tons (590ft x 77ft x 19½ft), which was still well short of the 18,000-tonners previously envisaged. Speed was somewhat improved because the ship was longer: 31.75/31kts. The final step in the series (GW 52) had the after Bofors guns replaced by twin 3in/70s. That added weight (160 tons to the armament group) and also magazine and gun bay volume, which had to be protected. The ship was lengthened by 10ft to add extra space, but there was no attempt to equate the required space with ship length or with bridge volume. The result was a 14,500-ton ship (600ft x 78ft x 19¾ft) with a speed of 31.75/31kts. That must have seemed satisfactory, and far better than the earlier 18,000-tonners.

Having explored the upper end of the missile cruiser range, in May 1955 the missile cruiser designer tried what he called a 'rock-bottom' ship, GW 53: two 4.5in and thirty-six missiles, without any flag facilities, and also without the Type 984 radar (it had Type 982/983 instead). The powerplant was the pair of 35,000shp units used in GW 48 and 49. Compared to GW 35, the ship would be larger because protection would raise its centre of gravity; better stability (higher GM) was required, hence the ship would be beamier; the depth of the machinery spaces would be greater; the machinery would be longer and heavier. Rock-bottom was not so rock-bottom: 9,850 tons (500ft x 71ft x 17¾ft, hull depth 45ft). As a consequence, speed was only 29.25/28.25kts, below the required figures.

Visiting Bath on 17 May 1955, First Sea Lord Admiral Mountbatten saw sketches GW 50, 51A and 52A (versions with a displacement margin added). He had sketches prepared for a Sea Lords' meeting on 26 May: a 6in cruiser with 3in/70 guns (GW 52A), a 6in cruiser with Bofors guns but no 3in (GW 51A), and a 5in cruiser to the Staff Requirement (GW 50A), as well as a GW fast escort. DNC also showed how the US Talos missile could replace Sea Slug in the cruiser. The Sea Lords' meeting on 26 May 1955 approved development of GW52A so that it could be presented to the full Board.

GW 58 was an upgraded version (dated June 1955) incorporating DNC's desired 4 per cent growth margin and further corrections to E-in-C weights (a jump of 200 tons had effectively finished GW 52). Power was pushed up to 120,000shp on four shafts, offering a better machinery arrangement. The design series had now returned to GW 25C, except for two Bofors mountings. However, it had much less protection (1½in vs 4in), better stability (a goal of 8.5ft rather than 6.75ft GM), and missiles carried a deck lower, at a cost in accommodation space. The 4 per cent margin was applied to length as well as to weights, so the ship grew to 624ft (the constructor hoped that beam could remain at 78ft). Draft would be adjusted to displacement, the constructor estimating 20¼ft, and 15,400 tons. Detailed arrangement calculations showed a 240ft missile magazine atop 220ft machinery spaces. Expected displacement was 15,800 tons (625ft x 78ft x 20¼ft, with a speed of 32.1/31.1kts.

GW 58 was clearly a likely basis for a final cruiser design, because unlike all others to date it was developed in sufficient detail to give confidence that the data were reliable. For example, there was an attempt to estimate the weight of the missile-moving machinery Vickers-Armstrong was then developing. This was also the first design for which damaged stability calculations were carried out. This study was given to Controller on 15 June 1955, and sketch Staff Requirements written around it (TSD 2305/55) were issued in July 1955. Speed should be 31kts 'deep and dirty' and endurance 4,500nm at 20kts 'operational and dirty' (six months out of dock). Armament was two twin 6in, two twin 3in/70, two twin L70 Bofors, and one twin launcher with forty-eight missiles. An unusual feature was provision for variable-depth Asdic as an alternative to the usual self-defence sets (Types 176 and 177). Study GW 60 (September 1955) met these requirements (it was essentially GW 58

THE MISSILE AGE

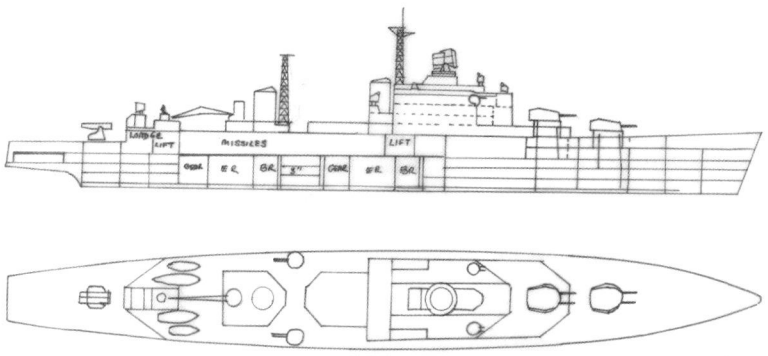

This missile cruiser sketch was dated June 1955. The ship would have displaced 15,400 tons deep (625ft pp, 637ft loa x 78ft x 20ft 3in). Armament would have been one twin Sea Slug launcher (forty-eight missiles), two twin 6in Mk 26, two twin 3in and two twin Mk 11 Bofors (40mm/70 rather than the wartime 40mm/60). Main machinery spaces would have been protected by 1½in sides covered by 1in decks; missile stowage, AIO, magazines and steering gear would all have had 1½in sides and 1½in decks. The ship was rated at 32kts deep and clean (30.75kts deep and 'dirty' [six months out of dock]) on 105,000shp (four shafts); endurance in operational condition would have been 4,500nm at 20kts. Accommodation capacity was eighty officers and 970 ratings. Subsequent versions had much the same configuration, although they were somewhat larger, with more missiles and, often with two rather than one Type 901 missile-guidance radars aft. The big radar atop the bridge was Type 984. Other search radars were Type 992 (with the GDS 3 target-indication system) and Type 978 (surface search), both on the foremast (which was topped by a Tacan dome). Gun fire control was by six MRS 3, each with a Type 903 radar (two for each calibre of gun). The small dish forward of the Type 901 aft was for missile telemetry. The two machinery units (boiler room, engine room, gearing room) were separated by a space containing the 3in magazines and the stabiliser machinery. The space immediately forward of the missile launcher was the loader, atop the check-out room. The spaces in the bridge just below the twin 40mm gun are the Operations Room above the Weapons Direction room. (Norman Friedman)

with provision for the variable-depth Asdic). It included space for the Y200 powerplant then being developed by E-in-C and the Yarrow-Admiralty Research Department (YARD). The resulting ship displaced 15,800 tons (645ft [pp] x 78ft). DNO wanted either four twin 3in/70 or two twin plus two sextuple L70 Bofors; GW 64 and GW 69 met this requirement. GW 64A was DNC's estimate of the effect of adding six more missiles. DNO also tried a kind of revolver stowage (GW 82) in a double-ended missile ship, but this effort had no effect on later work.

Given the GW 60 design, the Sketch Staff Requirement was revised and reissued in February 1956. DNO's request for four twin 3in/70 (in place of two twin 3in/70 and two twin Bofors) should be considered. Provision should be made for 25 per cent special (nuclear) Sea Slug warheads. The number of air-intercept positions should be increased. Complement should be increased by about ninety, and the variable-depth Asdic might be dropped. First Sea Lord wanted provision for the huge 'Stage 1¾' missile (see below), but that was clearly impractical. A rough estimate (GW 84, March 1956) suggested that the ship with four twin 3in would displace 18,300 tons (680ft x 82ft x 22ft). Controller was alarmed enough to ask DTSD to write the Staff Requirements to limit displacement to 16,000 tons. Unfortunately much of the content of the ship had not yet been completely developed, so size and weight kept growing: the twin 6in, the main machinery, the size of Sea Slug itself (which was now 21ft long with 67in span). The tube stowage, three high and two wide, which had been used for some time, was no longer practicable; instead missiles would be stowed on two levels of 27ft-wide tubes one high and four wide. DNO liked this type of stowage and GW 84 was based on it. A new GW85 study taking these increases into account (and providing for increased complement) would have displaced 17,000 tons (645ft x 79ft x 21½ft) with only the original two twin 3in/70.

DTSD still hoped to hold displacement (deep) to 16,000 tons. He called a meeting on 23 April 1956 to set major characteristics within that limit. DGD offered several armament schemes, all with sixty-four or forty-eight missiles and two twin 6in (and two or four twin 3in/70), with one or two missile guidance channels (Type 901 radars). He strongly recommended a version with sixty-four missiles, two guidance channels and four twin 3in/70. DNC estimated that this ship (GW 89) would displace 17,800 tons (665ft x 79ft). Other studies were GW 85 to GW 88, ranging from 16,960 tons (GW 86A: forty-eight missiles, two twin 3in/70, two guidance channels, and no Bofors guns) to 17,250 tons (GW 88: sixty-four missiles, two channels, two twin 3in/70, and two twin Bofors). A further study, GW 89B, was enlarged to provide sufficient accommodation: 18,415 tons (700ft x 79ft). Finally, to pare the ship down, GW 90 was sketched with one 6in mount (15,932 tons, 610ft x 79ft).

At the meeting, the Staff concluded that 16,000 tons was too little and displacement would probably exceed 18,000 tons. The ship could be somewhat reduced by using methods currently planned to reduce the complement of the missile destroyer (the 'County' class). Reducing Staff Requirements could cut the ship somewhat further, but 17,000 tons was an absolute minimum. Variable-depth sonar was abandoned, and minimum accommodation was set at 90 officers and 1,000 ratings. A new series of studies was produced, the modified version of GW 89 (90 officers, 1,100 ratings) now displacing 18,500 tons (700ft x 79ft).

DNC continued to produce new sketches in attempts to improve arrangements: GW 91 was GW 85 with its 3in mounts moved aft to decrease bending moment (but that caused blast problems with the missiles-guidance radar, also aft). GW 92 interchanged the missile launcher and 'B' 6in gun mounting.

Late in May the Staff made another attempt to hold displacement to 16,000 tons: a double-ended missile ship, with four twin 3in and four or six twin 4.5in guns. Studies were GW 93, GW 94 and GW 95, the first two with sixty-four missiles altogether (the third had sixty-eight). GW 93 and 94 had no 4.5in, but GW 95 had four. GW 93 would have displaced 15,848 tons (610ft x 78ft), and GW 95 16,529 tons (630ft x 79ft). DNC and his assistant met with DCNS on 30 May 1956. After DNC explained why the ship had grown, DCNS stated the major features he would support as a minimum: a Sea Slug system with one guidance channel (two preferable, for later ships) and forty-eight missiles; two twin 6in, two twin 3in/70 (each replaceable by the US Tartar missile), two 40mm L70, Type 984 radar with four intercept positions, a speed of 32/31kts, and an endurance of 4,500nm at 20kts. Accommodation should be provided for 90 officers and 980 ratings. DTSD should incorporate these requirements into a new paper for the Board. DTSD's paper should also include a ship reflecting the optimum cruiser (as DGD understood it) as GW 96, and also the GW 90 attempt to pare the ship back.

Then the project collapsed, so quickly that in August 1956, after all sorts of design discussions in June, DDNC stated that the future of the cruiser was so problematic that there was no point in considering forming a Design Liaison Group. On 27 August Controller put the planning dates for the cruiser back a year. The first of three cruisers would be ordered, under this revised schedule, in April 1959, to complete in October 1963, the last in April 1962 to complete in October 1966. The Staff was still arguing details of the ship. New draft

Staff Requirements for a 64-missile ship with four twin 3in were issued in November 1956, allowing for accommodation for 95 officers, 290 senior ratings, and 770 junior ratings. DNC was briefed on the GW 96 study on 22 November. By this time the design had been arranged so that other missiles could fit its stowage space: sixty-four Sea Slug or 124 US Terriers, or sixty-four US Talos, or twenty-two Blue Envoy ('Stage 1¾' missiles) with wing tips removed and boost fins hinged down, or ten Blue Envoy with complete wings but with boost fins hinged down. Replacement of Sea Slug by an army missile (Red Shoes/Green Flax) was being proposed (and fought down by the Royal Navy). The Ship Design Policy Committee considered fitting a new long-range missile instead of Sea Slug (e.g. twenty-five assembled and seventy broken-down Green Flax). Controller suspended work in January 1957. The missile cruiser was dead, only the 'County' class missile destroyer surviving. Without the Type 984 radar, it could hardly operate independently; the destroyers used their data links to obtain required data from carriers equipped with Type 984.

While Sea Slug cruiser work proceeded, the US Navy offered its Talos long-range missile to the Royal Navy. GW 61 was a study dated 20 May 1955 of GW 52 converted to a Talos ship, on the lines of the *Cleveland* class missile cruiser conversions then being made. As with Sea Slug, the US Navy favoured above-decks stowage, in this case in a box 130ft long, 50ft wide, and 14–16ft high. The missile itself was about twice as heavy as Sea Slug, and much longer, because it used a tandem booster. In GW 61 the missile hangar would be lengthened to 150ft because uptakes would have to pass through it. As in the all-British designs, the hangar/magazine would be covered with 1¾in plating. The ship would lose half a foot of metacentric height. Alternatively, Talos could replace the two 6in mounts forward instead of being installed aft. Not surprisingly, that would be a very elaborate redesign, and the loss of about 0.75ft of GM, which alone would probably rule it out.

Talos was broadly equivalent to the big land-based 'Stage 1¾' missile then being developed for British national air defence. It was expected to weigh 5 tons (compared to 3 tons for Talos), to be 25ft long, and to have a 7ft wing span. In June 1955 a ship to fire such missiles was sketched. Little was known of possible requirements, but it was thought that the ship should carry a total of thirty missiles, ten of which should be launchable in rapid succession. That could be achieved by placing four or five launchers abreast (fortunately they did not have to traverse), each fed by a line of missiles, one high. The control radar would be a larger version of Type 984. For self-defence, the ship might have six ASW torpedo tubes with twelve torpedoes. The ship would have fleet speed (31kts 'deep and dirty'). It was hoped that an unprotected ship armed with these missiles might displace no more than 10,000 tons. GW 59 was sketched to this end: it appeared that 10,500 tons would suffice, and that speed (on 70,000shp) would be 29.75/29kts. Given the relatively undefined missile, such figures were necessarily rough. Studies incorporating the '1¾' missile were GW 75 and GW 77 to 81.

Minimum or convoy ships were designed as well. A missile escort ship was briefly given priority just after the missile cruiser, at the end of 1954, then stopped after being presented to a January 1955 Naval Staff meeting. Later GW 54 and GW 55 were convoy escorts converted (on paper) into missile ships.

The Escort Cruiser

Despite the demise of the big missile cruiser, and the ageing of the gun cruiser force, the need for new cruisers remained. In March 1960 Director of Plans argued that cruisers were needed because carriers could not provide adequately for their own AA and A/S defence and because the mainly A/S escorts could not provide adequate anti-aircraft defence for themselves.[22] Moreover, given widespread British naval commitments, there was a real need for ships capable of operating independently – i.e. for cruisers. The two key technologies were the missile, which provided a measure of independent anti-aircraft capacity, and the helicopter, which provided something other than a carrier with a significant long-range ASW capability. In the past, air ASW had meant carrier-based fixed-wing aircraft. British carriers were relatively small, so their ASW aircraft ate significantly into their strike or fighter capacity, so moving that capability off the carrier and onto a helicopter ship could substantially improve carrier striking or air-defence power.[23] This possibility recalled the inter-war idea of using cruiser (and battleship) aircraft to reinforce the fleet carriers. Independent operations required high speed. The required combination of anti-aircraft missiles beyond what a ship needed for self-defence, an ASW helicopter force, command and control capacity, and speed all demanded a substantial ship displacing at least 10,000 tons. Given the decline of any surface threat, much or all of the traditional cruiser gun battery could be traded for an area air-defence missile system and helicopters, to meet the two primary threats: long-range air attack and submarines.

D of P wanted a cruiser armed with Sea Slug Mk II, with a bombardment weapon, and with nine Wessex 'single package' ASW helicopters (the minimum for adequate ASW services to a force). It should have sufficient operations room and accommodation for a senior officer, to back up frigates and destroyers. It might also be able to ferry a few VTOL aircraft (the P.1127, forerunner of the Harrier, had just flown). A sketch design showed a 10,500-ton ship. D of P wanted three such ships to be continuously available, so four should be ordered as soon as possible, a fifth being added to cover periods when one of the first four was undergoing a long refit. He thought the first two could be ordered in 1962/3 and the next two in 1963/4. The ships would be built instead of the projected last four missile destroyers, DLG 7–10, whose construction could be deferred until after 1970. There was already interest in single-purpose helicopter carriers, both large (nine helicopters, to defend a task force or convoy) and small (four to six helicopters, to protect a smaller convoy or to provide a support or hunter-killer group). Replies to a questionnaire showed considerable support for the escort cruiser, although DTWP (Director of Tactics and Weapons Policy) wanted nine big helicopters (presumably the future Sea Kings) or twelve of the smaller Wessexes. DAW warned that the minimum useful number of helicopters on a ship might be as many as eighteen. Eight helicopters on the cruiser would add two or three Buccaneer bombers on the carrier – a significant addition for a small carrier like HMS *Centaur*, but much less important for a large one like those likely to be in service when the escort cruisers appeared. He preferred to keep smaller numbers of helicopters aboard carriers, as a last-ditch defence. ASW helicopters became more important as the threat shifted to nuclear submarines, against which evasive tactics were useless. In addition, the escort cruiser could carry a substantial force of troops from point to point, land them, and support them with gunfire.

DTWP pointed to the carrier replacement programme then taking shape. The Staff favoured keeping anti-air missiles off the carriers and on board accompanying escorts. The Escort Cruiser would fill this requirement, assuming that its air defence function was emphasised. For example, its missile launcher might be placed aft instead of forward. He suggested that the covering fire role, which justified a twin 4.5in gun for the Escort Cruiser, could be deleted to save manpower. Overall, the carrier and escort cruiser programmes should be coordinated.

The new cruiser concept developed in the context of the continuing

THE MISSILE AGE

decline of Cold War priorities, anti-submarine warfare being seen as less important than more traditional naval roles. In this context a helicopter carrier might at first blush seem less important than other types of ships. However, to the extent that it increased carrier striking power in the context of local submarine opposition, as in the Indonesian Confrontation later on, it was very relevant to the new emphasis.

The helicopter carrier was first discussed by the Fleet Requirements Committee, whose recommendation for them was endorsed by the Ship Characteristics Committee. On 7 April 1960 the Board discussed the case for introducing helicopter ships, and also their broad characteristics. DCNS pointed to the growing difficulties of maintaining helicopters aboard aircraft carriers, and the growing role of helicopters in anti-submarine warfare. A helicopter carrier would be flexible, and could be used in other roles. It would also make the fleet in general more flexible, because in some circumstances it could be used independently of carriers or frigates. Two helicopters was the usual tactical unit, helicopters being stowed below (to protect their fragile rotor heads from spray) when not flying. They typically reacted to long-range contacts and hence could not replace screening ships. Typically two were held at readiness on deck (Condition 2), with two more in Condition 3. If the first two flew off, the second pair had to be brought up and a third pair prepared. That made the minimum helicopter complement six aircraft. Because stability was important for helicopter operations, larger ships were preferable to smaller ones – an eight-helicopter ship was better than two four-helicopter ships. A total of eight helicopters would support the minimum tactical unit of six. An eight-helicopter ship would displace at least 5,000–6,000 tons. It was noted that that the next larger acceptable helicopter ship would carry sixteen helicopters, about half the total planned Royal Navy force. Such concentration would seriously limit fleet flexibility, so it was ruled out. Ship characteristics should include fleet speed (defined as 25kts), though escort speed (28kts) would be desirable. Endurance should be as great as possible. The Fleet Requirements Committee considered five alternative armaments: (a) four Seacat launchers; (b) one twin 3in/70 and two Seacats; (c) a full (US-supplied) Tartar plus two Seacats; (d) the 'small ship' Tartar (eighteen rather than forty-two missiles) and two Seacats; and (e) Sea Slug (twenty-four missiles, eight ready-use) plus two Seacats. Tartar was ruled out on expense, timing, and uncertainty of US plans to develop the small-ship version, while the version with the 3in gun was preferred to the all-Seacat version because the gun offered some surface fire and bombardment potential and might be more valuable in Cold or limited war roles. Studies were made of ships armed with the 3in gun and with Sea Slug, both with the same limited flag facilities as in missile destroyers. Grade II flagship facilities were rejected as too expensive because of the considerably increased accommodation involved. Estimated capital costs (less first outfit) were £12 million for the Sea Slug ship and £10.5 million for the ship with the 3in gun. First cost would be higher for the Sea Slug ship because it would include the missiles, and the missile system had a higher annual running cost. Both the Fleet Requirements Committee and the Ship Characteristics Committee considered the added cost of Sea Slug fully justified, because it added to fleet air defence as well as to fleet anti-submarine defence. Also, in a limited war involving negligible submarine opposition, or for a police action, the helicopter ship could support an assault by 200–300 men (the Royal Navy had pioneered vertical assault).

Supporting this analysis were two descriptions of sketch design studies, one for a Sea Slug ship (6,200 tons, 485ft long), the other for a 3in gun ship (5,900 tons, 460ft long). Each was powered by two steam turbines (total 36,000shp), and each offered an endurance of 3,500nm at 20kts or 5,000nm at 12kts. The Sea Slug ship, but not the 3in ship, had the same analog computer combat system (CDS with DPT) as the first-generation Sea Slug destroyers. Search radars were the same: Type 965 for long-range air search, Type 992 for weapon direction and Type 278 for height-finding for helicopter control and also for Sea Slug missile control.

Controller wanted the helicopter ships to have the sustained endurance and self-supporting abilities of cruisers, but accepted that these advantages were prohibitively expensive. The ships would therefore be built to destroyer standards, which in turn would increase the need for maintenance ships. Six ships were envisaged; First Sea Lord suggested building three so as to get cruiser standards. The Board directed that further consideration be given to the cost and other implications of building to cruiser standards, taking into account the need for more maintenance ships if they were built to destroyer standards.

Meanwhile the Board nervously watched the British cruiser force run down. The '88 Plan' (for a manpower ceiling of 88,000) approved in 1957 allowed for an operational force of four carriers (three air groups), three cruisers, forty-six destroyers and frigates (plus trials and training ships), and thirty-one submarines. Further ships were provided to maintain the operational force through refits, so a five-cruiser force was envisaged (three *Tigers*, *Swiftsure* and *Belfast*). As money became tighter, there was already interest in scrapping *Swiftsure* and *Belfast*, leaving two or even only one operational cruiser by 1966/7. Yet, given the growing emphasis on Cold and limited war, the role of ships capable of operating independently should be widening, not shrinking. There had to be some way to create a more modern equivalent of the classic cruiser. Once the projected helicopter cruiser or carrier had both area anti-aircraft (Sea Slug II) and area anti-submarine (helicopter) capabilities, it was clearly just that ship.

On 30 May 1960 the Board considered a memorandum suggesting that cruisers might be replaced by helicopter ships. As these would have virtually no surface-attack capability, the Board directed studies of something more like a cruiser. Alternatives were construction of five helicopter cruisers with twin 4.5in guns, construction of five helicopter cruisers with two twin Bofors each, construction of five helicopter destroyers, construction of six missile destroyers to cruiser standards (and no helicopter ships), and simply retaining the *Tigers* and *Belfast* and deferring the last four missile destroyers. There was, it turned out, insufficient design capacity either for the helicopter destroyer or for the missile destroyer upgraded to cruiser standards. The case for including helicopter ships in the fleet was considered so strong that the options not including them were pointless. That left the helicopter cruiser. At this stage the two alternatives considered were helicopter ships with and without guns, both built to cruiser standards. Both alternatives would have the new automated command and control (ADA and the TIDE [Link 11] data link). Both alternatives were designed to embark eight Wessex III helicopters, to make 28.25kts, and to have endurances of 4,500nm at 20kts and 6,000nm at 12kts. Both were armed with one Sea Slug (twenty-six missiles) and two Seacats for close-in defence. Without the gun the ship could displace 8,600 tons (33,000shp) and would cost £14.25 million (exclusive of first outfit). Alternatively, a 9,100-ton ship (40,000shp) could add a twin 4.5in gun (cost £15.25 million).

On 28 July the Board considered a memorandum prepared to answer the question of perpetuating the cruiser under modern conditions. It favoured a helicopter ship built to cruiser standards. DCNS described its value under both Cold and limited war conditions; it would also greatly increase carrier effectiveness in air-defence and strike roles. The Board directed that Staff Requirements and outline drawings be prepared, to come before the Board in 1961. As the First Lord proposed, henceforth the ships would be 'escort cruisers' rather than 'helicopter ships'. To offset their cost, the four planned missile destroyers could be deferred.

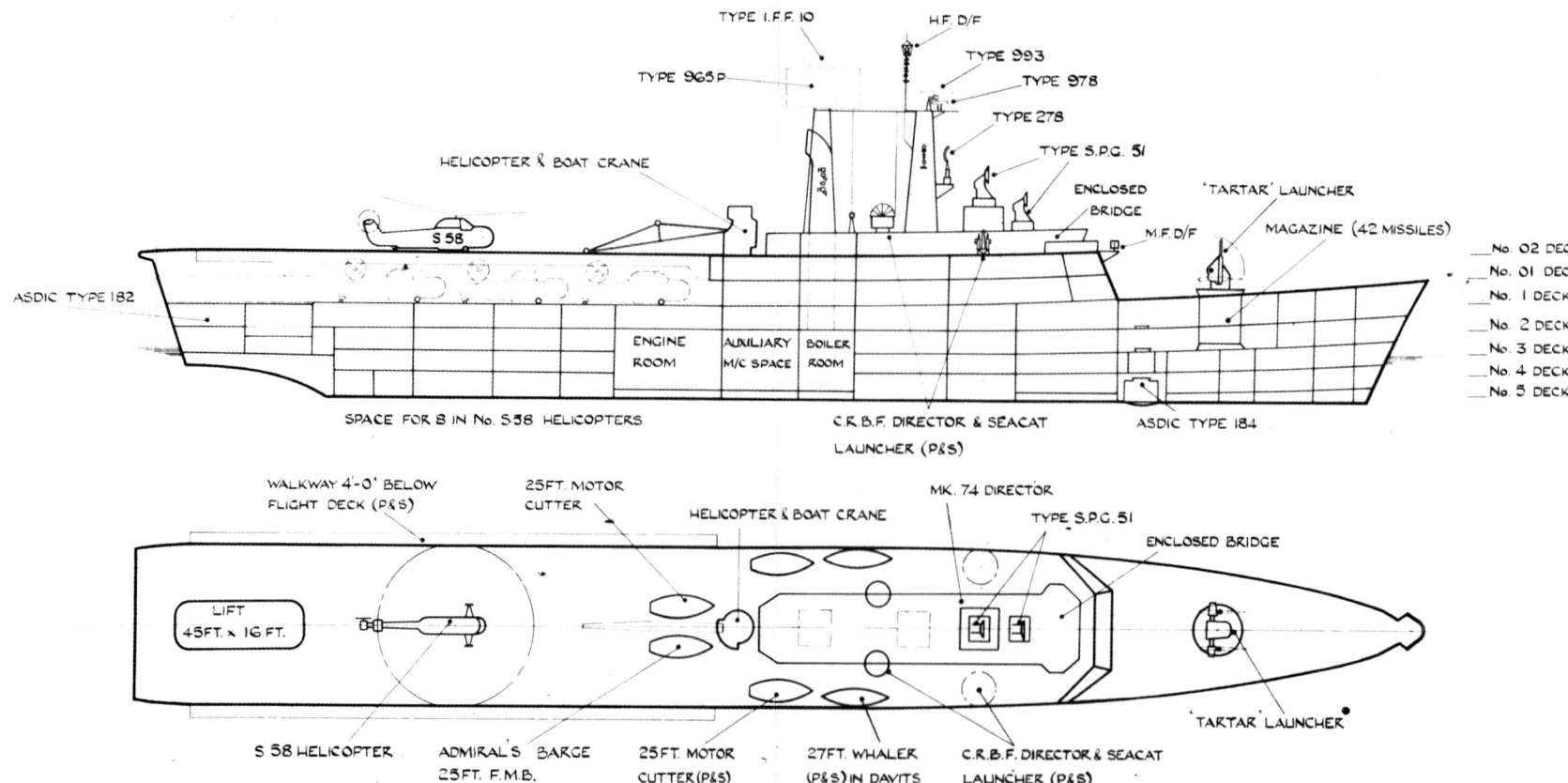

The main objection to the ships from outside the navy would be the lack of manpower, and the plan to build escort cruisers had to include manpower savings elsewhere. DCNS pointed out that, unlike a conventional cruiser, an escort cruiser would have a complement not much different from that of a missile destroyer.

Sketch designs were produced. In the late 1950s there were designs for helicopter carriers displacing up to 19,000 tons (with twenty-two helicopters). Attention then turned to ships armed with twin 3in/70 guns or Tartar or Sea Slug missiles, capable of 26kts 'deep and dirty'. Three series of such studies were reported to the Ship Characteristics Committee (successor to the Ship Design Policy Committee) in October 1960, Series 6, 9 and 21. Series 6 envisaged a complement of 50 officers and 400 ratings (as there was no firm Staff Requirement). Complement was a useful parameter because by this time Royal Navy planning was dominated by the approved total number of officers and ratings. Thus in 1960 the navy was working to the '88 Plan' in which it could have 88,000 personnel (an improvement on the '75 Plan' and '80 Plan' of 1957). With this complement, Study 6C was a 5,400-ton ship with a waterline length of 430ft (eight Wessex, one twin 3in/70). Study 6D was a ship of similar size armed with the US Tartar missile. DG Manpower then estimated that accommodation for 61 officers and 534 ratings would be needed for a missile destroyer-type flagship, so the sketches were revised to 6E and 6F, both 460ft long (5,900 tons).

Series 9 was developed on the theory that so large a ship should have a missile armament. With no decision to adopt the US Tartar, Sea Slug was the only option. Study 9C was 485ft long (6,400 tons) with twenty Sea Slugs (twelve ready-use). It carried special (i.e. nuclear) weapons. The powerplant was two-shaft 36,000shp steam turbines. Study 9D followed the Fleet Requirements Committee recommendation for greater hangar height to accommodate the future HSS-2 helicopter

Above: Preliminary Study 6D (January 1960) was an early version of the small escort cruiser (more a helicopter carrier) armed with the US Tartar missile instead of Sea Slug (Sea Dart did not yet exist). On 5,400 tons she could accommodate eight S58 (Wessex) helicopters, but she had no margin of growth to accept the planned Sea King. Dimensions were 430ft deep waterline, 459ft overall x 55ft; she would have had a two-shaft steam powerplant (36,000shp) for better than 26kts deep and dirty, and an endurance of 3,500nm at 20kts. The launcher was the standard US type (Mk 11) with forty-two missiles, as in US *Adams* class missile destroyers. There would not have been a computer combat system.

(22,000lbs). That increased displacement to 6,600 tons. Study 9E (6,730 tons) increased deck area by moving the superstructure into a starboard island; it could operate four helicopters simultaneously.

Series 6 and 9 were to destroyer standards, while Series 21 was designed to cruiser standards, including the ability to operate as a Grade II flagship. Endurance was 4,500nm at 20kts. Study 21D was a Series 9 ship with splinter protection around vital spaces. That boosted displacement to 8,530 tons (length 535ft). Two machinery units (total 40,000shp) would maintain the required 26kts 'deep and dirty' speed. Study 21H2 had the island superstructure of 9E and also had a twin 4.5in gun above the flight deck facing aft. Missile stowage increased to forty-four (somewhat more than in a 'County' class destroyer) with twelve ready-use rounds. Freeboard forward was increased to keep the launcher, in the bow, dry. Displacement was 9,500 tons on the same length of 535ft. Study 21J2 had two twin 4.5in, but otherwise matched 21H2. Displacement rose to 9,700 tons. Study 21K had the guns mounted forward and the Sea Slug aft as in the missile destroyer, with the flight deck between them. Missile capacity was reduced to twenty-eight and the internal arrangement was awkward. Displacement increased to 9,850 tons (560ft). Study 21L2 was similar to 21J2 but had

THE MISSILE AGE

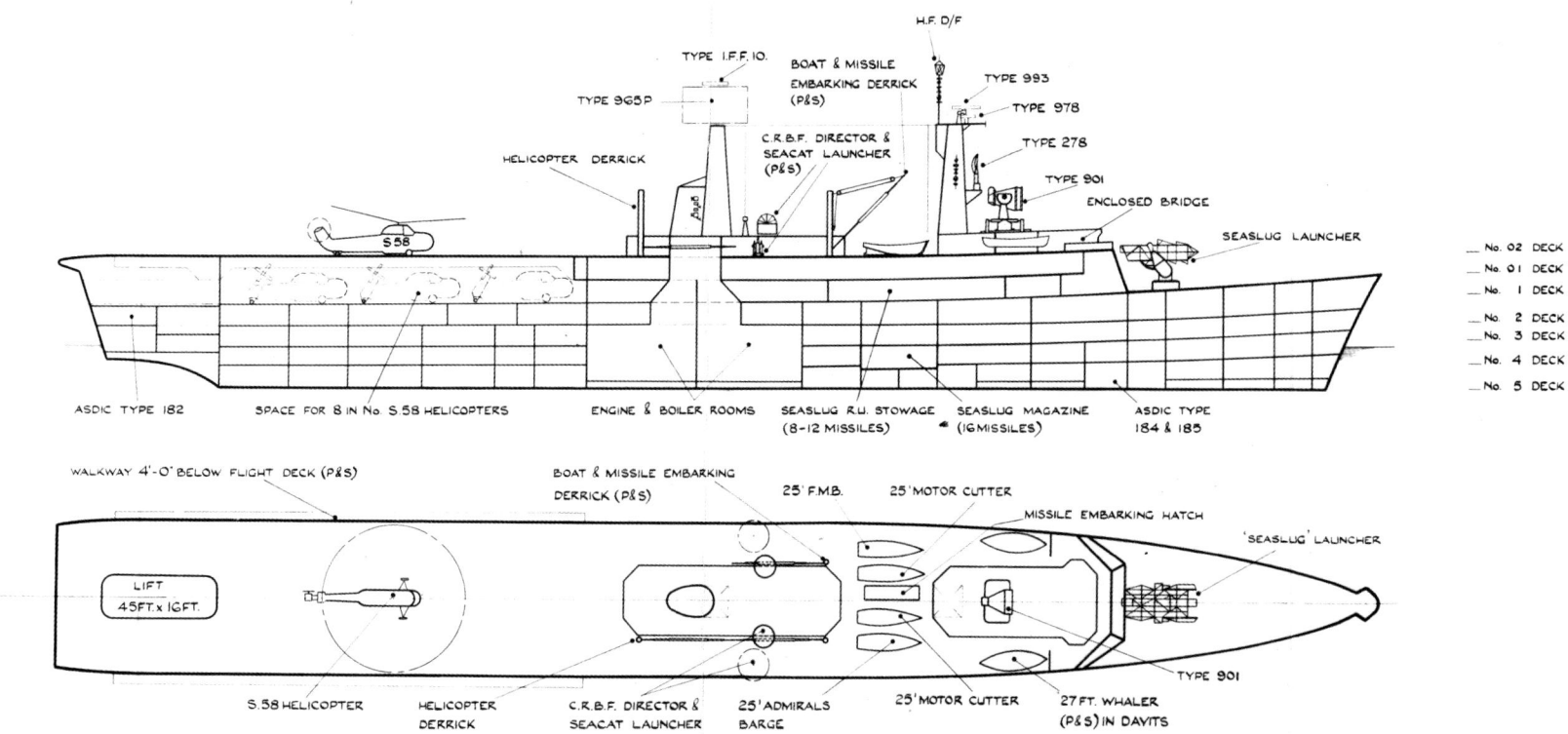

Above: Preliminary Design Study 9C (March 1960), for the escort cruiser, was a Sea Slug variant not too different from the contemporary Italian semi-helicopter carrier/cruisers. She would have been 485ft long on the waterline; beam and displacement depended on hangar height (58ft, 6,400 tons for 16.5ft, 59.3ft, 6585 tons for 18.5ft). The lower hangar was suited only to the S58 (Wessex) helicopter, but a height of 17.5ft or 18.5ft would also suit the coming HSS-2 (Sea King). The ship would be armed with a twin Sea Slug with twenty-four to twenty-six missiles and the control system then being installed in the similar-size 'County' class missile destroyer. For her helicopters, the ship would carry 150 Mk 43 or Mk 44 lightweight torpedoes. She would have both a jammer (Porky) and an intercept system (Cooky). Accommodation would be provided for sixty-one officers and 530 ratings.

greater power (60,000shp) for higher speed (29.5kts clean, 28.5kts 'dirty'). Displacement rose to 10,250 tons (550ft). There was at least one further study, not reported to the Board: 21M1.[24]

In September 1960 First Lord told the Minister of Defence about the planned Escort Cruiser programme. On 20 October the Board considered a joint proposal by VCNS and DCNS to build five escort cruisers, already approved in principle by the Board, with an analysis of the financial and manpower issues involved. The first two escort cruisers would replace the conventional cruisers *Swiftsure* and *Belfast* in the current '88' fleet plan. On completion of the second two, two *Tiger*s would be reduced to operational reserve. Four guided-missile destroyers (already deferred) would be further deferred until after 1970, cutting the overall fleet by one hull, which in turn could be replaced by one additional *Leander*-class frigate. The escort cruisers offered much the same fleet air-defence capability as the destroyers, one Sea Slug launcher each. A planned helicopter carrier programme would be dropped, and the four DC *Daring*s would be refitted rather than modernised. Delaying the missile destroyers would make it possible to arm them with the more

capable Sea Slug II missile. The changes would save £55 million in the long-term plan for the period before 1970, but more would be spent after that date. At this time the desired cruiser force was five ships. By 1967/8 it would consist of the three *Tiger*s and the first two escort cruisers. The escort cruiser as now envisaged was a 10,000-tonner with 45 officers and 550 ratings, presumably the 21L2 design.

In December a formal proposal was made to include the ships in the coming five-year plan. The concept was presented to the Minister of Defence in May 1961; the Chiefs of Staff endorsed the five-ship building programme in December 1961. At the start of the 1962 Long Term Costings the ten-year programme showed five ships, the first two to be ordered in 1964, the next two in 1965, and one in 1975. By this time estimated displacement was 13,500 tons. Ships would be armed with one Sea Slug II system (twin launcher) and twin 4.5in guns, and they would carry eight or nine large helicopters in the Chinook (or Sea King) class. Complement had grown to 70 officers and 900 ratings, and estimated cost was £21.6 million. Because of its greater endurance, the Chinook was easier to operate alongside fixed-wing carrier aircraft than earlier helicopters. The Naval Staff now considered that the carrier had to have some ASW helicopters on board, as she should not have to depend on accompanying units for such cover. It also seemed that a single Chinook or would be equivalent to two Wessex in either the ASW or the commando (amphibious) roles. For ASW, five Chinook could be equated to eight Wessex. On board a carrier, one Chinook would replace one strike or airborne early warning aircraft.[25] When cuts had to be made in the 1962 long-term programme, they included elimination of the nuclear warhead that made Sea Slug II an effective anti-ship weapon. Manpower had to be cut just as that required per escort cruiser grew. Also, a new air-defence missile (which became Sea Dart) was emerging. The fifth ship was removed from the programme, and the escort cruiser

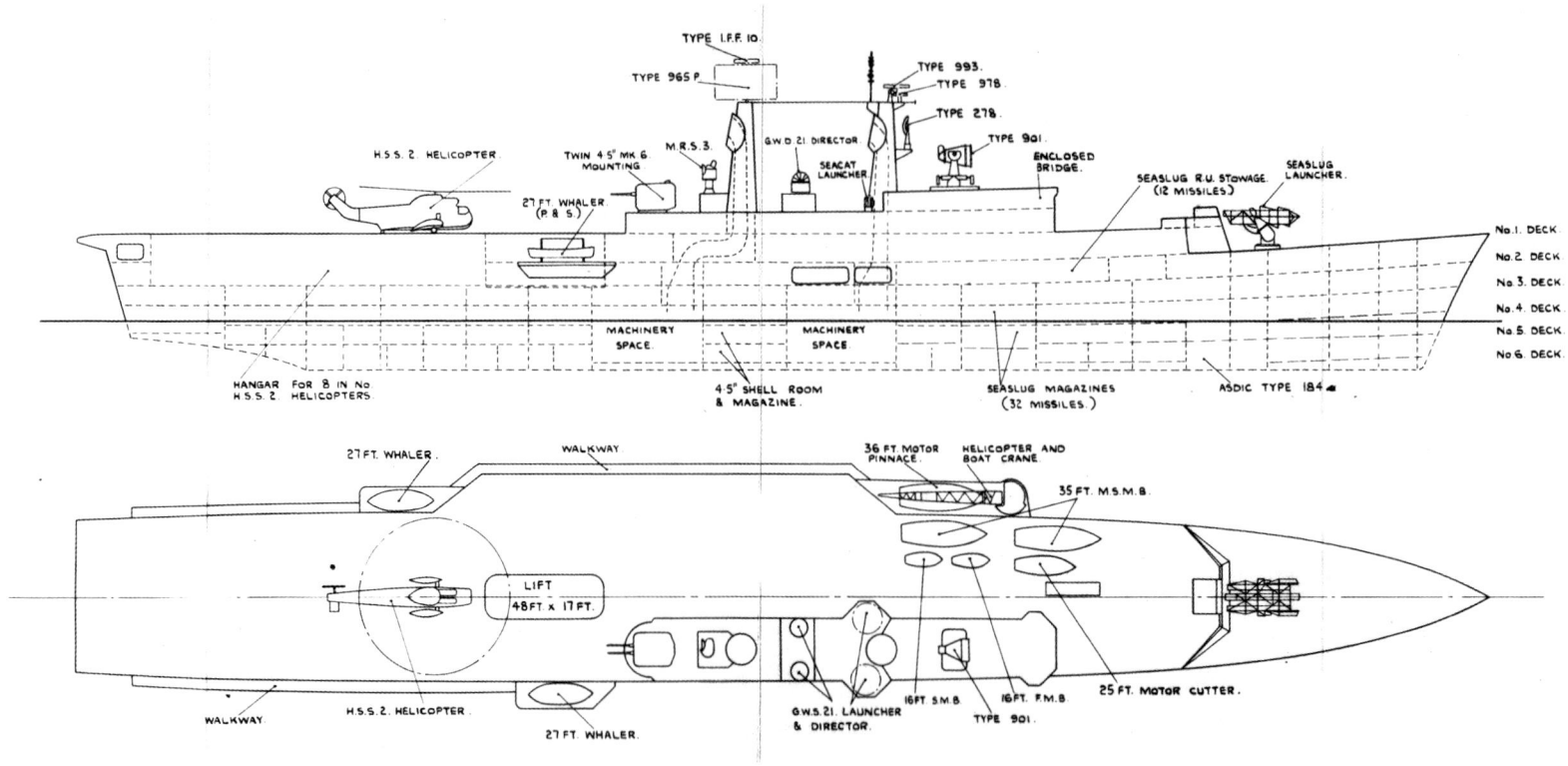

Above: Study 21H2 (October 1960) was a much-enlarged ship (9,460 tons) with a through deck, approaching what was eventually adopted in the *Invincible* class. Dimensions were 535ft (deep waterline) 573ft (overall) × 65ft × 19.7ft. There was enough internal volume for more missiles than a 'County' class missile destroyer could accommodate (forty-four). The ship would have both a computer combat system (ADA) and a TIDE (Link 11) data link. She would have a 40,000shp steam powerplant (two shafts) which could propel her at better than 27kts deep and clean (26kts-plus deep and dirty); endurance would be 4,500nm at 20kts or 6,000nm at 12kts. Accommodation would be provided for seventy-five officers and 635 ratings. There was sufficient space for eight HSS-2 (Sea King) helicopters.

design rethought.

The modified design, which was included in the 1963 Long Term Costing, would have displaced 10,500 tons and carried five Chinooks (roughly equivalent to eight Wessex). It would be armed with the new air-defence missile, with the Ikara long-range ASW missile, and with a pair of short-range anti-ship missiles (the French wire-guided SS 11). She would have the new Anglo-Dutch 'Broomstick' three-dimensional radar and a digital combat system (ADA/TIDE). Speed would be 27kts. Complement would be cut back to 70 officers and 700 ratings, and cost to £16.8 million. On the basis that the first Sea Dart system would not become available until 1969, the planned ordering dates were deferred, so that the first ship would be ordered in 1966 for completion in 1970. As compensation, two of the last four planned missile destroyers were brought forward. The deployment of ASW helicopters in ships other than carriers was also deferred. The projected dates required that the Board approve Major Characteristics in June 1963 and Staff Requirements in December, which dates could not actually be met. The first three ships would replace the three *Tigers*.

The escort cruiser role was recast to providing ASW helicopters in sea areas which could not be covered by the few fleet carriers (as it was no longer considered essential to free those ships of their organic helicopters). The shrinking air group of the escort cruiser reduced its potential in this role. On the other hand, the Cabinet was already cutting the carrier force to two operational ships, so there would be many such areas. The cruiser role in air defence also came into question, because Sea Dart could fit on board a smaller ship (it was conceived to replace the standard twin 4.5in gun mount, although that proved impossible). The case for the escort cruiser declined, VCNS declaring that it was not absolutely critical. Moreover, the Staff Requirement for an escort cruiser was linked to the carrier programme. Until 1964 that seemed uncertain. When a new long-term programme (including one rather than two carriers) was approved, it was possible to insert escort cruisers into the 1964 Long Term Costings. A decision was needed by February 1964.

Meanwhile it was considered vital to gain experience with large ASW helicopters from fleet units other than carriers. C-in-C Home Fleet suggested informally that the *Tigers* be converted to carry helicopters. An alternative proposal to build more helicopter training ships like HMS *Engadine* was considered and rejected.[26] As of September 1963 the three alternatives were either to continue with the escort cruiser programme, to cancel the escort cruisers in favour of *Tiger* conversions, or to postpone the escort cruisers while converting the *Tigers*. Ships designed for the purpose would be better, and their additional anti-aircraft missiles would be welcome, it being unlikely that the MoD and the Treasury would approve providing that firepower in the form of extra *Bristol* class destroyers if the escort cruisers were not built. Moreover, the escort cruiser programme had been accepted, if not formally approved by MoD and the Treasury. However, no heavy helicopters at all would become available before 1971 at the most optimistic, and five Chinooks was not much to get from so expensive a ship. The project would overload the design staff, so some other project would suffer (a pencilled comment: 'cannot touch for 2 years'). Converting the *Tigers* would bring the heli-

THE MISSILE AGE

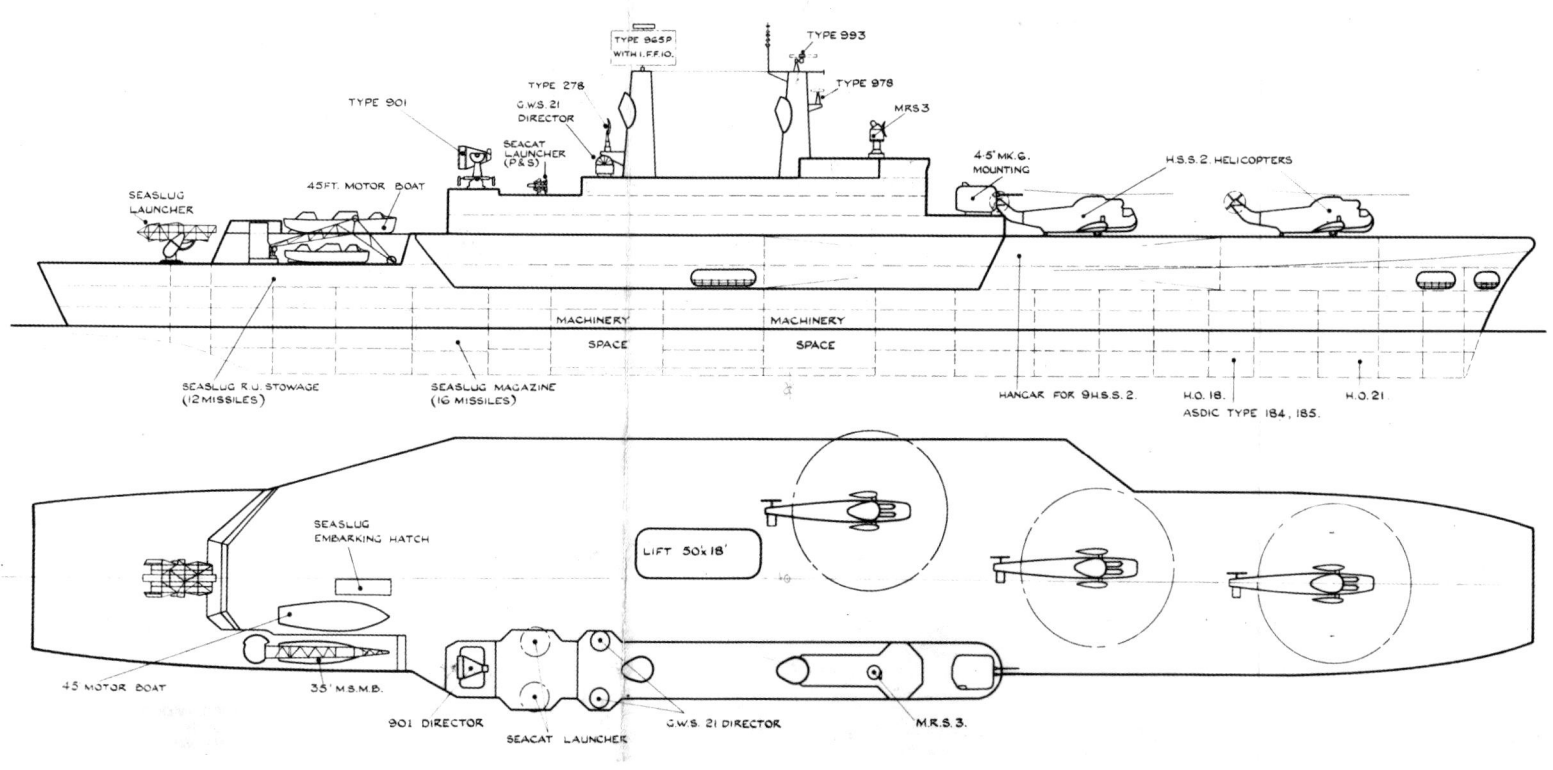

Above: Escort Cruiser Design 21M3 was dated March 1961. Deep displacement would have been 11,800 tons. Dimensions: 570ft deep waterline, 603ft oa x 72ft x 20.2ft. She would have had a 45,000shp (two-shaft) steam powerplant, giving her a speed of over 27kts deep and clean (over 26kts deep and dirty); endurance at 20kts would have been 4,500nm. The ship would have carried nine HSS-2 (Sea King) helicopters, with one 50ft x 18ft lift. Other armament would have been a twin Sea Slug launcher (twenty-eight missiles), using a GMS 2 Mod 1 control system, two quadruple Seacats (GWD 21 directors), and one twin 4.5in mount with an MRS 3 director. The ship would have had a digital combat system (ADA) and the TIDE (Link 11) data link; radars would have been Types 901, 993, 978, 278, 965P and 262 (for Seacat). ESM would have been the then-standard UA-8/9/10. Asdic would have been Type 184. Accommodation was 104 officers and 941 ratings.

copters to sea at least five years earlier and would save a great deal of money (although it would leave three old hulls instead of four new ones). The *Tiger*s would gain a useful role after 1966 and their lives would be prolonged. The strain on design capacity would be reduced and the dockyards given more work during a lean period. On the other hand, the fleet would lose anti-aircraft capability and the *Tiger*s could not accommodate nearly as many helicopters as could escort cruisers. Given their age, the *Tiger*s would need further expensive refits to keep running for more than two or three more years. In effect converting the *Tiger*s would only defer the need for new helicopter ships, and past experience showed that MoD and the Treasury would soon forget the saving involved. It was not clear how long the escort cruiser could or should be postponed, because that depended on further factors such as future amphibious (helicopter) carrier policy.

As of September 1963 Plans Division preferred to convert the *Tiger*s at once and defer (but not abandon) the escort cruiser. Conversion would give the ships at least six more years of useful life. Three preliminary *Tiger*-class conversion studies were made, of which a scheme which

accommodated four Wessex helicopters and retained both the forward twin 6in gun and all the 3in guns was preferred.[27] The idea was discussed by the Fleet Requirements Committee and then by the Board in October 1963, and the design further developed, to allow a firm decision in January 1964, so that work could begin early in 1965. The helicopter deck was built up from the ship's stern, a hangar (stowing the helicopters side-by-side) replacing the after superstructure and the after 6in turret. Spaces under the flight deck included stowage for the helicopters' Mk 44 torpedoes and also a maintenance space. The ship's Type 960 radar was replaced by the narrower-beam Type 965M, which was now standard (and was needed for air control), and the Type 277Q was replaced by a similar (more maintainable) Type 278 height-finder, the type used in missile destroyers. Director-General Aircraft pressed for roll-stabilisation, pointing out that otherwise flight operations would be severely limited. Compared to destroyers and frigates, the cruisers were particularly affected by pitch because their flight decks were so far from the centre of pitch, hence subject to the greatest distances of rise and fall. DGS countered that there might not be space for stabilisers. The Ship Characteristics Committee approved fitting stabilisers (of the type used in the 'County' class missile destroyers). The Board formally approved conversion at its meeting on 26 January 1964, the order of conversion being *Blake*, *Lion* and then *Tiger*. The first would be taken in hand early in 1965, the others at twelve-month intervals. In November 1964 Director of Naval Air Warfare pointed out that recent exercises had shown that large helicopters enjoyed a considerable advantage over the smaller Wessex, so he asked Director General Ships whether the *Tiger*s could be adapted to the larger aircraft. Director General Ships thought the hangar would have to be extended aft onto the flight deck, so that it could take only one rather than two helicopters at a time. Moreover, it was far too late to change the design, if ships were to be converted as planned. However, it was inexpensive to strengthen the deck to handle

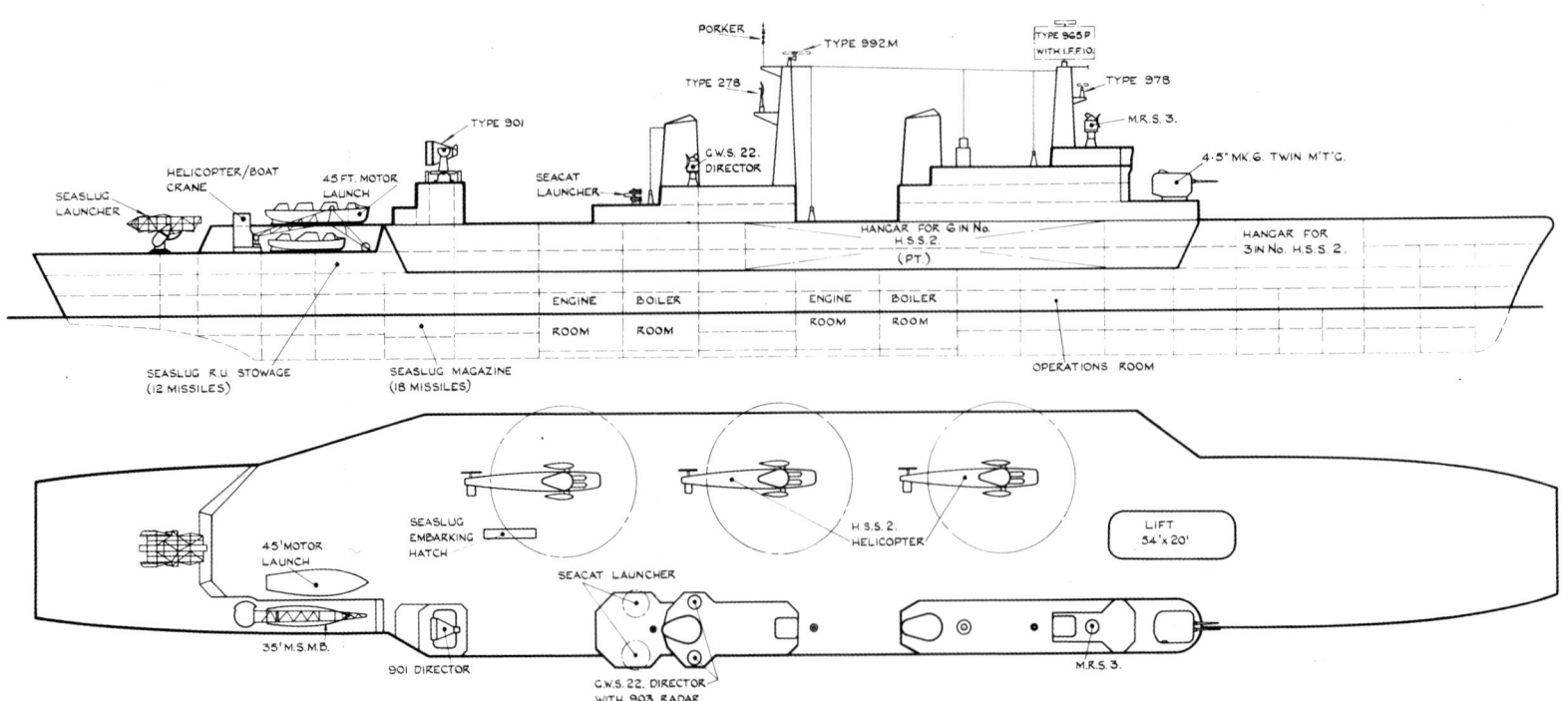

Above: Escort Cruiser Design 21M8 was a larger through-deck cruiser. This design was dated September 1961. When the cruiser was revived in the wake of the cancellation of the carriers, this design was apparently the starting point for what became HMS *Invincible*. Deep displacement would have been 13,550 tons, carrying the same aircraft and weapons as 21M3 (but with thirty Sea Slugs using the full GWS 1 Mod 1 system of a 'County' class missile destroyer) and a better Seacat control system (GWD 22), and also presumably carrying more Mk 44 ASW torpedoes (144). Instead of the Type 993 target-indication radar, the ship would have had the far more capable Type 992. Like 21M3, she would have had ADA and TIDE. Dimensions were 620ft (deep waterline) 650ft (overall) x 73ft; the four-shaft steam plant (presumably using frigate machinery) would have produced 60,000shp for a speed of 28kts deep and dirty. Endurance would have been 5,000nm at 20kts or 5,800nm at 12kts. Accommodation would have been 106 officers and 970 ratings.

the heavier helicopters.

Four Wessex I helicopters were earmarked for *Blake*, and plans called for forming two flights of Wessex III helicopters for the *Tiger*s. However, largely for manpower reasons, the 1965 review assumed that the cruisers could not be commissioned until the carriers were run down, in which case the aircraft would become available. The new helicopters were therefore deleted from the 1966 Long-Term Costings. In September 1966 the Board decided that the conversions should accommodate the more modern (and much larger) Sea King. That was possible because the helicopter hangar was on the same level as the flight deck. Instead of extending onto the flight deck, it was extended forward, the two waist 3in mounts being replaced by much more compact Seacat missile launchers and their directors. The Board decided to complete two ships: *Blake* in October 1968 (two years late) and the other in July 1969. *Lion* had been reduced to care and maintenance rather than immediately beginning conversion. The Board allowed for a third conversion later, if it was approved.[28]

Once begun, the project fell afoul of the 1965 Defence Review. *Blake* was taken in hand at Portsmouth in April 1965 but suspended in August pending the outcome of the review. She was suspended until March 1966 because yard capacity had to concentrate on the refit of the carrier *Victorious*, and then continued at reduced speed due to labour shortages and urgent yard commitments; the specification was changed early in 1967. *Blake* was further delayed by a serious fire in January 1969, completing only in April 1969. Once completed, she suffered a series of machinery failures and had to be taken in hand at Portsmouth for twenty-eight weeks to rectify defects. *Tiger* paid off in December 1966 and was placed in preservation in January 1967 for a refit beginning that July at Devonport for completion in late 1969. In April 1971, cost having overrun by £3.16 million on an original estimate of £5.4 million; later the overrun was £8.25 million. At that time completion date was May 1972 (later it was July 1972). The ship was delayed partly by unexpected deterioration while in the dockyard. *Lion* was due to begin her refit in January 1966, so a decision on her future was urgent. She was running afoul effort to refit the carriers *Bulwark* (amphibious) and *Hermes*. The 1970 Supplemental Statement on Defence Policy announced that she would not be converted at all. The experience of conversion was so bad that it generated an official enquiry. *Lion* never emerged from reserve and *Blake* was placed on the Disposal List on 3 December 1980.

Defending the conversion programme in October 1966, Director of Plans pointed out that if the carriers were phased out without replacement, the *Tiger*s would be the only big ships to be 'put into the breach, not only to fill the ASW/troop-lifting role but also to pave the way for a replacement big ship capable of undertaking a similar role. Once this role is allowed to lapse, it will be difficult to resurrect it.'

Early in 1967 there was a proposal to eliminate the 3in/70in guns these ships in favour of Seacat; Assistant Director of Naval Tactical and Weapons Policy defended the gun as a mature, reliable system offering (with its new fuse) significant surface-attack potential, which the Seacat missile lacked. The ships ended up with two Seacats (twenty-eight missiles in all) and one twin 3in/70 in 'B' position. In February 1975 the Board rejected a proposal to replace the remaining 6in mount with four Exocets. The *Tiger* conversion in effect killed the specially-built escort

THE MISSILE AGE

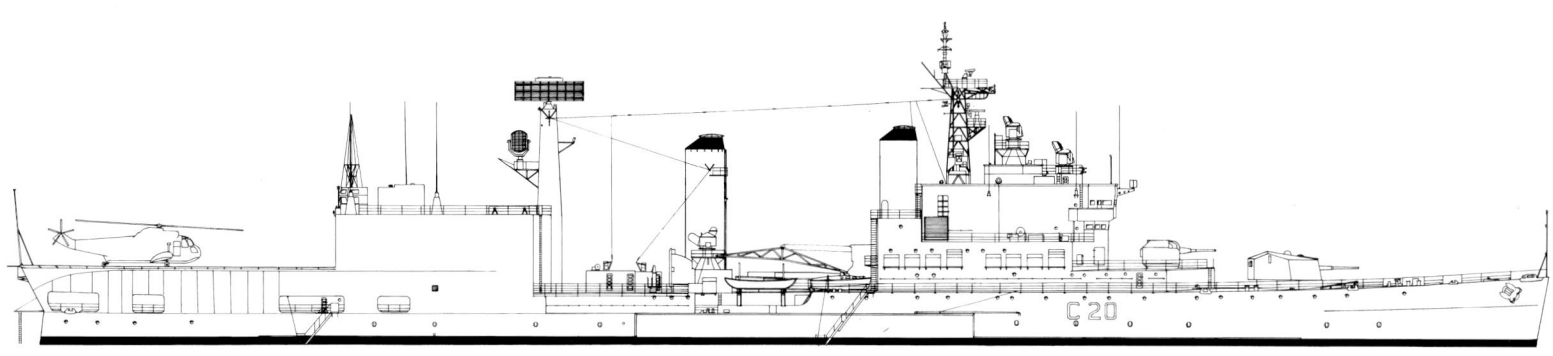

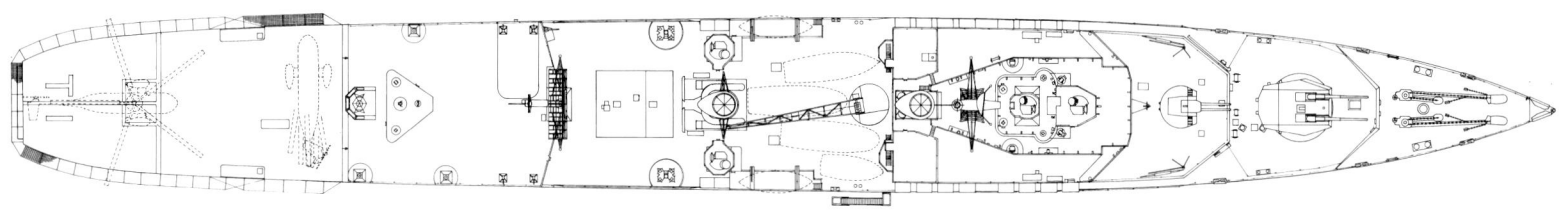

Above: Tiger as an escort cruiser/helicopter carrier, 1972. During conversion she received four sets of fin stabilisers. She retained her sonar. *Blake* was similar, but had the elaborate HF whip installation on her forecastle just abaft the capstan instead of atop the 'pri-fly' atop the helicopter hangar where on *Tiger* it could be folded horizontally to port. *Blake* retained her original funnel height, but on *Tiger* both funnels were raised. There were numerous other small differences in detail between the two, especially with regard to communications antennas and ventilation duct arrangements. Only three of the four helicopters could be stowed in the hangar, two fore and aft and one athwartships in the forward part of the hangar (nose to port). The hangar roof above the rotor hub was raised to permit maintenance and blade replacement on the rotor hub. Note that the boot-topping obscures the fore and aft extension of the external belt armour. The steering gear spaces at the stern were also armoured, as well as the Seacat reload magazine between the launchers (thickness not known). Of the four retained MRS 3 Mk 1 directors, the two on the forward superstructure had revised operator cabs, while the pair assigned to Seacat control retained their hemispherical operator station covers (*Blake* always had the revised configuration on all four directors). This drawing was based on a sparsely-detailed class configuration drawing issued in July 1967; many of the details shown were added based on photographs. (A D Baker III)

cruiser for the time being, and the project was dropped in the turmoil following cancellation of the new carriers and an accompanying re-think of the shape of the fleet.

The Command Cruiser and *Invincible*

The escort cruiser was conceived as part of a fleet built around carriers as its striking force, but in 1965–6 that basis was destroyed when, successively, the carrier replacement programme was cancelled and then the East of Suez mission which had largely (though not completely) justified it was abandoned. The carriers and East of Suez did not vanish instantly, but the logic of any future programme now had to be rethought. The Royal Navy continued to argue for a global maritime mission, but the key naval mission was now the defence of NATO sea communications in the eastern North Atlantic. The Royal Navy argued successfully in 1966–7 that frigates alone could not suffice, even for anti-submarine warfare. It was possible to distribute air-defence and ASW weapons among relatively inexpensive surface combatants only if they operated with a command ship – which should be built to cruiser standards.[29]

In July 1967 the Cabinet approved the shape of the future surface fleet, which would be built around three types of ship: the Type 42 Sea Dart destroyer, the Type 22 ASW frigate (with Type 21 as interim), and a command cruiser. The Naval Staff made the case for a command ship, whether or not she might carry helicopters. It was described as the core of the planned Maritime Contingency Force (MARCONFOR), the British naval contribution to NATO, which would offer a combination of anti-submarine, anti-air and anti-surface capabilities. Such a force, created *ad hoc*, could dominate a selected sea area, conduct offensive anti-air and anti-submarine operations, offer seaward support for amphibious operations (sealing off the beach from enemy seaborne attack), and could offer distant and close support of sea lines of reinforcement/supply to the UK and to Western Europe. Plans envisaged two to three such forces, each consisting of a command/helicopter ship (then designated CCH), ten surface combatants and nuclear submarines. This was in addition to British amphibious forces.

The command cruiser would provide the command and control capability which otherwise would reside in a carrier, now being given up. In addition to whatever ships might be operating with the command ship, she would control co-operating land-based aircraft (as the Royal Navy had been doing for many years). This was particularly important (at least in justifying the character of the ship) because part of the rationale for giving up carriers in the new NATO context was that the mission of their air groups in maritime warfare could be taken over by land-based aircraft. This sort of command and control required, above all, elaborate computer systems with large numbers of operators – and hence considerable internal volume in a ship.[30] By 1969, the first two programmes were well under way, and it was time to define the command cruiser and to approve the three-ship programme. In 1967 the command ship was

envisaged as a 10,000-tonner – essentially the earlier escort cruiser – which would cost £30 million.

In effect the converted *Tiger*s were the bridge between the strike carriers (twelve Sea Kings) and future capability. They had the necessary sea-keeping and support facilities – if only for six Sea Kings each – and HMS *Blake* had sufficient command and control for her own aircraft. Much as had been argued in 1960, the bare minimum helicopter outfit for a force was six Sea Kings, and that was now equated with at least a 12,000-ton ship. By the time the ship entered service, the Sea King would (it was imagined) be nearing the end of its life. That eventually justified 'stretch', in the form of capacity for nine rather than six helicopters.[31]

Since the ship would already need enough command and control for her helicopters, she might as well be extended to accommodate force command and control. She was too important to rely on other ships for her defence, so she should have her own area air-defence weapon. Moreover, placing Sea Dart on board the command cruiser (which would be part of an integrated force in any case) reduced the number of relatively expensive Type 42 destroyers required, without requiring extra hulls. The MoD Operational Requirements Committee approved this single-package idea. It asked for three parallel design studies: an austere Study 21 (reflecting capabilities put forward to the Chancellor of the Exchequer) offered Sea Dart, command and control, and six Sea Kings. Study 22 increased to nine helicopters, with an option for twelve. Study 23 made additional provision for bottom-bounce sonar on a 'for but not with' basis, as this sonar might become available at half-life in the late 1980s. Provision for VSTOL aircraft was added later. Study 21 implied a half-deck solution, about the size and shape of the *Tiger*. If the number increased to something between six and nine helicopters the most economical solution had a through-deck, because of the numbers of aircraft and the need to hangar them. Because it seemed unlikely that provision

THE MISSILE AGE

probe, identification (e.g. of surface ships), limited air defence and organic strike. At times it might be desirable to carry VSTOL aircraft instead of some helicopters. The MoD Operational Analysis unit was conducting a study of maritime strike, including the possible role of VSTOL aircraft.

With nine aircraft, Study 22 would cost £35 million in 1968 prices, and it would cost another £1 million to add three more aircraft. The Naval Staff therefore pressed for a twelve-helicopter version of Study 22. The ship had to have enough hangar space for her whole aircraft complement, because with so few, wastage due to exposure would be a serious problem. The ship was expected to displace 18,000 tons and to cost £37.5 million at 1968 prices (later the price was readjusted). The earlier 10,000-tonner approximated to Study 21. Hopefully nearly everything in the ship would be more or less off the shelf. Even the command-and-control system would be adapted from that in HMS *Bristol*. Three ships were needed to keep two operational, a step down from the three-cruiser fleet envisaged a decade earlier. As proof that austerity was not worth much, a scaled-down ship with a through deck but only six helicopters and point rather than area air defence (Sea Wolf rather than Sea Dart) was expected to cost £32.8 million – and the MoD analysts agreed that it was not worthwhile. Cuts to reduce cost included reducing stores endurance from sixty to forty-five days and fuel endurance from 7,500nm to 5,000nm.

In effect the cruiser was successor to the big missile destroyer *Bristol*, which had been justified as a command ship even after the carrier she was designed to escort was cancelled. The Sea Kings were the one element of the carrier air group which remained once the emphasis shifted away from carrier-based strike. This was not too different from the escort cruiser, the main difference being abandonment of shore bombardment (Sea Dart had some anti-ship capability). The key questions were how many Sea Kings the ship would operate, and also whether she should be adapted to the emerging Harrier VSTOL attack aircraft. Helicopters could operate from a platform at the stern of the ship, but a Harrier should be able to make a rolling take-off, hence needed a carrier-like 'through' deck.

The 'command cruiser' morphed into the light carrier *Invincible*, whose large superstructure suggests disproportionately large command spaces (actually they are mostly buried in her hull). She was also a direct descendant of the escort cruiser. With the revival of large British carriers, and with the appearance of frigates (Type 23s) which carry single-package Merlin helicopters, it is no longer so clear that a specialised escort cruiser is needed, and the *Invincible*s may thus have ended the cruiser story as they revived the British carrier force.

for a big bottom-bounce sonar (the British envisaged a 120ft transducer, 8ft deep) would be successful, Study 23 was abandoned.

US Navy experience was cited to show that it took two big helicopters to contain and track (and attack) a nuclear attack submarine. To maintain two on task for ten days required at least seven helicopters, in effect killing Study 21. That number would not allow any flexibility, for example to prosecute a second contact or conduct screening or reconnaissance. These arguments could justify much larger numbers of helicopters; the Staff took twelve as a reasonable compromise. Moreover, the through deck offered the greatest potential for accepting helicopter growth over the lengthy (25-year) life of the ships. Thus the through-deck option, which made VSTOL operation possible, was justified without reference to VSTOL aircraft. Yet extended discussions with the RAF and within the Sea Air Warfare Committee had identified a variety of attractive roles for a small number of embarked VSTOL aircraft:

Above left: The converted *Tiger*s were a half-way step towards a fully air-capable cruiser. Although it must have seemed to many that calling the *Invincible*s 'through-deck cruisers' was nothing more than a sop to anti-carrier fanatics in and out of the Royal Navy, in fact in their emphasis on command and control and in their original heavy area-defence missile armament these ships were cruisers whose main battery was ASW helicopters, supported by Sea Harriers, rather than guns. Later they evolved into something closer to a light carrier. The through-deck cruiser idea recalls US interest in 'flight deck cruisers', which combined substantial air capability with conventional cruiser weapons, about 1930. No Royal Navy equivalent seems to have existed, although in the 1930s the Royal Navy certainly was very much interested in using cruiser and battleship aircraft for strike and even fighter functions. HMS *Invincible* is shown in September 1991, after she had surrendered her area-defence missile battery and added more carrier capability in the form of more aircraft supported by, among other things, more capacity for aviation fuel.

Adventure is shown on 10 April 1942 after refit.

APPENDIX 1
FAST MINELAYERS

During the First World War, several cruisers were converted as fast minelayers, and proved useful enough for Controller to request a specially-built ship in July 1918.[1] The first requirement was high enough speed to slip into nominally German-controlled waters overnight, i.e. 30–35kts seagoing. DNC suggested basing the ship on either *Cavendish* (the *Hawkins* class cruiser being completed as the carrier *Vindictive*) or the new 'E' class, preferably *Cavendish* with a slight bulge (he wanted to try for 32kts). Economical speed would be no less than 20kts, which was appreciably higher than the speed typically aimed at. Radius of action should be at least 2,000nm at 20kts, and 1,000nm at full speed. To get into enemy-controlled waters, the ship had to have a relatively shallow draft, not more than 12–13ft (DDOD(M) considered draft the least important item, however, and DNC decided to adopt the same draft as in *Cavendish*). The ship should carry at least 400 moored mines, preferably all internally, on one mining deck (existing minelayers used canvas screens to conceal the mines they carried on their upper decks). The mining deck, the ship's main deck, had to be at least 430ft long to accommodate this many mines. The need to conceal the minelaying mission also shows in a requirement that the ship resemble a modern warship as much as possible. Devoting the whole main deck to mines would force officers onto the upper deck and ratings onto the lower deck.

How high the mining deck could be (the mine drop should not exceed 12ft) depended on the nature of the stern wave at speed; therefore, a cruiser stern was desirable. Mines would be stowed on four sets of rails, as straight as the curve of the ship would allow, with four

traps (for launching) and without any interconnections (points). The ship would have four embarking hatches on each side, each with its own stump mast and derrick, for quick loading. Hatches should lead directly onto the rails, so that mines and their sinkers could be loaded directly onto the ship. Mines would be moved aft to launch by an endless chain or conveyor belt, i.e. should not need bogies (this idea was inspired by US systems then operational for the Northern Mine Barrage). Armament would be six 4.7in or eight 4in guns (the latter option was chosen) and two 3in anti-aircraft guns.

A sketch design showed almost no protection, because mines accounted for so much tonnage. The big *Cavendish* devoted 616 tons to armament, while each mine weighed 1,850lbs, so 400 of them accounted for 330 tons. Allowing another 1¼ tons per mine for rails, etc brought that to 500 tons, and winches and other mine-related equipment might add 150 tons more, a total of 650 tons before allowing for the gun armament (140 tons). If the minelayer were based on *Cavendish*, it would be about the same size (11,375 tons deep compared to 11,630 tons). The endurance requirement was difficult because at maximum speed the ship would be burning twice as much fuel per nautical mile as at 20kts.

Alternatively, the minelayer could be based on the smaller and faster 'E' class cruiser, with its lighter machinery (this version had 80,000shp machinery rather than the 40,000shp of the *Cavendish*-based design). In an 'E' class cruiser, armament plus protection amounted to 1,055 tons; in addition to minimal protection (90 tons in the earlier design), the ship needed only 790 tons for mines and gun armament. Apparently this too was unsatisfactory, because within a few days Controller decided to abandon the large ship and go instead for a Design G based on the 'D' class cruiser: 420ft x 42ft x 15ft deep, 4,100 tons, making 30kts deep or 31kts light with mines on 52,000shp. To get down to this size, the mine load had to be halved to 200 (250 if possible) and armament cut to four 4in or 4.7in guns and one 3in HA gun. Mines would be carried on the upper deck. Radius of action would be as before, using 950 tons of oil (compared to 300 tons in a 'D' class cruiser). As before, the ship would be unprotected. A Legend dated 22 August 1918 showed a normal displacement of 4,100 tons, and a deep displacement of 4,850 tons, but there may have been some question as to whether the hull could carry the engines, and a revised statement showed 4,772 tons normal. A key issue in the design was whether E-in-C could provide 13,000shp in each of four boilers. Cost would be £480,000 and building time twenty-two months.

The project lapsed about in August 1918, perhaps because the end of the war was coming into sight, and was certainly less than twenty-two months away. However, the idea of the cruiser minelayer survived, to be revived in July 1920. Controller revived the 1918 requirements (at least 250 mines, to be carried internally on one mining deck) but specified sea speed and radius corresponding to those of a light cruiser of the same date and construction. The specified battery was a good anti-aircraft armament. Controller expected a displacement of 4,000 to 7,000 tons. DNC told his cruiser designer, Lillicrap, to base the ship on an improved 'D' class cruiser with bulges, to achieve 28kts in deep condition, and 29kts with half-oil; anything more would require a much larger ship (the PWQ Committee wanted 30kts deep, and 6,000nm at 14kts).[2] A first cut showed a displacement of 4,900 tons, but that soon grew to 5,400 tons, including 300 tons of oil fuel (1,300 tons fully loaded) and eight 3in anti-aircraft guns. Further analysis showed grossly insufficient stability, so beam had to be increased considerably.[3] As before, the weight of mines and associated equipment (375 tons) dwarfed that of the gun armament (70 tons). The ship was unprotected apart from bulges. She would have had 1,600 tons of oil to achieve the desired (cruiser) radius of 6,000nm. The 1,000nm and 2,000nm figures chosen in 1918 had been associated with a North Sea rather than a Far Eastern war. DNC ordered the use of 'D' class machinery.

In October, after attending a minelayer conference, DNC asked whether mines could be stowed on two decks and also whether it might be better to convert an existing cruiser. A *Chatham* could be converted to oil fuel, its upper deck covered over and used for mines. As much armament as possible would be retained. Candidates were *Southampton*, *Lowestoft* and *Birmingham*, all of which were currently serving on foreign stations (South America and Africa). Four guns would be removed, leaving three forward and one aft on the forecastle deck. The submerged torpedo room would be eliminated and converted to oil stowage, giving a total of 1,300 tons (for 5,000nm). About 200 mines could be stowed on the upper deck, fed by rails to two traps. The forecastle deck would be extended aft to cover this mine deck and the ships' sides built up. Conversion would cost about £231,000, compared to over £900,000 for a new ship. CNS turned down the idea late in October 1920, and attention turned back to the new minelayer.

The Staff considered that the 6,800-ton ship met the stated requirements, but the Staff asked for more mines. DNC offered to stow them on the lower deck, but that was considered objectionable.[4] It proved possible to rearrange the mine deck to stow twenty more mines (300 in all). Increasing the mine load reduced oil fuel capacity, but it was clear that machinery space was available (Royal Navy practice was to stow oil fuel below the platform deck, so added machinery volume could not be freed for oil). On 7 February 1921 DNC proposed adding a diesel, both to gain cruising radius and also to gain experience with cruising diesels. At a Controller's conference E-in-C agreed with the idea, and Controller concurred. The diesel would nearly double endurance to about 13,000nm at 13kts, one steam boiler being kept lit for auxiliaries. With the diesel, the ship would displace something over 7,000 tons, compared to the earlier 6,800 tons.

Both 6in and 4in guns were suggested as alternatives to the 3in HA battery, the choice falling on six 4in HA. A conference on 8 March 1921 decided that armament would be four 4.7in HA guns (otherwise installed on board the *Nelson* class battleships and the aircraft carriers *Courageous* and *Glorious*, with 200 rounds per gun) and two of the multiple pompoms just specified by the Naval Anti-Aircraft Committee. Armament weight increased from 110 to 140 tons. The design grew further.[5]

The new minelayer was HMS *Adventure*, the first post-First World War cruiser, albeit a special-purpose ship. As completed she had the first British transom stern. On trials, the transom created unusual water conditions, the wake suddenly spurting out above 24kts, and also created a suction effect which drew mines towards the stern after they were released, and the ship had to be rebuilt in 1931–2 with a more conventional rounded stern (which added 19ft to her length). This experience was reviewed when a transom stern was chosen (as a way of reducing resistance) for the *Fiji* class cruisers.

A second cruiser minelayer was included in the proposed 1924/5 programme, but it was not ordered.[6] At his conference on 12 November 1923 Controller decided not to fit a diesel; the ship would presumably have been a repeat *Adventure* in other respects (Controller specified the same 40,000shp powerplant). That made sense, as *Adventure* was experimental and would not be completed until 1926. At the next meeting (on 26 November) DNC said that as there were considerable alterations to *Adventure* required on account of different machinery arrangements and other requirements from various departments, the ship could not be ready 'at the early date given'. In December, DNC said that no progress on the new minelayer could be made until work on the new cruisers (the *Kent*

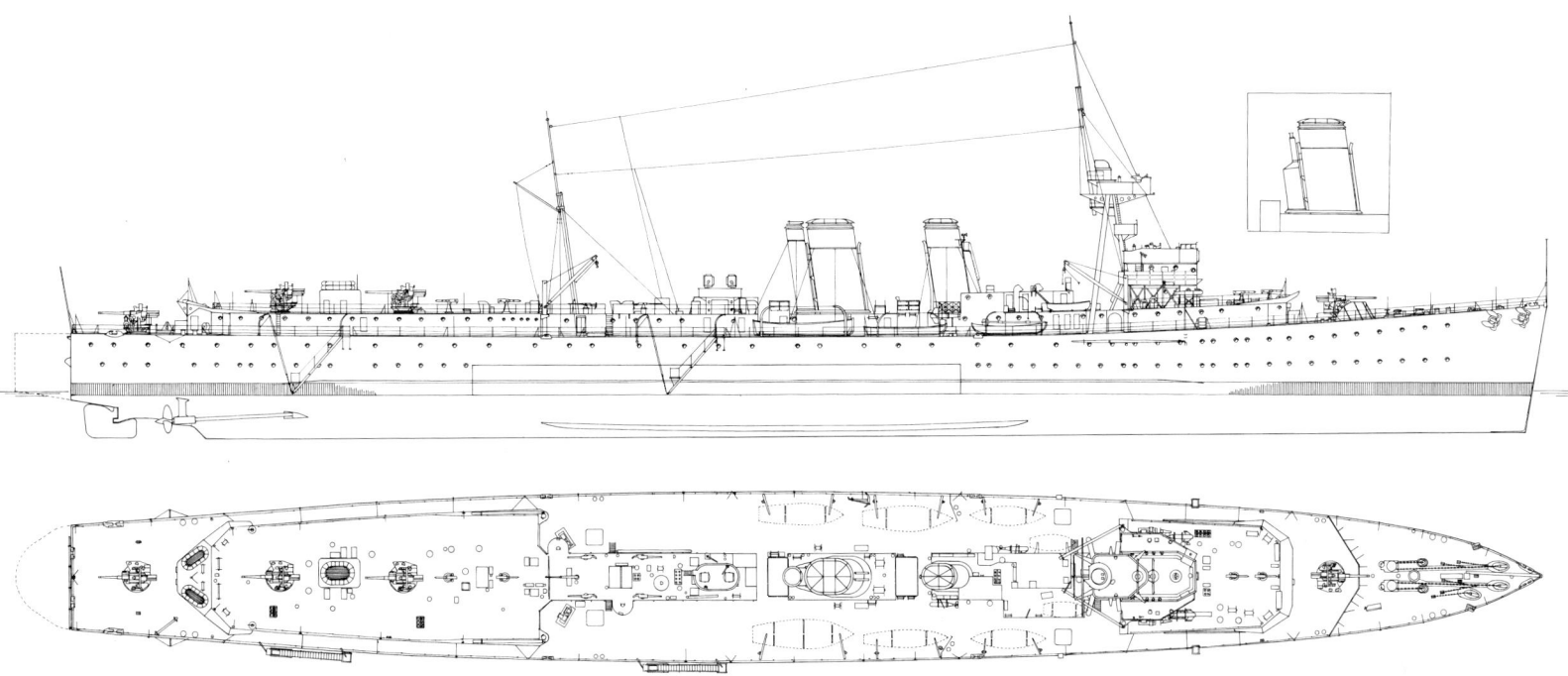

Above: Adventure as fitted on completion, January 1927. Dashed lines show how the stern was lengthened in 1932, overall length increasing from 520ft to 539ft. The inset shows the original diesel exhaust stack, replaced by the end of 1926. The diesels were removed altogether about 1938. The weather deck abaft the vertical spray shield at the bow was probably planked but is not shown as such in the drawing. The 1927 'as-fitted' plan shows a folding spray shield extension, but it appears either not to have been fitted or not to have been retained. Note that the forward mine-loading hatches served the outer mine rails and the after hatches the inner pair. By 1942 the single pompoms had been replaced by one octuple pompom in 'B' position, and the 4.7in director on the upper foremast platform had been replaced by a standard HA director carrying Type 285 radar. The foremast had been cut down, the mainmast restructured as a tall tripod, and nine single Oerlikons added. Design speed on steam power was 28kts, and by the middle of the Second World War her best speed was 27kts. At about the same time Vickers designed a cruiser-minelayer for Spain (Design 1040, 7 July 1923), to carry the same mine load (400) as *Adventure*. She had the usual Vickers cruiser battery of two twin and two single 6in guns, two 4in HA and two sets of 21in torpedo tubes. She would have displaced 5,750 tons normal (430ft pp, 458ft loa x 54ft 6in x 34ft x 16ft), making 25kts on 23,000shp (radius 5,000nm at 15kts). This design was probably related to an earlier unnumbered design, dated 19 June 1923, for a 4,150-ton (400ft x 50ft 5in x 24ft x 13ft) cruiser armed with three 6in/50, two 105mm HA and 400 mines; power would have been 20,000shp for 25kts (there was also a 15kt, 2,750-ton version armed with two 6in/50). No such ship was built (the Spanish Navy did later build some minelaying gunboats). The only inter-war equivalent to HMS *Adventure* was the French *Pluton*. In addition to the design for Spain, Vickers offered Turkey a small 'torpedo cruiser' (actually a minelayer, the title probably referring to the fact that mines were originally called torpedoes), Design 930 of 25 August 1927. She would have displaced 2,500 tons on trial (380ft x 37ft 9in x 23ft x 10ft 10in) and would have made 36kts on 52,000shp (five oil-fired boilers). Armament would have been sixty mines plus three 6in guns, two 3in HA, four single pompoms and two triple 21in torpedo tubes. Like the Spanish ship, this one did not materialise. When the Turkish Navy ordered four destroyers from Italian builders in 1930, two of them were equipped to lay mines. (A D Baker III)

Below: HMS *Adventure* as completed, showing the transom stern which caused such problems. She was given an unusually powerful anti-aircraft battery because aircraft would her main enemies during mining sorties. By 1937 the AA rearmament programme called for her to receive one octuple pompom and two 0.5in machine guns, plus one set of HACS III*C. However, that was not done for some time.

BRITISH CRUISERS

Adventure is shown on 10 April 1942 after refit. Note the round stern introduced to solve the problems created by the transom. The single antenna on the mainmast is for the rotating version of the Type 286 radar just fitted. An earlier 1942 photograph shows the Type 285 atop her HA director (as here) but a DF coil atop her pole mainmast.

class) had eased off; he thought he could have a design in February 1924. E-in-C and DEE could both be ready in mid-February. DNC renewed this pledge in January 1924. No design emerged, because the second minelayer soon disappeared from the approved building programme.

The main improvement to HMS *Adventure* before 1939 was the addition of a pair of quadruple 0.5in machine guns by April 1939. By October 1941 she had been fitted with an octuple pompom in 'B' position; installation was presumably delayed because of a shortage of such mountings, and because it was assumed that the ship would operate mainly at night (she did have a high priority for installation of an early air-warning radar). By October 1942 her quadruple 0.5in machine guns had been landed, and she had seven Oerlikons. By April 1944 she had been upgraded to four twin power and ten single Oerlikons. In 1943 proposals were made to rearm her.[7] The initial proposal was to remove the 4.7in guns and lighter weapons (but not the pompom), and to mount four twin RPC 4in in their place. Two quadruple 0.5in guns might have been retained. Another possibility was three twin upper-deck 4.5in mounts. Neither refit was carried out. She was reportedly converted into an emergency repair ship for landing craft to support the D-Day operation, but the official list of British and Commonwealth warships continued to describe her as a large minelayer even in October 1945, when she was laid up disarmed.

By 1936 there was interest in a replacement for HMS *Adventure*. In October 1936 DNC laid out requirements: high speed (38kts), a heavy gun battery (six 5.25in, two quadruple pompoms and two quadruple 0.5in machine guns, but no torpedoes). The ship should carry 360 mines.[8] This time cruiser protection was wanted (approximately 3in side and 2in deck for magazines ([later 4in side and 2in deck were chosen], but apparently little or nothing over the machinery).[9] DNC estimated the ship would be 620ft x 63ft and would displace 9,000 tons. Design estimates began with a detailed calculation of the required length of the mining deck. A rough Legend suggested that the ship could be built on 8,945 tons at a cost of about £2,110,000.[10]

The single fast minelayer was dropped in favour of building four smaller ones, each to carry 100 mines.[11] A draft Staff Requirement was circulated in February 1937, by which time some preliminary work had presumably already been done. The ships would rely on their speed to evade surface attack; the Staff wanted 40kts in standard condition (hence 37–38kts in deep condition). Endurance should be appreciably greater than (later amended to 'not less than') that of the 'Tribal' class, as the ships might have to carry out strategic mining operations subsidiary to main fleet operations (as was done by HMS *Abdiel* after Jutland). The target was 6,000nm at 15kts when six months out of dock. The Staff envisaged a ship of about 2,000 tons, as seaworthy as a modern destroyer. Mines would be laid at 20kts, with 120ft spacing. Armament would be primarily HA, allowing four guns to engage a single aircraft, i.e. four 4in HA. One quadruple pompom, with a wide arc of training on each side, was also wanted. Stabilisation was written in as a requirement. Later the Staff Requirement was amended: the ship should be able to carry fifty more mines at a cost in speed.

This project, which became the *Abdiel* class, began as design K33, in the same series as the *Fiji*s (K32 was probably the larger minelayer). Her hull form was initially based on that of the slightly smaller 'Tribal' class destroyer; at about 2,600 tons she was expected to be 400ft (wl) x 39ft x 11ft (depth 24ft amidships), carrying 100 mines on a 234ft mining deck. Work on the preliminary design was well advanced by late June 1937, E-in-C providing machinery space dimensions.

DNC submitted a sketch design in April 1938 for a 2,600-tonner. Although the ship was not much larger than a destroyer, she had considerably more powerful machinery, two-shaft turbines fed by four boilers.

FAST MINELAYERS

HMS *Welshman* is shown as completed in 1941, with a fixed-antenna Type 286 radar on her foretop.

There was no space for the sort of alternating engine and boiler rooms characteristic of cruiser designs. Instead, the boilers were in two adjacent rooms, the uptake from the after boiler of the forward boiler room and that of the forward boiler in the after boiler room being trunked together to form the middle of three funnels. In the initial design power was 70,000shp – a cruiser powerplant. The ship was rated at 39.75kts in standard condition and 37kts deeply loaded. Endurance was 5,500nm at 15kts. The ship had two mine rails extending forward to the fore end of the forward boiler room, with endless chains to move the mines to the stern. The mine rails ran in passages alongside the machinery spaces. The fire-control system for the two twin 4in guns was that of the rearmed 'W' class destroyers ('Wairs'), with a transmitting station below decks. It housed a Fire Control Box, which contained both HA and LA computers.

Controller (Rear Admiral Henderson) was unhappy with the armament of only two twin 4in; he wanted a third added on the after superstructure (which he understood would cost 60 tons and a quarter-knot). Alternatively, the 4in guns could be replaced by powered weather-proof twin 4.7in guns (as in 'L' class destroyers), probably at the cost of an increase to 2,720 tons (and the loss of about half a knot). A third possibility was to add two quadruple torpedo tubes (2,680 tons). Henderson thought HA fire was overrated; as many others in the Royal Navy pointed out, although the 50° 'L' class mounting could not elevate as far as a 4in HA gun, it could deal with an aircraft flying at 6,000ft or below up to the moment of bomb release, and it lost aim on an aircraft at 16,000ft for only 13½ seconds before the moment of bomb release. A very fast minelayer would be a poor air target in any case. It might have a much greater need for good LA armament, and it might find lucrative torpedo targets during minelaying runs. But ACNS pointed out that Controller had misunderstood the meaning of 'offensive' in the role of the ship. The minefields were offensive, but the ship was expected to avoid action, hence her high speed. ACNS liked Controller's suggested six 4in battery, which could engage dive bombers, and which did not require power operation, and having four guns bear aft would be particularly useful for a ship trying to escape enemies. The 4.7in guns cost too much speed and could not engage dive bombers. Despite the relative failure of the 4in HA gun, on balance the three-mount battery was the better choice. It was also better than fitting torpedo tubes. If the ship were surprised while laying mines, she would turn and run, in which case good torpedo firing solutions probably could not be developed. First Sea Lord was inclined to go for six 4in and one set of torpedo tubes.

As to whether the extra twin 4in gun should be forward or aft, ACNS observed that placing it forward and the pompom aft would give the latter the widest possible arcs, placing the two guns forward being slightly better (if spray could be avoided). In that case perhaps the after superstructure could be eliminated, the pompom and its director being placed on bandstands. That was impossible. As with HMS *Adventure*, the mining deck took up considerable internal volume, some of which was gained back in the superstructure. The after 4in gun had to be moved aft to provide more space for the Captain.

DNC considered carrying the fifty extra mines on the upper deck, but that gave too great a drop (about 20ft). It turned out that the two main mine rails could be curved to allow four lines of mines abaft the machinery on the mining deck. The outer rails could be extended forward into the accommodation spaces. Total capacity was increased to 120 mines. DNC offered two twin 4in aft and one forward; if he placed the additional 4in mount forward, he would have to raise the bridge. DNC understood that two four-man directors were required, so with one of them aft, placing the extra 4in gun there required that the pompom be relocated; he proposed placing it between the centre and after funnels. DNC pointed out that fifty more mines would add weight, and that they could be carried only if a reduction to 39.5kts could be accepted.

In July the forward position of the extra 4in mount was approved. DNO pointed out that in that case the after director was not needed, freeing a position for the pompom much better than the one between the funnels. There was sufficient space for an octuple pompom in the same position, but this possibility was never realised. The bridge was raised to clear 'B' 4in mount. It was essentially the angle-sided bridge adopted in the 'Tribal' class and later destroyers, but with a large gun crew shelter for the crew of 'B' mount built out from it. These features figured in the final design reflected in a Legend dated November 1938. Now the ship displaced 2,650 tons rather than 2,600 tons in standard condition. To maintain the 39.75kt speed in that condition, power was increased to 72,000shp; speed at deep displacement was 36kts.

The Board approved the Legend and drawings of the three 1938 ships, *Abdiel*, *Latona* and *Manxman*, on 1 December 1938. A fourth ship, *Welshman*, was included in the 1939 programme, and two more (*Apollo*

BRITISH CRUISERS

Above and below: Abdiel, lead ship of the new fast minelayers, is shown on 8 May 1943. The big cranes visible in the view from astern were used to load mines through hatches onto the mine deck below the upper deck. The combination of space and high speed made these ships ideal for running high-value cargoes into besieged places like Malta. Conversely, their limited endurance made them ineffective for Pacific warfare as it was developing in 1944–5.

FAST MINELAYERS

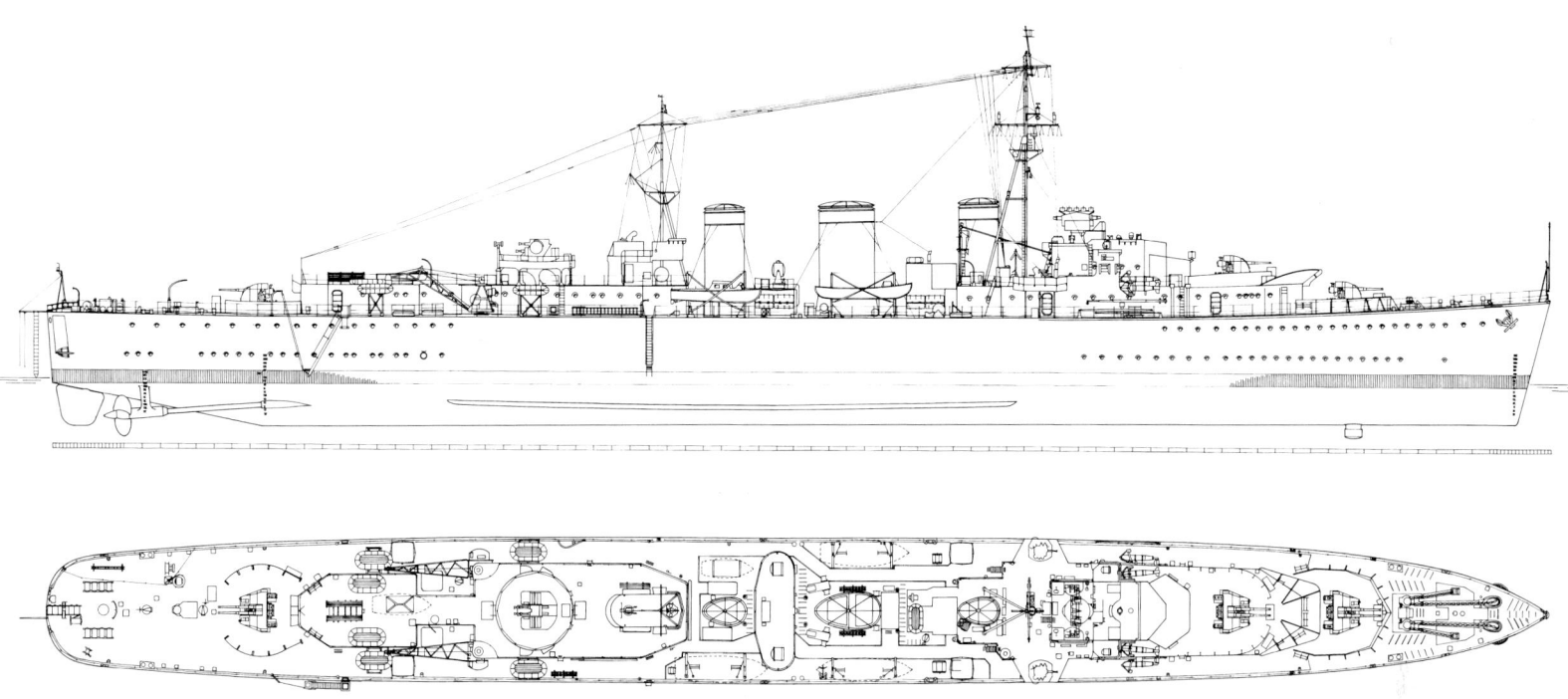

and *Ariadne*) were included in the 1941 programme. In these two ships 'B' twin 4in mounting was replaced by a second multiple anti-aircraft weapon, both of which were Hazemeyer twin Bofors instead of pompoms.[12] A proposal to include additional ships in the 1943 programme was rejected.[13]

As war approached, on 31 August 1939 DNC asked for a more austere minelayer, to carry 150 mines with the original armament of two twin 4in and one quadruple pompom, but with speed cut to 30kts.[14] His designer pointed out that, as in the other designs, the sheer size of the mine decks dominated any cruiser minelayer design. Experience with destroyers suggested that the powerplant needed (about 30,000shp) could be squeezed under the main deck, freeing enough space for four lines of mines. *Abdiel* required a total rail length of 380ft from the after bulkhead; in this ship only 190ft might be needed. Lillicrap thought that the ship might displace about 1,800 tons (350ft x 56ft x 10ft), with a 26,000shp powerplant. Endurance would match that of the fast minelayer, 5,500nm at 15kts. Cost would roughly match that of the fast minelayer. If the 30kts referred to deep condition, the ship would make 32.5kts in standard condition, and she would need 32,000shp (1,950 tons). Alternatively, 30kts in standard condition might equate to 28.25kts in deep condition. Nothing came of this estimate.

Latona is shown as completed. The accommodation ladder is shown in both deployed and stowed positions; the 'Mediterranean' ladder is shown deployed. It was stowed athwartships just abaft the after funnel. Early design sketches showed tilted rather than upright funnels, and 'B' gun in the superfiring position aft. Contrary to some publications, these ships were never armed with 4.7in guns (nor was there any interest in such armament; the main threat they faced was from the air). *Latona* had a very short life: completed on 4 May 1941, she was lost to Italian bombs off the Libyan coast on 25 October. (A D Baker III)

The mine deck of HMS *Latona* shows just how much space even 100 mines occupied, and, by extension, why these ships were effective carriers of high-value cargo to Malta and to Tobruk. For much the same reason the US Navy adapted its large-minelayer design (*Terror*) to carry first nets and then amphibious vehicles. (A D Baker III)

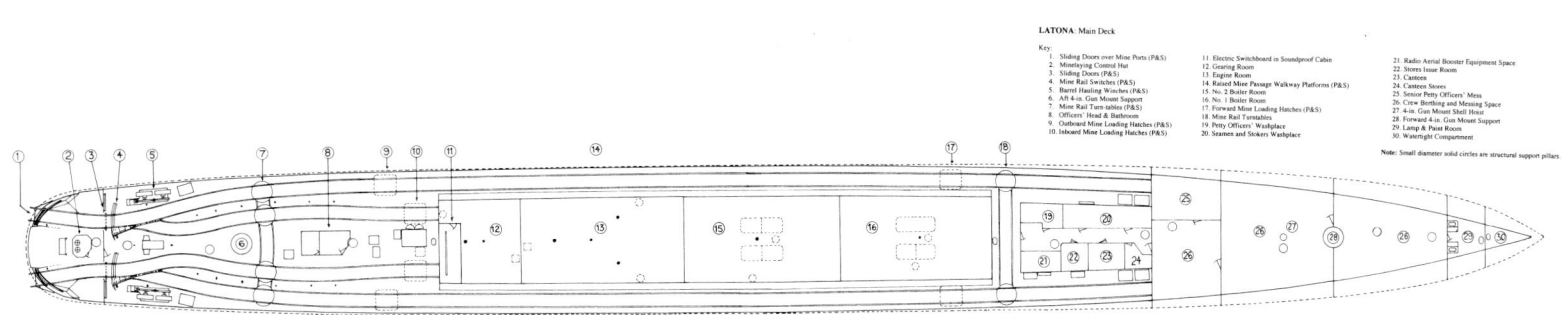

BRITISH CRUISERS

Above and below: Apollo was lead ship of the repeat Abdiel class. She is shown on 26 February 1944. Note the elimination of 'B' twin 4in gun in favour of a twin power Oerlikon. She also has the self-contained Hazemeyer twin Bofors guns with their on-mount Type 282 radars.

Above and below: Apollo was recommissioned in 1951 for the Korean War emergency, remaining in service aftewards for a decade. She is shown on 21 October 1954, stripped of her light weapons and provided instead with the new twin 40mm Mk 5 guns. She had laid minefields in support of the Normandy invasion (to protect the invasion force from German light forces and U-boats); she also laid ASW minefields in 1945. *Manxman* was also recommissioned (for the Mediterranean), but went back into reserve in 1953. She was recommissioned in 1956 as flagship of Mediterranean Fleet flotillas, and served at Suez before going back into reserve. For this service her after 4in gun was landed and a large deckhouse fitted. The third surviving ship, *Ariadne*, was refitted for recommissioning at the time of the Korean War, but that was not done. Similarly, she was brought out for Suez, but she could not be manned, and so was put back into reserve. She had the distinction of serving with the US Seventh Fleet (deployed May 1944), laying a field of 146 mines off Wewak, and then was nominated for use as an assault troop carrier. She was converted for this role and landed troops (presumably she served as a very fast APD) at Leyte Gulf. She was then converted back, returning to the United Kingdom in time to lay a number of ASW fields with HMS *Apollo*.

NOTES

1. Introduction

1. For an example of this practice, see Captain Peter Hore, RN, *Sydney, Cipher, and Search: Solving the Last Great Naval Mystery of the Second World War* (Seafarer Books, Rendlesham: 2009). Captain Hore describes the way in which the Admiralty and the Australian naval intelligence centre tried to locate the German raider operating in the Indian Ocean. This was the German *Kormoran*, which sank (and was sunk by) HMAS *Sydney*.
2. See P J Cain and A G Hopkins, *British Imperialism 1688–2000* (Longman, London: 2000 [second edition; first edition 1993]). The position of the United States within the informal empire but also as a force attempting to disrupt the formal empire gives some idea of the complexity of informal empire. Much of the formal empire was obtained to support the trading requirements of the informal empire; places like Hong Kong were valuable as trading ports, not in themselves. The British (or at least some of them) seem to have been unique in the nineteenth and early twentieth centuries in accepting the modern idea that investment and return were what counted, not physical control; hence many modern claims that conquest does not pay. Of course the British (or at least some of them, in government and the City) well understood that control of some territory made it more attractive for informal-empire partners to work with the British.
3. Staff development up to 1929 is described in the Naval Staff Monograph (Historical), 'The Naval Staff of the Admiralty: Its Work and Development' (ADM 234/434).
4. Unfortunately the Notebook collection is incomplete. The greatest apparent gap is in the series of wartime light cruisers. The collection also seems to lack almost all references to cruiser reconstruction or major modernisation; for example, there is no reference to the *London* reconstruction. Reconstruction designs may have been assigned to constructors at the dockyards performing them, and the Notebook series seems to include only work done in the central DNC establishment in London, and later in Bath (with some material from the London headquarters). It includes virtually nothing from constructors assigned to the fleet.

2. Protecting Trade

1. ADM 116/1013A, Case 3749 Vol. 2, consisting of notes on various designs. Controller's note of 13 January 1908 is numbered CN.024/1908. The general design concept was approved on CN.9153/07, but 'several subsequent discussions showed there was a general feeling of building rather a better class of vessel, in view of its principal role being to meet the German 3rd class Cruisers'.
2. The Cover shows two alternatives for the 'New *Boadicea*,' a preferred 420-footer (420ft x 43ft x 14ft 9in, 4,000 tons, 19,000shp) and a 410-footer (410ft x 43ft x 14ft 9in, 20,000shp), both capable of 25kts, and both armed with twelve 4in guns (twice the *Boadicea* battery) and two 18in deck torpedo tubes. An alternative version of the 420-footer had four 6in guns instead of twelve 4in. Protection was limited to a ½in deck with 1in slopes over the engine room and a 4in conning tower, as in *Boadicea*. Both designs showed 650 tons of fuel in normal condition, compared to 450 tons for *Boadicea*.
3. On 17 January 1908 Watt offered Designs A and B. A was armed with twelve 4in guns, B with two 6in and eight 4in guns (each had two submerged 18in torpedo tubes). As in the later design, each had much more fuel than *Boadicea* (650 tons normal, 1,600 tons fully loaded, compared to 450 tons and 1,050 tons). Required speed was 25kts, and power was given as 21,000shp. Displacements were given as 4,325 tons and 4,400 tons respectively, somewhat below Controller's estimate. Estimated costs were £400,000 and £410,000.
4. The 6in capped AP shell could penetrate 4in KC armour at 6,000yds and 6in at about 3,500yds under optimum conditions. Because the gun was not very effective against armour, the planned capped common (higher capacity) shell had not yet been developed as of August 1908; it would probably penetrate 4in KC armour at 4,000–4,500yds. The most effective shell against unarmoured sides was the HE (Lyddite). As of August 1908 the main role of these cruisers was to fight torpedo craft and cruisers of about their own size, so DNO (Captain Roger Bacon) proposed arming them only with Lyddite (75 per cent) and pointed common shells. Controller suggested that, as the ships were intended for foreign as well as home service, they should have some shrapnel shells as well, presumably for anti-personnel use (Imperial policing). The approved outfit was then 70 per cent Lyddite, 25 per cent common and 5 per cent shrapnel (approved August 1908).
5. These became Designs I, II and III, respectively; unfortunately the Legend is undated. However, a comparison of the original twelve 4in design (A) with these three (B, C and D) was dated 17 January 1908. Design I (two 6in) was a 4,400-tonner (21,000shp). Design II (four 6in) was a 4,600-tonner (21,500shp). Design III was a 4,800-tonner (22,000shp). All were expected to make 25kts, and all had a total fuel capacity of 1,600 tons (coal and oil), of which 650 tons were carried at normal displacement (*Boadicea*'s capacity was 1,050 tons, 450 tons of it carried at normal displacement). Ammunition was the normal 150 rounds per 6in and 4in gun.
6. A Legend for Designs B and E was dated 23 January 1908. The ship was lengthened (430ft x 46ft x 15ft 3in, 4,650 tons rather than 4,400 tons) and power was increased from 21,000shp to 21,500shp.
7. These ships, like earlier turbine cruisers, had three boiler rooms, all adjacent, arranged fore to aft, with engines abaft them. *Bristol* (Curtis turbines) had two adjacent engine rooms, each with a condenser on one side (to port of the forward engine room, to starboard of the after one). HMS *Newcastle* of this class, with Parsons turbines, had three engine rooms, all abeam of each other, with condensers abaft them. Similar arrangements were planned for the Parsons and Curtis cruisers of the 1909–10 programme. All had four funnels, the middle two wider than those at the ends. Each boiler room had

uptakes at both ends, the after uptake of the forward room being combined with the forward uptake of the middle room, and the forward uptake of the after room being combined with the after uptake of the middle room.

8. The papers of the 1909 Imperial Conference are in ADM 116/1100B. Given the German threat, on 22 March New Zealand offered a modern battleship, which became the battlecruiser HMS *New Zealand*. On 29 March the Canadian House of Commons passed a resolution calling for the establishment of a naval service. On 4 June 1909 the Australian premier offered a battleship for Empire Defence. Given the New Zealand offer, the Canadian decision and ongoing Australian offers of assistance, on 30 April the British government invited the Defence Ministers of the four Dominions and of the Cape Colonies (South Africa) to an Imperial Conference. In the Australian case, the effect of the 1909 Conference was dramatic. As late as 1908, the Australian government was interested in a coastal navy: six 'River' class destroyers, nine 'C' class submarines, and two depot ships. See document 45, a draft letter dated 20 August 1908 to the Australian premier, in Nicholas A Lambert, *Australia's Naval Inheritance: Imperial Maritime Strategy and the Australia Station 1880-1909* (Royal Australian Navy, Canberra: 1998; No. 6 in the series of *Papers in Australian Maritime Affairs*).

9. The difference was two guns (abreast) on the forecastle rather than one.

10. A 6 May 1915 note by E-in-C estimates that, if they burned only oil fuel, the boilers in the middle boiler room would produce an additional 3,500shp.

11. Folio 15 of Cover for 'Atlantic cruiser'. This memo is dated 8/5 (1914) and signed by S V Goodall. It replied to a May 1914 telegram from the Australian Naval Board: the Commonwealth had decided to build two additional 'Town' class cruisers at Sydney dockyard and wanted to order material for one immediately. 'As machinery of *Brisbane* is now obsolescent, it is proposed to substitute a two shaft arrangement with Parson's impulse reaction turbines. Boilers of *Brisbane* type have proved to be of insufficient capacity to obtain full results using Australian coal. It is therefore proposed to fit ten boilers in lieu of twelve. Four of these to burn oil fuel only ... Hull and armament same as *Brisbane*.'

12. With these figures radius of action would be slightly superior to that of *Birmingham* when burning Welsh coal, but appreciably less if burning Australian coal; E-in-C suggested carrying more oil at the expense of coal. When the ship was lengthened, oil capacity increased to about 560 tons.

13. DNO had recently recommended that all light cruisers be fitted with an alternative director position and also with a rangefinder specifically for torpedo control. Although this initiative had not yet been approved, the necessary arrangements were provided in the new cruiser. A provisional armament statement dated 1918 showed nine 6in (200 rounds per gun), one 3in HA (300 rounds), three 2pdr pompom anti-aircraft guns (800 rounds each), and eight Lewis guns in twin mountings (5,000 rounds per pair). There were two submerged 21in torpedo tubes for seven Mk IV or Mk IV* torpedoes.

14. Vickers Estimate Book 3 (Brass Foundry) lists a Vickers design (706) dated 10 July 1914 for a second-class cruiser for Spain which may be *Reina Victoria Eugenia*. She is described as similar to *Birmingham* class, 440ft (pp), 460ft (wl) x 49ft 6in (50ft over armour) x 25ft 7⅜in moulded depth x 15ft 9in draft, 5,500 tons (normal), armed with nine 6in guns, four Maxims, one 12pdr and two submerged torpedo tubes, 25,500shp for 25.5kts. At this time Vickers effectively owned the Spanish shipbuilding industry. Armstrong was probably acting as sub-contractor.

15. Vickers Estimate Book 3 (Brass Foundry) lists Vickers designs of light cruisers for Spain: designs 730 to 733 of 10 March 1915, and designs 731 (modified) and 741 of 7 June 1915. All were armed with six 6in guns and either two 21in or four 18in torpedo tubes. They differed in displacement: 4,500 tons (730), 4,700 tons (731), 5,000 tons (732 and 733, with four 21in tubes), 4,650 tons (730 mod), or 5,100 tons (741). *Mendez Nunez* as built displaced 4,650 tons. Designs 730 and 731 were about the size of the *Birmingham* class, 440ft (pp) x 45ft 3in x 24ft 10in (depth) x 14ft, with four-shaft turbines (43,000shp for 29kts or 37,500shp for 28kts, depending on whether all boilers were oil-fired). They had 2in side plating. Designs 732 and 733 were of similar size but heavier, with coal- and oil-fired boilers. The modified version of 731 had two twin deck tubes and 47,000shp machinery. It appears to have been the design chosen.

16. Policy laid out in Captain Ballard's (and others') remarks on the 'peace cruiser' in the Cover for that design and for the 'Atlantic cruiser'. Ballard's comment that the foreign service cruiser was already being provided for Australia makes it clear that he had the 'Town' class in mind. One justification for the small colonial cruiser was that it could be built in large numbers, to deal with expected attacks by numerous German armed merchant cruisers converted from medium-sized merchant ships. In rejecting this idea, Captain Ballard referred to the conclusions of a Special Committee appointed nearly a year earlier (i.e. about February 1912), whose advice the Admiralty Board had accepted. It claimed that the best and cheapest form of protection would be to provide British merchant ships with guns on a large scale; 'arrangements to that end are already far advanced.'

17. The Cover includes a Legend dated July 1913 for this B3 design, the culmination of the design process. The ship would have been 500ft (pp) x 52ft x 18ft 6in (fwd) and 21ft 6in (aft), displacing 7,400 tons at normal load, with freeboard 26ft forward, 12ft amidships and 13ft aft (with a long forecastle); on 30,000shp she was expected to make 26kts when loaded. Fuel would have been oil only: 500 tons normal, capacity 1,200 tons. She would have carried the usual 150 rounds per gun, plus twenty torpedoes. An unusual feature was four anti-aircraft guns. Deck armour would have been limited to 1½in over the steering gear. Barbettes would have been 4in, and gunhouses 4in and 3in. The note of 4 August from Churchill to DNC asked specifically for two 7.5in fore and aft and six 6in (centre line or *en echelon*), the whole to fire on either beam (B4 could fire only five 6in on either beam), with four anti-aircraft and four anti-destroyer guns and four 21in torpedo tubes ('rather a long ship') with a speed of 26–27–28kts, using either coal and oil fuel or oil fuel only, with maximum 4in armour, and displacement of 6,500 tons. Cost should be £500,000 if oil only, or £550,000 burning both coal and oil. This note is Folio 2 in the Cover. The design showed anti-aircraft but not anti-destroyer guns. Magazine capacity was 200 rounds per 6in gun.

18. The same Cover giving details of the big fast commerce protection cruiser includes notes of a conference on 14 August 1912 to define the characteristics of a vessel for service on foreign and colonial stations. It in turn reflected the conclusions of a committee (Sir Francis J S Hopwood, Rear Admiral Sir Edmond J W Slade, Captain G A Ballard [Director of Operations Division], Mr V W Baddeley [Asst Secretary for Finance Duties], and Cecil Perham [Secretary]) responding to the First Lord's minute of 25 April 1912. It was expected to report which ships on colonial duties would have to be

replaced during the next eight years and propose replacements. Such ships would undertake ceremonial and other visits and police patrols, 'but without any appreciable equipment as a fighting machine'. However, a 'peace ship' was of no value; the Foreign Office and the Colonial Office repeated their 1907 advice that they needed 'warships, looking the part, flying the White Ensign, carrying several effective guns and capable of disembarking landing parties'. However, a modern light cruiser was far more than what was needed. The Committee therefore argued that it wanted an economical ship which would nonetheless have a useful wartime function, e.g. in enforcing blockades and in other duties in remote areas. She would replace the small cruisers, sloops and smaller vessels on foreign stations 'which are rapidly wearing out'. The War Staff pointed out that a slow (17–18kt) ship would be 'a gift' to enemy cruisers, to which the Committee answered that in much of the world the enemy would have no cruisers (German overseas cruiser operations in 1914 suggested the opposite). Higher speed would be attractive but too expensive. To support the Committee, the August conference laid down tentative characteristics. Present at this conference were A C Lord, Admiral Slade, and DNC (d'Eyncourt). It was apparently intended to meet requirements framed by a Committee on the Design of Cruisers for Foreign and Colonial Stations, which envisaged an economical ship specifically for peacetime operations. Admiral Slade gave the requirements as: 2,000 to 2,500 tons, draft less than 14ft, speed 17 to 18kts on service (trial speed 20kts), to burn coal and/or oil, with a straight stem and a two-mast fore and aft rig with sails (for additional endurance, and as insurance if the engines broke down very far from a port). Armament would be six 4in (four on the broadside, one each on poop and forecastle) and no torpedo tubes. There would be a protective deck. The ship would carry stores for four months and would have maximum radius of action at economical speed, the figure given being 5,000nm. She would have turbine machinery. Desired unit cost was £100,000. DNC offered two guns abreast on poop and forecastle, and two in the waist, to give the maximum number the best command, and to provide more ahead and astern fire. To provide a double bottom and to minimise draft, he had to give up the protective deck above the boilers. Instead he offered a 2½in side (including hull plating, hence contributing to overall strength) extending from 2ft 3in below the waterline to the upper deck over the machinery; he considered this better than a 2in deck. The ship would have four boilers in two boiler rooms, the boilers on the centreline with stoking spaces at the ends of each boiler room. That made for a longer boiler space, but all coal would be stowed abreast the boilers, offering protection. Based on experience with recent third class cruisers, DNC expected to require 400 tons of coal and 100 tons of oil for 5,000nm at 10kts. He commented that the ship could accommodate more economical machinery, such as geared turbines. Power (5,000shp) would suffice for 20kts at 1,750 tons displacement. Estimated cost (as of 6 September 1912) was £115,000. The ship was smaller than Admiral Slade had expected, 300ft (pp) x 35ft x 10ft 9in (fwd) 12ft 9in (aft) and 1,750 tons; DNC referred to a deep draft of 13ft 6in. Deck protection was limited to 1in over magazines and ½in elsewhere, with nothing over the machinery. A further conference (on 4 October) called for enlarging the ship to about 2,000 tons and a length of 320ft to increase armament to two 6in guns (at the ends) and four 4in, the belt being carried aft to protect the steering gear, and three rather than two funnels (presumably to make the ship more impressive visually). A new Legend was accordingly developed (330ft x 38ft x 10ft 9in/12ft 9in, 2,000 tons, with the same machinery). The Committee turned in its report on 1 February 1913, the main difference from the design being a desire for 8,000nm endurance. The new War Staff was unenthusiastic. Its chief, Admiral H B Jackson (later First Sea Lord) commented that the Royal Navy currently used sloops, gunboats, and obsolete Second and Third Class Cruisers for such duties – it used ships designed for war for peacetime duties. He considered it a false and uneconomical policy to build inferior fighting vessels 'in these days of rapid advance in naval engineering and architecture'. Only ships with very special roles, such as surveying or river work, should sacrifice fighting power. Even so, it seemed wise to develop a design for a 'peace sloop', which would be laid up in wartime. The idea survived until about August 1913 as a design for a replacement sloop for the Pacific.

3. Destroyer-Killers

1. On trials, *Undaunted* made 28.47kts at 3,480 tons (i.e. slightly light) on 38,252shp. *Aurora* made 28.93kts at 3,495 tons on 41,786shp. Data from DNC First World War Cruiser History. This class was designed by Stanley V Goodall, later DNC. Quoting from an unpublished lecture Goodall delivered to US constructors in 1918, D K Brown wrote that Goodall knew the ship could never make the desired 30kts, but he and DNC believed Churchill would be out of the Admiralty before the ships were completed. The reference to propeller experiments was intended to cover this reality. Churchill was not gone in time, and there was considerable embarrassment. No propeller trials were ever run. The time scale was so tight that the hull form had to be chosen before any tank tests could be carried out: the next class was considerably better. 'The Design of HMS *Arethusa* 1912,' *Warship International* No.1 (1983).
2. Early in April 1912 Churchill asked Controller to begin one of the Super *Actives* as soon as possible so that this radical type of ship could be tested before the others were complete. He added that it was essential not to underspend in 1912–13; 'it would be much better to spend more than we have asked for and obtain a supplementary estimate [appropriation], all of which would relieve 1913-14. I must therefore see the design of a Super *Active* this week [written 3 April] ...'
3. According to a 20 November 1912 note by DNO to Third Sea Lord, the modification was in response to (undated) verbal instructions. DNC wrote that 'personally I always favoured this arrangement – I do not attach very great importance to the mixed-armament theory of objections in these small craft but armed as at present designed they are merely large destroyers and could not face a cruiser with one 6in gun. It is true they could probably save themselves by running away, but this is a *most* undesirable thing to do on many occasions of a cruiser service, although under certain circumstances the power to do so is necessary, where the conveyance of information may be of more value than to fight a successful action.' DNC resisted on the grounds that the change would be considerable and that it would add a total of about 86 tons, costing a loss of metacentric height (about 3in) and an increased trim by the stern (about 9in), resulting in the loss of about half a knot of speed. The change was formally approved at a conference on 7 January 1913. The 6in guns were provided with 150 rounds per gun; the 4in had 200.
4. The DNC cruiser design history attributes the *Falmouth* experiments to a committee on light cruiser armament presided over by Captain Grant.
5. The existing twin tube design had to be modified because its racer

(training circle on deck) was too large. Stowage had to be arranged for three more torpedoes. With space for only one further torpedo in the torpedo body room, two had to be stowed in the tubes, a practice previously avoided. It was not yet accepted that nothing but torpedoes already in tubes could be ready for quick use. Total outfit was ten torpedoes. Modified tubes could simply replace the singles, or twins could replace the two aftermost 4in guns (which were relocated to the former single-tube positions forward of the new twin tubes). The latter seems to have been adopted.

6. Atlantic Cruiser cover, Folio 19. Controller (Rear Admiral F C T Tudor) pointed out that neither the *Arethusa* nor the follow-on *Calliope* could be altered, but that DNC should take the problem into account in new designs. The second *Arethusa* Cover includes a report of her experience at Heligoland Bight, where she was hit by German shells but not badly damaged: 'It is satisfactory to know that our shells are so much superior to the German shells in destructive effect. Ours would be still better if all Lyddite [HE] shell were fused with No. 18 detonating fuse, which I believe is unfortunately not the case.' The ship did suffer considerable damage to electrical leads, which might better be carried below the waterline.

7. The *Calliope* class Cover does not indicate when or why this change was made. However, this logic was laid out in DNC's December 1914 comments on the proposed change to all-6in armament.

8. Designed revs for full power were 650rpm in direct-drive ships, 340rpm in *Champion*, and 480rpm in *Calliope*, *Champion* (two-shaft Parsons pinion geared turbines) being the most successful. *Champion* made 29.5kts at 41,188shp (337rpm) at 3,850 tons. There was a noticeable absence of vibration, and machinery spaces were roomier and access better than in direct-drive ships. *Caroline* (direct-drive) made 29.1kts at 40,780shp (586rpm). As a measure of efficiency, on ⅜power (16,000shp), a direct-drive ship (*Centaur* class) burned 260 tons in 24 hours, but a geared-turbine ship (*Ceres* class) burned 200 tons. At full power a direct-drive ship (*Arethusa* class) burned 550 tons, a geared-turbine ship (*Calliope*) 435 tons.

9. CAB 37-122, issued to the Cabinet in December 1914, comparing the war programme with what would have been ordered under the 1914/15 and 1915/16 programmes. Actual cruiser orders amounted to six ships. Cuts in the 1914/15 building programme saved about £1.5 million out of the usual £16 million. The usual number of capital ships (four) and torpedo craft was unchanged. Presumably new construction was being traded off against some other expense, such as new facilities (Churchill's Cabinet paper did not say).

10. At this time generators were rated in amps (of current) rather than, as later, in kilowatts (power). *Arethusa* and *Calliope* class cruisers had two 500-amp generators; DEE wanted two 700-amp in the new ships. He also proposed that the after capstan be electric. DNO pointed out that four searchlights would require 600 amps plus at least another 150 amps to handle fluctuations in searchlight power. A thousand amps would give no margin for ventilating fans, for the alternators which powered current British wireless sets, and for other necessary motors. As of July 1914 trials were being conducted to determine whether even larger searchlights were needed.

11. Nicholson found his ship 'remarkably steady' and a great advance on all previous light cruisers. He feared that raising weights in his ship would reduce that steadiness. Presumably he meant stiffness, because raising weights would have made for a longer, slower roll due to reduced metacentric height (GM).

12. The modified sketch design had been submitted to the Third Sea Lord by 5 December 1914, according to a note by DNC; the Cover dates comments by Fisher and Churchill to a few days later. DNC found the all-6in proposal attractive because the entire battery could be fought in heavy weather (and was relatively dry), because it offered the five-gun broadside, and because it made director firing easy to fit. DNC added that the existing mixed armament was best if 'as prescribed, the ships are to be employed as Flotilla Cruisers or attached to Battle Squadrons, where one of their roles would be the attack of enemy's destroyers, I am strongly of opinion that the 6in gun is not a rapid or handy enough weapon for this purpose.' The DNC history describes a debate over 6in, 4in, and mixed battery continuing until the *Calliope* design was finally approved by the Board on 5 May 1913. It is not reflected in the Cover, but Covers of this era often omitted discussions prior to final submission to the Board. The argument favouring an all-6in battery may be what Admiral Jellicoe referred to when the question was raised again in 1914–15. Arguments cited in the DNC history in favour an all-4in battery were that it sufficed to defeat destroyers, the ship's intended targets; that with 6in guns the ship's commander might be tempted to engage a superior enemy instead of performing the vital scouting function; that the evils of a mixed battery on a small ship outweighed any advantage of heavier guns; and that 6in guns would be ineffective on board a small ship. The mixed battery was chosen because the two 6in aft could deal with a pursuing cruiser. The ships suffered the same spray problems as the *Arethusa*s, and were fitted with sponsons for their waist 4in guns.

13. The forward gun in *Arethusa* was 98ft from the fore perpendicular, and that of *Calliope* was 105ft back. This was reduced to 97ft in *Centaur* and further reduced to 96ft in *Calypso*. Presumably the latter figures were acceptable in view of *Arethusa*. Of earlier cruisers, *Fearless* had her fore gun 92ft from the bow, and *Melbourne* (*Birmingham* class) 109ft back.

14. Churchill asked for a proposal for six 60pdr on the centreline on 8 December 1914. The next day DNO wrote that the only existing design was for a QF gun (semi-automatic) using a brass cartridge case. The Royal Navy was experiencing difficulties with its new 4in QF gun. A QF gun had to be tested in prototype form first, whereas a conventional gun with bagged charges could be ordered from a design. DNO therefore called for a designs of a 5in gun with a service breech mechanism (i.e. bagged charges) with percussion firing. Even then the armament firms would need time to make a new type of gun and mounting. DNO badly wanted two guns firing right ahead. The same day DNC submitted a sketch of a cruiser with six 60pdr on the centreline, presumably with two guns firing forward and two superimposed aft.

15. Ships were renamed on 4 October 1916.

16. Presumably the Germans had to fill the centrelines of their cruisers with far more massive coal-burning machinery. Guns could be mounted over engine rooms, but their supports could not be allowed to penetrate the stokeholds of coal-fired boilers.

17. In submitting the design, DNC referred to 'yesterday's directions for a modified "C" class cruiser with 6 x 6in guns' but that may have been Controller's response to DNC's initial proposal. DNC's remarks on the logic of such a design suggest that the initial request was broader. At this stage the ship was 440ft (wl) x 44ft 6in x 14ft 2in, 4,500 tons; *Calliope* was 3,750 tons (*Arethusa* was 3,500 tons) and the all-6in version displaced 4,120 tons. The beam was 2ft wider than in *Calypso*, and the hull was a foot deeper. Machinery was unchanged (40,000shp). As revised, the ship was 445ft x 45ft 6in x 15ft 5in (fwd) 15ft 3in (aft), and would displace 4,650 tons, i.e. nearly a thousand tons more than *Calliope*. Oil capacity in normal condition was 329 tons rather than 300 tons (maximum 1,000 tons).

18. This probably reflected experience at Jutland. A report from the Captain of HMS *Conquest* (Roger Backhouse, later Controller and then First Sea Lord) in the third *Arethusa* Cover (pp 145–8) suggested that all future ships have 1in decks over their magazines and ½in bulkheads around them, with similar protection to lower conning tower and auxiliary wireless office (the main wireless office on the main deck should have a ½in screen around it). Backhouse also wanted a 1in HT forward machinery space bulkhead like the after one. He also strongly recommended 1in decks for bridges and control positions, to stop splinters from shells bursting below them. The ship took only one or two (Backhouse thought one) direct hits. This shell struck the forward edge of the flash screen aft (separating the two 6in guns) and passed through to detonate against a 6in gun support. The explosion blew holes in both the superstructure deck above and the upper deck below and jammed the 6in gun, damaging nearby equipment. Fragments flying forward pierced as many as five light (7½lb) bulkheads. They also penetrated the lower deck and even the platform deck. Some of the holes were large (one was 3–4ft^2). The fragments were stopped by the 1in machinery bulkhead, which was dented. Backhouse thought this bulkhead saved the engine room from damage as well as limiting the effect of the burst on the lower deck. The burst destroyed everything within a radius of 10–12ft and caused damage at distances up to 60ft. The fire from its flash burned men in the 6in space on the platform deck and in the 6in shell room.
19. Folio 23 in the Atlantic Cruiser Cover rather than in the 'D' class Cover.
20. DNC pointed out that the current bridge had been designed after full discussions with DNO, Superintendant of Charts (S of C) and DNE and also with Commodore (T). DNC also pointed out that the Officer of the Watch, particularly at night, would never leave the compass platform, so he had to have the instruments for night attack close by (gun, torpedo and searchlight control); the committee's vision of an entirely open navigating platform was nonsense. The problem was solved in the new battlecruisers *Renown* and *Repulse* by raising the fore part of the manoeuvring or standard compass platform higher than after part, with no gunnery or other armament instruments placed in it, though they were near enough for immediate control, but cruisers had nothing like enough space for this.
21. According to a letter from a constructor on board HMS *Comus* dated 24 October 1916, these weights included an extra generator, the third 6in gun, paravanes and their attachments, fifty extra men and kits, extra 6in ammunition (200 rather than 150 rounds per gun), oil fittings, handing rooms, and protective mattresses (5 tons). A list of possible weight removals followed. It was headed with the comment that, if tubes were substituted for the after 4in guns, it would be better to replace the remaining 4in guns on each side with one 6in, as the value of the remaining two 4in on each broadside would be almost negligible, and certainly so at longer ranges, given their small splash (difficult to spot) and the additional time of flight. Proposals included replacement of the existing coal-fired kitchen with an oil-fired one, eliminating the 15-ton culinary coal bunker at upper deck level.
22. In May 1915 the four *Arethusa* class cruisers of the Harwich Force (*Arethusa*, *Aurora*, *Penelope* and *Undaunted*) were all given downward-sloping flying-off tracks for fighters intended to shoot down Zeppelins. Although the Zeppelins were often sighted, they were not engaged, and the tracks were all removed by August 1916. These platforms do not figure in the Covers, but Alan Raven and John Roberts, *British Cruisers of World War II* (Arms and Armour Press, London: 1980), pp 30–1 show photos of a Deperdussin monoplane on board *Aurora*, on 30 October and 4 November 1915, and of a short Seaplane stowed aboard *Aurora* (presumably to be hoisted in and out, as it did not need a ramp) in March 1915. This was a national defence measure rather than the later attempt to use fighters to destroy Zeppelins on naval reconnaissance missions.
23. For example, the list of Admiralty Board decisions for the week ending 29 September 1917 includes a decision to fit HMS *Cordelia* with a flying-off platform during her current refit; her conning tower was to be removed and, if time permitted, a ¾in splinter-proof structure substituted. The only other change was substitution of two twin 21in torpedo tubes for her two after 4in guns. ADM 167/54. Raven and Roberts, p 48, refer to a 1917 decision to fit all the ships of the *Caroline*, *Calliope* and *Cambrian* classes, the prototype being *Cordelia* (refitted at Hawthorn Leslie between September and October 1917). After *Cleopatra* was inclined with her fourth 6in gun early in 1918, DNC asked that ships which already had flying-off platforms have them removed when the fourth 6in gun was added. Between April and August 1918 the additional gun was installed on board *Constance*, *Conquest*, *Calliope*, *Cordelia*, *Champion* and *Canterbury*.
24. Mixed-battery 'C' class ships with flying-off platforms at this time were *Calliope*, *Caroline*, *Cordelia* and *Comus*, with *Constance* about to be fitted.
25. Figure given by Cooper, Farnborough and the Fleet Air Arm, quoting an official report by R H Nailer, *Flying-Off Platforms 1917–34*. From 1917 onwards flying-off platforms were installed on board (by classes) *Weymouth* and *Yarmouth*; *Dublin*, *Melbourne*, *Sydney* and *Southampton*; *Birkenhead*; all seven surviving *Arethusa*s; *Calliope*, *Caroline*, *Comus*, *Constance* and *Cordelia*; *Caledon* and *Cassandra* (with hangars alongside the bridge); *Carlisle* (with hangar under a raised bridge); *Caledon* (revolving platform); *Dragon* and *Dauntless* (hangars under the bridge); *Delhi*, *Despatch*, *Dunedin* and *Dragon* (revolving platform); and *Emerald* and *Enterprise* (revolving platforms); the thirty-second platform was the one aboard HMS *Arethusa*, sunk before she could be fitted with the later fixed platform. According to Raven and Roberts, the supports for the revolving platforms were retained in 'D' class cruisers until 1927–8 because they were to have been used as supports for catapults (apparently the plan to install catapults on board these ships was abandoned at that time).
26. Unfortunately the 'E' class Cover has been lost (a few papers not relevant to its origins were collected in a replacement Cover). Controller's query is in the second *Calliope* class cover, together with Tyrwhitt's ideas on greater forward firepower. The Board note, but not its content, is from 1917–18 Board Minutes and Memos (ADM 167/54). The account of the design is mainly from the DNC history of warship design and construction, the cruiser part of which was dated October 1918 (ADM 1/8547/34).
27. D'Eyncourt Design Notebook, National Maritime Museum, p 53. They were given on the same page as details of a proposed submarine cruiser armed with two 6in guns. Details of the two designs were: *see opposite page*. Side armour includes hull plating, as in earlier light cruisers. Note that 'E' class weights are for the Legend, not the deep load, condition. 'D' class data are as designed, before the decision was taken to arm them with triple instead of twin torpedo tubes. As completed, 'D' class displacement was 4,750 tons, increases being 5 tons for general equipment, 5 tons for armament (triple tubes), 35 tons for protection (magazine protection), and 100 tons for the hull. The additions to equipment, armament and protection consumed the 45-ton margin. DNC's Notebook included the abortive 5,100-ton modified 'D' with 295 tons of armament, 955 tons of machinery, 350 tons of oil, 440 tons of protection, 2,725 tons of hull, and 50

	A	B	'E' class May 1918 Design	'D' class
Length overall	465ft	552ft	570ft	471ft
Length waterline	530ft	510ft	535ft	445ft
Beam extreme	43ft	52ft	54ft 6in	45ft 6in/46ft (Mod D)
Load draft fwd	14ft 3in	13ft 6in	15ft	13ft 3in
aft	16ft 3in	18ft	18ft	15ft 3in
Displacement	4,500 tons	6,700 tons	7,550 tons	4,650 tons
Freeboard fwd	25ft 6in	30ft 0in	30ft 0in	22ft 6in
Freeboard aft	14ft 3in	16ft	17ft 0in	13ft 3in
Freeboard minimum	9ft 9in	12ft 6in		12ft 3in
Shaft horsepower	60,000	80,000	80,000	40,000
Speed deep	31kts	32.75kts	32kts	29kts
Speed light	32.5kts	34kts	33kts	
Oil	900 tons	1,250 tons	1,600 tons (650 tons Legend)	1,000 tons
6in guns	4 (150 rounds)	6 (200 rounds)	7 (240 rounds fwd guns, 215 others)	6 (200 rounds)
3in HA	2 (150)	–	–	2 (300 rounds)
4in HA	–	2	2 (200 rounds)	–
Triple TT	2	4	4	4
2pdr pompom	–	–	3	–
Maxim guns	–	–	1	1
Side amidships	2in (3in fwd ER)	3in	3in (machinery; 2in magazines and oil)	3in
Side fwd	1in	2in and 1½in	1½in	2in and 1½in
Side aft	1in	2in and 1¾in	1in and 2in	2in and 1¾in
Deck over steer	1in	1in	1in (and on engines and magazines)	1in
Deep Condition:				
General Equipment	315 tons	350 tons	360 tons	280 tons
Armament	160 tons	330 tons	355 tons	275 tons
Machinery	1,110 tons	1,460 tons	1,590 tons (incl RFW and Engineers' Stores)	945 tons
Res Fresh Water	105 tons	140 tons		
Engrs' Stores	30 tons	40 tons		
Oil	900 tons	1,250 tons	650 tons	300 tons (Legend)
Culinary Coal	15 tons	20 tons		
Hull + Protection	2,250 tons	3,740 tons	700 tons (protection) 3,820 tons (hull)	2,430 tons 375 tons
Margin	15 tons	70 tons	75 tons	45 tons
	4,900 tons	7,400 tons	7,550 tons	4,650 tons

tons for margin. Length would have increased to 487/460ft and beam to 47ft 6in. Commenting on a 1921 Japanese attempt to buy a fast light cruiser in the United Kingdom, NID likened it to the initial 'E' class sketch design (A) 'which was found to involve so many sacrifices to get the desired speed, that it was necessary to about double the displacement to obtain the desired features'. The Japanese project began when the total tonnage budgeted for in the 1920–1 programme left about 4,000 tons, which they wanted to use for a light cruiser built in the United Kingdom. They told their London Naval Attaché to make inquiries. Like the recently-built *Tenryu* and *Tatsuta*, it would be designed to lead destroyers. It should displace not more than 3,700 tons, mounting four 5.5in/50 (120 rounds per gun), one 3in/40 HA (200 rounds), and four twin 21in torpedo tubes (two pairs on each side; twelve torpedoes), with 2½in side and 1in deck armour over the machinery. Speed should be 33kts using mixed coal and oil fuel, with a radius of 5,000nm at 14kts. NID doubted that the Japanese could get the desired speed using coal-burning boilers; they would probably have to accept a lightly-built hull with little protection. Although no cruiser was bought in the United Kingdom, this was presumably the origin of the cruiser *Yubari*. Reports of this cruiser project included a report that the Japanese had recently bought an air flask from Armstrong Whitworth for a 24in torpedo, which would probably have a range of 12,500yds at 29kts – probably the first Western report of this calibre in the Japanese navy.

28. The discussion of protection survives (as a carbon copy, marked 'for 'E' class Cover') as Folio 21 in the 'Atlantic Cruiser' cover. It began with DNO's 14 March 1918 remarks, suggesting that the design was circulated at about that time. However, it is not at all clear to what extent (if any) the May 1918 design had been modified.

29. The version for the Board is described in the volume of Board Minutes and Memoranda for 1917–18 and is also reproduced (with a drawing) as ADM 1/9223.

30. The Cover on Foreign Cruisers from 1915 onwards includes as Folio 16 a copy of an extract from the 26 July 1918 issue of *Engineer*, dated 6 August in file. It claimed that the latest German light cruisers, to be named *Mannheim* and *Köln*, were a completely new design, said to be 520ft x 49ft x 17½ft, 6,300 tons, 33kts, with two 8.2in BL (21cm) at ends, six 5.9in QF, four TT. Oil-fired boilers, nearly 12,000nm at cruising speed. Some light side armour, probably about

4in thick, 320ft long with the usual protective deck. It claimed that the ships were armed with 8.2in guns because that was the next most powerful gun above the then-standard 5.9in in German service, the 6.7in (17cm) having been a notorious failure. No such ships seem to have been planned; *Köln* being leader of a class armed with eight 5.9in. Later it emerged that the Germans were interested in a much larger cruiser armed with 6.7in guns. Raven and Roberts, p 102, describe the two enlarged 'E' class designs. A substituted 7.5in guns in 'A', 'B', and 'X' and 'Y' positions for 6in and omitted the waist guns and 'Q' gun of the 'E' class. It would have had much the same performance, and would have displaced 7,700 tons. B was lengthened to provide space for a fifth 7.5in gun between the funnels: 570ft (pp) 607ft (wl) x 58ft x 15ft 9in, 8,850 tons, the extra length increased fuel capacity to an estimated 2,000 tons (from 1,600). Both designs showed four 4in HA guns and the four triple torpedo tubes of the 'E' class. By this time it seems to have been clear that the 7.5in gun was too large for a cruiser, because it could not be hand-loaded. These designs were not developed in sufficient detail to appear in D'Eyncourt's Notebook describing his designs as DNC.

31. Commenting on the final report of the Fire Control Requirements Committee (March 1920) DNC stated that the test arrangements for *Enterprise* had already been proposed on paper G.719/20. Substitution of quadruple for triple torpedo tubes had already been provisionally approved.
32. ADM 1/8397/365, first docket in file (S.0796/1914).
33. Date given by DNC First World War cruiser history. Apart from the design alternatives, taken from the D'Eyncourt Notebook (p 43), details are taken from the DNC First World War cruiser history. DNC tried four versions of the same hull (560ft pp x 64ft x 16ft fwd/18ft aft, 9,100 tons, with freeboard 25ft forward, 21ft amidships, and 15ft 6in aft. Deep draft varied: 19ft 3in in Design A, 19ft 4in in B, 19ft 5in in C, and 19ft 8in in D. Two other designs (E and F) used a slightly larger hull, 565ft (605ft overall) x 65ft x 16ft 3in/18ft 3in, 9,750 tons, freeboard 24ft 9in forward, 20ft 9in amidships, and 15ft 3in aft, deep draft 19ft 4in. All versions had 60,000shp machinery for 30kts, with 880 tons of fuel at load draft and 500/1,750 tons of oil fuel. In all the versions, coal-burning boilers accounted for ⅙ of total power. Because they had different armaments, the versions varied in complement, from 440 in Design A up to 470 in E and F. Protection was on typical light cruiser lines: 3in side to the upper deck (but 2in to the forecastle deck), extending down to 3ft below the waterline, with 2½–1¼in side armour forward and 2½–1¼in aft, a 3in conning tower with a 2in tube, a 1in deck over the steering gear, and 2in and 1in ammunition tubes. Torpedo tubes were submerged in each version of the design. The design chosen used the same hull as alternatives E and F, but with a load draft of 19ft 3in and a slightly reduced fuel load. Instead of eight 7.5in guns and four 3in, it had seven 7.5in and ten 12pdrs, four of them HA. Details of alternatives, other than dimensions and horsepower and speed, were: *see table below*.
34. ADM 1/8424/171. As with the 'E' class, unfortunately the Cover for these ships was lost. The Minute described the (non-cruiser) programme recently approved, which included thirty-six sloops (mainly minesweepers) as first priority, followed by twenty-four destroyers, three leaders, thirteen minesweepers, and one small floating dock.
35. *Principal Questions for DNO 1918*, p 16, Naval Historical Branch. The reference was to a Minute by Controller dated 12 December 1917 citing an earlier query by First Sea Lord. DNO stated that in September 1917 Portsmouth Dockyard was asked to design a 6in gun capable of firing at any elevation from −5° to +90°. Existing land-based mountings were too cumbersome for shipboard use, and the 6in QF used had too low a muzzle velocity (2,150ft/sec). DNO envisaged a new high-velocity QF gun. CSOF, Vickers and Elswick were all being asked for designs of 4.7in, 5.5in, and 6in anti-aircraft guns, of which the 4.7in weapon eventually armed the *Nelson*s. DNO pointed out that a dual-purpose mounting would not be as effective in either HA or LA fire as a specially-designed gun, but the idea survived, eventually producing the 5.25in gun and mount. A particular problem for LA fire was the increased height of the mounting (so that it could elevate), which would make loading more difficult. DNO concluded that 6in dual-purpose guns could not form part of the armament of the next light cruiser ('E' class), but that it would be wise to strengthen HA positions so that they could accept such guns if and when they became available.
36. The 1923 Admiralty Memoranda volume contains a discussion (item 1712, 26 July 1923) of plans to convert *Frobisher* and *Effingham*. The alternatives were either to convert the present after boilers to burn oil at a rate corresponding to their existing oil and coal consumption; or to replace the four coal-burning boilers with two large oil-burning

	A	B	C	D	E	F
Fuel (load)	880 tons	820 tons	760 tons	500 tons	1,000 tons	1,000 tons
Coal	500 tons	500 tons	500 tons	500 tons	500 tons	500 tons
Oil (full)	1,750 tons	1,750 tons	1,750 tons	1,600 tons	1,750 tons	1,750 tons
9.2in BL	–	–	–	2 (150 rounds)	2 (150 rounds)	–
7.5in BL	–	–	–	–	–	8 (150 rounds)
6in BL	10 (200 rounds)	12 (300 rounds)	14 (300 rounds)	8 (200 rounds)	8 (200 rounds)	–
3in HA	2 (300 rounds)	2 (300 rounds)	2 (300 rounds)	2 (300 rounds)	2 (300 rounds)	4 (300 rounds)
Maxim	4	4	4	4	4	4
21in TT	2 (7 reloads)	2	2	2	2	2
General Equipment	400 tons	400 tons	400 tons	400 tons	400 tons	400 tons
Armament	400 tons	460 tons	520 tons	680 tons	600 tons	600 tons
Machinery	1,880 tons	1,880 tons	1,880 tons	1,880 tons	1,900 tons	1,930 tons
Fuel	880 tons	830 tons	760 tons	500 tons	1,000 tons	1,000 tons
Protection	850 tons	850 tons	850 tons	900 tons	920 tons	940 tons
Hull	4,600 tons	4,600 tons	4,600 tons	4,650 tons	4,750 tons	4,780 tons
Margin	90 tons	90 tons	90 tons	90 tons	100 tons	100 tons
	9,100 tons	9,100 tons	9,100 tons	9,100 tons	9,750 tons	9,750 tons

ones similar to those at the fore end of No. 1 boiler room. The first alternative was rejected because it would add 96 tons, and the ships were already over the 10,000-ton limit. The second was expected to save 9 tons, while increasing oil capacity by 196 tons. Endurance would increase by about 500nm at cruising speed. Further weight would be saved by reducing complement (eliminating stokers). As both ships were badly congested, saving men was welcome. Both ships were still under construction. To avoid delay, Director of Dockyards proposed removing two large boilers from *Effingham* and place them on board *Frobisher*, replacing four small boilers. Four new boilers would be ordered for *Effingham*. DNC observed that, had the ships been designed for all-oil fuel in the first place, they would have been more economical and satisfactory; as it was, structural arrangements did not allow the best advantage to be taken of the coal bunker spaces freed by conversion. At this time (item 1731 of 1923 memoranda) it was estimated that on completion *Effingham* would displace 10,120 tons; *Frobisher* would be similar. DNC was trying to shave weights to bring the ships within the limit. Director of Plans argued that, because the ships had been laid down in 1916, they were outside the Treaty regulations, the additions over 10,000 tons being automatic increases due to construction (ships normally gained about 80 tons per year).

37. DNO discussion of *Hawkins* class armament arrangement in *Principal Questions for DNO 1922-23*, beginning p 2621, Naval Historical Branch. The initial DNO paper was dated 6 February 1922. It referred to a conversation with Captain R G H Henderson, the ship's former CO (later Controller). DNC claimed that in conversation with a member of his staff Henderson said that the after gun, in the worst position, could be fought on about 95 per cent of the days the ship was likely to be at sea.

38. The Naval Anti-Aircraft Gunnery Committee was asked to recommend an alternative anti-aircraft battery. On a weight for weight basis it could barely substitute two 4in guns for the four 3in HA. However, if all ten 3in guns were replaced, there was enough weight for a far more effective HA battery of six 4in. Ideal positions were one gun each side of the conning tower platform, one gun on the platform between the funnels, one each side abaft No. 4 7.5in gun, and one on the existing HA platform. The Committee emphasised the poor sky arcs of the existing guns. The recommendation was forwarded to DNO in May 1920, but nothing came of it. In April 1921 DGD wrote to ACNS that it was pointless to replace the four 3in, because two of them could fire star shell (that could not be done if there were only two of the heavier 4in). The rigging problem was solved by an alternative 'fighting' rig. The 7.5in main battery could not deal effectively with surfaced submarines, but no LA secondary armament was worth fitting: submarines were being given deck and conning tower protection which nothing short of a 4.7in gun could defeat. On the other hand, it did not seem worthwhile either to add four 4in HA and two 4.7in LA guns. It seemed unlikely that the cruiser would really find herself engaging surfaced submarines. The four 3in LA guns were useless, and the 3in HA guns could fire at surface targets.

39. Minute in *Principal Questions for DNO 1923-26*, beginning p 3113 (all DNO Minutes were sequentially paginated), Naval Historical Branch.

40. Lillicrap Constructor's Notebook 5, notes dated 17 February 1930.

41. ADM 229/18.

42. ADM 229/22, describing a conference on 3 October 1939.

4. War Experience

1. According to Stephen Roskill, *Naval Policy Between the Wars* Vol I (Collins, London: 1968), p 214, the rule was formulated in response to a Admiralty query of 12 August 1919 as to what period of immunity from war the Admiralty could expect for reconstruction. On 15 August the Cabinet directed that in preparing the 1920–1 Estimates the services assume no major war for the next ten years. The rule was extended several times during the 1920s and was made self-perpetuating in June 1928 at the behest of Winston Churchill, then Chancellor of the Exchequer. It was abrogated in 1932 after the Japanese invaded Manchuria and directly threatened British interests in China. The rule was generally used not to veto new construction but rather to cut purchases of expendable items, such as ammunition or quartz sonar (Asdic) transducers, which would be needed in wartime. In 1929, however, Churchill directly attacked the central Admiralty assumption that Japan was the most likely future enemy, deploying the Foreign Office to deny any such threat.

2. The first memorandum on oil fuel stocks (21 May 1921) seems to have appeared in the 1921 Admiralty Board Memoranda (ADM 167/64). It emphasised the value of the fleet as a deterrent to Japan and also as a means of maintaining British neutrality in the 'not unlikely' event of a war between the United States and Japan. A war fleet to fight Japan would include twenty-six light cruisers (re-estimated as thirty-seven in 1922; this figure was mentioned in the 1923 Admiralty memo on the required capacity of the Singapore base, item 1710 of the 1923 Board Memoranda). Fleet passage from the UK to Singapore would take forty days, assuming three days of 'tension' before the outbreak of war. The fleet would proceed as a unit at 16kts (capital ships steaming from Suez to Aden at 15kts). The fleet would stop only to refuel. Under the 'Ten Year Rule' the specified oil fuel stocks were not built up; they were among the items covered by the deficiency spending of the 1930s.

3. The many explanations in surviving papers are not credible partly because they did not match up over time. It seems clear that the Admiralty did not want to admit that the seventy-cruiser figure was associated with war in the East, probably for fear that the government might disown that scenario (as Winston Churchill induced it to do in 1929).

4. For example, in May 1928 First Sea Lord Admiral of the Fleet Sir Charles Madden laid out the disposition of cruisers he envisaged once the big 10,000-tonners were in service with the main fleet (the five *Kent*s went to China upon completion). Eight 10,000-ton cruisers (*London* class plus four later ones) would form the scouting line ('A-K' line) of the main fleet, presumably operating with its four battlecruisers. Another ten cruisers (eight 'D' class and two 'E' class) would work with the fleet's destroyers. The rest of the cruisers were on foreign stations, where they would contribute to trade protection in wartime: the four *Hawkins* class on the East Indies station; *Norfolk*, *Dorsetshire* and three 'C' class on the North America station; *York* and *Exeter* on the New Zealand station; and two 'C' class at the Cape. Another four 'C' class would be in reserve. By this time British cruiser squadrons generally contained four rather than five ships.

5. The fire-control committee report is ADM 116/2068. The Committee presented its interim report on 10 August 1919 and its final report on 9 March 1920. It proved impossible to provide the desired strong representation of fleet officers. Instead the fleet was represented by the Director(ate) of Naval Artillery and Torpedoes (DNA&T). Captains of HMS *Excellent* and HMS *Vernon*, the gunnery

and torpedo schools, were also consulted. Stanley V Goodall of the DNC Department signed the report. Commanders represented DNO and DTM.

6. The DCT was first proposed by Commander F Elliott (HMS *Benbow*). The report traced the DCT idea back to the *King George V* class battleships, which had a revolving armoured control tower carrying a rangefinder, control officer, and rate keeper. It failed because of inefficient training and hunting (servo transmission) and limited view; arrangements to bring the tower onto the target were, according to the report, very inefficient compared to those available in 1919. The tower was attractive because personnel remained in the same positions whatever the target bearing (otherwise they had to walk around to face the target). 'Now that director firing is universal, and inclinometers are also being introduced, it follows that considerable duplication of "aids to spotters" is avoided if all the important control personnel can be contained in the same revolving structure, able to align their optical instruments parallel at any moment.' For the Committee, the issue was whether gyro training control could eliminate own-ship movements in azimuth to allow effective spotting and tracking of the target. This type of gyro had recently been fitted to turrets in HMS *Valiant*. Existing capital ships and cruisers had a fore top for fire control with a separate control tower lower down plus turret positions. The report cited HMS *Hood* and recent light cruisers as examples of unacceptably-complicated aloft fire control arrangements.

7. Paper from DNO (G.0775/1920 in ADM 116/4041). The same file contains a longer paper on the armaments of future capital ships and light cruisers, a memorandum of 4 March 1920 by DNO.

8. RN practice was to fire alternating salvoes for quicker results from spotting; a ship could fire two half-salvoes in the time it would take to fire a single full salvo. The US Navy preferred full salvoes, and its large broadsides made a considerable impression on the Grand Fleet in 1918 (but did not change British practice).

9. DNC suggested that the Committee wanted to counter the new US *Omaha* class then being built, only four of whose eight 6in could bear on each broadside. He did not say, probably because no one in the Royal Navy knew, that they were being modified with a twin 6in mount at each end, i.e. with the same eight-gun broadside the Committee wanted.

10. Given its enthusiasm for aircraft on board cruisers, on 19 November 1919 the Committee proposed that the *Hawkins* class all be capable of carrying aircraft. HMS *Vindictive* was to be converted back into a cruiser. The Committee proposed that, like *Vindictive* (as a cruiser), the other ships should all have a hangar forward in place of No. 2 gun and a light steel deck abaft the after funnel to stow two two-seat seaplanes. The hangar would accommodate four Sopwith Baby seaplanes or four fighters. Nothing came of the proposal.

11. Lillicrap Notebook 4, p 6. Lillicrap was one of two cruiser preliminary designers active in the 1930s; the existence of the other designer is evident because Lillicrap was not involved in a few major designs, such as that of the *Southampton* class. Unfortunately no Notebooks for the other designer seem to have surfaced.

12. ADM 1/8586/70 is the summary report (27 March 1920). DNC referred to this report in commenting on the Fire Control Requirements report. The Committee, headed by Vice Admiral Richard F Phillimore, was formed in response to a Admiralty letter of 16 June 1919. The Committee was to take into account the increased effectiveness of shells against armour and also the increased effectiveness of anti-torpedo protection. It was to report on what ships should form the main fleet upon mobilisation. Evidence taken is in ADM 116/2060.

13. It seems neither DNO nor DNC was aware of the role of poor British magazine practices in the Jutland disasters.

14. ADM 7/943 is the Minutes of the committee, April 1919–June 1921.

15. The ideal system, as described by the 1921 report, would employ a stabilised director, set to the required elevation by a fuse range sight, using target vertical and horizontal angular velocities to allow for target motion. The initial sensor would be a height finder, the expectation being that the fire-control officer could estimate its course and speed when the aircraft was first sighted, at the same time that its height was measured. The computer would calculate vertical and horizontal deflections and fuse length (i.e. slant range) from height, angle of sight and relative course and speed of the target and of the wind. Fire control could be made more effective by measuring rather than guessing target vertical and horizontal angular speeds. The British army employed such a rate-measuring (tachymetric) system, but the naval problem was substantially more complicated than that ashore due to ship motion. DNO argued that no system of measurements could properly replace estimated Angle of Presentation (aircraft altering course would frustrate any tachymetric system). Unknown to the Admiralty, the US Navy employed tachymetric fire control, and the 1932 anti-aircraft committee strongly endorsed that approach.

16. Ships had rangefinders and fuse indicators. The rangefinder measured slant range to the aircraft, which was converted into height, and the vertical deflection estimated. The vertical deflection was doubled to allow time for loading and fuse-setting, and the corresponding fuse delay set. Guns fired a quick burst, whose duration was about the time of flight, then stopped so that aim could be corrected for the next burst (much as ships fired salvoes at surface targets and spotted the splashes). A ladder of fuse settings was used against a crossing target. All of the manual steps – such as reading off range, reading off fuse setting, setting the fuse – slowed the process and introduced errors. They made sense only for slow air targets not far from a ship.

17. Because anti-aircraft engagements lasted a very short time, no fire-control system could deduce aircraft motion from successive ranges, the way surface systems deduced a target ship's motion. Instead, a measured range was combined with some instant measure of aircraft motion. The British made what turned out to be an unfortunate choice: the fire-control officer estimated speed (actually, the rate at which the direction to the aircraft changed). The US Navy of the time relied heavily on gyros, so it used a gyro (which defines a line in space) to measure aircraft motion, a method called tachymetric (literally, rate-measuring). The Royal Navy of the 1920s probably regarded any such technique as insufficiently sailor-proof. However, as early as 1919 it certainly recognised that the director sights had to be stabilised (the Committee discussed a simple two-man HA director with stabilised layer's and trainer's sights). On the eve of the Second World War DNO (in the last pre-war issue of the DNO publication *Progress in Gunnery*) declared tachymetric methods superior, and announced that they would be adopted in a few years – but the war intervened. During the war the Royal Navy tried to obtain an approximation of tachymetric operation by adding gyros to its HADT, but that apparently was not enough. There may also have been problems with data transmission from HACP to guns, the gun crews moving their weapons in using follow-the-pointer methods.

18. The four-gun requirement apparently first appeared in a submission to the Committee by Captain E Altham RN on 25 April 1919, who

described his experiences under German air attack off the Belgian coast. In his view (and the Committee's) the guns had to be centrally (director) controlled. Altham's views were supported by Colonel Simon RE (Royal Engineers), who had been involved in the wartime anti-aircraft defence of London. To Simon what mattered was the moral effect of barrages, which would cause pilots to veer away. Moral effect depended on the intensity of the barrage, and it would be of little effect with fewer than four guns of at least 3in calibre. 'By day a majority of pilots, unless they happen to have considerable experience, will turn from a well-placed barrage seen in front of them and at the same level or just below.' Simon preferred fire to hit, if the target could be seen.

19. Board Minute 1537; the rearmament programme is described in *Principal Questions for DNO 1922-23*, Minute 451 (p 2804), Naval Historical Branch.

20. Torpedo bombers typically attacked at 1,000yds or closer in. In February 1920 the Committee remarked that that longer launching ranges probably would not be adopted in future, because extra torpedo weight, which a higher-performing aircraft could support, would best be used for more explosive. Fire would probably be opened at 3,000yds. The defence would rely on cumulative hits. Unlike a 4in gun, the weapons would not have flat trajectories, and would depend on simple sights. The Committee also considered the 'wall of water' defence that several navies tried, in which LA guns fired into the water at ranges between 1,000 and 3,000yds. The contemporary British tactic was to attack in a group of six aircraft, which worked into a position ahead of the target, then split into threes to attack from both bows (so that evasion became difficult or impossible).

21. The idea for the multiple pompom came from Captain C V Usborne RN, president of the committee, in 1919. This 'multiple Pom-Pom mounting' (later called the Mk M) should carry the maximum number of 2pdrs on a total of guns and mounting not to exceed 2 tons 12 cwt (the weight of a 3in HA gun), to achieve not less than sixty rounds per minute per barrel. It was to elevate and train manually at 15°/sec while firing, with maximum elevation of 45° and maximum depression of 10° – consistent with the emphasis on torpedo bombers and attacking boats. Control would be either by hand or by power, in the latter case using a continuously-running electric motor. The mount would be locally- or director-controlled (director control could be waived if it proved too difficult). If no more than six barrels could be accommodated, a further design was wanted with double the weight (5 tons 4 cwt). The lighter weight was intended to fit a destroyer, which was typically armed with a single 3in HA gun. The specification is in the minutes of the Naval Anti-Aircraft Gunnery Committee. In November 1920 trials were held to decide whether the mounting should be arranged so that fire from its guns converged at some set range between 1,000 and 3,000yds. At this time the mounting was expected to carry six to ten pompoms. The weapon as built used a new type of pompom gun (but with the existing ammunition and the same 1,900ft/sec muzzle velocity) and its octuple mount weighed far more than expected: 15 tons. In 1932 the experimental quadruple mount was expected to weigh 6¼ tons. Both versions were clearly unsuited to destroyers and to the surviving First World War cruisers. Rate of fire was better than required (ninety rounds per gun per minute) and maximum elevation was 80° rather than 45°. In 1928 a Mk M Pom Pom Committee recommended a further increase to 120 rounds per gun per minute. The specification required at least two minutes of continuous fire without reloading. The design selected offered 1½ minutes, but ammunition trays could be reloaded while it fired, for about 2½ minutes of continuous fire at ninety rounds per gun per minute.

22. ADM 1/8685/151.

23. Admiralty Board Memoranda for 1926. The Royal Navy seems to have been unique in planning to fly fighters and even bombers from its surface combatants.

24. Performance: E.I.H (nine made), 8,000lbs at 50kts; E.II.H (eight made) 8,000lbs at 50kts; E.III.H (ten made): 8,000lbs at 57kts; E.IV.H (three made). Of the latter, one was installed on board HMS *Effingham* but a second, intended for HMAS *Perth*, was never installed. There was also an S.I.H: 8,000lbs at 49kts or 6,000lbs at 54kts. Light catapult: S.II.L, 5,500lbs at 50kts. Data from Geoffrey Cooper, *Farnborough and the Fleet Air Arm* (Midland Publishing, Hersham: 2008). Because launch acceleration was limited to 2.5G, end (launching) speed depended on the length of the catapult. British catapults telescoped to greater length to achieve higher end speeds: 50kts for 43ft, 52kts for 46ft, 54kts for 49ft, 57kts (actually 56.5kts) for 53ft. The heavy vs. light distinction applied more to the energy available, hence the weight of the aircraft. In February 1931 it was estimated that the light catapult (46ft stowed length) weighed 38 tons, compared to 54 tons for the *Leander* type heavy catapult (53ft stowed length), in each case plus 19 tons of structure.

25. Performance: D.I.H: 8,000lbs at 56kts, upgraded to 12,000lbs at 56kts. D.IV.H: 12,000lbs at 60kts, upgraded to 15,000lbs at 60kts.

26. Dates of catapult removal (by classes) prior to *Fiji* class: *Emerald* (March–April 1944), *Enterprise* (February 1944); *Australia* (date not known), *Berwick* (May–August 1942 refit), *Cumberland* (date of removal not known), *Devonshire* (May 1943–March 1944 refit), *Kent* (July–November 1942, may already have had catapult removed, leaving a seaplane and a crane), *London* (December 1942–May 1943), *Norfolk* (March–May 1943), *Shropshire* (November 1942–June 1943), *Suffolk* (December 1942–April 1943), *Sussex* (by December 1943); *Ajax* (mid-1941), *Achilles* (April 1943–May 1944 refit), *Leander* (June 1941, replaced late 1941, removed late 1943), *Orion* (July–August 1941); *Hobart* (completed October 1942), *Perth* (received catapult previously aboard *Ajax* in July 1941, sunk with it on board); *Birmingham* (April–August 1943 refit), *Glasgow* (June 1944–May 1945, when 'X' turret was removed), *Liverpool* (August 1942–July 1945, when 'X' turret was removed), *Newcastle* (October–November 1942), *Sheffield* (March–June 1943); *Belfast* (August 1944–May 1945 refit). Ships not listed were sunk with their aircraft on board. These data are from Raven and Roberts, Appendix 2.

27. ADM 1/8653/266, printed paper on Empire naval defence (February 1921), p 15.

28. ADM 1/8653/266: Light Cruiser Construction Programme, PD 01914/1923, a docket including 1921–3 material. The paper, dated February 1921, was titled *Empire Naval Policy and Co-Operation*.

29. It would be impossible to provide against both; the best the Royal Navy could do would be to match the strongest other naval power. 'The worst situation with which the British Empire could be faced, from a naval point of view, would occur if Japan seized the opportunity of aggressive action in the Pacific at a time when the situation at home was threatened from another quarter, and reinforcements capable of dealing with the whole of Japan's main force could not immediately be spared ... Happily this extreme case is improbable ...' (p 11). This would be exactly the case if Japan seized the opportunity presented by a war with the United States. Presumably it was assumed that war with the United States was grossly improbable (at

30. PD 01620/21 of April 1921. Dreyer's remarks were dated 7 May 1921.
31. These big cruisers were apparently inspired by the *Hawkins* class (the US standard gun calibre closest to 7.5in was 8in). The US wartime naval staff in London, led by Admiral Sims, accepted the Admiralty view that the *Hawkins* class were freaks, and that something like an 'E' class cruiser was ideal. The General Board war planners, focussed on the Pacific after the Armistice, preferred the larger cruiser. The London staff came home hoping that their organisation, the Office of the Chief of Naval Operations, would take over programme planning from the General Board. US Navy preference for large cruisers was a major consequence of their defeat.
32. A thickness of 2in was chosen on the basis of recent test firings against the German light cruiser *Nürnberg*. There should be increased thickness over magazines.
33. Dreyer associated the new DCT with medium ranges, hence wanted a spotting top and director tower carried as high as possible. This idea went nowhere.
34. Remote-controlled explosive motor boats had been a prominent feature of the German defences of the Flanders coast, but it is not clear why the British thought the US Navy was so interested in such weapons. A few years later the Royal Navy worked on a sort of ultimate distance-controlled boat, a semi-submersible controlled from an aircraft, intended to penetrate enemy harbours.
35. Surviving Constructors' Notebooks do not appear to include any more detailed basis for these conclusions. No relevant designs appear in the d'Eyncourt Notebook. DNC's comments were dated 30 June 1921.

5. Treaties and Heavy Cruisers

1. This clause was included in the proposal examined by the US Navy's General Board beginning on 12 September 1921. General Board papers on the Washington Treaty in William V Pratt Papers, Naval War College. Admiral Pratt was the US naval adviser at the Washington Conference, and he was Chief of Naval Operations during the 1930 London Conference; in effect he was the US Navy's arms-control expert (and exponent). In the pre-Conference proposal cruisers were included in a more general category of auxiliary surface combatants, and not considered as a category on their own. As listed in 1921, the US Navy had 213,550 tons of cruisers built and building and the Royal Navy had 358,620 tons, but the US figure was dominated by obsolete armoured cruisers (only 75,000 tons of new *Omaha* class cruisers were underage). Because the auxiliary category also included destroyers, the US Navy in theory balanced the tonnage of British cruisers with the enormous tonnage of new US destroyers, so the total of cruiser and destroyer tonnage was 592,940 tons for the US Navy and 510,390 tons for the Royal Navy. The General Board proposed totals of 500,000 tons for each navy (and 300,000 tons each for Japan, France and Italy). It also proposed that cruisers (like capital ships, whose replacement age was later increased to twenty years) become overage at fifteen years and destroyers and flotilla leaders at twelve. Replacements for overage ships were exempted from the ban on new construction, and each navy was allowed to build new ships to provide an underage fleet of the proposed size. Overage ships could be replaced on a ton-for-ton basis, but no keels could be laid until the older cruisers were fifteen years old. The proposed treaty included a ban on cruiser guns of more than 8in calibre (initially the General Board rejected any limit on battleship tonnage, but the limit on cruiser guns was intended to prevent capital ships from being built in the guise of cruisers). The limit on cruiser guns does not occur in the General Board paper on the treaty, but it does occur in a copy of the draft treaty in a volume of pre-Conference papers in the Pratt papers.
2. British positions are taken from ADM 1/8630/142, the collected Minutes of the British Empire Delegation, 13 November 1921 through 31 January 1922. These were meetings 48 to 73 inclusive.
3. Capped shells were most effective when they hit nearly perpendicular to the target's armour. Remarks are in ADM 1/8653/266, the docket on light cruiser design 1921–3. In June 1923 DNO asked for a staff ruling on the requirements to be met; his query is DNO Minute 499 (p 3194) in *Principal Questions for DNO 1923-26*, Naval Historical Branch. In 1920 the requirement for 7.5in AP shell had been laid down: it had to be in a fit conditions to burst after perforating a 4in plate at 30° to the normal at 1,100ft/sec (equivalent to a range of 15,000yds; DGD minute, 9 December 1920). Trials in 1918–19 showed that this requirement could be met by a 7.5in SAPC shell; but the same shell could not penetrate a 6in C (cemented) plate at 20° at 1,729ft/sec (5,900yds). DNO said that he understand these requirements had been kept low because existing light cruisers did not have armour more than 3in thick. He reported that current 7.5in AP trials showed that one round nearly succeeded in penetrating a 5in C plate at 30° at 1,602ft/sec (equivalent to 6,500yds range). DGD asked Naval Intelligence for details of the side armour of the new Japanese 10,000-ton cruisers, but nothing was available (data were available for other foreign 10,000-ton cruisers). DNC estimated that unless armour was concentrated over a very small area, leaving most of the ship unprotected, maximum thicknesses were 4in side and 2in deck. The maximum given for any foreign ship was 4in side and 2½in deck. DGD suspected that, compared to British practice, the Japanese would add deck protection at the expense of side protection. They actually used a 3.9in belt and a 1.4in deck in their new *Myoko* class. DGD concluded from the 7.5in shell tests that any 8in APC (i.e. capped) or SAPC shell could penetrate 4in side armour at the required ranges, so effort should concentrate on deck penetration. If (as incorrectly expected) the Japanese sacrificed side for deck armour, the latter could be as much as 4in thick over the vitals, probably non-cemented (NC). He therefore proposed as a Staff Requirement that the new 8in shell penetrate 4in NC plate at about 20,000 to 25,000yds (i.e. at a 50° angle at 1,100ft/sec). Recent trials showed that the latest British APC and SAPC shells were at least as good as specially-designed ones at oblique angles (60° to normal). He therefore proposed that the shells be capable of penetrating (in a fit state to burst) a 6in C plate at 30° angle at 1,450ft/sec (equivalent to 7,000yds). DNC argued that a 4in plate could indeed defeat an 8in shell arriving at an angle of descent of 40° (i.e. striking a deck at 50° to the normal). However, it was most unlikely that so thick a deck could be provided over both magazines and machinery; it would probably be limited to magazines. Even 3in armour would suffice. Moreover, the 1,650ft/sec shell would strike side armour at an angle of descent of about 7° (i.e. about 83° to the normal) and would be unlikely to penetrate even a 1in deck at such a range. There was a real possibility that proof-testing shells at very oblique angles would cause ricochets which would ruin the tests. DNO generally agreed with DNC, and therefore suggested reducing the target thickness to 3in, but keeping 4in as a goal. As more information was received, the situation

became less difficult. In November 1924 DGD remarked that data on foreign cruisers showed that all of them sacrificed protection either for speed or for armament; semi-official reports of US and French designs showed no armour at all. The Japanese reportedly carried twelve (actually ten) 8in guns, for which they had sacrificed protection, and the Italians had sacrificed protection for speed. Some earlier 6in cruisers were actually better armoured. DGD therefore returned to his earlier proposal, that shells be designed to penetrate a 4in side and a 2½in deck. He thought that side penetration would be relatively easy at all ranges out to those at which the shell struck at 40° to the normal. However, a 2½in deck would be a more difficult proposition. He wanted penetration at 20,000yds (30° angle of descent, 1,150ft/sec). DNO pointed out that the velocity actually required would be 1,450ft/sec and the actual critical velocity 1,325ft/sec, though at the angles envisaged the shell might fail to 'bite' the armour. Failure against 2½in deck plate was particularly disturbing in view of reports (which were wrong) that the later Japanese cruisers had such a deck. One solution was to reduce 8in muzzle velocity so that the shell would fall at a steeper angle, for better deck penetration. Discussion continued through 1926; manufacture of shells for the *Kent* class had to begin in June 1926. DNO recommended a high-capacity (5 per cent) SAPC shell for all later 8in cruisers. They would be the only anti-ship shells carried, replacing the earlier CP and HE shells. An additional consideration was blast and splinter effect when the shells hit the large <u>unarmoured</u> parts of foreign cruisers. Undue attention to penetrating relatively thick armour might reduce general damaging effect (this applied to a choice between 5 and 6 per cent shells). In the end, ACNS decided to hedge against a possible change of policy by foreign navies (which certainly occurred in the United States, France and Italy) towards better protection by adopting the 5 per cent shell. CNS (First Sea Lord) approved this decision on 9 June 1926. All British heavy cruisers received the 5 per cent shell.

4. In July, DNO wrote that sketch designs of twin turrets would probably cost about £2,000 each unless there was an immediate prospect of orders. A detailed version of the preferred design would cost about £15,000. The budget contained no surplus money, and it was 'not safe' to ask for more. He therefore proposed a preliminary step, to develop a general specification so that turret designs could be requested in November or December if money was available (by then there would be a sketch design of the gun). Money for a prototype gun could be included in the 1923/4 Estimates.

5. At Jutland uncovered charges had been piled up in the turrets and in the working chamber half-way down the barbettes, creating a powder train from turret to magazine.

6. The deck thickness was based on built-up construction; presumably it could have been reduced had the deck been in a single thickness. The character of the side armour (face-hardened or homogeneous) would depend on whether an enemy used capped or uncapped shells.

7. Lillicrap Constructor's Notebook 4.

8. For some reason this reference to Admiralty Memo 285-B (which seems not to have survived) appeared only in September 1922 remarks by Director of Plans. Pound drew up tables showing ships due for replacement under the two standards, with two ships due in 1923 and ten in 1924 under the eight-year standard. If the existing total was to be maintained, a total of thirty-three ships would have to be built in 1923–32: one in 1925, seven in 1926, six in 1928, two in 1930 and five in 1932. Allowing a fifteen-year life, replacement could begin with two in 1926 and two in 1927.

9. Admiralty Board memo from Plans Division, 11 June 1926, on replacement of cruisers, destroyers, submarines, and twin-screw minesweepers, 1926; given the Ten Year Rule, the objective was a fleet at full strength by 1 April 1936, including seventy cruisers. One advantage of choosing this particular age was that replacement construction could be spread out. At this time the Royal Navy hoped for an annual programme of three cruisers, nine destroyers, and six submarines, leaving it in 1936 with five overage cruisers. On this basis one cruiser could be scrapped in 1927 and two in 1928.

10. Director of Dockyards pointed out that the two 'E' class cruisers could not be completed using 1923/4 funds, but could only be advanced while *Effingham* and *Frobisher* were completed. Completing all four cruisers would delay the carriers *Eagle* and *Hermes* as well as leaders, destroyers and submarines left over from the wartime programmes.

11. Lillicrap Constructor's Notebook 4, instructions dated 2 November 1922.

12. Lillicrap's notes include a table of length to depth ratios, which were generally about 16.4 to 17.3 in cruisers, but 14.88 for *Hawkins*, 14.6 for *Adventure* and 11.9 in the new battleship. The lower the ratio, the greater the hull depth. DNC approved a higher stress, 10–11 tons per square inch. Lillicrap later listed two more advantages of the deep hull: it made the adoption of longitudinal framing under the top deck easier, as height was available for good transverse bracket connections; and it made for good ventilation to the mess decks, important for a ship operating in the tropics.

13. The minelayer was 500ft x 59ft x 14ft (6,480 tons), so he scaled to 10,120 tons (580ft x 68½ft x 16¼ft). He estimated speed on the basis of the resistance curves already developed for the minelayer. Resistance (effective horsepower per ton) was measured on a scale of speed divided by the square root of length, so Lillicrap could estimate that the minelayer speed equivalent to 34kts cruiser speed (at a length of 580ft) would be 31.5kts (30.6kts, equivalent to 33kts for the cruiser). For 30.5kts the minelayer needed 34,800ehp. Hulls of the same shape needed the same effective horsepower per ton at the same scale speed, so Lillicrap scaled up to the cruiser on the basis of tonnage. He then converted effective horsepower into the shaft horsepower the ship actually produced by estimating a propulsive coefficient. Scaled-up figures were 101,200shp for 33kts and 116,800shp for 34kts; Lillicrap modified these figures slightly. This was standard practice for preliminary designers in the DNC Department, as illustrated by various Constructors' Notebooks.

14. DNC did <u>not</u> mention the *Hawkins* class lines, which were based on those of HMS *Furious*, similar to *Courageous*.

15. DNO pressed for a dual-purpose 8in mounting in a comment on a proposed design for a 6in HA mount: *Principal Questions for DNO 1922-23*, DNO Minute 454, p 2831, Naval Historical Branch. The first item in the file was dated November 1922. Designs had been proposed by Vickers, by Elswick and by Woolwich; they emphasised HA performance at the expense of LA. DNO observed that there was no current requirement for such a mounting, and asked whether it would not be more far-sighted to abandon such mountings in favour of ones which were suitable for both kinds of fire. Such a design would be very different, because for LA fire it would have to be arranged for a relatively small angle of elevation for loading. The reverse would be true of an HA mounting – in either case the time to come to the required elevation would be minimised. However, providing a mounting with two alternative loading positions was rejected as far too complicated. All-angle loading was also rejected as too complicated. It entailed long swinging loading arms, power ramming on the mounting (to be usable at any elevation), and tilting

and swinging trays for shell and powder. True HA operation required a QF gun, partly so that shell and charge could be loaded together and partly because the alternative would entail providing water to clean out the breech after firing (which would be unacceptable for HA firing). A LA gun could use either BL or QF ammunition, but the Royal Navy had elected to use BL. The limit for a fixed round was apparently the new 4.7in destroyer gun; a 6in QF gun would have to use a separate cartridge case. A dual-purpose gun would probably have to be QF because it would have a simple breech for a higher firing rate (and also because the cartridge case would make ramming easier and surer). A QF gun would also be lighter than a BL gun. The Royal Navy did not adopt a 6in QF gun until after the Second World War, and even then it encountered problems of reliability in making such large cartridge cases. For the moment, DNO favoured a dual-purpose 6in mounting in which LA fire would receive priority. DNO's analysis killed the proposal for a dual-purpose secondary battery for the battleships then being designed (which became the *Nelson*s). They were accordingly armed with 6in LA secondary guns in twin mounts and with separate 4.7in anti-aircraft guns. DNO recognised that, given the limitations in the Washington Treaty, future light cruisers would be armed with 8in rather than 6in guns, and that it was most unlikely that they would add 6in anti-aircraft guns. He did think that if a dual-purpose 6in mount could be developed (for battleships) 'it would form the armament of a very attractive class that would serve the Fleet well, both against surface attack and against air attack'. DGD considered it unlikely that a heavy anti-aircraft mount could follow the motion of an aircraft nearly right overhead, when the bearing of the aircraft changed most rapidly. Nor was fire control likely to be effective in that case. Most anti-aircraft firing occurred at angles below 50°, so total time under fire at elevations over 70° would be very small. DGD therefore suggested that a future dual-purpose gun should elevate only to about 75°. Allowing for vertical deflection, that would suffice for elevation angles up to 70°. A few guns would be needed to fill the blind spot immediately overhead. To this end a new gimbal mounting was to be developed. DGD therefore proposed that the 8in guns of the new light cruisers be dual-purpose weapons elevating to 75°. DNO concurred (20 December 1922), adding only that in the 8in mount LA fire should be given first consideration, and that nothing should be done to reduce the rate of fire at low angles. For this reason DNO asked for triple or twin mountings with a fixed (relatively low) loading elevation. He originally called for 50° maximum elevation; firms were asked to report whether it would be much more difficult to arrange for a greater maximum elevation.

16. Machinery weight was based on E-in-C's estimate of space and weight required. In this case it was based on 'E' class machinery. E-in-C proposed three boiler rooms to house his ten 10,000shp boilers (two each 42ft long, one 26ft long) and two 54ft engine rooms, for a total machinery length of 226ft, compared to 221ft for the much less powerful machinery in a *Hawkins*. Typically powerplant weight was divided into an 'E-in-C weight' proportional to power (in this case, 1,826 tons) and an auxiliary portion whose weight depended on factors such as generator power, for turrets and searchlights, in this case 241 tons; Lillicrap rounded the total of 2,067 tons up to 2,100 tons. To take into account DNC's acceptance of higher stresses, he divided hull weight for a known ship into strength and other elements, then handled them separately. In this case he used the most recent cruiser design, the cruiser-minelayer *Adventure*. Hull weight estimation was apparently difficult; Lillicrap's Notebooks show several quite different techniques. Typically the next stage was for junior constructors to develop hull weight in detail.

17. This Design 2 (there is no record of Design 1, but it was probably X/Y/Z) was approved on 15 October 1923 (presumably by DNC, for submission to the Board).

18. As submitted in October 1923, the Legend offered 1,025 tons of protection, achieved mainly by shaving total machinery weight (including auxiliaries) to 1,850 tons. Machinery had 1in bulkheads at the ends, plus the deck and side. Main magazines and handing room sides had 4in armour, with a 3in deck overhead (2½in of NCD and ½in of D steel). Shell rooms, which were adjacent, had 1in decks and sides and ends, because they were far less likely to explode if hit. The midships (anti-aircraft) magazine had 3in sides, 2in ends, and a 2in deck. Steering gear was protected by a 1½in turtle deck and a 1in bulkhead.

By this time it was clear that turret weight would grow, so DNC allowed 1,050 tons. Equipment was 675 tons, and hull weight was given as 5,400 tons. No margin was allowed. Page 75 of d'Eyncourt Notebook, National Maritime Museum. When the Legend was submitted, E-in-C machinery weight (i.e. without auxiliaries) was 1,700 tons, but E-in-C hoped to reduce that to 1,570 tons, using faster-revving propellers (300rpm), which DNC considered acceptable.

19. HMAS *Australia* and *Canberra* were completed with the raised funnels

20. A sketch dated 19 October 1923 showed two boiler rooms (each 45ft long) and two engine rooms (each 48ft); there was some hope that the boiler rooms could be shortened to 45ft each. It appeared that 80,000shp could be accommodated on the same 186ft length. Estimated weight was 1,850 tons, compared to 1,550 tons for the same output in the lightweight 'E' class plant. However, in December E-in-C thought he could cut machinery weight (wet) to 1,700 tons. This figure did not include new compressors wanted for new-generation torpedoes, which would add another 4½ tons.

21. Lillicrap Constructors' Notebook 7, in text apparently written much later for part of a paper. He dated the decision to 1925, but it was probably considerably earlier. The Italian navy was considered the leader in ultra-lightweight machinery, designed for high overloads (in the large cruisers *Zara* and *Fiume*, the design figure was reported as 76,000shp, but a designed-in overload of 95,000shp was reported). Naval Intelligence (NID) reported a 1930 Italian statement on lightweight machinery: 'There is a vast difference between a cruiser which works near its base in the Mediterranean and a ship that has to steam to the other end of the world and to be more or less self-supporting as is the case with British ships, which ships must naturally therefore be of a more robust type and of less speed.' The previous year the first Italian treaty cruiser, *Trento*, reportedly experienced trouble due to excessive forcing of her boilers, i.e. of trying to get too much power out of them. On the other hand, her 150,000shp turbines seemed to be much smaller than the 80,000shp turbines of HMS *London*. In 1931 NID reported that the standard practice of paying one million lire for every knot above contract speed had recently been abandoned because to get this bonus firms forced the engines too much on trials. In 1932 NID reported failure of the machinery of the two heavy cruisers built in Italy for Argentina.

22. The main items in the action load were two 50-ton pumps (for the bulge), one 5-ton feedwater pump, two 4in submersible pumps, eight 25in fans for the engine room, and seventy ship-ventilating fans. Turrets used hydraulic power.

23. The ships were completed without transducers or other gear. As of April 1926, with the *Kent* class under construction, tests had been

conducted in the Atlantic and in home waters, and it was considered desirable to gather operating data in the China Sea and in the Pacific. No Asdic-equipped destroyers were operating in those areas, but all the new cruisers were to go to the China station. The problem was considered particularly urgent because underwater sound (as in Asdic) was also being developed for communication, particularly between submarines, which would have an important wartime role there. HMS *Suffolk* was therefore ordered to be equipped with Asdic. Installation had been deferred pending a determination as to whether sufficient weight was available within the 10,000-ton limit.

24. In mid-1924 DNO pointed out that the ships would probably operate on trade routes in wartime, their opportunities to replenish ammunition being few and far between; they therefore needed maximum ammunition stowage. He wanted 300 rounds per gun (outfit and reserve), the reserve presumably being held ashore. DGD pointed out that the new 8in gun would fire about as fast as existing 6in guns – and 6in cruisers typically found their 200 to 230 rounds per gun inadequate in wartime. ACNS approved the 100 rounds per gun peace allowance, another 50 rounds per gun being held on each station to be embarked under 'abnormal' conditions. Controller cited the terms of the Washington Treaty. If it were found that the 10,000 tons would be exceeded if the ships carried more than 100 rounds per gun. The outfit should be 100 rounds and 200 rounds reserve, and that this action should be kept *most secret* (his emphasis), the instructions for embarking the extra 50 rounds 'should be in the form of sealed orders to the Captain only to be opened on receipt of the [war] warning telegram'. The bays for the extra 50 rounds should be removed and retained with the reserve ammunition until needed. Any decision was deferred until it became clear how critical weight would be. DNC (now Berry) wrote that he doubted there would be a problem; ships would be designed to take all 150 rounds, with reductions to be made only if they proved overweight. The episode suggests just how difficult it was to design a ship within the 10,000-ton limit. Similarly, the peace outfit for 4in anti-aircraft ammunition was 150 rounds per gun, the war outfit 200.

25. According to the 1927 report of HMS *Vernon* (torpedo school), ideally the cruiser torpedo should have the same 750lb warhead as the battleship torpedo, with long range (but not as long as the battleships' 20,000yds) at 35kts, with the ability to be angled and with deep running and 'W' (pattern-running) gear. Mk VII was a single-speed torpedo, originally oxygen-enriched (57 per cent oxygen), but converted to air operation early in the Second World War. It was replaced by the Mk IX as the opportunity arose, only nine being fired during the war according to John Campbell, *Naval Weapons of World War II* (Conway Maritime Press, London: 1985), p 84. Range with oxygen was 16,000yds at 33kts (with natural air, 7,800yds at 35kts). The natural air figure is from a note in the *London* Cover; Campbell, p 84, gives a range of 5,700yds. In 1927, when heavier torpedo batteries were being considered, Mk VII was rated at 16,750yds at 35kts with a 750lb warhead. The Mk V of the *Kent*s was a conventional-air heater torpedo with 25kt and 35kt settings, with nominal ranges of 8,000 and 14,000yds (as given in test specifications in the 1929 *Torpedo Manual*). In 1939 the means of gaining greater range was higher-pressure air, called HA air. The Mk IX** HA torpedo offered 10,500yds at 40kts and 16,000yds at 35kts and a duplex (magnetic) pistol which Mk VII lacked. Many British cruisers carried Mk IX torpedoes converted to natural air.

26. PD 01912/23 of 27 October 1923 in ADM 1/8672/227, docket on Emergency Cruiser Programme. This paper referred to justification of the seventy-cruiser goal on other papers, not included. As of 1924 the Admiralty goal was to build up the seventy-cruiser force by 1929, presumably in accordance with the Ten Year Rule.

27. As described in Admiralty minutes and memoranda for 1923, the peace fleet comprised three battlecruisers, a large aircraft carrier, three *Hawkins* class cruisers, four 'D' class cruisers, four other cruisers (probably also 'D' class), eighteen 'V' class destroyers, twenty-one 'L' class submarines, two destroyer depot ships, three submarine depot ships, five sloops, fifteen gunboats, and twelve auxiliaries. The war fleet to operate in the East would comprise twelve battleships, three battlecruisers, four carriers, thirty-seven light cruisers, eighty-one destroyers and leaders, forty-two submarines, and numerous subsidiary ships including armed merchant cruisers. Presumably the war fleet included the peace fleet (the Royal Navy had only four battlecruisers, for example).

28. Roskill, *Naval Policy Between the Wars*, I, p 428.

29. Roskill, *Naval Policy Between the Wars*, I, pp 419ff. Roskill does not mention the cruiser connection, but it is evident in DNC's explanation of the 6in cruiser design he presented to the Board in 1926. He referred to the 1924 arms-control project. The Admiralty docket is ADM 1/8683/131. It includes the usual comment that the Admiralty would welcome a reduction in cruiser numbers as long as it retained enough for the scouting line of the main fleet and for trade protection. It stated that the attempt at cruiser limitation at Washington was abandoned 'because the other signatories were not prepared to recognise our special needs in cruisers'. ACNS offered to cut cruisers to, say, 7,000 tons and 6in guns (presumably he had the 'E' class in mind). DNC (W J Berry) wrote on 25 April 1924 that a new 6in cruiser design should carry eight 6in guns in pairs, have a speed of 32–33kts, and an oil capacity of 1,800 tons, with a standard displacement of 7,500 to 8,000 tons, with nothing but the lightest protection, and destroyer leader machinery, with something more than half the endurance of the 10,000 ton cruisers. DCNS pointed out that the 8in limit had been adopted at the request of the United States, and that it was unlikely to be changed.

30. Lillicrap Constructor's Notebook 4, entry dated 23 September 1924: 70 tons could buy a 3in deck over the 8in magazines (23 tons for 1in). The plating had to extend all the way to the ship's side, so that no shell could pass through an unprotected deck to hit the magazine side. That increased weight to 50 tons per inch of armour, so only 2in could be provided.

31. The Legend for 'Light Cruiser A (light cruiser carrying aircraft)' was sent to the Board on 29 October 1923. It may be the ship involved. The Legend in the Notebook uses HMS *Hermes*, a carrier (sometimes described as an aircraft-carrying cruiser) for comparison. The ship was 530ft (pp), 581ft (overall) x 76ft (extreme, at flight deck) x 15ft (fwd), 18ft (aft), 9,750 tons standard, with 40,000shp machinery (26.5kts at load draft) and a fuel capacity of 2,000 tons and a complement of 720. Armament was given as seven 4.7in HA (240 rounds each) and eight single pompoms (800 rounds each). Protection would have been limited to side armour: 100lbs (2½in) over machinery, 80lbs (2in) over magazines and bomb rooms, and 60lbs (1½in) over steering gear. In contrast to a cruiser, the ship had very limited armament (275 tons), but her equipment weight (855 tons) clearly included aircraft. Machinery was somewhat heavy for a cruiser (1,060 tons) with limited power, and the hull weighed 7,000 tons because it included hangar and flight decks. Armour amounted to only 460 tons. D'Eyncourt Notebook p 74, National Maritime Museum.

32. Lillicrap shaved the 320 tons of deck protection to 230 tons, the 96 tons of side armour to 24 tons, and end bulkheads from 17 tons to

14 tons, presumably because machinery protection was the least important. Weakening machinery protection required some increase in the armour over the midships magazine, which otherwise benefitted from that armour (84 tons rather than 63 tons), but Lillicrap still saved 245 tons on machinery protection. The torpedo battery cost a total of 58 tons, including compressors and spare torpedoes. The turtle deck over the steering gear weighed another 40 tons. Eliminating both gave a total saving of 343 tons, enough to make the triple-turret ship feasible in weight terms.

33. The testing tank at Haslar used models to estimate the power needed to drive a hull of a given displacement at a given speed. This was effective horsepower (ehp). Only part of the power a ship's engines produce (shaft horsepower for a turbine ship) actually goes into propelling the ship; the ratio of ehp to shp is propulsive efficiency (typically about 50 per cent). Thus, given a Haslar ehp figure, a designer would double it to estimate the power output he needed, although in practice he might need a bit more or less.

34. Protecting magazines with 3in deck, 4in side and 3in bulkheads would cost 182 tons; similar protection to handing rooms (which shared bulkheads with the magazines) would cost another 310 tons, and providing the same protection to the 4in magazine would cost another 66 tons. This combination would be 70 tons over the 10,000-ton limit, counting the increased machinery weight. However, it might be possible to shave 50 tons from the machinery weight (to 2,300 tons plus 140 tons for machinery) and to save 22 tons by shaving gunhouses from 1in to ¾in. At least on a weight basis, the design was feasible.

35. Lillicrap Constructor's Notebook 4, 10 November 1924.

36. Lillicrap's 8,500-tonner was 535ft x 58ft x 16½ft, carrying about the same protective weight as a Kent (1,090 tons). She was expected to make 31.8kts. She needed 2,400 tons of oil to make the desired 8,000nm at 12kts, but could stow only 2,200 tons below the platform deck. However, 'peace tanks' could hold another 400 tons.

37. Lillicrap's 7 April 1925 notes show a Legend adding up to 6,795 tons after discussing these designs with DNC. However, the next day he received the bad news from E-in-C that a more realistic machinery weight would be 1,550 tons. He compensated partly by reducing protection from 1,090 tons to 990 tons at DNC's behest. He estimated that the ship would be 530ft pp (560ft oa) x 54ft x 16ft (depth 31ft). Fuel stowage below the platform deck would be 1,635 tons, compared to 1,500 tons in the 6,795-tonner, for an endurance of 6,000nm rather than 5,000nm at 12kts.

38. E-in-C offered 72,000shp on 1,450 tons (E-in-C weight alone). This scaled up to 1,612 tons for 80,000shp, which was short of the Kent figure (as designed; Kent had not yet been completed). E-in-C agreed to 1,610 tons exclusive of auxiliaries such as generators.

39. Estimated dimensions at this point were 565ft x 54½ft x 16½ft.

40. PD 01813, cited in the 1926 analysis of cruiser suitability, ADM 1/9272.

41. DNC found it impossible to provide space amidships for a second catapult. Apparently the British rejected the US practice of mounting two catapults side by side in the waist, preferring the centreline position.

42. Staff comments dated 11 November 1925 on a sketch design offered by DNC on 20 October 1925.

43. Protection shown was 4in sides and 2½in crowns for magazines with 2½in and 1½in ends, plus 3in sides, 1½in deck, and 2½in and 1in bulkheads for the machinery. The 4in side and 2½in crown protection extended over the shell rooms, which in previous cruisers had been more lightly protected (1in sides and crowns). The extra shell room protection had been requested because there was a fear that the ship might be blown up by an enemy shell bursting in a shell room and detonating the shells there. This had not previously been considered a major problem, and it may have been connected with the decision to use relatively high-capacity 8in shells instead of lower-capacity 8in armour-piercing (AP) shells.

44. Weight was saved by simplifying the structure at the bottom of the hoist; the gunhouse was almost identical to that in Kent. Some additional protection was provided to the hoist.

45. 540ft pp, 575ft oa x 57ft x 17ft, compared to 535ft pp, 565ft oa x 57ft x 17ft in the December 1925 Legend. The change was ordered early in March: 5ft more pp, and 5ft more overhang aft. The extra length was intended to maintain speed without any increase in power.

46. Given the extra height, the bridge structure was redesigned, an extra deck being worked in. The sea cabins were placed on the first deck above the shelter deck rather than on the shelter deck as in the past.

47. The load on the turret would be considerable, the aircraft's wings would block the view from the bridge, and the aircraft and catapult might be affected by blast. One way to solve the muzzle blast problem was to mount the catapult to fire backwards along the turret top – assuming there was sufficient space between turret and bridge (needed in any case to limit blast damage to the bridge). In February 1927 DNC asked that the roof of one turret of the new 10,000-ton cruiser *Berwick* be loaded to represent a catapult and aircraft, as a test of whether those weights would affect turret training (the tests were actually carried out aboard HMS *Suffolk* in March 1928). The existing 40hp training engine sufficed when the ship was on an even keel, but two such engines would be needed if she were rolling heavily.

48. The Cover gives no indication of when this decision was reached. In March 1929 the builders (Palmers) requested particulars of aircraft and catapult gear, and a paper was prepared announcing that they would not be needed. By this time installation of a rangefinder in 'B' turret was being considered, because there would be no catapult to interfere with it. The ship received her main catapult (an E.II.H*s cordite unit) after delivery.

49. Chatfield was Third Sea Lord (Controller); Field was DCNS; Dreyer was ACNS. As DGD, Dreyer had probably been most instrumental in urging the 8in cruiser.

50. Roskill, *Naval Policy Between the Wars*, I, p 499.

51. *York* Cover and also Foreign Cruisers Cover (a better version). Ships reportedly carried sixty rounds per 8in gun in standard condition (110 when fully loaded); they had eight 4.7in dual-purpose secondary guns. Machinery and magazines were protected by a 3in belt extending from 2ft 6in to 3ft 6in below the waterline to a deck above the machinery; the magazines had 3in crowns, but machinery had no deck protection other than structural deck plating. The armour was said to be superior to Krupp. Radius of action was given as 10,000nm at 14kts. The Japanese had saved weight by using high-elasticity Manganese steel. DNI (Rear Admiral W W Fisher) signed the basic paper, dated 11 October 1926. ACNS (Admiral Dreyer) made the various suggestions in a 13 October 1926 comment on the DNI paper. He thought that the 'E' class had been designed to carry sixteen torpedo tubes but had been reduced to twelve, which was the opposite of what had happened. The British credited the Japanese with four 10,000-ton cruisers under construction under the 1924 and 1925 programmes (*Myoko* class), and another four in the new construction programme which was being disputed by the Japanese treasury. They became the *Atago* class.

52. ADM 1/9272, 'Gun Armament of Future Cruisers,' a paper written by ACNS dated 31 August 1926.
53. On the basis of 100 rounds per gun, Controller estimated that a two-gun 8in turret cost 218 tons, a twin 7.5in, 173 tons; and a twin 7in, 139 tons. Thus the most extreme decrease, to 7in calibre, would save 79 tons per turret, or a total of 237 tons in a B Cruiser or 316 tons in an A Cruiser. Ammunition weight was based on an estimate of 211lbs for the 7.5in shell and 172lbs for the 7in, not on the 200lbs of the standard 7.5in shell. DNC pointed out that there would be additional savings due to smaller magazines and shell rooms, small savings in hull weight (smaller turret supports), and perhaps even savings in complement; these would probably amount to 200 tons for an A Cruiser with 7.5in guns and 360 tons for an A Cruiser with 7in guns, or 150 tons and 270 tons respectively for a B Cruiser. To calculate armour penetration, DNO assumed that all three guns had the same muzzle velocity, 2,775ft/sec. At 90° inclination (the worst case) 5in vertical armour would keep out a 256lb 8in shell at about 17,500yds; 4in armour would be effective at slightly over 21,000yds. At a much sharper 50° inclination (which might apply to a chase), 5in armour would be effective at about 11,000yds, 4in at about 15,000yds, and 3in at about 18,000yds. Deck armour defined the outer edge of an immune zone: beyond 25,000yds for 3in, about 23,000 for 2½in, and about 21,300yds for 2in. Against 7.5in shell at 90°, 5in was immune beyond about 11,000yds, 4in beyond about 13,200yds, and 3in beyond about 19,000yds. At 50° inclination both 4in and 5in were effective inside 10,000yds, which was the inner edge of the expected fighting zone; 3in was immune at 13,000yds. However, shells fell more steeply at longer ranges, so deck armour was less effective: the outer edge of the immune zone was about 22,200yds for a 3in deck, about 21,500yds for a 2½in deck, and somewhat over 18,000yds for a 2in deck. As might be imagined, the 7in gun was even worse against vertical armour: 5in was immune inside 10,000yds, 4in at about 10,200yds, and 3in at about 14,500yds. Against deck armour, the shell would penetrate a 3in deck at about 22,500yds, a 2½in deck at about 21,000yds, and a 2in deck at about 18,000yds. Discussing the different critiques of cruiser design, DGD mentioned that 'war in the Far East [has] been laid down as the basis on which our preparations are to be made'.
54. Based on their analysis, the British thought the Japanese *Myoko*s had 1–2in side armour, and a 120,000shp powerplant (for 35kts). In fact they had 3.9in side armour and developed 130,000shp, these figures being possible because they displaced 10,980 tons rather than 10,000 tons in standard condition.
55. The previous DNO had said that it was more difficult to design a triple 8in turret than the triple 16in, due to restricted space. The output per gun might be reduced by a quarter due to structural arrangements, so in 1924 DNO estimated that the output of a triple turret compared to a twin would be 9:8 rather than 12:8 (DNC disagreed). These estimates were probably made when DNC considered the cruiser with four triple turrets described above.
56. Calculation from Lillicrap Constructor's Notebook 4. At 16kts a ship burned 1.6lbs/shp/hr; for 20kts the ship would need another quarter-ton per hour for every additional 10,000shp boiler. At 28kts, the ship burned about 1.1lb/shp, and adding steam for full power added ¼ ton per hour for each additional boiler. Consumption at 18kts was 1.5lb/shp/hr, and steam for full power added ¼ ton for every additional boiler. For a *Kent*, the run-out would require 1,195 tons, the sortie another 170 tons, and the battle another 420 tons, a total of 1,785 tons. Endurance was calculated on the basis of 90 per cent oil usage, hence the 2,000 tons. Dreyer thought his new definition would require about 2,250 tons of oil, 350 tons less than that demanded by the Staff Requirement and 950 tons less than *London* class capacity.
57. Details from Lillicrap Constructor's Notebook 4.
58. Notes dated 9 October show 575ft (pp) 610ft (wl) x 60½ft x 18ft (hull depth 34ft) for Design X, compared to 540/575ft x 57ft x 17ft (32ft) for *York* and 595/630ft x 66ft x 17ft for *London*. Shortening the hull saved 610 tons compared to *London* (5,460 tons) but cost 510 tons compared to *York* (4,340 tons).
59. Yet another version was larger (615ft x 62ft x 18ft), with *London*-class armament except for six instead of four 4in. This version had increased magazine and shell room protection (5in sides and 4in crowns).
60. Proposed dimensions were 570ft x 54½ft x 16ft (changed to 16ft 6in), hull depth 31ft. Lillicrap estimated 4,000 tons for the hull (confirmed by scaling up the B design hull), 1,640 tons for E-in-C machinery, 130 tons for auxiliary machinery. 900 tons was available for protection. Data dated 4 December 1926. On 1 October, probably as a starting point for designs with more power, E-in-C estimated that a 100,000shp plant would require 52ft boiler and engine rooms (44ft and 48ft respectively for *Kent*); 90,000shp would require a 48ft boiler room and a 50ft engine room. Lillicrap chose 90,000shp because he had to add only 10ft to the ship's length.
61. 530ft x 54ft x 16ft; 7,500 tons. In contrast to the 7,800-tonner, armour over the shell rooms was reduced to 1in (side and crown). DGD wanted a thicker (1½in) deck over the machinery. The 125 tons required would be gained partly by reducing side armour over the machinery by an inch (70 tons). Eliminating the box protection saved 320 tons, but the more complete platform deck cost another 125 tons, for a net saving of 195 tons. Lillicrap had to add 1in bulkheads to the magazines, because shells could plunge through the unprotected ends of the ship.
62. As in *Surrey* as then conceived, half the machinery (the central unit comprising after boiler room and forward engine room) would have heavier armour, in this case 5in on 1in C sides covered by 2¼in on ½in NCD deck. The rest of the machinery had the usual 1in side and 1½in deck. The thick side offered immunity against 8in fire beyond 7,000yds at up to 50° target angle (58° in the A Cruiser) and immunity against deck protection inside 20,000yds. In December 1927 DNC suggested thickening the upper parts of the magazines at the expense of the lower parts protected mainly by being further under water. In this way the ship could have a quarter-inch more on its roof (total 3¾in) and 5in upper sides (2in lower sides) instead of the earlier uniform 4in, without any increase in weight. That offered another 1,000yds of protection to the magazine crown. The 5in side gained 20° in inclination for protection at all ranges against 8in fire; it reduced the minimum range by 3,500yds. First Sea Lord approved ACNS' suggestion that the magazine crowns be shaved to 3½in, the weight saved being transferred to the 2in parts of the magazine sides.
63. Habitability depended largely on volume available for personnel, but it was very difficult to calculate volume and to relate it to overall ship dimensions. The usual surrogate for crew space (and for oil stowage) was hull length, so a design sacrificed habitability by having its length cut. This was quite sensible; the short *York* was described as badly congested.
64. DNC provided 3in NCD armour worked on a ⅜in structural deck on the magazine crowns, but he had to take about ¼in off the side armour, both abreast the midships machinery and on the magazine sides. That left 5¾in side armour (NCD) on 1in C armour on the sides. The machinery deck armour was 2¼in on ½in NCD. The

desired 2in machinery space bulkheads extended 14ft below the deck (5ft below the lower edge of the side armour).

65. The final version of the design had 1,925 tons of protection. The design provided 2,200 tons of oil rather than the 3,200 tons of the earlier cruisers, but that provided the endurance originally desired (8,000nm at 12kts).
66. The bridge structure had to accommodate considerable equipment for fire control and for night control. Along each side, fore to aft, were a captain's sight (for target designation), a gyro repeater, a searchlight sight, and a star shell sight (with a star shell calculator nearby). In addition to the DCT for own-ship main battery control, the ship was designed to work with others in concentration fire (as in the battle of the River Plate in December 1939). To that end she had to work with a master ship, receiving her firing instructions and measuring the offset to the master ship so that she could adjust her own fire-control solution accordingly. She therefore had two 9ft PIL (Position in Line) rangefinders. Wholly separate from the PIL system was a pair of 12ft torpedo-control rangefinders used both for that purpose and to support tactical plotting. In earlier cruisers these rangefinders interfered with the view from the standard (magnetic) compass itself, making it difficult for officers to take bearings. In *Exeter* these rangefinders, mounted on large boxes to limit vibration, hindered access to the wing gyro compass repeaters on the bridge. Along the angled forepart on each side of the bridge was a chart table, and down the centreline of the bridge was the compass platform, a raised section carrying a standard compass with a pelorus forward of it. The torpedo control position was on the lower bridge, behind a prominent slot in the side of the structure. At its fore end was a 12in signal light. Abaft that light were a trainer's sight, a B sight, a firing pistol, a clear range indicator, and a torpedo order instrument (behind plating). The upper steering position was at the fore end of this level. Blast shutters protected the instruments. The after part of the bridge, abaft the DCT, supported the stalk of an HADT, finally provided forward, as long desired. *Dorsetshire* and *Exeter* were both fitted experimentally with roofed bridges (the roof fitted to *Exeter* was a prototype for one planned for the *Leander*s). In *Exeter* the roof covered the compass platform and the fore part of the bridge, its sides angled forward. The area between roof and bridge front was covered by glass windows forming a windshield, which extended beyond the end of the roof to the end of the angled fore part of the bridge. In June 1932 the CO of *Exeter* wrote to the Admiralty that experience in both ships convinced him that the weather protection involved outweighed any disadvantage due to a more limited view (this would change drastically due to air attacks during the Second World War). Overall, the captain much liked the bridge, which he considered well designed and convenient. He complained only about the wind baffles, which seemed useless, even though (he was told) they had been designed after wind-tunnel experiments.
67. ADM 1/9301, dated 2 July 1929. This was DNC's submission of the new design.
68. The measures Lillicrap proposed suggest how much could be done. He eliminated the 100 tons of protection added for the 1929 ship. Cutting belt armour from 5in on 1in to 4in on ½in (a total reduction of 1½in) would save another 130 tons; cutting to 3in on ½in would save 215 tons. Cutting the deck from 90lbs on ½in (110lbs) to 80lbs on 1in (100lbs), would save 55 tons. Cutting magazines crowns etc from 3in to 2½in (good to 21,000yds) would save another 25 tons.
69. Lillicrap Constructor's Notebook 6. Requirement TD 340/35 was dated 2 December 1935.
70. Notes in Lillicrap Constructor's Notebook 6 refer to TD 126/32, presumably the Staff Requirement, and to D.0618/33, containing instructions from CNS. The originals do not appear in the *Kent* class Cover.
71. Data from inclining experiments are given in various Constructors' Notebooks. DNC tried to estimate weight growth due to age. On initial inclining, *Berwick* was the lightest (13,428 tons deep). In July 1931 *Cornwall* displaced 13,624 tons deep, having gained 115 tons in three years. *Kent* was the heaviest (13,520 tons deep). In March 1931 her deep displacement was 13,744 tons, having gained 184 tons. Since her original standard displacement was 9,850 tons, it was now 10,034 tons. In March 1934 *Kent* displaced 13,742 tons deep.
72. ADM 229/19.

6. The 1930 London Treaty and Its Cruisers

1. ADM 1/8765/313, the Minutes of the Naval Planning Committee for 1928–9 and 1931–2.
2. Each salvo had to contain at least one deliberate miss so that it could be spotted, and cover a large enough area that the gunner would not unnecessarily change ranges. Firing at enemy destroyers required a good volume of fire, the minimum being four 8in or six 6in. The paper on small 8in cruisers included a summary of the expected results of engagements between various kinds of cruisers, assuming each was protected against fire from its own calibre of gun, and on gunnery practices conducted in 1922–7, the results being reduced by 30 per cent to simulate action conditions. On this basis, the small 8in cruiser would put a 6in cruiser out of action in fifty-eight minutes at 15,000yds, suffering 25 per cent loss of efficiency (but not counting the effect of hits in the vitals). At 9,000yds the 8in cruiser would put the 6in ship out of action in eighteen minutes, losing 38 per cent efficiency. By way of comparison, at 15,000yds a cruiser armed with eight 8in guns would put two 6in ships out of action in forty minutes, but at the cost of two-thirds of her efficiency. A third 6in cruiser would turn the scales in favour of the 6in ships. A cruiser with eight 8in would knock out the four-gun cruiser in twenty minutes at the cost of 30 per cent of her efficiency, but would be knocked out by two such ships in thirty-two minutes (one would lose 90 per cent of her efficiency, the other 20 per cent). In reality, most 8in cruisers were hardly immune to 8in fire, so these figures were deceptive. They also probably envisaged an unrealistically high rate of 8in fire.
3. ADM 1/8765/313 contains Chatfield's memorandum is dated 1.6.28 (i.e. 1 June), but given the dates of other papers in the file it seems more likely to have been written in January 1928. Chatfield proposed a three-type fleet: full 8in cruisers, convoy cruisers and inexpensive 6in cruisers for fleet work, with enough full cruisers to face down the enemy's 8in cruisers. On this basis he envisaged twenty A Cruisers, twenty-five convoy cruisers and twenty-five fleet cruisers.
4. Appendix III to the Staff Requirements (copy in ADM 1/8765/313) laid out the reasoning for the gun. A single hit must be sufficient to stop a destroyer; a large calibre was desirable to attack aircraft carriers and minelayers. Against destroyers the ship needed volume of fire, which meant both numbers of guns and rate of fire per gun. Mountings should have good 'A' arcs. Mountings could be open (with one, two or three guns) or could be turrets (two or three guns). Open mounts with more than one gun had to be power-operated, and were inferior to turrets with similar numbers. Twin 6in turrets

existed, and had recently been developed to achieve six to seven rounds per gun per minute. The Germans had adopted triple 6in mounts in their latest light cruisers, but the Royal Navy could not adopt such mountings without trials, so they were not considered. The conference decided that the ship should have 6in directors fore and aft, and that there should be consideration of allowing for divided fire.

5. Policy was that not fewer than two HA guns should bear on any part of the sky, hence the requirement for four such guns, mounted two to a side but capable of firing across the deck. A policy had also been adopted that 6in LA guns in battleships, and the 8in guns of cruisers, should be usable against aircraft. However, at the conference to determine Staff Requirements, Controller rejected a proposal for simple arrangements to allow the main armament (as in the A Cruisers) to fire at aircraft: there was no such thing as a simple HA control system. The ships would have to have the complete HACS 1 system, but it probably would not be worthwhile for a ship carrying only a few HA shells (and changing over from one type of shell to the other would not be quick). This was hardly worthwhile for a ship carrying 4in HA guns. The conference decided to require 50° elevation (more if possible) of the 6in mountings, but no space would be sacrificed to accommodate a third fire-control table (computer) in addition to 6in low angle and 4in high angle. Should some simple 6in HA fire-control arrangement be developed, it would be worth considering, but nothing should be done for the present. The earlier plan to mount four single 4in guns and two quadruple 0.5in machine guns was accepted; there was no interest in the 2pdr multiple pompoms of the larger cruisers.

6. DTM favoured quadruple tubes (work on a quintuple mounting was not promising) with two torpedoes per tube. The conference on the ship's Staff Requirements agreed on two quadruple tubes. DTM favoured the new Type J destroyer torpedo, which was relatively simple, with a single setting (11,000yds at 35kts) and which did not require enriched air, hence did not require a massive plant unsuited to lively ships. The conference agreed.

7. The basic requirements were to protect magazines so that the ship could not be blown up by similar or smaller ships, and to protect the machinery spaces as well as possible to preserve mobility. At the least, magazines should be protected between 10,000–16,000yds against 5in fire and engine- and boiler-room decks should be immune to 5in fire beyond 15,000yds. Magazines should be immune to 4.7in fire beyond 7,000yds, and engine and boiler rooms beyond 14,000yds. Since the engine- and boiler-room sides were only about a third of the target presented by the deck over them, the sides could be reduced as necessary to provide sufficient deck protection. In these figures the inner edge of the immune zone was set by side armour and the outer by deck armour. The logic of the ranges used was that although 6in splashes could be seen at 18,000yds, hitting was unlikely outside 16,000yds. Because the machinery space was so large, and because hits there could not blow up the ship, it was acceptable for armour to be thinner (this logic had applied to the large cruisers). Hence immunity inside 15,000yds (considered the 'outer limit of the important fighting range against other cruisers') was specified. For magazines, immunity at 10,000yds was considered equivalent to immunity at 8,000yds at 60° inclination, and to shorter ranges at greater inclinations. Magazine sides were probably completely protected by water below 7,000yds and between that range and 10,000yds the tops of the magazine boxes presented small targets to 6in fire. It was considered impossible to keep 4.7in shell out of machinery below 14,000yds when fighting beam to beam (0° inclination), but at such ranges the cruiser would probably be trying to close any destroyers attacking her in a day action, and hence would be at a greater inclination, perhaps 60°.

8. The 7,000nm at 16kts equated to a modified version of the endurance formula previously proposed: 200 hours at 16kts (3,200nm) with steam for 20kts; eight hours at 16kts with steam for full speed; ten hours at 24kts with steam for full speed; and twelve hours at 18kts with steam for full speed, all on the basis of being out of dock for four months, with a 25 per cent margin for contingencies. Although calculated on the basis of fleet operation, this was considered sufficient for trade route work (otherwise not analysed). This is from the Staff Requirement for the 6,000-ton light cruiser included in ADM 1/8765/313.

9. Initially the depth charges were to be held at bases and issued only in war, but in 1934 it was decided to keep them on board ships which would have been sent East in the event of war, on the theory that sending them back to base for war equipment would entail too great a delay. The same reasoning applied to splinter mattresses to be used in wartime.

10. The Cover gives no indication of when the change was proposed by E-in-C.

11. Weight increases in *Leander* were due to increased shaft horsepower (machinery weight 1,395 tons vs. 1,345 tons), to increased side armour over the machinery (but reduced deck armour: total 845 tons vs. 780 tons), and to fitting of a heavy catapult and flagship accommodation (hull weight 3,551 tons vs. 3,570 tons, equipment weight 498 tons vs 485 tons, armament [including the seaplane], 55 tons vs 40 tons). The 1931 ships already had heavier machinery (1,420 tons) and grew due to added equipment and hull weights (845 tons and 3,586 tons, respectively) and to the weight of the heavier seaplanes.

12. In May 1933 DNC tried to check its weight by comparing that figure with the weight of the twin 6in secondary mounts in the battleship *Nelson*. Initially it was expected that the guns themselves would weigh about as much as in the battleship (total 72.1 tons), the mountings adding up to another 330 tons (compared to 355.1 tons in the battleship). In fact the guns weighed only 55.4 tons, but the mountings weighed 379.5 tons. Had this sort of underestimate applied throughout the ship, she would have come out 1,400 tons heavy.

13. Practice varied between reporting design and actual displacements. Except for the *Arethusa*s, the difference was small, and far smaller than that practised by other navies. After the Second World War, British and US constructors reviewed Japanese practice and stated that overage was accidental rather than deliberate. A draft version of the review notes is in the foreign ship series of US Preliminary Design papers at NARA II. By way of contrast, a German document giving stated, actual design and completion weights was introduced at the Nuremberg war crimes trials as evidence of deliberate cheating, hence an intent to conduct aggressive war. According to its official cruiser history, the Italian navy deliberately designed its *Zara* class for 12,000 tons rather than 10,000 tons, to secure adequate protection with eight 8in guns. That seems to have been the sole example of Italian cheating. G Giorgerini and A Nani, *Gli Incrociatori Italiani 1861-1964* (Ufficio Storico Della Marina Militare, Rome: 1964), pp 496–8.

14. For example, on 4 October Lillicrap discussed with DNC a 470ft (wl) (approximately 450ft between perpendiculars) x 45ft x 13ft 6in ship, for which he was about to do a rough layout. He expected to use four of the new 17,000shp Thornycroft boilers (rather than 13,000shp destroyer leader boilers); the ship would make 35kts on 60,000shp.

The ship was about the size of a 'C' class cruiser (425ft between perpendiculars, with a length to depth ratio of 17, which probably could not be exceeded, so hull depth would be about 27ft, i.e. freeboard would be 13ft 6in). Adopting destroyer engine-room practice could dramatically shrink machinery length: the leader *Codrington* needed only 52ft for 39,000shp, whereas *Surrey* needed 90ft for 60,000shp. Lillicrap thought he could manage on 85ft (less than proportionate to *Codrington*), with two 44ft boiler rooms.

15. This seems to have been worked out in considerable detail. Dimensions were 470ft (wl) x 45ft x 12½ft (forward) 14½ft (aft); freeboard was 26ft forward, 13½ft amidships, and 13ft aft. Since the ship had open mounts rather than turrets, there was no difficulty in using triple shafts. The ship had six boilers (presumably the 13,000shp destroyer leader type) for 72,000shp at high revs (350rpm) and a speed of 36kts. Oil fuel stowage was 850 tons (5,000nm at 12kts). In addition to the five 6in guns she had two 4in HA guns and one quadruple torpedo tube (no catapult or aircraft). Protection was limited to boxes over magazines (2in crown and side), shell room (1in crown and side), fire-control room (*sic*, not transmitting station, perhaps because it contained multiple computers: 1in) and steering gear (1in). Protection amounted to 120 tons. E-in-C machinery amounted to 1,168 tons, auxiliaries adding another 55; the hull was expected to weigh 2,250 tons, general equipment 330 tons and armament 260 tons (Lillicrap rounded the total, 4,180 tons, up to 4,200 tons).

16. As reported on 22 January 1930, this ship was 525ft (wl) x 52ft x 14ft 6in (forward) 16ft 6in (aft) with freeboard 28ft forward, 14ft 6in amidships, and 16ft 6in aft. She had the same powerplant as the 5,600-tonner, but the greater length equated to more oil fuel: 1,200 tons, sufficient for 5,500nm at 12kts. Except for the additional turret, she had the same armament as the 5,500-tonner, and the same level of protection (430 tons). Armament (without the aircraft and its catapult etc) weighed 645 tons, compared to 500 tons in the 5,600-tonner and 260 tons in the 4,200-tonner. General equipment increased to cover the larger crew (430 tons), and hull weight increased to 3,150 tons.

17. Dimensions reported to DNC were 430ft (wl) x 42ft x 11ft (fwd) 13ft(aft), with freeboards of 25ft forward, 14ft amidships, and 15ft aft. As in the 4,000-tonner, the ship had a six-boiler 72,000shp powerplant with three shafts, running at slightly higher revs (360rpm); expected speed was 38kts. Oil capacity was 750 tons (another estimate was 800 tons). Armament weighed 240 tons, not far from that in the 4,000-tonner. The hull was simply scaled from the 470ft of the 4,000-tonner, and required power was initially calculated as 60,000shp for 36kts. Lillicrap then realised that the same 72,000shp plant he was using for other designs could just provide a speed of 38kts. E-in-C weight would be 1,165 tons. Without power turrets, auxiliary weight was only 55 tons. Endurance was given as 5,500nm at 12kts. DNC asked for a more detailed investigation on 4 April 1930, the ship to armed with four or five single 6in guns, with endurance of at least 5,000nm. The 3,000-ton displacement was not to be considered a rigid limit. Lillicrap thought that he might be able to reduce beam to 41ft. As before, he offered five 6in (200 rounds per gun), two 3in HA (200 rounds per gun), and two quadruple torpedo tubes, but this time he added a pair of quadruple 0.5in machine guns, Lewis guns, and the depth charges standard on British cruisers. He retained the 72,000shp plant, and tried alternative lines based on the flotilla leader *Codrington* and on the pre-1914 light cruiser *Forward*, of which *Codrington* lines were likely to be better (based on comparisons with other hull forms, not new tank tests).

18. An upper-deck 6in mounting weighed 20 tons, but a twin weighed 90 tons, adding 50 tons directly (for the 5.5in Lillicrap estimated 15 tons and 70 tons). However, that was not all. The short-trunk twin mounts also needed twelve men for each ammunition lobby and eighteen for the turret itself, a total of thirty per pair of guns. In *Emerald*, each single gun was served by nine men (the twin in *Enterprise* needed fifteen). The single had two men for ammunition supply, so two singles would need a total of twenty-two men. Mounting weight could be cut somewhat by halving the thickness of the shield (total weight would fall to 80 tons per mount). Magazines and shell rooms had to be protected. General equipment weight increased, too. The ship grew to 4,500 tons (480ft x 46ft x 13ft) and speed would be about 35kts on 72,000shp. The longer hull added enough oil tankage to make up for the losses due to the new magazines. This study was reported early in February.

19. As Chancellor of the Exchequer, Winston Churchill urgently wanted to cut defence spending in hopes of helping the British economy recover. In November 1927 he suggested that none of the six 1927/8 and 1928/9 cruisers be laid down. The Board agreed; Roskill, *Naval Policy Between the Wars*, I, p 555 suggests that was partly due to uncertainty about what sort of cruisers were wanted. At this time current expenditure was running above the Estimates, so any new construction would have required a supplemental appropriation – which, presumably, would not easily have been forthcoming. In the end two of the 1927/8 cruisers were cancelled, leaving HMS *Exeter*. The British financial situation continued to deteriorate. Churchill demanded further cuts, arguing that they were justified by the favourable political situation. Churchill cited Foreign Office claims that Japan was unlikely to cause any problems over the near future to attack the central justification for British naval planning. Even so, for 1929/30 the Board secured Cabinet agreement for three 10,000-ton cruisers. Labour won the May 1929 British election, with Ramsay MacDonald, who had previously championed arms control, gaining office as a minority Prime Minister. His presence in itself guaranteed that the Admiralty would be pressed to accept any apparently reasonable agreement. For the navy, probably the worst outcome of the London conference was that the Government accepted a drastic reduction in destroyer numbers, which the Admiralty had predicated on abolition of submarines: the destroyers were cut even though the submarines were not. Cruiser cuts were of far less moment, although it did not seem so at the time.

20. There was little hope of building up to seventy cruisers, let alone seventy underage ones, by 1936. On 18 January 1930 the Admiralty sent a memo to the British Empire delegation: it did not retreat from the seventy-cruiser requirement as a minimum, 'not taking the United States into account'. Not only could fifty be accepted only for a strictly limited period, but it was acceptable only provided other sea powers limited their own programmes, and 'provided that in our number there is a proper proportion of new construction suitable for extended operations, that is, they must be comparable to the types being built by other Powers, they must be capable of defeating any armed merchantman or raider, they must have sufficient radius of action to carry out their tasks, and must be habitable in all climates'. Statement repeated in connection with the future cruiser programme in ADM 167/86.

21. Stephen Roskill, *Naval Policy Between the Wars* Vol II (Collins, London: 1967), pp 40–1.

22. On the 339,000 tons allowed to the Royal Navy, 192,200 tons of 6in cruisers could be divided among the three types. To meet the fifty-cruiser requirement the Royal Navy had to build thirty-five ships

armed with 6in guns, which could ultimately be divided among the three types: *Leander* (7,000 tons), intermediate (4,500 tons), and fleet scout (3,000 tons). Without the intermediate ships the tonnage allowed for twenty-one 7,000-tonners and fifteen 3,000-tonners, but 'the fifteen is too many of such a weak type, and it is more probable that we would want only 6 or 8 at most'. Hence the Royal Navy needed the intermediate type, though actual numbers could not be set. 'Years ahead may bring many changes but I think the above confirms our view that an intermediate class of 4,500 tons maximum is a probable type. Work done on the 3,000 ton Scout class is however all to the good, through now less pressing than the next size larger.' Controller memo, 9 August 1930, copied in Lillicrap Constructor's Notebook 5.

23. As noted in an Appendix to the February 1932 Board Memorandum on the cruiser programme in the light of large foreign cruisers, in ADM 167/86. From 27 May 1930 C-in-C Mediterranean was Admiral Chatfield, who had left office as Controller in 1928.

24. However, in connection with the question of building more large cruisers to deal with the new large foreign cruisers, a Board memorandum of February 1932 on the building programme described these ships as useful only with the fleet, adding that 'the number of small cruisers that could usefully be included in the Fleet in time of war has been the subject of careful study'. This memo (in ADM 167/86) posited twelve such cruisers 'but it would be unwise to project beyond 1933, when we shall know more about cruiser types which are being built by other countries.'

25. Lillicrap Constructor's Notebook 5 shows a June 1930 study, presumably to help the Naval Attaché in Romania estimate what could be done to meet a Romanian request for a cruiser with eight 6in guns. Naval Attaché (Head of the British Naval Commission, Rear Admiral Reginald G H Henderson – later Controller) wrote that the Romanians wanted two or three cruisers; he wondered whether they might be happy with an improved 'E' displacing about 800 tons more, with more deck armour and all-oil fuel. Lillicrap sketched an 8,000-ton 32kt cruiser but also a 5,700-tonner – showing how much could be done on a very limited displacement. The sketches resembled a *Leander*, and they give some idea of what could be provided in an eight-gun ship. The 8,000-tonner was 570ft (wl) x 55ft x 16.5ft, mounting four twin 6in, four 4in HA and two triple torpedo tubes, with an 80,000shp powerplant giving 32kts. Magazine boxes and machinery all had 4in sides and 2in decks, somewhat more than contemporary British light cruiser protection. Estimated cost was £1.9 million. On 5,700 tons (520ft x 52ft x 14ft, 65,000shp for 32kts), Lillicrap offered the same armament but no armour at all over the machinery. Magazines could be protected by 3½in sides and 2in decks, and shell rooms by 1in sides and decks. The ship would cost £1.5 million.

26. Lillicrap Constructor's Notebook 5 describes the development of this design in detail. Controller's request for the design was dated 7 August 1930 (copy in Lillicrap Notebook). Lillicrap's notes written the next day include the cut in 4in guns.

27. Power was 60,000shp. A three-shaft arrangement was rejected because it was difficult to work in the after magazine. E-in-C adamantly opposed destroyer machinery both because he did not want to lose reliability and because he wanted the ship to develop full speed in the tropics.

28. He developed a series of alternative designs 470ft or 490ft long: A, B, X, Y, Z, P and Q, all with the desired three twin 6in guns (with three 4in HA guns in Designs A and Q, two in all the others). All had the same protection, 3in side and 2in crown over magazines, 1in shell room, 3in (total) side and 1in deck over machinery, 1in over transmitting station side and deck, and 1½in over steering gear. Except for Designs Z and Q (490ft long) all were 470ft (wl) x 47ft 6in x 14ft. The other variable was number of boilers, hence machinery box length: four boilers in Design A, six in the others. Design B showed that the extra protection weight required for a longer machinery box (620 tons rather than 560 tons) more than made up for lighter armament weight; on 60,000shp the ship would make 31.8kts rather than 32kts. Design P used 86,000shp cruiser machinery, the heaviest in the series (1,540 tons [E-in-C weight, since auxiliaries were the same in all designs]) to drive a 470ft ship at 34kts, but the extra machinery box length cost so much oil that the ship would make only 4,000nm at 15kts. Design Q was therefore lengthened to 490ft. The longer hull was easier to drive (82,000shp, 1,465 tons) to achieve the desired 34kts, and the shorter machinery box and longer hull accommodated enough oil for 5,000nm at 15kts. Design X used 68,000shp destroyer machinery to make 33kts (machinery weight was the same as in Designs A and B, 1,070 tons). Design Y used 72,000shp cruiser machinery (1,285 tons) and sacrificed endurance (4,500nm at 15kts) but made 33kts. Design Z showed that 64,000shp cruiser machinery (1,140 tons) could drive a longer hull at 33kts, and the extra length accommodated enough oil for 5,300nm at 15kts, the greatest endurance in the series. These ships were about the size of a 'D' class cruiser (465ft 6in (wl) x 46ft x 14ft 6in (mean)ft, 4,850 tons, 40,000shp, 5,950nm at 10kts). The 82,000shp Design P would probably displace 5,500 tons. The table of these designs went to DNC on 22 October 1930, for use by the Sea Lords at a meeting on 23 October.

29. Dimensions at this stage were 480ft (wl) x 47.5ft x 14ft, and an estimate based on other cruiser designs showed that the ship should make 32.5kts on 64,000shp.

30. Analysis of eleven possible hits on the machinery showed that the new arrangement was superior; it always retained at least a third of its total power. The analysis did not take into account the additional protection to the after boiler room given by the longitudinal bulkheads.

31. The larger boilers would operate at 350psi and at a superheat of 630° F. This degree of forcing saved 85 tons. At this stage the boilers and turbines were rated at 15,500shp each. It would be somewhat uneconomical to keep one large boiler lit for harbour services, so DNC suggested providing a diesel generator. Slightly later DNC and E-in-C favoured a 16,000shp boiler, which was adopted.

32. As pointed out by D K Brown, *Nelson to Vanguard: Warship Design and Development 1923-1945* (Chatham Publishing, London: 2000), p 73. The designers rightly calculated that flooding the wing space would cause only a small list. However, flooding one or two of the big adjacent compartments would dramatically reduce the ship's stability, so that the small amount of asymmetric flooding would have a much greater effect. A ship so damaged would capsize quickly – as several cruisers did. In a footnote to his edition of the official DNC history of wartime construction, Brown pointed out that this subtlety would not have been obvious before the era of computer calculation, presumably because it was a dynamic rather than a static effect.

33. The design offered slightly better protection: the deck over the transmitting station and the low power supply was 1in NC over ⅜in D.1 plating, equivalent to the 1⅜in D.1 approved for repeat *Leanders*. Careful review of weights showed a saving of 50 tons, for a displacement of 5,450 tons. The ship was 480ft long on the waterline.

34. Lillicrap Constructor's Notebook 5.

35. Probably the ship would need no more than 11,000–12,000shp. That would scale to 210 tons, but instead 400 tons (four boilers, two shafts) was used; machinery spaces would probably be only 70ft long (two 18ft boiler rooms, and 34ft for engines). Lillicrap used the hull weight of the 5,000-tonner and *Leander* armament to get a total unprotected displacement of 3,540 tons. He took *Surrey* protection as sufficient to resist 8in fire: the machinery required 5½in side armour (over a ½in skin), 5½in (on ½in) bulkheads, and a 2¼in (over ½in structure) deck. The belt would be about 10ft deep, so machinery protection would come to about 410 tons. Magazines and shell rooms would be similarly protected (5¾in sides, 3in decks), a total of 400 tons if the lower part of the magazines had 2½in armour, as in other cruisers, plus another 60 tons if the whole side were 5¾in thick. Because the machinery space was so short, the belt would not really protect the ship's stability and buoyancy. Lengthening it to 200ft would add about 800 tons of armour, displacement rising to about 6,000 tons. Protection included 1in over the ammunition hoists (a ring atop the magazine, and a ring above the deck) and over an ammunition lobby. The ammunition lobby set the minimum hull depth at 'Y' turret, where it was immediately above the magazine.

36. Lillicrap Constructor's Notebook 6, pp 9ff.

37. Lillicrap was apparently not assigned to this study; he worked instead on studies of small capital ships which were probably intended to help form British policy at the (abortive) 1932 Geneva Conference. The figure comes from the second *Amphion* Cover, folio 9, paper dated 5 July 1933. The original call for design studies, of the nine- and ten-gun ships, was made by Controller in a paper dated 8 February 1932. A later Minute on the same paper (which has since been lost) called for ordering one or two triple turrets for installation in 1932/3 programme ships, giving the nine- or ten-gun armament. Later the weight of the triple was fixed at 125 tons. Additional protection was dropped because of the weight limit; DNC thought the *Amphion*, at 7,250 tons, was the largest acceptable ship. He gave the displacement of the ten-gun ship as 7,550 tons, of the nine-gun as 7,300 tons, and of the eight-gun (three turrets) as 7,200 tons. The seven-gun ship seems to have been added for this 1933 paper. Speeds of the ten- and nine-gun ships were given as 31.8kts and 32.4kts, using the *Amphion* powerplant.

38. Lillicrap Constructor's Notebook 6, p 67 includes a DNC request on 16 November 1932 for an *Arethusa*-size cruiser with twin turrets fore and aft and the other two guns in single weather-deck mountings. That would save weight on magazine protection and handing rooms. Lillicrap answered that such a ship would be unable to carry an aircraft and catapult, which were essential. He also thought it unlikely that three-gun ahead fire would be acceptable, and that it would be difficult to provide sufficient HA guns. Putting the single mounts on the broadside, to give a total of five in some directions, would not be acceptable. The only acceptable arrangement would be single guns in 'B' and 'X' positions. Much of the weight saved would go into blast screens to protect the crews of the open mounts, but enough would be left over to provide another ½in on the belt (2¾in on ½in structural plating).

39. *Leander* had four turbo-generators, two on the platform deck immediately forward of the forward boiler room and two in the gearing compartment. With the modified arrangement, the forward generator compartment had to be omitted and the two generators relocated to the wings of the after boiler room. They would be diesel-driven (one set of turbo-generators was placed in each engine room). This was the same arrangement as in the new *Arethusa* class. Later, answering Controller's request that extras be ruthlessly cut, DNC commented that British electric generation capacity exceeded that of US ships, but the example he gave (*Omaha*, four 100kW turbo-generators) was badly out of date, closer to the First World War 'E' class.

40. ADM 268/52.

41. Like the 1921 Naval AA Gunnery Committee, the 1932 Committee took into account the only existing experience of air attack, during the First World War, although it was well aware that technology had changed radically since then. According to the 1932 report, British ships operating off the Belgian coast and in the Heligoland Bight were often attacked, but the only success was a hit on the light cruiser HMS *Attentive*. The Battlecruiser Force and 1st Cruiser Squadron were attacked, without success, on 1 June 1918. A German destroyer had been sunk by bombing at Zeebrugge, and another destroyer severely damaged while underway. British aircraft harassed the German *Goeben* and *Breslau* when they sortied in January 1918, and made many bombing attacks on *Goeben* when she was stranded for six days afterwards. A Turkish shore AA battery forced the attackers to such altitudes that they scored only two hits in 250 tries. Aircraft were far more successful against submarines and merchant ships.

42. The Committee cited rapid growth of aviation in the US, British and Japanese navies; the development of French long-range flying boats for Mediterranean operations; and the reported training of Italian bombers for long-range flights over the sea, plus the recent development of torpedo bombers by the Italian air force. On the other hand, attack over the sea was a specialised art, not quickly mastered by an air force not interested in the problem. Air Ministry representatives and DNAD estimated the threat posed to a fleet transiting 100 miles from enemy air bases ashore. They assumed that three aircraft could be rearmed and refuelled in fifteen minutes, which they said was the usual capacity of a modern air base. In that case the fleet might be subject to an initial wave attack (eighteen aircraft per wave), waves following each other quickly. It would take the enemy force two to four hours to regroup, after which it might attack with thirty-six, twenty-four or eighteen aircraft every three, two or one hours respectively. This analysis led the Committee to conclude that the problem was two or more waves of about eighteen aircraft each in quick succession.

43. The US aircraft approached at 8,000–10,000ft, diving vertically at the target and releasing bombs at 3,500–4,000ft (the aircraft could not level out from a lower altitude). The press reported that US attempts had been disastrous, and the Air Ministry thought it unlikely that an aircraft could be designed that could sustain the stresses involved and also lift a heavy-enough bomb. It preferred to work on a sight useful to a glide bomber, approaching the target at a 45° or smaller angle. The 1932 report discounted American claims that near-vertical dive bombing made possible 40 per cent hits compared with 4 per cent for horizontal bombing.

44. By this time work on radio-controlled targets was well advanced, so the next step to an anti-ship missile seemed obvious. 'It is difficult to visualise this form of attack being profitably employed against any target other than large areas.' According to a footnote, 'The Air Ministry regard this idea as exceptionally secret and would prefer that it is not generally promulgated.' The report went on to point out that *piloted* explosive aircraft, which could hit manoeuvring ships, 'cannot however be ruled out. It is reported that, sooner than accept defeat, ramming other aircraft is a recognised principle among Japanese pilots.' (Ch 2, para 22) Did anyone remember that when the Kamikazes appeared?

45. According to the report of the 1932 Committee, but not to the Cover or to any other design material. This would tally with the

decision to give their main batteries 60° elevation. Plans called for not fewer than thirty HE anti-aircraft shells per gun. The peacetime outfit per gun was 180 CPBC (i.e. anti-ship) plus twenty HE with provision for time fuses, the extra ten HE shells being provided in wartime either in vertical stowage round the foot of the shell bins or in place of the ten practice rounds per gun.

46. It seems to have been clear that many ships would be badly cramped, so the report showed five alternative arrangements for twin 4in guns: (1) three twins on each side amidships; (2) two on each broadside and one on the centreline abaft them, saving one mount; (3) two on the centreline and two in the waist forward of them; one on the centreline, two in the waist, roughly as in the 'E' class; (5) two, *en echelon* so that four might bear together on some arcs, hence better than four single mounts. Note that the standard arrangement of later British cruisers, two mounts on each side, was not included.

47. Controller (Backhouse) memo copied in Lillicrap Constructor's Notebook 5. Controller's paper on the triple turret has not been found.

48. This explanation was given in *Progress in Gunnery*, but Campbell, *Naval Weapons of World War II*, p 36, attributes it to an attempt to reduce shell interference. In the Mk XXII mounting the cordite hoist was totally separate from the shell supply. The cordite hoist was outside the mounting, linking the handing room in the hold to a small (protected) compartment outside the barbette on the deck immediately below the turret. From here it was passed into the mounting and then handed up into the gunhouse.

49. Board Memoranda for the first half of 1932, ADM 167/86, p 2928 (memo dated 25 February 1932) is Controller's memorandum for the Board on the problem presented by the big American and Japanese cruisers, which were of 10,000 tons and 8,500 tons respectively.

50. ADM 167/87, memorandum dated 22 December 1932.

51. Unfortunately this study has not been found in the Constructors' Notebooks, and there is no trace of it in Board Memoranda. Bessant Constructor's Notebook 3, p 58 contains details: 545ft (pp) 570ft (wl) x 58ft (extreme) x 16ft 6in, 50,000shp for 30kts. These provisional data were dated 24 July 1933. Armour data were not given (Bessant was calculating the resistance of a scaled-up *Leander* hull). This design was designated KVII, in a series which probably began with the ten-gun Improved *Leander*: there were two designs each for ten- and nine-gun ships, and two for the eight- and seven-gun *Arethusa*s. However, K designations seem not to have been applied to these designs, at least in the Lillicrap Notebooks, so the missing first six K designs may have been later approaches to developing a more powerful cruiser.

52. Memorandum on cruisers by First Sea Lord, 25 July 1933, in ADM 167/89.

53. The undated Staff Requirement was inserted into the *Southampton* Cover as loose sheets. A note indicates that it was found, and added, on 17 October 1941. These sheets are marked Folio 121.

54. For each aircraft, DNAD wanted an 18in torpedo plus three 500lb bombs, six 250lb bombs, and twelve 20lb bombs, corresponding to a torpedo plus three aircraft-loads of bombs. In 1936 DNAD wanted the torpedo plus two loads of the heaviest bombs carried, one of the next heaviest, and one of anti-submarine bombs. That meant two loads of 500lb or 'B' bombs, one of 250lb bombs, and one of anti-submarine bombs. DNAD defended these loads on the grounds that only 5 per cent hits were expected, and that the ship might be weeks from a fleet base when she fought. The A/S and light bombs were considered useful with the fleet for disturbing patrols; on trade routes the 250lb bomb was more useful. Below a certain point aircraft which could carry heavy bomb loads were no longer worthwhile, so if there as not enough space for good bomb stowage it might be better to supply ships with seaworthy seaplanes rather than high-performance ones.

55. Constructor's Notebook J L Bessant III.

56. In May 1934 E-in-C suggested a modification in which the boilers in the after boiler room would be placed side by side instead of in tandem, their shapes modified so that the outer shafts could pass under them. DNC rejected the idea on structural grounds. That was unfortunate, because the outboard voids associated with the tandem boilers allowed dangerous flooding.

57. War complement was set at 731. About 610 sleeping billets could be arranged in the normal places. If in addition turret spaces, ammunition lobbies, workshops etc were used, as approved for *Leander*, another 100 could be provided. However, accommodation would be worse than in later ships of the *Leander* class. The ship might have to be lengthened to provide the full 7.5 per cent margin over the calculated war complement. The peace complement was 706, including twenty-three officers (twenty-nine in War Complement), nine warrant officers (nine), and six midshipmen (six). Using *Leander* practices the ship could accommodate the full war complement, but the margin would be twenty-three rather than the desired fifty-two. The authorised margin had not been provided in any 6in cruiser to date. The worst problem was insufficient sleeping space, which was measured by the space between slinging billets (for hammocks). The alternative to providing more space was to sling hammocks closer together (reducing the space between billets from 21in to, say, 18in, the old destroyer scale; capital ships and the *Kent*s had 24in, *Leander*s 21in, and *Arethusa* and recent destroyers 20in; but the Medical Director General strongly opposed anything below 20in and preferred 24in). Another possibility, which was rejected, was to merge several small messes, e.g. to form one for all chief petty officers, including Engine Room Artificers. Various measures, such as reducing washrooms, which had already been accepted for the *Leander*s, were adopted.

58. The ship could now dock in No. 2 Dock at Gibraltar, in No. 5 at Malta, and in the Admiralty Dock at Hong Kong, all of which had been referred to in a CNS paper.

59. The belt and the deck covering it were extended 46ft forward, for a total protected waterline length of 232ft (40 per cent of waterline length), compared to 167ft (31 per cent) for D. The main part of the belt was reduced from 5in to 4½in and the magazine side protection somewhat reduced. It was now possible to place the transmitting station, main switchboard room, low power supply rooms, and auxiliary W/T offices, telephone exchanges, and lower steering position on the platform deck behind 4½in belt armour. The belt was covered by 1¼in decks and 2½in bulkheads. Magazine boxes had 4½in sides tapering to 3in at lower edge, decks 2in, bulkheads 2½in. Sides, ends, and decks of 'A' and 'Y' shell rooms were 1in. Turret ring bulkheads and recesses for shell and cordite hoists had 2in sides, 1in on ends, and decks made up to 1in as necessary. Shell and cordite hoists were in 1in tubes, with 3in armour rings where they passed through the armour deck. Protection to steering gear: 1½in sides, 1¼in deck. Bullet-proof protection to the bridge: 20lb side, 15lb roof; this was on the compass platform, plotting office, remote control office, wheelhouse, and after control positions. All boiler room and engine room fans were under main belt protection except after engine-room supply fans.

60. In the October statement 4in guns were credited with 200 rounds per gun, but that was increased to 250 in November, and the outfit

per 2pdr barrel was increased from 1,400 rounds to 1,800. The November statement also showed six double Lewis guns, with 2,000 rounds per barrel. The 6in guns were given 200 rounds each.

61. Controller (Rear Admiral Charles M Forbes) pointed out that in the original design, aircraft had accounted for the weight of a 6in turret, but offered an offensive load of only 4,500lbs, compared to 67,300lbs for the turret, and had a scouting value which depended largely on the weather.
62. Dimensions: 572ft (pp) 600ft (lwl) x 61ft x 16ft 6in; 8,835 tons std. 72,000shp = 32kts. Endurance 7,000nm at 16kts. Armament four triple 6in, three twin 4in, three quad 0.5in machine guns, two triple TT, one heavy catapult and five TSR aircraft. Apparently at first plans called for four twin 4in, but reducing to three saved on the required complement. At about this time DNC stated that there was space for only three aircraft. Armour over magazines and bomb rooms: 5in NC on ⅜in D.1 on lower 3ft 6in of sides; 2in NC on ⅜in D.1 crowns, 3in NC on ⅜in D.1 ends. Shell rooms: 1in D.1 sides, crown, ends. Machinery: 5in NC on 1in D.1 sides, 1¼in D.1 crown, 2½in NC on ½in D.1 ends. TS, LP supply, HA calculating position: 1in D.1 sides and crown. Steering gear 1½ to 1¾in D.1. Turrets: 1in NC shields and rings, 1in D.1 supports; 1in D.1 ammunition lobbies, etc. Bridge: ⅜in BP.
63. Approved by DNC on 30 October 1933.
64. Alternatively, the second HADT could be added aft; that was rejected to avoid smoke interference. Placing the two directors forward offered a more self-contained arrangement of guns and control positions on each side, as well as better arcs of training on after bearings.
65. The belt and 50lb deck were extended aft over the 4in magazine (a distance of 33ft), which therefore did not need their own 4½in side armour. The existing 2in crown was reduced to 1½in which, with the 1¼in deck immediately over it, was equivalent to the 2in against bombs. The extended belt and deck protected the main W/T office on the platform deck at the waterline. DNC pointed out that 4½in C armour could resist the 6in 112lb shell beyond 7,000yds at 30° to normal (5in C would resist the same shell at 6,000yds). The 1½in D.1 deck would be immune at ranges below 16,000yds (the Staff Requirement). DNC added that it would cost 42 tons to add ¼in to the belt, and 84 tons to add ½in (to reach 6,000yd immunity). Thickening the 1¼in deck to 1½in over the machinery would cost 69 tons. That addition would raise the minimum altitude at which the deck would resist a 250lb bomb only from 2,000ft to 2,800ft; as heavy bombs would presumably be dropped from greater heights, this was not worthwhile. Thus DNC preferred simply to extend the belt and deck aft. To do that while retaining the existing 2in deck over the 4in magazine would cost 42 tons, and the magazine would still be open to 6in gun fire above 16,000yds through the wing portions of the lower deck (which would have to be thickened to 2in, giving a total of 75 tons). Thickening the belt to 4¾in would bring the total to 120 tons, absorbing the entire margin.
66. Ordered on 21 November 1935 by DNC in accordance with verbal orders from Controller. Bessant Constructor's Notebook 3 includes a report dated 6 November 1935 on moving the forefunnel back by angling the smoke pipe back above the upper deck. At most the funnel could be moved 9ft back in this way, using supports from the forecastle deck. That would have cost weight and upper deck space. Anything more would have required lengthening the ship.
67. She was chosen because she would entail the least disturbance. *Sheffield* duplicated *Newcastle* because they were building at the same yard. The modification seems to have been proposed in March 1935.
68. J L Bessant Notebook IV (308/4) dated 1934–6.
69. ADM 1/27412.
70. This total had no binding force. It was adopted to allow for growth beyond the 9,000 tons of the original ships.
71. Surviving Constructors' Notebooks do not show Designs KXI through KXIII. Presumably the developed *Southampton* design was KXI. It seems unlikely that KXV was the first attempt at a cruiser with quadruple turrets, so it may have been KXII or KXIII.
72. The original quadruple-turret design was 589ft pp/616ft wl/623ft 6in oa x 63ft 4in x 17ft 3in x depth 33ft 6in; the new one was 579ft/606ft/613ft 6in x 63ft 4in x 17ft 3in x 33ft 6in, 10ft shorter but with the same standard displacement of 10,000 tons. In both cases, speed was 32.5kts at standard displacement and 31kts at deep displacement. The only difference in armament was substitution of triple for quadruple 6in guns, in both cases with 200 rounds per gun. In addition, both designs showed four quadruple pompoms and two quadruple 0.5in machine guns (the change to octuple pompoms was not shown), plus two triple torpedo tubes. The new Legend showed the belt tapered to 3in over magazine sides (forward magazines had box protection). In both versions the belt was closed by 2½in C bulkheads, but in the revised version the deck over the belt was 2in NC instead of 1¼in D. In the earlier version magazines behind the belt had 2in crowns rather than the 1¼in deck elsewhere. Turret protection in both versions was 4in C face, 2in NC sides, and 2in roofs. Turret support rings were 2in to 1in NC. Bullet-proof plating on the bridge was ½in deck, ⅜in sides. Steering gear protection (in both versions) was 1½–1¼in D over steering gear.
73. Magazine crowns, whether in boxes or behind the belt, had 3in rather than 2in armour. Forward magazine sides were 4½in (like the belt), tapered to 3in at their lower parts. Shell rooms were above the magazines, for additional protection. The steering gear was given a 2in deck and 1½in D sides and ends. Low-power cables carrying data from the transmitting station and HA calculating positions to the DCTs and HACS were enclosed in trunks of 20lb bullet-proof plating, apparently an innovation.
74. This amounted to protection for exposed personnel including gun crews; radar, RPC for the pompoms; stiffened upper deck; modified transition plating between the forecastle and the side; a modified forecastle deck forward; a modified sheer strake; a new transverse bulkhead at 104 station aft; and Oerlikons.
75. ADM 1/9355.
76. Lillicrap Constructor's Notebook 6 describes the design process, beginning with the assignment of Designs P through T by Director of the Tactical Division, 13 April 1934:

	P	*Q*	*R*	*S*	*T*
	4,500 tons	*X*	*4,500 tons*	*4,500 tons*	*X tons*
Armament					
6in	6 x I	as P	as P	X	2x III*
4in HA	4 or 2 x quad 2pdr				2 quad pompom
Speed (std)	X	33kts	33kts	33kts	33kts
Protection	Mags vs 6in ER, BR vs 5.1	As P	X	As P	As P**
	No aircraft in any of these				

*Turrets both forward, superimposed.
**Turrets splinter-proof.

Sketch designs for the 4,500-tonners were scaled down from *Arethusa*, with similar machinery spaces and weights. Based on the 'M' class, to protect magazines against 6in fire required 4½in sides (tapering to 3in) and 2in decks. In 1928 the Tactical Section had called for machinery protection against 4.7in (destroyer) fire above 4,000yds. That required a 3in side (effective at 4,000yds at a 30° angle to the normal, and at 5,000yds at any angle) and 1in deck (proof to 15,000yds). Against the 5.1in gun, then being considered for future destroyers (and capital ships, as secondary armament), the 3in side was proof above 9,000yds (7,500yds at a 30° angle). For Design P, Lillicrap added up hull (scaled from *Arethusa*), armament, general equipment, protection, and machinery weights not proportional to power, and subtracted from the allowed 4,500 tons to get what he could have for the powerplant: 840 tons, which scaled down from *Arethusa*'s 64,000shp plant to give 47,500shp, which in turn gave a speed of 30.75kts using curves of known hulls and scaling. For Designs R and S, with their required speeds, he went the other way, finding that they needed all of *Arethusa*'s 64,000shp. Given armament and machinery weights, he subtracted to find what was left for protection. It was not much. Taking *Arethusa* magazine protection (weaker than required), he was left with enough for either a 1¾in belt (plus ½in of hull steel) and a ⅜in deck or a ¾in belt and 1in deck. Design Q had to exceed 4,500 tons, because adding up the basic hull, machinery as in Design R (to make 33kts), protection as in Design P, and general equipment accounted for 4,450 tons. The heavier ship needed a bit more power, and that made her heavier still, with more protection to cover somewhat larger machinery spaces. Design T was derived from Design Q using triple turrets weights from the new 'M' class (*Southampton* class). Compared to *Arethusa*, armament weight was about 20 tons less, but protection was 120 tons more, e.g. for the ammunition lobbies. The ship also needed more generator power.

Controller then asked for an additional design, for a small fleet cruiser not to exceed 3,800 tons, with five or six 6in guns and two quadruple torpedo tubes, to make 38kts light and 35kts deep. Endurance was 6,000nm at 16kts. If practicable, the ship should have a 1in belt and ½in or ¾in deck over the machinery, with side protection over the magazines. A six-gun design should have two twins forward and two singles aft, i.e. for maximum chasing firepower. A five-gun design would have three guns forward (one twin and one single or three singles, two *en echelon*) and two singles aft. Work began on 9 May 1934. Instead of scaling up from the 3,000-tonner sketched in 1930, Lillicrap again scaled down from the much larger *Arethusa*, presumably because she represented more modern design practice. He found that her hull form was too resistful; he was forced back to the 3,000-tonner, which had been based on destroyer hull form. Machinery had to follow destroyer leader rather than cruiser practice. Even so, it proved impossible to reach the desired 38kts. This sketch design was submitted on 2 July 1934, becoming the basis for a further alternative, Design U.

ADM 1/8828 includes TD Memorandum 126, 'Memorandum on Small Cruisers,' dated November 1934, which includes the table of design alternatives, with approximate Legends. Design data were:

	P	Q	R	S
Displacement	4,500 tons	5,000 tons	4,500 tons	4,500 tons
Single 6in	6	As P	As P	3
4in HA	4	As P	As P	2
Speed (std)	30.75kts	33kts	33kts	33kts
Magazines*	4–3in/2in	As P	3in/2in	As P
Machinery	3in/1in	As P	1¾in/⅜in	As P
Endurance	6,000nm at 15kts	As P	As P	As P
Cost	———————£1.1 million———————			

	T	U	V
Displacement	5,600 tons	3,500 tons	1,830 tons
Single 6in	2 x III	5	5 x II 4.7in
4in HA	2 pompom	–	–
Speed (std)	33kts	38kts	36.25kts
Magazines*	As P	None**	None
Machinery	As P	None	None
Endurance	As P	As P	5,300nm at 15kts
Cost	£900,000	£480,000	

*Given as side/deck.
**No armour, but 1in side plating abreast machinery and ⅜in deck.

77. Magazine sides reduced from 4in to 3in, belt over machinery from 3in to 1¾in and deck from 1in to ⅜in or, alternatively, to ¾in side and 1in deck.
78. C-in-C Mediterranean wanted twelve 4.7in HA guns on 3,500 tons with a speed of 26kts, with maximum possible protection, Asdic and good searchlights. A quick estimate suggested that protection would be the usual 3in sides and 2in crowns for magazines and 2in sides and 1in deck over machinery.
79. According to a paper summarising views on the small cruiser, in ADM 1/8828, C-in-C Mediterranean envisaged the cruiser force in the war zone as five 8in, four *Leander*s, four *Arethusa*s, and eight new-type AA cruisers. The staff considered this too weak to meet Japanese reconnaissance forces. C-in-C Mediterranean wanted a policy of two AA cruisers to every five *Southampton*s, until eight were available.
80. ADM 1/9384.
81. This was the largest calibre allowable for destroyers under the 1930 treaty, and it was used aboard French destroyers. The Royal Navy considered it for future destroyers, but dropped it after unsuccessful trials. In 1932, when new battleship designs were being prepared as a basis for the British official position at the 1932 Geneva conference, the twin 5.1in was proposed as their dual-purpose secondary gun. The Staff Requirement for the 5.1in gun was issued in March 1932. It was to fire ten rounds per gun per minute, with elevation limits of +70° and −10°, and with a maximum surface range of 18,000yds. Ceiling was to be 10,000ft at a plan range of 16,000yds. Time of flight to this range should be less than 45 seconds. A battleship should have nine or ten guns on each side. To achieve the desired rate of fire, the gun had to fire fixed ammunition (no separate cartridge case) and the round had to be within the 110lb limit for man-handling. That in turn implied a 70lb shell; the performance requirement set a muzzle velocity of 2,500ft/sec. The 70lb shell implied 5.1in calibre. It is not clear why early versions of the *King George V* design showed 4.5in rather than 5.1in guns. The 5.1in in turn gave way to the 5.25in selected as the battleship secondary gun.
82. ADM 1/9384 (and also the *Dido* Cover) includes TD 83/35 of

January 1936, 'Staff Memorandum: The Small Fleet Cruiser and RA(D)'s Flagship: Examination of Design 5.25-B'. The memorandum assumed that the choice lay between five 6in guns and ten 5.25in, declaring the latter clearly superior, the 80lb shell being very well suited to a ship of this size, sufficient to deal with a small cruiser. Greater numbers offered a much better chance of hitting an evading destroyer. The powerful anti-aircraft armament would be a considerable fleet asset. The extra 250 tons, to provide alternating engine and boiler rooms, was well worth while. At this stage combined HA/LA DCTs were envisaged. The small fleet cruiser would work in the screen, for shadowing, for supporting British flotillas against enemy destroyers, and, under cover of more powerful forces, for offensive raiding and patrol operations with light craft. DNC was asked to consider two more possibilities, Designs C and D. Design C was Design B with two triple torpedo tubes, with 3in rather than 2in deck protection over magazines, and with a light catapult. Flag and plotting facilities would be as in a cruiser flagship rather than RA(D) flagship. Design D was as Design C except for a 3¼in deck over magazines. DTD wanted to know the effect on Design C of a heavy catapult, or 400 rather than 300 rounds per gun, and the effect if one set of machinery spaces was given 3¼in deck protection rather than the 1in envisaged. It is not clear from the printed version whether the increased deck protection in alternative D was 2¼in over the machinery rather than the magazines. The RA(D) ship was compared to two US designs recently reported by NID, A and B (neither of which can easily be identified). A was a 29–30kt 3,500-tonner armed with eight 5in guns, thirty-two 0.5in machine guns and four triple torpedo tubes, a catapult and three aircraft, and 3in belt and 2in deck protection. B was a 29–30kt 5,000-tonner armed with two triple 6in guns, thirty-two AA machine guns and four triple torpedo tubes, the same catapult and aircraft, and 3in to 4in belt and 2in to 3in deck armour. 'The USA ships have well-protected turrets, somewhat better decks, and include a Torpedo armament and aircraft, but it is recognised that the figures are not conclusive.' It was estimated that a 29.5kt version of the 4,750-ton RA(D) design would displace about 4,200 tons. The table also included the slightly earlier T design (5,600 tons, 35kts, two triple 6in), a 29.5kt version of which would displace about 4,900 tons. There is no indication that DNC produced Designs C and D, although he did head off calls for 3in deck protection.

83. Lillicrap Constructor's Notebook 6 shows how the two alternatives and their successors worked out. The requirement for the RA(D) ship, TD 83/35, was dated 1 June 1935. Lillicrap once more began by scaling down the *Arethusa* and her machinery, and adopting the earlier arrangement of undivided boiler rooms. Taking armament weights into account, he had enough protection weight to provide the 3in sides and 2in deck magazine protection of the larger ship, leaving enough for 3in side and 1in deck over the machinery. On this basis he could achieve 33kts on a 4,500-ton hull 470ft long. With alternating engine and boiler rooms the ship would displace 4,700 tons (at 475ft). Another calculation showed that reducing speed from 33kts to 31.5kts could save 200 tons, enough to increase the HA battery to two twins and to add 1in to the magazine sides. With alternating boiler and engine rooms, the resulting ship would displace 4,500 tons (475ft long). If the ship had the earlier undivided boiler rooms, the weight could go into another inch of belt over the machinery and another quarter-inch of deck over the machinery: 4,500 tons (470ft). Alternatively, with the original armour and the single 4in guns, the ship could be cut to 4,100 tons (455ft). On a 4,700-ton, 475ft hull, with undivided boiler rooms (63,000shp for 33kts), Design B had enough weight to protect machinery but not magazines against the new 4.7in shell at 5,000yds: 3½in side and 1in deck over machinery, 3in side and 2in deck over magazines – whose sides would be protected almost completely by water at that range. Armament weight was based on 250 rounds per gun plus 50 star shell per ship. Particulars of Designs A and B went to DNC on 28 June 1935.

84. On 8 October DNC, presumably at Controller's request, asked for the cost of a 5,000-ton ship with ten 5.1in HA/LA guns in twin mounts, capable of 32–33kts. Lillicrap took *Arethusa* as his basis, with about the same hull and protection. He estimated that the ship would cost £1.07 million. On 24 October DNC asked for a 4,500-tonner on the lines of the RA(D) ship armed with five twin 5.25in guns and two quadruple pompoms, to make 32kts. The ship would carry neither torpedoes nor aircraft. A marker in Lillicrap's Notebook at this point is '*Dido* and other cruiser designs'. The 5.25in gun had not yet been completely designed, and on a ship this size a slight change in armament weight could have considerable effect. Lillicrap began with the 4,500-ton (470ft x 48ft x 13¼ft) ship, recalculating power for 32kts (say 54,000shp). Available tonnage for protection gave him 3in sides and 2in crowns and ends for magazines, leaving enough for 2in side and 1in deck over the machinery (a re-estimate added 1in ends to the machinery). Lillicrap noted that the available length was too little, but his Notebook does not include any calculation of what was needed. This ship had adjacent boiler rooms, and greater length would be needed if the more survivable alternating engine and boiler rooms were adopted. the sort of lengthwise arrangement sketch which became common a few years later. Lillicrap discussed the design with DNC on 2 November, and in view of difficulties in providing enough length (probably for the five centreline 5.25in mounts), he proposed an alternative in which extra length was used to provide alternating engine and boiler rooms (which he called the Modified *Leander* machinery layout). Lillicrap justified adding 15ft to accommodate the new machinery arrangement and recalculated hull weight. He ended up with a 4,750-ton (485ft x 49½ft x 13¾ft) ship requiring 55,000shp to make 32kts. More weight was available for protection, so he could provide 3in side armour (and the earlier 1in deck) over the machinery. Lillicrap reported this design and the 4,500-tonner to DNC on 5 November, and they were forwarded to Controller as Designs A and B. The 485-footer became the basis for the *Dido* design.

85. At its meeting on 28 November 1935 the Board decided that the 1936/7 cruiser programme would comprise two *Southampton*s, the RA(D) flagship, and two further cruisers of a type to be decided depending on how the London Naval Conference came out. At that time the conference seemed close to collapse, and it seemed unlikely that any reduction beyond 10,000 tons would be achieved. Nothing smaller than a *Southampton* should be built until it was clear what foreign powers were doing. It was certainly not clear that the proposed 8,000-ton limit would be ratified. On 25 February 1936, however, First Sea Lord compared the British underage cruiser force with the combined forces of Germany and Japan, the two likely future enemies. The Royal Navy enjoyed parity in 8in cruisers built and building, superiority in large 6in cruisers (8,500 tons and above: ten to six) and in medium 6in cruisers (5,500–8,500 tons: ten to six), but inferiority in small 6in cruisers (nine to fifteen). Chatfield therefore decided that further construction of large cruisers should be suspended in favour of small ones, about 5,000 tons, any of which might be suitable as RA(D) flagship. The two cruisers of undecided type became two more *Dido*s. They offered an additional advantage

in that they were less expensive than the ships originally included in the programme, as it was reported to the Defence Requirements Committee. As the international situation worsened, in June the DPR (the Sub-Committee [of the Committee of Imperial Defence] on Defence Policy and Requirements) formally requested ways to accelerate the building programme. It was told that in each of the next three years seven rather than the planned five cruisers could be ordered. The 1936/7 programme was the first affected. ADM 167/95, Board memoranda for 1936. The acceleration memo is DPR.88. The new programme was brought before the Board at its meeting on 24 June 1936. The fourth and fifth *Dido*s were part of the Second Supplementary Estimate of July 1936.

86. The main points agreed at a meeting on 12 February 1936 in the First Sea Lord's office, with First Lord (CNS), Controller and DCNS present were that they should be prepared to go to 5,000 tons; that the ship should have 300 rounds per gun; that she should have two triple torpedo tubes; that magazine crowns should be thickened from 2in to 3in; that the forward gun might be on the forecastle deck; and that the questions of having octuple rather than quadruple pompoms and a thicker deck over machinery should be considered. The proposed additions (other than a thicker deck over machinery) added up to 152 tons, the largest items being extra ammunition (33 tons for 5.25in, 48 tons for pompom ammunition and pompoms). Lillicrap thought the ship would have to be lengthened 10ft to accommodate larger magazines, adding another 80 tons. That brought her to about 5,000 tons (495ft x 49½ft x 13¾ft). At this point the 4,500-ton Ship A was still in contention. Lengthening her by 10ft would bring her to about 4,750 tons (480ft x 49ft x 13½ft) or even 5,000 tons if she was given 6in more draft. That in turn would require more horsepower and more machinery weight (and length, to be protected). Another possibility was to lengthen the belt 60ft and thicken it from 3in to 4in. On 8 February Lillicrap and another constructor, Jackman (who had investigated hull weights in greater detail) produced a table of alternatives for Controller (submitted 19 February): *see below.*

Controller did not see any great benefit in a 5,000-tonner, and preferred A1 but with 3in magazine crowns. ACNS preferred B2 (alternating machinery spaces). The 3in belt could be penetrated by the new 6in shell (112lbs) at 16,000yds at 90° or at 13,500yds at 60°. Penetration by the new 4.7in destroyer shell was doubtful at 40,00yds. Comparable figures for a 4in belt were 11,000yds and 9,500yds; it was immune to 4.7in fire. A 3in deck would keep out a 1,000lb bomb dropped from 5,000ft and a 500lb bomb dropped from 9,000ft. There was little to choose from between a 1in and a 1¼in deck. The file on the RA(D) cruiser includes small-scale drawings of cruisers with adjacent and alternating engine and boiler rooms labelled P (495ft x 50ft x 13?ft, 5,150 tons) and Q (506ft x 50½ft x 14¼ft, 5,300 tons), which were supplied to support the decision for or against this arrangement, with BD guns.

87. Machinery was covered by a 3in belt and 1in deck and ends. Magazines had box protection: 3in sides and roof and 2in ends. Other key spaces, including shell rooms and transmitting station, had 1in sides and crowns. These thicknesses in Controller's memo to the Board do not quite match those in the Legend submitted at the same time. The Legend shows turret protection of ½in bullet-proof plating rather than the 1in planned earlier.

88. Only on 5 May did a meeting in DNO's office choose a mounting design from among several offered by Elswick. It was an upper-deck mounting with a 1in shield weighing 75 tons including ammunition

	A	B	A1	B1	A2	B2
LWL	470ft	485ft	470ft	485ft	485ft	495ft
Dispt	4,500 tons	4,750 tons	4,620 tons	4,830 tons	5,000 tons	5,000 tons
Speed	32	32	32	32	32	32
5.25in twin	—————————————————————5—————————————————————					
RPG	250	250	300	300	300	300
2pdr	2 x IV	2 x IV	2 x IV	2 x IV	2 x VIII	2 x VIII
TT	–	–	2 x III	2 x III	2 x III	2 x III
Magazine side	—————————————————————3in—————————————————————					
Crown	2in	2in	2in	2in	3in	3in
Machinery Belt	—————————————————————3in—————————————————————					
Deck	1in	1in	1in	1in	1¼in	1in
Mach	Adj	Alt	Adj	Alt	Adj	Alt
shp	54,000	55,000	65,000	56,000	58,000	56,000
Legend						
Hull	2,130 tons	2,230 tons	2,130 tons	2,230 tons	2,230 tons	2,330 tons
GE	400 tons	400 tons	400 tons	400 tons	400 tons	400 tons
Armt	430 tons	430 tons	490 tons	490 tons	540 tons	540 tons
E-in-C	920 tons	960 tons	940 tons	975 tons	990 tons	975 tons
Dyn etc	70 tons	70 tons	70 tons	70 tons	70 tons	70 tons
Pro	4,540 tons	4,740 tons	4,620 tons	4,815 tons	5,000 tons	5,000 tons

In machinery arrangement, Alt means alternating ER and BR. Otherwise engine and boiler rooms are adjacent.
In A2, alternative protection over machinery was a 4in belt and 1in deck.

hoists. At this point the heavier BD mounting was introduced; it seemed that the choice lay between four BD and five upper-deck mountings, the former saving 6 tons by eliminating one set of magazines and their protection. Expected revolving weights were 56 tons and 70 tons. Making all five mountings the heavier BD type would add 170 tons. It was pointed out that the extra weight could be balanced by reverting to adjacent engine and boiler rooms, but that was not acceptable; instead, the ship was allowed to grow to 5,300 tons. An investigation into the stability of a ship with the BD mounting showed the need for increased beam, so that was made 50ft 6in. Figures balanced well enough for a 506ft x 50½ft, 5,300-ton ship, and powering estimates suggested that it needed 58,000shp (32,000ehp with propulsive coefficients typical of small light cruisers). The modifications planned were reported to bidders on 3 February 1937 (the extra beam had not yet been chosen), and all the modifications were approved by Controller on 8 March. ACNS was gratified that the magazines and shell rooms were now protected against 6in plunging fire at 19,000yds and below, and against 250lb dive bombs. The Board formally approved these changes on 15 April 1937.

89. When the decision to proceed was made in April, 1936, it was £1.3 million. When the Board approved the 5,300-ton design, it was £1,340,000. The 5,450-ton design was expected to cost over £1.6 million.

90. Cruisers and larger ships were provided with a closed lower bridge below the compass platform. In August 1938 Director of Navigation (D of N) attacked this practice, arguing that the compass platform was and should be the primary conn, and that simply having a closed bridge reduced the effort to make the compass platform entirely efficient. 'After many years of handling H.M. Ships, D of N would not permit a closed bridge to be used for this purpose in any vessel he may command. The "lookout" is inevitably more restricted and subject to "blind" spots. Sound signals are less likely to be heard and the difficulty of estimating their direction would be much greater. The risk of the bridge personnel becoming drowsy at night is immeasurably greater. During dark hours, the faintest light, e.g., the glow from the pelorus repeater, would make the windows opaque from within, except for external objects having their own illumination ... the use of a boxed cubby-hole from which to control the ship entails a degree of risk which in D of N's opinion should not be accepted.' DNE argued that in some (unspecified) situations a ship should be conned from a closed position. It was decided that the existing arrangement, in *Manchester* and later large cruisers, was too elaborate; the upper steering position (two decks below the upper bridge) should be combined with the existing closed bridge. No change would be made in *Manchester*, *Liverpool* and *Gloucester*.

91. The 5,300-ton design showed only one DCT on the bridge (both DCTs were HA/LA at this stage). Initially the design showed only a single transmitting station (and two HA control positions). ACNS pointed out that there was a design requirement that the armament be split for HA fire; it might as well be split for LA, too.

92. Controller announced this change at a meeting on 3 September 1939; he hoped to accelerate completion not only of these two ships but also of the two following them. They were to be adapted as RA(D) flagships because they would be the only two of this sub-type, because the lack of gunpower was more acceptable in an RA(D) flagship, and because RA(D) would now be able to fly his flag in a more up-to-date ship. *Aurora* and *Galatea* would be released from RA(D) flagship roles to form a homogeneous squadron with their sisters *Arethusa* and *Penelope*. DSD agreed, subject to the ships being arranged so that ultimately they could be rearmed with 5.25in guns (which was never done). Projected completion dates for *Scylla* and *Charybdis* were June and May 1941; actual dates were June 1942 and December 1941. The armament statement on 20 November 1939 showed 400 rounds per gun. The ships retained their pair of quadruple pompoms and their torpedo tubes. In October 1941 plans called for them also to have a single 4in star shell gun, fifteen depth charges (rather than the original six), two single power-operated pompoms (S.B. Mk VIII), and six Oerlikons (rather than the four planned earlier). Endurance was given as 5,500nm at 16kts. On full-power trials on 3 December 1941 *Charybdis* made 30.774kts at 62,440shp at 6,800 tons.

7. The Slide Toward War

1. This account is based on Notes on Naval Air Defence dated 2 January 1937, circulated by DNC Stanley V Goodall to his sections. Copies are in the *Fiji* Cover and also in the General Battleships Cover. This paper laid out various kinds of air attacks: high bombing, dive bombing, torpedo attack, and close-range attack (strafing) by day, and individual high-level bombing, dive bombing, and torpedo bombing by night. High-level bombers were expected to salvo their bombs (on command from the flight leader) from 10,000 to 14,000ft altitude, the maximum in a group being eighteen (typically divided into flights of three to five, flying at up to 200kts speed). By this time the Royal Navy certainly accepted that US-style dive bombing was feasible, since a dive-bombing attack was expected to begin with aircraft approaching at 10,000–15,000ft, diving in steps to 6,000ft, then making final dives at a 40–70° angle, releasing their bombs at 1,500–2,000ft at a speed of 200–300kts. Torpedo bombers would likely approach at 10,000ft in formations of no more than six, diving to attack at 6,000yds range (45° dive, maximum speed 200kts) and dropping their torpedoes at a height of 30–200ft and a range of 800–2,000yds, the maximum time between flattening out and dropping being five seconds. As in previous analysis, it was assumed that up to eighteen aircraft could crowd around one target (in this account, for simultaneous drops and successive attacks at thirty-second intervals). The maximum number of strafers attacking one target was set at six, at speeds of 150–250kts. The account of night attacks stressed how difficult it would be to illuminate the attacking aircraft, a theme carried over from previous evaluations of the air threat. Drawings in General Cover 1 show the expected ADO position in a *Southampton*. The ship had a pair of combined ADO and star shell sights at the after end of her bridge, separate from the compass platform forward of the DCT. ADO would stand at the port sight, his assistant (AADO) at the starboard one. ADO was responsible for distributing anti-aircraft fire. He and AADO were both lieutenants. A smaller cruiser, such as a *Dido*, would not have dedicated ADO sights, instead ADO and AADO would stand to either side of the chart table, which was alongside the ship's 3m navigational rangefinder. At least in the large cruiser, each ADO sight was surrounded by communication devices which made it an ADP: links to the HADT and to the pompoms and machine guns, with a selector switch to the LADT. There was also a gyro compass repeater and a master cease-fire push. The 1937 report advocated a sight recently developed on board the cruiser *Devonshire*. Ideally they were paired, each pair covering a 60° sector from horizon to zenith; the maximum time on duty should have been an hour.

2. ADM 167/91, Memos for 1934. The key point raised in the Board memo was that the ships had been designed in view of war experience with light ships, on which basis their machinery was unprotected. Since then foreign navies had built large numbers of fast super-destroyers and, in effect, armoured cruisers. Memo dated 18 October 1934, describing the planned upgrade.
3. The surviving 'E' class Cover gives details. By 1942 *Enterprise* had had her No. 7 gun replaced by a pompom. When modernisation was ordered in September 1942 it included restoration of that gun and removal of the two after sets of torpedo tubes. She would be given new masts and new plotting arrangements in her bridge structure, as well as a new air-defence position, new radar and a new Type 132 Asdic. By this time *Emerald* had already had her two after sets of tubes removed, and she had two quadruple pompoms. In 1943 the allowance for both ships was seven 6in, three 4in HA, two quadruple pompom and six twin Oerlikons, one E.I.H catapult, and two quadruple torpedo tubes.
4. Controller (Rear Admiral Henderson) announced the programme, which he described as acceleration of fleet AA rearmament, in a memo dated 18 March 1936; the decision was reached at a meeting with CNS (i.e. First Sea Lord Admiral Backhouse). Additions and alterations need no longer await the need for large repairs. Some reserve anti-aircraft equipment had already been ordered to make up for deficiencies, under the 1935 programme. This memo and the list of planned upgrades are in General Cover 1. ADM 116/4037 gives policy statements concerning the programme. Progress was reported to the Board on a quarterly basis, the reports being included in Board Memoranda. The Defence Policy and Requirements Subcommittee (of the Committee on Imperial Defence) approved the Class II anti-aircraft rearmament programme (as DPR.196) in July 1937 subject to Treasury sanction. Class II included producing or modifying guns to arm merchant ships.
5. The Class I programme kept changing. Thus the report of the 6th Meeting of the AA Rearmament Committee, in 1938, included the decision that a Board decision on the design for 'C' class rearmament was still pending, and that equipment for these ships would not be needed during the first half of the year. Approval was being sought to rearm *Amphion* (to become HMAS *Perth*) between December 1938 and April 1939 (*Apollo* had already been rearmed). HMS *Achilles* could not have her fire-control gear modified until she returned from New Zealand in 1940 (that did not happen), and an Australian decision on HMAS *Sydney* was pending. *Emerald* and *Enterprise* would not be rearmed until September and June 1940 respectively. *Hawkins* and *Frobisher* could not be taken in hand before 1 March 1939. No time would be saved by rearming them like *Effingham* without materially altering the 'C' class conversion programme, in which case fire-control gear, the limiting factor, would become available earlier.
6. As described in July 1937, the approved set comprised eight 4in guns, one HACS, two quadruple pompoms and two quadruple 0.5in machine guns. Approval was sought for eleven such sets for cruisers plus seven for merchant ships. When the merchant ship-conversions were discussed, ACNS proposed two rather than one HACS for them, but DNC provided only one, in view of a later production decision. DTSD pointed out that the two prototypes had two HACS each, and that a ship with so powerful a battery should be able to engage two targets simultaneously. The only exceptions were sloops with eight such guns, which clearly lacked space for the second system. DTSD accepted that some of the ships would be scrapped before the second control system became available.
7. Priorities given in MFO 1382/1939, in ADM 116/4037.
8. A July 1937 paper refers to DNC's sketch designs. An unusual feature of the design was that each mount was provided with multiple hand-ups from ammunition hoists terminating on the deck below, the mount being served by the nearest hand-up, depending on how it was trained.
9. The choice was either to replace two of the three 60kW steam (reciprocating) generators with two 200kW turbo-generators and adding motor-generators to convert DC to AC for the armament; or adding two 200kW diesel generators specifically for the armament and adding more generators to meet the increased DC load; or replacing one 60kW turbo-generator with a 200 to 250kW AC turbo-generator with a good overload rating, to supply armament directly and, through a motor-generator, DC power to the ring main, plus a second AC turbo-generator (200kW) for the armament alone. There was some question as to whether a 200kW diesel generator would fit within the ship's deck height. The choice was to replace one of the 60kW generators with a 200kW AC unit and to install a new 200kW AC generator, with two 60kW motor-generators suitable for conversion from DC to AC or from AC to DC (electric motors are reversed electric generators).
10. Armament would have been four 5in/38 Mk 30 with two Mk 37 directors, two Oerlikons on the bridge wings (as in *Curacoa*), two 0.5in or Oerlikons aft, and two pompoms amidships as were being considered for 'C' class AA ships with two directors.
11. Dimensions: 546ft (wl) x 61ft 6in x 17ft 3in. Bessant Constructor's Notebook 4. Unfortunately Bessant concentrated on the development of the *Belfast* class after making the estimates described. The Notebooks of the responsible senior constructor, Kennet, have not survived. W G John, whose Notebooks have survived, took over the 8,000-ton cruiser project in October 1936.
12. Bessant Constructor's Notebook 4.
13. Lillicrap Constructor's Notebook 6. Unfortunately Lillicrap was not assigned to develop the design alternatives for this ship, and no comparable Notebook has survived. ADM1/9402 contains the designs submitted, but not the reasoning behind them.
14. W G John Constructor's Notebook 11 shows estimated displacement for K25G* of 8,210 tons, up from 8,150 tons a few weeks earlier. She would require about 74,000shp for the desired 31.75kts in standard condition, and deeply loaded would make 30.25kts; machinery weight was 1,440 tons, and armament weight was 1,050 tons. K31 was credited with 66,000shp (31kts at standard displacement, 29.6kts deep), the reduced weight of her machinery making up for her heavier 6in battery. Required endurance for both ships was 7,000nm at 16kts, six months out of dock. K31 particulars given to the towing tank (Haslar) in December 1936: 550ft (wl) 526ft (pp) x 59.5ft (wl) 60ft at upper deck x 16.5ft (depth 33ft on centreline, 32½ft at side).
15. The two designs are compared in W G John Constructor's Notebook 11. Vickers offered eight twin 5.25in, the usual two multiple pompoms and two quadruple 0.5in machine guns, and two triple torpedo tubes. Protection would have included a 3in belt covering machinery and magazines, covered by a 2in deck. Dimensions were 540ft wl/535ft pp x 57/60ft upper deck x 52ft 6in depth mld x 16ft 6in mean. Machinery: 75,000shp (300 rpm) for 32kts. Vickers admitted that the ship was 700 tons heavy, and John estimated that she would actually displace 8,967 tons. His analysis was dated 13 October 1936.
16. K31 could make the desired 30kts deep if her deep displacement were held to 9,770 tons. To do that oil fuel had to be reduced from 1,890 tons to 1,220 tons, giving an endurance of 5,600nm at 16kts

with a clean bottom. Alternatively, she could be credited with a speed of 30kts when 18 per cent deep, having burned 670 tons of oil. Normal oil-fuel capacity, set by the requirement for 30kts in deep condition, would be 1,500 tons, further tankage being labelled a reserve.

17. The preferred four-shaft arrangement produced 66,000shp using four boilers (300rpm, 11ft propellers 1,180 tons). The alternative was a 72,000shp two-shaft arrangement using five boilers (240rpm, 15ft propellers, 1,435 tons). The five boilers would have been arranged in tandem, because otherwise the shafts would have had to be too far apart, so that they could clear the boilers.

18. According to a typed report dated 9 November 1936 (W G John Constructor's Notebook 11), calculated metacentric height in standard condition was 2.08ft, compared to 3.27ft for a *Southampton* and 2.79ft for the Modified *Leander*. With a beam of 59ft 6in, the new ship would have a metacentric height of 2.09ft at 8,150 tons and 1.90ft at 8,300 tons. With 60ft it would be 2.34ft at 8,300 tons.

19. Meeting in DNC's office, 3 November 1936, cited in *Fiji* class Cover.

20. Sketch and data in W G John Constructor's Notebook 11. Although not labelled as such, this was probably K32 (K33 was the *Abdiel* class fast minelayer). A very crude sketch showed the 4in magazine between the forward boiler room and the forward engine room, the machinery otherwise divided as in recent cruisers; the magazine of No. 3 turret would have been 61ft forward of the forward boiler room. Unusually (perhaps uniquely) the sketch was used to estimate the required length: a total of 95ft for the magazines etc under the three forward gun mounts (the forward end of this length coincided roughly with the barbette of No. 1 turret), a 34ft forward boiler room, 24ft for the 4in magazine, 41ft for the forward engine room, 52ft for the after boiler room (boilers in tandem), 48ft for the after engine room, 34ft for the magazine under No. 4 turret, with shell room above it. At the after end of the ship, 117ft was allowed from the after boiler room bulkhead to the rudder post (presumably indicating allowable propeller shaft length), and 24ft from the rudder post to the stern at the waterline. Waterline length seems to have been calculated on the basis that the centre of buoyancy had to be shifted to keep the ship on an even trim, rather than on the basis that dryness required that the forward turret be a particular distance from the bow.

21. W G John, who prepared the estimates for the new ship, was responsible for analysing *Gorizia*.

22. The US Navy suddenly decided to add underwater armour to the *North Carolina* class at about the same time, to deal with the same sort of threat.

23. According to W G John Constructor's Notebook 11, Controller (Admiral Henderson) said that he had to think about this possibility, but that it should not be mentioned to E-in-C. Hearing about the exchange, DNC asked what would be the length of such a ship, and what reduction in armament would be necessary to bring it to 8,300 tons. Mr. John started with a 33.5kt ship and asked how much power he needed: as a first cut, 99,000shp, not far from the 96,250shp he got by boosting the 72,500shp of the new design by 30 per cent. He tried lengths of 550ft, 560ft and 570ft (waterline), and concluded that it did not pay to lengthen the ship. He could not approach E-in-C for possible machinery box dimensions, but noted that Yarrow was advocating a boiler which would actually reduce boiler room length by 6ft while increasing output. He estimated that he would have to save 340 tons in a 550ft ship or 387 tons in a 560ft ship. Simply reverting to twin 6in guns would save 321 tons, so the ship could retain her side and deck armour. More detailed calculations would show additional savings due to smaller magazines and shell rooms. The ship could have two triples and two twins if her belt armour were reduced from 3½in to 3¼in and deck armour from 2in to 1¾in (but 2in could be retained over the 6in magazines). The torpedo tubes would be deleted. These steps would save 333 tons. The other 7 tons could come out of the 4in magazine crowns. Another possibility was to replace the two triple torpedo tubes with one quadruple (but it is not clear where it would have gone). John then developed a series of alternative schemes. Scheme X deleted all aircraft and torpedoes, while Scheme Y deleted the aircraft but retained one set of quadruple tubes, and had the deck shaved to 1¾in. Scheme Z left the torpedo tubes on board, but eliminated the aircraft and shaved deck armour to 1¾in (upper deck and magazine crowns). Then he tried more possibilities: deleting the torpedoes and taking another ¼in off the decks (Scheme R); reducing torpedoes to one quadruple mount and reducing decks and belt by ½in (Scheme S). By late April, Controller was discussing the idea with E-in-C. A lesser increase in power might be wanted; John estimated that the ship could make 32.5kts on 90,000shp, or 32.1kts on 80,000shp. John then developed five versions (A to E) of the 8,300-ton ship with 85,500shp (18 per cent increase) machinery sufficient for 32.5kts, differing in what was sacrificed. Scheme A sacrificed aircraft and torpedo tubes, but retained full protection. Scheme B sacrificed the aircraft and had 1½in decks over the machinery (but retained 2in over magazines). Scheme C retained the aircraft but sacrificed the torpedo tubes, and had 1¾in decks over magazines and 1¾in and 1½in over machinery. Belt armour was reduced from 3½in to 3¼in. Scheme D had aircraft and torpedo tubes but 3in belt armour and the reduced decks of Scheme C. Finally Scheme E had the belt over the magazines increased to 3¼in, with a 1¾in deck over magazines and 1½in over machinery. W G John Constructor's Notebook 11 includes a set of Legends dated 14 May 1937 for alternative designs A, A', B, B' and B2 with increased power. The A designs were the same length as the earlier design, 526ft pp, but the B designs were lengthened to 538ft pp. Beam was the 61ft 6in adopted for the 1937 cruiser, except that B2 was 61ft beam. Torpedo tubes were eliminated, except in B2. Displacements were about 8,200 tons. Power was boosted: to 83,000shp and to 86,000shp in A and B respectively, to 80,000shp in A1 and B1, and to 76,000 shp in B2, with various combinations of side and deck armour. Controller rejected the sacrifice in features such as protection associated with the 83,000 to 86,000shp powerplant, and asked instead for 80,000shp.

24. Either the boilers could be forced to give more power on roughly the original size, or squeezed down while giving 72,500shp at the higher forcing rate. E-in-C offered both alternatives in a memo dated 1 June 1937. The plant already accepted used a 180ft machinery box (1,280 tons). The new forcing rate was 61, compared to a previously-accepted 52. At the higher rate, a 72,500shp plant could give 84,000shp in a 182ft machinery box (1,430 tons). Alternatively, a 63,500shp plant could be forced to 72,500shp at the high rate (169ft machinery box, 1,250 tons).

25. The original armour provided immunity against the new 6in shells (112lb) at normal inclination (90°) between 12,000 and 19,000yds. In Scheme A a thinner deck reduced the outer edge of the zone to 15,000yds but increased speed by 0.75kt. A' had similar armour but less power (0.5kt advantage). B had a thinner belt and the thin deck, so the immune zone was only 14,000–15,000yds, the speed advantage being 1.25kts. Scheme B' offered the same narrow immune zone but with reduced power (advantage 1kt). Reduction in

the deck over the machinery rendered all the modified ships open to a 250lb dive bomber hit on their machinery. DCNS suggested accepting smaller boilers for the 72,500shp, and rated the Schemes A', A, B', and B in fighting efficiency.

26. Armament weight was reduced from 1,050 tons to 1,010 tons by omitting torpedo tubes, and armour weight was reduced from 1,340 tons to 1,330 tons. At this point B2 was credited with 76,000shp.

27. On October 1927 trials, the ship had a critical speed (about 24kts). At about this speed the wake shot out from the transom 'like a needle spray bath'. When the ship accelerated from a lower speed this effect was seen at 23.8kts, and when power was reduced the effect disappeared at 24.1kts. Tank tests showed that the water shot out from below the transom to form a hump a little astern, leaving a vee of comparatively quiet water between the transom and the hump (i.e. in effect acting as though the stern were longer). A Haslar report of 5 July 1938 mentioned earlier model tests with a modified 'C' class hull (reported in June 1920) and with a modified *Kent* class hull (reported in September 1923).

28. The control position atop 'X' turret carried two officers (spotting and rate officers) and three ratings (two range takers, one telephone operator), plus a 22ft FM duplex rangefinder and spotting glasses for the officers. The DCT contained a 15ft FM7 rangefinder (ships were later fitted with 22ft rangefinders). It carried three officers (control officer, spotting officer and rate officer) and twelve ratings, including trainer and layer whose functions were duplicated by those controlling 'X' turret, a telephone operator, and two range takers. The DCT also had two ratings to operate its inclinometer (not duplicated in the 'X' turret position), a cross-levelling operator, and two wireless operators, plus two ratings at the PIL (Position in Line, for concentration firing) position. Two 9ft rangefinders on the compass platform supported the ship's tactical plot (the DCT carried a shorter PIL rangefinder). Data from a summary of the ship's fire-control system as of April 1938, in *Fiji* Cover 2.

29. Based on estimated performance of French and German shells exceeding that of British; it turned out that there was little or no evidence. DCNS said that the British were granting their enemies the best case and themselves the worst. Controller argued that a *Fiji* was well enough protected, and that a cruiser would rarely fight at the 90° inclination at which its armour seemed inadequate. The conference also compared *Fiji* protection with that of *Southampton*. DNC much preferred *Fiji*, particularly against air attack. At short ranges the belt might be penetrated, but the trajectory would be so flat that shells would pass over the magazines deep in the ship.

30. CAB 102/536, L Errington, 'Naval New Construction Requirements September 1939 – December 1941,' one of a series of unpublished Cabinet war histories prepared about 1946.

31. The deletions were relatively painless, but they did not completely cover the weight of the tubes and torpedoes: two 24ft boats (replaced by one 25ft boat of a new type), one after ladder, the sheet anchor and hawse pipe (already eliminated in 1938 and 1939 ships), side lining from all spaces other than cabins and officers' living spaces, and the awning over the forward superstructure. DNC could find no further painless deletions, and ACNS demanded torpedoes as 'of the greatest importance'.

32. The occasion was the dockyard conference prior to a refit. Addition of Types 284, 283 (three sets), 271 and two Type 282 had been arranged. What compensation should be arranged? The ship was scheduled for two Oerlikons on the quarterdeck, but the dockyard had insufficient labour to effect the approved compensation, stripping quarterdeck wood planking. Now two more Oerlikons (or 2pdr power singles) were wanted. After the conference, first priority went to radar (Types 284, 285 and 271). Next was close-in armament with splinter protection. The ship's captain pointed out that removing two searchlights reduced the ship's defence against night destroyer attack. Removing wood decking reduced habitability, which was worsened by the encroachment of RDF spaces. The ship already had four Oerlikons and two 0.5in, and the Captain wanted two more Oerlikons or, preferably, two single 2pdrs.

33. For a four-gun ship, these additions added up to 2,360ft-tons (weight multiplied by height above the keel), taking into account the effect on stability rather than just weight. The catapult and aircraft offered 1,220ft-tons. The centre gun and gear in 'X' or 'B' turret offered just 300ft-tons. Two Oerlikons on the quarterdeck cost 180ft-tons. Barrage directors for 6in guns cost 570ft-tons (equivalent to all three guns and gear of 'Y' turret or to five searchlights and their structures, at 110ft-tons each). Splinter protection for the HA armament was the largest potential addition, at 990ft-tons. Replacing pompom directors with predictors would cost 260ft-tons. Eliminating the port crane and lowering the boats would gain 930ft-tons. DGD's proposals amounted to 1,250ft-tons, so dropping them would leave 170ft-tons to be found. That would equate to removing part of the main battery or aircraft or giving up the third searchlight.

34. The ship's captain disliked Type 273 because it bore only out to Green and Red 115, and even so limited the training of 'B' turret under action conditions to 90° either side. Type 272M was approved as a replacement.

35. ADM 116/3921.

36. In 1937 Japan had not signed the treaty, but there was hope that she would do so, and it seemed (at least to the British) that Italy would soon sign (she did not). Thus Japan was the most likely builder, if Chile went ahead. DCNS doubted that the news that such a ship was being built for Chile would have much effect on the United States or the Soviet Union, but that it was being built in Japan would have enormous impact. He advocated bringing pressure on Chile. That did not work. Director of Plans thought the Chileans were interested mainly in quick delivery; the British could build faster than anyone else. They could offer two *Fiji*s. Given the poor performance of the Italian-built ships in Argentine service, Director of Plans doubted the Chileans would buy Italian ships this time. The Chilean specification called for six 8in guns (maximum range 29,000yds), firing at up to two to five rounds per gun per minute, averaging one hit per minute on an 8,000-ton cruiser at 11,000yds. Other weapons desired were eight 4in, two quadruple pompoms, two 0.5in machine guns and two triple torpedo tubes. The ship should have a catapult and a double hangar with space for three aircraft. Desired speed was 31.5kts, with a range of 8,500nm at 16kts. Desired protection was a 4in belt (5in over magazines) with a 2in deck (3in over magazines); magazine immune zones against 6in and 8in fire were 7,000–24,500yds and 13,000–22,500yds respectively. DNC thought these figures feasible. The preferred Admiralty argument was that a *Southampton* or *Fiji* was better, achieving 1½ hits per minute at 11,000yds albeit not as well protected (immune zones for a *Southampton* were 8,000–20,000yds and 15,000–18,000yds). Unfortunately having 8in guns to face the Argentines' 7.5in was probably more important than anything else. In October 1937 DNC cited Chilean protection requirements much closer to *Fiji* characteristics. Controller understood that builders in other countries had been allowed to bid whether or not construction itself was legal, and in February 1938 he suggested that Vickers and John Brown be allowed to bid with the understanding that they might not be

permitted to build the ships. In February 1938 the Chilean attaché provided fresh details of what was wanted: 8,300 tons, six 8in (two triples), secondary AA armament, 30kts, heavy armour, triple torpedo tubes, three aircraft and a catapult, and radius of action 8,000nm at 14kts. These may have been the original figures, inspired by the characteristics of HMS *Exeter* and *York*.

37. FO371/21436.
38. British Naval Attaché, Santiago, 4 April 1939, in ADM 116/3921. He claimed that the ships were being built to Chilean specifications, using Yarrow boilers. Presumably they were the 8,700-ton cruisers *De Zeven Provincien* and *Eendracht*, which have always been described as replacements for the ageing *Java* and *Sumatra*.
39. W G John Constructor's Notebook 12; reports from this exercise are also in the DNC Private Office papers in ADM 229.
40. She was 730ft x 86ft x 25ft, armament accounting for 2,990 tons and protection for 4,960 tons.
41. DNC told John to reduce belt and deck to 7in and 3in respectively (4in deck over magazines). In detailed calculations, bulkheads were 5in thick. Given the dimensions (700ft x 86ft x 22½ft) and a depth of 43.5ft, DNC wanted strength calculations leading to detailed hull weight. There was also a version of the 700-footer with a bulge, but with the same vertical belt as the others. Vertical rather than sloped belt armour was consistent with the recently-completed *King George V* battleship design, of which this was, in a way, a reduced-scale version. DNC's detailed interest in the project suggests that it was more than a quick calculation to assess possible hostile programmes.
42. In detailed calculations, the barbettes of the 9.2in ship were given 10in sides but 9in ends.
43. The Japanese 20,000-tonner appears in notes in DNC's private office files (ADM 229 series). The connection with Japan is also suggested by an adjacent page estimating armour weight for the new Japanese *Tone* class cruiser. The other design work in this Notebook, at this time, is concerned with weight growth and stability loss in the *Fiji* class.
44. This was the earliest in a list of 'cruiser designs since 1939 developed by Section 7 [cruiser section]' in a post-war *Fiji* class cover. Inclusion presumably means that it was developed in detail and was not merely a quick sketch ordered by DNC. In the list it is dated August 1939.
45. Both ships were given 580ft x 66ft hulls. Design A would have displaced 10,385 tons, Design B 10,030 tons. Most of the difference was machinery, 1,445 tons vs 1,120 tons.
46. A table dated 8 June 1939 showed resistance of various thicknesses of cemented (C) armour against 8in Mk VIII guns at different inclinations:

	90°	70°	60°	50°	40°
5in	15,600yds	14,600yds	13,400yds	–	–
6in	12,800yds	12,000yds	10,900yds	9,200yds	6,600yds
7in	10,800yds	10,000yds	9,800yds	–	–
8in	9,000yds	8,200yds	–	5,400yds	–

A 2in deck could not be penetrated inside 21,000yds, and could keep out a 250lb bomb from 5,000ft. A 3in deck would be effective out to at least 26,500yds, and it could keep out a 250lb bomb from 11,000ft and a 500lb bomb from 7,000ft. A 3½in deck would be required at maximum 8in range, 29,000yds. Required thicknesses were considered to be 6in belt and 2in deck. A note added in pencil stated that a German 11in shell could be kept out by a 2in deck at 23,000yds.

47. Drawings of this type are extremely rare in Constructors' Notebooks, so these may indicate how seriously the project was being taken.
48. The next sketch design he did (dated 29 August) reflects the deepening emergency: it was for the 4.5in HA armament of a 'Bank Line' ship as an anti-aircraft escort. Armament would have amounted to four twin 4.5in, two quadruple pompoms, and two quadruple 0.5in machine guns, with a single HADT and its HACP. John's figures included 10 tons for radar (RDF).
49. The secondary armament, aircraft arrangements and endurance were all chosen by DNC. Speed was set at 33/32.5kts because the stated speed of the German ships was 32kts.
50. DNC's typed and dated instructions are taped into the Notebook.
51. W G John Constructor's Notebook 12 includes a horsepower calculation for a 33.5kt ship, 700ft x 56ft x 23½ft, 20,700 tons standard. The hull would have been bulged to 58ft below the waterline.
52. Turrets would have been protected more lightly. The 9.2in turret had a 6in face, 5in sides and 3½in roof; the 8in turret had 5in face and 2in sides and roof or, as an alternative, the same armour as the 9.2in. Presumably the 9.2in was the same gun being used for coast defence. As used in cruiser calculations, it fired a 410lb shell.
53. Comments signed by Rear Admiral T S V Philips, DCNS, 29 November 1939, in Heavy Cruiser Cover. According to a paper by Director of Plans, the 15in gun capital ship could be completed in spring 1944, an 8in cruiser in summer 1944, a new 14in or 16in capital ship in winter 1944 (if ordered at once), and a 9.2in cruiser not until early 1945. At this time characteristics of the new Japanese cruisers of the 1937 programme were unknown, but they were suspected of carrying guns of over 8in calibre, and of displacing 12,000 to 16,000 tons (none of which was correct). In view of the 'characteristic Japanese lack of originality', they were presumably based on German designs, and were probably up-armoured 'pocket battleships' displacing about 14,000 tons. It was also thought that the Japanese had sacrificed endurance for protection and firepower (actually they had sacrificed protection).

8. War

1. According to F A Kingsley (ed), *The Development of Radar Equipments for the Royal Navy 1935-45* (Macmillan, London: 1995), cruisers with Type 286 were (with installation dates and some removal dates): *Achilles* (New Zealand: before February 1943 to May 1944); *Arethusa* (before July 1941 to April 1942), HMAS *Australia* (December 1940), *Birmingham* (before February 1941 to June 1941), *Black Prince* (April 1942 to August 1943), *Capetown* (before March 1942), *Ceres* (before August 1943 to before April 1944), *Dauntless* (before March 1942), *Despatch* (before July 1943), *Dido* (before May 1941), *Dunedin* (New Zealand: December 1940), *Durban* (August 1942), *Emerald* (before February 1941 to April 1943), *Glasgow* (July 1940 to August 1942), *Leander* (New Zealand: before November 1941), *Manchester* (November 1940), *Newcastle* (before March 1941 to November 1941), *Norfolk* (before May 1941), *Orion* (before May 1941), HMAS *Perth* (before May 1941), and *York* (before May 1941). For the RNZN ships and probably for HMAS *Australia* this is a mistaken identification of the indigenous New Zealand air-search radar. Cruisers with Type 290: HMAS *Adelaide* (before February 1943 to before June 1944), *Aurora* (May 1941 to June 1942: Type 290X), HMAS *Australia* (before March 1943 to before August 1943), *Caradoc* (before October 1942), *Cardiff* (before February 1943), *Ceres* (before

April 1944), *Dorsetshire* (May 1941: Type 290X), HMAS *Hobart* (before February 1942 to October 1942), *Leander* (New Zealand: mid 1942), and *Newcastle* (November 1941 to November 1942). Type 286 and 286M had fixed antennas (Outfit ATN), the central pair of dipoles transmitting, two pairs on the sides receiving. Type 286P and PQ had a rotating antenna (Outfit ATQ) whose dipole supports look like an 'X' in a photograph. These versions had a T/R switch. The same antenna was used by Type 290, which had a more powerful transmitter. The two indigenous New Zealand radars were a gunnery set (SWG, initially operating at 73cm) and a 150cm SW (warning) set. The latter used two short Yagis mounted side by side (presumably one for transmission, one for reception) on a rotating stem, high enough to clear a ship's funnel. By November 1941 both were on board the cruiser *Achilles*, and either were about to be or had been fitted to the cruiser *Leander*. Production of further sets, to be installed when ships called at New Zealand ports, was underway. Both were entirely indigenous: according to a US attaché, 'except for the similarity of function, and the use of a few valves [tubes] of the same type, the above mentioned sets differ in nearly every respect from any cognate English model'.

2. The barrage director was a simple eye-shooting device using a ring sight. It had two settings for convergence of fire, near (2,000yds) and far (4,000yds), range and time of flight being preset on the associated automatic barrage unit (ABU) in the ship's main battery fire-control system. Normally the ABU would fire the weapons automatically, the director operator pointing the guns (via magslip transmission) at the aim point, leading the target using his ring sight. However, he could also fire manually, using a pistol. The ranging radar fed its data into the ABU. At least in the last version (Mk III) the Type 283 radar used beam-switching for a limited degree of blind pointing, and a second operator could point the guns on that basis. The Mk III handbook (BR 1527) is in NHB.

3. A chart dated 14 November 1940 showed the great variety of Type 284 configurations already in use. The small pre-DCT directors of the cruisers *Devonshire*, *Shropshire* and *Sussex* could not accommodate Type 284. Note that there were separate transmitting and receiving elements, presumably the 3 x II top and bottom in version 12A. *Norfolk* was not included in the table, although she was fitted with Type 284 in September 1941. Her sister-ship *Dorsetshire* was not fitted. At this time the arrays were coded according to how many units (blocks of one or two reflectors [I or II] except as noted) were mounted, e.g. 12B (*London*) had 3 x II across the top of the DCT, one single on each side, and 2 x II on the bottom. *Fiji*s with 15ft rangefinders and *Kent* had 12B. *Fiji*s with 22ft rangefinders had 12A (3 x II upper tier and 3 x II lower). Until she had the new mast, *Kent* had 3R + 4 F: 4 fishbones on top of the director and II plus I reflectors (not fishbones) on the face of the DCT in front of the rangefinder. *Suffolk* had 12F: two fishbone arrays, three wide and two high, side by side, one transmitting and one receiving. All others had reflector arrays: 11A in *Southampton*s and *Belfast*s (3 x II on top, I each side, and II and I below); *Dido* had eight (2 x II below, plus, two, atop the other on each side of the DCT); the *Sydney* and *Arethusa* classes and *Ajax* had seven (II plus I on the bottom of the DCT plus 2 x I stacked on each side); *Leander*, *Orion*, *Neptune* and *Achilles* had 5A (II plus I on the bottom tier, I each side); and *York* and *Exeter* had 5B (II plus I somewhat higher, I each side brought onto the face of the director). In fact at least *Leander* and *Achilles* had a New Zealand-developed surface-gunnery radar.

4. Many cruisers were fitted with the FV1 radar detector, first installed on board the monitor HMS *Erebus* in 1941 (initially to locate German coastal radar positions). It operated on the Germans' usual search radar frequency, the 45–125cm (VHF, hence the V in the designation) band. Each of two or four antennas was a vertical dipole inside a semi-cylindrical wire mesh reflector (dipoles could also be mounted horizontally, to pick up horizontally-polarised signals, but this seems not to have been done in practice). The initial installations comprised two such antennas, one on each beam, the ship turning to determine direction of the signal (when the ship was pointed at the source of the signal, neither antenna received anything). This was acceptable because at that time ships turned to determine the direction using fixed Type 286 radar arrays. The four-antenna unit was credited with an accuracy of 20°. The aerials were connected to the P29 radio receiver using a six-position switch: the receiver could be connected to any one aerial, or to the entire array acting as an omni-directional antenna, or to some alternative antenna, such as a cone. Typically FV1 was connected to a Type 91 jammer, which had its own double-cone receiving antenna (using the sixth position of the switch); this antenna was much the same as the antenna associated with the Type 253 IFF transponder (but, unlike that antenna, was not for transmission). The four FV1 antennas were mounted on the spurs of the foretop or maintop or on the bridge structure. The double cone (antenna ARM) was rotatable to lie either horizontally or vertically, depending on the polarization of the incoming signal (Type 91 was a spot [single frequency] jammer and therefore needed signal data so that it could be set properly). When fitted, the double-cone was used to search for signals, since the four FV1 aerials were less sensitive when used in parallel. Type 91 transmitted its jamming signal via the four antennas associated with FV1. ADM220/1907, a file on radio warfare 1941–5, describes initial work on FV1. According to the post-war Admiralty history of radio warfare (1949), FV1 was unsatisfactory for radar direction-finding; it was described as nothing more than an early experimental set (which belies its widespread employment). This history claimed that a useful noise (radar) intelligence set was produced only in 1944, to gather radar intelligence off the French coast. The first invasion in which radar jamming was used was Sicily; a few ships were fitted with jammers there. More were available for Salerno. Type 91 was tunable and monitorable. It operated at 200 to 600 MHz (50 to 150cm) with an output of 8–22 watts. The improved Type 91M had a wider frequency range (90 to 600 MHz), and was the standard wartime British large-ship and shore jammer. When requirements were set for the Normandy invasion in November 1943, no other equipment had past the prototype stage. For this invasion further jammers were developed: an automatic swept-frequency jammer (needing no operator monitoring the enemy signal), Naval Carpet II (Type 653/1 and /2) producing about 100 watts at 350 to 750 MHz, parallelled by American Carpet I (Type 654: 5 watts, 480–700 MHz, a spot or barrage jammer): American Rug (Type 656, 8–20 watts, 200 to 550 MHz, a pre-set spot or barrage jammer): Pimpernel/Carpet I (Type 657): British Mandrel (Type 658: 100 to 152 MHz, 10 watts): American Mandrel (Type 659: an aircraft jammer): Carpet III (Type 655), British Moonshine (Type 660), Peter (Type 661, apparently abortive), American Carpet IV (Type 663: pre-set spot or barrage jamming at 350 to 1400 MHz, 5 to 40 watt output); and Carpet IIA (Type 664: non-tropicalised) In addition, the Royal Navy used the US TDY trainable jammer (360 to 800 MHz). Type 662 seems to have been an abortive countermeasures set developed by the Admiralty Signal Establishment. Type 652 was the airborne Jostle. Most of these sets are listed in an appreciation of D-Day radar countermeasures in a SHAEF report on D-Day communications in the

Top Secret Naval Attaché file (RG 38, 1944–7) in NARA II. Missing numbers are from a 1944 list of British radio nomenclature (ADM 220/1645). Most of the D-Day sets were automatic, and were fitted on board craft as small as LCTs. Note that in the fall of 1944 three Eastern Fleet ships had Naval Mandrel (presumably Type 658), according to an Eastern Fleet report on Japanese naval radar in the US Naval Attaché (Secret) series in RG 38, NARA II. In addition to the types listed, Types 650 and 651 were wartime British radio jammers to counter German guided missiles. In March 1944, in the run-up to D-Day, battleships, monitors, and cruisers were listed for installation of jammers. At that time available cruisers with FV1 were listed as: *Emerald*, *Diana*, *Frobisher* (being fitted), *Hawkins* (being fitted), *Ajax*, *Glasgow*, *Belfast*, *Mauritius*, *Bellona*, *Argonaut* and *Black Prince*. They were also to be fitted with some standard airborne jammers. In April 1944 plans were made for ships' staffs to modify Type 91 as a CARPET jammer. The SHAEF report mentioned above contains a table giving numbers of jammers of various types present on D-Day. It lumps battleships and cruisers together, and it also lumps together British and US ships. Totals for the battleship/cruiser category were: Type 91, twenty-one sets; TDY jammer (US), nineteen British and five American ships; Type 650, twenty-six (all British); Type 654, twelve (all British); no other jammers in major British ships.

    The 1943 FV1 handbook (CB 4268, later BR 1557, originally issued as a CAFO) is in the RG 38 (ONI) library of foreign technical publications at NARA II. This file also contains the Type 91 handbook (CB 4226). The designation FV2 was initially applied to the four-antenna version of FV1, but later dropped. Also, initial versions used a P19 receiver. FV3 was described as a 'rush job' for coastal forces; it may have been the satisfactory D/F set mentioned above (FV7 was an improved late-war version). Other units in the FV series were intended for carriers working with VHF-equipped aircraft. The fixed FV1 had relatively low gain, but could operate as a wide-open set likely to pick up short enemy transmissions, e.g. from U-boats. US wartime development emphasised long interception range, because the most important targets were approaching enemy search aircraft. The US Navy therefore developed spinning antennas with considerable gain, albeit with only an intermittent view of the target. In 1944–5 the Royal Navy, presumably looking towards Pacific operations, developed a series of US-style UHF intercept sets, RU1 through RU4, each using a single dipole inside a reflector. Only RU4 was completed, and a few were in service at the end of the war. Post-war, because it was most interested in transient submarine transmissions, the Royal Navy developed wide-open intercept sets in a new U series (e.g., UA-4) derived to some extent from FV1. The difference was that instead of being able to switch from one antenna to another, they collected data from all the antennas simultaneously, using them for instantaneous direction-finding (a separate omni antenna was used to measure the parameters of the intercepted signal).

5. Signal Division (Admiralty), *Radio Warfare* (S.D. 1080/47), issued 1949, in RG 38 ONI library of foreign technical and related publications, NARA College Park. Date of issue was 29 August 1949; this was one of the post-war series of historical technical monographs. RU4 used a reflector-backed dipole spinning at up to 120rpm. RU4 was particularly valued for its ability to detect and track aircraft radars. The others in the series were RU1 (130 to 500 MHz), RU2 (500 to 1,000 MHz), and RU3 (1 to 3 GHz). As of August 1945, RU4 was expected to be in full production at the end of the year, along with RU1. The other two would follow. Cruisers and battleships were to have one RU1 and one RU4, plus lower-frequency sets.

6. Cdr J O H Gardner RN, 'The Evolution of the Modern Gun Mounting: Pt III, The Remote Power Control of Guns', in *Papers on Engineering Subjects* (predecessor to the *Journal of Naval Engineering*) (Sept. 1946) (copy in the Brass Foundry NCD series). Other sources were articles in the 1946 *AGE Journal*. Gardner pointed out that hydraulic drive (which the Royal Navy generally favoured) was lighter for a given power output, produced more torque (hence needed less power to accelerate its own mass), and could last much longer; electric drive was simpler, quieter, cleaner, and easier to produce (i.e., better adapted to mobilised civilian industry), and offered less fire risk. By 1946 British practice was to limit hydraulic RPC to heavy mountings which already used hydraulic power, and to use electric RPC for everything from 4in guns down. RP 10 and RP 40 both exerted control via hydraulics. In RP 10 a pump pushed fluid into a driving system; the magslip controlled the throttle on the pump to control the system. In RP 40 control was exerted by changing the amount of fluid running around a closed circuit by controlling the pump driving the fluid (the angle of a swash-plate pump was controlled by the magslip signals). RP 50 was all-electric using a metadyne which controlled the electric motors on the mounting. Metadyne control in particular required a considerable additional space (the metadyne had been developed by Metropolitan Vickers for the London Underground system). Magslip signals were electronically amplified to control two generators (training and elevation), each of which controlled a motor on the mounting. Estimated weight for a pompom installation (as calculated in 1940) was 2,600lbs below decks plus 1,100lbs on the mounting (or 3,400lbs plus 1,500lbs on a 4in mounting). Both RP 10 and RP 50 were applied to pompoms.

7. Even the boats on board cruisers could be inadequate. In 1942 the captain of a *Fiji* pointed out that nearly half the crew were typically granted shore leave after fourteen days at sea. Leave might be given for only four or five hours, but it took five trips to land 260 liberty-men. Once, when only 150 wanted liberty, all boats were not in service, and the last trip left the ship two hours after leave was piped. This would become a matter of grave concern if sailing orders were received towards expiration of leave. The boats themselves were unreliable, and the stowage left them at the mercy of 4in blast and buffeting from the sea.

8. ADM 229/29, DNC private office papers.

9. According to a wartime US naval attaché report (in RG 38 NARA Naval Technical Intelligence series for 1940–6, Confidential series), the UP was specifically intended to deal with dive bombers. The attaché described it as nothing more than a stop-gap, presumably until more 2pdr pompoms and other light weapons became available. The launcher carried its rockets in horizontal rows of four, and fired them in ten-round salvoes. Rockets had a ceiling of 1,500–2,000ft. US attachés first saw it ashore in Dover in June 1940. In January 1941 the attaché reported that defects in the rocket and projector had brought about a decision to cease development of the 7in weapon and to concentrate on 2in and 3in versions to defend merchant ships and small warships (these weapons survived as PAC [Parachute and Cable] weapons, e.g., equipping LSTs). Deficiencies included crudity of control and long flight time (plus time to stream the cable). The last ship fitted was HMS *Prince of Wales* (February 1941). The US file includes an Air Ministry report of August 1940 on AD (Airfield Defence) Mk II, the land version of the UP. The shipboard weapon was sometimes designated AD Mk I. Cruisers

listed with UP projectors in the April 1941 ship data book (CB 1815) were *Norfolk*; *Birmingham*, *Glasgow*, *Sheffield* and *Newcastle*; and *Aurora* (one rather than the two in the other ships). The last ships to retain UP mountings were probably *Birmingham* and *Newcastle*. No UP mountings were listed in the October 1941 edition, although some were still installed (nor were they in the October 1940 edition). In addition, Raven and Roberts (Appendix 2) also list *Arethusa* (two mountings, by July 1941), *Norfolk* (two fitted during March–May 1941 refit after damage), and *Sussex* (two fitted by mid-1942). Details are from the US file because it has been impossible to obtain British official references to the UP weapon.

10. On 30 July 1940 DNC asked that, as a first action concerning the *Fiji*s, a rough estimate be prepared of the displacement and speed of a ship with *Dido* armament and protected by 60lb side, 2½in decks, and boxed magazines The belt would be reduced from 130/140lbs to 60lbs and extended from the after end of the citadel to the stern. The upper deck would be 2½in (100lbs) on a structural deck; the lower deck would be 2½in worked structurally, protection being extended aft in way of the lengthened belt. Magazines would be boxed (60lbs NC sides and 40lbs NC tops). More weight would have been added than removed, and removing machinery low in the ship could not compensate for adding weight high in the ship. Having done the calculations, W G John (Notebook 12, July 1940) suggested retaining the existing 2in armour on upper and lower decks, which would solve the stability problem. Having calculated that a ship with *Dido* armament and *Fiji* hull form could not have satisfactory stability with upper decks armoured throughout, John offered an alternative retaining *Fiji* side armour but gaining 2½in armour on the upper deck (the lower deck remained 2in), and with boxed magazines. It would be somewhat larger than a *Fiji*: 10,500 tons standard (595ft x 65ft x 17½ft [hull depth 32ft 3in], with a transom stern), making the same speed as *Belfast* (32.5kts) on 80,000shp.

11. As listed in September 1940, added protection was 30lb DKM plating on the ship's side between platform and lower decks, over the magazines; ring bulkheads and trunks of turrets increased from 20lbs to 40lbs D plating; protection in way of steering gear brought out to the ship's side to be continuous with DKM plating; additional splinter protection to cables, bridges, and way of pompoms, and over ready-use magazines; and plating on bulkheads at stations 83 and 99 (forward and after bulkheads to forward engine room) was increased to 40lbs (1in). In August 1942 DNC decided not to thicken analogous engine-room bulkheads in the *Tiger* and *Minotaur* classes because they had better deck protection.

12. Originally the gun was to have elevated only to 30°, but it was then re-thought as a dual-purpose weapon with a maximum elevation of 70°; it could then be used during the day. That entailed further changes: magazine stowage for ammunition and some form of fire control. The constructor involved commented that 'it seems a pity in a ship mounting 8 5.25in HA guns to complicate things for one single 4in HA gun'. Controller decided that the gun would be limited to ready-use ammunition, but that did not solve the problem of additional complement – in a ship which had lost a bridge level forward.

13. 'Y' turret was considered too low to control the whole of the 6in battery, and the local director sight in it was not set up to direct other guns. In divided control, one group would consist of 'Y' turret only. That hardly justified the clock, with its wiring and changeover switches, in the transmitting station. DNO proposed to substitute a simplified Fire Control Box (used in sloops for anti-aircraft and surface fire) in 'Y' control position to allow control from the HA/LA director tower aft. The box would be able to transmit range and deflection to 'A' and 'B' turrets if the transmitting station were knocked out (e.g., by flooding), the entire armament being controlled from the HA/LA director tower through 'Y' turret operating in local control mode. There would not be provision in the 'Y' turret control position for spotting and rate officers, although the turret could be controlled from this position if all other control positions were out of action. The Fire Control Box would communicate its range and deflection data by emergency telephone.

14. Another argument favouring a squared-off upper bridge was that the jackstaff was not visible from the existing bridge. Those conning the ship used their view of the jackstaff as a way of seeing in exactly what direction the ship was steaming in relation to other ships. A squared-off bridge front was an alternative way to sensing the ship's precise course. Abolition of the wheelhouse was controversial, as the quartermaster at the wheel could respond to general commands and thus reduce the load on captain and navigator. For example, he could be told 'get that ship at 15° on port bow, and keep her there'. At a conference on 9 December 1940 on bridge design there was considerable interest in providing a protected fighting position on the bridge (the French had conning towers there).

15. All *Fiji*s had been given plating on the fore bridges and cable trunks. The modified ships were given 20lb screens to their 4in guns, 20lb screens and deck to their pompoms, 15lb decks for machine guns, 15lb screens and deck for pompom directors, 20lb screens for torpedo tubes, 20lb for HACS supports, 15lb bulkheads and decks for RDF offices (Types 281, 282, 284 and 285), 15lb screen and deck for surface lookouts, 15lbs for the emergency conning position. The switchboard rooms moved from forward to an amidships sided position to reduce the length of vulnerable control cables. Circuit breaker rooms were staggered to make the ring main less vulnerable. The protective deck was extended forward and aft to provide a more gradual transition in the thickness of the strength deck.

16. George Moore, *Building for Victory: The Warship Building Programmes of the Royal Navy 1939-1945* (World Ship Society, Gravesend: n.d.) describes the almost constantly-shifting Second World War programme. See also CAB 102/536. In February 1941 First Lord set the 1941/2 cruiser programme at three modified *Fiji*s to be ordered at once plus four heavy cruisers to be ordered in the autumn (later deferred to March 1942 before the project died). Because the battleship *Vanguard* enjoyed special priority (at the behest of the Prime Minister), the cruiser *Bellerophon*, building at the same yard, was effectively suspended. For the 1942 programme the Naval Staff wanted seven modified *Fiji*s, one in effect replacing the single 8in cruiser which had survived in the previous year's programme. Controller agreed to six, but four were cancelled in August 1942 and another in November, *Hawke* being substituted. In December 1943 *Minotaur* and *Superb* were offered to Canada as a free gift. The offer was accepted on 7 February, but in April *Superb* was replaced by *Uganda*. The Canadians accepted the ships in a message in July 1944. No cruisers were included in the 1943 programme, but the 1944 programme included five of a new type.

17. In December 1941, *Fiji* was credited with a metacentric height (GM) of 1.5ft at light load and 3.4ft at deep load. *Uganda* was the same, because the weight and moment saved by removing 'X' turret had been exactly balanced off. The three ships ordered under the 1941 programme (63ft beam) were credited with 2.1ft light and 3.9ft deep. The ships just ordered would have a foot more beam (64ft): 2.6ft light, 4.5ft deep.

18. Under a new policy announced in September 1945, installation of

single power-worked 2pdrs (Mk 18*) and Boffin mountings ceased, although these mountings were not to be removed. They were to be superseded by Mk 7 power-worked single Bofors guns. Oerlikons not yet removed should remain. Until Mk 7 mountings became available, Mk 3* single mountings would be fitted in positions already earmarked for single hand-worked mountings, but existing Mk 3 would not be replaced.

19. The DCT had to be able to turn 360° in either direction, presumably to follow fast surface targets; that required a longer and larger cable run below it.
20. Type 277Q should replace Type 277 (requiring a new mast) and Type 293Q should replace Type 293; Type 274 should be fitted with a stabilised aerial and with the Type 931 supplementary gunnery set. A standby warning radar (Type 268) should be fitted. An AEW receiver would be installed. The 1944 standard wireless and radio countermeasures outfits should be installed. A third gyro should be fitted (requiring yet more space). Spaces should all be air-conditioned to the latest standard.
21. *Fiji* class Cover 6, Folio 83.
22. As a very preliminary guess, DNC estimated that if the after machinery, after turret and all 4in guns were removed, space and weight might be available to install GAP and one or possibly two CR/MR mountings (the future twin 3in/70). If the after set of main machinery were retained, it seemed 'very problematical' that a GAP battery could replace the after turret and after 4in guns, 'even assuming the most favourable developments'. The quadruple torpedo tubes would have to be surrendered. Removing the armament and after machinery would make possible conversion into a GAP experimental ship, but more investigation would be needed after requirements became firm.

9. Wartime Cruiser Design

1. For the earlier project, see the author's *British Destroyers* (Seaforth Publishing, London: 2009). This one was dated 5 February 1940 in Lillicrap Constructor's Notebook 7. It was a fast (35kt) low-freeboard ship with a protected hull and six bow torpedo tubes (eighteen torpedoes), gun armament being limited to one quadruple pompom. Lillicrap considered it actually a non-submersible submarine using ballast tanks to reduce buoyancy (hence silhouette) when attacked. Ballast tanks might also be needed for stability. It proved impossible to meet the desired characteristics on less than about 4,900 tons (the Legend weights totalled 4,965 tons) with dimensions of 400ft x 47ft x 17ft, with 8ft freeboard (design development had begun at 3,500 tons). The machinery box would be 260ft long. With little space in the hull, much of the considerable accommodation needed for the large crew serving the machinery would have to be to submarine standards, in the superstructure above the protective deck. The ship would have a 2in curved (turtle) upper deck extending about 5ft below the waterline. Sides and top would be 30lbs for splinter protection. Initial estimates suggested that something like an *Abdiel* powerplant would suffice, but the sketch design submitted on 26 February 1940 showed a two-shaft 120,000shp plant. About 650 tons of oil would give an endurance of about 3,700nm at 12kts. The bluff submarine-like forward end would present particular problems at high speed. Details of design logic are in W G John Constructor's Notebook 12. He went so far as to resurrect data on the Victorian torpedo ram *Polyphemus* (the First World War torpedo cruiser project was 'New *Polyphemus*'). His sketch showed a submarine-like hull containing six 20,000hp boilers in four rooms (end rooms with one boiler each, two centre rooms each with two boilers in tandem), uptakes from each pair of boilers being trunked together into tall funnels. The funnels were connected by a flying deck, the fore end of which contained a splinter-proof conning tower. The after end of the hull resembled that of a submarine, and the cross-section was submarine-like.
2. Controller asked DNC for two alternative designs, armed with either four twin 4in or four twin 4.5in, on 29 February 1940. The designs are described in Tozer Constructor's Notebook 1. They would also have two octuple pompoms. The initial 3in deck was soon changed to 3½in. Speed should be as high as possible. In the design developed, side armour followed the current practice of covering machinery to the upper deck (to protect boilers) and magazines fore and aft of that. Thickness was 1½inches. Approximate dimensions were 470ft (pp) 490ft (wl) x 54ft 6in. Displacement at 14ft 3in mean draft was 5,700 tons. This seems to have been the 4.5in version. Work continued as late as June 1940. Tozer Constructor's Notebook 1 includes calculations for Design K, presumably with three twin 5.25in, displacing 6,380 tons (standard) and 7,850 tons deep. Design M had *Abdiel* class machinery, and Design N had 'L' class (destroyer) machinery. Endurance was calculated for 1,200 tons of oil at 16kts. Tozer Constructor's Notebook 3 includes calculations for a follow-up, *Fiji* modified as an A/A ship, with four twin 5.25in, two octuple pompoms, and a thick deck. Provisionally the total depth of the new design would be 1.75ft greater than for a *Dido*. The ship would also have the usual pair of quadruple 0.5in machine guns and a pair of torpedo tubes. Legend details in the text are from a book of Legends maintained by DNC Department and held, at least at one time, in the National Maritime Museum; I am indebted to late David Lyon, at the time head of the Draught Room, for access during the 1970s, when I photographed it.
3. This entailed 40lbs over the ship's side between upper and forecastle decks and blast screens on the forecastle deck; 40lbs over ammunition hoists (screens on the upper deck); protection to pom-pom ready-use magazine (60lb side, 20lb top, 40lb bottom, all NC); and 40lb tubes to the HACS supports.
4. Details in Harrington Constructor's Notebook 4, dated 2 February 1940. The lists of weights added and removed do not include radar, a small weight but with considerable effect on stability. However, the same Notebook gives an extensive account of weights and moments for Type 281 a few pages later. The modified *Belfast* also occurs in W G John Constructor's Notebook 12, with the note that beam is to be increased by 2ft 6in after a discussion with Lillicrap. On 19 December it was decided to add degaussing and also the D.IV.H catapult used in the *Fiji* class. A typed note in W G John Constructor's Notebook 12 lists changes in the ships and notes that as inclined *Belfast* and *Edinburgh* displaced (presumably on average) 10,550 tons, hence the modified ship would displace 10,866 tons. GM in standard condition was expected to increase from 1.5 to 2.4ft, a measure of how well top-hamper was being reduced.
5. When he visited Bath on 8 January 1940, DNC agreed with Lillicrap that effective bulge protection would be impossible, hence that machinery and other spaces should be subdivided as much as possible, with boilers in separate boiler rooms and extra bulkheads in engine rooms. Lillicrap planned to try keeping boilers below the lower deck. Subdivision lengthened the citadel from 190ft to 250ft and also demanded a larger engineering crew. These considerations pushed the length of the 15,000-ton ship to 670ft.

6. Lillicrap, quoting DNC on his visit to Bath on 8 January 1940.
7. W G John Constructor's Notebook 12, memo dated 23 August 1940, referring to work completed while he was on leave (in a 6in cruiser file, now lost) and to further instructions. He began with a 580ft (wl) x 68ft x 17½ft (33½ft depth) ship, presumably the one recently sketched. The standard displacement given had been reduced by fifty rounds per 6in gun and by 100 rounds per 4in gun, from 10,950 tons.
8. W G John Constructor's Notebook 13 includes a comparative table. The nine-gun ship (A) would have been 585ft (wl) x 70ft 6in; the twelve-gun (B), 615ft x 72ft 6in; *Belfast* was 606ft x 63ft 4in, displacing 10,540 tons. Both modified ships had 4½in belts (like *Belfast*) and 3in decks (2in in *Belfast*). Both had the six twin 4in guns of *Belfast*, but four quadruple pompoms instead of her two octuples. All three ships had 80,000shp powerplants. Speeds were 29.5/31kts (A), 29.25/30.75kts (B) and 31/32.5kts (*Belfast*). John also provided some higher-speed alternatives. To achieve 30.75kts deep, Design A had to be enlarged to 605ft x 72ft (12,270 tons) with 96,000shp. At standard displacement speed would be 32.25kts. Design B could have similar performance on 108,000shp (630ft x 74ft, 13,930 tons standard). In each case, the powerplant used four boilers. Alternatively, an eight-boiler plant like that of the earlier 9.2in cruiser studies could be used, but in that case a ship capable of 30.75kts at deep displacement would be even larger: 685ft x 85ft, 15,830 tons, 110,000shp. That was not too far from the 21,500-ton cruiser (720ft x 84ft) armed with nine 9.2in or twelve 8in guns, with an eight-boiler, 160,000shp powerplant offering a speed of 31.25/33.5kts. A comparative Legend dated September 1940 showed a modified *Fiji* with its 4in magazine abaft the after boiler room, and with the six twin 4in and four quadruple pompoms: 568ft (pp) 580ft (wl) x 70ft 6in, 11,280 tons, 89,000shp to make 31kts in standard condition. This ship would have *Belfast* side and deck armour (4½in and 2in).
9. A typed table in W G John Constructor's Notebook 13 shows three sets of heavy cruiser data added 11 October 1940, the other data being dated October 1939 and 22 January 1940. The 10,500-tonner of 1939 used *Fiji* machinery (72,500shp with 80,000shp when forced). The January 1940 designs used higher power to achieve higher speed: 96,000shp for a 12,500-ton ship (610ft x 72ft) with thicker deck armour (3in and 2½in instead of 2in) and the same 5in belt as in 1939 or 125,000shp for a 15,300-ton (670ft x 77ft 6in) ship similarly protected with six rather than four twin 4in guns, and slightly faster (33/31.5kts). Two of the October 1940 designs reverted to the 5in/3in protection of the 1939 design, but had four quadruple pompoms instead of two octuple ones. Both had 80,000shp *Belfast* powerplants. The nine-gun version displaced 11,950 tons (590ft x 70ft), and was expected to make 30.75kts in standard condition. The eight-gun version displaced 12,300 tons (610ft x 72ft) and was expected to make 31kts in standard condition. A third design had 3in decks (2in plus 1½in plating) and nine 8in guns. That boosted armour and protection weight from 2,300 tons in the first October 1940 nine-gun design to 2,800 tons. The hull matched that of the eight-gun ship (610ft x 72ft), but displacement increased to 13,100 tons. It took 96,000shp to achieve 32kts in standard condition.
10. Design X was 10,300 tons (580ft x 66ft), with 90,000shp for 30.75/32.25kts. The nine-gun version was 9,245 tons (550ft x 64ft); the same powerplant would drive it at the same speed (reduced length balanced off reduced tonnage). Design Y was 14,050 tons (635ft x 75ft) with 100,000shp for 30.5/32kts. The nine-gun version of Design Y was 12,750 tons (605ft x 73ft) with 100,000shp for 30.75/32.25kts. Z was initially 625ft x 74ft x 19ft, 13,500 tons, which John thought he could cut down to 615ft x 75ft x 19ft, 13,200 tons. Unlike *Belfast*, Y had four twin 4in guns and four quadruple pompoms rather than two octuple (X had two quadruples). All designs showed four rather than two quadruple 0.5in machine guns. All had the D.IV.H heavy catapult.
11. He also wanted the speed, in half-oil condition, of X, Y, and Z six months out of dock in the tropics: 29.5kts, 28.75kts and 29kts. Design Z was 13,500 tons (615ft [wl] x 78ft) and required 96,000shp.
12. The *Admiral Graf Spee* argument and the comparison with the improved 6in cruiser are from CAB102/536.
13. The Board seems to have become interested in 8in cruisers early in 1941, because W G John Constructor's Notebook 13 includes three sets of estimates marked 'work done at Admiralty, Whitehall, 13th and 14th January 1941' (p 61). The twelve-gun Design Z seems to have been used as a basis for a design with three twin 8in, one with three triples, and one with two triples. The ship with three triples was considered similar to Design Z: 635ft long (14,150 tons), 100,000shp (30.5/32kts), with 4in side and 2in and 3in decks. Raising the armour to the upper deck would increase displacement to 14,780 tons. The same change would increase Design Z displacement to 14,030 tons.
14. According to Moore, *Building for Victory*, p 144, the names initially chosen were *Benbow*, *Blake*, *Effingham* and *Hawke*, *Benbow* being changed to *Albemarle* and *Effingham* to *Cornwallis* by the new Ship Names Committee early in 1941. These names were approved by King George VI.
15. DNC ordered the first of these designs on 27 February 1941, according to a typed note in W G John Constructor's Notebook 13 (p 71): nine 8in, 30.5kts deep, 12,000nm at 16kts; 4½in belt with 2in deck over machinery and 4in over magazines; close-range AA armament of six twin 4.5in guns, four quadruple pompoms, and Oerlikons; one heavy catapult and two aircraft (as in the *Fiji*s). The forecastle deck would be the strength deck, carried right aft. The upper deck (i.e. the deck below the forecastle or weather deck) would be the armoured deck, except right aft. Gunhouse protection would be a 4½in face and 4in roof. A sketch of a 671ft (wl) x 79ft x 20ft ship shows one twin 4.5in gun alongside the bridge structure and two more alongside the superstructure between the two funnels, their magazine between the two machinery blocks. The sketch shows the superstructure deck (on which 'B' turret sits) extending aft to just forward of 'X' turret. Note that the belt is of constant height along its length, instead of rising to cover the forward boiler room. The version with four twin 4.5in eliminated one of the after pair on each side, but retained the pair near the bridge, presumably to reduce vulnerability. A drawing dated 7 March 1941 showed 637ft (pp) 651ft (wl) x 79ft x 20ft. Legends from the DNC Legend Book (Folio 18): *see overleaf*.
16. The Staff also wanted 1in bulkheads. Turret faces were to be 4½in, with 2in sides and roofs if possible. Protection generally followed that of the previously-designed heavy 6in cruiser.
17. Requirements were resolved at a Controller's meeting on 26 March after DNC produced a series of Legends. According to notes in W G John Constructor's Notebook 13, DDTSD plumped for a high bow for seakeeping, citing the old 'D' class. It was agreed that the ships' high freeboard was satisfactory. German ships had high bows but very low freeboard, hence might not be relevant. DNC would consider experiments. DNC was later instructed to provide considerably more

	I	II	III	IV
4.5in twin BD mountings	4	–	–	–
4in twin mountings	–	8	6	4

All versions had nine 8in, four quadruple pompom, four quadruple 0.5in or Oerlikons, and two triple torpedo tubes (DTM accepted that they would be 23ft 6in above water). All had 100,000shp powerplants.

	I	II	III	IV
Standard Displacement	16,100 tons	16,500 tons	16,700 tons	16,800 tons
Length pp	636ft	656ft	643ft	630ft
Length wl	650ft	670ft	657ft	644ft
Beam	80ft	80ft	80ft	79ft
Mean draft	20ft 6in	20ft 6in	20ft 6in	20ft 6in
Speed Standard	32.25kts	32.5kts	32.25kts	32.5kts
Speed Deep	30.75kts	30.5kts	30.25kts	31kts
Radius at 16kts	12,000nm	12,000nm	12,000nm	11,800nm
Oil Fuel	3,500 tons	3,500 tons	3,500 tons	3,400 tons
General Equipment	670 tons	700 tons	685 tons	660 tons
Aircraft	160 tons	160 tons	160 tons	160 tons
Machinery	2,050 tons	2,050 tons	2,050 tons	2,050 tons
Armament	2,000 tons	2,000 tons	1,970 tons	1,860 tons
Protection	4,000 tons	4,180 tons	4,060 tons	3,940 tons
Hull	7,220 tons	7,410 tons	7,315 tons	7,130 tons

sheer than in earlier classes, the limitation on LA fire (minimum 2° elevation right ahead) being accepted. DDSTD initially favoured a 4.5in secondary battery, but weakened following discussion. Plans wanted eight twin 4in, but doubted that four HACTs could be accommodated. Controller favoured 4in guns. On 8in ammunition, an important factor in weight (at 296lbs per shell), DDTSD wanted to increase to 170 rounds per gun; DNC considered 150 the most possible in the available length. Going to 200 rounds would require 30ft more length (20ft if this applied only to the forward turrets). Cordite was the limiting factor, not shells. At this meeting the decision as to whether to order the ships was deferred to September, but drawing work was to proceed. Staff Divisions were to prepare papers and drawings for a Board decision. The design with eight twin 4in was 670ft x 80ft x 23ft 6in (depth 43ft) and would have displaced 16,835 tons.

18. The tentative Staff Requirement called for four heavy high-velocity automatic weapons of new design, possibly twins, of 50–60mm; and four medium automatic guns (Oerlikons or equivalent). The Staff also wanted two triple torpedo tubes and fifteen depth charges, the latter at that time standard on board British cruisers.

19. The design with four twin 4.5in (A or I) was 636ft (pp) 650ft (wl) x 80ft x 20ft 6in (freeboard 22ft 6in) (16,100 tons), with 110,000shp engines for 32.25/30.75kts. It originally had 150 rounds per 8in, but that was reduced to 100; and it had 200 rather than 400 rounds per 4.5in. Design II (sixteen 4in) was 656ft (pp) 670ft (wl) (16,500 tons); Design III (twelve 4in) was 643ft (pp) 657ft (wl) (16,200 tons); and Design IV (eight 4in) was 630ft (pp) 645ft (wl) (15,000 tons). All had the same 110,000shp engines, for speeds of 32.25/30.75kts, 32.25/30.75kts, and 32.5/31kts. Stowage for 4in ammunition was cut from 350 to 200 rounds per gun. A sketch of the 670-footer dated 23 March 1941 shows the forward 4in magazine ahead of the forward machinery block, separated from it by a generator (dynamo) room.

20. Discussions of the requirements are in ADM 1/12758, and, in shorter form, in the early papers of the Future Building Committee.

21. If the gun loaded at a fixed angle and then elevated to fire, bagged charges would suffice. DNO preferred them, as there would be no need to get rid of used cartridge cases, and more ammunition could be carried on a given weight. Loading at high angles added considerable complication and weight, so considerable weight could be saved by limiting the gun to 55° elevation. DNO wanted intermediate weapons (he was urging development of a 3½pdr) to reach aircraft at angles of sight over 50° flying at heights beyond the reach of 40mm guns. DNO pointed out that the existing twin 6in already elevated to 60°. For 1943 DNO could offer either the existing triple 6in modified to elevate to 60° or the existing 70° 5.25in mounting. The main modification envisaged was to allow for quick changeover between LA and HA shells: a two-bucket shell hoist would replace the existing endless chain. Larger ammunition capacity would also be needed. DNO provided weights of possible alternative armaments: three triple 6in, five twin 5.25in, five twin 6in (60°), four triple 6in and six 5.25in or twin 6in. Three triple 6in, as in a *Tiger*, weighed 420 tons, compared to 450 tons for five twin 5.25in.

22. The undated Legend for Design A is Folio 25 in the DNC Legend Book: 12,000 tons, 630ft (wl) x 70ft 6in x 17ft 6in (fwd) 19ft 6in (aft) standard and 23ft 0in deep; freeboard 31ft (fwd) 24ft 6in (amidships) 18ft (aft) at standard displacement. Armament was nine 6in (400 rounds per gun, far more than in existing ships), twelve Hazemeyer twin Bofors (twenty-four barrels), twelve twin Oerlikons, and two quadruple torpedo tubes, and the usual fifteen depth charges. Protection matched that of a *Fiji*: 3¼in side and 2in deck over machinery, 3½in and 2in deck over magazines, 2in turret roofs, 2in–1in turret sides. Power was 80,000shp (four shafts) for 31.75kts standard and 30.25kts deep; 3,750 tons of oil fuel gave an endurance of 6,000nm at 24kts when six months out of dock in the tropics ('deep and dirty'). Weights: general equipment, 740 tons; machinery, 1,600 tons; armament, 1,250 tons; protection, 1,940 tons; hull, 6,230 tons; margin, 240 tons. The DNC Legend Book does not include legends for Designs B to L.

23. Gawden Constructor's Notebook 1 shows later calculations concerning Design A alongside Designs G, H, and J for 5.25in ships. Gawden's stability figures for Design A suggest dimensions of 643ft x 63ft. Light displacement was 12,250 tons (deep 16,290 tons). No trace of Designs B through F seems to have survived. It seems likely

that the designs were designated according to a list of possibilities laid out by DNO in July 1942: (a) three triple 60° 6in; (b) five twin 5.25in; (c) five twin 60° 6in; (d) four triple 60° 6in; (e) six twin 5.25in; and (f) four twin 60° 6in. These were roughly in weight order, from 420 tons for the three triple 6in to 552 tons (plus fixed hoists) for six twin 6in. By way of comparison, the *Tiger* armament of three triple LA 6in and five twin 4in plus their directors weighed 520 tons, slightly below the weights of (d) and (e) (560 tons and 540 tons). This association would explain why Gawden's first design other than A was G, incorporating a new weapon, the triple 5.25in. DTSD laid out the requirement for endurance, but specified neither close-range armament nor protection. Both seem to have been proposed by DNC, who argued that the most difficult protection requirement to meet was against bombing; a 2in deck would keep out only the smallest German AP bomb in a diving attack. Nothing thicker was practicable.

24. ADM 116/5150 for 1942–3 and ADM 116/5151 for 1944–5. Each volume contains both papers and minutes of meetings.
25. Characteristics were:

	Three twin 5.25in	Three triple 5.25in
Length	525ft	575ft
Oil Fuel	2,000 tons	2,500 tons
Speed	31.25–29.5kts	31.75–30kts
Endurance/20kts	6,000nm	6,000nm
Standard Displacement	6,800 tons	10,500 tons
Protect, RDF, etc as	Latest *Dido*	Latest *Fiji*

Formal Legends have not survived. Traces of some of the designs considered can be found in various Constructors' Notebooks. Thus Gawden Notebook 1 includes tankage and armour weight calculations for Design G (four triple 5.25in guns, 71ft beam), October 1942 (dimensions were also given as 610ft x 69ft). Side armour would have matched that of a *Fiji*, with 3½in over magazines and 3¼in over machinery; decks were 2in thick, and bulkheads 1in. An alternative Design J was 615ft x 69ft, with similar armour and armament. The design with three twin mounts was K; a modified K2 had two more Bofors and four more twin Oerlikons. The design with three triple 5.25in mounts (L) was worked out in November 1942; Gawden shows dimensions of 575ft x 68ft x 18ft 3in (hull depth 32ft).

26. A later meeting (on 30 November 1942) decided that there was too little difference between the 5.25in and the 6in to make a 6in cruiser of any type worth while. Given the probable difficulties of designing a 6in dual-purpose gun, 'it was hardly worth pursuing the 6in type'. Later it was pointed out that 5.5in seemed to be the upper limit for a dual-purpose mounting, as anything larger would have to be fully powered.

27. A slightly later paper by DDNC(L) on the disadvantages of speed pointed out that three *Dido*s had been torpedoed, *Naiad* and *Hermione* abreast the machinery spaces and *Phoebe* (twice) abreast the forward magazines, of which the first two had sunk after rapid flooding of machinery spaces. Two main machinery compartments of a *Dido* were 16 per cent of her buoyancy. There is no indication in this paper that DDNC(L) was aware of the flawed machinery arrangement.

28. Passmore Constructor's Notebook 1 gives dimensions of Design M as 506ft x 55ft x 29ft x 15ft 6in; Design P had the same dimensions but 56ft beam. He gives Design N2 as 550ft (538ft pp) x 64ft x 32ft x 16ft 9in (8,650 tons). Design R was the much larger cruiser with 6in and 4.5in guns. Legends of Designs M1, N1 and N2 were included in a presentation for the Board dated 30 June 1943, when N2 was being chosen as preferable alternative (Board Memos, ADM167/118). M1 was 520ft x 55ft x 29ft x 16ft 3in (7,150 tons); N1 was 550ft x 62ft x 32ft x 16ft 3in (8,200 tons), and N2 was as above. All were designed for 29kts at standard displacement and 28kts at deep load (not given). The 28kts at deep load was equivalent to 25.5kts deep load six months out of dock in the tropics. Power was, respectively, 44,000shp, 44,000shp and 48,000shp, in each case on four shafts. Endurance at 18kts when six months out of dock was 6,200nm, 6,000nm and 7,700nm, respectively. M1 had three and N1/N2 had four twin 5.25in (400 rounds per gun), each with its own HA director. The designs showed, respectively, six, six and eight Busters (twin Bofors) and eight, ten and twelve twin Oerlikons, plus, in each case, two quadruple torpedo tubes. In the N2 design, the Busters were forward and aft on the centreline, plus six sided, and at least half of the twin Oerlikons were hand-worked so that the ship could keep fighting if she lost power (this requirement may have been added when the associated Staff Requirement was written). All had 3in belts over machinery with 1in bulkheads, 1in (1½in for N2) magazine fore ends, and 1in shell room after ends. Decks were 2in (1¼in in M1) over machinery and 2in over magazines. Turrets were ½in. Protection requirements: the belt over the machinery should defeat the ship's 5.25in shells at 11,250yds (90° inclination) as well as German 5.9in shell (17,000yds) and Japanese 5.9in shell (12,500yds, 90° inclination). The deck over the belt should defeat 5.25in shells at all ranges, and German 5.9in inside 27,000yds (or Japanese 5.9in inside 22,500yds); it should also defeat a 500lb SAP bomb dropped from 2,500ft at an air speed of 300ft/sec (180kts). Magazines should be similarly protected, with splinter protection for turrets and steering gear.
Weight data:

	M1	N1	N2
General Equipment	490 tons	530 tons	530 tons
Machinery	1,000 tons	1,010 tons	1,040 tons
Armament	745 tons	920 tons	970 tons
Protection	830 tons	1,020 tons	1,090 tons
Hull	3,925 tons	4,560 tons	4,850 tons
Margin	140 tons	160 tons	170 tons
Displacement	7,130 tons	8,200 tons	8,650 tons

29. The 44,000shp plant of N1 was replaced by a 48,000shp plant; endurance increased from 6,000 to 7,700nm at 18kts. Displacement increased from 8,200 to 8,600 tons. The modified ship could have eight rather than six Busters and twelve rather than ten twin Oerlikons. This N2 design was generally considered satisfactory. DGD would have liked the funnels closer still, but DNC retorted that a single hit should not take out both uptakes.
30. DGD compared three alternatives to the Mk 24 planned for the *Tiger* class:

	A	B	C	Mk 24
Rate of fire per gun (rpm)	8–9	10–12	10–12	5–6
Maximum elevation	60	60	80	60
QF or BL	BL	QF	QF	BL
Total weight per turret (tons)	215–220	225–240	240–255	162½
Roller path diameter (ft)	19	22	24	19

All of these guns were envisaged as LA weapons with HA capability. Design A was essentially a modernised Mk 24, achieving a higher rate of fire by adopting a fixed loading angle (hence enjoying a reduced firing rate if it had to elevate to high angle every time it reloaded). High rate of fire was attractive because a faster-firing gun could hit a fleeting target, and also for anti-aircraft fire. Maximum elevation had to be reduced by about 10° to allow for deceleration required for RPC, so effective maximum elevations were 50° and 70° for A, B and C. Only C could be considered an effective anti-aircraft gun. On the basis that a ship with Mk 24 turrets would displace 14,200 tons, DGD estimated that with Design A it would displace 14,700 tons, with B, 15,100 tons, and with C, 15,300 tons. A Design C ship would be about 20ft longer than one with Mk 24. DGD recommended Design C for the 1944 cruiser. DNO estimated that it would take four years to develop Design B or C to the point where turrets would be available for installation, compared to 2½ years for Design A (only one design could be developed at a time). At this time three 1944 cruisers were expected to complete in 1948 and two in 1949. It would be possible to complete the first three with Mk 24 and the last two with Design C. DGD rejected Design A on the grounds that if it were pursued it would be impossible to begin work on a fully-modern 6in turret before 1948. He recommended accepting some delay and completing all five ships with Design C. In August 1944 the Sea Lords chose Design B, which became the Mk 25.

31. Minutes of the Future Building Committee meeting on 24 August 1944. It is not clear whether any designs had yet been prepared.

32. A table in the DNC Legend Book (Folio 31) lumps Designs Q and R together. Both designs showed eight 500kW generators, twice the equipment of a modified *Fiji*: four turbo-, four diesel-generators. Dimensions were 630ft(wl) x 74ft x 19ft 6in (fwd) 21ft 6in (aft) at standard displacement (14,200 tons); hull depth was 43ft. Power output was 100,000shp for a speed of 32.5kts standard and 31.5kts deep clean, equivalent to 29kts six months out of dock in the tropics. Endurance was 6,500nm at 20kts at full load under trial conditions, equivalent to 4,200nm at 20kts when six months out of dock in the tropics. Complement was 1,050, compared to 870 for *Belfast*, which displaced 11,700 tons standard as bulged and rearmed. Weights (Design R): general equipment, 780 tons; machinery, 1,900 tons; armament, 1,820 tons; protection, 2,200 tons; hull, 7,220 tons; margin, 280 tons.

33. Data from DNC Legend Book (Folio 31B): *see below*.

34. The document refers to Japanese 5.9in shells, but Japanese cruisers had 5.5in guns. Bulkheads at the ends of the main belt were 100lbs NC, and 40lb DW bulkheads were fitted at both ends of the machinery units. Gun houses had 4in faces and 2in tops and sides, as in *Belfast*.

35. 655ft (wl) x 76ft x 20ft 11in, depth to upper deck 36ft. Standard displacement was 15,350 tons (18,708 tons deep, 17,283 tons in half-oil conditions); complement was 1,050 as a private ship. Armament weight was 2,049 tons. She would have carried 2,850 tons of oil fuel, for an endurance of 7,580nm at 20kts (reduced to 4,660nm when six months out of dock in the tropics). This ship met draft Staff Requirements dated 14 July 1944 (final ones were issued in August), which did not explicitly demand dual-purpose 6in guns. They called for 200 rounds per gun, but the figure 300 was pencilled in.

36. Displacement increased to 15,560 tons, beam to 76ft 3in, and draft increased from 21ft fore and aft to 21ft 3in fore and aft. The 250 rounds per 6in gun ammunition allowance was increased to 270.

37. There had already been some cancellations. This paper largely shaped the post-war British carrier fleet, in that it called for deferring construction of the four *Gibraltar*-class fleet carriers and also of four of the eight *Hermes* class (it did, however, call for continued work on all three *Ark Royal*s, only two of which were built). By this time many lesser ships, including forty-two submarines, had already been cancelled.

38. An unhappy ACNS asked for a comparison of Mk 25 with US turrets. There was no exact comparison, because the standard US triple turret was LA only, and the US dual-purpose mount was a twin. The maximum elevation of the US triple was 41°, but Mk 24

	Q/R	V	W	Belfast *rearmed*
Length (wl)	630ft	606ft	606ft	606ft
Beam	74ft	73ft	73ft	66ft 4in
Displacement (standard)	14,200 tons	13,300 tons	13,350 tons	1,700 tons
Draft fwd	19ft 6in	19ft 6in	19ft 6in	18ft 3in
Draft aft	21ft 6in	21ft 6in	21ft 6in	20ft 9in
Freeboard fwd	31ft 6in	31ft 6in	31ft 6in	29ft 0in
Freeboard amids	22ft 6in	22ft 6in	22ft 6in	14ft 3in
Freeboard aft	16ft 9in	16ft 9in	16ft 9in	16ft 9in
Draft deep (mean)	4ft 0in	24ft 0in	24ft 0in	22ft 6in
shp	100,000	100,000	100,000	80,000
Speed std	32.5kts	32.5kts	32.5kts	31.25kts
Speed deep dirty	29.5kts	29.25kts	29.25kts	27.5kts
Fuel oil	2,650 tons	2,650 tons	2,650 tons	2,250 tons
At 20kts clean	6,500 tons	6,900 tons	6,900 tons	6,100 tons
At 20kts dirty	4,200 tons	4,300 tons	4,300 tons	4,300 tons
Complement	1,050 tons	990 tons	980 tons	870 tons
6in guns	12 (200 rpg)	12 (200 rpg)	12 (200 rpg)	12 (200 rpg)
4in guns	–	12 (400 rpg)	–	8 (375 rpg)
4.5in guns	12 (400 rpg)	–	8 (400 rpg)	–
Quad pompom	–	–	–	4
Buster	10	9	10	–
Oerlikon	28	28	28	26
Torpedo tubes	8	8	8	6

elevated to 60°. US turrets were somewhat more heavily armoured, and they carried more shells. DNO calculated weights assuming British armour thicknesses and shell stowage. His table showed:

	Mk 24	US triple	Mk 25	US twin
Revolving structure	163 tons	175 tons	204 tons	192 tons
DNO fixed fittings	10	25	27	25
Ammunition	38	71	80	74
	220 tons	271 tons	311 tons	291 tons
Extra armour		9 tons		16 tons
Extra ammunition		24		25
If above omitted (i.e. to British standards)		238 tons		250 tons

Presumably the much greater weight of ammunition in the US triple could be attributed to the use of cased rather than bagged ammunition. The British shell was also lighter (112lbs rather than 130lbs). DNO pointed out that the new US twin DP mount was heavier than the triple mount in the *Defence* class (Mk 24). There were dramatic differences in design. The US triple carried all three guns in one sleeve, only 4ft 4in apart. DNO had recently investigated a plan for a quadruple 6in gun (which apparently did not appear in contemporary cruiser designs) with guns paired in two cradles. Trials suggested that the minimum distance guns had to kept apart so as not to interfere with each other was 6ft (similar trials were carried out before designing the triple turrets for the *Nelson*s). Alternatively, firing could be delayed; the British delayed the centre gun, the US Navy the two wing guns. American practice was being reviewed to see whether the minimum distance was excessive. In any case, using a common cradle made it impossible to correct for the muzzle velocities of individual guns, increasing the spread of the turret; and the failure of one gun could put the turret out of action. However, the common cradle made for a smaller roller path and saved one or two men as gunlayers, and allowed one or two machines to be omitted. DNO noted that the American dual-purpose twin mount had independent guns. US turrets also had higher trunnions than British, partly because of their narrower roller paths, but also in order to increase the fixed loading angle to provide a slightly higher rate of fire at high elevation. However, the blow on the deck due to firing was greater, so a heavier structure was needed under the turret. The US twin 6in trained at a much higher speed (25°/sec; 12°/sec in the Mk 25, 10°/sec in the US triple and in the Mk 24) than any of the other turrets, because it had been conceived primarily as an anti-aircraft gun (even in the Mk 25 anti-aircraft fire was a secondary consideration). However, the British LA triple elevated fastest, because the gun had to be re-elevated after loading. Both the Mk 25 and the US twin loaded at all elevation angles. DNO pointed out that loading was completely automatic in the British Mk 25, whereas the US twin required some manhandling both in the shell stowage space and in the gunhouse. He also noted that the US Navy expected to use two weights of shell (lighter for anti-aircraft), compared to one weight for the British.

39. Guns were 6ft apart, the minimum acceptable to prevent undue interference. The trunnions were 11ft above the roller path, nearly 4ft higher than in Mk 25, considerably increasing firing forces. A revolving cartridge holder at the gun held nine rounds, which could be fired in twenty-two seconds (twenty-six rounds per gun per minute), followed by a lull while the revolver was refilled. The gunhouse crew would number about fifteen. Protection at this stage was 4in face and 2in sides, roof, and back. Estimated revolving weight was 184 tons, more than a twin 8in (the constructor took 190 tons as a good estimate). The battery would be controlled by two LRS 1 (for divided control) and three MRS 1 controlling three turrets in separate controlled fire.

40. Unfortunately the Cover does not include DGD's paper.
41. Rate of fire was ninety rounds per gun per minute. DNO estimated that each round would weigh 40lbs, and that the self-contained version would carry 600 rounds per gun. He expected it to weigh 25 tons, with a crew of sixteen.
42. As tabulated in Perry Constructor's Notebook 10, they were:

I	5 turret	Z 1	620 x 70 x 19.5 x 13,150 tons
II	4 turret	Z 2	620 x 71 x 19.5 x 13,100 tons
III	5 turret	Z 3	635 x 73 x 20 x 14,000 tons
IV	5 turret	Z 4	655 x 76 x 21 x 15,450 tons
V	3 turret	Z 5	625 x 70-6 x 19-9 x 13,100 tons
VI	3 turret	Z 5A	630 x 71 x 20 x 13,350 tons
VII	5 turret	Z 4A	645 x 75-5 x 20-9 x 15,080 tons
VIII	5 turret	Z 4B	655 x 76-5 x 21 x 15,470 tons
IX	4 turret	Z 2A	635 x 73-5 x 20-25 x 14,160 tons
X	3 turret	Z 5 B	630 x 71 x 20 x 13,350
XI	4 turret	Z 2B	630 x 73 x 19-9 x 13,880 tons
XII	5 turret	Z 4C	645 x 75 x 20-3 x 15,070 tons
XIII	4 turret	Z 2C	630 x 73 x 20 x 14,090 tons
XIV	5 turret	Z 4D	620 x 74 x 21 x 14,380 tons
XV	5 turret	Z 4C	645 x 75 x 20-6 x 15,045 tons

43. *Minotaur* was not included in the 1944 Cruiser Cover, the comparison with *Worcester* being in the general post-war cruiser cover (without any lead-in showing the basis of the design). No separate Cover for this design has survived. The switch in designs, late in 1946, is evident in a volume of DNO notes for Controller (February 1947–November 1948) in Naval Historical Branch.
44. She was 645ft (wl) x 75ft x 20ft 9in (std) 24ft 0in (deep); hull depth was 44ft, and deep displacement was 18,415 tons. She was expected to make 31.5kts deep (clean) on 100,000shp with four shafts. Endurance 'deep and clean' was 6,000nm at 20kts. Design ZA was 13,070 tons standard (616ft x 73ft x 20ft/25ft 6in; 16,760 tons deep load). Because she was shorter, she needed more power to make the required 31.5kts: 110,000shp. Design ZB was 15,210 tons standard (616ft x 74ft x 20ft 3in/24ft 0in, 17,960 tons deep load) and would require 120,000shp. In comparable terms, *Worcester* displaced 15,210 tons standard (664ft x 70ft 7¾in x 21ft 6in/25ft 0in, hull depth 43ft 9in, deep displacement 18,000 tons). She was expected to make 32kts 'deep and clean' on her 120,000shp, or 32.5kts using the 10 per cent overload.
45. G E Moore, 'Post-War Cruiser Designs for the Royal Navy 1946-1956', in J Jordan (ed), *Warship 2006* (Conway Maritime Press, London: 2006). The basic design (D) was the *Minotaur* chosen the previous year: five twin 6in, eight twin 3in, 645ft (wl), 18,380 tons. The difference in tonnage was that this time deep load rather than standard displacement was used. Alternative P showed that substituting quads for twin 3in and removing one twin 6in required another 15ft of length and an increase to 19,250 tons. Eliminating torpedoes saved 25ft (length 635ft) and reduced displacement to 18,500 tons, about that of *Minotaur*. Alternative Q was a step further up, another twin 6in being replaced by two quad 3in (total eight mountings): 675ft, 19,750 tons (or, without torpedoes, 665ft, 19,550 tons). Alternative R showed what happened if all the 6in were restored and eight quad 3in demanded: 710ft, 21,000 tons

(without torpedoes, 700ft, 20,500 tons). Alternative S showed that cutting back to four quad mountings but demanding all five twin 6in would cost 660ft and 19,250 tons (without torpedoes, 660ft, 19,000 tons, presumably because the 3in guns did not interfere with the torpedoes, so no extra length was involved). A series of alternatives with eight or four twin 3in was appended, R1 of the series corresponding to *Minotaur* but larger: 660ft, 19,500 tons.

46. ADM 116/5966. The ten-year period was set by the Chiefs of Staff in a 1947 paper on future defence policy. They estimated five years in which war was unlikely and another five in which the probability of war would increase; after the ten years it would increase sharply. Thus 1956/7 became a horizon for force modernisation. The abortive building plan is in ADM 167/129, Board Minutes and Memoranda for 1947, as part of a lengthy long-range plan.

10. Post-war Cruisers

1. There were two arguments. One was that it would take until 1957 for the Soviets to make up for wartime devastation. The other was that Soviet nuclear development would set the timetable. By 1948 the British regarded the United States as the key to any war, so the Soviets would not fight until they had enough bombs (often estimated as 100) to disable the United States. The 'year of maximum danger' was a set interval after the Soviets exploded their first bomb, the date of which was estimated as 1952. The US government adopted the British approach and also the 1957 deadline (after plumping for 1955 for a time), but once the Soviets exploded their bomb in 1949, the US advanced the 'year of maximum danger' to 1954. The British did not, possibly partly because no earlier date was even remotely affordable.
2. This plan included a missile ship (not a cruiser) in its 1956/7 and 1959/60 programmes, and the modernisation of four cruisers: *Royalist*, *Belfast*, *Swiftsure* and *Superb*.
3. According to Paffett Constructor's Notebook 2, p 183, about 20 June 1952 DTSD decided that the three missile ships formerly considered (A, 30kts; B, 20kts; and C, 12kts) should be fused into an 18-knot ocean convoy escort armed with one Sea Slug launcher and two twin 3in/70. This was the preface to a design study, using a merchant-ship hull.
4. The policies described here, including the decision for four cruisers and then for missile cruisers, are taken from the post-war draft Staff History held by the Naval Historical Branch and from my *The Post-war Naval Revolution* (Conway Maritime Press, London: 1986), the British programmatic sections of which were taken from Admiralty documents in the PRO. There is some question of interpretation, as it is not clear that references to guided-missile cruisers in the Staff History do not sometimes refer to all-gun large cruisers. The first major expression of the new deterrent-based strategy was the Global Strategy paper, a report dated 12 June 1952 by the Chiefs of Staff to the Defence Committee. The Chiefs of Staff reported on cuts after mobilisation in a paper dated 31 October 1952. Controller put the Navy's case in 1954.
5. At a meeting in May 1952 DTSD said the Naval Staff strongly favoured torpedo armament, apparently to fire the new 'Ferry' guided anti-ship torpedo. The only available position was the quarterdeck, on which either four fixed tubes firing aft or one remote-controlled quadruple tube could be placed. Fixed tubes would save weight and did not have to be manned in action. However, there was no experience of firing torpedoes into or across a ship's wake (and it was not clear how 6in blast would affect the tubes). To test the idea, two single tubes were to be taken from the frigate *Relentless* and mounted on board the trials cruiser *Cumberland*, one firing fore and aft and the other athwartships. About thirty to forty trial shots were required. The Ship Design Policy Committee decided to complete the ships without tubes, which could be installed afterwards if desired (the necessary trials went ahead). This option was never exercised. The star shell gun was dropped because it offered no other capability, although it did offer illumination (for target identification) at a greater range than the existing rocket flare. Presumably illumination was not very important in a ship optimised for anti-aircraft performance. However, the subject was considered important enough for fleet commanders to be polled. The rocket flare was their favourite.
6. MRS 3 used what the US Navy called a linear-rate computer. It projected ahead the observed rates at which the target was moving. Over time those rates changed, but the computer was not wired to calculate the change, so in effect the system was limited in range. LRS 1 presumably did not share this limitation. It saw no service as a gun fire-control system, but was the basis of the missile-control system used by the Sea Slug missile.
7. In 1948 MRS 3 was envisaged as control system for the new 3in/70 gun; as primary control system in destroyers which could not accommodate LRS 1 (to replace Mk VI director/Flyplane); and to control dual-purpose mountings split from primary control in new cruisers and destroyers. It would control 4in guns in AA frigates, and also 6in, 5.25in and 4.5in mountings when split from primary control in cruisers, carriers, battleships, destroyers and AA frigates. It required AC power (hence a motor-generator for power conversion in DC ships). Like other British (and US) fire-control systems, MRS 3 had a below-decks computer, space requirements for which could pose problems in older ships.
8. See the author's *British Destroyers and Frigates: The Second World War and After* (Chatham Publishing, London: 2006).
9. Except for the two *Tiger*s still being built in private yards, modernisation would be done at two Royal Dockyards (Portsmouth and Devonport), so nine years (to the 1957 target) gave eighteen yard-years of work. At this time only the ten *Dido* class had dual-purpose guns, so they enjoyed the highest priority for modernisation (they needed new fire controls). Worthwhile modernisation would take two years, and work on the *Dido*s would leave eight yard-years for the 6in ships – for a total of four, plus the two *Tiger*s in private yards. Priority for new construction was carriers and escorts, so it was unlikely that new cruisers would be built before 1957.
10. DNO was beginning to work on a new medium-calibre gun, at that time expected to be a 5in/70, which he hoped would become available in 1955. He expected the Mk 26 mount to be available for ship-fitting in 1952 if development proceeded at a normal rate, but spending on it had been cut in plans for the 1948/9 programme. The decision had been made to use US guns and ammunition, so that large supplies would be available in wartime, and it seemed that a proposal to incorporate the US 6in/47 into the existing twin mount design would take more time. All of these dates turned out to be delusionary; the 5in gun was abandoned in 1954, having formed the basis of a cruiser-destroyer concept. A meeting in November 1947 (chaired by Deputy Director of Naval Ordnance [DDNO]) decided that since modernisation was planned before 1955 (also a delusion), only the 6in should be considered. The projected fire-control systems were Flyplane, for 5.25in and 6in guns, and MRS III, for existing anti-aircraft weapons and divided control of dual-purpose guns. Flyplane already existed, and could be provided 3½ years after a contract was

placed. MRS III was expected to be available in 1954. The November meeting covered modernisation of the whole cruiser fleet. Flyplane was an anti-aircraft computer, so it was proposed in combination with an Admiralty Fire Control Clock for surface control.

11. A ship with three HACPs could take two Flyplanes, a ship with two HACPs (as in the original *Fiji* class) could take only one. In this sense the Improved *Fiji*s with three turrets were equivalent to *Tiger*s, but they could not take as much topweight.

12. There was no space in the machinery box for additional generators. Alternatives were one 1,000kW generator in space vacated by 'B' turret shell room, to serve both turrets; or one 500kW forward and one aft in place of the after 150kW diesel generator (finding space for a 50kW emergency diesel generator aft); or to replace both 150kW diesel generators with 500kW units. They could be steam or gas-turbine powered. This was one of the first British proposals for gas-turbine power. The two 500kW generators would not be connected to the ring main, but in the last proposal would also be emergency generators. In all cases the existing 500kW turbo-generators would be retained. However, to additional steam generators near the turrets would have long, hence vulnerable, steam lines. Another possibility was to replace the existing generators with more powerful ones. To further complicate the situation, the new weapons used AC rather than the usual DC power.

13. Key features were Type 274 radar, Type 932 radar (for splash-spotting), automatic cross-levelling (to compensate for roll), and improvement of the fire-control table (computer). All four requirements were met in *Ceylon*, *Nigeria*, *Newcastle*, *Birmingham* and *Uganda* (HMCS *Quebec*). Of the other ships, *Superb* and *Swiftsure* both had Type 932, as did HMCS *Ontario*. Automatic cross-levelling was on board *Swiftsure*, *Kenya*, *Sheffield* and *Ontario*. The computer improvement had been done in *Kenya* and *Sheffield*. The ships listed were those with the best surface fire-control systems, *Nigeria* otherwise being quite limited.

14. In 1950 *Euryalus* still had Type 79B, the first wartime air-search radar; *Nigeria* had Type 279, its immediate successor. The others had one version or another of the standard wartime Type 281. Most ships had Type 277 surface/air-search sets. Exceptions were the two New Zealand ships (*Black Prince* had a US SG1, and *Bellona* Type 268), *Dido* (Type 268), *Sirius* (Type 268), *Diadem* (Type 268), and *Nigeria* (SG1). Remarkably, *Royalist* had no surface search set at all, but she did have the Type 293 target-indication radar. Several ships, including *Nigeria*, lacked Type 293. *Newfoundland* and *Superb* were being fitted with Type 960 air-search radar; *Swiftsure* had Type 281BQ instead because of lack of time during a refit. *Mauritius*, *Kenya*, *Jamaica*, *Gambia*, *Ceylon*, *Ontario* and *Uganda* all had sufficient space and weight to take Type 960, and *Bermuda* was scheduled for installation.

15. See the author's *British Destroyers and Frigates: The Second World War and After* for attempts to design or build a FADE.

16. These alternatives were listed in a sheet dated 14 November 1950, giving deep load displacements and metacentric heights for each. The original scheme with four twin 4.5in would involve an increase in complement of ninety men, hence was presumably not practicable. The ship would have displaced 11,670 tons deep (GM 4.22ft). All the other alternatives would have been possible without increasing complement. With two twin 4.5in, the ship would displace 11,548 tons (4.46ft, no trim); with six single 4.5in, 11,554 tons (4.56ft, trim 4in by bow); and with four twin 4in, 11,522 tons (4.57ft, 5.2in trim by bow). The emergency outfit included a barrage director and GDS 2*. Ratings required for the modernised battery, not including STAAGs, were 172, compared to 264 for the battery with four twin 4.5in guns. The problem was thus to eliminate ninety men. Note that the 6in Mk 26 battery required fifty-two ratings, the Mk 24 battery, ninety-two, reflecting the greater automation of the new mounting. A twin 4.5in mount required thirty men. Four MRS 3 Mod 1 required the same number (twenty-eight men) as four Mk 6M directors and their four FPS 5 computers below decks.

17. Modernisation, as planned in March 1953, would have added back the third twin 4in on each side, controlled by two MRS 3. The close-range battery would have been at least six Bofors Mk 11. Torpedo tubes would be landed. The ship would be fitted with TIU 3/Type 992 and the bridge enclosed. At this time plans called for replacing the 4in guns in *Swiftsure* and *Superb* with three twin 3in/70 (one on the centreline), and for installing the maximum possible number of L70 Bofors. Unfortunately the latter would have had to be converted to all-AC power, at an enormous cost. Detailed calculation showed that the goal of bringing 4in control up to *Tiger* standard in *Belfast* (with outstanding approved alterations and additions) entailed about 200 tons of overweight, which could be balanced by removing 'X' turret or multiple measures including deleting two of the twin 4in guns (keeping the original four) and eliminating the torpedo tubes. These measures offered only a total of 58 tons of compensation. Much more electrical power was needed. A revised Staff Requirement called for three triple 6in, four twin 4in (with two MRS 3) and six Bofors Mk 11. DTSD disliked the loss of the turret. DPT reception and a sixteen-track display were included and then dropped. In February 1954 plans called for retaining all four turrets plus four twin 4in, plus the six Bofors Mk 11s. The agreed Staff Requirement for what was now called limited modernisation (March 1954) showed eight twin L70 Bofors. It must have seemed ludicrous that DEE said that the modernisation would be as elaborate as that of *Tiger* (replacing TIU 3/992 with TIU 2/Type 293 radar helped). As of June 1954 it was estimated that the ship would displace 11,600 tons (light), compared with 11,800 tons before; and 14,716 tons deep compared with 14,826 tons, the current figures having been taken from a recent Inclining Experiment. They compared with 11,920 tons light in April 1945, 13,776 tons in average action condition (1954: 13,665 tons), and 14,933 tons deep load. A Legend dated January 1956 showed the four triple 6in turrets, six (rather than four) twin 4in (this may have been an error), and six twin L70 Bofors. Extreme beam was given as 67ft (a January 1955 Legend showed 69ft), and deep displacement as 14,895 tons. As of May 1956 the planned refits of *Swiftsure* and *Superb* had been scaled back to fitting new close-range weapons (six twin L70 Bofors) but also DPT, and possibly the new Type 965 radar (then called WAIR, which DND badly wanted). Like *Belfast*, both would have had closed bridges and lattice masts.

18. About August 1948 attack on, and defence of, trade became the second main task, the support of destroyer striking forces becoming the first of the secondary tasks. The June 1951 version of the Staff Requirement omitted trade attack and protection altogether, as well as reconnaissance (a secondary role in the earlier Staff Requirements). However, the central task of providing close air-defence was extended to include convoys as well as carrier task groups.

19. It also asked how many could be mounted in a *Dido* if one or all 5.25in mountings were removed. One twin 3in/70 could replace a single 5.25in turret. They could replace all five original turrets (and the pompom in 'Q' position if it had replaced a twin turret). In the modified *Dido*s, Nos. 2 and 3 mountings would have to be at the same level, unless new structure were added (plans for the original *Dido*s were later changed to show Nos. 2 and 3 on the same level).

Nothing could be done with the pompom positions amidships. Each 3in/70 mounting would need its own MRS IV or V director, which could be accommodated (some would have to be sided). Space sufficed for 400 rounds per gun. The approved addition of two 300kW generators would provide sufficient power. Estimated light displacement would decrease from 5,970 tons to 5,850 tons, but stability would suffer (GM would be reduced from 1.6ft to 1.4ft). For the modified *Dido*s, comparable figures were 5,880 tons falling to 5,760 tons, GM reduced from 2.1ft to 1.9ft. A more detailed estimate for the early *Dido* with bridge and 'Q' turret lowered one deck was 5,809 tons (GM 1.72ft).

20. Both versions showed two STAAGs and six single Bofors, plus two quadruple torpedo tubes. The mixed-battery ship was expected to displace 9,551 tons standard and 11,705 tons fully loaded (deep). The all-3in ship would displace 9,191 tons standard and 11,345 tons at deep load. Estimated speed in deep condition (mixed-battery) was 31.5kts (31kts in the tropics); deep load speed was 30.75kts (30.26kts), the corresponding figures for six months out of dock were 29.75kts and 29.25kts. Speeds for the all-3in ship were given as 31.75/31.25kts standard, 30.9/30.4kts deep, and 29.9/29.4kts six months out of dock in deep condition. The average 3in ammunition load was 624 and 628 rounds per gun respectively.

21. Each 6in mount would have a Mk 6 director with Flyplane computer below decks, with one Admiralty Fire Control Table forward for LA fire, and one Admiralty Fire Control Clock aft for LA fire. If necessary, DGD would accept a fire control clock instead of the table, and dispense with the clock aft. Each 3in/70 would have an MRS 3 director. Gun direction would be by a TIU III with Type 992 radar (TIU 2/293 as interim if TIU III were not ready in time). If the 3in/70 was not ready in time, 4in guns could be mounted temporarily. Data from a note by DGD on 17 February 1948.

22. At a Staff meeting in September 1948 to discuss the Staff Requirement, Director of Plans regretted the limited armament offered, compared with large new Soviet destroyers which would have eight 5.9in guns, eight 3in guns and eight torpedo tubes (no such ships were actually built) and with the existing *Kirov* class cruisers (nine 7.1in and eight 3.9in guns). DGD pointed out that rate of fire and accuracy were more important than numbers of guns 'and there was reason to suppose that the *Tiger*s would be superior to their possible adversaries'. DTSD (chairman) pointed to the ships' anti-aircraft role. Director of Plans 'did not feel reassured' in view of the vulnerability of only two gun mounts and the question of reliability; a single breakdown or hit would halve the ship's main battery. The constructor responsible for the cruiser section (H S Pengelly) soon commented to DNC that with the planned armament the limit imposed by the relatively small hull had been reached, and even exceeded. Even if space could be found for a third Mk 26 in place of the forward 3in/70, it would add about 200 tons and would reduce stability unacceptably. If the new high-performance gun was essential for the primary role of the ship – anti-aircraft defence – then the reduction in number of guns and mounts could not be avoided. VCNS reaffirmed this conclusion. The two-turret arrangement did prompt DGD to demand somewhat thicker turret armour. Mk 26 had been conceived for the five-turret *Minotaur*; losing one turret was not nearly so serious as losing one of two turrets. *Minotaur* thicknesses were: 3in (originally 4in) face, 2in sides, 1½in roof, 1½in rear and 1in floor. Thicknesses proposed for *Tiger* were reduced: 2in face, 1½in sides, and 1in roof, rear and floor. Late in 1948 DGD demanded *Minotaur* thicknesses (with a 2in protective ring). The extra inch on the face was described as weight to balance the mount. DNC had pressed for minimum weight (splinter protection only), both for stability and to make it easier to train the turret at maximum speed.

23. In September 1948, estimated action load was 2,000kW, so the desired total generator capacity, exclusive of emergency diesel power, should be 4,000kW (with 3,000kW considered acceptable). There was a 'reasonable chance' that the four 500kW turbo-generators in the machinery spaces could be upgraded to 750kW, to give a total of 3,000kW rather than 2,000kW (this was DEE's initial AC power plan). It would probably be necessary to upgrade diesel generator power to 700kW (say, two 350kW), which was impracticable. However, the 150kW diesel generator aft could probably be upgraded to 200kW. DEE stated that switching to AC would not involve extra weight, but it was not at all clear that a 3,000kW AC installation would be no heavier than the existing 2,000kW DC one. Conversion to AC power was likely to be expensive, although in the end an AC system would be more efficient and more suitable. DEE later proposed installing, if possible, two 1,000kW AC generators in place of two of the 750kW units, to provide a more nearly adequate reserve of electric power. By November, E-in-C wanted all four generators in the machinery space to be 1,000kW (DNC commented that when they went from 350 to 500kW they were told it would be very difficult). By this time all machinery, including the original quartet of 500kW DC generators, had been installed, but E-in-C pointed out that no doubt the generators could be used elsewhere. DEE claimed that the change to AC would save 75 tons (DNC's cruiser designer, Pengelly, did not believe that). The estimated weight saving was later reduced to 65 tons, which could be set against an anticipated 100 tons excess (over Legend figures) in the DC design. DEE also pointed out that existing DC switch gear could not handle the loads now envisaged. At its sixth meeting (12 November 1948) the Ship Design Policy Committee decided that these and all future ships should have AC power, but given the implications of the decision, it wanted approval from First Lord and First Sea Lord. In fact First Lord was not informed when the order was given (March 1949) to cease work on DC drawings in favour of AC, on the theory that he would not want to be informed prior to consideration by the full board. Then that decision was reversed. The same meeting affirmed the *Tiger* design with Mk 26 and 3in/70 guns. The larger implication for cruiser modernisation was that DEE would want any cruiser rearmed with Mk 26 turrets to be rewired for AC, at a considerable price. The formal decision in favour of AC power was delayed because its cost was difficult to calculate, yet the increased expenditure required Board and Treasury approval. The change was probably set by January 1949. In March 1949 DNC estimated costs for the Treasury. The total cost of the planned DC installation was £1.6 million, of which £480,000 had already been spent. The AC installation would cost £1.4 million. Stripping out the existing installation would cost £40,000, and modifying machinery would cost another £80,000, but £200,000 of the DC machinery cost could be recovered. The net increase was £200,000, a seventh of the cost of the AC installation alone.

24. After the Board approved CDS for cruisers, approval was sought to buy a prototype for installation on board *Superb*; as of 1954 the requirements for CDS/DPT were being referred to the Air Defence Working Party for review in light of the new strategic concept of deterrence.

25. There were also delays in supplying key equipment. In July 1951 delays in providing the GDS 3 gun direction system would delay the ships by five to seven months, pushing expected completion dates to January 1956 (*Tiger*), May 1956 (*Blake*) and July 1956 (*Defence*). This was before the one-year deferment.

26. These guns appear in the July 1953 version of the Staff Requirement, to be controlled by MRS 8, ultimately to be replaced by MRS 3 Mod 2. The L70 weapon was intended to replace the wartime L60 (i.e. 40mm/60). It offered greater effective range. For a time, the L70 featured in all new British warship designs, but it was superseded by the Seacat missile before entering service (the British Army did buy the L70).

27. Enclosures to the Ship Design Policy Committee dated 31 December 1948 from DNC (C S Lillicrap). The papers involved were 'The Functions and Status of the Cruiser' (SDPC(48)24), 'The Shape of the Cruiser of the Future' (SDPC (48)33), and 'Protection of Convoys by Cruisers' (SDPC(48)33). They appear to have been written by DTSD. He proposed visionary developments (for 1948) such as atomic power and flush decks with armament sunk into them (to deal with nuclear fallout and blast), but DNC chose to limit himself to developments already under way. DNC's drawings are in the General Post-war Cruisers Cover.

28. Sketch I was 14,500 tons standard/17,500 tons deep (645ft x 74ft x 23ft deep). Sketch II would carry forty-eight missiles on 14,000 tons standard/17,000 tons deep (530ft x 73ft). The missile launcher and stowage displaced the two after twin 6in and also the two after twin 3in. A structure forward of the twin launcher carried a missile-assembly space, and had the Sea Slug director on top. The missiles themselves were shown stowed horizontally below the waterline in two compartments with a checkout compartment between them. Sketch II can be compared to the abortive 1956 missile cruiser, which had a similar main battery. Sketch III was 10,500 tons standard/13,000 tons deep (585ft x 68ft x 22ft). The drawing showed a *Daring*-style forefunnel with the lattice foremast wrapped around it, and a conventional after funnel, widely separated from it. Sketch I/II had the single funnel. Sketch IV was 12,250 tons/14,750 tons (600ft x 70ft x 23ft), with separate engine and boiler rooms forward and aft, separated by the 3in magazine amidships. It was expected to make 30kts on 90,000shp (four shafts) when 'deep and clean'. This design showed fixed quadruple torpedo tubes in the raised area amidships. Funnels were as in Sketch III rather than the single funnel of Sketches I and II. Sketch V was a version of IV with two combined engine and boiler rooms forward but with separate boiler and engine rooms aft. It had three rather than four shafts. It would displace 12,000 tons standard/14,250 tons deep (580ft [wl] x 70ft x 23ft; 90,000shp would give 31kts in deep clean condition).

29. ADM 1/22160, a file on gun armament for the emergency cruiser programme.

11. The Missile Age

1. Paffett Constructor's Notebook 2, which begins with a reference dated 28 March 1951 to Perry's Sketch Design I. Unfortunately, Perry's last Notebook in the Brass Foundry collection ends in 1946 (Paffett refers to Perry's Book 11, which has not apparently survived). Paffett refers directly to the 'Large Cruiser of 1960 – Sketch I', so presumably that is the Perry design in question. His drawing of armour distribution showed a box over the forward magazines, and conventional belt and deck protection (with a high part to cover boilers) abaft that. The belt is 130lb (3¼in) NC, but the 60ft long box forward has only 60lb (1½in) NC on its side. End bulkheads are 130lb NC. Decks are 60lb (12lb structural, the rest protection).

2. In January 1952 the DACR (direct action close range) weapon was code-named Zenith. There were four possibilities: D.10 (30 tons), Scarecrow (39 tons), V.A. Crow T. (36 tons), and Marquardt (27 tons), each of which included 11¾ tons of ammunition.

3. Paffett wrote that he decided to try a 6in design because the virtues of the two heavy guns were still being debated. However, he also referred to TSD 00105/51, the preliminary Staff Requirement (which has apparently been lost). Hull weights were scaled from those of *Belfast*, corrected for the flush deck planned for the new ship.

4. Estimated weights for the weapons in contention, as of June 1951, were 225 tons for the twin 5in, and 280.5 tons for the twin 6in, 99.8 tons for the 3in/70, and 30 tons for the close-range weapon. In each case, weight included ammunition. Type 984 radar was expected to weigh 60 tons.

5. Paffett worked out armour for a 625ft ship (2,095 tons available): 3in side and 2in deck over the whole citadel. That was actually very tight, and he had to shave to provide it (he managed to increase armour weight to 2,250 tons by cutting endurance to 5,820nm at 20kts, or 4,580nm 'deep and dirty').

6. Paffett's comments on CR 6 suggest that 3in guns would have to be on the centreline, because otherwise the ship could not make full use of these weapons. In that case CR 8 would have four centreline positions, compared to no more than five (two 3in, three 6in) for CR 4, since in CR 4 two of the 3in would be sided, as in *Tiger*. CR 8 length was given as CR 4 less a 6in, plus a 3in, which might equate to 625ft or 600ft, which was very long for an 11,000-ton ship. A estimate in November 1951 gave dimensions as 600ft x 65ft, with endurance 4,500nm 'deep and dirty' at 20kts, capable of 34.25kts 'deep and clean'. On this basis, adding up known weights (hull, equipment, machinery, armament, fuel oil, reserve feed water and a 200-ton Board margin), only 300 tons was left for protection. To get minimal protection, Paffett proposed going to 12,000 tons, in which case speed would drop to 33.6kts 'deep and clean'.

7. Some of the missing numbers represented alternatives not worked out in detail. CR 9 was a version of CR 8 with three twin 5in instead of two twin 6in and two twin 3in. CR 11 was a version of CR 10 with three twin 5in instead of the 6in/3in battery. CR13 was a version of CR 12 with three twin 5in. All of these ships had two quadruple torpedo tubes. CR 16 was *Minotaur* brought up to date (Paffett's Notebook gives no details at all).

8. W G John Constructor's Notebook 15 mentions a revised version of this TSD 2230/52 dated 8 August 1952 in a memo of 16 January 1953 describing the current study (22), a 435ft, 4,770-ton ship armed with two twin 5in and two twin L70 Bofors, with six-channel GDS 3 (Type 992 radar) plus air-warning and fighter-control radar (Types 982 and 983). The ship also had a single 'stripped' 4.5in star shell gun, six fixed torpedo tubes on each beam, single Limbo ASW mortar, and the Camrose anti-torpedo torpedo. On 30 July 1952 DNC had directed designers to use destroyer practice throughout. The memo proposed a 4,750-ton ship with endurance reduced from the desired 3,500nm at 22.5kts to 3,050nm. Note that at this point the twin 5in mounting was expected to weigh 270 tons.

9. Armament was four twin 5in, four Bofors Mk 12, and eight torpedo tubes. The ship would have a single Limbo ASW mortar, in effect left over from her rather distant cruiser-destroyer ancestry. John's Notebook also shows a May 1953 study of a 15,400-ton cruiser armed with two twin 6in, four twin 3in, and four twin Bofors Mk 11, a slightly lighter battery. This ship would have been 650ft long.

10. W G John Constructor's Notebook 15 gives 7,450 tons for a 520ft x 54ft ship with three sets of unit machinery and 6,700 tons for a 480ft x 52½ft ship with two machinery units. Both versions devoted

700 tons to protection. Later John gave 6,100 tons as the displacement of a cruiser with two twin 5in guns. Protected to *Dido* standards the ship would displace 7,900 tons.

11. W G John Constructor's Notebook, 25 June 1953; John was also asked for a fast escort, on which he had been working.
12. T J O'Neill Constructor's Notebook 1, as he resumed work on the CR series. This note was dated 11 November 1954.
13. These tables are in W G John Constructor's Notebook 15, and surrounding documents suggest they were produced early in July 1953, about the time when the cruiser-destroyer was abandoned. The Notebook includes more detailed sheets describing the ships further. The big cruisers and the missile ship all had four-unit machinery. The 18,200-ton and 16,850-ton ships both had 2½in side and 1½in deck armour; radars were Types 960, 982, 983, 974 and 992. Both had four twin Bofors L70 as close-in armament. The 18,200-ton ship had twelve MRS 3 fire-control systems, the 16,850-tonner (and the 16,000-tonner), ten. The 13,000-tonner had three-unit machinery, hence its reduced speed (it also had eight rather than ten MRS 3 systems). All the new-design cruisers had two quadruple torpedo tubes. None of the cruisers had Type 984 radar. The 8,000-tonner was 550ft x 57ft 6in x 16ft (hull depth 34ft 6in). The Y102 powerplant offered a speed of 30.5kts 'deep and clean' (29.5kts 'deep and dirty') and an endurance of 3,600nm at 20kts. Armament was three twin 5in with MRS 3 Mod 1 directors, one sextuple Bofors (one MRS 3 Mod 2), two 40mm twin with TOM, GDS 3, and four triple torpedo tubes. It had 1½in sides and 2in crowns for magazines, and 2.5in sides and 1in deck over machinery. The ship had accommodation for 675.
14. This time the Type 984 aerial and office were taken as 35 tons, and the CDS system as another 30; GDS and Type 992 added 32 tons.
15. T J O'Neill Constructor's Notebook 2, confusingly dated prior to Notebook 1, but devoted to missile ships. This entry is dated 6 September 1954 (O'Neill's work on gun cruisers was dated November 1954).
16. This version showed four tandem bays, each with six missiles on three levels. Each level fed into a lift and traverser feeding missiles into two checkout spaces abreast the loaders which would place them on the twin launcher. Missiles failing checkout would be stowed in these spaces.
17. Dated 3 December 1954 in T J O'Neill Constructor's Notebook 2.
18. Associated equipment space would be about 3,000ft^2 (about 1,500ft^2 beyond what was already available), and the extra ten officers and sixty ratings would add another 3,000ft^2. Given a waterplane coefficient (0.82), the designer could estimate the effect of adding 20ft of ship length: 1,100ft^2 on each of three decks, a total of 3,300ft^2 – about half of what was needed. Three bridge decks, 20ft x 50ft each, offered another 3,000ft^2, which would be enough. The result was 550ft (pp) x 70ft x 18¼ft (hull depth 45ft), 11,300 tons.
19. GW 43 had two 6in guns but no 3in/70; she also had four twin Bofors. She would have made 32/31kts on 14,000 tons. This design seems not to have been worked out in any great detail.
20. In deep stowage, as used in the test ship *Girdle Ness*, missiles were stowed one atop the other in cradles. When needed, they were grabbed one by one by a crane overhead, moved to an opening in a watertight bulkhead, and rammed through. By way of contrast, tube stowage entailed continuous positive control of the missile, which was easy to move fore and aft using lugs to hold it to the stowage rails. Its railed supports had to move sideways (traversing) to move it off the fore-and-aft direction, but that was not too difficult to arrange. Sea Slug was also relatively easy to fin before launching; the need to stow it fully-assembled complicated stowage system design. The capacity of later 'County' class missile destroyers was greatly increased, nearly to that planned for the missile cruisers, by stowing some missiles partly assembled and assuming they would not be fired with the assembled ones. DNC disliked the considerable open volume of deep-stowage magazines and pointed out that they could not be vented satisfactorily. However, their worst feature was surely that the missile on its grab was not under full control unless a very complicated rigid grab was used.
21. Comments dated 4 April 1955 in T J O'Neill Constructor's Notebook 3.
22. ADM 1/27685, nominally on the role of helicopter ships in the '88 Plan' (29 March 1960).
23. At this time frigates and missile destroyers typically deployed small helicopters without dipping sonars, hence essentially extensions of the ship's own sonar system. The big 'single package' helicopters were limited to otherwise fixed-wing aircraft carriers to, according to D of P, 'their mutual embarrassment and the detriment of the carrier's main role'.
24. Detailed calculations are in Chilcott Constructor's Notebook 1, 9 February 1961. These are mainly deck area calculations, hence do not provide much insight into design characteristics. Millman Constructor's Notebook 17 includes a space analysis for Study 21H3. Stickings Constructor's Notebook 2 includes capacity calculations for 21M15 and for 21N6.
25. A paper on the 1963 Long Term Costings in the *Tiger* Cover makes the interesting point that the 50,000-ton carrier then planned was only large enough to carry thirty strike aircraft, four airborne early-warning aircraft, and two search-and-rescue helicopters; adding ASW aircraft would push up the size and cost of the ship. It already seemed that displacement would go to 52,000 tons to accommodate two ASW Chinooks. On 54,000 tons, the most that could be driven by the planned three-shaft powerplant, two more Chinooks could be accommodated. To the extent that the carrier was cancelled because it was so expensive, conceivably greater reliance on the escort cruiser concept could have saved the project.
26. According to a history of the escort cruiser dated 6 September 1963 in the third *Tiger* Cover, no fewer than three formal proposals had come from the fleet over the past eight months: to convert *Belfast* to carry helicopters, to convert the *Tiger*s, and to convert the LST *Lofoten* 'as an interim off-shore garage' until the FOST support ship (*Engadine*) was completed in six Wessex, with limited maintenance for eight days. Speed was 15kts, and endurance was 5,000nm at 15kts. Were it to be turned into a fleet unit, it would need both better maintenance facilities and better communications. The ship was unarmed and lacked command-and-control facilities. The one FOST ship in the programme was due to complete at the end of 1965.
27. This was Scheme Z, offering flight deck space for two Wessex (with rotors spread) and hangar stowage for four, with forty-eight Mk 44 torpedoes for the helicopters. Scheme X offered deck space for one Wessex and a hangar for three, and Scheme Y offered deck space for two and stowage for four, both waist 3in/70 mounts being landed. In Scheme Z the helicopters were stowed under the flight deck, hence the guns could be retained. Ultimately a form of Scheme Y (hangar forward of the flight deck) was adopted in which the flight deck was sponsoned out over the stern and the guns retained; the complication of a hangar under the deck (hence a lift) was eliminated.
28. An undated Legend of *Blake* as converted to operate Sea Kings showed a Seacat system (forty-two missiles) and two twin 3in/70. Standard displacement was given as 11,280 tons (full load 12,190 tons). As inclined upon completing conversion, *Blake* displaced

12,440 tons deep; 11,510 tons in average action condition; and 9,920 tons light.
29. DEFE 24/386, Command Cruiser papers 11 August–31 December 1969.
30. In theory the four last 'County' class missile destroyers could trade armament for command and control spaces, but they were due to be discarded in the 1980s, when the command cruiser would reach mid-life. Once they were gone, that capability would have to go somewhere else. Accommodating the facilities in Type 42s would be difficult (a 1969 paper gave a price but then argued that it would be uneconomical to split up the limited number of Type 42s, and that in any case the desired level of command and control could not be accommodated in so small a hull.
31. As explained in an 11 December 1968 paper written to help support the Staff Requirement the following year. DEFE 24/1385. Another paper in this docket pointed out that the command at sea requirement had to be justified in the face of a trend towards exercising command of maritime operations from shore, using higher-capacity communications. At about the same time proposals were being made for the future automation of the Northwood maritime command centre. Points to be raised with the MoD Operational Requirements Committee included non-NATO scenarios and details of the limited command facilities aboard all other ships. Recent studies had shown that there had to be an Officer in Tactical Command afloat, well provided with command facilities. The navy's position was that the three ship types were parts of an integrated package which was ineffective without the cruiser, its seaborne command and control and heavy-helicopter element. The Staff Target (the basis for the later Staff Requirement) was submitted for Board approval in August 1968.

Appendix: Fast Minelayers

1. Lillicrap Constructor's Notebook 3, p 319.
2. Lillicrap Constructor's Notebook 4 contains an estimate for a 30kt ship based on an 'E' class hull, to carry 250 mines: 510ft x 32ft x 16ft 6in, 6,700 tons (with 450 tons of oil). It needed 24,000ehp to make 30kts. The design, based on a 'D' class cruiser, had large trimming tanks fore and aft. Estimated dimensions were 450ft x 46ft (wl) 52ft (extreme) x 15ft, for a displacement of 4,900 tons without protection. The extreme beam was underwater, the ship having 4ft-deep bulges on each side.
3. This ship was 495ft x 52ft (wl) 58ft (ext) x 15ft (6,760 tons).
4. Design history from a memorandum dated 25 April 1921 in Board Memoranda for 1921, ADM 167/64. Unfortunately the *Adventure* Cover has been lost, the Lillicrap Notebooks providing most of the design history.
5. As of 1 April 1921 Lillicrap was considering a 7,550-ton ship, 510ft (pp) x 59ft 10in x 15ft 6in, which would need 52,500shp to make the desired 28kts. He then cut back to 500ft x 59ft x 15ft 3in (7,150 tons), the size chosen, for which 55,000shp was needed. As built the ship had the standard wartime 40,000shp light cruiser plant, hence was somewhat slower than desired, about 27.5kts. The cut-back dimensions are the ones cited in the April memorandum.
6. Some notes on its design are in the summaries of Controller's conferences in the *Kent* class Cover.
7. Details were taken from Brinton Constructor's Notebook 5.
8. At this time *Adventure* was credited with 340 H (First World War) mines or 280 Mk XIV (current mines). Mines would be carried in four rows, two on each side of the ship, the mining deck being 464ft long (about 110ft longer than in *Adventure*).
9. W G John Constructor's Notebook 11, instructions dated 20 October. John estimated that the ship would need 165,000shp. DNC also asked John to work up a design for an aircraft-carrying cruiser, perhaps inspired by the Swedish *Gotland* (and not too different in concept from the Japanese *Tone*), with three turrets forward, as in HMS *Nelson*, and one aft. She would carry three Walrus amphibians. Nothing seems to have come of this idea, which seems to have repeated a design estimate offered by Lillicrap in 1936.
10. Mines would weigh 558 tons, armament 436 tons, a very different proportion than in *Adventure*. Protection would have accounted for 390 tons. Machinery weight (2,587 tons) was based on comparisons with fast Italian cruisers. Later in the design, beam was reduced to 56ft, and desired mine load increased to 400 mines.
11. The specified mines were Mk XIVs, whose bogies used a broader-gauge track than the old H2, which in the late 1930s was still by far the most numerous mine in British stocks. Ships would therefore have a third rail (in storage) for every mine track, so that they could quickly be converted to lay the older mine.
12. As refitted in July 1945 *Ariadne* had three US twin Bofors instead of the two Hazemeyers, two of these guns being mounted aft to superfire over 'X' twin 4in. During the same refit, five single Bofors replaced the earlier ten Oerlikons. *Apollo* had fourteen Oerlikons, eventually replaced by six single Bofors.
13. On 2 December 1942 ACNS(W) asked the Future Building Committee to examine pros and cons. He pointed out that the ships had proved useful, though perhaps more as blockade-runners (particularly in the Mediterranean) than as minelayers. In later stages of the war they might serve both as minelayers and to supply advanced bases. Naval Assistant to Controller pointed to their low endurance at high speed, which ill-suited them to the long distances of the Far East. From a production point of view they were a nuisance, as they are a special type (large, of light construction) with special boilers, engines and propellers. They took disproportionately long to build and broke into any sort of production line. Slips for *Ariadne* and *Apollo* were being absorbed by an intermediate Aircraft Carrier (*Hermes* class) and two destroyers in January 1943; adding minelayers in 1943 would cut a carrier from the 1942 Supplementary programme and delay one in the 1943 programme by about six months, as well as delaying large destroyers. It was most uneconomical to build this sort of ship for use as a cargo carrier. Director of Plans pointed out that one ship (*Latona*) had been lost and another (*Manxman*) severely damaged by torpedo; these ships were subject to more than ordinary hazards. Offensive minelaying would certainly be useful against Japan, considering the large expanses of mineable water in the Far East. However, there were already minelaying destroyers, and the number available seemed sufficient for a balanced fleet. ACNS(W) gave up, but regretted that, 'as we will want more once there is a probability that Japan is our only enemy.' During the following year two more ships were lost: *Abdiel* to a ground mine and *Welshman* to a torpedo.

    Some design work was done, as Brinton Constructor's Notebooks 4 and 5 both contain calculations for either repeat *Abdiel*s or for a new-design minelayer. Calculations for the repeat ship began in August 1942. Initially the main improvements would have been RPC for the 4in guns and, probably, additional electric power.
14. Lillicrap Constructor's Notebook 6, p 251.

BIBLIOGRAPHY

Published Works

Bastock, John, *Australia's Ships of War* (Angus and Robertson, Sydney: 1975).
Brown, D K, *The Grand Fleet: Warship Design and Development 1906-1922* (Chatham Publishing, London: 1999).
_____, *Nelson to Vanguard: Warship Design and Development 1923-1945* (Chatham Publishing, London: 2000).
_____ (ed), *The Design and Construction of British Warships 1939-1945: The Official Record, Vol 1, Major Surface Vessels* (Conway Maritime Press, London: 1995).
_____, 'The Design of HMS *Arethusa* 1912', *Warship International* No. 1 (1983).
Campbell, John, *Naval Weapons of World War II* (Conway Maritime Press, London: 1985).
Chesneau, Roger (ed), *Conway's All the World's Fighting Ships 1922-1946* (Conway Maritime Press, London: 1980).
Gillett, Ross, *Australian and New Zealand Warships 1914-1945* (Doubleday Australia, Lane Cove: 1983).
_____, *Australian and New Zealand Warships Since 1946* (Child & Associates, Brookvale: 1988).
Giorgerini, Giorgio, and Nani, Augusto, *Gli Incrociatori Italiani 1861-1964* (Ufficio Storico Della Marina Militare, Rome: 1964).
Gray, Randal (ed), *Conway's All the World's Fighting Ships 1906-1921* (Conway Maritime Press, London: 1985)
Groener, Erich (revised and expanded by Dieter Jung and Martin Maass), *German Warships 1815-1945, Vol. 1: Major Surface Vessels* (Conway Maritime Press, London: 1990).
Hall, Christopher, *Britain, America and Arms Control 1921-1937* (St Martin's Press, New York: 1987).
Hore, Peter, *Sydney, Cipher, and Search* (Naval Institute Press, Annapolis: 2009).
Howse, Derek, *Radar At Sea: The Royal Navy During World War II* (Macmillan, Basingstoke: 1993).
Kingsley, F A (ed), *The Development of Radar Equipments for the Royal Navy, 1935-45* (Macmillan, Basingstoke: 1995).
_____ (ed), *The Application of Radar and Other Electronic Systems in the Royal Navy in World War 2* (Macmillan, Basingstoke: 1995).
Lacroix, Eric and Wells, Lincoln, *Japanese Cruisers of the Pacific War* (Naval Institute Press, Annapolis: 1997)
Lenton, H T, *British and Empire Warships of the Second World War* (Greenhill Books, London: 1998).
Marriott, Leo, *Catapult Aircraft: Seaplanes That Flew From Ships Without Flight Decks* (Pen and Sword, London: 2006).
Moore, George, *Building for Victory: The Warship Building Programme of the Royal Navy 1939-1945* (World Ship Society, Gravesend: n.d. [2002]).
_____, 'Post-War Cruiser Designs for the Royal Navy 1946-1956', *Warship 2006*.
Morris, Douglas, *Cruisers of the Royal and Commonwealth Navies Since 1879* (Maritime Books, Liskeard: 1987).
Pfennigwerth, Ian, *The Australian Cruiser Perth 1939-1942* (Rosenberg, Dural: 2007).
Raven, Alan, and Roberts, John, *British Cruisers of World War II* (Arms and Armour, London: 1980).
Roskill, Stephen, *Naval Policy Between the Wars* Vol I (Collins, London: 1968) and Vol II (Collins, London: 1976).
Watson, Raymond C, Jr., *Radar Origins Worldwide: History of its Evolution in 13 Nations Through World War II* (Trafford, Victoria [British Columbia]: 2009).

Official Handbooks

(In the Public Record Office except as indicated; NHB is the Royal Navy Historical Branch, Portsmouth, and NARA is the US National Archives, in this case at College Park)

Director Handbook: Application of RP 40 to Director Control Towers in King George V, Fiji, Dido, and Earlier Classes, 1950 (BR 912(23)) (NARA).
Director Handbook: General Information, 1935 (BR 912(1)) (HMS *Excellent* Library).
Director Handbook: Instruments Fitted in King George V, Fiji, and Dido Classes, 1943 (BR 912(17)) (HMS *Excellent* Library).
Guard Book and Index For the High-Angle Firing Manual, 1942 (CB 3085) (NARA).
Guard Book and Index for Pamphlets of the Handbook on the High Angle Control System Mks I – IV (and Associated Equipment), 1940 (CB 4056 (GB&I) (NHB).
Handbook for 6-inch BL Mk XXIII Gun on Twin Mk XXI Mounting 1932 (CB 1895A).
Handbook for 6-inch BL Mk XXIII Guns on Triple Mk XXIII Mounting 1939 (BR 962).
Handbook for Outfit Type FV1 1943 (CB 4268) (NARA).
Handbook for Type 91, 1943 (CB 4226) (NARA).
Handbook for Pom Pom Directors Mks II and III, 1940 (OU6365/40) (NARA).
Handbook for the Barrage Director Mk III 1946 (BR 1527) (NHB).
Handbook of Fire Control Tables, Clocks, and H.A.C. Systems: Introductory Pamphlet on Fire Control Tables, 1939 (CB 4048(1)) (HMS *Excellent* Library).
Handbook on the Use of RDF For Gunnery Purposes 1943 (CB 4112) (NHB) (Guard Book and Index for Pamphlets).
Handbook on the High-Angle Control System: General, 1940 (BR 919(A)) (HMS *Excellent* Library).
Handbook of Radio Warfare 1945 (CB 04404) (NARA).
List of His Majesty's Ships, Showing Their Armaments (published half-yearly, 1914-1918).
Naval Staff Monograph (Historical): The Naval Staff of the Admiralty: Its Work and Development (September 1929: CB 3013).

Particulars of War Vessels and Aircraft (British Commonwealth of Nations): Half-Yearly Returns (CB 01815B, various editions, April 1939 through October 1945).

Particulars of War Vessels and Aircraft (British and Foreign) Quarterly Return (half-yearly beginning in 1924) (OU 6107) (April 1923 through October 1938; CB 1815 [i.e., confidential rather than for official use only] from April 1928) (NHB).

Progress in Gunnery (CB 3001 series: various editions through 1939).

Radio Warfare (S.D. 1080/47: volume of *Technical Staff Monographs 1939-1945*) (NARA).

Unpublished Material Consulted at the National Maritime Museum

D'Eyncourt design notebook
Armstrong design portfolios 3 and 4
Vickers papers on export designs including:
Vickers Design Books 2 and 8
Vickers Design Book for Designs 958-1352
Vickers Design Book for Designs 1013-1118
Vickers Estimate Book 3 (1911-1920)
Vickers Estimate Book 5
Vickers Estimate Book for 1925-32
Vickers Weight Book 1 for Warships
Constructors' Notebooks (See Endnotes)

Ships' Covers (most have several volumes)
Foreign Cover (210)
Bristol Class (240)
Dartmouth Class (253)
Melbourne (257)
Birmingham Class (272)
Brisbane and abortive Canadian cruiser (276)
Arethusa Class (286)
Calliope Class (303)
'Atlantic Cruiser' (319)
Cambrian Class (327)
Chester and *Birkenhead* (352)
Vindictive (356) (most of *Hawkins* Class cover was lost)
Calypso Class (370)
Cairo Class (402, includes AA rearmament)
'D' Class (404)
Foreign Cruisers 1930-36 (426)
Kent Class (430)
London Class (442)
RAN 'County' Class (444)
York (450)
Dorsetshire Class (452)
Exeter (463)
Northumberland Class (470)
Leander Class (479)
Repeat *Leander* Class (494)
Intermediate Cruiser (499)
Arethusa Class (508)
Neptune Class (509)
Amphion Class (518)
Southampton Class (529)
General Cover (554)
Dido Class (555)
Belfast Class (566)
Fiji Class (567)
Repeat *Dido* Class (572)
Dido Class 1938 (592)
Fast Minelayers (594)
Heavy Cruiser (624)
Delhi Rearmed (656)
Hawkins Class (678; only for war refits)
E Class (679; limited data)
Cruiser 1944 (729)
Tiger Class (777)
Dido Modernisation (785)
Missile Ships (789)
General Cruiser Cover (790)
Liverpool Modernisation (817)
Belfast Modernisation (839)
Foreign Cruisers 1926 (no number)

DATA LIST

Except as noted, these are design data, from Legends (generally from the Covers). The number in parentheses after numbers of guns is rounds per gun. TT are torpedo tubes. Complement is as the ship was designed or completed, not as she was later operated; numbers could vary considerably. Prior to the Washington Conference (1921), displacements were given as 'legend' or 'normal' or 'Navy List' and deep load.

Standard displacement was a somewhat artificial figure. Where available, for post-1921 ships loads, including equipment, are divided into those contributing to standard displacement and extras (including extra machinery weight, due to machinery stores) included only in deep load. RFW is Reserve Feed Water. Note that some armament weight (e.g., additional ammunition) was included only in deep displacement. Aircraft, when carried, were included in armament weight.

The 1921 and 1930 Treaties gave pre-Treaty ships standard displacements equal to their earlier 'normal' or 'Navy List' displacements, which included fuel and reserve feed water, and thus considerably exceeded standard displacements calculated according to the new treaty procedures. The effect of this procedure was to down-play the jump in cruiser (and, for that matter, battleship) size when new ships were built, as in their case standard displacement excluded fuel and reserve feed water, and was closer to light ship displacement.

Actual, as opposed to design, data are from Inclining Experiments, indicated as IX, with their dates.

Inclining experiment data for cruisers, particularly during the Second World War, are from Constructors' Notebooks, which describe the experiments (and often provide earlier figures for comparison). These figures rarely included the standard displacement, which was an artificial figure. The half-oil condition was sometimes called the average action condition. Constructors evaluated the effect on stability of all proposed modifications, including small ones such as adding water coolers for drinking water. Thus Constructors' Notebooks, particularly those compiled by A J Neal in 1943–53, also describe various proposed modifications, which are mentioned in the notes below. Unfortunately it is not generally clear whether proposed modifications were actually made.

A few general proposed modifications are best mentioned here:

1. Installation of Action Information Centres (AIC). This was difficult because the British AIC differed considerably in concept from the US CIC. The US Navy had a single space built around a radar plot, while the Royal Navy emphasised the need for different users to have full access to a plot. Thus the Command on the bridge (compass platform) needed direct sight (via a viewing glass looking down) of a (largely surface) plot. Aircraft direction required a different kind of plot, in an Aircraft Direction Room. A third kind of plot, in a Target Indication Room, was used to allocate targets to directors and weapons. A flagship needed another plot, for the admiral trying to co-ordinate a force. In the British view, the multi-purpose US CIC was crowded to the point where it could not function effectively. A critique of US ASW performance in the post-war Exercise Mainbrace suggested that the US system provided so little information to the Command on the bridge that ASW performance was badly degraded. Conversely, installation of an AIC entailed major reconstruction of a ship's bridge. This was the reconstruction of bridges in the cruisers *Birmingham* and *Newcastle* in 1950–2, in which topweight was reduced by using light alloys for internal partitions.
2. Replacement of existing HA directors by Mk VI, which was expected to provide blind-fire capability using its Type 275 radar. Estimates were prepared for the heavy cruisers and for the *Leander* class and HMAS *Hobart*, generally in connection with landing 'X' turret.
3. Air-conditioning. Fitting of individual air-conditioners was apparently ordered about October 1944, probably in connection with planned deployments to the Pacific (the notebooks do not say as much, but they include calculations and an account of overall air-conditioning policy).
4. Replacement of torpedo tubes by additional light anti-aircraft guns (generally Bofors guns) in the British Pacific Fleet, formally authorised on 2 June 1945. Torpedo tubes were restored post-war in at least some ships. It is not clear how many ships were modified, and reports in this file are ambiguous.
5. In 1943 surviving 'C' and 'D' class (non-AA) cruisers were ordered modified for shore bombardment, presumably initially to support the D-Day landings. The most important change was installation of a special indirect-fire director, i.e., a director which could aim the ships' 6in batteries at a designated shore target which could not be seen from the ship. In May, the ships were ordered refitted but not modernised. Alterations to gun armament were to be limited to substituting twin Oerlikons for existing singles and for single pompoms (2pdrs) in ships on ocean escort duty. In the 'D' class, the centreline (after) 4in HA gun was to be replaced by as many single and twin Oerlikons as possible. The two sided 3in ('C' Class) or 4in ('D' class) HA guns were retained to fire star shell. By July, plans called for landing the torpedo tubes and two 36in searchlights in these ships and adding the indirect-fire director. In the 'C' Class, seven single Oerlikons were to be replaced by twins, and four more twin Oerlikons added (two of them on the boiler room vents). In the 'D' class eight singles were replaced by twins and four twins (the most for which there was space: two of them were on the centreline) added. Installation of Type 970 radar (the shipboard version of the aircraft H2S, used for landing-craft navigation close inshore) was considered but rejected, and the ships no longer needed air-warning radars. Boats were reduced to one motor cutter and one whaler; all un-needed davits were landed. Ships involved were *Ceres*, *Capetown* and *Dauntless*. An earlier proposal for *Ceres* (April 1943) envisaged something like what was done to HMS *Danae*; all torpedo tubes, the after controls, the 3in guns and the single pompoms (and some boats) would be landed, and single 4in gun with barrage director, two quadruple pompoms (with directors and radars), and two twin Oerlikons added, as well as a 44in searchlight and a D/F office. Asdic and radar Types 291 and 271M (or 276) would have been installed. This project presumably paralleled the 'D' class reconstructions, and it died with them. Estimated deep displacement of the modernised ship was 5,458 tons. According to the official ship data book (CB 1815), *Capetown* retained her two single pompoms. Both ships had fourteen single Oerlikons in 1945.
6. The three surviving anti-aircraft cruisers in 1944–5, *Caledon*, *Colombo*

and *Delhi*, had (or were planned for) a major radar refit in which the normally land-based GCI (ground-controlled intercept) radar as well as the new Type 277 (in place of the existing Type 273) were installed. Ships also had army-type single hand-worked Bofors installed with their predictors. For *Caledon* these were to be additional to her existing battery, for *Colombo* in place of two twin Bofors Hazemeyers, and for *Delhi* in addition to her existing battery. Calculations were dated 10 July 1944. In January 1945 calculations were made on rearrangement of close-range anti-aircraft weapons on board *Colombo*: two twin power Oerlikons were removed, and two single Bofors guns added on the forward superstructure deck and two on the after superstructure deck. In 1945 both *Colombo* and *Caledon* were credited with four single Bofors in addition to their other armament (four twin Bofors and, respectively, four and six twin Oerlikons and two and three singles). Note that the data card for *Caledon* shows her with (presumably an initial AA armament) of one twin and two single Bofors, and four twin and three single Oerlikons. It seems unlikely that the radar upgrade was carried out. *Delhi* apparently did not receive the single Bofors.

The constructors' notes suggest that most or all modifications made when ships were refitted in the United States were analysed and approved not by constructors on the staff of the British Admiralty Delegation in Washington, but rather by the same Section 9 constructors (particularly Neal) responsible for analysing changes planned in the United Kingdom. These notes also show that the British constructors were unaware of changes made to ships in Australia and in New Zealand, because when considering further changes to ships (e.g., HMAS *Hobart*) they refer not to the most recent modifications but to the ship's state in 1939.

N/A indicates data not available.

1. 'Town' Class

CLASS	BRISTOL (LIVERPOOL)	DARTMOUTH
LBP	430-0	430-0
LOA	453-0	453-0
BEAM	47-0	48-6
HULL DEPTH	26-7⅝	26-10½
FREEBOARD AMIDS	11-9	N/A
DRAFT		
FWD	14-0⅝	N/A
AFT	16-3⅛	15-6 mean
MEAN DEEP	15-3	N/A
NORMAL	4,825	5,275
SHAFTS	4	4
SHP	21,500	22,000
SPEED NORMAL	25	25
COAL	1,353 capacity	1,290 capacity
OIL	256 capacity	269 capacity
ENDURANCE	5,830/10	5,610/10
COMPLEMENT	310	534 in 1922
WEIGHTS:		
HULL	1,910 (2,104.3)	2,150.75
MACHINERY	1,115 (1,014.57)	1,057.58
ARMAMENT	225 (283.20)	309.59
EQUIPMENT	245 (327.50)	333.34
PROTECTION	420 (501.73)	481.93
FUEL	600 (570.0)	800
RFW	23.60	44.11
NORMAL	4,700 (4,825)	5,275
ARMAMENT:		
GUNS		
6 IN	2 x 6in/50 (150)	8 x 6in/45 (150)
4 IN	10 (150)	N/A
MAXIM MG	4 (5,000)	4
TT	2 x 18in (7)	2 x 2in (7)
PROTECTION		
BELT	N/A	N/A
DECK	2in – ¾in	2in – ¾in

LIVERPOOL and DARTMOUTH weights are from Vickers light cruiser book, recording actual weights.

CLASS	CHATHAM	BIRMINGHAM	CHESTER
LBP	430-0	430-0	430-0
LOA	457-0	457-0	456-6
BEAM	49-0	49-10	49-10
HULL DEPTH	26-0	26-0	N/A
FREEBOARD			
FWD	21-7	21-11 as built	22-1
AMIDS	10-7	18-10	19-0
AFT	12-4	12-8	12-10
DRAFT			
FWD	14-9	14-10	15-3 mean
AFT	16-9	16-10	N/A
MEAN (DEEP)	18-0	17-4 (17-5)	N/A
NORMAL	5,400	5,440 (5,372)	5,185
SHAFTS	4	4	4
SHP	25,000	25,000 (27,900)	31,000
SPEED NORMAL	25.5	25 (26.17)	26.5
COAL	1,240	1,270 (1,164)	N/A
OIL	260	230 (224)	1,161
ENDURANCE	4,460/10 (1922)	4,540/10	N/A
COMPLEMENT	390	395 (548 in 1922, flag)	317
WEIGHTS:			
HULL	2,330	2,297 (2,914 with armour)	2,460
MACHINERY	1,075	1,075 (1,048)	1,165
ARMAMENT	335	358 (361)	370
EQUIPMENT	300	300 (299)	145
PROTECTION	660	680	520
FUEL	650	680	525
BOARD MARGIN	50	50	N/A
NORMAL	5,400	5,440 (5,372)	5,185
ARMAMENT:			
GUNS			
6 IN	8 (150)	9 (200)	N/A
5.5 IN	N/A	N/A	10 (250)
MAXIM MG	4 (5,000)	4	N/A
TT	2 (7)	2	6
PROTECTION			
BELT	3in	3in	3in
DECK	⅜in	⅜in	⅜in

NOTES: *Chatham* weights are from the *Melbourne* Cover. *Birmingham* weights as designed and as completed are from the Cover for *Birkenhead* and *Chester*. *Lowestoft* (*Birmingham* Class) IX data 18 Jun 1914: light ship 4,564 tons, deep load 6,228 tons.

2. ARETHUSA Class and Successors

CLASS	ARETHUSA (PENELOPE)	CAROLINE/CALLIOPE
LBP	410-0	420-0
LWL		
LOA	436-0	446-0
BEAM	39-0	41-6 ext
HULL DEPTH	24-7 mld	24-7 mld
FREEBOARD		
FWD	21-0	22-0
AMIDS	10-6	11-6
AFT	11-6	12-6
DRAFT		
FWD	13-0	12-6
AFT	14-0	14-6
MEAN (DEEP)	15-7	15-6
NORMAL	3,512	3,750
SHAFTS		
SHP	40,000	40,000
SPEED NORMAL	29.5	28.5
OIL	840 capacity	175
ENDURANCE	1,400/half-power	1,650/10 (1922)
COMPLEMENT	270	368 (1922)
WEIGHTS:		
HULL	1,860 (1,606.08)	1,860
MACHINERY	875 (908.6)	875
ARMAMENT	198 (123.27)	232
AMMUNITION	(74)	N/A
EQUIPMENT	253 (245.43)	253
PROTECTION	333 (301.14)	333
FUEL	210 (240)	175
RFW	21 (22.48)	22
NORMAL	3,750 (3,521)	3,750
ARMAMENT:		
GUNS		
6 IN	2 (150)	2 (150)
4 IN	8 (300)	8 (300)
MAXIM (MG)	1 (8,000)	1 (8,000)
TT	2 x II (10)	2 (7)
PROTECTION		
BELT	3in	3in
DECK	1in	1in

NOTE: Weights for *Arethusa* and *Caroline* are Vickers Barrow detailed estimates for the 1914/15 cruisers. Weights in parentheses for *Arethusa* are actual weights of HMS *Penelope* (from Vickers file). *Aurora* IX 19 Oct 1914: light ship 3,196 tons, legend displacement 3,587 tons, deep load 4,204 tons. *Caroline* IX 26 Nov 1914: light ship 3,486.5 tons, deep load 4,630.95 tons, which the constructor involved considered very high in view of the design figures.

CLASS	CENTAUR
LBP	420-0
LOA	446-0
BEAM	42-0
FREEBOARD	
FWD	22-0
AMIDS	11-6
AFT	12-6
DRAFT	
FWD	12-6
AFT	14-6
MEAN DEEP	15-6
NORMAL	3,750
SHAFTS	
SHP	40,000
SPEED NORMAL	30
OIL	900
ENDURANCE	1,500/half-power
COMPLEMENT	322
WEIGHTS:	
HULL	1,800
MACHINERY	834
ARMAMENT	180
EQUIPMENT	221
PROTECTION	335
FUEL	300
BOARD MARGIN	20
NORMAL	3,750
ARMAMENT:	
GUNS	
6 IN	5 (150)
4 IN	N/A
3 PDR AA	2 (300)
MAXIM MG	1 (8,000)
TT	2
PROTECTION	
BELT	3in
DECK	1in

NOTE: *Centaur* data are taken from d'Eyncourt Notebook in NMM.

DATA LIST

CLASS	CALYPSO (CASSANDRA)	CERES (CURLEW)
LBP	425-0	425-0
LOA	450-0	450-0
BEAM	42-3 mld	43-0
HULL DEPTH	24-7 mld	24-7
DRAFT MEAN	14-4¾	14-5¼
DEEP MEAN	18-9	N/A
NORMAL	4,238	4,250
DEEP	4,910.85	4,939.3
SHAFTS	2	2
SHP	40,000	40,000
SPEED NORMAL	29	29
OIL	312 capacity	313
ENDURANCE	2,050/half-power	2,100/half-power
COMPLEMENT	438 (1922)	432 (1922)
WEIGHTS:		
HULL	1,845 (2,659.51)	2,062.64
MACHINERY	848 (896.23)	897.19
ARMAMENT	243	217.68
EQUIPMENT	246	305.57
PROTECTION	320	313.93
FUEL	365	313
RFW	30 (28)	30
NORMAL	3,897 (4,238)	4,250
ARMAMENT:		
GUNS		
6 IN	5 (150)	5 (204)
3 IN	2 HA (300)	2 HA (252)
3 PDR AA	N/A	2 (2pdr)
MAXIM MG	1 (8,000)	1
TT	4 x II	4 x II
PROTECTION		
BELT	3in	3in
DECK	1in	1in

NOTE: *Cassandra* designed weights are from Vickers data book on light cruisers, presumably are as built. Actual weights are from Vickers Weight Book No. 1. *Curlew* data are from Vickers Weight Book No. 1. In addition to oil, ships carried culinary coal, e.g. 15.65 tons in *Curlew*. Armament is from the *Calypso* class Covers (armament statement for *Calypso*). In 1918, the 3in rounds per gun aboard *Ceres* class cruisers were: 160 time-fused for HA, 40 night tracers (anti-Zeppelin), 100 anti-submarine (common), and 52 practice. At that time Rear Admiral Light Cruisers wanted to eliminate the anti-Zeppelin rounds and the practice rounds, adding 74 star shells. The weight saved would go into 6in rounds. Instead of the existing 113 CPC, 66 HE, 9 shrapnel, and 14 practice, he wanted 113 CPC and 99 HE.

CALEDON Class IX data:

	Light Ship	Half-Oil	Deep	Standard
CARADOC 27 Feb 42	4,098	4,856	5,323	4,112

(with 116 tons permanent ballast and 22 tons of 6in ammunition at the guns)

	Light Ship	Half-Oil	Deep	Standard
CALEDON 19 Dec 43	4,143	4,868	5,322	(as AA cruiser)

CERES and later 'C' Class IX data:

	Light Ship	Half-Oil	Deep	Standard
CARLISLE 15 Jan 40	4,175	N/A	N/A	N/A
COVENTRY 27 Apr 40	4,288	N/A	N/A	N/A (RDF fitted)
COLOMBO 27 Feb 43	4,192	4,950	5,408	N/A

Caledon and *Colombo* were converted into anti-aircraft cruisers, and in June 1944 they and *Delhi* were the only surviving operational ships of that type. Both were earmarked to receive Mk III barrage directors (the only other old cruisers so designated were the modernised *Danae* and *Dragon*).

CLASS	DANAE (DESPATCH)	EMERALD
LBP	445-0	535-0
LOA	472-6	570-0
BEAM	46-9	54-6
FREEBOARD		
FWD	29-0	30-0
AMIDS	11-9	14-6
AFT	12-9	17-0
DRAFT		
FWD	13-9	15-0
AFT	15-9	18-0
MEAN (DEEP)	16-11	18-6
NORMAL	4,970	7,550
SHAFTS	2	4
SHP	40,000	80,000
SPEED NORMAL	29	33
SPEED DEEP		About 32
OIL	1,065 capacity	1,600
ENDURANCE	2,300/half power	
COMPLEMENT	27/427	34/560 (*Enterprise*)
WEIGHTS:		
HULL	2,620	3,820
MACHINERY	1,030	1,590
ARMAMENT	320	355
EQUIPMENT	300	360
PROTECTION	400	700
FUEL	300	650
NORMAL	4,970	7,550
ARMAMENT:		
GUNS		
6 IN	6 (200)	7 (240/215)
4 IN HA	2 (130)	2 (200)
POMPOM	2 x I (800)	3 x I
MAXIM MG	2 (5,000)	1
TT	4 x III	4 x III
PROTECTION		
BELT	3in	1in
DECK	1in (magazines)	1in

NOTES: Data for *Despatch* are from a Legend prepared in 1921 (ADM 1/9235). The enclosed 6in gun in *Diomede* added 15 tons to armament weight. 'E' Class data are from a Legend dated 9 May 1918 (ADM 1/9223). It indicates 240 rounds for each of the two foremost 6in guns, and 215 for each of the others.

IX Data for 'D' Class:

	Light Ship	Half-Oil	Deep	Standard
DRAGON	4,250	5,075	5,678	N/A
DURBAN 8 Oct 21	4,430	5,150	5,760	N/A
DESPATCH 20 May 22	4,535	5,275	5,885	N/A
DURBAN 12 Nov 27	4,705	5,420	6,035	N/A (95 tons ballast)
DRAGON 18 Jan 30	4,492	5,239	5,853	N/A
DESPATCH 5 Feb 32	4,588	5,416	5,946	N/A
DELHI 21 Sep 26	4,668	5,420	6,030	N/A
DAUNTLESS 24 Jun 39	4,552	5,324	5,793	N/A (47 tons semi-permanent ballast)
DELHI 10 Jan 42 (as rearmed with 5in guns)	4,676	N/A	6,051	N/A
DESPATCH 11 Jul 42	4,739	5,529	6,055	N/A (permanent ballast)
DURBAN 13 Aug 42 (shows addition of 195 tons of permanent ballast after this IX)	4,895	5,687	6,222	N/A (95 tons ballast)
DELHI 25 Mar 43 (shows addition of 200 tons of permanent ballast after IX; figures also given as 4,979, 5,832 and 6,392 tons)	4,904	5,738	6,279	N/A
DIOMEDE 16 May 43 (after refit as training ship, No. 1 gun mount replaced with standard 6in gun; consideration was given to installing FV1/Type 91 at this time)	4,881	5,667	6,190	N/A
DANAE 6 Jun 43	4,679	5,463	5,996	N/A (as modernised)
DRAGON 8 Aug 43	4,674	5,448	5,969	N/A
DAUNTLESS 12 Sep 43	4,620	5,811	5,836	N/A

NOTES: After *Delhi* was ordered rearmed with US 5in guns, alternative schemes of rearmament for other 'D' class cruisers were developed about April 1941: one with five twin 4in (500 rounds per gun), two quadruple pompoms (with directors), and four quadruple 0.5in machine guns; and one with four twin 4.5in guns. Four twin 4.5in plus two HA directors came to 130 tons, five US 5in plus two Mk 37 came to 113 tons, and five 4in twin plus two HA directors were 82 tons. Plans for radar modernisation of *Delhi* (Type 293 forward, Type 281B on the mainmast, Type 91 jammer) were developed about June 1944, but apparently never executed.

Initial details of the 'D' class modernisation, which ultimately applied only to *Danae* and to *Dragon*, were settled at a conference on 10 June 1942. Plans initially called for landing Nos. 3 and 4 6in guns (but apparently not the torpedo tubes) as well as the three single 4in HA guns. *Danae* would retain her original 6in director, but later ships might have new DCTs (none was converted in this form). The twin 4in mount would be hand-worked, controlled by a barrage director. The ships would all have two quadruple pompoms and four twin Oerlikons (P&S on the fore side of the bridge and aft). Planned radars were Type 290 on the foremast, Type 273M, two Type 282 for the pompoms, one Type 283 for the 4in gun (no fire-control radar for the 6in guns), a Type 251 beacon, and a Type 252 IFF transponder. One Duplex AV rangefinder would replace the bridge rangefinders. Compensation would include 50 tons of ballast. In fact one 6in gun (No. 3) and the torpedo tubes and 4in HA guns were landed. During a 28 April – 7 May 1944 refit *Danae* received missile-jamming equipment: two Type 650 masts and aerials, a US-supplied CXFR aerial on her mainmast (with transmitter in her Type 282 office), and Type 655. She also received countermeasures receivers, Types QD and QH3. In July 1944 an abortive study examined further improvement of the ship's light anti-aircraft by removing another 6in gun and adding either a quadruple pompom or four twin Oerlikon Mk V at superstructure level, or four singles on the superstructure and two on the upper deck.

The Polish Navy wanted to modify *Dragon*; one Constructor's Notebook contains a note from DNC to the ship's new commanding officer (20 January 1943) reminding him that it had already been found necessary to fit the ship with 200 tons of ballast, and that more could not be added because freeboard was already a minimum. This answered a letter from a Captain Rymenowicz of the Polish Navy headquarters, which wished to replace No. 4 6in gun with a US-type twin Bofors, as that was the only suitable midships position. The Engineering Captain of the Polish Navy wanted to add more twin Oerlikons and, apparently, a second twin Bofors. These letters seem to have led to the decision to modernise *Dragon* to match *Danae*. As inclined in August 1943, *Dragon* had a twin 4in gun with a 4in barrage director aft. Her light battery amounted to five twin and four single Oerlikons: twins P&S on the superstructure abreast the bridge, P&S on pompom seats (until pompoms are available), one single P&S on the pompom director seats (until pompoms were fitted), singles P&S abreast the Type 273 radar on the superstructure level, and one twin Oerlikon on the quarterdeck with a blast screen. Two new diesel generators had been installed to carry the additional electrical load. Torpedo tubes had been landed. The ship had Type 291 radar and Type 86 short-range radio. The inclining experiment report does not indicate whether the 6in gun had been removed, but that must have been done to free the space for the pompoms. The report does indicate that Type 91 was planned, but not yet installed at this refit.

As inclined in September 1943, *Dauntless* had had her torpedo tubes removed, as well as her single Mk II* pompoms. She had eight single Oerlikons: P&S on the lower bridge, P&S forward of the fore funnel, P&S abaft No. 4 6in gun on the superstructure deck, and P&S on the quarterdeck. Single Oerlikons also replaced the after 4in HA gun and the two single pompoms. This was part way towards the planned upgrade. Radars were Types 271 and 291.

BRITISH CRUISERS

3. HAWKINS Class

CLASS	RALEIGH
LBP	565-0
LOA	605-0
BEAM	65-0
FREEBOARD	
FWD	24-9
AMIDS	20-9
AFT	15-3
DRAFT	
FWD	16-3
AFT	18-3
MEAN DEEP	19-3
NORMAL	9,750
SHAFTS	4
SHP	60,000
SPEED NORMAL	30
COAL	800
OIL	1,580
ENDURANCE	5,640/10
COMPLEMENT	37/672 (*Frobisher*)
WEIGHTS:	
HULL	4,900
MACHINERY	1,950
ARMAMENT	560
EQUIPMENT	430
PROTECTION	810
FUEL	1,000
BOARD MARGIN 100	
NORMAL	9,750
ARMAMENT:	
GUNS	
7.5 IN	7 (150)
12 PDR QF	6 (300)
12 PDR HA	4 (300)
TT	7
PROTECTION	
BELT	3in

NOTE: A calculation of weights to go off when *Frobisher* was converted into a hulk (1947) to serve the HMS *Defiance* establishment gives her armament as a training ship: three 7.5in guns (forecastle deck forward, superstructure deck [on a platform], and upper deck aft), one 6in P.Mk IX ('B' position), one 4in HA Mk III (upper deck aft), five Oerlikons (four on the superstructure deck, one on the upper deck aft), and one quadruple Mk VIII torpedo tube.

4. 'County' Class

CLASS	KENT	LONDON
LBP	590-0	595-0
LOA	630-0	630-0
BEAM	68-4 ext	66-0
FREEBOARD		
FWD	33-0	32-0
AMIDS	27-3	26-6
AFT	29-0	28-0
DRAFT		
FWD	15-3	16-0
AFT	17-3	18-0
MEAN DEEP	20-6	21-3
STANDARD	10,000	9,840
SHAFTS	4	4
SHP	80,000	80,000
SPEED STD	31.5 light	32.25
SPEED DEEP	30.5	31.25
OIL	3,424	3,222
ENDURANCE		9,900/12
COMPLEMENT	50/736	48/745
WEIGHTS:		
HULL	5,600	5,450 (5,190)
MACHINERY	1,830	1,826 (1,730)
ARMAMENT	1,000	1,004 (1,245)
EQUIPMENT	570	570 (640)
PROTECTION	1,000	960 (940)
STANDARD	10,000	9,840 (9,745)
FUEL	3,424	3,222
RFW	180	165
EQUIPMENT	100	100
ARMAMENT	106	111
DEEP LOAD	13,810	13,438
ARMAMENT:		
GUNS		
8 IN	4 x II (130)	4 x II (150)
4 IN	4 x I (200)	4 x I (100)
POMPOM	2 Mk M (1,000)	2 Mk M (1,000)
TT	2 x IV (9)	2 x IV (9)
PROTECTION		
BELT	Magazines Only	1in
DECK	1⅜in	1⅜in

NOTES: Mk M pompom was the octuple; it was included in the design and in the Legend weights, but was not ready in time for installation. Some data are from the DNC history. Figures in parentheses for *London* are estimated weights based on material weighed as it was placed on board, from Day Constructor's Notebook 4 (June 1928).

IX data for KENT Class:

	Light Ship	Half-Oil	Deep	Standard
KENT as designed	10,000	N/A	13,752	N/A
KENT 16 May 28	9,555	N/A	13,520	N/A
KENT 4 Jun 38	10,239	12,499	14,197	N/A
BERWICK 15 Oct 38	10,356	12,619	14,297	N/A
SUFFOLK 9 Feb 41	10,608	12,858	14,486	N/A
SUFFOLK 27 Mar 40			14,556	N/A
CUMBERLAND 16 Apr 42	10,590	13,052	14,672	N/A
SUFFOLK 9 Feb 41	10,608.5	12,858	14,486	11,014
BERWICK 15 Aug 42	10,798	13,236	14,911	N/A
KENT 18 Oct 42	10,696	13,023	10,696	N/A
CUMBERLAND 1951	9,417	11,530	12,890	N/A
AUSTRALIA 2 Jul 45	10,240	12,565	14,253	N/A

NOTES: *Berwick*: Plans for removal of 'X' turret (February 1944) envisaged adding two pairs of Mk VII (quadruple) pompoms, a Type 284 radar for the after 8in director, and modernising the ship's other radars (installing Types 277 and 293). This was not done.

Cumberland: plans for removal of 'X' turret (January 1944) envisaged alternatives with and without installation of an AIC and Mk VI directors. The approved scheme (which was not carried out) did not include this additional improvement, but it did include radar modernisation (Types 277 and 293, and a second Type 284 for the after 8in director). This plan called for adding four more Mk VI octuple pompoms (total six) and five more twin Oerlikons (total ten), at the cost of one single Oerlikon (total reduced to nine). When the ship was under his command in August 1945, C-in-C East Indies proposed adding four single Bofors, presumably as part of the anti-Kamikaze programme: one each P&S on the former catapult deck in place of two single Oerlikons, and one each P&S on the quarterdeck abreast the Captain's Hatch.

Kent: The proposed removal of 'X' turret (November 1943) was initially simply to add light anti-aircraft weapons, including four quadruple pompoms. However, schemes for installing an AIC were proposed in January 1944. Detailed lists of alterations (for the turret removal) were made in May 1944, including shortening her funnels (a feature not included in contemporary plans for *Norfolk*). Plans for the AIO were still being made that September, including ones to modernise the ship's radar (including installation of Type 277). However, nothing was done, and late in 1944 plans were made to convert *Kent* into an accommodation ship. All of her 8in guns would be removed, and eight octuple pompoms mounted. Space would be added for 80 officers and 130 ratings, with a ward room and crews' mess on the upper deck. Estimated light condition after conversion was 10,229 tons.

Suffolk: Early calculations for the refit to remove 'X' turret (which was never carried out) were dated January 1944; plans then included an AIC and Mk VI directors. At this time the ship had two octuple (MK VII) pompoms, five twin and four single Oerlikons (she had only three singles in April 1945). It appeared that she could be fitted with six quadruple pompoms, fourteen twin and four single Oerlikons. A more refined proposal (April 1945) called for four octuple pompoms (Mk VI) and eight twin and four single Oerlikons. However, at the same time plans were drafted to convert her for trooping, her hangar converted to house 150 men.

HMAS *Australia*: After being damaged at Lingayen Gulf, HMAS *Australia* was patched in Sydney and then sent to Devonport for refit. At Sydney her 'X' turret was landed (she was the only *Kent* class cruiser to have her 'X' turret removed). With the turret were landed two single Oerlikons on top and two single Bofors on the after controls. Her two cranes had been landed, boat derricks being installed. Her funnels had all been shortened by 5ft. As inclined in the UK in July 1945 the ship had two octuple pompoms (Mk VI), one abreast the after control on each side (but no pompom directors on board), one single Bofors on the centreline of her quarterdeck (with seatings for two more, which were not fitted), and eight single Oerlikons (P&S on the upper deck fore side of the forward superstructure, on sponsons P&S at the fore end of 'B' gun deck, P&S on 'B' turret, and P&S on the upper deck at the after end of the after superstructure). Both of her 8in directors carried US-supplied FC radars. She also had a Type 273 surface-search set, a two-aerial Type 281, a US-supplied SG surface-search set on her foremast, and Type 285 on her 4in directors. The British constructors offered two possibilities. One was to add four more Mk VI pompoms (total six). Her earlier nine single Bofors could be restored, and two of her four single Oerlikons landed (apparently the British initially confused single Oerlikons with single Bofors in her reports). However, this idea was soon superseded by plans to install a close-range battery of the original two pompoms plus three quadruple and two twin Bofors and four single Oerlikons. The ship would have a quadruple Bofors on her quarterdeck and two others abreast her bridge, and two twins in place of 'X' turret. A later proposal added four single Bofors (two abreast the catapult structure and two between the 4in guns). In addition, an AIO would be installed, and the ship's radars modernised (fitting Types 277 and 281BQ and three barrage directors). Estimated light ship displacement was 10,639 tons.

IX data for LONDON Class:

	Light Ship	Half-Oil	Deep	Standard
LONDON as designed	9,340	N/A	13,438	N/A
SUSSEX 1 Mar 29	9,500	N/A	13,220	N/A
LONDON 12 Feb 41	10,820	13,007	14,578	11,014
LONDON Mar 43	10,877	N/A	N/A	N/A
SUSSEX 4 Aug 42	10,560	12,722	14,282	N/A
SHROPSHIRE 13 Jun 43	10,577	12,757	14,263	N/A
DEVONSHIRE 12 Mar 44	10,673	12,917	14,450	N/A
SUSSEX 11 Mar 45	10,556	12,760	14,273	N/A
DEVONSHIRE 1 Apr 47	9,421	11,772	13,304	N/A

(as training ship)

NOTES: As rebuilt, *London* was given 3in armour over her side and her fan chambers. The planned reconstruction of *Sussex*, stopped by the outbreak of war, would have involved 4½in side armour and 3½in over her fan chambers.

Devonshire: In 1943 DG of AAW devised 'ideal AA armaments' for ships being refitted. For *Devonshire* he proposed, in June 1943, to reduce the 4in battery to two single star shell guns. She could then have four octuple and the two existing quadruple pompoms if she retained 'X' turret, or eight octuples if (as was done) 'X' turret was landed. The four-turret scheme also included eight twin and two single Oerlikons (sixteen twin and two single in the three-turret scheme). Instead, all the 4in guns were retained when 'X' turret was landed. The scheme adopted at this time also initially called for fifteen twin and no single Oerlikons (this was not the final scheme). Presumably the retention of the 4in guns was part of the reversal of the view that anything larger than a pompom was useless for anti-aircraft fire. All schemes included removal of the ship's torpedo tubes. Existing quadruple 0.5in machine guns were removed at this time. A further rejected scheme envisaged replacing the 8in director tower with Oerlikons. It was rejected because it involved too much work, and because it added only one twin Oerlikon. *Devonshire* did have 'X' turret removed during a 1943–4 refit. Planned improvements included raising the 4in directors by 7ft. The ship's close-range battery was somewhat reduced to compensate for fitting an admiral's plot on the signal deck. As inclined in March 1944, *Devonshire* had had two diesel generators added on her main deck. Her two tactical rangefinders had been landed to save weight. She had six Mk VII (quadruple) pompoms (P&S on the superstructure abreast the forward boiler room vent, P&S on the superstructure abreast the after controls, and P&S in way of the former 'X' turret); twin Oerlikons P&S forward of the lower bridge, P&S raised 9ft 6in above the superstructure abreast the forward boiler room vent, P&S on the superstructure abreast the after end of the bridge, P&S raised 9ft 6in above the superstructure abreast the centre funnel, P&S between the 4in guns raised 9ft 9in above the deck, P&S on the upper deck on the fore side of the 27ft whalers, P&S in way of 'X' turret raised 6ft 9in above the superstructure, and on the quarterdeck; and singles P&S on 'B' and 'Y' turrets, P&S on the superstructure deck abreast the fore end of the bridge, P&S on the superstructure abreast the blacksmith shop, and P&S in the former barrage director positions aft. Pompom directors were P&S abreast the fore funnel over Type 282 offices, P&S on the after control, and P&S forward of 'X' turret position. The barrage directors were P&S on the lower bridge and on the centreline forward of 'X' turret position. Radars were Types 277, 293, 281B, 282, 283, 284 and 285. According to the IX report, Type 277 office structure but not the aerial had been fitted; the ship still had her Type 273 (which was to be removed before the refit was complete). The main mast carried FV1/Type 91, and there was a D/F coil on the fore side of the director platform. In October 1946, *Devonshire* was credited with one single Bofors, plus four quadruple pompoms, one twin power Oerlikon and one single Oerlikon. That may reflect the beginning of stripping for her training role (in April she had no Bofors, but fourteen twin power Oerlikons and four singles).

London: Also in 1943, the effect of removing 'X' turret from *London* was calculated as part of a larger modernisation including installation of an AIC (Ops Room, Radar Display Room, Aircraft Plot and Target Indication Room, and air-conditioning plant [3 tons]). and Mk 6 directors. 'X' turret would have been replaced by a quadruple pompom and two twin superimposed Oerlikons, plus a barrage director. Two more quadruple pompoms would have been added between the funnels. Additional Oerlikons would have been mounted. None of this was done, but the project was apparently revived in October 1944. By that time it included removal of the ship's torpedo tubes. It still included the AIC, but apparently not the Mk 6 directors. This version envisaged installation of four octuple pompoms: two in place of 'X' turret and two abreast the after 8in DCT (the ship already had two such mounts, so she would emerge with six: forty-eight barrels). In place of her existing four twin and fourteen single Oerlikons, she would have five twins and two singles (singles would be removed from the quarterdeck, from the top of 'X' turret, from forward of the mainmast (P&S), from abreast the searchlight support (P&S), from abreast the fore funnel (P&S), and from the signal deck (P&S), and two twins would be removed from P&S of the main mast). Three twins would be added: P&S in way of the former 'X' turret and on the quarterdeck centreline (plans for six others were abandoned). Radars would have been modernised, including the addition of a Type 651 missile jammer (the ship already had the FV1/Type 91 combination). At about this time it was decided to install Type 274 on the forward director. Again, nothing was done. *London* never received an AIC. At the end of the war *London* had single Bofors atop her 'B' and 'X' turrets (and two other single mounts). By 1947 plans called for landing the two turret-top single Bofors, as well as other guns. Note that in December 1943 *not* fitting the planned AIC was expected to make it possible to add ten Oerlikons. In April 1946 *London* was credited with two octuple pompoms, four single Bofors, and eight twin Oerlikons (plus four single Oerlikons by October 1947). In April 1949 she was credited with four single and four single power Bofors (presumably Boffins replacing four twin power Oerlikons), plus four twin Oerlikons. Note that she had retained her torpedo tubes.

Shropshire: The ship was inclined (for the first time since 1929) to measure her stability after a refit prior to being transferred to the Royal Australian Navy. A memorandum dated 17 December 1942 on newly-ordered changes observed that that although the stability of *Sussex* and *London* had been accepted as a wartime measure, to enable essential fighting equipment to be installed, 'it is appreciably less than would normally be accepted, and efforts are consequently being made to improve the stability of these two ships.' For *Shropshire*, the largest improvement would be achieved by eliminating the aircraft and catapult, the next largest being shortening of each funnel by 6ft. However, the Australian Navy Board wanted to retain the catapult. Some, but hardly all, of the difference could be made up by landing the torpedo tubes. In the end, the aircraft and catapult had to be landed (the torpedo tubes remained). As inclined in June 1943, *Shropshire* had had her aircraft and catapult landed. She had two barrage directors abreast her bridge and two more aft, one per 8in turret. Her foremast carried Types 281 and 291 radars, an odd combination. She had a single Oerlikon on 'B' turret, twins P&S on the 4in deck level, two twins abreast her fore funnel, two twins on the centreline on the fore side of her Type 273 radar, singles P&S on the aft side of the pompoms, a single on 'X' turret, a twin on the quarterdeck (with a blast screen), and three single portable Oerlikons. She had Type 132 Asdic and the FV1 radar countermeasures system.

Sussex: In September 1943 the ship asked for more Oerlikons. DNC found enough weight to replace the ten singles on board with twins, but could not allow anything else. The main source of weight compensation was the aircraft and catapult, others including the two 12ft rangefinders on the bridge and the Type 273 'lantern' (to be replaced by a Type 277 radar). In December, removal of 'X' turret was approved, freeing more space and weight. DNC prepared various alternative schemes in May 1944. The ship then had two octuple (Mk VI) pompoms. One possibility was to add four quadruple pompoms (Mk VII); in addition the ship would have thirteen twin and six single Oerlikons. Another was to use octuples in place of the four new quadruple mounts (total of six), with seven twin and no single Oerlikons. Yet another possibility was to add two octuple and two quadruple pompoms (total of four octuples), with thirteen twin and five single Oerlikons. In each case the torpedo tubes were landed as compensation. If the torpedo tubes were retained, the ship could have the six multiple pompoms and eleven twin and two single Oerlikons. The big refit included radar modernisation and addition of missile countermeasures (Type 651 and the US CXFR) plus Loran (fitted to many British cruisers late in 1944). On completion of her refit the ship was inclined (11 March 1945). Her 'X' turret and torpedo tubes had all been removed, and the petrol tank for her aircraft removed through her side. Light armament was six Mk VI (octuple) pompoms (RPC, with directors), four twin and six single Oerlikons. The pompoms were P&S abreast her fore funnel, abreast her main mast, and on her superstructure deck in way of 'X' turret position. Twin Oerlikons were P&S on the superstructure abreast the bridge and on the superstructure (on supports) abreast the centre funnel. The singles were P&S on 'B' turret, P&S on the lower bridge level on supports on the fore side of the bridge (they could be replaced by twins when available), and P&S on supports forward of 'X' gun position (they too could be replaced by twins when available). Radars were Types 281B (mainmast, with Type 941 interrogator above it), Type 277 (with Type 242 interrogator), Type 293 (with Type 242 interrogator), Type 251M beacon (on foremast, with equipment in Type 293 office), and Type 243Q on the cross-trees. Countermeasures were Type 91 on the mainmast and CXFR (in Type 91 office). The list did not mention the Type 651 missile jammer. The ship had a D/F coil on the fore side of the bridge, and a Y (radio intercept) office in the bridge at 'B' gun deck level. Note that *Sussex* retained her torpedo tubes.

CLASS	NORFOLK	NORTHUMBERLAND
LBP	595-0	570-0
LOA	633-0	600-0
BEAM	66-0	64-0
FREEBOARD		
FWD	32-3	30-6
AMIDS	26-6	24-0
AFT	28-3	17-6
DRAFT		
FWD	16-0	17-0
AFT	18-0	19-0
MEAN DEEP	21-0	21-6
STANDARD		9,975
SHAFTS	4	4
SHP	80,000	60,000
SPEED STD	32.25	30
SPEED DEEP	31.25	29
OIL	N/A	2,200
ENDURANCE	9,300/16	N/A
COMPLEMENT	57/672 (*Dorsetshire*)	N/A
WEIGHTS:		
HULL	N/A	4,700
MACHINERY	N/A	1,435
ARMAMENT	N/A	1,200
EQUIPMENT	N/A	625
PROTECTION	N/A	1,900
MARGIN	N/A	125
STANDARD	10,000	10,000
FUEL	N/A	2,450
RFW	N/A	146
MACHINERY	N/A	–
EQUIPMENT	N/A	80
ARMAMENT	N/A	128
DEEP	N/A	12,664
ARMAMENT:		
GUNS		
8 IN	4 x II (150)	4 x II (100)
4 IN	4 x I (200)	4 x I (200)
POMPOM	2 Mk M (1,000)	2 Mk M (1,000)
TT	2 x IV (9)	2 x IV (9)
PROTECTION		
BELT	1in (machinery)	5½in on ½in
DECK	1⅜in	2¼in on ½in

Because *Norfolk* was a modified *London*, no Legend was drawn up for Board approval. Data are given here for comparison with the abortive *Northumberland*, the 1928 'A' Cruiser.

IX Data:

	Light Ship	Half-Oil	Deep	Standard
NORFOLK 22 May 43	10,944	13,086	14,600	N/A

Removal of 'X' turret was approved in January 1944. At this time the ship had two Mk VI pompoms (octuple), no twin Oerlikons, and thirteen singles (one atop 'B' turret, two atop 'X' turret, singles P&S of the fore funnel, two P&S of the after funnel, a single to port abreast the crane, P&S aft near the Type 273 radar, and one on the quarterdeck). Plans included installing AIC. Alternatives were to replace the ship's Mk VI pompoms with RPC Mk VIIs, four more being added (P&S abreast the middle funnel and aft in way of 'X' turret); or to retain the Mk VIs and add the four Mk VIIs, which was favoured. In that case the ship could also have seven twin and two single Oerlikons or, differently arranged, eleven twins and four singles. If no AIC were installed, the ship could gain six Oerlikons (two in the bridge structure, two twins forward of the fore funnel, one twin on the catapult platform to port, and one twin on the upper deck aft, to starboard). In November 1945 further radar modernisation (Type 281BQ instead of Type 281B, Type 277P instead of Type 277, Type 293P instead of Type 293) and the addition of ten single Bofors was proposed, the cost being eleven twin and nine single Oerlikons. In April 1946 *Norfolk* was credited with six quadruple pompoms, nine single Bofors, and eleven twin Oerlikons. She had the full AIC. *Norfolk* apparently retained her torpedo tubes, but they were gone by October 1947.

CLASS	YORK	EXETER
LBP	540-0	540-0
LOA	575-0	575-0
BEAM	57-0	58-0
FREEBOARD		
FWD	30-6	29-9
AMIDS	15-0	15-0
AFT	17-6	16-9
DRAFT		
FWD	16-0	16-0
AFT	18-0	18-0
MEAN(DEEP)	20-6	20-4
STD	8,418	8,621
SHAFTS	4	4
SHP	80,000	80,000
SPEED STD	32.25	32.25
SPEED DEEP	31.25	31.5
OIL	1,900	1,923
ENDURANCE	N/A	7,850/12
COMPLEMENT	43/589	44/480
WEIGHTS:		
HULL	4,954	4,338
MACHINERY	1,755	1,770
ARMAMENT	901	990
EQUIPMENT	491	523
PROTECTION	1,017	1,020
STANDARD	8,418	8,621
FUEL	1,969	1,923
RFW	165	165
EQUIPMENT	64	56
ARMAMENT	33	33
DEEP	10,649	10,798
ARMAMENT:		
GUNS		
8 IN	3 x II (150)	3 x II (150)
4 IN	4 x I (200)	4 x I (200)
POMPOM	2 x octuple (1,600)	2 x octuple (1,500)
TT	2 x III	2 x III
PROTECTION		
BELT	3in	3in
DECK	1½in	1½in

NOTE: *York* weights are as finally designed. *Exeter* weights are from the official DNC history (table on page 118). These are not the ones in the approved Legend, dated 25 February 1928, subsequently considerably revised. Neither ship was completed with the planned octuple pompoms. IX data:

	Light Ship	Half-Oil	Deep	Standard
EXETER 1 Mar 41	8,850	10,360	11,300	9,009

Exeter IX is after her post-River Plate major repair.

5. LEANDER Class

CLASS	LEANDER
LBP	522-0
LOA	554-3
BEAM	55-2
FREEBOARD	
FWD	29-10
AMIDS	15-4
AFT	17-10
DRAFT	
FWD	15-2
AFT	17-2
MEAN DEEP	19-11
STANDARD	7,448
DEEP	9,452
SHAFTS	4
SHP	72,000
SPEED STD	32.5
SPEED DEEP	31
OIL	1,745
ENDURANCE	8,000/12
COMPLEMENT	42/550 (later ships 46/562)
WEIGHTS:	
HULL	3,875
MACHINERY	1,504
ARMAMENT	731
EQUIPMENT	507
PROTECTION	871
STANDARD	
FUEL	1,785
RFW	134
EQUIPMENT	63
ARMAMENT	22
DEEP	9,452
ARMAMENT:	
GUNS	
6 IN	4 x II (200)
4 IN	4 x I (200)
0.5 IN MG	3 x IV (2,500)
TT	2 x IV
PROTECTION	
BELT	3in on 1in
DECK	1¼in

NOTES: IX Data:

	Light Ship	Half-Oil	Deep	Standard
LEANDER as designed	7,448	N/A	9,452	N/A
LEANDER 18 May 33	7,080	8,455	9,350	N/A
ACHILLES 30 Sep 33	6,907	8,250	9,140	N/A
LEANDER 24 Apr 37	7,352	8,701	9,559	N/A
ORION 17 Jul 37	7,286	8,614	9,462	N/A
AJAX 13 Jul 40	7,259	8,626	9,512	N/A
AJAX 23 Oct 42	7,379	8,768	9,653	N/A
LEANDER 5 Aug 43	7,556	8,909	9,769	N/A
ACHILLES 21 May 44	7,478	8,865	9,738	N/A
LEANDER 31 Dec 45	N/A	8,909	9,770	
LEANDER 15 Jul 46	7,541	8,948	9,809	N/A
ACHILLES Apr 47	7,495	N/A	N/A	N/A

Note that *Achilles* light GM was 2.74ft in 1933, but only 1.62ft in 1944.

Achilles was modernised in 1943–4, her single 4in guns finally replaced by twins (with two sided HADT instead of the earlier centreline one), and four pompoms installed. 'X' turret was landed. She emerged with five twin and six single Oerlikons and modern radars. Hence the May 1944 IX. Fitting of Type 651 missile jammers was approved in September 1944; two 36in searchlights were removed from abreast the funnel as compensation. In September 1945 *Achilles* reported that during a refit in March–April 1945, while part of the British Pacific Fleet, she had landed her torpedo tubes and added five Bofors Mk III (single hand-worked US guns). The single Oerlikons P&S on the superstructure abreast the bridge were moved to the 4in blast screens, and their position extended to take two single Bofors on each side abreast the bridge (the fifth was on the quarterdeck). The torpedo parting space was temporarily converted to a mess deck for the additional complement required by the increased armament and by radar personnel. Apparently this action had *not* been vetted by DNC. DNC replied that the Bofors had to be surrendered when the torpedo tubes were restored. However, *Achilles* was an RNZN ship, and in March 1946 the New Zealand Navy Board proposed to compensate for the restored torpedo tubes by removing only one Bofors and two twin Oerlikons from the forecastle. The issue became moot when the ship was returned to the Royal Navy in exchange for a new *Dido*. As inclined in May 1944, *Achilles* had had her torpedo tubes and 'X' turret landed. She had four Mk VII pompoms (P&S abreast the empty catapult position, and P&S in way of 'X' turret); twin Oerlikons P&S on the forecastle abreast her bridge, P&S on the catapult platform, and P&S on the roof of the officers' galley, and on 'Y' turret; and singles P&S on 'B' gun deck and on the quarterdeck. Two 4in HADT had replaced the former single one. Barrage directors had been fitted on the centreline on 'B' gun deck abaft 'B' turret and on the centreline aft between the after pair of pompom directors over the look-out position. Radars were Types 276, 293 and 281B, plus gunnery sets. She also had FV1/Type 91, a US-type TBS radio, and British Types 86 and 87. A fighter control plotting office, radio telephone office, and chart house had been built on the lower bridge, and the roof on the upper bridge had been removed. In April 1946 *Achilles*, on loan to the RNZN, was credited with the five single Bofors, plus seven twin and four single Oerlikons – and her torpedo tubes.

Ajax: The British Admiralty Delegation in Washington requested approval of the planned rearmament on 25 March 1943. Pompoms would be replaced by two US-type quadruple Bofors. DNC approved fitting four twin and one single Oerlikons, working from that point to a total of three quadruple Bofors, eleven twin and two single Oerlikons. That involved removing the single Oerlikon on the after controls as well as the two quadruple pompoms (replaced by Bofors). A third quad would be mounted abaft the funnel. Oerlikon positions were: two twins on the quarterdeck, twins on 'B' and 'X' turrets, a twin on the after controls, two singles on the upper deck, a twin either side abreast the catapult platform, and two twins on the superstructure deck forward. Boats, cranes and torpedo tubes would have been removed as compensation. In July 1945 it was approved to fit two RPC quad pompoms in place of the Bofors, together with British directors.

Leander was modernised in the United States, her 'X' turret landed. Like the others, she received sided 4in directors instead of the earlier centreline HADT. Catapults, rangefinders, and quadruple 0.5in machine guns were all removed. She was expected to emerge with two quadruple pompoms, six twin Oerlikons, three barrage directors, and modernised radars (including Types 277, 293 and 281B, and the Type 91 jammer). Compensation items were her torpedo tubes and wood from her upper deck forward of her torpedo tubes. In March 1945 it was proposed to replace two twin Oerlikons Mk V with single power 2pdrs (Mk XVI), offering a net saving of 3.5 tons and heavier anti-aircraft projectiles. However, as plans evolved, she was expected to receive US-type quadruple Bofors. They apparently were not on board when she returned to the United Kingdom, to the point where in November 1945 it was proposed that twin Bofors be fitted instead. When the ship was inclined in July 1946, she had on board twin Bofors P&S abreast her after controls (with directors), plus a single Bofors P&S on 'X' gun deck and a single Bofors on a pedestal on the centreline of that deck, plus twin Oerlikons P&S on the forecastle deck abreast the bridge, and twin Oerlikons P&S abreast the after control. The ship had barrage directors on the fore side of the bridge and on the after superstructure over the control position. Radars were Types 277, 276 (on the foremast), and 79B, with Type 91/FV1 on the mainmast, plus the usual gunnery types (282, 283, 274 and 285). Aerials for the short-range Types 85M and 87 radios were on the W/T yard of the mainmast. Note that in October 1945 the official ship data list (CB 1815) ascribed two quadruple Bofors and three twin power and four single Oerlikons to *Leander*. A year later she was credited with the battery listed above.

5. Modified LEANDER Class/SYDNEY Class

NOTE: Legend data from Legend dated 27 October 1932 for 1932 Programme cruiser and from slightly later calculations for the Board. Armament includes 55 tons for seaplane, catapult and crane. Machinery weight is 1,310 tons of E-in-C weight plus 90 tons of refrigerating machinery, dynamo, and engineers' lubricating oil.

CLASS	SYDNEY
LBP	530-0
LOA	562-0
BEAM	56-8
FREEBOARD	
FWD	29-9
AMIDS	15-8
AFT	17-2
DRAFT	
FWD	15-3
AFT	17-3
AT DEEP: MEAN	19-5
STANDARD	7,197
DEEP	9,330
SHAFTS	4
SHP	72,000
SPEED STD	32.5
SPEED DEEP	31
OIL	1,837
ENDURANCE	7,000/16
COMPLEMENT	46/586
WEIGHTS:	
HULL	3,679
MACHINERY	1,393
ARMAMENT	746
EQUIPMENT	473
PROTECTION	906
STANDARD	7,197
FUEL	1,837
RFW	148
EQUIPMENT	73
ARMAMENT	23
DEEP	9,278
ARMAMENT:	
GUNS	
6 IN	4 x II (200)
4 IN	4 x I (200)
0.5 IN MG	4 x IV (2,500)
TT	2 x IV
PROTECTION	
BELT	4in (machinery)
DECK	2in (magazines)

NOTE: *Sydney* data from DNC official history, table on page 118 of various cruisers. Initial additions to all three ships included about 35 tons of protective plating. As inclined at Garden Island in November 1941, *Perth* displaced 7,209 tons light (7,080 tons in 1939) and 9,763 tons deep; GM was reduced from 2.2ft to 1.08ft. She was assigned 175 tons of permanent ballast, but shipped only 18 tons before being sunk. *Hobart* did ship the 175 tons (early 1942, estimated deep load 9,908 tons). As modernised *Hobart* displaced 9,420 tons (deep). This is official RAN information, not from RN Constructors' Notebooks.

7. ARETHUSA Class

CLASS	ARETHUSA
LBP	480-0
LOA	506-0
BEAM	51-0
FREEBOARD	
FWD	27-9
AMIDS	14-3
AFT	15-3
DRAFT	
FWD	13-3
AFT	15-3
MEAN DEEP	17-10
STANDARD	5,419
DEEP	6,896
SHAFTS	4
SHP	64,000
SPEED STD	32.25
SPEED DEEP	30.75
OIL	1,127
ENDURANCE	6,500/16
COMPLEMENT	500
WEIGHTS:	
HULL	2,581
MACHINERY	1,221
ARMAMENT	563
EQUIPMENT	493
PROTECTION	633
STANDARD	
FUEL	1,327
RFW	116
EQUIPMENT	16
ARMAMENT	18
DEEP	6,896
ARMAMENT:	
GUNS	
6 IN	3 x II (200)
4 IN	4 x I (150)
0.5 IN MG	2 x IV (2,500)
TT	2 x III
PROTECTION	
BELT	2¼in
DECK	1in

DATA LIST

8. SOUTHAMPTON Class

NOTE: Second series of figures for *Belfast* are as deduced from inclining experiment, according to a comparative table of cruiser data in Wood Constructor's Notebook 6 dated 25 March 1947.

CLASS	SOUTHAMPTON	GLOUCESTER	BELFAST
LBP	558-0	558-0	579-0
LWL	N/A	N/A	606-0
LOA	591-6	591-6	613-6
BEAM	61-8	62-4	63-4
FREEBOARD			
FWD	30-10	30-7	31-0
AMIDS	16-1	15-10	16-3
AFT	18-7	18-4	18-9
DRAFT			
FWD	16-2	16-5	16-3
AFT	18-2	18-5	18-3
MEAN (DEEP)			
STANDARD	8,947	9,400	10,069 (10,565)
SHAFTS	4	4	4
SHP	75,000	82,500	80,000
SPEED STD	32	32.3	32.25
SPEED DEEP	30.5	30.8	31
OIL	1,943	1,970	2,256 (Legend 2,400)
ENDURANCE			
COMPLEMENT	790	790	700 (Flag)
WEIGHTS:			
HULL	4,282		4,734 (5,138)
MACHINERY	1,492	1,570	1,498 (1,498)
ARMAMENT	1,162	1,205	1,366 (1,456)
EQUIPMENT	580	675	610 (612)
PROTECTION	1,431	1,480	1,861 (1,861)
STANDARD	8,947	9,400	10,069 (10,565)
FUEL	1,943	1,970	2,256 (2,256)
RFW	148	N/A	148 (150)
MACHINERY	N/A	N/A	9
EQUIPMENT	104	N/A	131 (149)
ARMAMENT	51	N/A	59 (59)
DEEP	11,193	N/A	12,672 (13,179)
ARMAMENT:			
GUNS			
6 IN	4 x III (200)	4 x III (200)	4 x III (200)
4 IN	4x II (200)	4 x II (200)	6 x II (200)
POMPOM	2 x IV (1,800)	2 x IV (1,800)	2 x VIII (1,800)
0.5in MG	2 x IV (2,500)	2 x IV (2,500)	2 x IV (2,500)
TT	2 x III	2 x III	2 x III
PROTECTION			
BELT	4½in	4½in	4½in
DECK	1¼in	1¼in	2 in and 3in

NOTES: Comparative displacements of *Southampton* and *Gloucester* classes:

	Light	Half-Oil	Deep	Standard
GLASGOW 20 May 45	9,696	11,404	N/A	9,902
SHEFFIELD Feb 46	9,630	11,344	12,404	N/A
BIRMINGHAM 16 Mar 48	9,464	11,136	12,174	N/A
BIRMINGHAM 19 Apr 52	9,504	11,187	12,234	N/A
NEWCASTLE 29 Feb 52	9,267	10,967	12,024	N/A
BELFAST Class:				
EDINBURGH 26 Feb 42	10,710	12,590	13,720	10,890
BELFAST Apr 45 (as modernised)	11,919	13,776	14,934	12,111

The proposal to modernise the anti-aircraft armament of *Liverpool* (September 1943) called for much the same work as on *Birmingham*, but with US quadruple Bofors guns in place of both 'X' turret (two quads) and the pompoms atop the hangar roof.

9. DIDO Class

CLASS	DIDO	BELLONA
LBP	485-0	485-0
LWL	N/A	506-0
LOA	512-0	512-0
BEAM	50-6	50-6
FREEBOARD		
FWD	24-1	27-6
AMIDS	11-5	13-6
AFT	12-7	14-1
DRAFT		
FWD	12-11	13-3
AFT	13-6	16-2
MEAN DEEP	16-10	N/A
STANDARD	5,450 (5,521)	5,770
SHAFTS	4	4
SHP	62,000	62,000
SPEED STD	32.25	32
SPEED DEEP	30.75	30.5
OIL	1,105	1,110
ENDURANCE	5,500/16	5,500/16
COMPLEMENT	487	508
WEIGHTS:		
HULL	2,521	2,645
MACHINERY	1,146	1,165
ARMAMENT	730	670
EQUIPMENT	406	430
PROTECTION	718	860
STANDARD	5,521	5,770
FUEL	1,105	1,110
RFW	99	N/A
MACHINERY	23	N/A
EQUIPMENT	39	N/A
ARMAMENT	49	N/A
DEEP	6,836	N/A
ARMAMENT:		
GUNS		
5.25 IN	5 x II (360)	4 x II (340)
4 IN	N/A	1 x I star shell (150)
POMPOM	2 x IV (1,800)	2 x IV (1,800)
0.5 IN MG	2 x IV (2,500)	2 x IV (2,500)
TT	2 x III	2 x III
PROTECTION		
BELT	3in	3in
DECK	1in	1in

NOTE: *Bellona* data are from a Legend dated 31 August 1940.

IX Data:

	Light Ship	Half-Oil	Deep	Standard
BONAVENTURE 27 Apr 40	5,430	6,385	6,940	5,530
CLEOPATRA 24 Feb 45	5,925	6,870	7,424	8,346
DIDO 30 Mar 46	5,903	6,862	7,418	6,020
SIRIUS post-war	5,785	6,750	7,925	N/A
SPARTAN (calculated)	5,455.5	N/A	7,206.8	N/A

10. FIJI ('Colony') Class

Weights in parentheses are for the ship as built (weighted by weight group).

CLASS	FIJI	UGANDA
LBP	538-0	538-0
LWL	550-0	550-0
LOA	555-6	555-6
BEAM	62-0	62-0
FREEBOARD		
FWD	30-0	29-6
AMIDS	15-6	15-0
AFT	17-6	17-0
DRAFT		
FWD	15-6	16-0
AFT	17-6	18-0
MEAN DEEP	N/A	20-0
STANDARD	8,253 (8,630)	8,640
SHAFTS	4	4
SHP	80,000	80,000
SPEED STD	32.25	32.25
SPEED DEEP	31.25	31.25
OIL	1,700	1,700
ENDURANCE	8,000/16	8,000/16
COMPLEMENT	738 (Flag)	754 (Squadron Flag)
WEIGHTS:		
HULL	3,819 (4,130)	4,190
MACHINERY	1,450 (1,430)	1,450
ARMAMENT	1,190 (1,220)	1,130
EQUIPMENT	550 (550)	550
PROTECTION	1,289 (1,300)	1,410
STANDARD	8,253 (8,630)	8,640
FUEL	1,700 (1,700)	1,700
RFW	128 (128)	N/A
MACHINERY	8 (0)	N/A
EQUIPMENT	93 (93)	N/A
ARMAMENT	172 (169)	N/A
DEEP:	10,354 (10,720)	N/A
ARMAMENT		
GUNS		
6 IN	4 x III (150)	3 x III (150)
4 IN	4 x II (150)	6 x II (150)
POMPOM	2 x IV (1,200)	2 x IV (1,200)
0.5 IN MG	2 x IV (2,500)	2 x IV (2,500)
TT	Space reserved	
PROTECTION		
BELT	3½in	3½in
DECK	2in	2in

NOTE: *Fiji* draft and freeboard are from an 11 October 1937 Legend for a standard displacement of 8,170 tons, later increased. Fuel capacity and deep load for speed exclude reserve fuel. SHP is for temperate climates. In tropics, shp is reduced to 72,500 and speeds to 31.5kts and 30.5kts, respectively. *Uganda* data are from a Legend for a Modified *Fiji* dated August 1940, <u>not</u> for the completed ship. Notes show stowage for 200 rounds per 6in, 250 per 4in, and 1,800 per 2pdr, but not at the Legend displacement. In *Uganda*, armament weight includes 160 tons for aircraft equipment.

IX Data

	Light Ship	Half-Oil	Deep	Standard
KENYA 23 Aug 40	8,464	9,823	10,673	8,580
MAURITIUS 17 Nov 40	8,526	9,886	10,736	8,642
GAMBIA 2 Jan 42	8,735	10,167	11,017	8,346
Modified FIJI Class:				
NEWFOUNDLAND 29 Oct 44	8,812	10,158	11,008	8,909
CEYLON 12 Jan 46	8,902	10,261	11,107	9,036
CEYLON Dec 48	8,781	10,140	10,991	8,912
NEWFOUNDLAND 29 Nov 52 (after large repair)	8,862	10,244	11,086	N/A

In April 1946, *Ceylon* had four single Bofors and *Newfoundland* had one quadruple Bofors mount. Both had three quadruple pompoms and five (*Newfoundland* six) twin power Oerlikons, plus two singles. Presumably the Bofors were Pacific War additions (but both ships retained their torpedo tubes).

For the 1950–2 large repair of *Newfoundland*, the approved light armament was two twin and one single Bofors on each side atop the hangar, one twin Bofors on the centreline amidships, and two twin Bofors P&S at the after end of the forecastle deck. All had external directors: two STD forward, one amidships and two aft. This battery was adopted in favour of one including the self-contained STAAG mounting (and retaining the existing Mk 4 directors for the 4in guns). The net saving on the close-range battery was 41 tons, but the adoption of two Mk 6 directors in place of the earlier three Mk 4s cost a net 13 tons. Estimated displacement, to be compared to the 1952 data above: light ship 8,957 tons, half-oil 10,314 tons, and deep 11,160 tons.

11. Improved FIJI Classes

Weights in parentheses are deduced from inclining experiment data.

CLASS	ONTARIO (rebuilt)	SUPERB	TIGER
LBP	538-0	538-0	538-0
LWL	550-0	550-0	550-0
LOA	555-6	555-6	555-6
BEAM	63-0	64-0	64-0
FREEBOARD			
FWD	N/A	N/A	24-1¼
AMIDS	N/A	N/A	10-5 ⅜
AFT	N/A	N/A	12- 5½
DRAFT	(DEEP)		
FWD	20-8	20-7	20-10¾ mean
AFT	21-8	21-7	N/A
STANDARD	8,878 (9,218)	8,960 (9,272)	9,330 (light)
SHAFTS	4	4	4
SHP	80,000	80,000	80,000
SPEED STD	31.8	31.5	N/A
SPEED DEEP	30.25 clean	30.25 clean	30.25 temperate
	28.25 tropics 6 mos.	28.25 tropics 6 mos.	N/A
OIL	1,851 (70 coal)	1,893 (48 coal)	2,015
ENDURANCE	6,800/15	6,800/15	3,750/20
COMPLEMENT			839 (Flag)
WEIGHTS:			
HULL	4,553 (4,774)	4,576 (4,751)	4,883
MACHINERY	1,431 (1,431)	1,454 (1,454)	1,494
ARMAMENT	991 (992)	1,015 (1,012)	1,101
RADAR/RADIO	N/A	N/A	1,10
EQUIPMENT	558 (716)	560 (700)	799
PROTECTION	1,345 (1,345)	1,355 (1,355)	1,370
STANDARD	8,878 (9,218)	8,960 (9,272)	9,757
FUEL	1,902 (1,920)	1,902 (1,940)	2,015
RFW	123 (123)	123 (123)	N/A
MACHINERY	8 (8)	8 (8)	N/A
EQUIPMENT	73 (64)	73 (75)	N/A
ARMAMENT	143 (148)	143 (146)	N/A
DEEP	11,127 (11,481)	11,209 (11,564)	11,900
ARMAMENT:			
GUNS			
6 IN	3 x III	3 x III	2 x II Mk 26 (400)
4 IN	5 x II	5 x II	N/A
3 IN	N/A	N/A	3 x II Mk 6 (851)
POMPOM	4 x IV	4 x IV	N/A
	4 x I power	2 x I power	N/A
40 MM	N/A	4 x I	3 x II (1,500)
20 MM	8 x II, 6 x I	8 x II, 2 x I	N/A
TT	2 x III	2 x III	1 x IV if weight permits
PROTECTION			
BELT	3in	3in	3in
DECK	2in	2in	2in

NOTE: *Tiger* figures are from September 1954 estimates (ADM 1/25800); information in the Cover is not complete enough to provide later hence more accurate figures. The twin 40mm and the torpedo tubes were never fitted.

IX Data:

	Light Ship	Half-Oil	Deep	Standard
ONTARIO 14 Apr 45	9,009	10,521	11,481	N/A
SWIFTSURE 1944	8,769	10,380	11,240	N/A

12. Wartime Designs

CLASS	1940 Cruiser	1941 Cruiser
LWL	720-0	690-0
BEAM	84-0	82-0
DRAFT MEAN	24-0	21-0
HULL DEPTH	N/A	45-0
STANDARD	21,500	18,740
SHAFTS	4	4
SHP	160,000	120,000
SPEED STD	33.5	31.5
SPEED DEEP	32.25	30.5 clean
OIL	2,900	
ENDURANCE	9,000/16	6,000/24
WEIGHTS:		
HULL	9,200	8,100
MACHINERY	5,100	2,350
ARMAMENT	2,560	2,540 (170 Aircraft)
EQUIPMENT	9,10	740
PROTECTION	5,730	4,660
BOARD MARGIN		350
STANDARD	21,500	18,740
ARMAMENT		
GUNS		
8 IN	12 x III (200) OR 3 x III 9.2in (200)	3 x III (200)
4.5 IN	6 x II HA/LA (400)	N/A
4 IN	–	8 x II (250 HA, 100 LA)
POMPOM	2 x VIII	5 x VIII (1,800)
0.5 IN MG	2 x IV	Max 20MM
PROTECTION		
BELT	7in	4½in (machinery)
DECK	2¾in (machinery)	2in

NOTE: No Legend was produced for the 1941 cruiser. Weights were taken from W G John Constructor's Notebook 13; they represent the design as in March 1942, when it died. Some figures reflect Staff Requirements circulated in November 1941. Dimensions are mainly from requests for horse-power passed to the model basin. Through 1941 overall length was 670ft (656ft between perpendiculars) and beam was 80ft. Initially the secondary battery was four twin 4.5in, but the late version of the design showed the eight twin 4in above. Data are uncertain because the design kept evolving. W G John was in charge of cruiser design at the time. Data on the 1940 design are from a comparative table circulated in 1940, in the Cover.

CLASS	CRUISER 1944: DESIGN N.2	DESIGN M.1
LBP		
LWL	550-0	520-0
LOA		
BEAM	64-0	55-0
FREEBOARD		
FWD	29-3	26-9
AMIDS	15-3	12-9
AFT	17-3	14-3
DRAFT		
FWD	15-9	15-3
AFT	17-9	17-3
MEAN DEEP	20-6	19-6
STANDARD	8,650	7,150
SHAFTS	4	4
SHP	48,000	44,000
SPEED STD	29	29
SPEED DEEP	28	28
OIL	2,300	1,700
ENDURANCE	7,700/18 home waters 6 out	6,200/18
COMPLEMENT	ca. 675 (Flag)	ca. 600 (Flag)

DATA LIST

WEIGHTS:		
HULL	4,850	3,925
MACHINERY	1,040	1,000
ARMAMENT	970	745
EQUIPMENT	530	490
PROTECTION	1,090	830
STANDARD	8,650	7,130
ARMAMENT		
GUNS		
5.25 IN	4 x II (400)	3 x II (400)
BUSTERS (TWIN 40)	8 (1,440)	6 (1,440)
20MM (POWER)	12 x II (2,400)	8 x II (2,400)
TT	2 x IV (9)	2 x IV (9)
PROTECTION		
BELT	3in	3in
DECK	2in	2in

NOTE: Data are from an undated Legend in Pt 3 of the 1943 papers of the First Sea Lord (ADM 205/29). Recalling his unhappy experience when the 1914 *Arethusa* failed to make her required speed, DNC doubted that these slow cruisers would be successful.

CLASS	NEPTUNE	MINOTAUR
LWL	655-0	645-0
BEAM	76-0	75-0
FREEBOARD		
FWD	32-6	
AMIDS	23-0	
AFT	17-9	
DRAFT		
FWD	21-0	
AFT	21-0	
MEAN DEEP	24-6	
STANDARD	15,350	15,280
SHAFTS	4	4
SHP	108,000	100,000
SPEED STD	32	31.5
SPEED DEEP	29.5	–
OIL	2,850	2,100
ENDURANCE	7,500/20 Clean	6,000/20
COMPLEMENT	1,050	
WEIGHTS:		
HULL	7,850	7,240
MACHINERY	2,000	2,100
ARMAMENT	2,050	2,270
EQUIPMENT	810	1,060
PROTECTION	2,340	2,480
BOARD MARGIN	300	3,05
STANDARD	15,350	15,280
FUEL	2,850	2,100
RFW	N/A	180
DEEP:	N/A	18,415
ARMAMENT		
GUNS		
6 IN	4 x III (250)	5 x II
10.5 IN	6 x II (400)	–
3 IN	–	8 x II
BUSTER	10 (twins) (1,440)	–
20 MM	16 x II (2,400)	–
TT	4 x IV (17)	4 x IV (17)
PROTECTION		
BELT	4½in	
DECK	2in	

Data on *Minotaur* are from a comparative table dated August 1947, which also showed Designs ZA and ZB and USS *Worcester*. The design was not completely worked out, so when a new cruiser design was begun in 1950 it used data compiled for the *Neptune* design.

13. The Missile Cruiser

CLASS	GW 96A
LWL	675
LOA	687
BEAM	80
DRAFT MEAN DEEP	22-0
STANDARD	15,762
SHAFTS	4
SHP	110,000
SPEED STD	–
SPEED DEEP	31 (32 deep and clean)
OIL	2,450
ENDURANCE	4,500/20
COMPLEMENT	95/1,020
WEIGHTS:	
HULL	9,050
MACHINERY	2,258
ARMAMENT	1,898
EQUIPMENT	1,218
PROTECTION	1,244
STANDARD	15,762
FUEL	2,450
RFW	104
BOARD MARGIN	198
DEEP:	18,420 (changed to 18,450)
ARMAMENT	
SEASLUG	64
GUNS	
6 IN	2 x II (420, including 20 starshell)
3 IN	4 x II (900)
PROTECTION	
BELT	1½in (machinery)
DECK	1in

NOTE: Details have been taken from Mansbridge Constructor's Notebook 7 (which gives weights) and from the Cover. This was the final design before cancellation, presumably the one which would have been built. The armament group includes radars.

14. Fast Minelayers

Abdiel data from Legend dated 18 November 1938.

CLASS	ADVENTURE	ABDIEL
LBP	495-0	400-0
LWL		410-0
LOA	516-0	417-11
BEAM	58-0	39-0 (40-0 as built)
FREEBOARD		
FWD	30-3	24-0
AMIDS	18-11	18-0
AFT	19-11	19-3
DRAFT (STANDARD)		
FWD	14-0	10-0
AFT	16-0	12-0
DRAFT (DEEP)		
MEAN	17-6	13-8
STANDARD	6,850 (Legend)	2,650
SHAFTS		2

SHP	40,000	72,000
SPEED STD	28 (Legend)	39.75 (later given as 40.2)
SPEED DEEP		About 36 (later given as 35.2)
OIL	1,500 capacity	700
ENDURANCE	5,500/15	
COMPLEMENT		240 (est)
WEIGHTS:		
HULL	4,300	1,270 (1,324)
MACHINERY	1,200	950 (928)
ARMAMENT	140	130 (139)
MINES/MINING EQT	320	215 (257)
EQUIPMENT	320	100 (200)
PROTECTION	–	5
STANDARD		2,650
FUEL	500	749
RFW	N/A	46
EQUIPMENT	N/A	N/A
ARMAMENT	N/A	N/A
DEEP	6,850	3,643
ARMAMENT:		
GUNS		
4.7 IN HA	4 (200)	N/A
4 IN	–	3 x II (250)
POMPOM	1 x VIII (1,800)	1 x IV (1,800)
0.5 IN MG	–	2 x IV (2,500)
MINES	300	100 Mk XIV (150 emergency)

NOTES: *Adventure* weights and dimensions are from a Legend dated March 1921 (ADM 1/9228); unfortunately the Cover is missing. This version included the diesel engine but also included the planned (but not fitted) octuple pompom. Displacement given is the Legend or Navy List figure, since this Legend was constructed before the Washington Treaty. Load displacement includes a 70-ton Board Margin. As IX September 1940, she displaced 6,650 tons standard (mean draft 13ft 10¾in). The major items fitted since that time were SA gear (to protect against acoustic mines), Asdic, radar (Types 251, 271 and 286PQ [likely to be replaced by 79BM]), a D/F office on the signal deck, and protection to vital internal communications. New approved additions as of 18 March 1943 included twin Oerlikons Mk V (three on each side), RPC for the multiple pompoms, Asdic 128T, and an aircraft plotting office. A proposed rearmament (July 1943) envisaged replacing No. 4 single 4.7in gun with a quadruple RPC pompom. Another proposal called for replacement of each single 4.7in gun by a twin 4in. A proposal in August 1943 showed three upper deck 4.5in mountings, all forward. Another proposal was to add two quadruple pompoms and to replace the octuple mount aft with a third.

Abdiel data are from a Legend dated 1938. The figures in parentheses are from the DNC history, and are probably final design data. IX data for *Ariadne*, a repeat *Abdiel*: hull 1,475.42 tons, machinery 919.34 tons, armament 123.93 tons (141.96 tons in deep condition), equipment 123.40 tons (195.42 tons in deep condition), and mines etc 155.43 tons (234.63 tons in deep condition). As inclined on completion, standard displacement was 2,805.62 tons and full load was 2,963.23 tons. Brinton 5/6 Constructor's Notebooks include details of alternative designs for a new fast minelayer (1943). Design A would have displaced 3,572 tons standard (5,026 tons deep). Design 'B' (August 1943) would have displaced 3,458 tons standard and 4,901 tons deep, with two twin 4.5in between-decks Mk IV HA/LA (400 rounds/gun), four Buster (two centreline, two sided), eight twin Oerlikons (one at each end on centreline, three each side). Design C sacrificed power by adopting unit machinery (54,000 shp, with the same hull form and the same armament: 410ft x 44ft x 29ft x 11.5ft std/15.1ft deep). In August 1943 there was a further armament study (three upper deck twin 4.5in instead of two between-decks mounts, with two forward and one aft in place of the two after Bofors and the quarterdeck Oerlikon atop the after crew shelter). Owing to the height of 'B' gun, the bridge had to be raised. Another alternative was to add short rails aft for thirty more mines. These studies continued into November 1943.

LIST OF SHIPS

Ships are listed in order of programme and then alphabetically. Pennant numbers are not given, because they were never painted up before the NATO era, except for anti-aircraft cruisers. For these ships pennant numbers are given.

The first column of dates gives date of laying down with date of launch below it. The second column of dates gives the date of completion. The third date is the date of decommissioning, where known. I have counted the date of paying off into reserve as the decommissioning date, for pre-Second World War cruisers (ships in reserve had nucleus crews, and the date of paying off for disposal was generally different). Where known, a date of recommissioning is given under the date of completion. Decommissioning for refit generally is not counted.

Old light cruisers listed as recommissioning in 1939 were generally ships in reserve brought forward with reservist crews for the Royal review of the Reserve Fleet at Weymouth on 9 August 1939, then retained in commission on mobilisation that month.

Note: wartime fates of British- and Commonwealth-built ships under British operational control during the Second World War (ORP, RAN) are given, but not details of non-British built ships (Royal Netherlands Navy and Free French Navy) under British operational control.

Abbreviations

BU	Broken up (date arrived for scrapping)
C	Collision (peacetime)
Canc	Cancelled
Coll	Collision
Conv	Conversion (converted)
CLAA	Anti-air conversion; date is that of completion
C&M	Care and Maintenance
Cpl	Completed
CTL	Constructive Total Loss (write-off, but not sunk); cause in parentheses as in war losses
IN	Indian Navy (initially Royal Indian Navy)
ORP	Polish Navy Ship
PN	Pennant Number
RA	Rear Admiral
RAN	Royal Australian Navy
RCN	Royal Canadian Navy
RNZN	Royal New Zealand Navy
RPN	Royal Pakistan Navy
Reb	Rebuilt
Ren	Renamed
SNO	Senior Naval Officer
STT	Ship Target Trials post-1945
TS	Training Ship
VA	Vice Admiral
WL	War Loss
WL(A)	War Loss (Abandoned due to damage)
WL(B)	War Loss (Bombing)
WL(G)	War Loss (Gunfire)
WL(M)	War Loss (Mining)
WL(T)	War Loss (Torpedo)

Yards

Armstrong	Armstrong or W G Armstrong Whitworth (Elswick)
Beardmore	W M Beardmore (Dalmuir, near Glasgow)
Brown	John Brown and Co (Clydebank, near Glasgow)
Cammell	Cammell Laird
Chatham	Chatham (Royal) Dockyard
Clydebank	J and G Thompson, Clydebank (Glasgow), later John Brown
Cockatoo	Cockatoo Dockyard, Sydney
Denny	Wm Denny and Bros (Dumbarton)

Devonport	Devonport (Royal) Dockyard
Fairfield	Fairfield Shipbuilding and Engineering Co (Govan)
H&W	Harland & Wolff (Belfast)
Hawthorn	R&W Hawthorn Leslie (Hebburn)
L&G	London and Glasgow Shipbuilding Co
Palmer	Palmers Shipbuilding and Engineering Co (Hebburn-on-Tyne)
Parsons	Parsons Marine Steam Turbine Co (hulls subcontracted)
Pembroke	Pembroke (Royal) Dockyard
Scott	Scott's Shipbuilding and Engineering Co Ltd (Greenock)
Sheerness	Sheerness (Royal) Dockyard
Stephen	Alex Stephen and Engineering Co Ltd (Greenock)
Swan Hunter	Swan, Hunter, & Wigham Richardson Ltd (Wallsend)
Thornycroft	John I Thornycroft & Co Ltd (Woolston)
Vickers	Vickers (Barrow), originally Naval Construction and Armaments Co, later Vickers-Armstrong (Barrow)
VA Walker	Vickers Armstrong Walker yard
VA Tyne	Vickers Armstrong Tyne yard
White	J Samuel White & Co Ltd (Cowes)

'Town' Class

1908/9 Estimates (BRISTOL Class)

Ship / Builder	Laid down / Launched	Completed	Fate date	Notes
BRISTOL	23 Mar 09	17 Dec 10	Jun 19	Sale list May 20, sold for BU May 21
Brown	23 Feb 10			
GLASGOW	25 Mar 09	19 Sep 10	May 20	Stoker TS (disarmed Jan 25) 21–6, sale list Mar 26, sold for BU Apr 27
Fairfield	30 Sep 09			
GLOUCESTER	15 Apr 09	Oct 10	Apr 19	Sale list Mar 20, sold for BU May 21
Beardmore	28 Oct 09			
LIVERPOOL	17 Feb 09	4 Oct 10	Jun 19	Sale list Mar 20, sold for BU May 21
Vickers	30 Oct 09			
NEWCASTLE	14 Apr 09	20 Sep 10	Feb 20	Sale list Dec 20, sold for BU May 21
Armstrong	25 Nov 09			

1909/10 Estimates (DARTMOUTH Class)

Ship / Builder	Laid down / Launched	Completed	Fate date	Notes
DARTMOUTH	19 Feb 10	Oct 11	Jun 21	Torpedoed by UC-25 on 15 May 17, recomm after repair Mar 19, refit Sep 24–Sep 26, Flag of Reserve Fleet Portsmouth Apr 27–28. Temp accommodation hulk for DEFIANCE I Jan 30, paid off May 30, sold for BU Dec 30
Vickers	13 Feb 11	Sep 26	Jun 29	
FALMOUTH	21 Feb 10	Sep 11		WL(T) 20 Aug 16
Beardmore	20 Sep 10			
WEYMOUTH	19 Jan 10	Oct 11	Jul 21	Damaged by Austrian U-28 on 2 Oct 18, paid off Malta Jun 19, refit for service Mar 20. Flag of VA Nore Reserve Dec 25. Sold for BU Oct 28
Armstrong	18 Nov 10	Dec 25	Sep 27	
YARMOUTH	27 Jan 10	Apr 12	Oct 28	Nore Reserve Dec 20, Signals School Portsmouth 1922–4, Refit Dec 24–25, trooping 1925–6, Signals School 27, Flag of RA Submarines Falmouth Apr–Oct 28, sale list Nov 28, sold for BU Jul 29
L&G	12 Apr 11			

1910/11 Estimates (CHATHAM Class)

Ship / Builder	Laid down / Launched	Completed	Fate date	Notes
CHATHAM	3 Jan 11	4 Dec 12	1918	RNZN 11 Sep 20, ret to RN 1924, sold for BU Jul 26
Chatham	9 Nov 11	1920	Nov 25	
DUBLIN	3 Jan 11	Mar 13	1924	Sold for BU Jul 26
Beardmore	9 Nov 11			
SOUTHAMPTON	6 Apr 11	Nov 12	Aug 24	Sale list Aug 25, sold for BU Jul 26
Brown	16 May 12			

RAN CHATHAM Class

Ship / Builder	Laid down / Launched	Completed	Fate date	Notes
BRISBANE	Jan 13	Nov 16	1929	Recomm May 35 – 14 Sep 35 to carry crew to UK for SYDNEY; sold Jun 36
Cockatoo	30 Sep 15			
MELBOURNE	14 Apr 11	Jan 13	23 Apr 28	Crew to HMAS AUSTRALIA, sold Dec 28
Chatham	30 May 12			
SYDNEY	11 Feb 11	26 Jun 13	8 May 28	BU 1929, tripod remains in Sydney harbour
L&G	29 Aug 12			

1911/12 Estimates (BIRMINGHAM Class)
NOTE: Orders for LOWESTOFT and NOTTINGHAM were considerably delayed in an attempt to prop up Thames Iron Works by ordering from it; the firm went bankrupt, and the ships were ordered from other builders.

Ship / Builder				Notes
BIRMINGHAM Armstrong	10 Jun 12 7 May 13	3 Feb 14 Nov 23	1920	Sold for BU Mar 31
LOWESTOFT Chatham	29 Jul 12 28 Apr 13	21 Apr 14		Lightweight director on tripod as completed. Sold for BU Jan 31
NOTTINGHAM Pembroke	13 Jun 12 18 Apr 13	Apr 14		WL(T) 19 Aug 16

HMAS ADELAIDE

ADELAIDE Cockatoo	Jan 15 27 Jul 18	Aug 22 13 March 39	1928 26 Feb 45	Sold Jan 49

Ex-Greek Cruisers
NOTE: A 1921 attempt to sell both ships back to the Greek navy failed.

BIRKENHEAD Cammell Laird	27 Mar 13 18 Jan 15	May 15	May 20	Ex-Greek ANTINARVOS KOUNTORIOTIS. Porstmouth Reserve 1919, sold Oct 21.
CHESTER Cammell Laird	7 Oct 14 8 Dec 15	May 16	May 20	Intended Greek name may have been LAMBROS KATSONIS. Taken over before she could be launched and named. Nore Reserve 1919, sold Nov 21

ARETHUSA Class and Successors
1912/13 Estimates (ARETHUSA Class)

ARETHUSA Chatham	28 Oct 12 25 Oct 13	11 Aug 14		WL(M) 11 Feb 16
AURORA Devonport	24 Oct 12 30 Sep 13	5 Sep 14		RCN 1 Nov 20 (in comm reserve Mar 19–Aug 20; C&M Halifax 1 Jul 22, sale list 1 Jul 27, sold 1927. Fitted for minelaying May 17, laid 212 mines
GALATEA Beardmore	9 Jan 13 14 May 15	Dec 14	Mar 20	Sold Oct 21
INCONSTANT Beardmore	3 Apr 14 6 Jul 14	Jan 15	16 Feb 22	Paid off Oct 19, then with 1st Submarine Flotilla (briefly Atlantic Fleet Flagship); sold for BU Oct 24
PENELOPE Vickers	1 Feb 13 25 Aug 14	10 Dec 14	Jun 19	Sale list Jul 23, sold for BU Oct 24 Fitted as minelayer Nov 17, laid 210 mines
PHAETON Vickers	12 Mar 13 21 Oct 14	Feb 15	Feb 20	Had been paid off for 12-month refit, at end into reserve, sale list May 22, sold for BU Aug 22
ROYALIST Beardmore	3 Jun 13 14 Jan 15	Mar 15	Jan 20	Sale list May 22, sold for BU Aug 22. Fitted as minelayer Feb 17, laid 1,183 mines
UNDAUNTED Fairfield	21 Dec 12 28 Apr 14	29 Aug 14	Apr 19	Completed with upper works camouflaged. Out of reserve Feb–May 21 to carry drafts to Med Fleet; sale list Apr 22, sold for BU Apr 23

1913/14 Estimates (CAROLINE/CALLIOPE Class)

CALLIOPE Chatham	1 Jan 14 17 Dec 14	Jun 15 May 24 Sep 28	Jan 21 Jan 30	Nore Reserve 1926 but trooping and SNO ship Dec 27. Sold for BU Aug 31
CAROLINE Cammell	28 Jan 14 29 Sep 14	4 Dec 14	Feb 22	TS for Ulster Division RNVR Feb 24, extant in 2010; was admin centre for Londonderry escorts 1939–45
CARYSFORT Hawthorn	25 Feb 14 14 Nov 14	6 Jun 15 1924	Sep 23 Apr 31	Trooping 1924; flag of Devonport reserve 1927, trooping Feb–Jul 29, SNO ship Devonport Reserve Jan 30, relieved by COMUS. Sold for BU Aug 31
CHAMPION Hawthorn	9 Mar 14 29 May 15	20 Dec 15 May 25	Oct 24 Dec 33	Attached VERNON torpedo school 1919–24. Gunnery Firing Ship May 25, Signal School 1928. Sold for BU Jul 34
CLEOPATRA Cammell	26 Feb 14 14 Jan 15	1 Jun 15 1923 Jan 25	1921 1924 Dec 26 Mar 31	Comm in Nore Reserve Dec 27, SNO ship Sep 28 –Mar 31. Sold for BU Jan 31

COMUS Swan Hunter	3 Nov 13 16 Dec 14	21 May 15	May 30	SNO ship of Devonport Reserve Apr 31, fully decomm Dec 33, sold for BU Jul 34
CONQUEST Scott	3 Mar 14 20 Jan 15	1 Jun 15 Feb 22	Apr 19 Apr 28	Into reserve after refit to repair mine damage Jul 18, 1922 SNO 1st Submarine Flotilla Atlantic Fleet to Jan 27; 1928 comm reserve Portsmouth. Sold for BU Aug 30
CORDELIA Pembroke	21 Jul 13 23 Feb 14	3 Jan 15 Jan 20	Jun 19 Dec 22	Patrol off Irish coast 1920–2. Sold for BU Jul 23

1914/15 Estimates (CAMBRIAN Class)

CAMBRIAN Pembroke	8 Dec 14 3 Mar 16	May 16	Nov 29	SNO ship Nore Mar 31, fully decomm Jul 33, sold for BU Jul 34
CANTERBURY Brown	14 Oct 14 21 Dec 15	9 May 16	Dec 33	Gunnery school Portsmouth 1920–2. Nore Reserve Mar 31–Dec 33, but trooping from Dec 32. Sold for BU Jul 36
CASTOR Cammell	28 Oct 14 28 Jul 15	12 Nov 15 Jun 28	Sep 26 Jul 30	Irish Patrol 1922, then gunnery School Portsmouth 1924–5, then refit into Nore Reserve, but Trooping to China From Oct 27, recomm for China. Devonport Reserve Jul 30, paid off May 35. Sold for BU Jul 36
CONSTANCE Cammell	25 Jan 15 12 Sep 15	26 Jan 16	Mar 31	Portsmouth Reserve to Jul 35, sold for BU Jan 36

War Estimates (CENTAUR Class)

CENTAUR Vickers	24 Jan 15 6 Jan 16	Aug 16 8 Apr 25	Oct 23 Mar 32	Flag of Commodore (D) Atlantic Fleet 1925. Sale list 1933, sold for BU Feb 34 Signals School
CONCORD Vickers	1 Feb 15 1 Apr 16	18 Dec 16 May 24 Oct 28	Jul 23 Oct 27 Jan 33	Portsmouth after refit Oct–Nov 28, Sale List Nov 34, sold for BU Aug 35

War Programme (June 1915: 1915-16 Programme): CALYPSO Class

CALEDON Cammell	17 Mar 16 25 Nov 16	6 Mar 17 Jul 39	1931 Apr 45	Training duty in Reserve Fleet 31 Jul 31–Nov 32, SNO Reserve Fleet Devonport Dec 32 –23 Oct 36, trooping to Malta Dec 34, CLAA (conv Sep 42 – Dec 43. Sold 22 Jan 48 for BU; also STT
CALYPSO Hawthorn	7 Feb 16 24 Jan 17	21 Jun 17 Aug 39	1932	WL(T) 12 Jun 40
CARADOC Scott	21 Feb 16 23 Dec 16	15 Jun 17 Aug 39	17 Oct 34 Apr 44	Boys Training Ship while in reserve Devonport Aug 38–Jul 39 Base ship Colombo 1944; reduced to reserve status Devonport Dec 45. Disposal List Feb 46
CASSANDRA Vickers	Mar 16 25 Nov 16	29 Jun 17		WL(M) 5 Dec 18

Repeat Programme (March 1916)

CARDIFF Fairfield	22 Jul 16 12 Apr 17	25 Jun 17 Jul 38	14 Jul 33 3 Sep 35	Ex CAPRICE. TS Home 1940–5. Was Flag of Reserve Fleet Nore Command 1933, Flag Reserve Fleet Chatham 1934; attended Silver Jubilee Review Jul 1935 and then returned Chatham as Flag of Reserve Fleet. Also at Coronation Review for George VI, May 37. Into reserve for CLAA conversion (cancelled) Jun 38, then brought forward for Review of Reserve Fleet Aug 39, retained in commission. Converted to gunnery TS at Portsmouth, 20mm guns added, Oct 40

CERES Brown	11 Jul 16 24 Mar 17	1 Jun 17 1932 Aug 39	1931 1933 Oct 44	Accommodation ship 1944–5, sold for BU 5 Apr 46. Note Normandy duty (see CAPETOWN). May have been nominated for duty during Munich Crisis 1938. Accommodation Ship Feb 44, converted May 44 to support Normandy landing. At this time 6in main battery and pompoms removed, 20mm added; this may be when Type 290 radar replaced Type 286. Resumed use as Accommodation Ship Plymouth Aug 44. Accommodation Ship Portsmouth Dec 44 –Jan 46
COVENTRY Swan Hunter	4 Aug 16 6 Jul 17	21 Feb 18 Aug 39	1936	Ex CORSAIR. CLAA Dec 35; WL(B) 14 Sep 42. Trials ship Mar 37; may have tested 20mm Oerlikon some time between Apr and Dec 37. Not clear whether or not she was in commission 1936–9
CURACOA Pembroke	Jul 16 5 May 17	18 Feb 18		Tender to Portsmouth Gunnery School 1932–9; CLAA conv 1939–24 Jan 40, WL(C) 2 Oct 42
CURLEW Vickers	21 Aug 16 5 Jul 17	14 Dec 17		Chatham CLAA conv Jan 36, WL(B) 26 May 40 Briefly in reserve 1927, trooping.TS Chatham after CLAA trials, Aug 38– Aug 39. Some time in Jan–Jul 39 nominated instead of DUNEDIN for installation of Type 79Z radar, refit and installation completed 23 Sep 39.

Improved CALYPSO Class (1917 Orders)

CAIRO Cammell	28 Nov 17 19 Nov 18	14 Oct 19 1939	1936	Chatham CLAA conv 1938–30 Jun 39, WL(T) 12 Aug 42
CALCUTTA Vickers	18 Oct 17 9 Jul 18	Aug 19 Mar 28	Jan 27 May 31 Nov 33	Boys Training Ship Chatham Nov 33–1938, Chatham CLAA conv 1938–8 Mar 39. WL(B) 1 Jun 41
CAPETOWN Cammell	23 Feb 18 18 Jun 19	10 Apr 22 17 Jul 34 Jul 39	31 Dec 29 Sep 38 Oct 44	Accommodation ship Jan 44–1945, sold for BU 5 Apr 46. Conversion to CLAA deferred 1939 due to international situation. Taken in hand Dec 43 to fit Oerlikons and Type 273 radar. Recommissioned 17 Apr 44 and nominated Control Ship for convoys from beachhead area of Normandy. Was flag of Captain Southbound Sailings Control Ship and Depot Ship at Mulberry A. CERES was Captain Northbound Sailings. Then reduced back to Reserve as noted
CARLISLE Fairfield	2 Oct 17 9 Jul 18	16 Nov 18 Nov 39	Mar 37 1943	Ex CAWNPORE. CLAA Chatham 1939–24 Jan 40 CTL(B) 9 Oct 43, Base Ship for escorts, converted Dec 43–Mar 44, proposal to use at Aden rejected; at Alexandria 1943–5, BU Alexandria 1949
COLOMBO Fairfield	8 Dec 17 18 Dec 18	18 Jun 19 Aug 39	Dec 35 Jun 45	CLAA Devonport Aug 42–Jun 43. Sold for BU 22 Jan 48 Mobilised with Reserve Fleet personnel for Munich crisis 1938. SNO Reserve Fleet Chatham 1936. Attended Coronation Review 1937. Disposal List 1947

'D' Class (1916 Programme)

DANAE Armstrong	11 Feb 16 26 Jan 18	22 Jun 18 1936 Jul 39	1935 Dec 37	ORP CONRAD 4 Oct 44, ret 1946, sold forBU 22 Jan 48. Modernised Sep 42–Jul 43 as prototype for class. Visited Oslo Sep 45, remained under Polish control. Ret Sep 46, Disposal List 1947
DAUNTLESS Palmer	3 Jan 17 10 Apr 18	2 Dec 18 1930	1929 1935 1945	Badly damaged by grounding off Halifax 2 Jul 28. TS Home 1943–5. Sold For BU 13 Feb 46
DRAGON Scott	24 Jan 17 29 Dec 17	16 Aug 18 1939	1938	Polish Navy Jan 43. Badly damaged by Marder 8 Jun 44, blockship Normandy 9 Jun 44

LIST OF SHIPS

(1917 Programme)

Name/Builder	Laid down	Launched	Completed	Notes
DELHI Armstrong	29 Oct 17 23 Aug 18	7 Jun 19 Aug 39	Mar 38 Jun 45	CLAA New York Navy Yard May –Dec 41 (further completed at Portsmouth Apr 42. Badly damaged Split 12 Feb 45. To Sheerness, survey Chatham, permanent repair uneconomic. Disposal List 1947. NOTE: she was repaired sufficiently to steam home
DESPATCH Fairfield	8 Jul 18 24 Sep 19	30 Jun 22 Aug 39	Jan 38 Oct 44	Accommodation ship 1945, sold for BU 5 Apr 46. Major refit Portsmouth Oct 43–May 44 as Depot Ship, 6in battery, one 4in, torpedo tubes landed, sixteen single Bofors and two twin power Oerlikon added, then as HQ ship at Arromanches Mulberry, duties ended 24 Jun 44. Accommodation Ship Portsmouth Feb 45–early 1946
DIOMEDE Vickers	3 Jun 18 24 Apr 19	7 Oct 22 29 Jul 37	31 Mar 36 Oct 45	NZ Division 1925–36; crew to ACHILLES. Recomm 1937 for training. But some of time as TS In Reserve Fleet. Nominated Jan–June 38 for refit as tender to Signal School Portsmouth, Type 79Z radar to have been fitted for development trials. Refitted, but this deployment was cancelled Mar–June 39. TS ship Home 1943–5 after conv Rosyth Jul 42–Sep 43; unique A mount changed to standard type. Torpedo tubes removed during Rosyth refit Aug–Sep 44. Sold for BU 5 Apr 46
DUNEDIN Armstrong	Nov 17 29 May 19	Oct 19 11 Nov 37	1937	WL(T) 24 Nov 41. New Zealand division replacing CHATHAM 1924–31, after refit to Med Fleet. Ship's company to LEANDER 1937. Attended Coronation Review 1937, Boys TS until operational service 1939.
DURBAN Scott	Jan 18 29 May 19	31 Oct 21 Aug 39	1936 Dec 43	Breakwater at Normandy 9 Jun 44. Decomm 1943, was reduction to C&M. Refit as blockship: 6in retained, 24 Oerlikons added. Sunk as part of No. 5 Gooseberry in British assault area. Was SNO Corncob in Gooseberry sheltering small craft. Hulk hit by German torpedo 3 Aug
DAEDALUS Armstrong	Not laid down			Canc 26 Nov 18
DARING Beardmore	Not laid down			Canc 26 Nov 18
DESPERATE Hawthorn	Not laid down			Canc 26 Nov 18
DRYAD Vickers	Not laid down			Canc 26 Nov 18

'E' Class

Name/Builder	Laid down	Launched	Completed	Notes
EMERALD Armstrong	23 Sep 18 19 May 20	14 Jan 26 Jul 39	Jul 38 Dec 44	Refit 1933–4; sold for BU 23 Jun 48. SNO Reserve Fleet Rosyth Mar–Dec 45. Disposal List 1948
ENTERPRISE Brown	28 Jun 18 23 Dec 19	31 Mar 26 Aug 39	1937 Dec 44	Refit 1934–5; sold for BU 11 Apr 46. Recomm Apr 38 in Portsmouth Reserve. Refit for trooping approved Jan–Feb 45, refit Chatham Mar–Apr 45, three 6in Landed, trooping Jun –Dec 45, again reduced to Reserve until on Disposal List
EUPHRATES Fairfield	1918			Canc 26 Nov 18

HAWKINS Class

Name/Builder	Laid down	Launched	Completed	Notes
EFFINGHAM Portsmouth	2 Apr 17 8 Jun 21	Jul 25 1938	1932	Rearmed 1937–8, wrecked 21 May 40 in Norway
FROBISHER Devonport	2 Aug 16 20 Mar 20 1942	20 Sep 24 1932 2 Jun 47	1930 1939	Flag of VA Reserve 1930 –2, Cadet TS 1932–7, Gunnery TS 1937–9, rearmed 39–42, conv at Rosyth to TS Oct 44 – Mar 45, Cadet TS 1945–7, sold for BU 26 Mar 49 (Disposal List 1949)

415

HAWKINS Chatham	3 Jun 16 1 Oct 17	23 Jul 19 1932 1939	1931 1935 1940 Jul 44	Rearmament Portsmouth cpl 23 Dec 39. In South Atlantic 1940–1, Portsmouth refit Dec 41–May 42. Sep–Apr 44 Refit and conversion to relieve DAUNTLESS as TS, but training requirement canc May 44 and ship reduced to reserve. STT 1947, sold for BU 22 Aug 47.
RALEIGH Beardmore	4 Oct 16 28 Aug 19	Jul 21		Wrecked 8 Aug 22
VINDICTIVE H&W	29 Jun 18 17 Jan 18	21 Sep 18 1923 1936	1921 1930 1945	Ex CAVENDISH. Completed as carrier, Used for trooping Used for trooping while in reserve 1921–3, conv to cruiser Chatham 1923–5, conv to TS 1936–7, Cadet TS 1937–9, to repair ship 1939–40. Refitted as Destroyer Depot Ship May to Jun 44, used for Accommodation and Repair in Normandy Operation, Accommodation Ship Portsmouth Jan–May 45, paid off Jun–Aug 45. Sold for BU 14 Feb 46

'County' Class

1924/5 Estimates (KENT Class)

BERWICK Fairfield	15 Sep 24 30 Mar 26	15 Feb 28	1946	Refit 1937–8. Sold for BU 15 Jun 48
CORNWALL Devonport	9 Oct 24 11 Mar 26	8 May 28		Chatham refit 1936–7. WL(B) 5 Apr 42
CUMBERLAND Vickers	18 Oct 24 16 Mar 26	23 Jan 28 1949	1946 May 59	Chatham refit 1935–6. Conv to trials cruiser 1949–May 51. Trooping from Far East about Nov 45–Jun 46. Sold for BU 1959. PN C 57
KENT Chatham	15 Nov 24 16 Mar 26	22 Jun 28	16 Feb 45	Chatham refit 1937–8. Flag Gareloch reserve 1945, sold for BU 22 Jan 48
SUFFOLK	15 Nov 24	31 May 28	1946	Trooping 1945–6. Sold Portsmouth 16 Feb 26 for BU 25 Mar 48

AUSTRALIAN SHIPS

AUSTRALIA Brown	26 Aug 25 17 Mar 27	14 Apr 28 16 Jun 47	1946 31 Aug 54	Sold for BU 25 Jan 55
CANBERRA Brown	9 Sep 25 31 May 27	10 Jul 28		WL(G) 9 Aug 42

1925/6 Estimates (LONDON Class)

DEVONSHIRE Devonport	16 Mar 26 22 Oct 27	18 Mar 29	6 Oct 53	Refit Liverpool May 43 –Mar 44 ('X' turret landed), trooping to Jan 46. Conv to TS Sep 46–April 47, Cadet TS 1947–53, sold for BU 16 Jun 54. While TS, landed armed parties Grenada Mar 51. Oldest warship at Coronation Review 1953
LONDON Portsmouth	23 Feb 26 14 Sep 27	31 Jan 29	1949	Rebuilt Chatham 1939–41. Refit Chatham 1946–7 for Far East, badly damaged Chinese gunfire during AMETHYST rescue 21 Apr 49 (23 hits). Arrived Chatham 8 Sep 49. Sold for BU 3 Jan 50
SHROPSHIRE Beardmore	24 Feb 27 5 Jul 28	12 Sep 29	1949	Refit Nov 42–Jun 43, to RAN 25 Jun 43, sold for BU 16 Jul 54
SUSSEX Hawthorn	1 Feb 27 22 Feb 28	19 Mar 29	Feb 49	Refit Sheerness Jun 44 –Apr 45 ('X' turret landed); East Indies 1945, refit for further service there 1946–7, Far East through 1949. Sold for BU 3 Jan 50

1926/7 Estimates (NORFOLK Class)

DORSETSHIRE Portsmouth	21 Sep 27 29 Jan 29	30 Sep 30		WL(B) 5 Apr 42
NORFOLK Fairfield	8 Jul 27 12 Dec 28	30 Apr 30	3 Nov 49	Refit Portsmouth Jan–Nov 44 ('X' turret landed). East Indies 1945–9. Sold for BU 3 Jan 50

1926/7 Estimates (YORK)

YORK Palmer	16 May 27	1 May 30 17 Jul 28		WL(A) 22 May 41

1927/8 Estimates (EXETER)

EXETER Devonport	1 Aug 28 18 Jul 29	23 Jul 31		River Plate battle: patched in Falklands, then refit Devonport 14 Feb 40 – 10 Mar 41. WL(G) 1 Mar 42

LIST OF SHIPS

1928/29 Estimates (NORTHUMBERLAND Class)
NORTHUMBERLAND
Devonport
Canc 14 Jan 30

SURREY
Portsmouth
Canc 14 Jan 30

LEANDER Class
1929/30 Estimates

LEANDER Devonport	8 Sep 30 24 Sep 31	24 Mar 33	1948	New Zealand division 1937–43, damaged at Kula Gulf 13 Jul 43, repaired Auckland Aug–Dec 43, then Boston Navy Yard to Aug 45, then Rosyth and Portsmouth to Mar 46; Med 1946–7 and then STT; sold for BU 15 Jan 49. Initial passage to New Zealand delayed a year by outbreak of Spanish Civil War. Departed after 1937 Coronation Review. In Corfu Channel incident 1946. Returned to UK to pay off Dec 47

1930/1 Estimates

ACHILLES Cammell	11 Jun 31 1 Sep 32	6 Oct 33	1947	To New Zealand Division and took NZ crew from DIOMEDE 31 Mar 36, Formally to RNZN on its formation Jul 41. At River Plate action Dec 39. Badly damaged 5 Jan 43 Guadalcanal, arrived Portsmouth and paid off 22 Mar 43, 'X' turret landed, radars modernised; recomm 23 May 44, required further repairs. To British Pacific Fleet on formation Nov 44. Auckland refit Feb–Mar 45, single 20mm replaced by twins. Ret to New Zealand Mar 46, left for UK Aug 46, ret to UK control. After refit, to India 5 Jul 48, str after 1976
NEPTUNE Portsmouth	24 Sep 31 31 Jan 33	23 Feb 34		WL(M) 19 Dec 41
ORION Devonport	26 Sep 31 24 Nov 32	18 Jan 34	1946	Fitted for STT June 48 (underwater explosions), Disposal List on completion, sold for BU 19 Jul 49. Had returned Devonport after two years in Med, 5 Jul 46

1931/2 Estimates

AJAX Vickers	7 Feb 33 1 Mar 34	6 Oct 33	16 Feb 48	Fought at River Plate. Refit Malta 1945, then South America cruiser, Then Med from Apr 46. Disposal List 1949

Modified LEANDER Class/SYDNEY Class
1931/2 Estimates

AMPHION Portsmouth	26 Jun 33 27 Jul 34	6 Jul 36		RAN PERTH 25 Jul 39, WL(G,T) 1 Mar 42

1932/3 Estimates

APOLLO Devonport	15 Aug 33 9 Oct 34	9 Oct 34	Aug 47	RAN HOBART 29 Sep 38. Sold for BU 1962.
PHAETON Swan Hunter	8 Jul 33 22 Sep 34	24 Sep 35		RAN SYDNEY as completed, WL (G) 19 Nov 41

ARETHUSA Class (1936)
1931/2 Estimates

ARETHUSA Chatham	25 Jan 33 6 Mar 34	23 May 35	Oct 45	Disposal list Jan 48, offered to Norway, not accepted; STT. Sold for BU 1950

1932/3 Estimates

GALATEA Scott	2 Jun 33 9 Aug 34	14 Aug 35		WL (T) 14 Dec 41

1933/4 Estimates

PENELOPE H&W	30 May 34 15 Oct 35	13 Nov 36		WL(T) 18 Feb 44

BRITISH CRUISERS

1934/5 Estimates

AURORA	23 Jul 35	12 Nov 37	17 Apr 46	To China as CHUNGKING 1948. Sale announced
Portsmouth	20 Aug 35			Nov 45. Began refit Feb 47

Southampton Class

1933/4 Estimates

NEWCASTLE	4 Oct 34	5 Mar 37	1950	Modernised Devonport
VA Walker	23 Jan 36	1952	Sep 58	Feb 50–May 52. Disposal List 1959. PN C 76
SOUTHAMPTON	21 Nov 34	6 Mar 37		WL(B) 11 Jan 41
Brown	10 Mar 36			

1934/5 Estimates

BIRMINGHAM	18 Jul 35	18 Nov 37	Dec 59	Modernised Portsmouth 1950–2. Sold for BU 1960.
Devonport	1 Sep 36			PN C 19
GLASGOW	16 Apr 35	9 Sep 37	Nov 56	Refit Portsmouth 1948, refit Chatham Apr 51,
Scott	23 Jan 36			for disposal Mar 58. PN C 21
SHEFFIELD	31 Jan 35	15 Aug 37	1959	Long refit 1949–51. Refit 1956–7, Operational
VA Walker	23 Jul 36			Reserve 1959–60, HQ Commodore Reserve Ships.
				Disposal List 1964, BU Sep 67. PN C 24

1935/6 Estimates

GLOUCESTER	22 Sep 36	31 Jan 39		WL (T) 22 May 41
Devonport	19 Oct 37			
LIVERPOOL	17 Feb 36	2 Nov 38	1953	Heavily damaged by Italian air attack during Operation
Fairfield		24 Mar 37		'Pedestal', 14 Jun 42, post-refit trials Aug 44, but C&M
				because shortage of manpower. Fully manned about May
				45, took Tripartite Commission to Germany to inspect
				German warships. To Med Oct 45. BU Jul 58. PN C 11
MANCHESTER	28 Mar 36	4 Aug 38		WL(T) 13 Aug 42
Hawthorn	12 Apr 37			

BELFAST Class

1936/7 Estimates

BELFAST	10 Dec 36	3 Aug 39	1963	Rebuilt Devonport 1955–9, refit Devonport 1963, then
H&W	17 Mar 38			reserve. Accommodation ship Portsmouth Aug 1965.
				Museum ship London 1971. Had been briefly recomm
				1963 for RNR exercises. Cut by 1968 Defence Review,
				removed from Reserve status but retained as
				Accommodation Ship. PN C 35
EDINBURGH	30 Dec 36	6 Jul 39		WL(T) 2 May 42
Swan Hunter	31 Mar 38			

DIDO Class

The starred ships were completed to the modified design.

1936/7 Estimates

DIDO	20 Oct 37	30 Sep 40	Oct 47	Into reserve after completing Chatham
Cammell	18 Jul 39			refit begun Oct 45. Flagship of Reserve Fleet
				Portsmouth 1952–6. BU 16 Jul 56. PN C37
EURYALUS	21 Oct 37	30 Jun 41	1947	Recomm to replace AJAX in Med after Rosyth refit.
Chatham	6 Jun 39	1948	Aug 54	BU 18 Jul 58. Returned home from 1944
				Commission 17 Feb 47. PN C42
NAIAD	26 Aug 37	24 Jul 40		WL (T) 11 Mar 42
Hawthorn	3 Feb 39			
PHOEBE	2 Sep 37	30 Sep 40	1952	Sold for BU 1956. PN C43
Fairfield	25 Mar 39			
SIRIUS	6 Apr 38	6 May 42	Apr 49	Briefly laid up 1947 due to manning shortage.
Portsmouth	18 Sep 40			BU 15 Dec 58. PN C82

1937/8 Estimates (three ships)

BONAVENTURE	30 Aug 37	25 May 40		WL(T) 31 Mar 41
Scotts	19 Apr 39			
CLEOPATRA	5 Jan 39	5 Dec 41	1954	Flag of Reserve Fleet
Hawthorn	27 Mar 40			Portsmouth 1954–6. Disposal List 1958. PN C33

HERMIONE Stephen *1938/9 Estimates*	6 Oct 37 18 May 39	25 Mar 41		WL(T) 16 Jun 42
CHARYBDIS Cammell	9 Nov 38 17 Sep 40	3 Dec 41		WL(T) 25 Oct 43
SCYLLA Scotts	19 Apr 39 24 Jul 40	12 Jun 42	Aug 44	Mined 23 Jun 44, uneconomic to repair, and planned conversion to BPF Escort Carrier Flagship cancelled. Disposal List 1948, STT, then BU 6 May 50
War Programme				
ARGONAUT Cammell	21 Nov 39 6 Sep 41	8 Aug 42	1946	Disposal List 1955, BU 19 Nov 55. PN C 61
*BELLONA Fairfield	30 Nov 39 29 Sep 42	29 Oct 43	1956	RNZN 1948–56, to reserve on return, Disposal List 1957, BU 5 Feb 59. PN C 63
*BLACK PRINCE H&W	2 Nov 39 27 Aug 42	20 Nov 43	1947	RNZN 1948–62, then Disposal List, BU Japan Mar 62. PN C81
*DIADEM Hawthorn	15 Dec 39 26 Aug 42	6 Jan 44	1950	To Pakistan 1956, refitted and ren BABUR Jul 1957. PN C 84
*ROYALIST Scotts	21 Mar 40 30 May 42	10 Sep 43	Jan 46	Rebuilt 1956, RNZN 1956-67, ret, Disposal List Jan 68. BU in Japan
*SPARTAN Vickers	21 Dec 39 27 Aug 42	10 Aug 43		WL(B) 29 Jan 44

FIJI Class ('Colony' Class)

Starred ships were completed to the modified design.

1937/8 Estimates

FIJI Brown	30 Mar 38 31 May 39	21 Aug 42		WL(B) 22 Jun 41
KENYA Stephen	18 Jun 38 18 Aug 39	27 Sep 40 Jul 48 15 Aug 55	Nov 47 1954 Sep 58	Emergency refit for Pacific Jul–Dec 45, Bofors replaced Oerlikons. Disposal List Feb 59. BU 29 Oct 62. PN C14
MAURITIUS Swan Hunter	31 Sep 38 19 Jul 39	1 Jan 41 1950	1947 1952	BU 27 Mar 65. Admiralty criticised for laying her up immediately after a refit lasting several months. PN C80
NIGERIA Vickers	8 Feb 38 18 Jul 39	23 Sep 40	1951	Flag South Atlantic 1946–50. Sale List 1952-3, to Indian Navy Apr 54, ren MYSORE and comm as flag Indian Navy Aug 57. Sold for BU 1980. PN C60
TRINIDAD Devonport	21 Apr 38 21 Mar 40	14 Oct 41		WL(B) 15 May 42

1938/9 Estimates

*CEYLON Stephen	27 Apr 39	21 Aug 42 30 Jul 42	1960	Modernised 1955–6. Sold to Peru 1960 as CORONEL BOLOGNESI (transf 6 Feb 60).
JAMAICA Vickers	28 Apr 39 16 Nov 40	29 Jun 42 1954	1952 2 Sep 58	Flag Reserve Fleet May 53 – Nov 54. Disposal List 1960, sold for BU 14 Nov 60. PN C44
GAMBIA Swan Hunter	24 Jul 39 30 Nov 40	21 Feb 42 1950	1948 7 Dec 60	RNZN 1943–7. Ret to RN 27 Mar 48. Disposal List 1968, BU 5 Dec 68. PN C48
*UGANDA VA Walker	20 Jul 39 7 Aug 41	3 Mar 43		To RCN on completion of major repair after hit by German glider bomb at Salerno 13 Sep 43. Recomm by RCN 21 Oct 44, retained name. To BPF, then back to Esquimalt and Canadian control 4 Aug 45. Ren HMCS QUEBEC. Disposal list about 1960, BU 8 Feb 61. PN C66, but RCN used a CCL number instead

BRITISH CRUISERS

1939/40 Estimates

BERMUDA	30 Nov 39	21 Aug 42	1947	Modernised 1956. Disposal List 1965, BU 26 Aug 65. PN C52
Brown	11 Sep 41	Oct 50	1963	
*NEWFOUNDLAND	9 Nov 39	20 Jan 43	Feb 47	TS 1947–50; reserve status Feb–Apr 47 and Jul 50–Feb 51; in hand for modernisation Devonport Feb 51. Sold to Peru Nov 59, ren ALMIRANTE GRAU Dec 59 on hand-over, ren CAPITAN QUINONIS 1973 when DE RUYTER took her name. Out of service 1979, sold for BU
Swan Hunter	19 Dec 41	14 Nov 52	Oct 59	

Improved FIJI Class

MINOTAUR	20 Nov 41	25 May 45	ca 1959	To RCN Jul 44, ren ONTARIO. Disposal List 1959. BU Nov 61
H&W	29 Jul 43			
SWIFTSURE	22 Sep 41	22 Jun 44	1955	Badly damaged in coll with DIAMOND 1953. Disposal List 1962, BU 17 Oct 62
VA Tyne	4 Feb 43			

TIGER Class

BELLEROPHON	–			Ex-BLAKE, ex-TIGER. Ordered 18 Dec 41, not laid down. Ren BLAKE and then BELLEROPHON on transfer to NEPTUNE Class. Canc 28 Feb 47
VA Tyne				
BLAKE	17 Aug 42	8 Mar 61	1963	Ex-TIGER, ex-BLAKE Work stopped (on all three) Jul 46, restart announced 15 Oct 54. Helo carrier conversion 1965–9. Arrived Chatham after Rosyth refit for Reserve Fleet. Standby Squadron Chatham, survey during Falklands War to decide whether to reactivate her and sister TIGER. Could not be done quickly enough, decided to BU; BU 29 Oct 82. PN C99
Fairfield	20 Dec 45	23 Apr 69	6 Dec 79	
HAWKE		1 Jul 43		Ordered 14 Oct 42, canc 15 Oct 45
Portsmouth				
LION	24 Jun 42	20 Jul 60	1964	Ex-DEFENCE. Approved for BU 1972. BU 24 Feb 75. Renamed 1957. PN C34
Scotts	2 Sep 44			
SUPERB	23 Jun 42	16 Nov 45	1958	BU 8 Aug 60.
Swan Hunter	31 Aug 43			
TIGER	1 Oct 41	18 Mar 59	1966	Ex-BELLEROPHON. Rhodesian talks held aboard 1966. Helo carrier conversion 1968–72. For disposal 1978. Left in tow for BU 23 Sep 86. PN C24
Brown	25 Oct 45	2 Jul 72	20 Apr 78	

NOTE: BLAKE was initially renamed TIGER, but the original name was restored after pressure by Admiral Sir Geoffrey Blake, Chairman of the Ships Names Committee. BELLEROPHON was then renamed TIGER to maintain the original class name. One ship (not named) was ordered from Stephens under the 1942 programme 7 April 1942 but then cancelled 10 August 1942. Planned 1942 orders for four ships (Fairfield, Hawthorn Leslie, Vickers and Cammell Laird) were not let. Names selected in May 1942 for the six 1942 cruisers by the Ships Names Committee were EDGAR, MAJESTIC, MARS, TRIUMPH, WARRIOR and THESEUS. Names selected for the (unordered, except for BELLEROPHON) NEPTUNE Class (1944 programme) were NEPTUNE, CENTURION, EDGAR, MARS and BELLEROPHON. The five initial builders (before BELLEROPHON was added) were John Brown, Hawthorn Leslie, Vickers Barrow, Swan Hunter and Harland and Wolff.

INVINCIBLE Class

ARK ROYAL	14 Dec 78	1 Nov 85		
Swan Hunter	4 Jun 81			
ILLUSTRIOUS	7 Oct 76	20 Jun 82		
Swan Hunter	14 Dec 78			
INVINCIBLE	20 Jul 73	11 Jul 80		
Vickers	3 May 77			

Minelayers

ADVENTURE	29 Nov 22	2 Oct 26	25 Jul 45	Refit Jan–Aug 40, mined 14 Aug 41, repairs Feb–Jun 41, Oct–Nov 41 (turbines), collision 3 Feb 42, repair Feb–Apr 42 and Type 128 Asdic fitted (mine detection), diesel removed Apr 43, Jul 1943 nominated as Depot and Repair Ship, conversion arranged for Liverpool Nov 43, taken in hand for conversion to Landing Ship (Engineering) 25 Nov 43, nominated for D-Day support, completed Feb 44, then Accommodation Ship Portland, then Portsmouth, D-Day and subsequent operations, in Mulberry B for support and repair, including of landing craft damaged by gale, back to Home Fleet Aug 44, C&M Nov 44, recomm 10 Dec 44, to Portsmouth Command, Disposal List Jul 47
Devonport	18 Jun 24			

1938–9 Programme

ABDIEL	29 Mar 39	15 Apr 41		WL(M) 10 Sep 43
White	23 Apr 40			
LATONA	4 Apr 39	4 May 41		WL(B) 25 Oct 41
Thornycroft	20 Aug 40			
MANXMAN	24 Mar 39	20 Jun 41	1947	Refit Sheerness Jun 46–Feb 47. Med Fleet Sep 51. Conv
Stephen	5 Sep 40	1951	1953	as Support Ship for minesweepers (Chatham)
		Feb 56	1957	Jul 61–Apr 63. Disposal List 1972 WL(T) 1 Feb 43
		Feb 63	Sep 70	
WELSHMAN	8 Jun 39	25 Aug 41		
Hawthorn	4 Sep 40			

War Programme

APOLLO	10 Oct 41	12 Feb 44	1946	Disposal List 1962,
Hawthorn	5 Apr 43	1951	1961	BU Nov 62.
ARIADNE	15 Nov 41	9 Oct 43	1946	Prepared for service at Suez, but insufficient manpower.
Stephen	16 Feb 43	1956		Never deployed, Disposal List 1965, BU 14 Feb 65

INDEX

Page numbers in *italics* represent illustrations.

Abbreviations
Adm = Admiral
Arg = Argentina
Cdre = Commodore
Cpt = Captain
def = definition
Fr = France
Ger = Germany
It = Italy
Jpn = Japan
R/A = Rear Admiral
RN = Royal Navy
Sov = Soviet Union
Sp = Spain
USA = United States of America
USN = United States Navy

'1960 Cruiser' design 295, 297–301, *plans 297–300*

A-type Cruiser (1929) 122–3, *plan 123*
Abyssinian crisis 125, 205–6, 228
Achilles, HMS 153, *238*
    see also *Delhi*
Active, HMS 38, 40–1
Adelaide, HMAS 29, *29*, *31*, 31–3, *32–3*, 67
Admiral Graf Spee (Ger) 258
Admiral Hipper class (Ger) 227–8
Admiralty organisation 13–16, 79
    Fleet Requirements Committee 315, 319
    Future Building Committee 261–5
    Post-War Questions Committee 79, 82, 85
Adventure, HMS 86, 99, 101, 234
Agato (Jpn) 264
AIO (Action Information Organisation) 238, 275, 289, 296, 304, 309–11
aircraft, shipboard 13, 55, 61, *61*, 61–2, 88–93, *89*, 105, 117, 161–2, 165, 174, 177, 179, 221–2, 228–9, 258
    aircraft-aft design 16
    catapults 9, 33, 64, 73, 74, 90, *90*, 91–4, *94*, *101*, 104, 106–7
    FDO (Fighter Direction Office) 249
    jet aircraft 281
    see also helicopters *and individual aircraft types*
Aircraft Direction Rooms (ADR) 296
Ajax, HMS *151*, 153, 234, *237*, 242, *plan 144*
    class 242
Alaska class (USA) 228, 256
Alexander, A V, First Sea Lord 156
Alynbank, HMS *233*
Amazon, HMS 108
Ambuscade, HMS 108

INDEX

Amethyst, HMS 21
Amphion, HMS 174, 177
   class 172, 178, 206, 215
   see also Perth, HMAS
Anglo-German Naval Agreement (1935) 227, 229
anti-submarine warfare 60–1, 91, 107, 274–5, 314–21, *315*
Apollo, HMS 243
   see also Hobart, HMAS
Arethusa, HMS 40–2, 47, 59, 123, 146, *160*, *162*, 163–4, 172, 177, 186, 189, 206, 243, *plan 38*, *plan 157*
   class 11, *11*, 29, 36–42, 60, 156–62, 165, 167, 173, 178, 205, 227, 264, 273
      see also individual ships
Arethusa, HMS pre-1914 264
Argentina 226–7
La Argentina (Arg) 227
Argonaut, HMS *188*, *190*, 192, 244, 278, 281, *plan 187*
Argus, HMS 91
Ariadne, HMS 243
Ark Royal, HMS 261
Armstrong Whitworth 87, 98, 312
Asdic 13, 162, 192, 214, 219, 260, 269, 313
Asquith, H H 38
Atlanta class (USA) 287
'Atlantic Cruiser' class (B3) 35, *plan 35*
Aurora, HMS *39*, 48–9, 60–2, 93, *158–9*, 205
Australia 10, 23–4, 27, 29, 33, 79, 104, 156, 165, 200, 206
Australia, HMAS *8–9*, *92*, 92–3, 104, *130*, *135–7*, 238, *240*, 242, *plan 125*

Backhouse, R/A Roger 121–2, 125–6, 155–6
Balfour, A J 56, 65
Ballard, Cpt 35
Bande Nere class (It) 155
Barrow shipyard 296
'B' bombing *def* 172, 181, 184
B class cruisers 108–13, *110*, 116–17, 155
   see also York, Exeter
Beardmore shipbuilders 50
Beatty, Adm Lord 60, 79, 104–5, 121
Belfast, HMS 123, 178, *180*, *182–3*, *184–5*, 185, 215, 219, 228–9, 233–4, 244, 254, 263–4, 275, 278, 281, *283*, 287, 289–90, 315, 317, *plan 179*
   class 178, 220–2, 235, 238, 255
      improved 263–4
      see also Edinburgh
   improved class 264
Bellairs, Captain R M 144
Bellerophon, HMS (ex-*Tiger*) 244, 249
Bellona, HMS/HMNZS 244, *245*, 254–5, *256*, 287
Beresford, Adm Sir Charles 37
Bermuda, HMS 224, *227*, 244, 248, *279*, 281, 287, *287*, 289–90, *plan 218*
Berry, William J 105
Berwick, HMS 9, *92*, 104, *128*, 233, 301
Bikini Atoll tests 277
Birkenhead, HMS 33, *33*, 62
   see also Chester
Birmingham, HMS *13–15*, 62, *178*, *239*, 242, 244, 278, *280–1*, 287, 289–90, *290*
   class 29–30, 35, 48, 65, 92, 242
      see also Adelaide, Lowestoft, Nottingham
Blackburn Roc aircraft 93
Blackburn Skua aircraft 92–3
Black Prince, HMS/HMNZS 244, *256*
   class 246
Blake, HMS 249–50, *302–3*, 319–20, 322
Boadicea, HMS 18, 21, *27*

class 18–21
Bofors guns *13*, 185, 192, 209, 213, 227, 242–3, 249–50, 264, 267–8, *278*, 281, 287, 289–90, 296, 305, 309–13
    'Buster' 243, 260, 264
    Hazemeyer 243, 260
Breslau (Ger) 62
    class 38
Bretagne class (Fr) 33
bridge design 56, *57*, 62, 103, 108, 113, 118, 122, 149, 155, 162, 191, 224, 248–9, 264, 277, 287, 289, 296
Brisbane, HMAS 26–7, 29, 156, 200
Bristol, HMS *10*, 21–2
    class 18–23, 26–7, 48, 62
        see also *Glasgow*, *Gloucester*, *Liverpool*, *Newcastle*
Bristol, HMS (destroyer) 318, 323
Brooklyn class (USA) 179, 202
Bulwark, HMS 320

'C' class 56, 60–1, 84, 90, 92–3, 116, 155, 173, 186, 188, 202, 205–6, 214, 229, 233–4
    see also *Caroline* class, *Calliope* class, *Cambrian* class, *Centaur* class, *Caledon* class, *Ceres* class, *Carlisle* class
Cairo, HMS 207, 209, 233
Calcutta, HMS 207, 209, 233
Caledon, HMS 50, 62, 89, *205*, 207, 209, 213–14, 243, *plan 52*
    class 60–1, 171
        see also *Calypso*, *Caradoc*, *Cassandra*
Calliope, HMS 47–8, 57, 60, *plans 44–5*
    class (later *Caroline* class) 29, 42–9
        see also *Champion*
Calypso, HMS 50, *50–1*, 57, 207, *plan 51*
    class 50–5
        improved class 55–6
Cambrian, HMS *58–9*, 59, 62, *plan 46*
    class 48, 59–60
        see also *Canterbury*, *Castor*, *Constance*
Cammell Laird shipbuilders 34, 50, 226
Camrose anti-torpedo weapon 305
Canada 23–4, 27, 79, 213
Canarias (Sp)
    class 114–15, 140, *plan 114*
Canberra, HMAS 92–3, 96–7, 104, *130*, 233, *plan 100*
Canterbury, HMS 48, 60, *plan 46*
Capetown, HMS 62, 207, 209, 214, 233, *plan 52*
Caprice, HMS 50
    see also *Cardiff*
Caradoc, HMS 50, 53, 57, 207
Cardiff, HMS 207, 214
Carlisle, HMS 62, 207, 209, 232, *plan 52*
    class 62
        see also *Cairo*, *Calcutta*, *Capetown*, *Colombo*
Caroline, HMS 57, 59, 61–2, *plan 43*
    class 59–60
        see also *Carysfort*, *Cleopatra*, *Comus*, *Conquest*, *Cordelia*
Carysfort, HMS 60, *plan 43*
Cassandra, HMS 50, 62
Castor, HMS 200, *plan 46*
Cavendish, HMS 69, 73
CDS see under fire-control systems
Centaur, HMS 48, *48*, 49–50, 53, 56, 60, 251, 287, 314, *plan 49*
    class 48–50, 54, 57
        see also *Concord*
Ceres, HMS 50, 56, 207, 214
    class 53, 60

see also Cardiff, Coventry, Curacoa, Curlew
Ceylon, HMS 222, 244, 246, 249, 278, 281, 287, 289, *289*, 290
Challenger, HMS 26
Champion, HMS 47–8, 56–7, 241, *plan 45*
Charybdis, HMS 192, 213, 233, 254
Chatfield, Adm of the Fleet Alfred 108, 113, 144, 173, 176–7, 190, 200–2
Chatham, HMS 62, *plan 27*
    class 24, 26–9
        see also Brisbane. Dublin, Melbourne, Southampton, Sydney
Chatham Dockyard 47, 74, 92, 126, 207
Chester, HMS 33, 62, 84
    class 33
Chester, USS 123
Chile 226–7
China 9, 11, 62, 79, 173, 275
Chinook helicopters 317–18
Churchill, Sir Winston 35, 38, 41–2, 48–9, 79, 229, 252, 273
Cleopatra, HMS 47, 60–1, 192, *192*, 244, 281
Cleveland, USS 250
    class 264–5, 309
coal fuel 19, 21–2, 29, 31, 34–5, 65, 67, 69, 78–9
Cockatoo Dockyard (Australia) 29
Colombo, HMS 207, 209, 213, 233, 243
'Colonial Cruiser' design (1,000-ton 8in cruiser) 79, 94–5
Command Cruiser design 321–3
Comus, HMS *plan 45*
Concord, HMS 48–9, 56, 60
Conquest, HMS 58, 59–61, 86, 155, *plan 45*
Constance, HMS 60, 62, 200
convoys 79, 86, 95, 284, 297, 314
Coolidge, President Calvin 113
Cordelia, HMS 62, *plan 44*
Cornwall, HMS 92, 103, 118, 123
Corsair, HMS 50
    see also *Coventry*
'County' class 104–5, 117, *118*, 123, 145, 241, 243–4, 264, 274, 278, 305, 310, 313
    see also Kent, London, Norfolk
Courageous, HMS 95, 101
Coventry, HMS 82, *203–4*, 207, 233, *plan 209*
Coventry Ordnance Works 42, 60
CR series designs 304–5, 309
CRBF *see under* fire-control systems
Cumberland, HMS 92, 103, 123–5, *133*, 251, 275, *276*
Cunningham, Adm Sir Andrew B 188, 262
Curacoa, HMS 61, *200–1*, 207, *232*, 233, *plan 52*
Curlew, HMS 50, *203*, 206–7, 231, 233

'D' class *see Danae* class
DA (Direct Attack) guns 300–1, 303
Danae, HMS *11*, *211*, *240*
    class ('D' class) 33, 54–6, 60, 62, 84, 86, 90, 92–3, 110, 116, 155, 157, 161, 173, 186, 188, 205, 207, 213–14, 229, 233–4
Daring class destroyers 295, 297, 300–1, 310, 317
Dartmouth, HMS 23
    class 22–5
Dauntless, HMS *54*, 55, 62, *210*, 214, *plan 55*
DCTs *see under* fire-control systems
D-Day 236
Defence, HMS 243, 249–50
Defence Requirements Committee (DRC) 173
Delhi, HMS 62, 213, *213*, 214, *plan 212*
depth charges *15*, 33, 86, 149, 209, 219

Des Moines class (USA) 269
Despatch, HMS 62
Devonport Dockyard 75, 121, 209, 277
Devonshire, HMS 75, 92, 118, 126, *135*, 230–1, 242, *plan 127*
   class 242
Diadem, HMS 244, *252–3*, 278, 281
Diana class (RN) 19
Dido, HMS *2–3*, 192, 222, 226, 242, 244, 278, 281
   class 178, 186–92, 216, 218, 224, 231, 233, 241–5, 249, 251, 258, 261–2, 264, 273–5, 277–8, 281, 284, 287
     redesigned *see Black Prince* class
diesel fuel 243, 250, 264, 269
Diomede, HMS 76–7, 78
direction-finders *81*, 236, 238
dive-bombing threat 167, 203, 209
Dorsetshire, HMS 92, 118–19, 121, 204
   class 107–8
Dover Patrol, The 60
Dragon, HMS 55, 62, 87, 89, 214, *plan 55*
Dresden (Ger) 19
Dreyer, Adm Frederic C 15, 79, 94–5, 98, 113, 117
Dublin, HMS 62
Dunedin, HMS 62, 214
Duquesne (Fr) 123
Durham, HMS 62

'E' class *see Emerald* class
Eagle, HMS 251, 287
Eden, Sir Anthony 125
Edinburgh, HMS 27, 178, *180–1*, 185, 233
Effingham, HMS 54, 69, *70*, 71–3, 75, 86, *plan 67*
Egerton, Captain G A 117
Emden (Ger) *10*, 19, 41–2, 79, 86
Emerald, HMS 63–4, 93, 150, 204–5, *208*, 244, *plan 63*, *plan 206*
   class ('E' class) 62–5, 84–6, 98–9, 103, 107–8, 113, 188, 204, *plan 63*
     Improved 'E' class 105
     see also Enterprise, Euphrates
Empress of India, HMS 161
Enterprise, HMS 11, 65, 82, 84, 85, 93, 205, 236, 244, *plan 207*
'Escort Cruiser' design 314–21, *plans 316–21*
Euphrates, HMS 62
Euryalus, HMS *81*, 186, 192, 244, 281
Evans, Adm 213
Excellent, HMS 82, 86, 172
Exeter, HMS 92, 94, 103–4, 113, 117–20, *119–20*, 122, 149, 204, 206, 233–4, *plan 118*

FADES 281
Fairey Flycatcher aircraft 90
Fairey IIID aircraft 90
Fairey IIIF aircraft 90–1
Fairey Swordfish aircraft 92, 174, 221
Fairfield shipbuilders 50
Falklands, battle of (1914) 78, 86
Falmouth, HMS 23, 42, 48
Fearless, HMS 38, 40
Field, Adm Sir Frederick 113, 142
Fiji, HMS 222, 224, 233, 248–9, 255
   class 16, 94, *94*, 126–7, 178, 191, 214–26, 229, 233, 235, 238, 242–3, 245, 249, 268–9, 273, 277–8, 281, 284, 289–90, 300, *plan 309*, *plans 216*
     improved and modified 227, 236, 243, 245–6, 254, 260–1, 273, 278, 289
     see also Bermuda, Ceylon, Gambia, Jamaica, Kenya, Mauritius, Nigeria, Trinidad, Uganda
Finland 227

INDEX

fire-control systems 11, 142, 260
    Bridge Operations Room (BOR) 296
    CDS (Combat Direction System) 275, 296, 315
    CRBF (Close Range Blind Fire) systems 250, 278, 281, 301
    DCTs (Director Control Towers) and positions 9, 13, 29, 42, 49, 56–7, 79–80, *81*, 82–3, 85–6, *102*, 103, 108, 111, 121, 125–6, *129*, *148*, 159, 165, 167, 173, 177–8, 184, 202, 204, 213, 221, 225, 229, 232–5, 234, 264, 269, 275, 278, 281, 284, 287, 289–90, 300, 303–4, 312
    fighter control 249, 264, 281, 293, 296
    Flyplane AA system 172, 275, 277–8, 284, 301
    HACPs (High-Angle Control Positions) and HACS (High-Angle Control Systems) 86, 121–2, *158*, 162, 168, 171–3, 177, 184, 191, 202–6, 209, 213–14, 221–2, 225, 227, *237*, 241–3, 246–8, 250, 275–7, 281, 284, 301
    HACTs (High-Angle Calculating Tables) 153, 191
    HADTs (High-Angle Director Towers) 86, 149, 159, 172, 177, 204, 206, 209, 221, 229, 249–50, 254–5, 278, 281, 289
    rangefinding directors 125, 165, *166*
    RPC (Remote Power Control) 185, 241, 248–50, 255, 284, 295
    STDs (Simple Tachymetric Directors) 243, 249, 275, 290
    Target Indication Units (TIUs) 275
First World War, the 8, 16, 18–19, 29–30, 35, 42, 60–2, 65, 76, 86–7, 116, 156, 171, 173, 202
Fisher, Adm Sir John 18, 23, 36–8, 40–2, 61
Fisher, Adm Sir William W 144, 186–7
Fleet Air Arm 93–4, 164
France 8, 94, 123, 155, 172, 202, 227
    in Second World War 243
Fraser, Adm Bruce 142, 249, 274
Frobisher, HMS 68, 69, *69–70*, *71–2*, *72–3*, *73–4*, 90, 103, 202, *plan 66*
    class 93
Fubuki class (Jpn) 186
Furious, HMS 67, 95
Furutaka class (Jpn) 72, 94, 111, 113, 151

Galatea, HMS *11*, *36–7*, 42, 60–1, 167, 206
Gambia, HMS *220*, 222, 226, 242, 244, 281, *286*, 289–90
Geddes, Sir Eric 79
generators 48, 243–4, 250, 264, 269, 277, 295
    turbo-generators 243–4, 262, 303
Geneva Conference (1932) 172–3
Geneva Naval Conference (1927) 113–17, 120, 142
Germany 86
    pre-First World War 8, 10–11, 19, 35–8, 41
    in First World War 54, 57, 60, 62, 65, 67, 76, 78, 87
    inter-war 82, 88–9, 94, 173, 227–9
    in Second World War 229, 235–6, 245, 252, 254, 258, 264, 297
Gibraltar 181, 220
Girdle Ness, HMS 251
Glasgow, HMS *18–19*, *176–8*, 233, 244, 278, 284, 287, 290, *plan 20*
Gloucester, HMS *20–1*, *175*, 181
    class 178, 214–15
    see also Liverpool
Gneisenau (Ger) 86
Goodall, Stanley C 16, 42
Gorgon, HMS 101
Gorizia (It) 220
Graf Spee (Ger) 95
Grasset, Adm 94
guns
    6in 16, 19, 21–2, 29, 31, 33, 38, 42, 47–50, 54–5, 59–62, 67, 72–4, 82, 84–5, 105, 113, 117, 142, 145, 149, 151, 153, 155, 157, 167, 172–4, 178–9, 184, 202, 217, 219–20, 225, 228, 246, 249–50, 254, 256, 258, 260–1, 263–5, 267, 276–7, 277–8, 284, 295–6, 300–1, 304–5, 309–11, 313, 320
        'all-big-gun' armament 49, 54–5, 59–60, 65
    8in 70–2, 79, 84. 95, 100–1, 103–4, 108, 113, 116–17, 123, 125, 144, 146, 150–1, 156, 172, 202, 228–9, 254, 256, 258–9, 261, 269
        compared to 6in guns 142, 144

8in Gun Cruiser 98–9
  see also Bofors guns, Lewis guns, fire-control systems, machine guns
'GW missile cruiser' designs 309–12, *plan 313*

HACPs *see under* fire-control systems
HACS *see under* fire-control systems
HADTs *see under* fire-control systems
Halsey, R/A Lionel 62
Harrier VSTOL aircraft 322–3
Harwich Force, The 48, 59–60, 62, 188
Hawke, HMS 249, 251
Hawker Osprey 91
Hawkins, HMS 65, *65*, 69, 71, *71*, 73, 86, 92–3, 99
  class 16, 31, 35, 50, 56, 65–75, 82, 84, 86, 94–5, 98–9, 101, 113, 145, 151, 156, 173, 200, 202, 229, 256, 261
    see also Cavendish, Effingham, Frobisher, Raleigh
Hawthorn Leslie shipbuilders 47, 50
helicopters 314–20, 322–3
Heligoland Bight, battle of (1914) 41, 48, 103, 107
Henderson, Adm Reginald 16, 126, 157, 179, 218, 221
Hermes, HMS 33, 91, 320
  class 293
Hermione, HMS 192
Hobart, HMAS (ex-*Apollo*, HMS) 168–71, 206, 243–4, *plan 163*, *plan 165*
Hood, HMS 33, 65, 67, 98
Hoover, President Herbert 156

Ikara missiles 318
Inconstant, HMS 42, 60–1
India 79, 287
Indianapolis, USS 103
Indonesia 275
Invincible, HMS *322–3*
  class 8
Italy 16
  inter-war 155–6, 202, 206, 226–8
  in Second World War 229

Jackson, Adm (First Sea Lord) 54
Jamaica, HMS *219*, 222, *223*, *225*, 226, 244, 281, *285*, 287, *288*, 289–90, *plan 224*
Japan 9, 72, 94, 120
  inter-war 76, 78–9, 86, 95–6, 99, 104–5, 113, 116–17, 125, 151, 155–6, 172–3, 186–8, 202, 206, 218, 221, 227–9
  invasion of Manchuria 173
  Sino-Japanese War, and the 11
  Second World War, in the 78, 225, 229, 258, 261, 264, 274, 301
Jellicoe, Adm of the Fleet Lord 53–4, 56–7, 59, 79, 94, 156
John, W G 219, 221, 228–9, 254
John Brown shipbuilders 50, 227
Jutland, battle of *83*, 101, 107–8

'K14' design *plan 222*
Kent, HMS 9, 92, *100–1*, 105–6, 123, 125, *128*, *131–2*, 242–3, *plan 124*
  class 9, 16, 71–2, 84, 91–2, 99–104, 107–8, 113, 117, 125–7, 155, 178, 204–5
    modernisation 123
    see also Australia, Berwick, Canberra, Cornwall, Cumberland, Suffolk
Kenya, HMS *214–15*, 222, 225–6, *241*, 242, 244, 278, 281, 287, 289–90
Keyes, Adm Roger 99
King George V class battleships 301
Kolberg class (Ger) 19
Köln class (Ger) 94
Königsberg (Ger) 19
Korean War, the 274, 278, 287, 296

INDEX

Krupps 227–8
Kuwait 11

Lambert, Cdre C F 56
'Large 1941 Cruiser' 254–60, *plan 254*
Leander, HMS/HMNZS 122, 142, 144–55, *145*, *147*, 151, *152–3*, 153, *154*, 155–6, 164, 167, 172–4, 177, 214, 243
    class 92–3, 122, 144–58, 162–4, 167, 171–4, 177, 204, 206, 215, 242–3, 264, 273, 281
        improved or modified class (*Amphion* class) 93, 161, 163–7, 173, 206
            see also Hobart, Perth, Sydney (all HMAS)
Lee of Fareham, Lord 98
Leipzig (Ger) 86
Lend-Lease Act (1941) 213
Lewis guns 33, 87, 185, 209, 246
'Light Cruiser A' design 105
Lillicrap, Sir Charles S 16, 85, 98–9, 101, 103, 105–7, 108, 117, 122–3, 149, 155, 157–9, 161–3, 228, 255–6, 258
Lion, HMS *294–5*, 319–20
Little, Adm Sir Charles 222
Liverpool, HMS 215, 244, 278, 281, 284, 287, 289
London, HMS 92, 104, *106*, 107, 116, 122, 126–7, *138–40*, 149, 233, 278, *plan 126*
    class 92, 104, 116–17, 122, 125, 204
        modernisation 178
London Conference and Naval Treaty (1930) 72–3, 116, 122–3, 142–99, 202, 206, 226
London Conference and Naval Treaty (1936) 123, 173, 200, 202, 226, 228
Lowestoft, HMS *30*, 31

'M' class *see Minotaur*
MacDonald, Ramsay 104, 122, 156
machine guns 75, 86–9, 104, 123, 145, 149, 164, 173, 176, 181, 184–5, 202, 204–5, 209, 219–20, 224, 228, 248–9
Madden, Adm Sir Charles (1st Baronet) 142, 155
Mainz (Ger) 86
Malaysia 274–5
Malta 181, 297
Manchester, HMS *12–13*, 178
Mauritius, HMS *216–17*, 222, 226, *235*, 244, 281, 287, 290
Maxim guns 19–20, 38, 50, 55
May, Adm William H 37
Mayo, Adm H T (USN) 94
Melbourne, HMAS *26*, 29
minelaying 57, 61, 104, 117
Minotaur, HMS 244, 249–50, 267–8, 297, 300–1, 303, *plan 267*
    class ('M' class) 177, 243, 249–50, 277
    see also Ontario
missiles 236, 238, 251, 274–5, 309–14
Mogami (Jpn) 227
Moore, R/A Archibald G H W 48
Mountbatten, Adm Lord 312
Mussolini, Benito 173
Myoko class (Jpn) 99
Mysore, INS *see Nigeria*

Naiad, HMS 192, 233
NATO 274–5, 289, 321
Nelson, HMS 100, 107, 120, 122, 179, 233, 241
    class 86, 99, 116–17
Neptune, HMS (1933) *plan 146*
Neptune, HMS (1944) 264–8, *267–8*, 295, 303–4, *plan 266*
    class 250, 273
Netherlands 227–8, 243, *plan 195*
Newcastle, HMS *19*, 242, 244, 278, 287, 289–90, *290*
Newfoundland, HMS 224, 244, 246, *247*, 278, 281, 284, 287, 289–90, *290*
New Orleans class (USA) 123

New Zealand 10, 79, 156, 200, 287
New Zealand, HMS 79
Nigeria, HMS 222, 224, 226, *227*, 233, 238, 242, 244, 278, 281, 287, *288*
Norfolk, HMS 9, *90–1*, *106–7*, *138*, 204, 238, 242–4, *244*, 245, *plan 127*
    class 92, 104
        see also Dorsetshire
Norfolk, USS 297
Northumberland, HMS 121
Norway 245, *plan 196*
Nottingham, HMS *30*
nuclear weapons and protection 274, 277–8, 287, 303–4, 314, 323
Nürnberg (Ger) 86

Oerlikon guns *13*, 33, 74, 162, 185, 192, 213–14, 224–6, 242–3, 246, 248–50, 260, 264, 269, *278*
oil fuel 19, 21, 29, 31, 33–5, 42, 65, 71, 73, 78, 84–5, 103, 108, 146, 157, 161, 219, 264, 301
Omaha, USS 95
    class 94, 111, 113, 145
Ontario, HMCS (ex-*Minotaur*) 249, 251, *291*
Orion, HMS *94*, *148–50*, 153, 242

Pakistan 287
Penelope, HMS 60–1, *159*, 177, 205
Pengelly, H S 304–5
Pensacola class (USA) 116
Perth, HMAS 234
Perth, HMAS (ex-*Amphion*, HMS) *32*, *80*, *164*, 165, *166*, 167, *167*
Phaeton, HMS *see Sydney*, HMAS
Phoebe, HMS *188*, 192, 233, 244, 278, 281, 284
pompoms 60, 73–5, 87, *87*, 88, 95, 98, 102, 104, 123, 155, 162, 171–2, 177, 184–5, 192, 202–3, 206, 209, 213, 220, 229, 232–3, 241–2, *242*,
    246, 248–50, 254, 259, 269, 281, 289–90, 301
Portland, USS 123
Portsmouth Dockyard 87, 121, 167, 207, 277, 289, 320
Portugal *plans 193–4*
Pound, Adm Sir Dudley 99, 121–2, 144, 262
Princess Irene, HMS 104

Quarto (It) 38
Quebec, HMCS (ex-*Uganda*) *291*
Queen Elizabeth class (RN) 38
Quincy, USS 65

radar *13–14*, 16, *129*, *135*, *150*, 162, 185, 202, 224, 231–41, 236, 238, 242–3, 246, 248–50, 264, 275, 278, 281, 287, 289, 297, 300–1, 303,
    309–10, 313
    microwave 234–6
radio 18, 33, *39*, *80*, 150, 153, 155, 189, 236, 249, 264–5, 281, 284, 289
    introduction of 8, 10
RAF 232, 273–4
Raleigh, HMS 69, 99, 103, *plan 66*
rangefinders 20, 125, 165, *166*, 232, 235
Reina Victoria Eugenia (Sp) 33
Resolution, HMS 90
River Plate, battle of the 258
rockets (UP) *135*, 192, 243, 254
Rodney, HMS 107
Roosevelt, President Franklin D 202
Rosyth Dockyard 74
Royalist, HMS *40–1*, 60–1, 244, 278, 281, 284, *284*, 287
Russia/Soviet Union 8, 79, 227, 252, 287, 297
    post-war 274–5, 278, 305, 309
'Russian' cruiser 252, 254

INDEX

SA (self-protection against acoustic mines) 162, 192
San Diego, USS 264
Sandys, Duncan 16, 275
Sardonyx, HMS 232
Saudi Arabia 11
Scharnhorst (Ger) 86
Scott's shipbuilders 50, 227
Scylla, HMS *189*, 192, 213, 244–5, 254, 277, *plan 191*
Seacat missiles 315, 320
Sea Dart missiles 317–18, 321–3
Sea King helicopters 314, 320, 322–3
searchlights (projectors) 16, 20, 29, 111, *148*, 177–8
Sea Slug anti-aircraft missiles 251, 274, 297, 305, 310–17
Sea Wolf missiles 323
Second World War, the 9, *12*, 16, 74–5, 161, 167, 172, 185, 202, 220, 224, 231–72, 297, 309
    designs 161, 252–72
Shakespeare, HMS 60
    class 65
Sheffield, HMS *17*, *178*, 231, 235, 244, *272–3*, 275, 278, *278*, 282, 287, 289–90, *plan 176*
Shropshire, HMS 87, 91, 126, *132*, *134–6*, 242, *plan 105*
Siam *plan 197*
Singapore 79, 99, 104, 117, 213
Sirius, HMS 192, 244, 281, 284
    class 287
Southampton, HMS *27*, 29, 56, 84, 177–8, 181, 184, 214, 225, *235*, 245, 258, 269, 278, 301, *plan 174*
    class 118, 123, 172, 178–80, 183, 188, 215, 218–22, 227, 235, 238, 243–4, 250, 277–8
        see also Birmingham, Glasgow, Newport, Sheffield
Soviet Union *see* Russia
Spain *11*, 226, 287
    see also Canarias class
Spartan, HMS *plan 246*
STAAG (Stabilised Tachymetric Anti-Aircraft Gun) 284, 289, 295–6, 301
Stettin (Ger) 19
Suffolk, HMS 92, 99, 107, 123–5, *129*, 233, *234*, 245
Superb, HMS 243, 249–51, *262–3*, 275, 278, 281, 284, 287, 289–90, *292–3*, *plan 251*
Surrey, HMS 121–3, 144, 149
    class 117, 120–2, 173
        'Y cruiser' 151
        see also Northumberland
Sussex, HMS 92, 126, 238, *242*
Sverdlov (Sov) 305, 309
Swan Hunter shipbuilders 50
Sweden 227
Swift, HMS 40
Swiftsure, HMS 244, 249–50, *258–9*, 277–8, 281, 284, 287, 289–90, 315, 317, *plan 249*
    class 226, 281
Sydney, HMAS (ex-*Phaeton*, HMS) *10*, *26*, 29, 41–2, 61, 86, 89, 92, *142–3*, *143*, *165*, 167, 206

Talos missile 314
Tartar missiles 313, 315–16
Tennyson d'Eyncourt, Sir Eustace 16, 34, 105
Thornycrofts 108
Tiger, HMS (1913) 87
Tiger, HMS (1944) 243, 249–50, 261, 268, 319–20, 322, *plan 294*, *plan 321*
    class 250–1, 264, 266, 273–5, 277–8, 284, 287, 289, 293–6, 301, 304–5, 315, 317–20, 322, *plan 293*, *plan 296*
    see also Bellerophon
torpedoes 29, 35, 41, 49–50, 53–4, 79–80, 82, 104, 127, 173, 188, 245, 252, 260, 295–6
    E ('Enriched Air') 104
    as threat and defence against 37, 91, 101, 107, 161, 261
    torpedo bombing 87, 90, 167, 171, 174, 177
torpedo tubes 19, 22, 33, 38, 42, 48, 50, 53–4, 57–9, 58, 60, 65, 75, 84–6, 95, 98–100, 104, 106, 110, 116–17, 124, 126–7, 155, 173–4, 178,

431

204, 221–2, 224, 228–9, 248, 250, 254, 260, 264, 267, 269, 277, 295, 300, 310, 314
   submerged 19, 22, 53, 58, 60, 67, 73–4, 117
'Town' class 19–21, 29–30, 33–4, 38, 49, 57, 82, 217, 242, 273, 278, 281, 284, 289–90, *def* 21
   see also *Bristol, Birkenhead, Birmingham, Chatham, Weymouth* classes
'Tribal' class destroyers 8, 186, 188–9
Trinidad, HMS 222
Tudor, R/A P C 48–9, 56
turbines 16, 19–20, 40, 48, 152, 297
   Brown-Curtis 21–3, 29, 40, 48, 56
   gas 311
   Parsons 21, 29, 33, 40, 47, 49
Turkey 287
Tuscaloosa, USS 65
Tyrwhitt, Cdre Sir Reginald 48, 59–60, 62

U-boats 76
Uganda, HMS 222, 225, 244, 246, *248, plan 247*
   class 249, 250, 289, 290
Undaunted, HMS 40, 42, 60, *plan 55*
USA 9, 16, 78, 94, 116
   in First World War 76, 78
   inter-war 79, 86, 95–6, 99, 104–5, 113, 122–3, 142, 150–1, 156, 167, 172–3, 186, 202, 206, 227
   in Second World War 192, 213–14, 228, 238, 242–4, 256, 258, 264, 267–8, 301
   post-war 274, 276, 287, 297, 313–15, 323

'V' class destroyers 206–7
Valiant, HMS 87
Valkyrie, HMS 60
Vanguard, HMS 229, 267
Venezuela 98, 226
Vernon, HMS 53
Versailles Treaty 94, 173
Vickers-Armstrong company 33–4, 42, 50, 59, 87, 98–9, 172, 218, 226–7, 254, 268, *plan 141, plans 193–9, plans 269–71, plans 306–8*
   Elswick Ordnance Company 98–9, 103, 111, 296
Victorious, HMS 320
Vindictive, HMS 73, 74, 75, *75*, 86, 90, *91*, 92

'W' class destroyers 206–7
Walrus flying boats *147*, 221, *plan 199*
Warspite, HMS 250
Washington Conference and Treaty (1921) 96–8, 104–5, 113, 117, 121, 125, 156, 202
Watts, Sir Philip 19–20, 22, 38, 41–2
Wessex helicopters 314, 316–20
Weymouth, HMS 22–3, *24–5*, 62, *plan 23*
   class 24, 29
      see also *Dartmouth, Falmouth, Yarmouth*
Whimbrel, HMS 243
Whitley, HMS 206
Wilson, Adm Sir Arthur K 37–8
wind tunnels 113
Wolf (Ger) 95
Worcester, USS 267–8

Yarmouth, HMS 23, *61*, 62
York, HMS 103–4, 108–13, *110–12*, 113, 117–18, 121, 204, 206, *plan 108*
   class 108

Zeppelins 61–2